Climate Change *and* Coastal Ecosystems

Long-Term Effects of Climate and Nutrient Loading on Trophic Organization

Marine Science Series

The CRC Marine Science Series is dedicated to providing state-of-the-art coverage of important topics in marine biology, marine chemistry, marine geology, and physical oceanography. The series includes volumes that focus on the synthesis of recent advances in marine science.

CRC MARINE SCIENCE SERIES

SERIES EDITOR

Michael J. Kennish, Ph.D.

PUBLISHED TITLES

Acoustic Fish Reconnaissance, I.L. Kalikhman and K.I. Yudanov
Artificial Reef Evaluation with Application to Natural Marine Habitats, William Seaman, Jr.
The Biology of Sea Turtles, Volume I, Peter L. Lutz and John A. Musick
Chemical Oceanography, Third Edition, Frank J. Millero
Coastal Ecosystem Processes, Daniel M. Alongi
Coastal Lagoons: Critical Habitats of Environmental Change, Michael J. Kennish and Hans W. Paerl
Coastal Pollution: Effects on Living Resources and Humans, Carl J. Sindermann
Ecology of Estuaries: Anthropogenic Effects, Michael J. Kennish
Ecology of Marine Bivalves: An Ecosystem Approach, Second Edition, Richard F. Dame
Ecology of Marine Invertebrate Larvae, Larry McEdward
Ecology of Seashores, George A. Knox
Environmental Oceanography, Second Edition, Tom Beer
Estuarine Indicators, Stephen A. Bortone
Estuarine Research, Monitoring, and Resource Protection, Michael J. Kennish
Estuary Restoration and Maintenance: The National Estuary Program, Michael J. Kennish
Eutrophication Processes in Coastal Systems: Origin and Succession of Plankton Blooms and Effects on Secondary Production in Gulf Coast Estuaries, Robert J. Livingston
Handbook of Marine Mineral Deposits, David S. Cronan
Handbook for Restoring Tidal Wetlands, Joy B. Zedler
Intertidal Deposits: River Mouths, Tidal Flats, and Coastal Lagoons, Doeke Eisma
Marine Chemical Ecology, James B. McClintock and Bill J. Baker
Ocean Pollution: Effects on Living Resources and Humans, Carl J. Sindermann
Physical Oceanographic Processes of the Great Barrier Reef, Eric Wolanski
Pollution Impacts on Marine Biotic Communities, Michael J. Kennish
Practical Handbook of Estuarine and Marine Pollution, Michael J. Kennish
Practical Handbook of Marine Science, Third Edition, Michael J. Kennish
Restoration of Aquatic Systems, Robert J. Livingston
Seagrasses: Monitoring, Ecology, Physiology, and Management, Stephen A. Bortone
Trophic Organization in Coastal Systems, Robert J. Livingston

Climate Change *and* Coastal Ecosystems

Long-Term Effects of Climate and Nutrient Loading on Trophic Organization

Robert J. Livingston
Professor Emeritus
Florida State University

CRC Press
Taylor & Francis Group
Boca Raton London New York

CRC Press is an imprint of the
Taylor & Francis Group, an **informa** business

CRC Press
Taylor & Francis Group
6000 Broken Sound Parkway NW, Suite 300
Boca Raton, FL 33487-2742

Printed on acid-free paper
Version Date: 20140911

International Standard Book Number-13: 978-1-4665-6842-6 (Hardback)

Library of Congress Cataloging-in-Publication Data

Livingston, Robert J.
Climate change and coastal ecosystems : long-term effects of climate and nutrient loading on trophic organization / Robert J. Livingston.
pages cm. -- (CRC marine science ; 35)
Includes bibliographical references and index.
ISBN 978-1-4665-6842-6 (hardback)
1. Coastal ecology--Mexico, Gulf of--Longitudinal studies. 2. Climatic changes--Mexico, Gulf of--Longitudinal studies. 3. Nutrient pollution of water--Mexico, Gulf of--Longitudinal studies. I. Title.

QH92.3.L58 2015
577.5'10916364--dc23 2014022789

Visit the Taylor & Francis Web site at
http://www.taylorandfrancis.com

and the CRC Press Web site at
http://www.crcpress.com

There are in fact two things, science and opinion; the former begets knowledge, the latter ignorance.

Hippocrates

Nature is wont to hide herself.

Heraclitus

There is nothing more frightful than ignorance in action.

Johann Wolfgang von Goethe

Facts are stupid things.
Trees cause more pollution than automobiles do.

Ronald Reagan

Some of the scientists, I believe, haven't they been changing their opinion a little bit on global warming? There's a lot of differing opinions and before we react I think it's best to have the full accounting, full understanding of what's taking place.

George W. Bush

This book is dedicated to the many highly talented people who contributed to the comprehensive and unique research effort over the past four decades. This includes the crews of thousands of field trips, completed without injury or loss of life. It is also dedicated to my wife and family, who participated in various phases of the work.

Contents

Part III
Trophic Response to Long-Term Climate Changes

Part IV
Impacts of Anthropogenic Nutrient Loading

List of Tables

List of Figures

Series Preface

Robert J. (Skip) Livingston is well known for his long-term ecosystem-based research on freshwater, estuarine, and marine systems in the southeastern United States. Skip's mission, when he first arrived in the Department of Biological Sciences at Florida State University in 1969, was to conduct long-term interdisciplinary research on coastal systems using a combination of well-crafted field and laboratory approaches. Focusing on the northeastern Gulf of Mexico river–bay systems—notably the Apalachicola, Choctawhatchee, Perdido, Escatawpa-Pascagula drainage systems; Escambia River–Bay system; Blackwater River and Mobile River estuaries; Winyah Bay; and Apalachee Bay (Ecofina and Fenholloway River estuaries)—Skip set out to assemble extensive comparative databases that would effectively assess system-level condition. The results were massive databases collected on a wide range of subject areas, including climate trends, physicochemical measurements, sediment characteristics, nutrients and other water quality parameters, toxic substances, phytoplankton, submerged aquatic vegetation (SAV), zooplankton, benthic invertebrates, fishes, and other biotic elements. There was also a considerable treatment of food webs/trophic organization and the study of natural and anthropogenic drivers of change in these systems, particularly the relationships of nutrient loading, nutrient dynamics, and eutrophication.

Skip has been extremely productive over the past 45 years. Since 1970, he has been the principal investigator on more than 100 projects and has authored 8 books, more than 170 scientific papers, and numerous technical reports. In addition, he has directed the academic programs of 46 graduate students. Skip has served as the director of the Center for Aquatic Research and Resource Management at Florida State University and is currently professor emeritus at the university.

This comprehensive book focuses on the northeastern Gulf of Mexico river–bay systems, notably Apalachee Bay, Apalachicoloa Bay, Choctawhatchee Bay, Ecofina River estuary, Pensacola Bay, and Perdido Bay. It is the culmination of long-term research on anthropogenic and natural driving factors in this region conducted by Skip and his colleagues at Florida State University. The main theme of the book includes climate change, climate change effects, and eutrophication. Skip has incorporated a climate database of more than 40 years and has provided a narrative on how climate change affects anthropogenic nutrient loading and biotic responses in the aforementioned systems, in particular Apalachee Bay, Apalachicola Bay, and Perdido Bay. The second half of the book is more analytical in nature in that it evaluates the effects of climate change in the region.

The data compilation is impressive by any measure, and thus the book will be useful as a reference volume for researchers working on riverine, estuarine, and coastal marine systems. It is an excellent addition to the scientific literature and adds substantially to the contributions of Skip's previous three books on this region (Livingston, 2000, 2002, 2005). I commend Skip for the tremendous amount of effort he has expended on the production of this book, as well as his other published books. The multifaceted nature of the book, inclusive of the huge amount of data compiled, is testimony to Skip's long and illustrious career as a leading aquatic scientist in the United States.

LITERATURE CITED

Livingston, R.J. 2000. *Eutrophication Processes in Coastal Systems: Origin and Succession of Plankton Blooms and Secondary Production in Gulf Coast Estuaries*. Boca Raton, FL: CRC Press.
Livingston, R.J. 2002. *Trophic Organization in Coastal Systems*. Boca Raton, FL: CRC Press.
Livingston, R.J. 2005. *Restoration of Aquatic Systems*. Boca Raton, FL: CRC Press.

Michael J. Kennish
Institute of Marine and Coastal Sciences
School of Environmental and Biological Sciences
Rutgers University
New Brunswick, New Jersey

Foreword

Seldom in the professional career of a scientist does one get the opportunity to publish a treatise that summarizes his or her life's research accomplishments that will have such a significant impact on the science of estuaries and global climate change. The ultimate goal of one's career as a scientist is to see his or her conclusions and theories accepted and acknowledged by colleagues and peers from diverse backgrounds and to see that knowledge applied, in this case, to the restoration and proper management of degraded coastal ecosystems.

Robert J. Livingston, emeritus professor of biological sciences at Florida State University, is one such individual. He and his team of research colleagues and doctoral and master's students have investigated various aspects of Florida's panhandle estuaries for the past 43 years and have accumulated and analyzed a wealth of biological, ecological, chemical, geological, climatological, physical, and trophic data from those coastal ecosystems. Dr. Livingston has published hundreds of professional papers, given hundreds of professional presentations, and has written a number of books as a result of his research. His research and published results have occasionally met resistance from those whose short-term research failed to detect the long-term changes to the coastal ecosystems they studied.

Dr. Livingston and his research team witnessed the unraveling of the *web of life* in Florida's panhandle estuaries. The Livingston Team documented the deterioration of Florida's Gulf Coast estuaries from Apalachee Bay on the east to Perdido Bay on the west. That impairment has resulted from a combination of the following factors: physical alterations (e.g., dredging of passes to the Gulf); biochemical alterations (e.g., kraft paper mill and domestic waste pollution); damming and dredging of upstream impoundments (e.g., in the Apalachicola–Chattahoochee–Flint River Basin); and global climate changes (e.g., droughts, reduced freshwater inputs, and increased salinity stratification [salinization] of estuaries), with (and resulting in) changes in the fauna and flora of the Gulf (e.g., decline of valuable oyster fisheries and changes in SAV). Dr. Livingston utilizes a unique method of comparing and contrasting trophic relationships among fauna and flora across various estuaries to document the natural and anthropogenic changes his research team documented.

Occasionally, a book is published that has the potential to change the way we think about estuaries and make us realize how dependent we are on them and how we have altered them in the recent past. This is one such book. Hopefully, it will help to initiate a recovery process for degraded estuaries along Florida's panhandle coast as well as elsewhere in Florida and the world's coastal zones. If not, humankind will suffer drastic consequences in the coming decades that few of us can imagine. Estuaries around the globe are being altered by human forces and now by human-induced global climate changes. There is now serious doubt that the highly vulnerable yet valuable estuarine and coastal ecosystems will survive in their naturally productive state due to worldwide climatological changes. We have much to be thankful for in terms of Florida's panhandle estuaries and our struggles to heal their ecological wounds; however, until and unless someone like Robert Livingston documents those *wounds*, we cannot initiate plans to insure their continued ecological health and productivity.

In this treatise, Dr. Livingston systematically documents the salient biological, ecological, climatological, and anthropogenic factors that have adversely impacted Florida's panhandle estuaries over the past 43 years. This includes detailed information concerning the deterioration of water quality; the reduced abundance of finfish, shellfish, and prey species; and the reduction of fishery resources in these once-productive coastal ecosystems. This is the most comprehensive review and explanation of the human alteration of estuaries and the resulting deleterious impacts on Florida's natural resources since the documentation of the demise of the Everglades ecosystem.

Dr. Livingston's treatise documents and explains the ongoing decline of Florida's largest oyster industry in the Apalachicola estuary in Franklin County, including previously undocumented impacts of global climate change, the resulting regional droughts in the Apalachicola–Chattahoochee–Flint

River Basin, and the mismanagement of regional water resources from that basin that have reduced the freshwater inflows to that estuary to the detriment of the various fisheries that depend on that ecosystem. The Livingston Team has significantly expanded our knowledge of the ongoing and now recognized impacts of global climate change in coastal ecosystems—not just the visible rise of sea level and the alteration of habitats, but also the invisible and difficult-to-discern changes in faunal and floral assemblages and the trophic alterations within various habitats.

This treatise is one of the most comprehensive studies of the negative impacts of climate change on the health and productivity of estuaries not only around the Gulf of Mexico but also around the world. Never before in this writer's memory has anyone or any team of estuarine scientists amassed such a wealth of data over a four-decade study period across such a diverse set of estuaries covering such a wide array of study topics. The task of accumulating and analyzing those data has been monumental.

The long-term record of the Livingston Team's estuarine research along the panhandle coast of Florida indicated that habitat differences among the subject estuaries have led to different responses to region-wide climate changes. These responses included altered food webs, disjunct energy relationships, reduced productivity, severely damaged fisheries, and the loss of keystone species. The data described in this book are not meant to be predictive of the future but represent empirical evidence that long-term climate changes have occurred in the northeastern Gulf of Mexico region and that such changes have adversely affected those formerly productive coastal ecosystems. Despite the scope and volume of this information, the general public has remained uninformed by the news media regarding the demonstrated adverse impacts of climate changes on rivers and estuaries of concern.

With the results of the Livingston Team's four-decade study of estuarine declines, natural resource managers and legislators and local environmental groups like the Apalachicola River Keepers can now begin to reverse the adverse impacts of human coastal development through increased preservation of estuarine drainage basins, reduction and eventual elimination of industrial and domestic pollution inputs, closure of passes across coastal barriers that disrupt the long-term natural salinity regimes, reestablishment of natural freshwater inflows that will foster the recovery of Florida's estuarine-dependent finfish and shellfish resources and fisheries, and creation of coastal preserves that will become future estuarine habitats as the sea levels continue to rise in response to global climate change. Shellfish managers and water resource managers should now cooperate more readily to restore ideal oyster-growing conditions in the Apalachicola estuary, which is historically the greatest source of oysters in the state of Florida.

The water management districts within the Apalachicola–Chattahoochee–Flint River Basin should now join with the states of Alabama, Florida, and Georgia to insure adequate supplies of freshwater inflows to the Apalachicola estuary and to reduce the human and agricultural effluents that now pollute the Tri-Rivers Basin. And finally, coastal zone managers at the local, state, and federal levels, including the US Army Corps of Engineers, should now examine the Livingston Team's data and conclusions to determine the possibility of closing the dredged inlets and passes across the coastal barriers that enclose Florida's degraded estuaries to help restore the natural salinity regimes within those salinity-stratified environments.

This estuarine treatise should serve as a scientific wake-up call that promotes the Florida Department of Agriculture and Consumer Services (Aquaculture Division), the Florida Department of Environmental Protection, the Florida Fish and Wildlife Conservation Commission, Florida's nongovernmental organizations, and the leaders of commercial and recreational fisheries to determine how best to reverse the declines in Florida's essential estuarine support systems for those fisheries and ancillary businesses that depend on healthy and productive estuaries. Those businesses include, but are not limited to, tourism, bait and tackle shops, charter boat operators, and seafood processing, wholesale, and retail businesses. It is time to set aside power politics and apply the power of science to insure the continued future of Florida's panhandle estuaries.

This comprehensive review and explanation of Florida's estuarine problems should be acknowledged and validated by the most astute and informed coastal scientists but may be eschewed by those bureaucrats and coastal zone (mis) managers, water resource (mis) managers, and coastal real estate and development interests who still believe that they can remake Florida's coastal estuaries and adjacent upland areas in their own misguided images. Many in the scientific community will try to rebut the conclusions of the Livingston Team with their short-term and shortsighted research results, inadvertently siding with the global warming deniers. Those who have data that support the conclusions of the Livingston Team will acknowledge Livingston's research results wholeheartedly; those who do not, will be quick to attack these long-term monitoring results.

This long-term comprehensive study of estuaries should serve as a model for future and continuing studies of estuaries in Florida and elsewhere, provided academic institutions and funding agencies have the will to address and reverse the damaging ecological impacts revealed by the Livingston Team's long-term studies. This treatise is recommended as a source book for all estuarine students, scientists, and coastal resource managers who claim an interest in the health and well-being of coastal ecosystems under their purview and management.

Finally, this estuarine treatise contains an up-to-date bibliography of estuarine publications and reports that will serve as a new starting point starting for future research and studies. It also contains the methods and materials that the Livingston Team used in assessing the health and trophic organization of Florida's estuaries. Methods and procedures have changed over the past 43 years, and new ones will be developed in the future; however, without some understanding of the historic trends in estuarine research reported by the Livingston Team, future research may not be comparable to that of the recent past.

Edwin W. "Ed" Cake, Jr., PhD
Chief Science Officer & Biological Oceanographer
Gulf Environmental Associates
Ocean Springs, Mississippi

Preface

The research effort on which this book is based has involved continuous analyses of a series of river-estuarine and coastal systems in the southeastern United States that encompassed the period from 1970 to 2012. These long-term studies were carried out using a combination of field-descriptive and experimental (lab and field) approaches. The research team for this work included field personnel, chemists, taxonomists, experimental biologists, physical oceanographers, hydrological engineers, statisticians, computer programmers, and modelers.

An interdisciplinary, comparative database was created based on consistent field sampling methods and detailed taxonomic identifications over the extended period of the study. The research program was designed to evaluate system-level responses to natural and anthropogenic nutrient loading and long-term climate changes. We concentrated on phytoplankton/benthic macrophyte productivity and associated food web organization (infaunal and epibenthic macroinvertebrates and fishes) in a series of river-dominated estuaries along the NE Gulf coast. Efforts were made to determine how human activities affect these processes against a backdrop of continuous and long-term responses to natural climatological cycles. The emphasis was on how river-estuarine systems differ in their response to natural and anthropogenic nutrient loading along with the effects of seasonal and interannual changes in climate conditions.

The field effort over the 43-year sampling period is outlined in Table P.1. It should be noted that all field sampling techniques (physical, chemical, and biological) were carried out based on written protocols (see Appendices). Most such protocols were developed at the beginning of the study. Sampling techniques in terms of spatial conditions were based on detailed habitat stratification in each coastal system. These methods were quantified early in the overall study period. There was a concerted effort to make sure that taxonomic identifications were made consistent with the ongoing efforts of a worldwide group of systematic experts. This adherence to quantitative sampling methodology and taxonomic verity along with the use of consistent laboratory analytic techniques allowed valid intersystem comparisons on which this book is based.

Complete descriptions of these ecosystems, data collection efforts and field/laboratory operations, protocols for all field data collection and experimental methods, and methods of statistical analysis and modeling have been published by Livingston (2000, 2002, 2005). The database included long-term data taken in the Apalachicola River estuary, Apalachee Bay, and the Perdido River estuary. Phytoplankton community analyses were carried out in the Perdido, Econfina, and Fenholloway estuaries. Such shorter-term analyses have been made of the Escambia, Blackwater, and Choctawhatchee bays as well. A 20-year program was carried out to determine the impact of a pulp mill on the Elevenmile Creek–Perdido Bay system.

The multidisciplinary database included physicochemical and nutrient loading data and information concerning detailed spatial and temporal distributions of (species-specific) phytoplankton, infaunal and epibenthic invertebrates and fishes collected in a series of coastal systems in the NE Gulf of Mexico. The intersection of variables such as nutrient loading, plankton responses, and the effects of plankton blooms on secondary productivity and coastal food webs was analyzed relative to long-term climatological changes. There was a particular emphasis on a comparative analysis of the aforementioned factors in the five major river-estuarine systems within the extended study area. Along with field sampling, there were associated field and laboratory experimental programs along with a major effort to define the trophic organization of the various coastal systems.

In a book that relates long-term climatological changes (which includes the droughts of 1980–1981, 1986–1988, 1999–2002, 2006–2008, and 2010–2012) to a series of multidisciplinary studies of a series of river-estuarine systems in the NE Gulf of Mexico, it is appropriate to acknowledge the periods covered by these research efforts. The general approach and field sampling methods were common to all study areas. These studies were subject to the same determination of

Table P.1 Field/Experimental Effort of the Florida State University Research Group from 1970 to 2012

System (Number of Years of Data)	MET.	PC	LC	POL	PHYT	SAV	ZOO	INF	INV	FISH	TR	NL	NR
Apalachee Bay (Econfina, Fenholloway)	50***	26*	12**	2	5*	15*	3*	2	20*	20*	20*	12*	1*
Apalachicola River–Bay system	90***	17*	3*	2	1	1	1	8*	14*	14*	8*	2*	ND
Choctawhatchee River–Bay system	30***	4*	4*	2	2*	1*	2*	4*	4*	4*	2*	2*	ND
Escambia River–Bay system	30***	1.5*	1.5*	1	1.5*	ND	1.5*	1.5*	1.5*	1.5*	1.5*	1.5*	ND
Blackwater River–Bay system	30***	1.5*	1.5*	1	1.5*	ND	1.5*	1.5*	1.5*	1.5*	1.5*	1.5*	ND
Perdido River–Bay system	50***	20**	20**	3	17**	3*	3*	20**	20**	20**	20**	20**	1*

Notes: MET, river flow, rainfall; PC, salinity, conductivity, temp., dissolved O_2, O_2 anomaly, pH, depth, Secchi; LC, NH_3, NO_2, NO_3, TIN, PON, DON, TON, TN, PO_4, TDP, TIP, POP, DOP, TOP, IP, DOC, POC, TOC, IC, TIC, TC, BOD, silicate, TSS, TDS, DIM, DOM, POM, PIM, NCASI color, turb., chlorophyll *a*, *b*, *c*, sulfide; POL, water/sediment pollutants; PHYT, whole water, net phytoplankton; SAV, submerged aquatic vegetation; ZOO, net zooplankton; INF, infaunal macroinvertebrates taken with cores, ponars; INV, invertebrates taken with seines, trawls; FISH, fishes taken with seines, trawls; TR, trophic analyses; NL, nutrient loading; NR, nutrient limitation experiments; ND, no data. The numbers of years of data taken are listed.

*monthly,

**monthly, quarterly,

***daily, monthly.

trophic organization. However, two of these studies were short-term efforts. The Choctawhatchee River–Bay project was carried out from 1985 to 1986. This included a period of relatively high river flows that preceded the drought of 1986–1988. The Pensacola estuary project was carried out from 1997 to 1998, a period of relatively high river flows with no major drought periods. These two projects had such short time spans that no conclusions could be drawn regarding responses to long-term climatological conditions. The Apalachee Bay study, which included intensive fieldwork from 1971 to 1980 and from 1992 to 2004, covered a period during the 1970s that had no major droughts. The latter field effort included intensive studies of drought conditions during 1999–2002. The Apalachicola studies, which were most intensive during the period from 1972 to 1991, included a period of no major droughts during the 1970s that was followed by a period of droughts during 1980–1981 and 1986–1988. The Perdido River estuary study was carried out from 1988 to 2007, a period that included the droughts of 1999–2002 and 2006–2007. Thus, these three projects were carried out over long enough periods so that the influence of climatological factors on physical, chemical, and biological indices could be evaluated.

Acknowledgments

There were many people who contributed to the research effort over the past 43 years. It is impossible to acknowledge all these people. Literally, thousands of undergraduate students, graduate students, technicians, and staff personnel have taken part in the long-term field ecology programs. Various people have been involved in research associated with the Center for Aquatic Research and Resource Management at Florida State University and Environmental Planning & Analysis, Inc.

The program has had the support and cooperation of various experts. Dr. A. K. S. K. Prasad studied phytoplankton taxonomy and systematics. Robert L. Howell IV ran the field collections and identified fishes and invertebrates. Glenn C. Woodsum designed the overall database organization and, together with Phillip Homann and Loretta E. Wolfe, ran database management and day-to-day data analyses of the research information. Dr. F. Graham Lewis III was instrumental in the development of analytical procedures and statistical analyses. Dr. David A. Flemer designed the overall approach for analytical chemistry.

The long-term field ecological program has benefited from the advice of a distinguished group of scientists. Dr. John Cairns, Jr., Bori Olla, Dr. E. P. Odum, Dr. J. W. Hedgpeth, Dr. O. Loucks, Dr. K. Dickson, Dr. F. J. Vernberg, and Dr. Ruth Patrick reviewed the early work. Dr. D. M. Anderson, Dr. A. K. S. K. Prasad, G. C. Woodsum, Dr. M. J. Kennison, Dr. K. Rhew, Dr. M. Kennish, Dr. E. Fernald, Dr. D. C. White, Dr. L. Beidler, and V. Tschinkel have reviewed various aspects of the research. Additional reviews have been provided by Dr. C. H. Peterson, Dr. S. Snedaker, Dr. R. W. Virnstein, Dr. E. D. Estevez, Dr. J. J. Delfino, Dr. F. James, Dr. W. Herrnkind, and Dr. J. Travis.

Over the years, we have had a long list of coinvestigators who have participated in various projects. These collaborative efforts in the field have included the following scientists: Dr. D. C. White (microbiological analyses), Dr. D. A. Birkholz (toxicology, residue analyses), Dr. G. S. Brush (long-core analyses), Dr. L. A. Cifuentes, (nutrient studies, isotope analyses), Dr. K. Rhew (microalgal analyses), Dr. R. L. Iverson (primary productivity), Dr. R. A. Coffin (nutrient studies, isotope analyses), Dr. W. P. Davis (biology of fishes), Dr. M. Franklin (riverine hydrology), Dr. T. Gallagher (hydrological modeling), Dr. W. C. Isphording (marine geology), Dr. C. J. Klein III (aquatic engineering, estuarine modeling), Dr. M. E. Monaco (trophic organization), Dr. R. Thompson (pesticide analyses), Dr. W. Cooper (chemistry), D. Fiore (chemistry), Dr. D. A. Flemer (chemistry), and Dr. A. W. Niedoroda (physical oceanography).

We have had unusually strong support in the statistical analyses and modeling of our data, a process that is continuing to this day. Dr. Duane A. Meeter, Dr. Xufeng Niu, and Loretta E. Wolfe performed statistical analyses through the years. These statistical determinations and modeling efforts have been aided by advice from the following people: Dr. T. A. Battista (National Oceanic and Atmospheric Administration), Dr. B. Christensen (the University of Florida), Dr. J. D. Christensen (National Oceanic and Atmospheric Administration), Dr. M. E. Monaco (National Oceanic and Atmospheric Administration), Dr. T. Gallagher (Hydroqual, Inc.), Dr. B. Galperin (the University of South Florida), and Dr. W. Huang (Florida State University).

The research has been integrated with an active graduate student program. The following people have achieved graduate degrees with Dr. Livingston as their major professor:

Columbus H. Brown, MS "The effect of photoperiodism on the respiration of the channel catfish"—1973; Stephen Brice, MS "The effects of methyl mercury on the channel catfish *Ictalurus punctatus*"—1973; Theresa Ann Hooks, MS "An analysis and comparison of the benthic invertebrate communities in the Fenholloway and Econfina estuaries of Apalachee Bay, Florida"—1973; Bruce D. DeGrove, MS "The effects of Mirex on temperature selection in the sailfin molly, *Poecilia latipinna*"—1973; Kenneth L. Heck, Jr., MS "The impact of pulp mill effluents on species assemblages of epibenthic marine invertebrates in Apalachee Bay, Florida"—1973; Gerard G. Kobylinski, MS "Translocation of Mirex from sediments and its accumulation by the hogchoker

(*Trinectes maculatus*)"—1974; F. Graham Lewis, MS "Avoidance reactions of two species of marine fishes to kraft pulp mill effluents"—1974; Aureal J. Tolman, MS "Effects of Mirex on the activity rhythms of the diamond killifish, *Adinia xenica*"—1974; Claude R. Cripe, MS "The effects of Mirex on a simulated marsh system"—1974; Michael Zimmerman, MS "A comparison of the benthic macrophytes of polluted system (Fenholloway River) and an unpolluted system (Econfina) in Apalachee Bay, Florida"—1974; Paul Muessig, MS "The determination of pathways of mercury concentration and conversion within specific organs of channel catfish (*Ictalurus punctatus*)"—1974; Christopher C. Koenig, PhD "The synergistic effects of Mirex and DDT on the embryological development of the diamond killifish, *Adinia xenica*"—1975; Susan Drake, MS "The effects of mercury on the development of the zebrafish"—1975; Allan W. Stoner, MS "Growth and food conversion efficiency of pinfish (*Lagodon rhomboides*) exposed to sublethal concentrations of bleached kraft mill effluents"—1976; Allan W. Stoner, PhD "Ecological relationships and feeding response of the pinfish, *Lagodon rhomboides*"—1979; Roger A. Laughlin, MS "Avoidance of blue crabs (*Callinectes sapidus*) to storm water runoff"—1976; George Gardner, MS "Behavioral reactions of pinfish to pulp mill effluents"—1976; Peter Sugarman, MS "Effects of bleached kraft mill effluents on activity rhythms of the pinfish, *L. rhomboides*"—1977; Bruce Purcell, MS "Effects of storm water runoff on grass bed communities in East Bay"—1977; James Duncan, MS "Short-term impact of clearcutting activities on epibenthic fishes and invertebrates in the Apalachicola Bay system"—1977; Peter Sheridan, PhD "Trophic relationships of fishes in the Apalachicola Bay system"—1978; Steven Osborn, MS "Ecological relationships of ophiuroids in Charlotte Harbor"—1979; Joseph Ryan, MS "Day-night feeding relationships of grass bed fishes"—1980; Holly Greening, MS "Spatial/temporal distribution of invertebrates in grass beds of Apalachee Bay"—1980; Bruce MacFarlane, PhD "Effects of variations in pH on topminnow physiology and behavior"—1980; Pat Dugan, MS "Long-term population changes of epibenthic macroinvertebrates in Apalachee Bay, Florida"—1980; Brad McLane, MS "Impact of stormwater runoff on benthic macroinvertebrates"—1980; Kathy Brady, MS "Larval fish distribution in Apalachee Bay"—1981; Duncan Cairns, MS "Detrital processing in a subtropical southeastern drainage system"—1981; Bruce Mahoney, PhD "The role of predation in seasonal fluctuations of estuarine communities"—1982; Will Clements, MS "Feeding ecology of filefish in Apalachicola Bay, Florida"—1982; F. Graham Lewis, PhD "Habitat complexity in a subtropical seagrass meadow"—1982; Ken Leber, PhD "Feeding ecology of decapod crustaceans"—1983; Kevan Main, PhD "Predator-prey interactions in seagrass beds: The response of *Tozeuma carolinense*"—1983; Kelly Custer, MS "Gut clearance rates of three prey species of the blue crab"—1985; J. Michael Kuperberg, MS "Response of the marine macrophyte *Thalassia testudinum* to herbivory"—1986; Susan Mattson, MS "The effect of post-settlement predation on community structure of epifauna associated with cockle shells"—1986; Joseph L. Luczkovich, PhD "The patterns and mechanisms of selective feeding on seagrass-meadow epifauna by juvenile pinfish, *Lagodon rhomboides*"—1987; Jon A. Schmidt, PhD "Patterns of seagrass infaunal polychaete recruitment: influence of adults and larval settling behavior"—1987; David Bone, PhD "Response of seagrass invertebrates to toxic agents"—1987; Carrie Phillips, MS "Influence of physical disturbance on infaunal macroinvertebrates: seagrass beds vs. unvegetated areas"—1987; Frank Jordan, MS "Trophic organization of fishes in the Choctawhatchee River"—1989; Jutta Schmidt-Gengenbach, MS "A study of the effects of pollutants on the benthic macroinvertebrates in Lake Jackson, Florida: A descriptive and experimental approach"—1991; Jeff Holmquist, PhD "Disturbance, dispersal, and patch insularity on a marine benthic assemblage: influence of a mobile habitat on seagrasses and associated fauna"—1992; Todd Bevis, MS "Effects of urban runoff on fish assemblages in Lake Jackson, Florida"—1995; S. E. McGlynn, PhD "Polynuclear aromatic hydrocarbons: Sediment and plant interactions"—1995; C. J. Boschen, MS "Effects of urban runoff on feeding ecology of fish assemblages in a north Florida solution lake"—1996; Julie Reardon,

MS "Ecology of phytoplankton communities in Lake Jackson"—1999; Barbara Shoplock, MS "Ecology of zooplankton communities in Lake Jackson"—1999.

Systematics and taxonomy are of central importance in a comprehensive ecosystem program. This attention to the accurate naming of biological subjects was something taught to Dr. Livingston from his major graduate professors (Dr. Carl Hubbs, Dr. C. Richard Robbins, Dr. Arthur A. Myrberg, and Dr. Charles E. Lane). The associated attention to natural history has benefited from the efforts of a long line of postdoctoral fellows and trained technicians: Dr. A. K. S. K. Prasad and Dr. K. Rhew (phytoplankton, benthic microalgae), Dr. F. Graham Lewis III (aquatic macroinvertebrates), Dr. R. D. Kalke (estuarine zooplankton), Dr. G. L. Ray (infaunal macroinvertebrates), Dr. K. R. Smith (oligochaetes), Dr. E. L. Bousfield (amphipods), Dr. F. Jordan (trophic analyses), Dr. C. J. Dawes (SAV), Dr. R. W. Yerger and Dr. Carter Gilbert (fishes), R. L. Howell III (infaunal and epibenthic invertebrates and fishes), W. R. Karsteter (freshwater macroinvertebrates), Dr. J. H. Epler (aquatic insects), and M. Zimmerman (SAV).

A number of phycologists have given their time to the taxonomy and nomenclature of phytoplankton species over the years: Diatom taxonomy—Dr. C. W. Reimer (Academy of Natural sciences of Philadelphia); Dr. G. A. Fryxell (the University of Texas at Austin); Dr. G. R. Hasle, (the University of Oslo, Norway); Dr. P. Hargraves (the University of Rhode Island); Professor F. Round (the University of Bristol, England); R. Ross, P. A. Sims, Dr. E. J. Cox, and Dr. D. M. Williams (British Museum of Natural History, London, UK); Dr. L. K. Medlin and Dr. R. M. Crawford (Alfred-Wegener Institute, Bremerhaven, Germany); Professor T. V. Desikachary (the University of Madras, India); Dr. J. A. Nienow (Valdosta State University, Georgia); Dr. P. Silva (the University of California at Berkeley); Dr. M. A. Faust (Smithsonian Institution); Dr. R. A. Andersen (Bigelow Laboratory for Ocean Sciences, Maine); Dr. D. Wujeck (Michigan State University); Dr. M. Melkonian (the University of Koln, Germany); Dr. C. J. Tomas (Florida Department of Environmental Regulation, St. Petersburg); Professor T. H. Ibaraki (Japan); Dr. Y. Hara (the University of Tsukuba, Japan); and Dr. J. Throndsen (the University of Oslo, Norway). Technical assistance with transmission and scanning electron microscopes was provided by K. A. Riddle (Department of Biological Science, Florida State University). A. Black and D. Watson (Histology Division, Department of Biological Science, Florida State University) aided in specimen preparation and serial thin sectioning.

The curators of various International Diatom Herbaria have aided in the loan of type collections and other authentic materials: Academy of Natural Sciences of Philadelphia (ANSP), California Academy of Sciences (CAS) (San Francisco), National Museum of Natural History, Smithsonian Institution (Washington, DC), Harvard University Herbaria (Cambridge, Massachusetts), the Natural History Museum (BM) (London, UK), and the F. Hustedt Collections, Alfred-Wegener Institute (Bremerhaven, Germany).

Others have collaborated on various aspects of the field analyses: W. Meeks, B. Bookman, Dr. S. E. McGlynn, and P. Moreton ran the chemistry laboratories. O. Salcedo, I. Salcedo, R. Wilt, S. Holm, L. Bird, L. Doepp, C. Watts, and M. Guerrero-Diaz provided other forms of lab assistance. Field support for the various projects was provided by H. Hendry, A. Reese, J. Scheffman, S. Holm, M. Hollingsworth, Dr. H. Jelks, K. Burton, E. Meeter, C. Meeter, S. Solomon, S. S. Vardaman, S. van Beck, M. Goldman, S. Cole, K. Miller, C. Felton, B. Litchfield, J. Montgomery, S. Mattson, P. Rygiel, J. Duncan, S. Roberts, T. Shipp, A. Fink, J. B. Livingston, R. A. Livingston (Reese), J. Huff, R. M. Livingston, A. S. Livingston, T. Space, M. Wiley, D. Dickson, L. Tamburello, and C. Burbank.

Project administrators included the following people: Dr. T. W. Duke, Dr. R. Schwartz, Dr. M. E. Monaco, J. Price, Dr. N. P. Thompson, J. H. Millican, D. Arceneaux, W. Tims Jr., Dr. C. A. Pittinger, M. Stellencamp, S. A. Dowdell, C. Thompson, K. Moore, Dr. E. Tokar, Dr. D. Tudor, Dr. E. Fernald, D. Giblon, S. Dillon, A. Thistle, M. W. Livingston, and C. Wallace.

Special thanks are extended to Dr. E. W. Cake (Gulf Environmental Associates, Ocean Springs, Mississippi) for his excellent review of the book and his comprehensive editing of the final version. I am indebted to Dr. Will Clements (Department of Fish, Wildlife and Conservation Biology, Colorado State University) and Dr. Joseph J. Luczkovich (Department of Biology, East Carolina University) for their outstanding reviews of the book. I am also indebted to Dr. M. J. Kennish (Editor, Institute of Marine and Coastal Sciences, Rutgers University) and John Sulzycki (Senior Editor, CRC Press/Taylor & Francis Group) for their efforts concerning the publication of our work over the years.

Author

Robert J. Livingston is currently professor emeritus in the Department of Biological Science at Florida State University (Tallahassee, Florida). His interests include aquatic ecology, pollution biology, field and laboratory experimentation, and long-term ecosystem-level research on freshwater, estuarine, and marine systems. His past research includes multidisciplinary studies of lakes and a series of river–estuarine ecosystems on the Gulf and Atlantic coasts of the United States. Over the past 43 years, Livingston's research group has conducted a series of studies in areas from Maine to Mississippi. Dr. Livingston has directed the programs of 46 graduate students who have carried out research in behavioral and physiological ecology with individual aquatic populations and communities in lakes, rivers, and coastal systems. Dr. Livingston is the author of over 170 scientific papers and has written or edited 8 books on the subject of aquatic ecology. He has been the principal investigator for more than 100 projects since 1970.

The primary research program of the author has been carried out in areas that include the Apalachicola drainage system, the Choctawhatchee River and Bay, the Perdido estuary, the Escambia River–Bay system, the Blackwater River estuary, the Escatawpa-Pascagula drainage system, the Mobile River estuary, the Winyah Bay system (including the Sampit River), Apalachee Bay (the Econfina and Fenholloway River estuaries) and a series of north Florida lake systems (Table P.1). A software has been developed over the years to aid in the analysis of the established long-term databases. This work has included the determination of the impact of various forms of anthropogenic activities on a range of physical, chemical, and biological processes.

Recent analyses of the long-term database have involved determinations of the effects of nutrient loading and associated plankton blooms on the trophic relationships of coastal infaunal and epifaunal macroinvertebrates and fishes. This has included questions concerning how such effects are related to long-term climatological changes.

Together with research associates and former graduate students, Dr. Livingston is currently in the process of publishing the long-term database.

PART I

Overview

CHAPTER 1

Introduction

The statement by Borja et al. (2010) concerning the status of long-term evaluations of coastal systems could provide an appropriate context for this book:

> There are very few examples of long-term monitoring data, including different biological elements (i.e., plankton, benthos, fishes, etc.) together with physical-chemical data from water and sediments, showing recovery trajectories after remediation or restoration processes in marine environments....

This assertion could apply to any and all so-called long-term research efforts in coastal systems. Of course, there are a number of major studies that have been carried out in such areas. Probably the most well-studied estuarine complex, the Chesapeake Estuary, has been intensively analyzed for decades. This is true in lesser degrees with regard to the various big-river systems in America and Europe. However, despite such research efforts, the actual definition of what makes an effective long-term, interdisciplinary research program has been largely overlooked in the scientific literature. Indeed, in some influential circles, such efforts have been the subject of ridicule and outright dismissal.

Temperate big-river estuaries and associated coastal areas are well known for their high rates of productivity and their complex, highly changeable habitat conditions due to their susceptibility to rapid shifts in physical habitat conditions. Located between upper freshwater systems and the open ocean, these estuaries have a broad range of habitats that are basic to the maintenance of complex food webs. These coastal areas are among the most complex ecosystems on earth because of the combination of high instability of the physical habitat and the relatively rapid response of the relatively shallow waters to climatological changes. To fully understand the ecosystem processes that underlie the incredibly high productivity that is the basis of major fisheries, there needs to be an understanding of the entire drainage basin along with an understanding of the complex relationships of the estuary with the open sea.

Physical (structural) components (depth, area, river inputs, openings to the open sea) combine with sediment and water-quality conditions to produce relatively high levels of primary productivity that are fundamental to the vast populations that inhabit many estuarine ecosystems. Marsh development, phytoplankton productivity, and the distribution of benthic plants provide both habitat and food for such populations. The key to any understanding of how a given estuarine system functions lies in the food web machinations whereby cyclic seasonal and interannual climatological conditions drive the complex interspecific processes among the ever-changing populations.

An ecosystem study of a given estuary must take into consideration the response of the various habitats to changes in the driving variables as they relate to the trophic components of the system. When it comes to an evaluation of the changes that might occur due to long-term trends related to climate change, the sheer complexity of the estuarine environment makes any estimation of present and future impact difficult to ascertain without a major research effort. This complexity is reflected in the scientific literature relating to climate change.

Chaalali et al. (2013) noted that, with respect to the consequences of climate change, the specific responses of estuarine ecosystems to such changes, along with identification of the drivers of such responses, are less understood than those in terrestrial ecosystems. Working in a southwestern European estuary, the Gironde, the authors examined climatic, physical, chemical, and biological parameters for the period 1978–2009. The authors found that the estuary experienced two abrupt shifts (~1987 and ~2000) over the last three decades, which altered the whole system. The timing of these abrupt shifts was related to alterations reported in both marine (e.g., in the North Sea, in the Mediterranean Sea, and along the Atlantic) and terrestrial (e.g., in European lakes) areas. However, it was not possible to determine the dynamic processes through which climate effects moved through the different compartments of the estuary. There was evidence that the dynamics of the estuary were strongly modulated by climate change at both regional and global scales.

Beaugrand and Reid (2003) noted long-term changes in phytoplankton, zooplankton, and salmon populations in relation to hydrometeorological forcing in the northeast Atlantic Ocean and adjacent seas. They found highly significant relationships between long-term changes in all three trophic levels and sea-surface temperature in the northeastern Atlantic, northern hemisphere temperature, and the North Atlantic Oscillation (NAO). Regional temperature was an important parameter that governed the dynamic equilibrium of northeast Atlantic pelagic ecosystems with possible consequences for biogeochemical processes and fisheries. Kroncke et al. (1998), working with macrofaunal samples that were collected seasonally from 1978 to 1995 in the subtidal zone off Norderney (one of the East Frisian barrier islands), found that interannual variability in biomass, abundance, and species numbers was related to interannual climate variability. Using multivariate regression models, they found that abundance, species number, and (less clear) biomass were correlated with the NAO and that the mediator between the NAO and benthos was probably the sea-surface temperature in late winter and early spring.

Cloern et al. (2010) noted that fish and plankton populations in the ocean fluctuate in synchrony with large-scale climate patterns. However, similar evidence is lacking for estuaries because of shorter observational records. Marine fish and invertebrates sampled in San Francisco Bay since 1980 showed large, unexplained population changes with record-high abundances of common species after 1999. Populations of demersal fishes, crabs, and shrimp covaried with the Pacific Decadal Oscillation (PDO) and North Pacific Gyre Oscillation (NPGO), both of which reversed signs in 1999. The authors indicated that synchronous shifts of climate patterns and community variability in San Francisco Bay were related to changes in oceanic wind forcing that modified coastal currents, upwelling intensity, and surface temperatures with an influence on recruitment of marine species that utilized the estuaries as nursery habitat.

Williams et al. (2010), working with 23 years of data taken in Chesapeake Bay (CB), used indicators to detect trends of improving and worsening environmental health in 15 regions and 70 segments of the bay. They evaluated the estuarine response to reduced nutrient loading from point (i.e., sewage treatment facilities) and nonpoint (e.g., agricultural and urban land use) sources. Despite various restoration efforts, ecological health-related water-quality (chlorophyll *a*, dissolved oxygen, Secchi depth) and biotic (phytoplankton and benthos) indices showed little improvement (submerged aquatic vegetation was an exception), and water clarity and chlorophyll *a* conditions actually worsened considerably since 1986. Nutrient and sediment inputs from higher-than-average annual flows after 1992 combined with those from highly developed coastal plain areas and compromised ecosystem resiliency were thought to be important factors responsible for worsening chlorophyll *a* and Secchi depth trends in mesohaline and polyhaline zones from 1986 to 2008.

Sobocinski et al. (2013) compared fish community composition data collected at similar seagrass sites from 1976 to 1977 and 2009 to 2011. Seagrass coverage at the specific study sites did not vary considerably between time periods. However, contemporary fish species richness was lower and multivariate analysis showed that such assemblages differed between the two data sets. The majority of sampled species were common to both data sets, but several species were exclusive to only one data set. For some species, relative abundances were similar between the two data sets, while for others, there were notable differences without directional uniformity. The authors concluded that observed changes in community structure might be more attributable to higher overall bay water temperature in recent years and other anthropogenic influences than to changes in seagrass coverage.

Allan and Soden (2008) used climate models that suggest that extreme precipitation events will become more common in an anthropogenically warmed climate. Satellite observations and model simulations indicated a distinct link between rainfall extremes and temperature, with heavy rain events increasing during warm periods and decreasing during cold periods. The observed amplification of rainfall extremes was found to be larger than that predicted by models. The authors noted that projections of future changes in rainfall extremes in response to anthropogenic global warming may be underestimated. Voynova and Sharp (2012) used models to predict that the number of extreme floods and droughts will increase with climate change, a prediction that is backed by recent analyses of satellite data. The authors analyzed 100- and 80-year records of the two major rivers of the Delaware Estuary. They found that about 20% of the very large and 50% of the extreme daily discharges occurred in the current decade (2001–2011). This was consistent with predictions of increased extreme weather conditions (inundation and drought) from climate change. Based on a 44-year agency monitoring record, the estuary was well mixed during summer periods with high bottom-water dissolved oxygen. In the summer of 2006, an extreme river discharge pushed the Delaware Estuary salinity gradient further downstream resulting in stratification that allowed for a rapid phytoplankton biomass increase with resulting oxygen depletion at depth. The significant increase in large and extreme floods in the last decade indicated that the typology of the Delaware Estuary was shifting as a result of climate changes.

Paerl et al. (2006) analyzed the effects of anthropogenic and climatic perturbations on nutrient–phytoplankton interactions and eutrophication in the waters of the CB and the Neuse River Estuary/Pamlico Sound (NRE/PS) system. Both systems experienced large recent increases in nitrogen (N) and phosphorus (P) loading, and nutrient reductions have been initiated to alleviate symptoms of eutrophication. These nutrient reductions have been strongly affected by hydrologic variability, including severe droughts and a recent increase in Atlantic hurricane activity. In both systems, the resulting variability of water residence time strongly influenced seasonal and longer-term patterns of phytoplankton biomass and community composition. Fast-growing diatoms were favored during years of high discharge and short residence time in CB, whereas this effect was not observed during high discharge conditions in the longer residence time. All phytoplankton groups except summer cyanobacterial populations showed decreased abundance during elevated flow years when compared to low flow years. Seasonally, hydrologic perturbations, including droughts, floods, and storm-related deep mixing events, overwhelmed nutrient controls on floral composition.

Mallin et al. (1993) using a 4-year (monthly) data set to investigate the effect of upstream physical forces on primary productivity of the Neuse River Estuary (North Carolina, the United States) found that the magnitude of estuarine primary production and the periodicity of algal blooms can be directly related to variations in upper watershed rainfall and its subsequent regulation of downstream river flow. Future changes in precipitation patterns for coastal regions may thus lead to substantial alterations in coastal primary productivity patterns. The authors noted that timing and duration of sampling programs are critical when assessing anthropogenic nutrient impacts on estuaries. In the Neuse Estuary, sampling during dry years (1988, 1991) would have indicated a system of moderate productivity, whereas sampling during wet years (1989, 1990) would have indicated a eutrophic system. Likewise, the effect of episodic rainfall and runoff events must be considered in mitigation efforts to reduce nutrient loading to coastal systems. Finally, the authors found that if precipitation decreases, coastal primary production may also decline, leading to possible trophic implications that would result in reductions of fisheries productivity.

Jackson et al. (2001) noted how top-down fishing effects influenced coastal ecosystem organization. The authors pointed out how the removal of suspension feeders affected water quality. Wiltshire et al. (2010), working in the North Sea (Helgoland Roads) since 1962, documented changes of phytoplankton, salinity, Secchi disk depths, and macronutrients along with a pelagic time series of zooplankton, intertidal macroalgae, macrozoobenthos, and bacterioplankton data. In the late 1970s, water inflows from the southwest to the German Bight increased with corresponding increases of flushing rates. Salinity and annual mean temperature also increased since 1962, the latter by an average of 1.67°C. This has influenced seasonal phytoplankton growth, causing significant shifts in diatom densities and the numbers of large diatoms (e.g., *Coscinodiscus wailesii*). Changes in zooplankton diversity were accompanied by macroalgal community increases of green algae and decreases of brown algae species after 1959. Over 30 benthic macrofaunal species have been newly recorded at Helgoland over the last 20 years, with a distinct shift toward southern species. These detailed data provided the basis for long-term analyses of changes on many trophic levels at Helgoland Roads.

Day et al. (2013) gave a comprehensive review of the various aspects of global climate changes on the Gulf of Mexico. These included changes in temperature, sea-level rise, rainfall and river flow, and the frequency and intensity of storms. One general prediction of future changes in rainfall and river flow is that there will be drier conditions in the outer tropics and subtemperate zones with wetter conditions in the higher latitudes. This prediction would lead to lower rainfall in various parts of the Gulf of Mexico with the possible exception of the Mississippi drainage where increased flows may result from increased rainfall in mountainous regions of the drainage basin. It was predicted that these rainfall and river flow changes will be accompanied by increased temperature and rising sea levels. These conditions would have a largely negative effect on fisheries production of the gulf estuaries. One complicating factor in the evaluation of such impacts will be the multifaceted current and future effects of an array of anthropocentric activities in most of the gulf estuarine systems.

Long-term resilience of estuarine and coastal areas represents an important element in the responses of such systems to climate changes. Lotze et al. (2006) reconstructed time lines, causes, and consequences of change in 12 once diverse and productive estuaries and coastal seas worldwide. They found that anthropogenic impacts have depleted >90% of formerly important species, destroyed >65% of seagrass and wetland habitat, degraded water quality, and accelerated species invasions. Twentieth-century conservation efforts achieved partial recovery of upper trophic levels but have so far failed to restore former ecosystem structure and function. Dolbeth et al. (2007), working in the Mondego Estuary, found a system that had been under severe ecological stress mainly caused by eutrophication and the occurrence of macroalgal blooms with decreases of the area occupied by *Zostera noltii* beds. After the implementation of mitigation measures, naturally induced stressors such as strong flood events induced a drastic reduction of annual production, not seen before the implementation of those measures. The authors found that the resilience of the populations may have been lowered by a prior disturbance history (eutrophication) and consequent interactions of multiple stressors.

The ecosystem approach has been accepted by estuarine and coastal researchers in conceptual terms for a long time. However, the integrative nature of such efforts has proven to be difficult to achieve in many cases. The repetitious development of nonquestions, the so-called mu effect (Dayton, 1979), is an integral part of the ongoing lack of adequate ecosystem-level data in coastal ecosystems. The term "mu," a Japanese word, refers to questions that cannot be answered because they depend on incorrect assumptions. Without adequate background information, it is not always possible to differentiate the effects of immediate human impacts from natural (long-term) variability of a given system. If a given project is dependent on untested assumptions, there is little chance of success. The need for a broad range of interdisciplinary variables at the beginning of any such effort is obvious; reduction of such variables should be based on time-tested statistical analyses of factors such as covariability and the lack of associations with key processes. Ecosystem studies should be designed with a minimum of unsubstantiated assumptions.

Once again, the issue of adequate sampling for unanticipated questions becomes a factor in the success or failure of many field programs. Duarte and Piro (2001) related the "fragmentation of science" to a proliferation of scientific information that continues to grow at a rapid pace but whose compartmentalization (i.e., increased emphasis on the proliferation of unrelated facts) is poorly suited to address issues that require evaluation of interannual variations in basic ecosystem processes. Impacts at such levels can only be answered through a thorough understanding of the natural, long-term variation of different coastal systems and the response of various complex biological processes to ongoing, long-term anthropogenic impacts. Comparative analyses of the response of different coastal systems to state variables represent an important part of the understanding of ecosystem-level processes in any given system.

COASTAL ECOSYSTEMS

Estuarine and coastal ecosystems encompass a spatial biological continuum from upland, freshwater drainage basins to the offshore marine areas. These ecosystems are among the most temporally variable of all those on earth, with cyclic variations measured from seconds to decades. Estuaries are characterized by a combination of habitats with common physical, chemical, and biological attributes that are marked by abrupt changes in space and time. Each such system represents a unique combination of these habitats so that interestuarine comparisons are not always easy to make.

Many estuaries are also highly productive. However, primary productivity differs widely both within a given system and among various coastal areas. Despite the fact that estuarine areas are composed of common habitat types, there is considerable ecological individuality from estuary to estuary. The biological attributes of each ecosystem, determined by adaptive responses of indigenous species from phytoplankton to fishes, combine to form unique combinations of food web processes. Seasonal and interannual cycles of controlling factors, together with intermittent droughts and storm events, provide the highly variable background for the system-specific characteristics of estuarine areas. To add to this complexity is the evaluation of the relative impacts of multiple anthropogenic activities relative to natural responses to factors such as climatological variables that drive productivity and food web processes.

Within individual drainage basins, the complex combinations of habitats form highly productive systems that are interconnected physically, chemically, and biologically. High nutrient levels, multiple sources of primary and secondary production, depth relationships, sediments quality, energy subsidies from wind and tidal currents, and freshwater inflows combine to establish the high natural productivity of these coastal areas. Although the same general patterns of change occur in many temperate estuaries, the unique combination of habitats in different systems leads to broad variations of trophodynamic processes and secondary productivity in different coastal ecosystems (Livingston, 2000, 2002, 2007a,b). Most such systems combine both autochthonous and allochthonous sources of nutrients and organic matter that form the basis of high secondary production in the form of sports and commercial fisheries. Most of the estuarine-dependent species in gulf coastal regions spawn offshore with young-of-the-year populations migrating inshore to nursery in the highly productive coastal waters. Consequently, different combinations of controlling factors, together with specific patterns of inshore–offshore migration of marine forms and movements of euryhaline species, contribute to the area-specific biological characteristics of any given inshore gulf estuary.

The basic characteristics of a given estuary are determined, in part, by physiographic conditions (depth, surface area, connections to the open ocean) and freshwater input of dissolved and particulate substances (inorganic and organic) (Livingston, 2000, 2002, 2005). Seasonal and interannual cycles of temperature, rainfall, and river flows add a dimension of recurrent changes of primary and secondary productivity. The unique combination of habitat features to such conditions determines the area-specific ecological responses that are tempered by myriad adaptations of estuarine species. Quantitative and qualitative features of the primary producers drive coastal areas. Inputs from freshwater and saltwater wetlands (allochthonous and autochthonous), in situ phytoplankton productivity, and submerged aquatic vegetation all contribute to the structure and productivity of coastal food webs.

With the exception of the few species that spend their entire life cycles in such systems (i.e., shellfish, infaunal macroinvertebrates, some fishes), many commercially important estuarine fishes and invertebrates migrate into inshore coastal areas as larval stages and juveniles after spawning offshore. Many commercially important species, such as oysters, penaeid shrimp, blue crabs, and various finfishes, are adapted to rapid habitat changes. These species utilize the abundant food resources of coastal systems while remaining relatively free from predation by the stenohaline (salinity-restricted) offshore marine forms. Thus, coastal systems are often physically stressed but exist as a highly productive sanctuary for developing stages of nursing, offshore forms. Many such populations are used directly and indirectly by humans. Gradients of physical variables, together with food availability and utilization, and predator–prey interactions provide the basic components of the highly complex food webs of inshore marine ecosystems.

The rapidly changing physical conditions and intermittent (seasonal, interannual) cycling of organic production are important determinants of the exact form of the biological organization of diverse coastal areas. Area-specific combinations of such variables, together with biological components that vary continuously, determine the final food web structure of a given estuary. Such structure is determined by complex interactions of nutrient loading, corresponding primary productivity, interacting habitat conditions, and biological modifying factors such as predator–prey interactions and competition. Biological interactions, in response to gradients of salinity and primary producers, determine the specific characteristics of any given coastal system. With increasing salinity, population distribution becomes more even, with a corresponding decrease in relative dominance and an increase of species richness and diversity (Livingston, 2002). Seasonal changes of predation pressure as well as interannual recruitment characteristics modify this general condition at any given time. In such a way, coexisting estuarine assemblages reach different levels of equilibrium that depend largely on the exact temporal sequence of key habitat and productivity factors.

The high adaptability of coastal species often prevents direct (linear) relationships with specific physical/chemical driving factors. In this way, nonlinear (i.e., biological) processes contribute to what may often appear as chaotic conditions even though there is often an underlying organization that is grounded largely in the trophic responses of existing species to a continuous series of physical/chemical interventions. There is a family of cyclic time periods that have a bearing on how coastal systems fluctuate. The most common such cycles involve the monthly changes of tides and seasonal changes of salinity and temperature over annual periods. More complicated interannual variation is important even though such long-term changes are seldom included in the planning of ecosystem-level research. It is the issue of interannual variation that is related to the potential impacts of climate changes.

ECOSYSTEM RESEARCH

There are various definitions of ecosystem research. In this book, I will define this term as a combination of interdisciplinary studies carried out within adequate spatial and temporal boundaries. The spatial boundaries of such studies comprise a suite of assumptions that are often not definitive within the context of the stated objectives. In many cases, little effort is made to qualify the habitat composition of the study area so that station placement becomes somewhat arbitrary.

When the temporal components of a given ecosystem study are taken into consideration, there is an even greater sweep of assumptions where the length of time and sampling frequency are often determined by the available funding and the predilections of the researchers in charge of the project. In most publications of the results, there is usually relatively little attention given to the spatial and temporal aspects relative to the objectives and results of said studies. Even less definitive is the attention given to the quantification of the field-sampling methods used in such studies, especially with respect to the biological aspects of the field effort.

In many cases of ecosystem-level efforts, there is little recognition of the need for a critical review of the suitability of the data on which such studies are based. Conclusions and projections based on untested assumptions are thus subject to serious qualifications that eventually preclude the expansion of short-term observations to generalizations concerning the temporal variability of the subject area. Interannual variability is often ignored in the data analysis even though there is often ample evidence, even in short-term studies, that the results may not be representative of trends within the context of multiyear variation. The lack of background information that results from this weakness explains the general inability to evaluate the impacts of major events such as an oil spill in the Gulf of Mexico or processes affected by climate change in coastal and marine systems. Without even a basic understanding of within- and between-system differences in terms of responses to natural cycles of driving forces, effective impact evaluations are often not feasible.

The definition of the spatial and temporal boundaries of a given ecosystem remains an issue to this day in many studies. The very purpose of ecosystem research remains debatable. However, by whatever definition, the ecosystem paradigm remains the only viable approach to many resource management issues and impact analyses in coastal areas. The development of ecosystem-level research is not consistent with the usual hypothesis tests that drive most such approaches to environmental science. The purpose of ecosystem science should be based on the development of a factual basis for the evaluation of the complex interrelationships of natural and anthropogenic driving functions with eventual answers to the development of management initiatives. With the growing importance of climate issues, the need for a more systematic approach to this subject becomes necessary.

The designation "long-term," so often applied in publications related to aquatic ecology, is actually a subjective appellation that has no precise definition. There is a wide range of interpretations of the term in the scientific literature of coastal areas. The complexity of the temporal element of coastal systems is often oversimplified by investigators who rely on inadequate observational data. The difficulty of applying a temporal context for many forms of field-based information has been ignored by the vast majority of papers and books on the subject. It is generally assumed that the results of the usual 2–3-year study need not be interpreted within the context of longer climatological trends even though results of a study carried out during a drought period may differ considerably from those encountered during a period of high rainfall and river flows. Consequently, the quality of the data in terms of adequate sampling effort for impact analyses and associated modeling leaves much to be desired given the high level of temporal variability in most coastal systems.

There is a long list of problems that typifies the usual scope and effectiveness of ecosystem operations. The trends of increased specialization and the lack of researchers trained in carrying out interdisciplinary studies have contributed to the relative dearth of meaningful ecosystem studies. The lack of coordination among the various disciplines in terms of sampling effort is the rule here. The use of different sampling stations and varying sampling periodicities by researchers in different disciplines is usually the rule rather than the exception. There is often no effort to quantify the effectiveness of a given sampling effort. In many papers, there is virtually no evaluation of the quality and the quantity of the data on which the results are based.

Over recent decades, taxonomy has become a forgotten discipline despite the fact that knowledge of the actual species involved in a given study should be a fundamental concern. The substitution of easily sampled factors (i.e., chlorophyll *a*) for species identifications (i.e., phytoplankton) has led to oversimplifications of the biological organization in a majority of so-called ecosystem-level studies. Midstudy changes in sampling methodology and/or taxonomic identifications often add to the reduced quality of the data generated.

The design of ecosystem-level studies thus requires a very different set of priorities relative to the usual execution of funded ecological research. The choice of sampling factors at the beginning of a given study is crucial to the success of a given project. The need for adequate, well-timed sampling procedures at various levels of physicochemical and biological levels of organization is important. The high level of habitat variability of coastal systems complicates the execution of adequate sampling that, in turn, adds to the uncertainty of the research results. Natural interactions between the trophic organization of coastal systems and long-term cycles of climatological factors such as temperature and precipitation are mediated by continuous adjustments of untold numbers of interspecific trophic processes. The complex predator–prey relationships in coastal areas defy simplistic assumptions that are often accepted in the formulation of contemporary models and predictions of long-term changes in such systems.

The qualifications of field-sampling methods play an important role in the success of a given study. Livingston (1987b), using 3 years of weekly sampling and 12 years of monthly sampling of physical–chemical and biological (infaunal and epibenthic macroinvertebrates and fishes) factors in the Apalachicola Estuary (Florida), carried out an investigation of the spatial and temporal aspects of field-sampling processes. The results indicated spatial/temporal scaling problems (i.e., fitting a given research question to the dimensions of habitat variability of the study area) in the highly variable system. Specific limitations of weekly, monthly, and quarterly sampling intervals were directly related to the efficiency of the sampling gear, the range of variation in the study parameters, and specific biological features (motility, recruitment, natural history) of the biota. It was concluded that there are families of spatial and temporal scaling phenomena that should be considered when establishing a given field-sampling program. The dimensions of variation change along spatial/temporal gradients of salinity, habitat complexity, and productivity. The limits of variation define the needed sampling effort for a given biological level. Without an adequate evaluation of such variation, representative samples cannot be taken. The resulting inadequate sampling effort often precludes reliable comparisons between and among different systems.

The loss of valuable estuarine resources is favored by the lack of adequate scientific databases that are consistent with the dimensions of the individual study areas. There is a continuum of scaling dimensions (and sampling problems) that ranges from small-scale experimental approaches to system-wide analyses. Misapplication of such scaling estimates has led to overgeneralizations of the results of what has been called "long-term" studies. Unless field-sampling programs are scaled to the dimensions of the research problem and the study area in question, there will be a continued proliferation of trivial studies at one end of the continuum and the progressive and related deterioration of estuarine resources at the other. When placed within the context of what has been termed "climate change," the adequate determination of the interannual variation of a given ecosystem becomes a crucial factor in the evaluation of current impact(s) and future changes. More serious attention should be given to the aspects of definable limits of variability if there is to be a credible determination of the impacts of altered climatological factors on coastal, estuarine and marine systems.

The purpose of ecosystem research should include gaining an understanding of processes that can eventually be used to answer questions that have not been asked at the beginning of the program. Most ecosystem research ends up being defined by a *post hoc* accumulation of information that is often taken in a disparate manner based on untested assumptions with regard to the nature of both spatial and temporal variabilities. Accordingly, often used methodology for ecosystem research, in terms of both data collection and analysis, remains illusory when attempting to answer system-level questions that occur after the development of a given program of study. This problem makes it necessary to approach the multidisciplinary aspects of ecosystem research in a wholly different way than that which is usually used in ecological research.

The reductionist approach remains the dominant force in the funding and publishing of ecological information. This approach is based on very different questions than those with regard to long-term, system-level changes in coastal areas. The reductionist scientist attempts to piece together independent bits of information by subdisciplines to explain interdisciplinary interactions of diverse factors that follow different timescales. As a product of the reductionist method, the environmental sciences remain compartmentalized in terms of both funding for research and the publication of results. This has led to outstanding failures of science to cope with even the most mundane of human impacts. By ignoring the relatively complex background responses of coastal ecosystems to natural variation through the multiple temporal cycles of driving factors such as meteorological conditions and associated changes in habitat, there is little basis for identification of complex anthropogenic activities when they occur.

The relative lack of consistent, multidisciplinary data sets in coastal systems has been accompanied by an increased emphasis on models for predictive purposes. Because of the paucity of ecosystem-level data, these models are often based on false assumptions and are usually not field verified. Many popular models ignore even the most basic long-term habitat changes, natural and anthropogenic. The generally accepted system of funding and publication of field research is thus antithetical to the development of integrated, multidimensional, and long-term research efforts. Long-term research is generally shunned by researchers whose futures depend on timely publications. As pointed out by Duarte and Piro (2001) regarding ecosystem studies, "the resulting publishing difficulties are often quite damaging for the career development of the committed scientists. Interdisciplinary research is, therefore, penalized by the inadequate scientific mechanisms of funding and evaluation."

A comprehensive ecosystem approach requires a very different set of initial research questions than that of the usual hypothesis-testing process. Unfortunately, this approach precludes the *modus operandi* that most environmental scientists (who are largely trained as reductionists) follow when faced with the prospect of interdisciplinary research. The usual patch–quilt approach (patching together the results of diverse but limited short-term studies) taken by various funding agencies is not only antithetical to an ecosystem approach but contributes to the ongoing loss of coastal resources. The absence of adequate scientific data provides uncertainty that is welcomed by political and economic interests that are not really concerned with protection of natural resources. In this sense, the scientific community inadvertently assists in the loss of natural resources.

There are various ways in which to conduct a comprehensive, long-term field program that maximizes the results at the end of the study. Livingston (2005) outlined a different way to design and implement an ecosystem study than that usually undertaken today. The process should begin with the development of reasonable research questions concerning system interactions. System-level questions should be based on the implementation of quantitative sampling with attention to interdisciplinary coordination of physical, chemical, and biological variables. The research program should be based on broad-based preliminary studies. The generation of background data should be broad enough to provide the background for future (hitherto unasked) research questions. The research should start with a maximal sampling effort so that covariables can be identified and simplified at an early point in the field effort. In this way, variables should be reduced based on data rather than added during the study.

The research should include comprehensive ecological monitoring along with comparable experimental work that bases its hypothesis-testing questions from information taken in the field. The distinguishing characteristics of an effective ecosystem-level program are thus diametrically opposed to the usual pattern of carrying out a series of uncoordinated, unrelated studies by specialists who labor independently on their own limited field without any regard for the processes that drive their results. There should be a plan to integrate the field findings with those of other disciplines through adequate statistical methodology instead of the usual patching together of the disparate results of the various disciplines to answer limited questions that are usually unrelated to issues at the ecosystem level. This should include commonality of the spatial and temporal aspects of the sampling effort among the different disciplines.

Coordination of the field effort will depend on someone who is in charge of such coordination. The initial research effort should start with the delineation of the spatial/temporal dimensions of the system in question. The evaluation of differences between natural variation and anthropogenic impacts depends on an adequate sampling of the key habitat parameters in space and time. This will involve a precise and comprehensive definition of the basic processes that underlie system function over varying temporal periods. Most coastal systems are composed of factors that are common to disparate systems. However, any given river-estuarine/coastal system has very distinct ways in which these factors interact, with corresponding differences in primary production and food web architecture. The necessity of using common protocols for the field-sampling procedures is obvious if the data taken are to be useful in a comparison of different systems.

The scope of ecosystem boundaries in coastal areas is often limited to the upland drainage basins on one end and connections to the open ocean on the other. This scope also is within the realm of more general climatological features (e.g., global warming) that need to be taken into consideration. Because of the uniqueness of the processes that entail different combinations of state variables that drive individual aquatic systems, the final product of any given comparative analysis should also include a detailed review of distribution in space and time of habitats that are common to the systems in question. To achieve a truly system-level research program thus requires sophisticated levels of planning and execution of the sampling effort that are consistent with the spatial and temporal limitations of the study in question.

The following synopsis of objectives should be considered when designing an ecosystem approach to the design of an ecological research program:

1. There should be a preliminary determination of realistic goals for the long-term research program.
2. There has to be one person who is familiar with all aspects of the program and who can keep all of the elements consistent with the stated research goals. Research execution by committee is self-defeating in long-term, multidisciplinary projects.
3. A multidisciplinary review board should be established that includes representation of the various scientific and engineering subdisciplines, sports, and commercial fishing interests and economists who specialize in resource issues.
4. There should be a complete review of available scientific information concerning the system in question prior to the initiation of the project.
5. There should be a comprehensive determination of habitat distribution in the subject system. This effort should be carried out in conjunction with the development of maps of point and nonpoint sources of pollution to the system.
6. A catalog of anecdotal information should be gathered based on interviews with people who have lived in the system for extended periods or who have in some way had unique personal knowledge of changes in the system (i.e., fishermen). This effort can be supplemented by a comprehensive long-core sediment analysis to determine the past history of the system.
7. A central computer system should be developed by a team that includes database management people, computer programmers, statisticians, and modelers for the development of a capability for the analysis of long-term, multidisciplinary data sets. This team should carry out a continuous analysis of the data with input to the principal investigator and the review team.
8. Meteorological databases should be developed that include long-term rainfall/river flow and temperature information for the production of models based on long-term trends of atmospheric conditions in the basin.
9. There should be a determination of permanent sampling stations for long-term monitoring of water and sediment quality. The monitoring program should be based on a review of the results of a preliminary sampling program that stratifies habitat distribution in the system.
10. Nutrient/pollutant loading models should be developed based on various elements of the initial monitoring program.

11. There should be a development of an integrated physical, chemical, and biological monitoring program centered on key populations and communities as indicators of water/sediment quality. This should include all elements of biological organization with an emphasis on the main sources (allochthonous, autochthonous) of primary production and associated food webs of the system.
12. Spatial/temporal food web changes should be integrated with physical/chemical and loading programs (common stations, synoptic sampling regimes) so that the long-term changes of the system can be followed.
13. There should be an integration of all databases in a series of modeling efforts that include hydrological models, 3D water-quality models, water/sediment models, population/community changes, and food web models to answer questions of long-term spatial/temporal changes of system in question.
14. Management questions should be outlined, based on interviews with policy makers, environmental managers, scientists, and knowledgeable laypersons.
15. Quality assurance/quality control (QA/QC) reviews should be continuous and comprehensive. There should be periodic reviews of the overall program to ensure that it is consistent with established project goals.

When viewed from the perspective of the published literature on coastal and marine ecosystems, there is an obvious dearth of long-term, well-integrated, interdisciplinary studies. Biological sampling is often minimized with the lack of quantification of sampling techniques, which is usually the rule. Taxonomy is fast becoming a lost art. There is little attention to the fundamental aspects of trophic organization at the ecosystem level. The tendency exists whereby a series of short-term, isolated, studies defined by a devotion to individual disciplines are substituted for an integrated ecosystem-level approach that is carried out in an interdisciplinary fashion over time periods that can address long-term changes of key driving factors. The popular patch–quilt approach precludes comprehensive statistical analyses while providing inadequate data for effective modeling. The general lack of interest in comprehensive ecological studies by the scientific community has led to a situation whereby evaluations of events such as major oil spills and changes in climate are hampered by a lack of reliable information.

STUDIES IN THE NORTHEAST GULF OF MEXICO

This book is based on a series of long-term ecological studies of a series of river–bay systems along the NE Gulf of Mexico. These systems have been described in detail by Livingston (2000, 2002, 2005) and include some areas that have been almost completely free of human influences. Eastern sectors (Apalachee Bay, Apalachicola Estuary) have remained in a relatively natural state relative to disturbed river estuaries to the west (Choctawhatchee Estuary, the Pensacola Bay system, Perdido Estuary) where human activities have been associated with various forms of adverse impacts on coastal productivity.

Our food web studies have been used to evaluate the long-term changes of these systems with the development of trophic units as characteristic forms of ontogenetic feeding patterns that are characteristic of the biota of the region (Livingston, 2002). The structural organization of the river and bay food webs has been defined and quantified in terms of spatial distribution along gradients of freshwater runoff and temporal aspects of habitat changes and associated predator–prey alterations due to long-term changes of climatological factors and anthropogenic changes in nutrient loading along the coast.

This book will address the changes of food web structure relative to long-term trends of climatological conditions. In addition, there will be comparative analyses of how the trophic organization of different river–bay ecosystems responded to variations of both anthropogenic impacts and natural driving factors in space and time.

CHAPTER 2

Methods

The collection methods in this study were based on a preliminary set of analyses that were quantified and standardized so that the physical, chemical, and biological samples taken over the study period were comparable among the various study areas. All field collections were subjected to statistical analyses carried out early in the program (Hooks et al., 1975; Livingston, 1976a, 1987a,b; Livingston et al., 1976a,b; Meeter and Livingston, 1978; Greening and Livingston, 1982; Flemer et al., 1998) to provide quantitative field data. Detailed protocols for the sampling methodologies are given in Appendix A. These methods have been published in the peer-reviewed scientific literature that includes the following: Livingston et al. (1974, 1976b, 1977, 1997, 1998a,b); Livingston (1975a,b,c, 1976a, 1979, 1980a, 1982a, 1984a,b, 1985a, 1987b, 1988b, 1992, 1997a); and Flemer et al. (1998). Water-quality methods and analyses and specific biological methods have been continuously certified through the Quality Assurance Section of the Florida Department of Environmental Protection (QAP #940128 and QAP#920101).

Taxonomic verification of the field biological samples was continuous with the highest level of consistency through time. The team assembled for these studies included workers who remained with the project for periods ranging from years to decades. The head of the field program operated over the active period of field collections. In this way, the data taken were consistent with the effort to make intersystem comparisons over periods descriptive of interannual variation. Statistical methods used over the 43-year period of study are described in Appendix B. These methods were developed with respect to the various experimental and field descriptive methods used in the various study areas. Modeling efforts were established in ways that were consistent with the questions and sampling efforts associated with individual projects.

There was a major emphasis on the long-term and intersystem changes of the trophic organization of the study areas. Livingston (2002) gave a detailed description of the methods used in this part of the study. Most published analyses of trophic relationships remain limited in scope, usually determining feeding patterns of individual species or groups of related species. Feeding habits of coastal organisms have been studied in habitats such as mangroves (Odum and Heald, 1972), seagrass beds (Brook, 1977; Kikuchi and Peres, 1977; Stoner, 1979a,b; Livingston, 1982a,b), coral reefs Hobson, 1978; river-dominated estuaries (Livingston, 1997a,b), and the open ocean (Tyler, 1972; Ross, 1977, 1978). Early studies (Darnell, 1958, 1961) defined the "opportunism" of coastal fishes within the context of ontogenetic progressions of "distinct nutritional stages." Subsequent work has delineated distinct ontogenetic changes of basic dietary patterns in a number of invertebrate and fish species (Carr and Adams, 1973; Adams, 1976; Kikuchi and Peres 1977; Laughton, 1979; Sheridan, 1979; Sheridan and Livingston, 1979; Livingston, 1980a, 1982a, 1984a, 1997d; Stoner, 1980a,b,c; Leber, 1983, 1985, 2003). These data have been used to construct ontological trophic units.

The species concept remains the fundamental unit of systematists and, as such, is useful as a descriptor of basic natural history. However, from early papers (Thienemann, 1918) to more recent discussions (Peters, 1977; Livingston, 2000), there has been an underlying discontent with the largely theoretical association of the species with a distinct and stable niche. Lindeman (1942) proposed a more bioecological species-distributional approach based in large part on the trophic relationships within the "community unit." Ontogenetic patterns of feeding add another dimension of complexity to the study of coastal food webs. Many trophic studies and models are based on species-specific trophic designations (Peters, 1977; Stoner, 1979a, 1980c; Livingston, 1980a).

In the overwhelming majority of ecological papers concerning trophic relationships in coastal systems, the central focus is on the species. Where the basic life-history characteristics of the individual species are the subject of concern, such emphasis is appropriate. However, when complex fish and invertebrate assemblages in shallow coastal areas are the subject of the study, the use of the species as a unit of measure represents an oversimplification of the trophic organization (Livingston, 1988a). This is especially true in attempts to associate process-oriented relationships with forcing factors such as meteorological events, habitat variables, and/or biological interactions. Based on the results of the analysis of the trophic processes of gulf coastal fishes and invertebrates, the use of species-level data in the quantification of habitat relationships was not possible because of the complexity of the life-history stages as they relate to feeding relationships. Instead, the database was reorganized according to cross species groupings of like trophic units, thus making a more realistic grouping of the similar trophic levels in the subject study areas based on interspecific, ontological trophic units.

A systematic analysis of the feeding habits of the dominant fishes and invertebrates of the eastern Gulf of Mexico was carried out from 1972 through 2007. This analysis was based on a biological field-sampling program that was carried out in a standard fashion for all analyses over the extended study period. The identification of multispecies assemblages of similar ontological food habits allowed a more comprehensive understanding of the underlying ecological processes by relating the relevant biological units of organization to process-oriented factors. In this way, the databases for our studies were reorganized based on ontological trophic relationships rather than species-level considerations. For questions related to process-oriented trophic organization, there was a good reason to look at subspecific units based on life-history stages rather than the species as a whole.

Each of the animal groups (infauna, epibenthic macroinvertebrates, fishes) was sampled in the field using quantitative methods (core data for infauna, otter trawls, and trammel nets for fishes and epibenthic invertebrates; Table 2.1). The numbers of samples taken in the field were determined by carrying out species accumulation curves for the different sampling methods; the data were quantified and the numbers of trawl tows and cores were determined so that over 80% of the total number of species would be represented at each sampling (Livingston, 1976a, Livingston et al., 1976b). The fish, infauna, and invertebrate (FII) data were quantified to numbers/biomass m^{-2} (Figure 2.1).

All animals were measured and weighed, and species-specific regressions were run to calculate biomass (dry weight, ash-free dry weight) from the numbers/size data. These regressions were then used for biomass transformations of all field collections. The total numbers, biomass, numbers of species, and other derived variables (including the trophic information) were then calculated as FII (fish, infaunal macroinvertebrate, epibenthic macroinvertebrate) indices. All such data were calibrated on a m^2 basis. Thus, the FII index was created for each system that represented the various indices based on numbers or biomass of all macroinvertebrates, infaunal invertebrates, and fishes in a square meter from the surface to 10 cm deep in the sediments (Figure 2.1).

As an example of the development of the trophic unit concept, a generalized outline of the determination of the ontogenetic feeding patterns of pinfish is given in Figure 2.2. This development has been outlined in a series of studies of the pinfish (*Lagodon rhomboides*), a numerical dominant in shallow inshore waters of the NE gulf (Stoner, 1979a,b, 1980a,b,c; Livingston, 1982a, 1984b, 2002; Stoner and Livingston, 1984). There was a well-ordered progression of changes in

Table 2.1 List of General Food Types Found in Fish Stomachs along with Codes Used to Describe Them in the Presentation of the Data

AM	Amphipods	HO	Holothuroidea
AN	Annelids	HY	Hydracarina
AP	Appendicularians	IE	Invertebrate eggs
AR	Animal remains	IL	Insect larvae
BA	Barnacles (ad, juv)	IN	Insects
BI	Bivalves (ad, juv)	IP	Insect pupae
BN	Barnacles (larvae)	IS	Isopods
BR	Branchiopods	MA	Microalgae
BZ	Bryozoans	MS	Miscellaneous
CA	Cambaridae	ML	Molluscan larvae
CD	Cephalochordates	MY	Mysids
CE	Cephalopoda	NE	Nematodes
CH	Chaetognaths	NM	Nemerteans
CI	Ciliophora	OL	Oligochaeta
CL	Cladocerans	OP	Ophiuroids
CN	Cnidaria	OR	Organic remains
CO	Copepods (calanoid, cyclopoid)	OS	Ostracods
CR	Crabs (ad, juv)	PL	Polychaetes (larvae)
CS	Crustacean remains	PO	Polychaetes (ad, juv)
CT	Chitons	PR	Plant remains
CU	Cumaceans	PY	Pycnogonida
DE	Detritus	RA	Radiolarians
DI	Diatoms	RP	Reptilia
DL	Decapod larvae	SA	Sand grains
EC	Echinoderms	SH	Shrimp (ad, juv)
EG	Egg cases	SI	Sipunculids
FE	Fish eggs	SP	Sponge matter
FL	Fish larvae	ST	Stomatopods
FO	Foraminiferans	SY	*Syringodium filiforme*
FP	Fecal pellets	TA	Tanaids
FR	Fish remains (ad, juv)	TH	*Thalassia testudinum*
GA	Gastropods	TM	Trematodes
HC	Copepods (harpacticoid)	TN	Tunicates

Note: "Plant remains" represents living plant matter and "detritus" represents dead organic matter. "Miscellaneous" (MS) represents all items that comprise <3% of the total mass.

food preferences of this species. In the unpolluted Econfina system, young-of-the-year recruits from offshore spawning grounds (<20 mm SL) were planktivorous, feeding mainly on calanoid and cyclopoid copepods and fish eggs. With growth (21–35 mm SL), there was gradual transition to benthic carnivory in the form of amphipods, mysids, harpacticoids, and other small animals. With growth to 60 mm SL, there was an increased preference for amphipods, shrimp, plant remains, and detritus in the pinfish diet. Plant matter consisted largely of microalgae. Fish ranging from 61 to 120 mm SL then fed on larger benthic invertebrates such as crabs, shrimp, and bivalve mollusks; plant matter changed with such growth from microalgae to pieces of *Syringodium* and *Thalassia*. Pinfish that exceeded 120 mm SL generally fed on plant matter in the form of *Syringodium* and *Thalassia* and were largely herbivorous in the Econfina system. This species thus passed through a planktivorous phase with a shift through benthic carnivory to two distinct stages of omnivory and ended in herbivory.

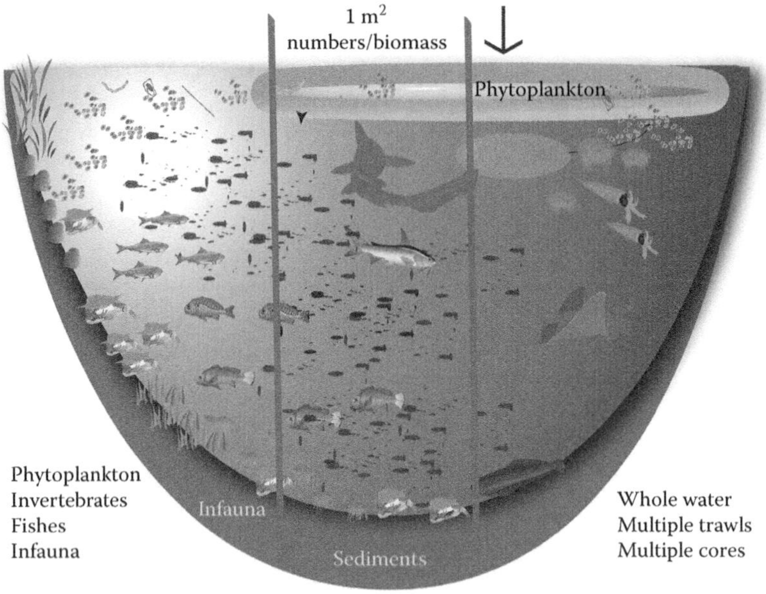

Figure 2.1 **(See color insert.)** Field sampling for phytoplankton, infaunal and epibenthic invertebrates, and fishes for studies in the eastern Gulf of Mexico. Detailed sampling protocols are given in Appendix A.

There was a spatial pattern to the differentiation of pinfish feeding habits between the polluted and unpolluted estuaries that was based, in part, on the distribution of grass beds. Plant biomass, dominated by *Thalassia* in the Econfina Estuary and *Syringodium* in the Fenholloway Estuary, showed divergent trends over the 7-year period of observation. The patterns of pinfish feeding were similar in both the inner and outer areas of the Econfina system over the study period. These patterns were quite different in the Fenholloway nearshore waters where submerged aquatic vegetation was reduced and even eliminated due to the impact of mill effluents. The reduction of pinfish in the polluted estuary was attributed in large part to the reduction of available habitat and food. There was a general decline of plant biomass in the Econfina system during the 1970s, while there were indications of a limited recovery in peripheral areas of the Fenholloway Estuary during this period.

In this way, the food web determinations were made for each of the estuaries under study in an effort to determine the relationships of the trophic progressions of the infauna, epibenthic invertebrates, and fishes to natural and anthropogenic driving factors.

All biological data (as biomass m^{-2} mo^{-1} of the infauna, epibenthic macroinvertebrates, and fishes) were transformed from species-specific data into a new data matrix based on trophic organization as a function of ontogenetic feeding stages of the species found in the subject bay areas over the multiyear sampling program (Figure 2.3). Infaunal macroinvertebrates were organized by feeding preferences based on a review of the scientific literature (Livingston, 2002). Ontogenetic feeding units (organized by size and by species) were determined for epibenthic invertebrates and fishes in the nearshore gulf regions that included the following: the Apalachicola River and Bay system (Sheridan, 1978, 1979; Laughlin, 1979; Sheridan and Livingston, 1979, 1983), the Apalachee Bay (Econfina and Fenholloway drainages, Livingston, 1980a, 1982a, 1984b, unpublished data; Stoner and Livingston, 1980, 1984; Laughlin and Livingston, 1982; Stoner, 1982; Clements and Livingston, 1983, 1984; Leber, 1983, 1985), and the Choctawhatchee River and Bay system (Livingston, 1986a,b, 1989b). It was determined that an adequate sample for a given trophic designation (size class, species, location, time of sampling) required processing at least 15 fish or invertebrate stomachs (Livingston, unpublished data). For example, 4129 blue crab (*Callinectes sapidus*) stomachs were processed from

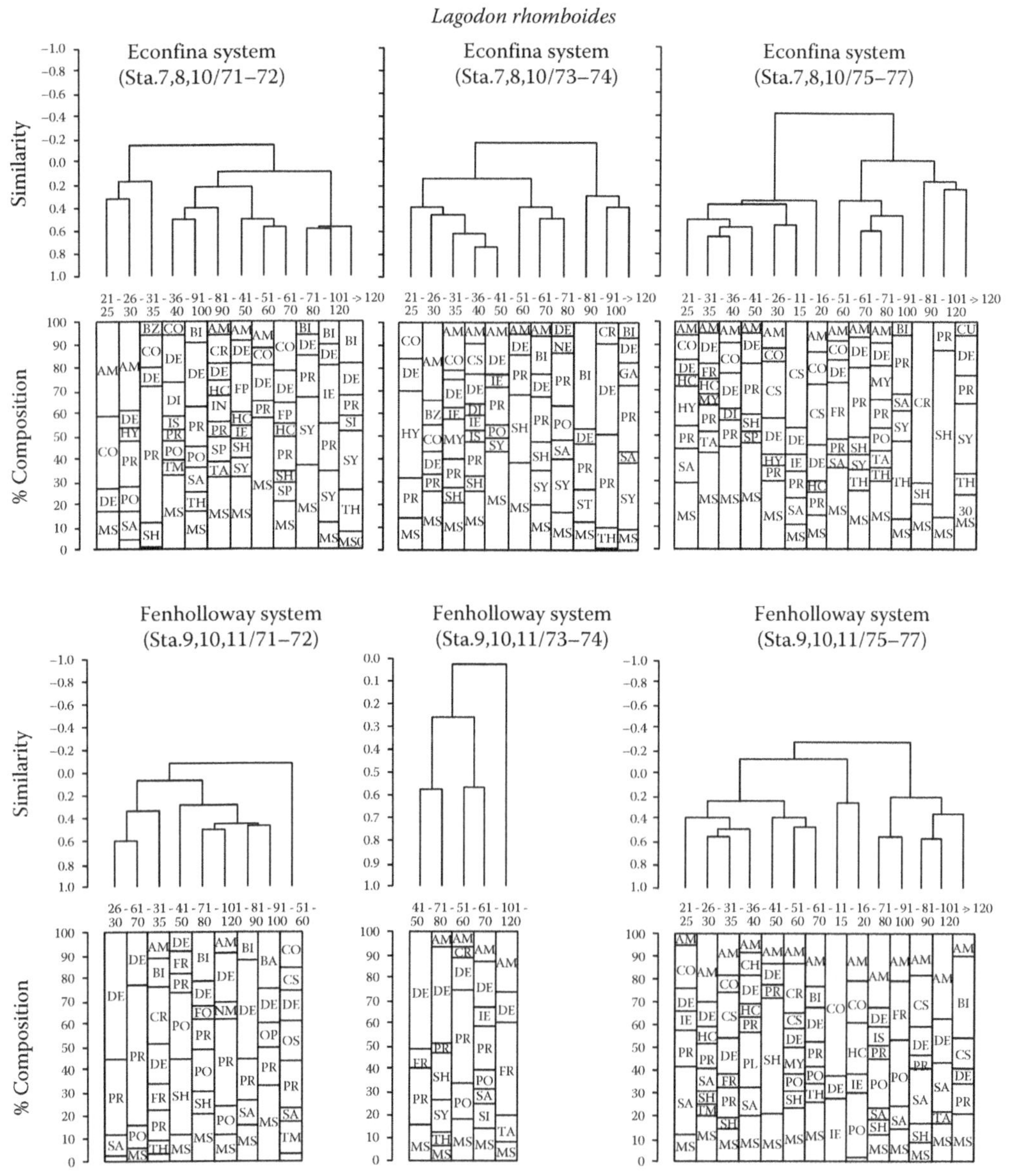

Figure 2.2 Ontogenetic changes in the diet of *L. rhomboides* taken in the Econfina and Fenholloway coastal areas from 1971 to 1977. Histograms represent the relative proportions of the major dietary components (dry weight), and dendrograms represent cluster analyses of the diet similarity among size classes. Codes for the food items are given in Table 2.1.

the Apalachicola system (Laughlin, 1979). Other representative species analyzed included penaeid shrimp (*Farfantepenaeus duorarum*; 1115 stomachs), pinfish (*L. rhomboides*; 4915 stomachs), spot (*Leiostomus xanthurus*; 4091), and pigfish (*Orthopristis chrysoptera*; 8634).

The trophic organization of infaunal and epibenthic macroinvertebrates and fishes in the gulf–bay systems is given in detail by Livingston (2002). The field data (infauna, epibenthic invertebrates, fishes) that were taken in all systems were reordered into trophic levels so that monthly changes in the overall trophic organization of the various systems could be determined. The long-term field data from the subject estuaries were used to establish a database based on the dry biomass and total

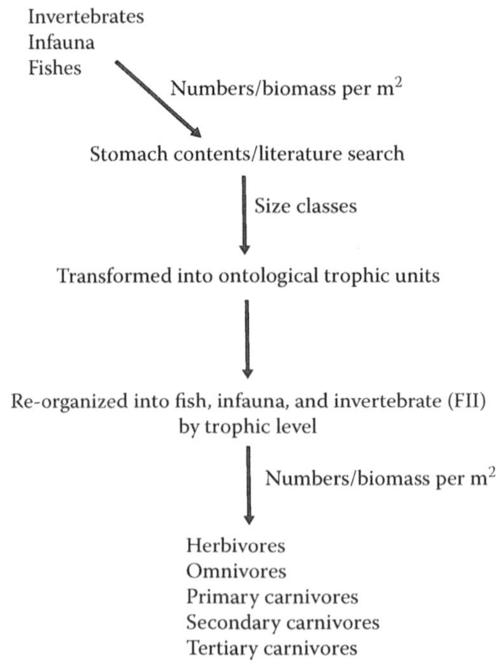

Figure 2.3 Steps taken to transform the FII database into a trophic reorganization of the field data.

ash-free dry biomass m^{-2} mo^{-1} as a function of the individual feeding units of the infauna, epibenthic macroinvertebrates, and fishes (Figure 2.3).

The data were summed across all taxonomic lines (FIIs) and translated into the various trophic levels that included herbivores (feeding on phytoplankton and benthic algae), omnivores (feeding on detritus and various combinations of plant and animal matter), primary carnivores (feeding on herbivores and detritivorous animals), secondary carnivores (feeding on primary carnivores and omnivores), and tertiary carnivores (feeding on primary and secondary carnivores and omnivores) (Figure 2.3). All data were given as dry biomass and ash-free dry mass m^{-2} mo^{-1} or as percent dry biomass and ash-free dry mass m^{-2} mo^{-1}. Details of this reorganization are given by Livingston (2002). Unless otherwise indicated, the data are presented as monthly means of stations or groups of stations in each system.

In this way, the long-term databases of the field collections of infauna, epibenthic macroinvertebrates, and fishes in all systems were reorganized into quantitative and detailed trophic matrices based not solely on species (Livingston, 1988a) but on the complex ontogenetic feeding stages of the various organisms in the subject estuaries. The data, now called the FII index, represented the biomass of all macroorganisms in a square meter of water and sediments (10 cm deep) at each station or station group per month or unit time in each bay system. These FII data were also transformed into the various natural history and trophic attributes of the joined databases.

A comprehensive diving survey was carried out in each of the study areas prior to the design of the sampling program. Physical, chemical, and biological samples were taken that, together with the diving results, were then used to stratify the habitats in the subject coastal areas. Based on the distribution of the individual habitat types, a sampling program was then designed to make sure that the different habitats were adequately sampled in a representative fashion. Within each major habitat, randomized, repetitive samples were taken that allowed a quantitative evaluation of factors in terms of depth and spatial distribution.

PART II

Long-Term Habit at Conditions

CHAPTER 3

Regional Background

The estuarine and coastal areas along the Gulf of Mexico coast are characterized by a series of habitats that are controlled to a considerable degree by the physiographic features of the adjoining freshwater drainage basins and associated gulf connections (Figure 3.1). The coastal zone and associated freshwater drainages of the north Florida area extend from the Perdido River and Bay system in eastern Alabama and the western panhandle of Florida to the small blackwater rivers on Apalachee Bay. Nine of the twelve major rivers and five of the seven major tributaries of Florida occur in this region. The upland watershed of the north Florida Gulf Coast, which is located in Alabama, Florida, and Georgia, includes about 135,000 km^2. The combined area of the inshore estuarine systems in the region approximates 2,150 km^2. The coastal zone in this area is characterized by saltwater marshes, sandy beaches, tidal creeks, intertidal flats, oyster reefs, seagrass beds, subtidal unvegetated soft bottoms, and a broad assemblage of transitional habitats.

Many of the inshore habitats of the major bays and estuaries are dominated by freshwater runoff from associated river basins and, to a lesser degree, groundwater contributions. The salinity regimes of these areas are variously affected by major river systems or, as is the case in Apalachee Bay, by a series of small rivers and groundwater flows. The combined drainage of springs and streams contributes about 1 billion gallons of freshwater per day to Apalachee Bay, which is the only study area not affected by a major river system (Livingston, 1990a,b, 2000). The primary rivers of the northwest Florida Panhandle (Perdido, Escambia, Choctawhatchee, Apalachicola) have their headwaters in Georgia and Alabama (Figure 3.1). A series of smaller streams along the panhandle coast include the Blackwater and Yellow Rivers of the Pensacola Bay system, the Chipola River (part of the Apalachicola drainage), and the Ochlockonee River on Apalachee Bay. Farther down the coast, a series of small streams (St. Marks, Aucilla, Econfina, and Fenholloway) with drainage basins in Florida flow into Apalachee Bay.

The Perdido Estuary (Figure 3.1) is a relatively shallow inshore embayment characterized by a dredged connection with the offshore gulf, Perdido Pass. The bay complex lies in an area of submergence on the north flank of the active gulf coast geosyncline. The Pensacola Bay system (Figure 3.1), a complex of three component bays with a restricted connection to the gulf, is dominated by sediments provided by stream discharge (mainly the Escambia River) and wave/current action from the gulf (Wolfe et al., 1988). Santa Rosa Sound has generally coarser sand-dominated sediments derived largely from offshore areas. The Choctawhatchee Estuary (Figure 3.1) occupies a Pleistocene river valley that has been recently enclosed by a barrier system. Until the late 1920s, the bay was connected with the gulf via a shallow pass and, as such, was periodically a fresh to brackish water lake. However, with the human creation of East Pass in 1929, Choctawhatchee Bay became a highly stratified estuarine system dominated by freshwater input from the Choctawhatchee River flowing over relatively salty gulf water.

The Apalachicola Estuary (Figure 3.1), bounded gulfward by three barrier islands, is an extension of the Apalachicola River. This barrier system is a shallow width-dominated coastal plain

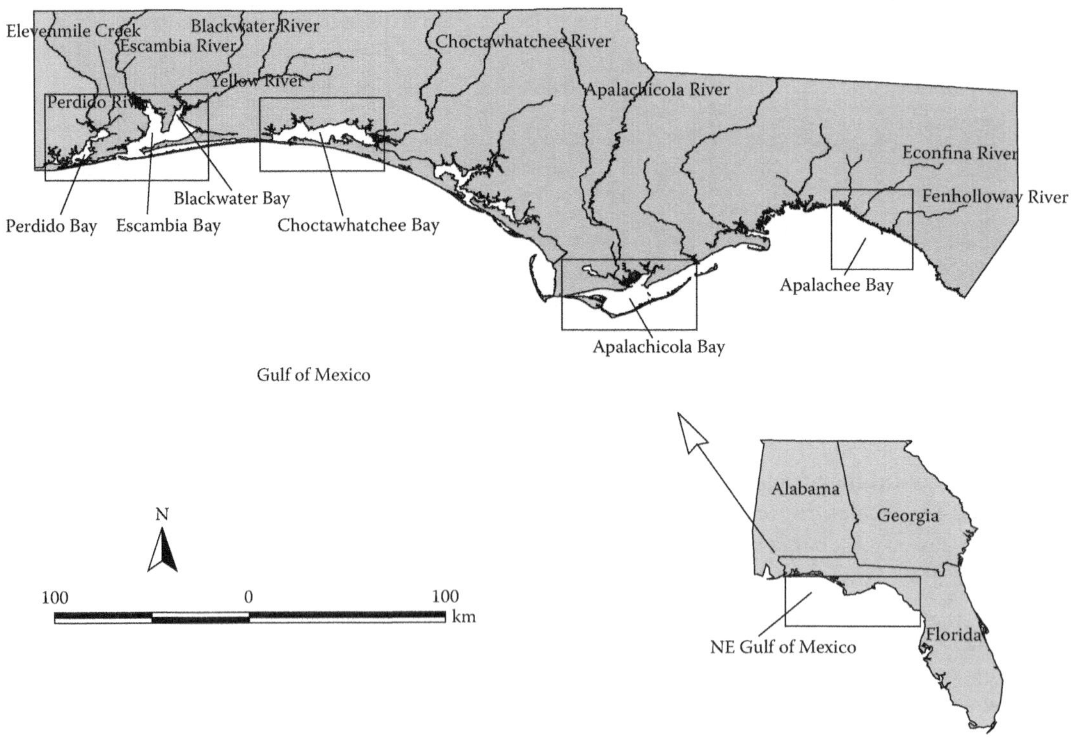

Figure 3.1 Rivers and bays in the NE Gulf of Mexico. The Florida Geographic Data Library (FGDL) provided the geographic data.

estuary (Livingston, 1984a). Sand, silt, and shell components that have accumulated over tertiary limestones and marls of alluvial origin with some recent modifications due to hurricanes dominate sediments in this estuary. The region along the upper gulf coast of peninsular Florida from the Ochlockonee River to the Suwannee River is characterized by a series of drainage basins that include the Aucilla, Econfina, and Fenholloway Rivers. These streams drain into Apalachee Bay (Figure 3.1). The smaller basins are wholly within the coastal plain as part of a poorly drained region that is composed of springs, lakes, ponds, freshwater swamps, and coastal marshes. The Econfina and Fenholloway River Estuaries both originate in the San Pedro Swamp. This basin has been affected, in terms of water flow characteristics, by long-term physical modifications through forestry and pulp mill activities. Apalachee Bay is part of a broad, shallow shelf area that occurs along the gulf coast of peninsular Florida.

Fernald and Patton (1984) and Wolfe et al. (1988) give comprehensive reviews of the physiography of the study area. The land region from Apalachee Bay to Perdido Bay is one of the least developed coastal areas in the continental United States. Except for the most populous areas such as Pensacola, the degree of urbanization and industrialization is low resulting in generally low levels of pollution especially in eastern parts of the study region (Figure 3.2). The Pensacola area, bordering the Perdido and Pensacola Bay systems, is the center of industrial development, whereas Destin and Panama City are mainly characterized by highly developed urbanization. Coastal regions farther east, from Apalachicola to the so-called Big Bend area (Apalachee Bay), have extremely low human populations with virtually no industrial or agricultural development. The primary economic activities in this region are forestry and sports and commercial fishing. There is a pulp mill on the Fenholloway River, the only major source of pollution on the largely unpolluted Apalachee Bay.

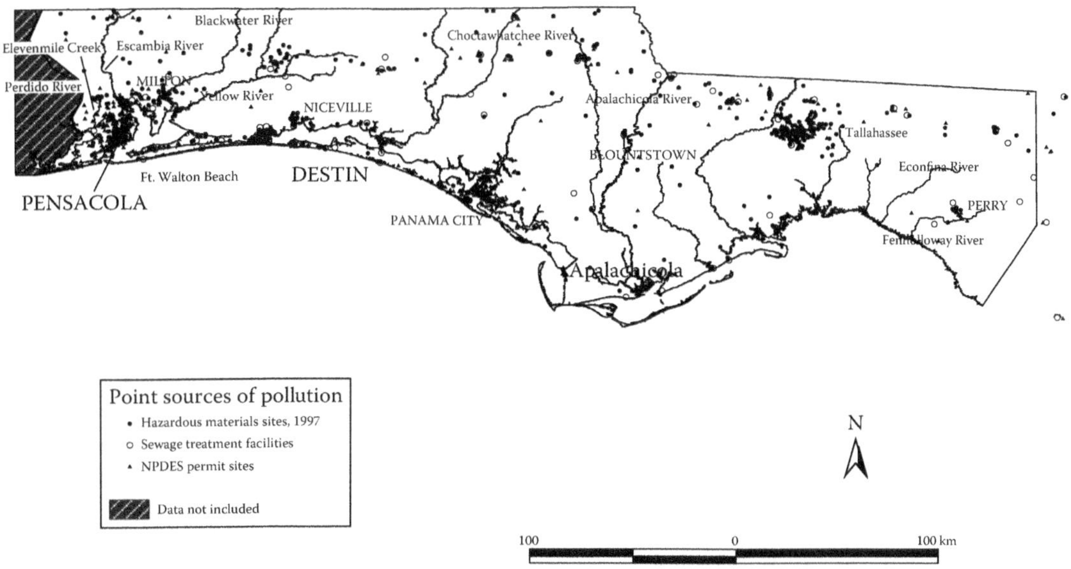

Figure 3.2 The Florida Panhandle showing the distribution of point sources of pollution. These sources generally follow the distribution of the human population of the region with coastal centers at Pensacola, Destin, and Panama City in the western panhandle region. (This figure is a composite of information provided by the National Oceanic and Atmospheric Administration, the Florida Department of Environmental Protection, and the ESRI Corporation.)

Much of the panhandle landscape is the result of stream and river flows and wave action that has acted on the land surface over the past 10–15 million years (Fernald and Patton, 1984; Wolfe et al., 1988). From the Pleistocene period to the recent (about 2 million years), there have been major changes in sea level that have greatly affected the land types in the panhandle region. During the periods of lowered sea levels in the Pleistocene, the present-day shallow areas were terrestrial resulting in an increase of 1.5–2 times the terrestrial acreage of the panhandle coastal regions. This has resulted in a current coastal physiography that is dominated by beach ridges (marine in origin), barrier islands, spits, cliffs, swales, sloughs, dunes, lagoons, and estuaries along a relatively flat upland configuration. Barrier islands start in the eastern end of the Apalachicola Bay system and include a series of such islands through Apalachicola Estuary and west to the Pensacola and Perdido Estuaries. Brackish water lagoons are common in the panhandle region and include St. George Sound (Apalachicola Estuary) and Santa Rosa Sound (Pensacola Bay system). On the eastern end of the panhandle coast, there is a massive development of coastal swamps and marshlands that rim Apalachee Bay. This area is characterized by the absence of barrier islands as a result of the shallow, sloping margins, the lack of wave action, and an inadequate supply of sand (Tanner, 1960).

The geomorphology of the Florida Panhandle is complex with strata that range in age from 450 million years ago (Paleozoic) to the present (Fernald and Patton, 1984). Wolfe et al. (1988) have reviewed the geology of the panhandle region. The surface geology of the Florida Panhandle is sedimentary, with limestones, organics, and clastics (gravel, sand, silt, clay) forming the primary sediment types. The western panhandle coast from the Perdido system to the Ochlockonee drainage is dominated by medium-to-fine sand and silt, whereas the Big Bend area (Apalachee Bay) is predominantly limestone or limestone/dolomite (Fernald and Patton, 1984). In the eastern half of the panhandle, ground surfaces are affected by surface limestones and solution activity. Only those predominantly marine deposits of the past 60 million years are important for groundwater resources. Sandy clays dominate northern portions of northwestern Florida or clayey sands deposited by ancient rivers, whereas sands deposited along the constantly changing shorelines characterize southern areas. The panhandle area has been emergent since the Miocene; however, about half of the open gulf portions of the beach areas have been eroding due to postulated sea-level increases (Wolfe et al., 1988).

The geology of the northern gulf coast of Florida is dominated by four structural features: the Gulf of Mexico Sedimentary Basin (west of Choctawhatchee Bay), the Chattahoochee Anticline (a rise at the eastern end of the panhandle), the Apalachicola Embayment (from Alligator Point south of Tallahassee, Florida, to the Choctawhatchee drainage), and the Suwannee Straits (in the Big Bend area of north-central Florida) (Tanner, 1960). The Apalachicola Delta lies at the center of the Apalachicola Embayment, which is a relatively shallow basin that is characterized by the ground surface exposure of Oligocene and Eocene carbonate rocks. These carbonate deposits substantially affect the physical and chemical properties of water in this region. Offshore of this area lies the North Gulf Coast Sedimentary Province, whereas the Florida Peninsula Sedimentary Province underlies the Apalachee Bay region.

The northwest Florida region is considered to be south temperate with mild temperatures and a humid atmosphere (Fernald and Patton, 1984; Wolfe et al., 1988). Mean annual temperatures in the region approximate 20°C, with some variation depending on elevation and proximity to the coast (Livingston, 1984a). The coldest months occur from December through February. Summer temperatures have mean values ~30°C with peaks reaching 32°C–33°C, usually in August. Coastal waters closely follow changes in air temperatures due in large part to the shallowness of the inshore areas and the highly active conditions resulting from tidal currents and wind-driven water movements.

The mean annual rainfall in the region approximates 152 cm, which ranges from 163 cm in western areas to 142 cm in the east. Rainfall along the northeast gulf coast of Florida is usually bimodal on an annual basis, with a major peak during summer–early fall months (June–September) and a secondary peak during winter–early spring periods (January–April). The Florida winter peak is shorter in duration and magnitude than the summer increases (Livingston, 1984a) although this pattern varies from year to year. A seasonal drought period usually occurs during October and November with a lesser one in April and May. Georgia rainfall has two peaks, summer and winter. The Georgia rainfall pattern is different from that of Florida in that both Georgia peaks are more or less equal. Periodic heavy summer rainfall is often associated with storm activity. Winter rains usually occur with frontal systems from the north and are generally of longer duration than the summer storms. There is also the heavy rainfall that is often associated with hurricanes whereby large amounts of precipitation and associated flooding occur during relatively short periods of time.

Gulf air circulation is mainly anticyclonic around a high-pressure region during periods from March through September (Wolfe et al., 1988). During the winter, air masses of continental origin move through the north Florida area. Winds tend to be highest during the months of January through March. In shallow estuarine areas, winds can have a major influence on tidal current patterns. Such effects are due to the orientation, physiography, and spatial dimensions of a given river–estuarine or coastal system. Southerly winds tend to augment the astronomical tides, whereas northerly winds can increase substantially the effects of a major tidal efflux (Livingston, 1984a). Hurricane occurrence in the northeast gulf is relatively frequent. During the period from 1885 to 1985, about 48 storms came ashore in the panhandle region. Relatively little hurricane activity has occurred along the northwest peninsula in Apalachee Bay (Fernald and Patton, 1984).

In major parts of the Florida Panhandle region, the groundwater and surface water bodies are often directly connected by rock channels (Wolfe et al., 1988). The karst topography that is so prevalent in this area is characterized by porous limestone that is penetrated by various solution channels and sandy substrates. The connections to the surface are typically in the form of sinkholes and springs, which can act as direct connections between the surface waters and groundwater aquifers. Rainwater drains rapidly into the surficial aquifer, which is composed mainly of coarse-to-fine sands intermixed with various forms of surface gravels. There is considerable variability in the freshwater fluxes in the various subbasins of the north Florida region (Livingston, 1984a, 1989b).

Most of the major river systems are composed of a series of tributaries that drain the various subbasins into a main stem. The subbasins often follow differing hydrological cycles (Livingston et al., 1977) than the main stem, depending on regional rainfall patterns and the highly variable physiographic conditions in the different regions. Some of the drainage basins of the various north Florida areas extend into Alabama and Georgia, and the temporal patterns and volumes of river discharge to associated estuaries depend on the cumulative climatological conditions and seasonally varying evapotranspiration rates in the respective watersheds. Winter–spring flooding in the alluvial streams is a product of reduced evapotranspiration rates during winter rainfall peaks along the heavily vegetated drainage basins (Livingston, 1984a). The reduced surface flows during summer periods are due to increased evapotranspiration in vegetated floodplains, even though the Florida rainy season usually peaks during this period. Surface flow to the gulf is minimal during the fall drought period. Major rivers such as the Perdido, Apalachicola, and Choctawhatchee have flow patterns that are more correlated with seasonal rainfall patterns in Georgia and Alabama (where high percentages of the respective drainage basins are located) than in Florida. Nearshore coastal regions are affected by both surface and groundwater runoff.

Astronomical tides in the western panhandle from the Apalachicola to the Perdido systems are diurnal, having relatively small amplitudes that vary between 0.37 and 0.52 m (Livingston, 1984a). Tides from Apalachee Bay to the Apalachicola Estuary are mixed and semidiurnal, having two (unequal) highs and lows each tidal day (24 h and 50 min). Tides in this region range from 0.67 to 1.16 m. The entire coastal region in the NE Gulf of Mexico area is characterized by low wave energy conditions (1.96–12.25 J m^{-2}) (Tanner, 1960).

Habitat conditions in the gulf coastal areas of north Florida are determined by different factors. The barrier estuaries of the panhandle section of Florida are associated with a series of rivers from the Perdido to the Apalachicola. Wind tends to be a major factor in the determination of water circulation and salinity in most of the shallow gulf coastal estuaries (Livingston, 1984a). In areas where river flow is a factor, frictional effects on the saltwater regimes are combined with salt-wedge circulation and geostrophic forces to affect the estuarine salinity distribution. The physical dimensions of a given river–estuarine basin comprise a key element in the determination of the distribution and nature of aquatic habitats. The size and depth of a given bay along with the location of passes to the gulf are important factors in stratification potential and salinity distribution. Thus, freshwater influxes and basin characteristics such as depth and areal dimensions determine the spatial and temporal habitat characteristics of the gulf coastal estuaries of north Florida.

The large river estuaries of the Florida Panhandle are dominated by freshwater input from the primary rivers of the region. River flow patterns determine important features such as salinity, color and turbidity, primary productivity, and the diverse blend of associated fisheries. Submerged aquatic vegetation (SAV) is restricted to areas less than 1 m deep in most of these estuaries. Much of the bay bottom in such systems is composed of unvegetated mud and oyster bars. High phytoplankton productivity and imported organic matter from the river contribute to the sedimentary microbial activities that form the basis of the detritivorous food webs (Livingston, 1984a).

In shallow estuaries such as the Apalachicola, wind and tidal disturbance of the shallow system enhance secondary production. The particulate organic carbon is then utilized by a seasonal succession of euryhaline dominants. This results in some of Florida's most productive inshore systems that are typified by the Apalachicola Estuary. Apalachee Bay, on the other hand, receives less surface freshwater runoff than the alluvial systems. Seagrass beds dominate the clear, nutrient-poor waters of Apalachee Bay with phytoplankton as a secondary source of organic carbon. Salinity is generally higher here; species richness and diversity are high, and relative population dominance remains low. In this case, the seagrasses have provided both the major habitat and the most important source of organic matter for the inshore areas.

Livingston (1990b, 2000, 2002, 2005) gave a comprehensive review of the habitats of the coastal areas of the northwest coast of Florida.

CHAPTER 4

Rainfall and River Flows

Long-Term Changes

River flow and nutrient concentration/loading data were compiled over 2-year periods for the various gulf river systems. An attempt was made to choose periods that were close to nondrought conditions in each system. A comparison of river flows is shown in Figure 4.1. The Apalachicola River system had the greatest flow rates of all the systems analyzed. Escambia and Choctawhatchee River flows were comparable. The Perdido River (including the Alabama Blackwater River and the Styx River) had the lowest flows of the major rivers of the region. In most cases, peak flows in these high-flow streams occurred during winter–spring periods, and seasonal drought periods were usually noted during fall periods. The Fenholloway River, the Econfina River, and the Elevenmile Creek (Perdido Estuary) had comparable, relatively low-flow rates with winter–spring peaks. The Econfina and Fenholloway Rivers had minor summer peaks reflecting the fact that their drainage basins were entirely within the state of Florida.

Regressions were run based on the association of Apalachicola River flows with flow rates of the various other river systems of the eastern gulf (Table 4.1). There were significant ($P < 0.05$) relationships between the Apalachicola River flows and those of the other rivers in the study area. However, the R^2 values varied widely. The association of the Apalachicola with the Econfina River was the lowest of all rivers surveyed, a result that could be expected since the Econfina drainage is restricted largely to local flow rates in Florida relative to the alluvial Apalachicola River that has its origins in Georgia and Alabama. The R^2 values were relatively high in the various big-river systems with gradients consistent with distances from the Apalachicola drainage; the Perdido River had the lowest such association with the Apalachicola River.

The region of study (Figure 1.1) is subject to certain commonalities with respect to rainfall that is usually bimodal on an annual basis, with a major peak during summer–early fall months (June–September) and a secondary peak during winter–early spring periods (January–April). There is a trend toward increased precipitation in coastal areas of the gulf coast of northern Florida when compared to that of inland areas. During dry years, however, the maximum rainfall occurs inland (Wolfe et al., 1988). There is, however, considerable spatial variability in rainfall patterns, with distinct differences along relatively short distances.

Some drainage basins (Perdido, Apalachicola) extend into Alabama and Georgia, and the temporal patterns and volumes of river discharge to associated estuaries depend on the cumulative climatological conditions and seasonally varying evapotranspiration rates in associated watersheds. Meeter et al. (1979) reviewed the long-term climatological changes in the Apalachicola River Basin. A spectral analysis of long-term trends of Apalachicola River flow and Georgia rainfall indicated drought–flood cycles of 6–7 years during periods up to the 1970s. Cross-spectral analyses of these two variables showed that Apalachicola River flow and precipitation in Georgia were in phase. This is similar to the pattern noted in the Perdido River. Winter–spring flooding in the large rivers is a product of reduced evapotranspiration rates during winter rainfall peaks along the heavily vegetated drainage

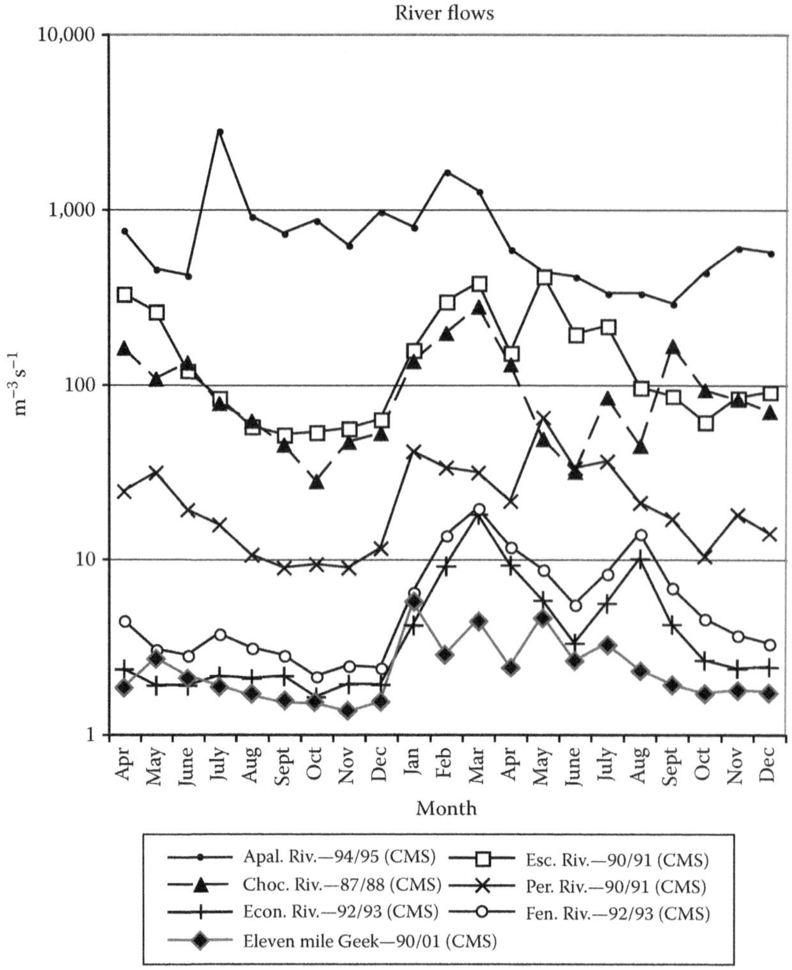

Figure 4.1 River flows (CMS) in drainage systems along the NE Gulf Coast. The US Geological Survey (Tallahassee, F.L.) provided river flow data.

Table 4.1 Regressions of Apalachicola River Flow with Other North Florida River Systems

	Econfina	Choctawhatchee	Escambia	Perdido
R^2	0.36	0.77	0.70	0.44
Probability	<0.0001	<0.0001	<0.0001	<0.0001

River flow data are provided by the US Geological Survey.

basins (Livingston, 1984a). Large river systems such as the Perdido and the Apalachicola have flow patterns that are thus more correlated with seasonal rainfall patterns in Alabama and Georgia, where high percentages of the respective drainage basins are located. The smaller streams along Apalachee Bay originate in Florida drainages and are associated with Florida rainfall conditions.

Rainfall in Georgia was compared to that in the Apalachee Bay drainage (Perry, Florida) over a period from 1957 through 2012 (Figure 4.2a). Percent differences with the long-term means of rainfall are shown in Figure 4.2b. Although the Georgia rainfall was significantly lower than that in Florida, the seasonal patterns were similar, and the long-term trends in both were downward after the drought of 1981–1982. The linear functions of the long-term declines were similar in both systems.

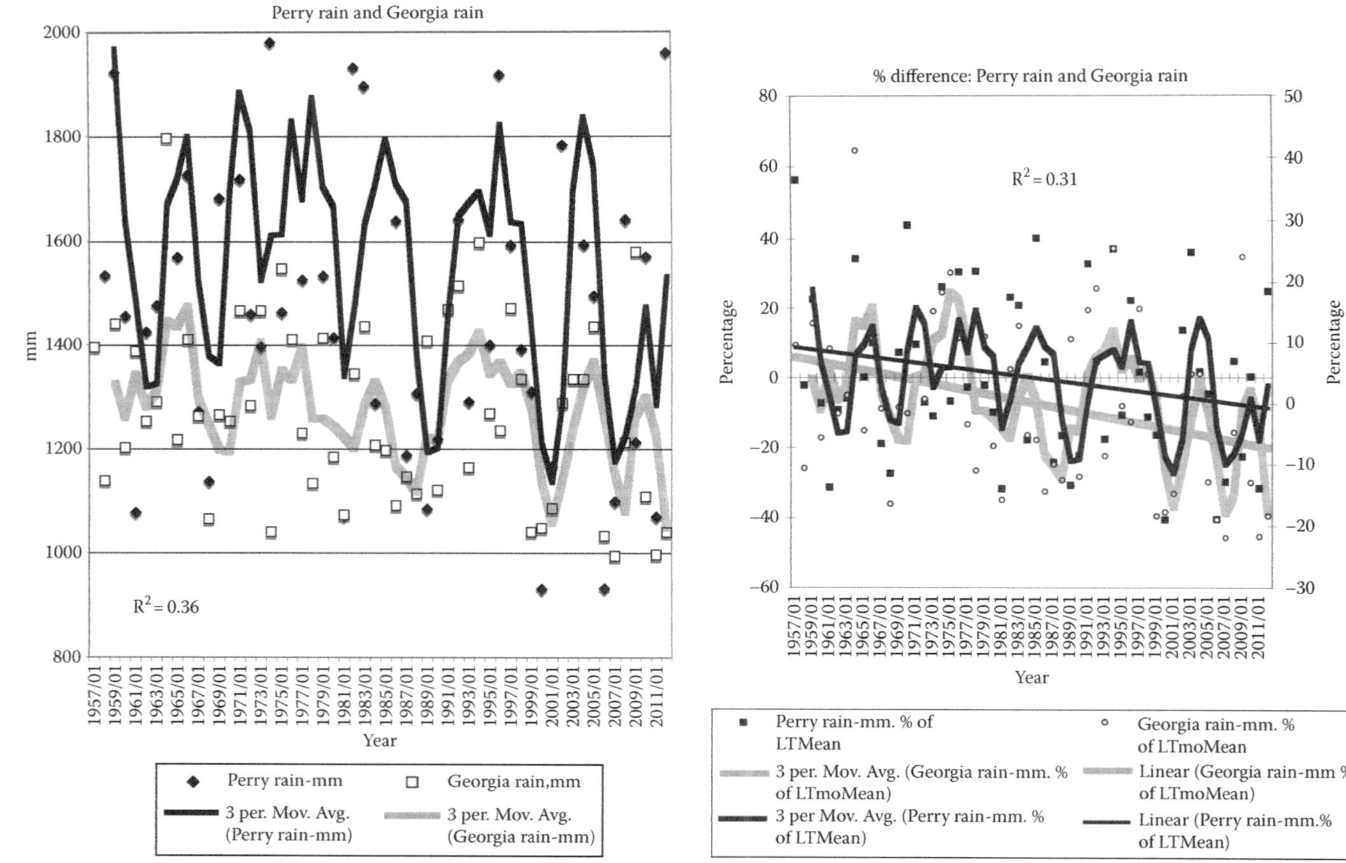

(a) (b)

Figure 4.2 (a) Georgia rainfall and Perry rainfall (annual averages) from 1957 through 2012. Rain data were provided by the National Oceanic and Atmospheric Administration. Three-month moving averages are given. (b) Georgia rainfall and Perry rainfall (percent differences from grand means) from 1957 through 2012. Rain data were provided by the National Oceanic and Atmospheric Administration. Three-month moving averages and linear regressions are also given. *(Continued)*

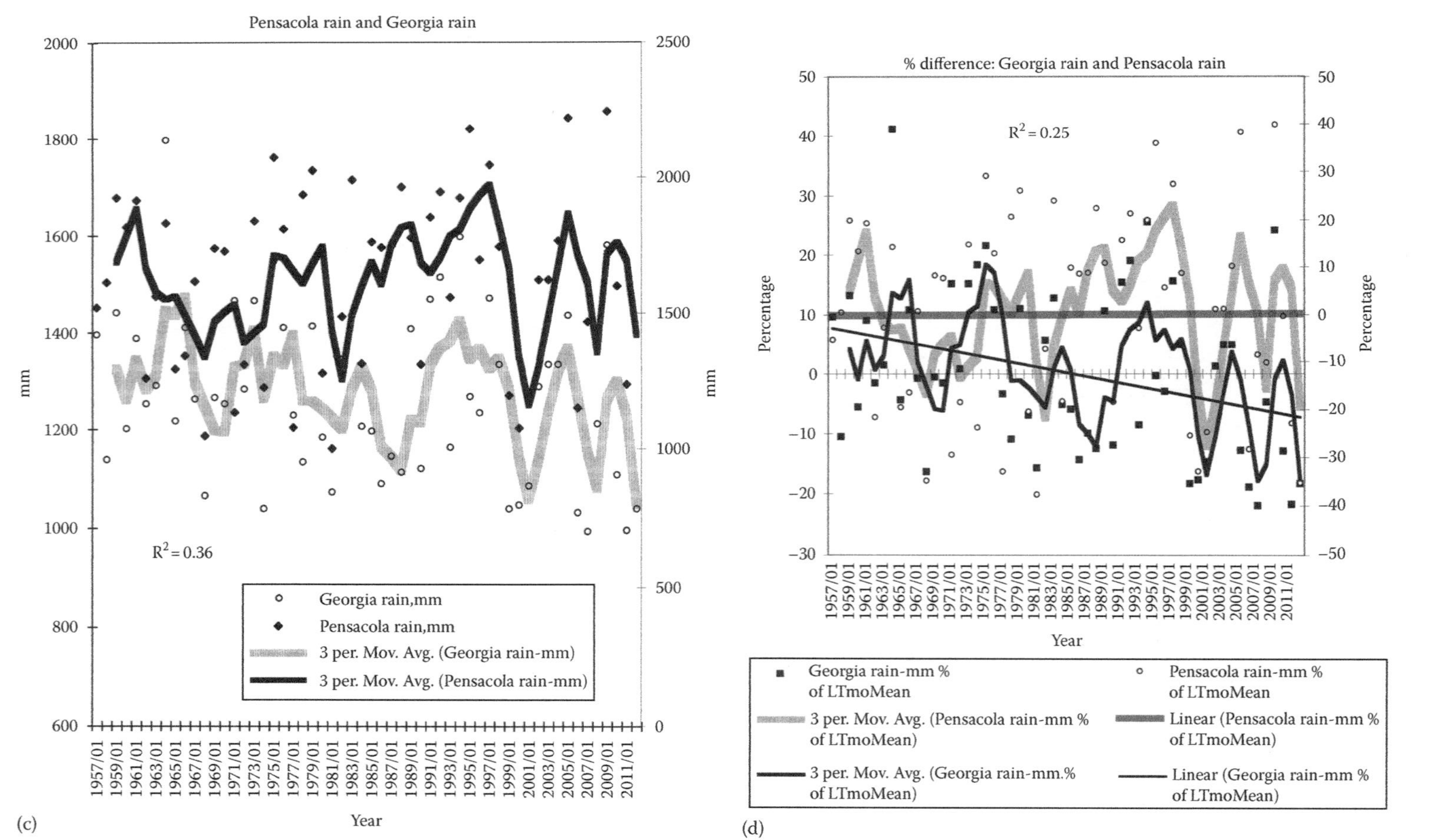

Figure 4.2 (*Continued*) (c) Georgia rainfall and Pensacola rainfall (annual averages) from 1957 through 2012. Georgia rain data were provided by the National Oceanic and Atmospheric Administration. Pensacola rain data were provided by the Florida Climate Center (Tallahassee, FL). Three-month moving averages are given. (d) Georgia rainfall and Pensacola rainfall (annual averages presented as % differences from the grand mean) from 1957 through 2012. Georgia rain data were provided by the National Oceanic and Atmospheric Administration. Pensacola rain data were provided by the Florida Climate Center (Tallahassee, FL). Three-month moving averages and linear regressions are also given.

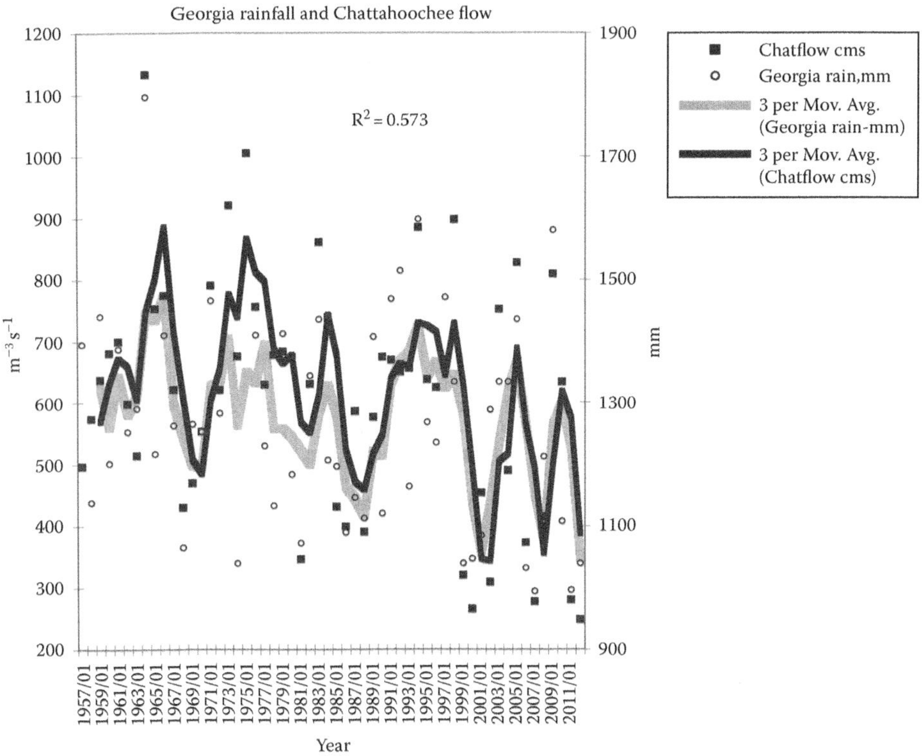

Figure 4.3 Georgia rainfall (annual averages, mm) and Chattahoochee River flow (annual averages, m^{-3} s^{-1}) from 1957 through 2012 as three month moving averages. Rain data were provided by the National Oceanic and Atmospheric Administration. River flow data were provided by the US Geological Survey, gauge number 02358000.

A similar analysis was run between the rainfall in Georgia and Pensacola, FL (Perdido Estuary, Pensacola Estuary) (Figure 4.2c and d). In this comparison, there was no real correlation between the rainfalls of the two systems over the entire period of analysis. The long-term patterns appeared to be similar after the early 1990s, but the overall linear relationships were different with no overt reduction of rainfall in the Pensacola system until the period following the drought of 1999–2002.

The Apalachicola River (Florida) is formed at the juncture of the Chattahoochee and Flint Rivers in Georgia. The Chattahoochee is a major river that flows from the Blue Ridge Mountains through Atlanta into the Apalachicola River that begins at the Jim Woodruff Dam at the Georgia–Florida border. A comparison of Georgia rainfall and Chattahoochee River flow (Figure 4.3) showed a closely associated trend ($R^2 = 0.573$) between these two factors. The declines in these variables started with the drought of 1980–1981 and showed a precipitous fall through 2012 in both the drought and the flood periods. The Chattahoochee River is the main source of freshwater to the Apalachicola River. This emphasizes the fact that Apalachicola flow rates reflect meteorological conditions in Georgia.

Overall, the long-term (temporal) rainfall and river flow patterns of Georgia and the Florida regions from Apalachee Bay to Apalachicola Bay were similar with less similar associations with the Florida drainages in the Pensacola region.

A detailed analysis was made concerning 82 years of the daily flow rates of the Chattahoochee River flows in Georgia. Daily flows are presented in Figure 4.4a and b. There was a series of low flows during the mid-1950s that was imbedded within a series of relatively moderate seasonal flow rates from 1930 to 1980. A brief drought was noted during 1981–1982 that was followed by a series of reduced flows during the 1986–1988 period. There was a pronounced period of low flows from 1999 through 2002 that was followed by extended droughts during the periods 2006–2008 and 2010–2012. The drought periods became progressively more pronounced from the early 1980s to the present time. The low-flow periods from 1980 through 2012 became more extreme with increased duration of the low-flow periods that was accompanied by reduced periods between the observed droughts. At the same time, periodic high flows were higher during the 1990s. These data indicated a definite change in the extent, duration, and frequency of droughts over the past three decades relative to the previous five decades.

A pattern of differences between day 2 and day 1 over the period of analysis is shown in Figure 4.4c. A positive number meant that the second-day flow was higher than the day before; a negative number meant the reverse of this pattern. There was an increase in the range of these differences from the end of the drought of the 1950s to the early 1970s when the peaks on both sides increased. By the late 1970s, both positive and negative differences reached peaks that were maximal during the 1990s. During periods of increased droughts during the new millennium, the interim periods between droughts were again typified by increased differences (both negative and positive) though such differences were not as pronounced as those during the previous decade. From an ecological standpoint, this overall increase in the extent of both positive and negative differences on a daily river flow basis increased the extent of both low- and high-flow patterns. In other words, there has been a gradual change in the pattern of increasing high and low flows at daily intervals.

If the Chattahoochee monthly flows were analyzed by differencing the intervals between the first year of each succeeding major drought from the previous initial drought year from 1955 to 2012, it would produce a series where the intervals were reduced almost by half with each succeeding drought (Figure 4.4d). The highly significant R^2 values clearly represent a nonrandom distribution of the differenced data. The series of intervals was structured, not random, with each succeeding initiating year of a major drought becoming shorter by about half over the 57 years of river flow data. This phenomenon is difficult to explain although it is clear that there is a certain structure to the increased frequency of droughts with time from 1955 to the present.

Another way of looking at these long-term changes is shown in Figure 4.4e, a graph of the averages of maximum and minimum Chattahoochee River flows. After the major drought during the 1950s, both forms of flows returned to the former distributions until the 1980s when these patterns changed at both levels. Low flows went down from the 1980s to the end of the sampling period. Peak high flows remained stable and even increased during the 1990s during this period, whereas peak low flows continued to decrease.

The data thus showed that, over the most recent 30 years of Chattahoochee River flows, there has been a trend of increasingly frequent and severe drought periods. This trend was due mainly to serial decreases of the minimum flow rates.

A regression of Chattahoochee River flow rates with Apalachicola River flow rates (monthly from 1950 through 2012) indicated a highly significant correlation ($R^2 = 0.94$) between the flow rates of the two rivers. This association of the Florida River with the Georgia River reflected previously noted associations of the Apalachicola flows with the patterns of Georgia (not Florida) rainfall.

The long-term (1930–2012) rainfall and river flow patterns of Georgia and the Florida regions from Apalachee Bay to Apalachicola Bay were similar with marked increases of the frequency and severity of droughts from 1980 to 2012. These droughts (1980–1981, 1986–1988, 1999–2002, 2006–2008, 2010–2012) were accompanied by periodically higher peak flows during the 1990s that increased the ranges between low and high flows. The low-flow periods from

1980 through 2012 became more extreme with time. This was accompanied by the increased duration of the low-flow periods along with reduced periods between the observed droughts. Rainfall/river flow conditions in the Pensacola and Perdido Bay systems differed somewhat from this pattern with the drought trends following a similar pattern to those farther east from 1999 through 2012.

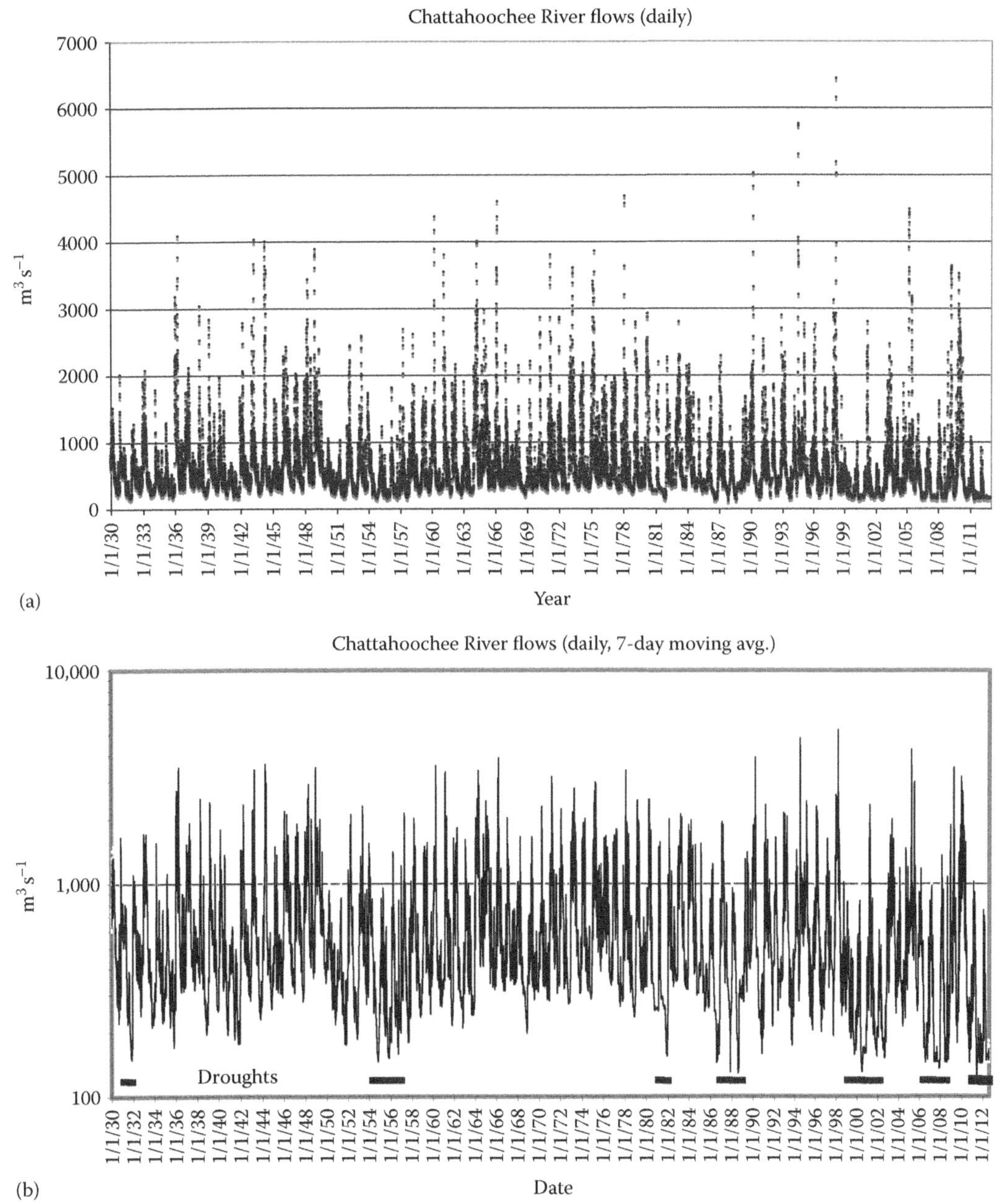

Figure 4.4 (a) Daily Chattahoochee River flow ($m^{-3} s^{-1}$) from 1930 through 2012. River flow data were provided by the US Geological Survey, gauge number 02358000. (b) Daily Chattahoochee River flow ($m^{-3} s^{-1}$) from 1930 through 2012 with 7-day moving averages. Drought periods are shown. River flow data were provided by the US Geological Survey, gauge number 02358000. *(Continued)*

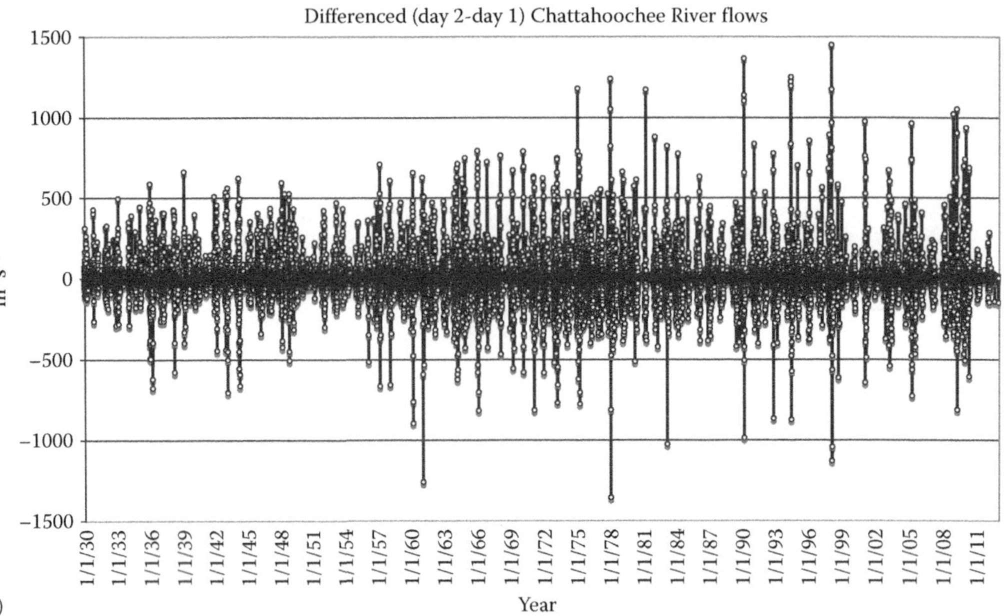

(c)

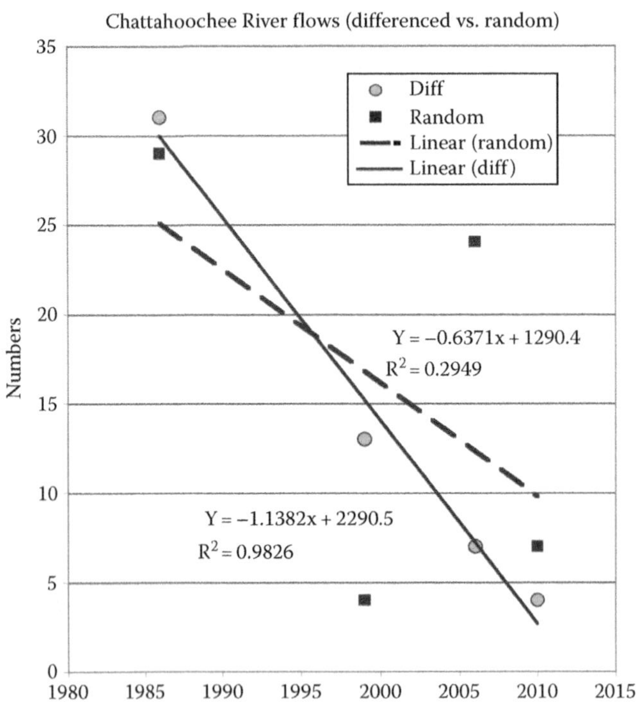

(d)

Figure 4.4 (*Continued*) (c) Daily Chattahoochee River flow (m^{-3} s^{-1}) from 1930 through 2012. Data were differenced (day 2 to day 1). River flow data were provided by the US Geological Survey, gauge number 02358000. (d) Chattahoochee River flows: data were generated by subtracting the number of year difference between the succeeding drought using the initial year of each major drought period from 1955 to 2012 (1986–1955, 1999–1986, 2006–1999, 2010–2006). Regressions were run showing formulas and R^2 values for differenced data and a set of random numbers. River flow data were provided by the US Geological Survey, gauge number 02358000. (*Continued*)

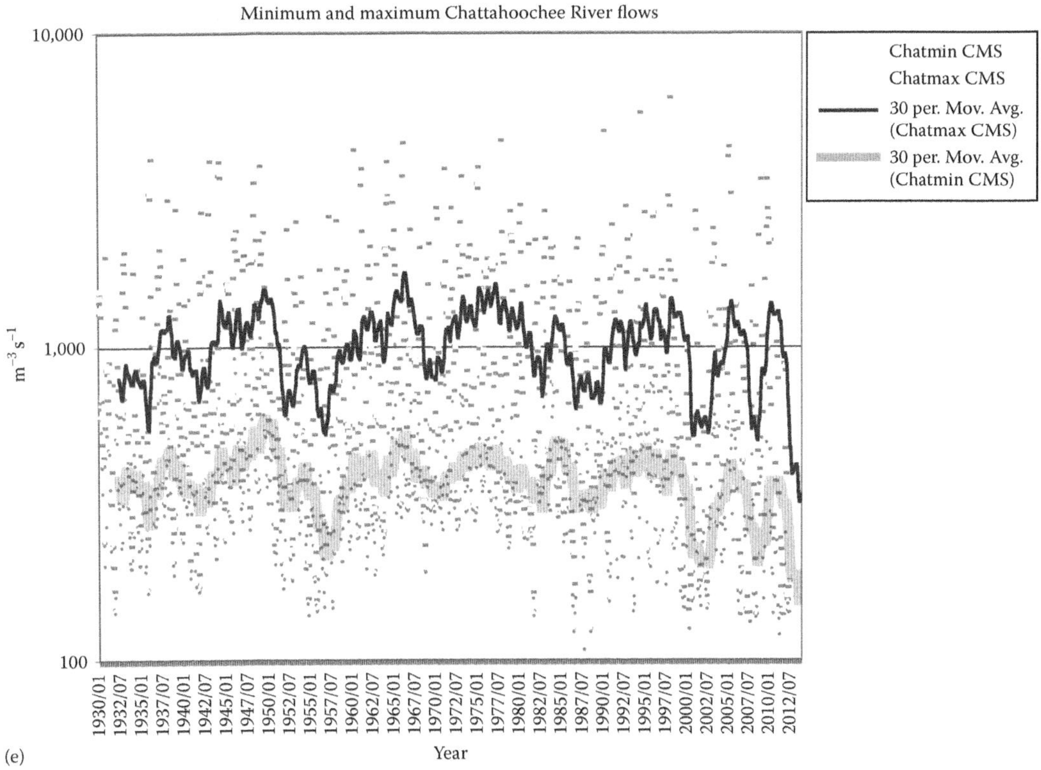

Figure 4.4 (*Continued*) (e) Monthly minimum and maximum Chattahoochee River flows (m^{-3} s^{-1}) with 30-month moving averages from January 1930 through December 2012. River flow data were provided by the US Geological Survey, gauge number 02358000.

The ecological responses of Apalachee Bay and the Apalachicola and Perdido Estuaries to these long-term changes of climate were measured within the context of differences of natural habitat conditions and the impacts of anthropogenic activities such as dredged openings to the gulf and nutrient loading.

CHAPTER 5

Nutrient Loading

Natural versus Anthropogenic Inputs

Anthropogenic nutrient enrichment has been a major theme in coastal ecology for over three decades (Livingston, 2002). Cloern (2001) outlined a modern conceptual model that included the following:

> a complex suite of direct and indirect responses including linked changes in water transparency, distribution of vascular plants and biomass of macroalgae, sediment biogeochemistry and nutrient cycling, nutrient ratios and their regulation of phytoplankton community composition, frequency of toxic/harmful algal blooms, habitat quality for metazoans, reproduction/growth/survival of pelagic and benthic invertebrates.

This approach has morphed into recent inquiries into the complex responses of coastal systems to anthropogenic nutrient loading that include interacting forms of toxic effects, habitat loss, responses of modern-day fishing pressure, and the complex effects of bloom species on planktonic assemblages, related biota, and the trophic organization of receiving areas (Livingston, 2000, 2002, 2005).

Seasonal and interannual attributes of the effects of nutrient loading on coastal systems are complex. Responses to nutrient loading and concentrations are modified by temperature, salinity, wind, tides, and sediment processes. Jouenne et al. (2007) noted that the timing and intensity of plankton blooms varied at various levels of temporal events. Other evidence indicated that seasonal patterns of microalgal succession existed with regular progressions of composition, biovolume, and diversity of planktonic assemblages that were influenced by various characteristics of the physical environment (Livingston, 2000). These effects resulted in progressions of phytoplankton community structure that were intricately involved in the trophic organization of a given system (Livingston, 2007a,b). Nutrient loading and responsive primary production represented processes that were central to our understanding of how highly productive estuarine and coastal systems function.

Hypoxic events are relatively common in coastal systems. Verity et al. (2006) related oxygen depletion in well-mixed estuaries in the southeastern United States to increased nutrient concentrations and increased decomposition of autotrophs below the pycnocline. The data indicated that hypoxia can occur from microbial respiration in well-mixed estuaries with strong vertical and horizontal mixing. Conley et al. (2007) reported recent (2002) hypoxic events in Danish coastal waters that were associated with high rainfall and nitrogen loading. The authors related such changes to those noted in the Chesapeake system (Hagy et al., 2004) where bottom dissolved oxygen (DO) was "modulated" by nitrogen runoff from adjacent land areas. These trends of low DO were also associated with bottom-water transport, wind events, and surface water temperature. However, decreased nutrient loading was not associated with increased DO during recent times. Borsuk et al. (2004), on the other hand, found that riverine total nitrogen (TN) played only a minor role in determining chlorophyll *a* concentrations in the Neuse River Estuary. River flow had a stronger effect on such productivity.

The relationship of plankton blooms and DO is complex and includes instances of both supersaturation and hypoxia. Hypoxia has been reported in most of the gulf estuaries (NOAA, 1997). Periodic hypoxia in some coastal systems can be the result of natural conditions (Seliger and Boggs, 1988). Episodes of naturally low DO ("jubilees") have occurred over relatively long periods in gulf coastal areas such as the Mobile Bay (Loesch, 1960; May, 1973). Lehrter (2008) noted that chlorophyll *a* was strongly related to freshwater flushing times. Estuarine phosphorus concentrations and photosynthetically active radiation were also good descriptors of chlorophyll *a* concentrations. Salinity stratification and associated restriction of vertical mixing in estuaries have been associated with hypoxic conditions. Schroeder et al. (1990) found that river flow was the dominant control mechanism of salinity stratification in Mobile Bay. Annual spring freshets flushed salt from the bay while

> the relative strengths of river discharge and wind stress changed the bay from highly stratified to nearly homogeneous and back on a variety of time scales ranging from daily to seasonal.

Stratification events and DO levels were tightly coupled with variations in freshwater discharge and wind stress in the Pamlico Estuary (Stanley and Nixon, 1992). Stratification appeared in a matter of hours, with hypoxia forming if the water is mixed only every 6–12 days.

Phytoplankton communities are affected by multiple processes (Cloern and Dufford, 2005). Anthropogenic activities resulting in increased nutrient loading, restricted water circulation, altered phytoplankton production, and trophic anomalies also influence the frequency and severity of such episodes (Seliger and Boggs, 1988; Breitburg, 1990). The magnitude of estuarine primary production and the periodicity of algal blooms can be directly related to variations in upper watershed rainfall and its subsequent regulation of downstream river flow (Mallin et al., 1993). Heil et al. (2007) noted that the quality of available nutrients drives the composition of phytoplankton assemblages. Paerl (2009) recently emphasized the importance of both phosphorus and nitrogen to the establishment of plankton blooms and the relevance of such dual limitation to the development of effective management programs. Flemer et al. (1998) found a similar pattern in nutrient limitation experiments in Perdido Bay.

Mean nutrient loading and concentration data were calculated for the major rivers as they entered the respective bay systems (Table 5.1). Nutrient loads were highest in the Apalachicola River and lowest in the Econfina River. Nutrient loading was high in the Fenholloway River (influenced by a pulp mill) relative to the reference Econfina River. Ammonia, nitrite/nitrate, and orthophosphate loadings in Elevenmile Creek (influenced by a pulp mill in the Perdido Estuary) were higher in 2004/2005 than in 1990–1991 due to increased loading from a pulp mill over the later period. Elevenmile Creek in both periods had higher loading of ammonia and orthophosphate than that in the Perdido River although the river loads of nitrite/nitrate were higher in the river than those in the creek.

Ammonia and orthophosphate concentrations were highest in Elevenmile Creek and the Fenholloway River due to loading of these nutrients from the respective pulp mills. Such concentrations were comparable among the remaining rivers despite the considerable disparities of flow rates. Nitrite + nitrate concentrations were higher in the Apalachicola River than the other rivers. The lowest such concentrations occurred in the Econfina River. Overall, these data indicated that higher river flows were associated with higher levels of nutrient loading in unpolluted rivers and that such loads were not generally associated with higher nutrient concentrations.

Monthly (21 months) mean data for the nutrient loading of the various river systems are shown in Figure 5.1a through c. Ammonia loading tended to peak during winter–spring months although such trends varied among the various rivers. This loading was highest in the Apalachicola River and in the creeks affected by pulp mill effluents. Orthophosphate loading also tended to increase during winter–spring months, but such loading was relatively low in the Apalachicola system relative to the Choctawhatchee River. Choctawhatchee orthophosphate loads were high compared to those in the large rivers. These nutrient loads were also high in rivers affected by pulp mill effluents.

Table 5.1 Twenty-One Month Mean Data for Flow Rates, Nutrient Loads, and Nutrient Concentrations in Rivers along the Eastern Gulf as They Entered the Respective Bays

Means (21 Months)		
	Apalachicola River (1994–1995)	**Escambia River (1997–1998)**
Flow ($m^3\ s^{-1}$)	767.2	156.0
NH_3 load (kg day^{-1})	1115.2	949.6
$NO_2 \pm NO_3$ load (kg day^{-1})	22418.6	2257.5
PO_4 load (kg day^{-1})	436.2	185.5
NH_3 conc. (mg L^{-1})	0.040	0.058
$NO_2 \pm NO_3$ conc. (mg L^{-1})	0.354	0.196
PO_4 conc. (mg L^{-1})	0.018	0.012
	Econfina River (1992–1993)	**Fenholloway River (1992–1993)**
Flow ($m^3\ s^{-1}$)	4.5	6.3
NH_3 load (kg day^{-1})	11.5	358.5
$NO_2 \pm NO_3$ load (kg day^{-1})	10.7	30.2
PO_4 load (kg day^{-1})	8.1	135.1
NH_3 conc. (mg L^{-1})	0.060	1.640
$NO_2 \pm NO_3$ conc. (mg L^{-1})	0.050	0.790
PO_4 conc. (mg L^{-1})	0.050	0.990
	Choctaw River 1987–1988	**Perdido River 1990–1991**
Flow ($m^3\ s^{-1}$)	97.8	22.8
NH_3 load (kg day^{-1})	467.3	121.6
$NO_2 \pm NO_3$ load (kg day^{-1})	3358.5	369.3
PO_4 load (kg day^{-1})	438.5	14.3
NH_3 conc. (mg L^{-1})	0.022	0.064
$NO_2 \pm NO_3$ conc. (mg L^{-1})	0.198	0.188
PO_4 conc. (mg L-1)	0.012	0.007
	Elevenmile Creek (1990–1991)	**Elevenmile Creek (2004–2005)**
Flow ($m^3\ s^{-1}$)	2.5	4.1
NH_3 load (kg day^{-1})	350.5	687.9
$NO_2 \pm NO_3$ load (kg day^{-1})	199.7	241.3
PO_4 load (kg day^{-1})	22.7	35.0
NH_3 conc. (mg L^{-1})	1.712	2.182
$NO_2 \pm NO_3$ conc. (mg L^{-1})	0.990	0.928
PO_4 conc. (mg L^{-1})	0.109	0.113

Nitrite + nitrate loads were dominant in the Apalachicola system relative to all of the other rivers. Nitrite + nitrate loading was relatively low in the creeks affected by pulp mill effluents.

Orthophosphate loading was highest in the Apalachicola, Choctawhatchee, and Escambia Rivers. This loading was also high in the Fenholloway River and Elevenmile Creek (1994–1995). The orthophosphate concentration data indicated comparability among the Econfina, Choctawhatchee, and Elevenmile Creek (1990–1991) systems at one level and the Escambia, Apalachicola, Perdido, Blackwater, and Yellow River systems at a somewhat lower level. Orthophosphate concentrations in the Fenholloway River and Elevenmile Creek (1994–1995) were one to two orders of magnitude higher than those found in the other systems. These comparisons indicate that nutrient loading was higher in the alluvial rivers than the nutrient-enriched Fenholloway River and Elevenmile Creek after increased orthophosphate loading (1994–1995). However, these small streams had significantly higher orthophosphate concentrations than the alluvial rivers. These differences between nutrient loads and nutrient concentrations were important in the assessment of algal bloom response in the various estuaries.

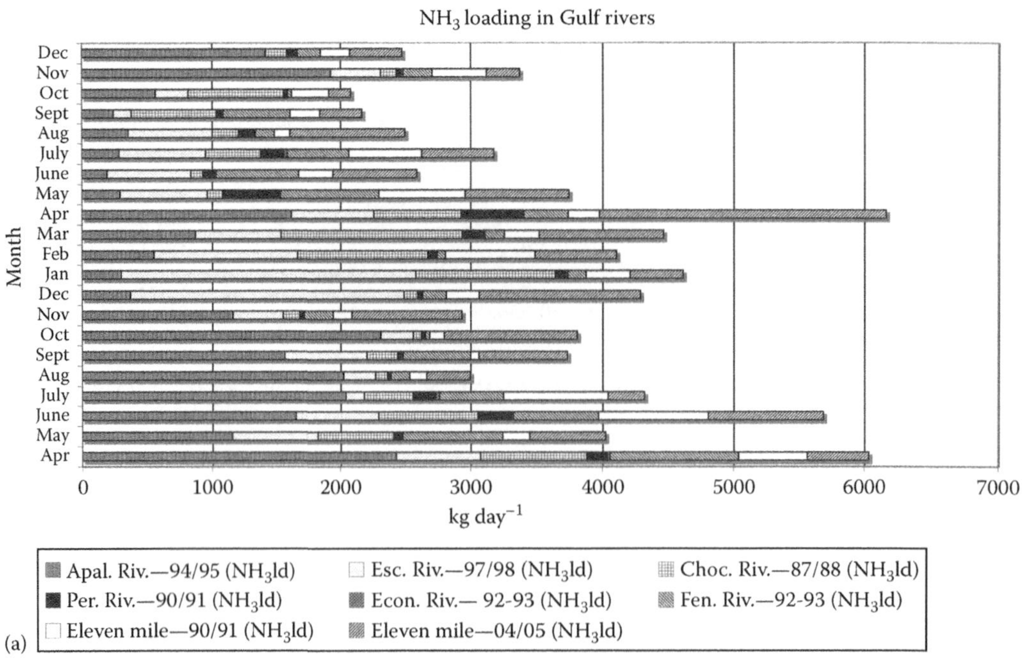

(a)

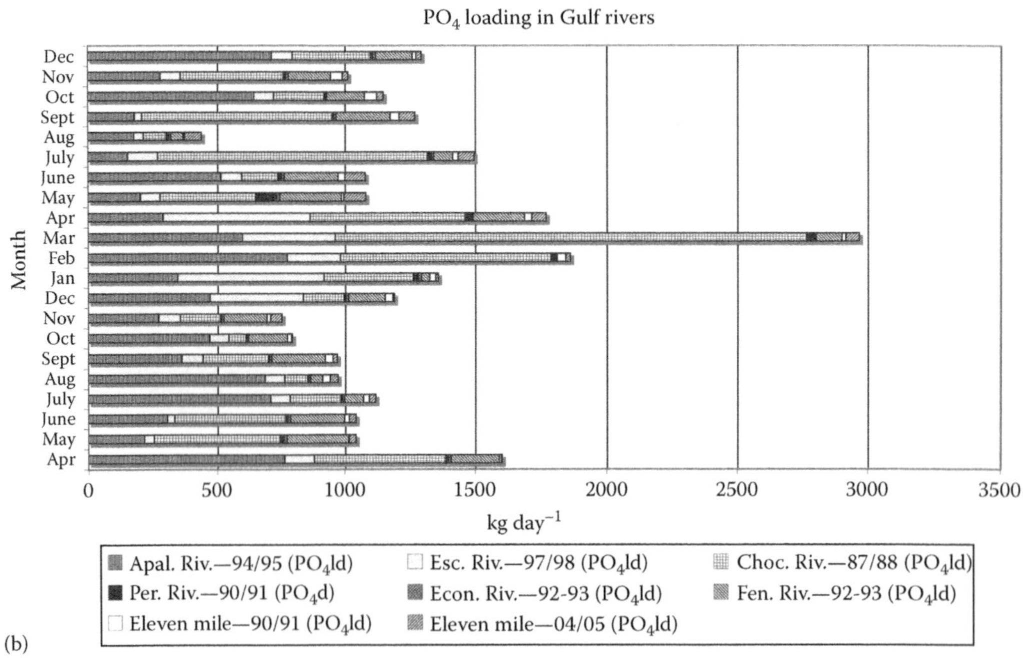

(b)

Figure 5.1 (a) Monthly mean (21 months) NH_3 loads (kg day^{-1}) in rivers along the eastern Gulf as they entered the respective bays. (b) Monthly mean (21 months) PO_4 loads (kg day^{-1}) in rivers along the eastern Gulf as they entered the respective bays. *(Continued)*

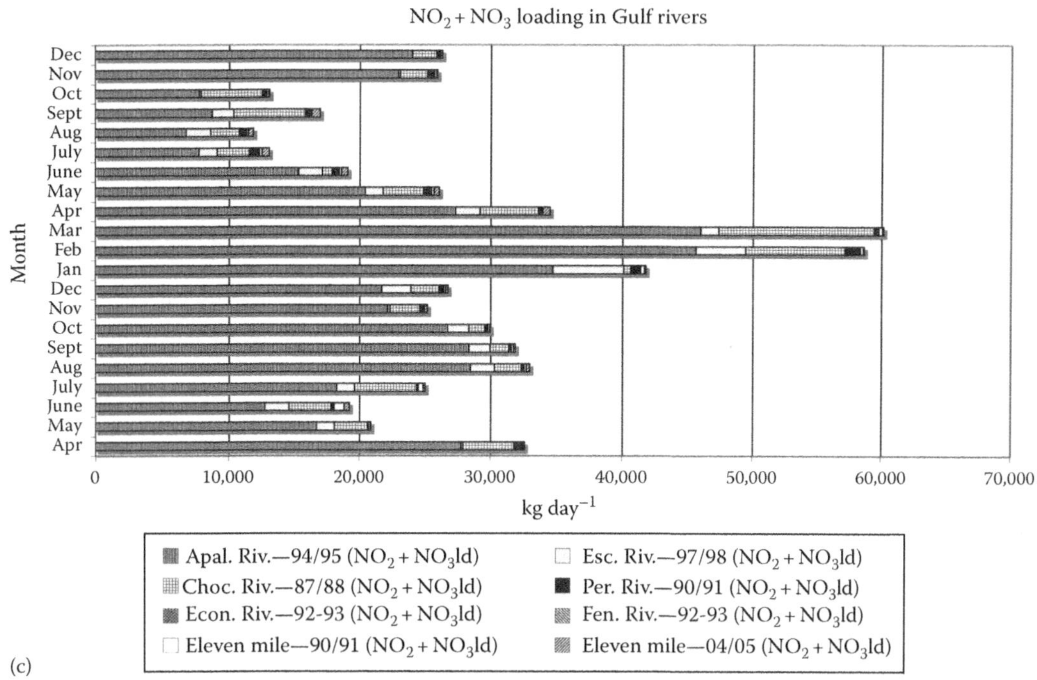

(c)

Figure 5.1 *(Continued)* (c) Monthly mean (21 months) $NO_2 + NO_3$ loads (kg day^{-1}) in rivers along the eastern Gulf as they entered the respective bays.

Ammonia loading was comparable among the various systems with the exception of the Econfina, which had relatively low ammonia loading. Ammonia concentrations were an order of magnitude higher in the Fenholloway River and Elevenmile Creek (1990–1991; 1994–1995) than the other rivers. Thus, as with orthophosphate, ammonia loading in the small, nutrient-enriched streams was comparable to that of the alluvial rivers, whereas ammonia concentrations in the Fenholloway River and Elevenmile Creek were much higher than those in the alluvial rivers. Nitrite/nitrate loading appeared to be directly related to river flow characteristics in most of the systems tested. Somewhat higher nitrate/nitrite concentrations occurred in Elevenmile Creek (1990–1991; 1994–1995); the lowest such measurements were found in the Econfina system.

The nutrient loading and nutrient concentration data indicate that there were differences between absolute nutrient loading and nutrient concentrations. Nutrient loading can be high as a function of higher river flow, whereas concentrations of nutrients vary primarily as a function of flow differences and anthropogenic sources such as pulp mill discharges, agricultural discharges, and urban runoff. The question concerning the differentiation of operable factors that initiate blooms (i.e., loading vs. concentration) is complex. However, overall loading is a key factor although such loading should be evaluated within the context of the flow characteristics of the delivering medium, the resultant nutrient concentrations, and the assimilative capacity of the receiving system.

A comparison between the relatively unpolluted Apalachicola Estuary and the Perdido Estuary indicates certain similarities. Iverson et al. (1997) found that dissolved silicate did not limit phytoplankton production in the largely mesotrophic Apalachicola Bay (Fulmer, 1997), as was the case with the Perdido system prior to the blooms (Livingston, 1992, 1997b,c; Flemer et al., 1998; Livingston et al., 1998a). In the Apalachicola system, orthophosphate availability limited phytoplankton during both low- and high-salinity winter periods and during summer at stations with low salinity. Nitrogen, on the other hand, was limiting during summer periods of moderate to high salinity (Iverson et al., 1997). Light and temperature limitation was highest during winter–spring periods, thus limiting primary production during this time (Iverson et al., 1997). High chlorophyll *a* levels during winter periods were attributed to low zooplankton grazing during the cooler months (Iverson et al., 1997). These conditions prevailed in the Perdido system during periods of moderate nutrient loading (1989–1991).

Iverson et al. (1997) noted that there had been no notable increase in chlorophyll *a* concentrations in Apalachicola Bay over the past two decades despite increases in nitrogen loading due to increased basin deposition of this nutrient. Perdido Bay, on the other hand, was subject to complex changes of both nutrient loading and nutrient concentrations due to ammonia and orthophosphate discharges from the pulp mill into Elevenmile Creek. The effects of the discharge of these nutrients appeared to be related to specific temporal patterns relative to the nutrient requirements of individual plankton species and to the complex temporal reactions of these species to combinations of these nutrients.

Nutrient loading in the Apalachicola River system remained relatively high without apparent hypereutrophication in the bay, whereas relatively lower nutrient loading in the Perdido system with higher concentrations of reactive inorganic phosphorus and ammonia in Elevenmile Creek ended in years of plankton blooms. This contrast indicates that high nutrient concentrations, as a product of high nutrient loading into a creek with relatively low flows, could have been an important factor in the initiation of plankton blooms in Perdido Bay. Nutrient loading per se was thus not the only factor in the generation of plankton bloom.

The drought/flood periodicity of river input is an important factor in the overall pattern of nutrient loading and eutrophication in the subject estuaries. As will be shown in the following, droughts were associated with blue-green algal blooms in Elevenmile Creek, whereas high river flows were associated with peak blooms of raphidophytes in Perdido Bay. The added nutrient loading from the Perdido River could also have been important in sustaining the plankton blooms once they were initiated by the high nutrient concentrations from Elevenmile Creek. Thus, the relationships of flow rates, nutrient loading, and the resulting nutrient concentrations appeared to be important in the initiation and sustenance of plankton blooms in the estuary. Climatological trends of rainfall/river flows, natural and anthropogenic nutrient loading, and the characteristics of the receiving estuaries were all involved in how the nutrients were processed. The details of such responses were related to differences in the receiving estuaries.

CHAPTER 6

Dredged Passes to the Gulf

Comparative Effects

The ecological implications of alterations of the physical components of coastal systems through various methods of dredging, filling, and channelization have not received as much attention by coastal researchers as other forms of anthropogenic impacts. The northern coast of the Gulf of Mexico is lined with bays that are enclosed by barrier islands that have been systematically altered by dredging activities. Dredged channels have been created in the Apalachicola, Choctawhatchee, and Perdido Estuaries that opened the naturally low-salinity bay waters in these areas to the entry of high-salinity gulf waters. The effects of these openings have been evaluated in the long-term studies of the NE Gulf coastal ecosystems.

Different studies have described various general conditions that result in anoxia and/or hypoxia in coastal regions of the United States. Salinity stratification has been shown to play a role in the development of benthic hypoxia in river-dominated estuaries. Episodes of low dissolved oxygen (DO) "jubilees" have occurred over relatively long periods in estuaries such as Mobile Bay (Loesch, 1960; May, 1973) and Chesapeake Bay (Smith et al., 1992). Periodic hypoxia in some estuaries could reflect natural conditions (Turner et al., 1987; Seliger and Boggs, 1988). It should be pointed out that anthropogenic activities resulting in increased nutrient loading, restricted water circulation, altered phytoplankton production, and trophic anomalies do influence the frequency and severity of such episodes (Seliger and Boggs, 1988; Breitburg, 1990).

According to Turner et al. (1987), oxygen depletion in the shallow bottom waters of Mobile Bay was directly related to the degree of water column stratification. Carbon sources for this effect were shown to be of recent origin and were not due to overwintering or "relic" sources. Schroeder et al. (1990) found that river flow was the dominant mechanism of salinity stratification in Mobile Bay and that wind stress was important to the salinity distribution only in the absence of major freshwater discharges. In other systems, it has been shown that, as water column stratification increases, vertical oxygen diffusion rates are reduced (Kemp and Boynton, 1980).

Smith et al. (1992) summarized the findings of 6 years of studies concerning the dynamics of the Chesapeake Bay system. It was determined that hypoxic and anoxic conditions in the bay occur during warm periods with reaeration of the lower layer waters strongly dependent on seasonal stratification due to spring runoff. Stratification was the dominant process in the decrease of vertical water and DO exchanges. The authors found that the deepwater hypoxic conditions in Chesapeake Bay resulted from cumulative effects of biochemical mechanisms that were modified by physical stratification. Nonpoint nutrient inputs supported phytoplankton biomass in excess of the oxygen assimilation capacity of the bay. However, the exact relationship of nutrient loading (and reductions of nutrient loading) to the lessening of hypoxic conditions remained undescribed because of unknown factors such as feedback loops and interannual changes in climatic factors. The long-term effects of physical alterations were not emphasized in most of these studies.

Numerous field studies have been carried out to evaluate the impact of hypoxia and anoxia on estuarine populations and communities. Pearson (1980) and Pearson and Rosenberg (1978) have carried out detailed studies on the effects of organic enrichment and low DO on marine systems. Van Es et al. (1980) demonstrated a direct response of meio- and microfauna in tidal flats along distinct gradients of organic enrichment and oxygen saturation. The authors could not isolate the impact of gradients of DO from the effects of other natural forcing factors such as salinity (absolute value and fluctuations), sediment composition, and relative emersion time. The interaction and covariance of important ecological features of an estuary often confound direct correlations between the oxygen regime and the biological organization of the system. However, in a series of studies (Santos and Bloom, 1980; Santos and Simon, 1980a,b), researchers showed that annual defaunation in Hillsborough Bay (Tampa Bay) was caused by hypoxia. A stochastic recolonization response of the soft-bottom macroinvertebrate community was demonstrated following the recurrent defaunation event.

There is considerable variability in the freshwater flows of the various drainages of the NE Gulf study area. Flooding often occurs during winter and early spring months in the various northern Florida River drainages. This winter–spring flooding of the alluvial streams is a product of reduced evapotranspiration rates during the winter rainfall peaks along the heavily vegetated drainage basins (Livingston, 1984a). The reduced surface flows during summer periods are due to increased floodplain evapotranspiration, even though the Florida rainy season usually peaks during this period. River flows to the gulf are lowest during fall drought periods. Streamflows in the large river systems such as the Perdido, Apalachicola, and Choctawhatchee dominate the salinity regimes of the associated estuaries, and along with the stratigraphy of the receiving basins, the freshwater flows determine the spatial/temporal distribution of salinity stratification in the receiving coastal areas.

Bays and estuaries with direct freshwater runoff have highly variable salinities that are dependent on rainfall, river flow, tidal effects, and wind conditions that interact with local physiographic features of the respective basins. Offshore coastal waters are often characterized by marine conditions with higher, temporally stable salinities. The smaller streams along the Big Bend area such as the Econfina and Fenholloway Rivers are associated with relatively limited estuarine areas relative to the large river systems to the west. Tidal effects vary from one area to another, although the general effect is to enhance mixing in the nearshore waters.

Habitat conditions in the gulf coastal river–estuaries of north Florida are determined by various factors. The barrier-enclosed estuaries are dominated by a series of rivers from the Perdido to the Apalachicola. These estuaries tend to be characterized by high water color. Wind is a major factor in the determination of water circulation and salinity in these estuaries. In areas where river flow is a factor, frictional effects on the saltwater regimes are combined with salt-wedge circulation and geostrophic forces to affect the estuarine salinity distribution. The physical dimensions of a given river–estuarine basin comprise a key element in the distribution and nature of aquatic habitats. The size, depth, and location of passes represent important determinants of the stratification potential and salinity distribution in a given estuary. The complex interactions of freshwater influxes, basin characteristics, and associated geostrophic forces thus determine the spatial and temporal habitat characteristics of the gulf coastal estuaries of northern Florida that, in turn, are associated with the primary and secondary productivity of these systems.

Sikes Cut, an artificial opening to the gulf in a barrier island that encloses Apalachicola Bay, was created during the mid-1950s by the US Army Corps of Engineers (COE) and is currently maintained by the corps. Due to the entry of gulf water, the southern part of the bay is characterized by salinities that resemble the open gulf. This area is an entry point of various oyster predators that move into the bay from the gulf. Menzel et al. (1966) found that the near-total demise of the once productive St. Vincent oyster bar resulted from increased mortality from predation following the opening of the cut. Livingston et al. (2000) noted that oyster mortality due to predation was highest at St. Vincent's Bar, Scorpion, Pickalene, Porter's Bar, and Sikes Cut. These areas are distant from river influence with high salinities and are also in closest proximity to the entry of oyster predators from the gulf through the cut. Despite the destruction of the prized oyster reefs, no action has been taken to correct this problem.

During the 1920s, citizens opened cuts into the barriers of the Choctawhatchee and Perdido Estuaries. Livingston (1997b) made a comparative analysis of the causal relationships of specific changes in the DO regime of the estuaries along the NE Gulf Coast of Florida with specific attention to the role of anthropogenic stratigraphic alterations in salinity stratification and associated hypoxia. Major parts of both of these estuaries were affected by the salinity stratification that accompanied the openings to the high-salinity gulf waters. As with the Apalachicola situation, the effects of the dredging were more or less permanent in the Choctawhatchee and Perdido Estuaries. The effects of the anthropogenic stratification of these estuaries were the subject of extensive field analyses, and an interbay comparison of the impacts of stratification on secondary productivity was instrumental in the determination of these effects.

METHODS

A salinity stratification classification was carried out based on the cooperative efforts of our research group with scientists from the National Oceanic and Atmospheric Administration (J. Klein and P. Orlando, pers. comm.). This transformation was carried out with long-term surface and bottom-salinity determinations in the various estuarine systems. Data were organized on an annual basis (by month, by station) for each of the subject estuaries according to the following criteria:

Highly stratified (*HS*): salinity difference (bottom–surface) ≥10 ppt
Partially stratified, strong (*PSS*): salinity difference ≥5 and <10 ppt
Partially stratified, weak (*PSW*): salinity difference ≥2 and <5 ppt
Vertically homogeneous (*VH*): salinity difference <2 ppt

A review was carried out using the long-term databases and selected scientific papers concerning the relationships of low DO to estuarine populations. Based on these analyses, we reordered the DO data taken in the each system according to potential impact on the biological components:

Anoxic to severely hypoxic (*ASH*): DO ≤ 2 mg L^{-1}
Hypoxic (*HYP*): DO > 2 and ≤4 mg L^{-1}
Biologically "neutral" (*BN*): DO > 4 mg L^{-1}

The ASH classification included DO levels that were actively detrimental to most estuarine fauna. The HYP classification includes DO levels that may be stressful to more sensitive organisms and thus can be associated with loss of certain faunal populations. The exact meaning of this transitional level of DO is, as yet, not well understood and probably has a stress value that remains relative to the stratigraphic characteristics of the area and the time of year. The data were analyzed according to methods described by Livingston (1997d).

STRATIFICATION COMPARISONS: DISSOLVED OXYGEN

Cross-bay comparisons of the relationship of stratification and bottom dissolved oxygen (BDO) have been published by Livingston (1984a, 1997d, 2000). The trends were tested with a series of statistical analyses to determine the relative importance of these variables on the bay systems in the NE Gulf of Mexico. Livingston (1997b) made a comparative analysis of the causal relationships of specific changes in the DO regime of the estuaries along the NE Gulf Coast of Florida with specific attention to the role of anthropogenic stratigraphic alterations in salinity stratification and associated hypoxia.

Apalachee Bay is part of one of the most extensive series of sea grass beds in the northern hemisphere (Iverson and Bittaker, 1986). Salinity stratification of the Econfina and Fenholloway River–Estuaries was maximal in the lower parts of the drainage areas. Offshore areas in both systems were not strongly stratified. The overall level of stratification was somewhat less in the Econfina Estuary than that shown in the Fenholloway Estuary. BDO was lowest in both estuaries at the respective river mouths. Those areas were characterized by higher stratification and, in the Fenholloway, the presence of pulp mill effluents.

The mean DO levels in offshore areas of the Fenholloway drainage were slightly lower than those in the Econfina areas. However, such levels were generally higher than 7 mg L^{-1} at all offshore stations and minima did not go below 4 mg L^{-1}. In most instances, the DO minima were comparable in the two offshore systems. Surface DO levels were also universally high in these systems. Hypoxic conditions at depth (HYP) were directly associated with mean depth ($R^2 = 0.402$, $p = 0.0003$), mean bottom salinity ($R^2 = 0.425$, $p = 0.0002$), and vertical stratification indices ($R^2 = 0.225$, $p = 0.01$), while such conditions were inversely related to mean surface salinity ($R^2 = 0.441$, $p = 0.0001$). Thus, despite considerable differences in physiography and river flow, the general relationships of depth, salinity stratification, and hypoxia at depth in the Econfina and Fenholloway Estuaries were similar.

The Apalachicola Estuary is relatively shallow compared to other estuaries along the NE Gulf Coast. Depth appears to be an extremely important factor in eutrophication processes of the subject bay systems. Early studies (Dawson, 1955; Gorsline, 1963) did not mention salinity stratification in the Apalachicola Bay system. However, more recent analyses indicate that the estuary is periodically stratified (Livingston, 1997d; Livingston et al., 2000).

In Apalachicola Bay, residual surface flows are directed primarily to Sikes Cut and West Pass (Livingston et al., 2000). The highest level of stratification in the bay occurred at the COE dredged and maintained Sikes Cut. The dredged pass opens onto the primary basin of Apalachicola Bay, whereas West Pass has restricted access due to oyster bar development (Livingston et al., 2000). At West Pass, tidal variability was predominant. Average bottom current velocities were highest at the passes and in the eastern region of the bay, with the highest average bottom velocities at Sikes Cut. This situation appears to have an important influence on the bottom current structure of Apalachicola Bay due to the density differentials brought about by the heavier gulf water penetrating through the cut. Stratification was relatively less well developed in western sections of the bay approaching Indian Pass where tidal and subtidal variation was prominent. The lowest levels of stratification were noted in East Bay, an area notable for having shallow depths and generally low salinities.

During periods of high river flow, there were increased salinity gradients in all directions, and the Apalachicola Estuary was generally stratified with the range of salinity gradients dependent on season and location. Using a combination of regression and analysis of variance with the long-term database, Livingston (1984a) showed that surface to bottom salinity-by-month interactions were significant at the dredged Sikes Cut and in the Intracoastal Waterway at the mouth of the river. Thus, stratification was significant over prolonged periods in two areas that were affected by dredging in the bay.

River flow, depth (dredging), and the entrances to the gulf (including Sikes Cut) were important variables in the long-term trends of stratification in the Apalachicola Estuary. However, there was a relatively low incidence of hypoxic conditions in the bay. The fact that the deepest and most highly stratified areas were found at the highly active river mouth and at the passes that were subject to highly oxygenated gulf waters could, in part, account for this finding. The relatively shallow Apalachicola Estuary, although periodically stratified, was in fact one of most well-mixed estuaries along the NE Gulf. This factor allowed increased utilization of mineralized nutrients from the bottom by the phytoplankton with periodic (i.e., drought) conditions enhancing production of benthic microalgae (Livingston, 1984a).

By comparison, in the deeper, highly stratified Choctawhatchee Estuary, there was little in the way of interchanges between bottom and surface areas. Choctawhatchee Bay has shallow shelf areas with relatively steep slopes that eventually level out to average depths of three meters in eastern sections and eight meters in western sections of the central portions of the estuary. Areas along the periphery of the bay were generally less than 2 m deep. The deepest parts of the bay were positively correlated with high stratification (HS, $R^2 = 0.335$, $p = 0.001$; PSS, $R^2 = 0.465$, $p = 0.01$). Extreme hypoxia (ASH) was highly correlated ($R^2 = 0.605$, $p = 0.001$) with stratified conditions in deeper areas of the bay. There was a general decrease of such correlation with decreasing stratification (PSS, $R^2 = 0.322$, $p = 0.001$; PSW, $R^2 = -0.294$, $p = 0.01$; VH, $R^2 = -0.553$, $p = 0.001$). Hypoxic conditions (HYP) showed a similar trend (HS, $R^2 = 0.451$, $p = 0.01$; PSS, $R^2 = 0.702$, $p = 0.01$; PSW, $R^2 = -0.384$, $p = 0.01$; VH, $R^2 = -0.552$, $p = 0.01$). Areas characterized by DO concentrations exceeding 4.0 mg L^{-1} showed an opposite trend as a function of the stratification indices (HS, $R^2 = -0.562$, $p = 0.001$; PSS, $R^2 = -0.672$, $p = 0.0001$; PSW, $R^2 = 0.407$, $p = 0.01$; VH, $R^2 = 0.627$, $p = 0.001$).

A regression of subhalocline hypoxic conditions (HYP) with mean depths in the Choctawhatchee Estuary showed a relatively high R^2 value ($R^2 = 0.636$, $p = 0.003$). Depth was a leading variable in the distribution of hypoxia in the bay, with salinity stratification as a necessary prerequisite for deteriorating water-quality conditions. There was thus a direct association of depth, salinity stratification, and the extent of hypoxia in Choctawhatchee Bay. The primary difference between Choctawhatchee Bay and Apalachicola Bay was the shallow depths in the Apalachicola system and the lack of hypoxia despite stratification at the passes and in areas directly receiving Apalachicola River flow. However, other factors were associated with these trends. Bayou stations located in urban areas of the Choctawhatchee Bay also had high levels of hypoxia with or without stratification indicating adverse effects due to urban runoff (Livingston, 1997b). High stratification (HS) and strong partially stratified (PSS) conditions were noted at all bayou stations with the exception of Hogtown Bayou. Dead-end canals surrounded by high levels of urbanization and marina development in the Choctawhatchee Bay (Old Pass Lagoon) showed the lowest DO at depth in the entire bay. In nearby areas where there was more exchange with the open gulf, hypoxic conditions at depth were minimal.

High-salinity water from the gulf moves up Pensacola Bay and into the deeper parts of Escambia Bay and East Bay (Livingston, 1999). As in most river-dominated estuaries, salinity variation was high in the upper and midportions of Escambia Bay. In both the Blackwater and Escambia Estuaries, river flows dominated the physical habitats. The lowest salinities were noted during winter–spring periods of high river flow. Peak salinities occurred during fall periods of relatively low river flows. North–south gradients of salinity were evident in both systems. Salinity stratification was maximal in deeper areas of the lower and midportions of Escambia Bay. High salinity stratification in Blackwater Bay/East Bay occurred in deeper parts of both systems.

Surface water DO was highest in upper Escambia Bay (Livingston, 1999). The lowest DO occurred in areas of the highest salinity stratification indices, usually at depth (i.e., below the halocline). Regressions of DO with other physical/chemical variables indicated that surface DO was significantly (but weakly) associated with surface salinity (i.e., near the mouth of the Escambia River). BDO was significantly (inversely) associated with bottom temperature and the salinity stratification index. The depth/hypoxia relationship occurred only in upper parts of both the Blackwater and Escambia systems. In Pensacola Bay, deeper areas were not associated with hypoxic conditions at depth due to active exchanges of bottom water from the gulf. In Blackwater Bay, bottom-water oxygen concentrations were highest in the upper bay and deeper parts of East Bay. Surface water oxygen increased with distance from the river mouth. This situation was similar to that in Escambia Bay, where the highest oxygen was noted in the midbay areas. Temporal relationships between BDO concentrations and salinity stratification indices indicated that bottom-water DO in the Pensacola Bay Estuary was controlled by salinity stratification in the upper and midparts of Escambia Bay and the Blackwater/East Bay system.

The analysis of the relationships of depth, salinity stratification, and DO distribution in Perdido Bay was carried out with data taken before the initiation of the plankton blooms. With the exception of the lower end of Elevenmile Creek (station P22), shallow waters above the halocline in Perdido Bay were characterized by DO concentrations above 4 mg L^{-1}. The highest levels of severe to moderate hypoxia at depth were located in Elevenmile Creek, which was affected by high dissolved organic carbon (DOC) (leading to high biological oxygen demand [BOD]) and ammonia released from the pulp mill located at the head of the creek (Livingston, 2007a). High levels of severe to moderate hypoxia were also found in estuarine parts of the Perdido Estuary. Relatively low levels of hypoxia occurred at depth in upper portions of the estuary. In the lower bay, subhalocline hypoxic conditions were found in just over 40% of the samples taken.

Spatial/temporal trends of salinity in Perdido Bay followed seasonal and interannual patterns of freshwater runoff. Winter to early spring periods represented the low-salinity times, whereas fall months of drought usually were characterized by the highest salinities of the year. Salinity stratification of the estuary essentially created two very different habitats that were depth related. The estuarine parts of the Perdido River and Elevenmile Creek showed the highest levels of salinity stratification with such stratification evident in over 80% of the samples. These areas represented the highest levels of stratification of all of the bays surveyed. Most areas of upper Perdido Bay were characterized by low levels of vertical stratification. These were the shallowest stations in the bay; areas less than 1 m in depth were usually vertically homogeneous in salinity distribution due largely to wind mixing. Overall, mean depth in Perdido Bay was highly correlated with high salinity stratification (HS) ($R^2 = 0.688$, $p = 0.001$) and was inversely correlated with vertical homogeneity (VH) ($R^2 = -0.468$, $p = 0.01$).

The distribution of hypoxia in Perdido Bay was comparable to DO conditions found in deeper portions of Choctawhatchee Bay. Extreme bottom-water hypoxia in Perdido Bay was highly correlated with high levels of stratification (HS, $R^2 = 0.749$, $p = 0.001$). There were inverse correlations of bottom hypoxia with decreasing levels of stratification (PSS, $R^2 = -0.210$, $p = 0.01$; PSW, $R^2 = 0.521$, $p = 0.01$; VH, $R^2 = -0.418$, $p = 0.003$). Vertically homogeneous salinity conditions showed an opposite trend (HS, $R^2 = -0.787$, $p = 0.01$; PSS, $R^2 = 0.275$, $p = 0.001$; PSW, $R^2 = 0.521$, $p = 0.001$; VH, $R^2 = 0.406$, $p = 0.01$). There was a negative association of severe hypoxia with surface salinity ($R^2 = 0.248$, $p = 0.007$). These trends were similar to those noted in the Choctawhatchee Bay Estuary. Both estuaries were subjected to increased salinity stratification by the creation of artificial passes to the gulf that are currently maintained by the US Army COE.

Detailed studies by Niedoroda (1992) and modeling by Gallagher (see Livingston, 1993a) indicated that the physiography of Perdido Bay, together with seasonal and interannual fluctuations of the Perdido River, determined the stratification potential and BDO regime of the estuary. Flushing rates of the bay played an important role in the generation of plankton blooms, which, in turn, were associated with increased surface DO and reduced BDO during warm periods. Saltwater intrusion into lower Elevenmile Creek was found at distances of 2–3 km from the mouth of the creek. The level of salinity stratification was quite significant in the creek, with considerable differences between surface and bottom salinities, thereby limiting mixing of the surface and bottom-water layers.

The modeling analysis indicated that the decrease of surface layer DO ranged from about 6.0 mg L^{-1} at mile 2.5 to between 3.0 and 4.0 mg L^{-1} at the creek mouth. This decrease was due primarily to the oxidation of the paper mill's wastewater carbonaceous BOD and ammonia. The lower layer DO levels of 1.0–2.0 mg L^{-1} were due almost entirely to the effect of Perdido Bay water quality and sediment oxygen demand as a consequence of stratification. However, the long-term studies in Perdido Bay indicated that salinity stratification exacerbated the effects of hypereutrophication through increased subhaloclinal hypoxia as phytoplankton biomass increased throughout the bay.

The history recorded in long cores sampled in Perdido Bay sediments (Brush, 1984, 1991) indicated changes that were possibly related to nonpoint sources such as timbering, agriculture, and municipal development, but very little if any changes could be specifically related to the paper mill activity in Elevenmile Creek. The historical data also indicated that salinity increases in the Perdido Estuary were a sharp deviation from the natural (i.e., freshwater) conditions of the estuary in the recent past (i.e., prior to the opening of Perdido Pass).

The modeling data, combined with the descriptive field analyses and the field/laboratory experiments, indicated that hypoxia in Perdido Bay prior to the plankton blooms was the product of highly complex chemical interactions that were mediated in large part by the hydrographic characteristics of the estuary. The establishment of a strong halocline was a central feature of the development of hypoxia in Perdido Bay. A series of screening models (Klein and Galt, 1986) were run to determine the importance of specific state variables in the hydrodynamics of Perdido Bay. The Klein models showed that meteorological events in the Gulf region were responsible for transport conditions that were consistent with the movement of nutrients and organic matter up the bay from postulated sources in the lower bay. These analyses all indicated that the opening and continued maintenance of Perdido Pass had extensive, adverse effects on habitats in deeper parts of Perdido Bay.

Based on rates of river flow and the physiography of the various large river–estuaries in the NE Gulf, Niedoroda (1999) found that circulation and mixing in Escambia Bay were intermediate between the more "vigorous" Apalachicola Bay system and the more restricted Choctawhatchee and Perdido Estuaries. Depth and stratification were highest in Choctawhatchee Bay; reduced wind mixing and high stratification led to similar reductions of mixing in the Perdido system. Stratification was relatively low in the Apalachicola and Apalachee systems and was intermediate in Pensacola Bay. Salinity stratification was closely associated with depth (usually occurring in depths >2 m); this stratification was closely associated with subhaloclinal hypoxia in both estuaries. The lack of vertical mixing also stabilized the sediments in these systems, leading to formation of liquid mud, which represented an impaired habitat for bottom-dwelling organisms.

The combination of periodic, warm-season hypoxia and liquid mud formation resulted in poor benthic habitat in the Choctawhatchee and Perdido Estuaries, a situation that was exacerbated by the radical alteration of the inshore salinity regimes by dredged passes to the high-salinity gulf waters (Livingston, 1997d). Thus, the dredged openings of the lower parts of both Choctawhatchee Bay and Perdido Bay impaired useful benthic production in areas with depths >2 m, which included considerable parts of these estuaries. A similar adverse impact on oyster production due to increased predation in high-salinity bottom waters entering through the artificial Sikes Cut in Apalachicola Bay has been documented (Livingston et al., 2000).

An interbay analysis of the salinity stratification and associated DO conditions was carried out with annually averaged data. Mean depths were highest in the Choctawhatchee and Perdido systems with East Bay of the Apalachicola Estuary being the shallowest of the various estuaries (Figure 6.1). Salinities (Figure 6.2) were highest in Apalachee Bay and Choctawhatchee Bay, whereas the most highly stratified systems were the Perdido and the Choctawhatchee Bay systems. The most anoxic conditions were in the highly stratified Perdido and Choctawhatchee Estuaries that were also subjected to the opening of the respective areas to the open gulf waters by anthropogenic (dredging) activities (Figures 6.3 and 6.4).

The effects of salinity stratification were not limited to the reductions of bottom-water DO. There was a strong connection between salinity stratification and sediment quality in the coastal systems along the Gulf Coast. The quality of sediments in a given estuary is an important component of habitat suitability for secondary production in terms of invertebrates and fishes. Resulting hypoxia inhibits nitrification and denitrification in the sediments that increases the algal production process and resulting bacterial decomposition and bacterial oxygen consumption. There was little evidence of year-to-year carryover of organic matter and associated nutrients in Perdido Bay (Livingston, 1992). The establishment of a halocline and subsequent hypoxia at depth is rapid, occurring within days or weeks of the stratification. The movement of salt water from the lower end of the bay is rapid (5–6 days, Livingston, 1994)

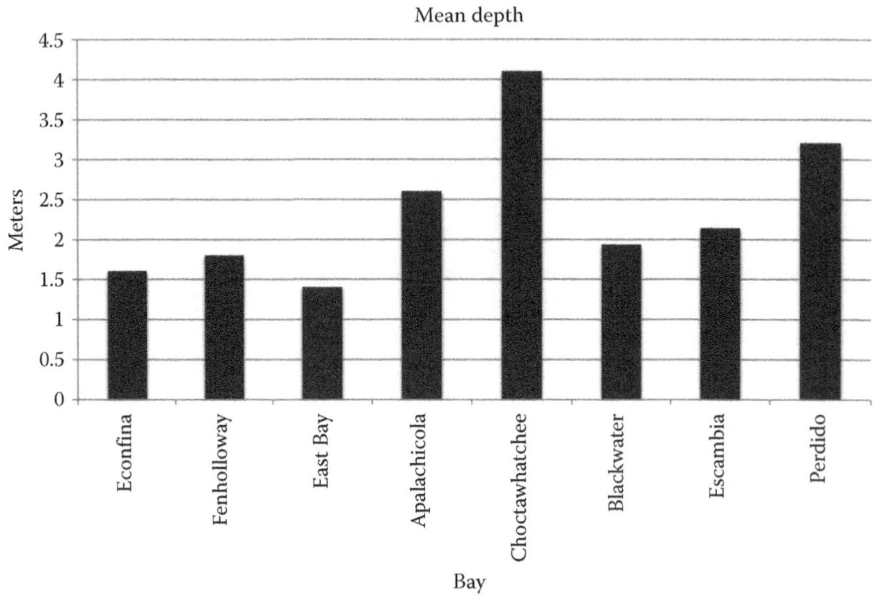

Figure 6.1 Mean depths of the various subject bay systems based on averaged data over the respective study areas.

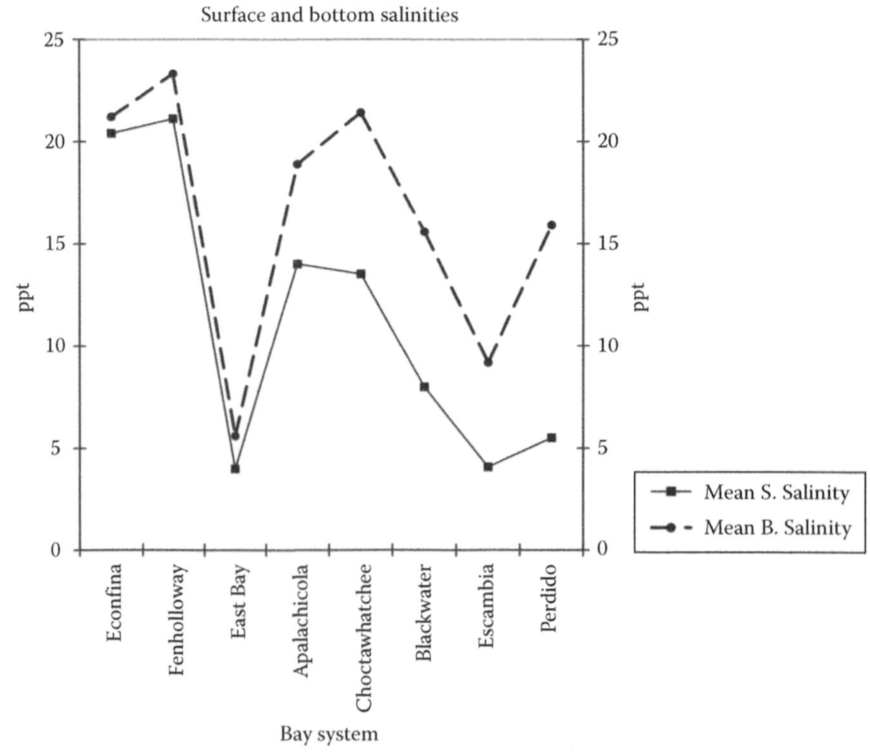

Figure 6.2 Surface and bottom salinities of the various subject bay systems based on averaged data over the respective study areas.

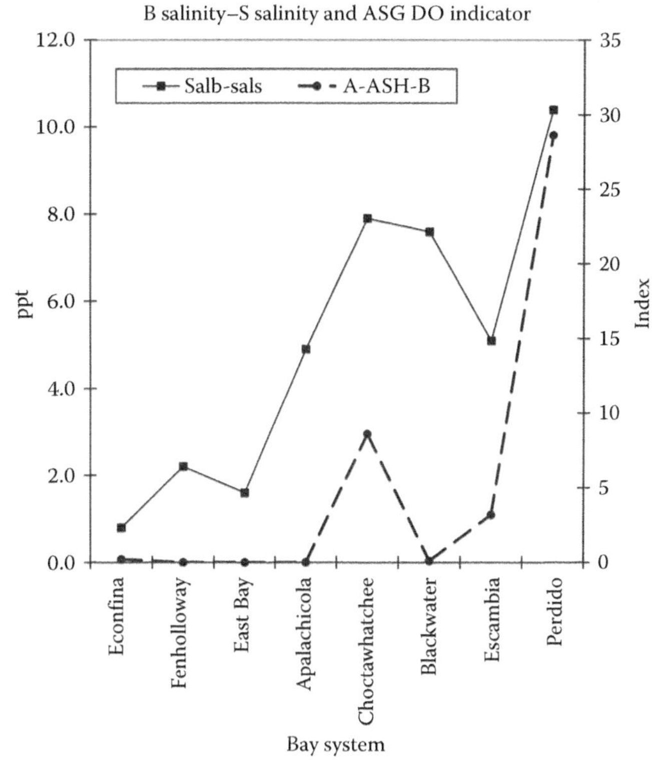

Figure 6.3 Bottom–surface salinities and DO indicators of the various subject bay systems based on averaged data over the respective study areas.

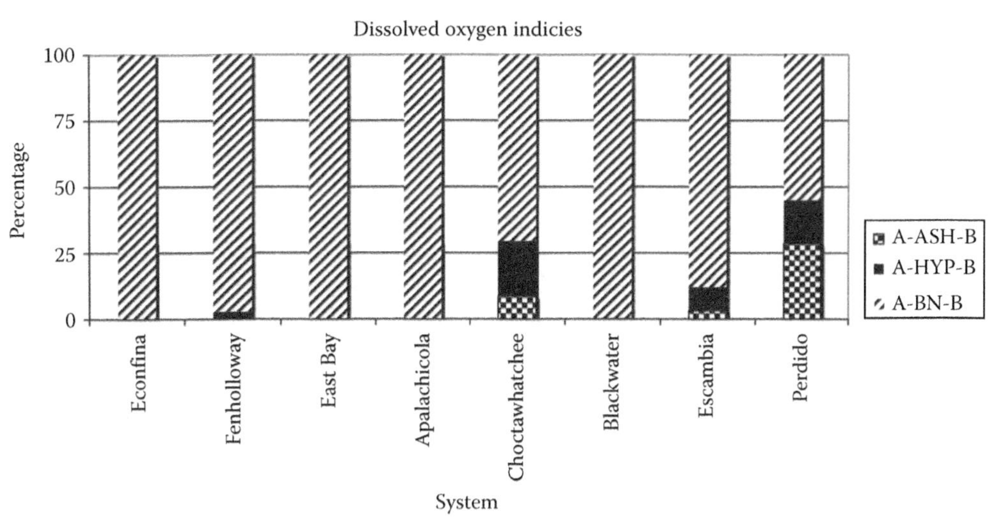

Figure 6.4 DO indicators of the various subject bay systems based on averaged data over the respective study areas.

as salt water moves along the bottom from the gulf to the head of the bay. Stratification inhibits reaeration and the reduction of DO occurs as the gulf intrusion moves northward.

In general appearance and consistency, the liquid mud in stratified areas of Choctawhatchee Bay was almost identical to that in like parts of Perdido Bay. There was a fine flocculent layer above the sandy substrate that was a gray-black color. In both systems, the salt wedge was almost always warmer than the surface layer. The flocculent sediments were easily disturbed by movements of divers, but no currents were evident in the liquid mud layer. The sediment orthophosphate levels in the lower end of Elevenmile Creek and the nearby bay areas tended to be the highest such concentrations in the Perdido Estuary. These levels were also higher than those in the Choctawhatchee sediments. The data indicated that Elevenmile Creek (and pulp mill discharges) had an effect on the orthophosphate in sediments of the immediate receiving area of the Perdido Estuary. The other parts of the lower Perdido River and upper Perdido Bay had levels of orthophosphate that were similar to those in the Choctawhatchee sediments.

BIOLOGICAL IMPACTS

There were significant relationships of the infauna, epibenthic invertebrates and fishes with various factors associated with salinity stratification in the Perdido Estuary (Table 6.1). The stratification index (botSal–surSal) had the highest and most prevalent statistical negative correlations with a range of biological indices. Depth and BDO were negatively and positively, respectively, associated with the biological indices.

This analysis indicated that the strong salinity stratification of deeper parts of Perdido Bay was not statistically associated with the distribution of the primary plankton blooms, but stratification was significantly and negatively correlated with a broad range of biological factors that included the trophic indicators.

Studies of the coastal systems of the NE Gulf coastal areas have indicated that the overall biological organization of these systems was affected by a combination of natural factors such as habitat and climatological variables and anthropogenic activities that varied in space and time. Deciphering the relative impacts of these factors has been difficult because of the complexity of interactions of multiple controlling agents and the temporal changes that occur over short and long periods. Because of long-term changes of climate variables such as temperature and rainfall/river flows, the effort to evaluate relative impacts has been hampered by the relative lack of long-term studies that have been carried out using consistent methods and adequate species identifications.

Another factor in this complex puzzle is that the processes that govern long-term impacts are often better understood when carried out as comparisons between and among similar coastal systems. Since many studies of the eastern gulf river–estuaries have been carried out over relatively short periods, the problems of extrapolation to longer periods have been largely ignored. Consistent and quantitative methods of collecting field data (particularly biological) have added another problem to the extrapolation of results to long-term generalizations. In short, comparisons among different coastal systems have not been routinely carried out in the gulf study area.

A principal component analysis (PCA) regression was run using data from the Perdido, Choctawhatchee, and Apalachicola Estuaries (Table 6.2). These estuaries had been subjected to significant dredging operations that opened these bays to high-salinity gulf waters. The dependent variables included the fish, infauna, and invertebrate (FII) indices (total biomass, herbivore biomass, omnivore biomass, C_1 carnivore biomass, C_2 carnivore biomass, C_3 carnivore biomass). Rather than using statistically significant independent variables in the PCA determinations, a standard set of key habitat variables was used to allow a comparative approach to determine what is meaningful in opening these estuaries to the gulf.

Table 6.1 Regressions of the Biological Indices Taken Bay-Wide in the Perdido Estuary and Averaged over the Period of Study Relative to Key Stratification Variables

1988–2007 Regressions	R^2 ($p < 0.05$)	All Stations				
Factor	**Depth**	**surSalinity**	**botSalinity**	**Surds**	**botDO**	**botSal–surSal**
Depth			0.39		−0.23	0.61
botSal	0.39	0.37				0.53
Inf n	−0.33		−0.26		0.28	−0.53
Inf b	−0.21				0.3	−0.34
Inf#taxa					0.52	−0.38
InfHerb b	−0.22		−0.19		0.3	−0.34
InfOmniv b			−0.33			−24
InfC1 b		0.4	−0.46	0.27	0.31	−0.21
Nvttu						
Nvt n	−0.2					
Nvt b						
NvtC_1 b						
NvtC_2 b	−0.2				0.24	
NvtC_3 b						
F b	−0.34					−18
FOmniv b	−0.28					−0.2
FC_1 b	−0.23					−0.16
FC_2 b	−0.17				0.27	−0.13
FC_3 b	−0.24		−0.23			
sqrtFIITu	−0.5		−0.16		0.44	−0.49
sqrtFII#taxa	−0.35				0.4	−0.37
FII b	−0.3				0.24	−0.34
FIIHerb b	−0.17				0.29	−0.29
FIIOmniv b	−0.27		−0.25			−0.26
FIIC_1 b	−0.22				0.25	−0.21
FIIC_2 b	−0.25				0.26	−0.16
FIIC_3 b	−0.24		−0.22			

Only significant ($p < 0.05$) factors are shown in the table. n, numbers; b, biomass; inf, infauna; nvt, epibenthic invertebrates; f, fishes; omni, omnivores; c, carnivores; tu, trophic units; C_1, primary carnivores; C_2, secondary carnivores; C_3, tertiary carnivores.

In terms of total FII biomass, all three bays showed direct relationships to sediment particle size and inverse relationships to % sediment silt (Table 6.2). Only Perdido Bay and Choctawhatchee Bay had an inverse relationship of this biological factor to salinity stratification and related indices such as water depth and bottom salinities. This finding is consistent with the fact that the adverse effect of Sikes Cut in the shallower Apalachicola Estuary was restricted mainly to southern parts of the bay, whereas the impacts of the opening of the Perdido and Choctawhatchee Bays were more general and affected major parts of the respective estuaries.

A similar pattern of response of FII herbivores and omnivores to sediment and stratification factors was noted in the three estuaries. Factors such as sediment quality and related levels of salinity stratification were overwhelmingly dominant with respect to the reductions of the primary trophic components of the Perdido and Choctawhatchee Estuaries relative to the more productive Apalachicola Bay. These results show that, relative to other key habitat variables, the anthropogenic salinity stratification of the Perdido and Choctawhatchee Estuaries was a dominant adverse influence on the herbivores and omnivores of these two systems, whereas these indices in the Apalachicola Estuary were not adversely affected by such stratification.

Table 6.2 PCA Regressions of FII Indices Regressed against Key Habitat Variables and Trophic Factors Taken in the Perdido, Choctawhatchee, and Apalachicola Estuaries

Eigenvectors	Principal Component Axis		
Variable	Perdido Bay	Choctawhatchee Bay	Apalachicola Bay
	(a) Total FII biomass m^{-2}		
RIV	0.076952	−0.004519	0.388992
SEDGS	−0.324567	−0.360979	−0.351406
SED%ORG	−0.285602	−0.258531	0.046878
%SAND	0.377293	0.362144	0.072876
%SILT	−0.350015	−0.351643	−0.299869
%CLAY	−0.243295	−0.350572	0.055532
DEPTH	−0.32612	−0.303711	0.067773
SECCHI	−0.191576	−0.107963	0.207013
BS-SS	−0.329942	−0.319692	−0.186916
BS-SS/D	−0.21885	−0.20049	0.11236
SDOAN	−0.079122	−0.045408	0.014954
SCOL	0.139485	−0.045841	0.04483
STURB	0.071349	−0.042598	0.008146
SCLRA	0.060388	0.020121	nd
BSAL	−0.317123	−0.21598	−0.170861
BTEMP	−0.018699	0.000941	0.124297
BDO	0.146337	0.224617	−0.029566
BDOAN	0.131011	0.271697	−0.011891
Analysis of Variance			
St error	0.06647995	0.00788937	0.215295
T fo HO	11.773	7.774	3.4524
Prob > ITI	0.0001	0.0001	0.0002
	(b) Total FII Herbivore Biomass m^{-2}		
RIV	0.076952	−0.334483	−0.000376
SEDGS	−0.324567	0.01002	−0.035011
SED%ORG	−0.285602	−0.014956	−0.235081
%SAND	0.377293	−0.014508	−0.237206
%SILT	−0.350015	0.351643	0.243825
%CLAY	−0.243295	0.088982	−0.245346
DEPTH	−0.32612	0.116178	−0.234526
SECCHI	−0.191576	0.347863	−0.144935
BS-SS	−0.329942	−0.319692	−0.293179
BS-SS/D	−0.21885	−0.249898	−0.271333
SDOAN	−0.079122	0.231	0.133237
SCOL	0.139485	0.045841	0.288845
STURB	0.071349	−0.330215	0.071727
SCLRA	0.06388	−0.020121	nd
BSAL	−0.317123	0.215882	−0.241806
BTEMP	−0.018699	0.282921	−0.241806
BDO	0.146337	−0.224617	0.169096
BDOAN	0.131011	0.163057	0.169096

(Continued)

Table 6.2 (*Continued*) PCA Regressions of FII Indices Regressed against Key Habitat Variables and Trophic Factors Taken in the Perdido, Choctawhatchee, and Apalachicola Estuaries

Eigenvectors	Principal Component Axis		
Variable	**Perdido Bay**	**Choctawhatchee Bay**	**Apalachicola Bay**
Analysis of Variance			
St error	0.10931612	0.04295233	0.09531973
T fo HO	13.109	5.531	3.654
Prob > ITI	0.0001	0.0001	0.001
	(c) Total FII Omnivore Biomass m^{-2}		
RIV	0.076952	−0.004519	−0.097561
SEDGS	−0.324567	−0.360979	−0.035011
SED%ORG	−0.285602	−0.258531	−0.235081
%SAND	0.377293	0.362144	−0.237206
%SILT	−0.350015	−0.351643	0.243825
%CLAY	−0.243295	−0.350572	−0.245346
DEPTH	−0.32612	−0.303711	−0.234526
SECCHI	−0.191576	−0.107963	−0.144935
BS-SS	−0.329942	−0.319692	−0.293179
BS-SS/D	−0.21885	−0.20049	−0.271333
SDOAN	−0.079122	−0.045408	0.133237
SCOL	0.139485	−0.045841	0.288845
STURB	0.071349	−0.042598	0.071727
SCLRA	0.060388	0.020121	nd
BSAL	−0.317123	−0.215841	−0.241806
BTEMP	−0.018699	0.000941	−0.241806
BDO	0.146337	−0.224617	0.169096
BDOAN	0.131011	0.271697	0.169096
Analysis of Variance			
St error	0.05352967	0.04014864	0.14310109
T fo HO	6.604	10.652	5.383
Prob > ITI	0.0001	0.0001	0.0001
	(d) Total FII C_1 Carnivore Biomass m^{-2}		
HWT	0.288595	0.028363	0.320159
OMWT	0.18503	0.212986	0.290606
RIV	0.061717	−0.016162	−0.061551
SEDGS	−0.295873	−0.353214	0.379645
SED%ORG	−0.260212	−0.252899	0.106607
%SAND	0.352705	0.352269	0.04464
%SILT	−0.328221	−0.341696	−0.156789
%CLAY	−0.221293	−0.342534	0.142433
DEPTH	−0.314165	−0.297075	0.163453
SECCHI	−0.167361	−0.095872	0.03563
BS-SS	−0.32469	−0.315622	−0.137528
BS-SS/D	−0.223365	−0.200767	−0.001958
SDOAN	−0.040896	−0.031957	0.052149
SCOL	0.118769	−0.058998	0.11839

(*Continued*)

Table 6.2 (*Continued*) PCA Regressions of FII Indices Regressed against Key Habitat Variables and Trophic Factors Taken in the Perdido, Choctawhatchee, and Apalachicola Estuaries

Eigenvectors	Principal Component Axis		
Variable	**Perdido Bay**	**Choctawhatchee Bay**	**Apalachicola Bay**
STURB	0.052759	−0.055779	−0.455783
SCHLOR	0.079583	0.016423	−0.15771
BSAL	−0.296078	−0.196183	−0.15775
BTEMP	0.008883	0.01203	0.103698
BDO	0.149709	0.216634	0.193698
BDOAN	0.143837	0.271864	0.09985
Analysis of Variance			
St Error	0.05131244	0.02800913	0.06651701
T fo HO	5.43	5.465	5.648
Prob > ITI	0.0001	0.0001	0.0001
	(e) Total FII C_2 Carnivore Biomass m^{-2}		
HWT	0.093771	0.182402	0.233463
OMWT	−0.28284	0.036376	0.238896
C1WT	0.201891	0.307526	0.111845
RIV	0.524781	−0.418584	−0.018452
SEDGS	0.154311	0.002646	−0.050399
SED%ORG	0.182813	0.039706	−0.255042
%SAND	−0.069693	0.011558	−0.259106
%SILT	0.106508	0.014621	0.258905
%CLAY	0.161685	−0.065611	−0.261216
DEPTH	0.2299	−0.095886	−0.250396
SECCHI	−0.027122	−0.119154	−0.169812
BS-SS	−0.138475	−0.118164	−0.24913
BS-SS/D	−0.044965	0.040505	−0.219288
SDOAN	−0.0389	−0.138758	0.022498
SCOL	−0.081277	−0.394891	0.263726
STURB	0.284	−0.013984	0.028534
SCHLOR	−0.282597	0.270767	0.263726
BSAL	−0.092751	0.279586	−0.237747
BTEMP	−0.384343	0.488718	−0.189994
BDO	−0.066085	−0.045841	0.170888
BDOAN	0.15304	−0.144768	0.144125
Analysis of Variance			
St error	0.07750323	0.05149559	0.1384877
T fo HO	5.873	7.554	3.442
Prob > ITI	0.0001	0.0001	0.002
	(f) Total FII C_3 Carnivore Biomass m^{-2}		
HWT	0.08457	0.182632	0.018889
OMWT	0.320527	0.119423	0.218611
C1WT	0.180244	−0.021985	−0.143086
C2WT	0.054566	0.207198	−0.072238
RIV	−0.0675	−0.323686	0.153413
SEDGS	0.197874	0.004391	−0.361196

(*Continued*)

Table 6.2 (*Continued*) PCA Regressions of FII Indices Regressed against Key Habitat Variables and Trophic Factors Taken in the Perdido, Choctawhatchee, and Apalachicola Estuaries

Eigenvectors	**Principal Component Axis**		
Variable	**Perdido Bay**	**Choctawhatchee Bay**	**Apalachicola Bay**
SED%ORG	0.199356	−0.01681	0.013326
%SAND	−0.104071	−0.011781	0.039889
%SILT	0.075001	0.337138	−0.029067
%CLAY	0.173159	0.080286	0.024706
DEPTH	0.060906	0.109495	0.035973
SECCHI	−0.067319	0.329414	0.161019
BS-SS	−0.092816	−0.128563	0.098739
BS-SS/D	−0.158997	−0.22758	0.160569
SDOAN	0.065724	0.23039	0.014921
SCOL	−0.156496	0.043762	0.028011
STURB	0.051154	−0.320463	−0.01903
SCHLOR	0.186876	−0.141725	−0.211505
BSAL	0.119778	0.202789	0.126314
BTEMP	0.43419	0.278259	0.176244
BDO	0.297979	−0.215968	0.419392
BDOAN	−0.098539	0.155152	−0.010259
Analysis of Variance			
St error	0.06162242	0.04673134	0.19852627
T fo HO	3.096	3.543	3.971
Prob > ITI	0.0022	0.0004	0.0007

Notes: Data were averaged over 2 years of sampling during moderate to high river flows. Variables included the following: HWT, herbivore biomass; OMWT, omnivore biomass; C1WT, primary carnivore biomass; C2WT, secondary carnivore biomass; RIV, river flow; SEDGS, sediment grain size; SED%ORG, sediment percent organics; %SAND, sediment percent sand; %SILT, sediment percent silt; %CLAY, sediment percent clay; DEPTH, water depth; SECCHI, Secchi depth; BS-SS, salinity stratification index; BS-SS/D, salinity stratification index/depth; SDOAN, surface dissolved oxygen anomaly; SCOL, surface color; STURB, surface turbidity; SCHLOR, surface chlorophyll *a*; BSAL, bottom salinity; BTEMP, bottom temperature; BDO, bottom dissolved oxygen; BDOAN, bottom dissolved oxygen anomaly.

Analysis of the C_1 carnivores (Table 6.2) indicated strong positive associations between this level of trophic organization with the FII herbivores of the Perdido and Apalachicola Estuaries. Once again, there were strong negative associations of the C_1 carnivores of the Perdido and Choctawhatchee Estuaries with sediment factors (inverse relationships with particle size and silt fractions) and salinity stratification and depth factors. These associations were largely absent in the shallow and relatively well-mixed Apalachicola Estuary where the main determinants were positively correlated with herbivore and omnivore biomass and larger sediment particles. The C_2 carnivores were not correlated with any of these habitat variables and were positively associated with C_1 carnivore biomass. The results were similar with the C_3 carnivores with the caution that the correlation coefficients were relatively low with this group in a way that would disqualify a strict interpretation of these results.

Livingston et al. (1997), working with the long-term Apalachicola Bay data, found that there appeared to be a dichotomy of response of the trophic elements with respect to controlling factors.

The herbivores and omnivores were directly linked to physical and chemical controlling factors that were associated implicitly with the primary productivity of the estuary. These observations were supported by the dynamic regression models in which these two feeding groups were coupled to the principal component axes representing the environmental variables. The river, which mediates such factors, was thus directly linked to the response of these trophic components. The primary, secondary, and tertiary carnivores, on the other hand, were associated more closely with biological factors; none of the carnivore groups were related directly to any environmental principal component after taking into account the effects of herbivores and omnivores (Livingston et al., 1997). This suggests that river flow and primary production were mainly associated with changes in the lower trophic levels and that the carnivores were associated primarily with other animal trophic components.

Depth and salinity stratification due to dredged openings to the gulf were highest in the Choctawhatchee and Perdido Estuaries. Stratification was relatively low in the Apalachicola and Apalachee Bay systems and was intermediate in the Pensacola Estuary. There was an adverse impact on oysters in parts of lower Apalachicola Bay that were affected by increased salinities due to enhanced predation in high-salinity bottom waters entering through the artificial Sikes Cut. Salinity stratification in the Perdido and Choctawhatchee Estuaries was closely associated with anthropogenic openings to the gulf and was statistically correlated with depth (usually occurring in depths >2 m); this stratification was statistically associated with subhaloclinal hypoxia. The lack of vertical mixing also stabilized the sediments in these systems, leading to the formation of liquid mud that represented an impaired habitat for bottom-dwelling organisms. The combination of periodic, warm-season hypoxia and liquid mud formation was statistically associated with poor benthic habitats and impaired useful secondary production in considerable parts of the Choctawhatchee and Perdido Estuaries. Dredged cuts had adverse effects on the trophic organization of major parts of these bays. There were different reactions of the various trophic elements to the stratified conditions. Overall, the adverse effects of the dredged openings of the affected estuaries were long term and pronounced, and the dredging factor should be included in any comparative review of the productivity of the gulf estuaries involved in the long-term studies in this book.

PART III

Trophic Response to Long-Term Climate Changes

CHAPTER 7

Climatological Impacts on Gulf Estuaries

The field-sampling effort (Figure 7.1) showed that the three long-term field efforts (Apalachee Bay, Apalachicola Estuary, Perdido Estuary) were run in such a way that there was an initial 10-year period of sampling that was not affected by a major drought. The Perdido study was included in this statement, because the first major drought occurred in 1999–2002 and that drought was followed by a second such event from 2006 to 2007. All three ecosystems were subsequently affected by the series of droughts that occurred over the succeeding years of the studies. This allowed an evaluation of the effects of the droughts on these estuaries. The short-term studies (Choctawhatchee Bay, Escambia Bay, Blackwater Bay) were carried out during no-drought periods. The relatively abbreviated analyses of these estuaries precluded an evaluation of climatic effects.

APALACHEE BAY

Background

The study area in the Apalachee Bay (Econfina and Fenholloway drainages, Figure 7.2) is part of a broad, shallow shelf area that occurs along the Gulf Coast of Florida. The coastal area from the Ochlockonee drainage to the southern reaches of the Apalachee Bay is characterized by the flat karst topography of the Gulf Coastal Lowlands. Small streams that include the Aucilla, Econfina, and Fenholloway Rivers flow into a coastal region dominated by low wave and tidal activity and relatively clear, high-salinity inshore waters.

The Fenholloway River and its associated estuary have been affected by long-term physical modifications through forestry activities that include a pulp mill that discharges into the upper parts of the stream. A field study was started in 1970 that compared the Fenholloway River and its offshore areas with the Econfina River and Estuary, a relatively natural blackwater stream to the north. The initial field monitoring was carried out from 1970 to 1980. A use attainability analysis, led by the State of Florida Department of Environmental Regulation, was performed from 1992 through 1993. The research effort was continued through a subsequent monitoring program from 1993 through 1999. Corresponding to a color removal program carried out by the Buckeye Technologies pulp mill in Perry, Florida, during fall 1998, a comprehensive field program was performed from 1999 to 2004.

The Econfina and Fenholloway drainage basins are wholly within the coastal plain as part of a poorly drained region that is composed of springs, lakes, ponds, freshwater swamps, and coastal marshes. The smaller streams along the Big Bend area are associated with relatively limited estuarine areas. The saltwater marshes of the Apalachee Bay region are highly developed, and the line of estuaries in this region has very high wetland/open water ratios compared to the alluvial estuaries

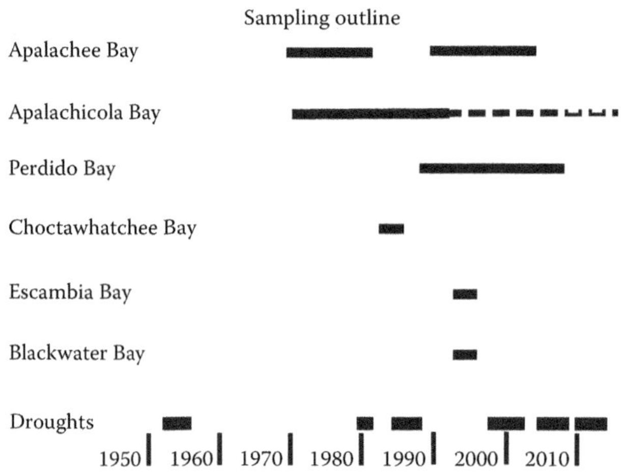

Figure 7.1 Outline of sampling effort of seven river–bay systems over a 43-year study program. The dashed line of the Apalachicola system represents reviews of fishery information. Field-sampling periods were compared to the distribution of droughts from 1950 through 2012.

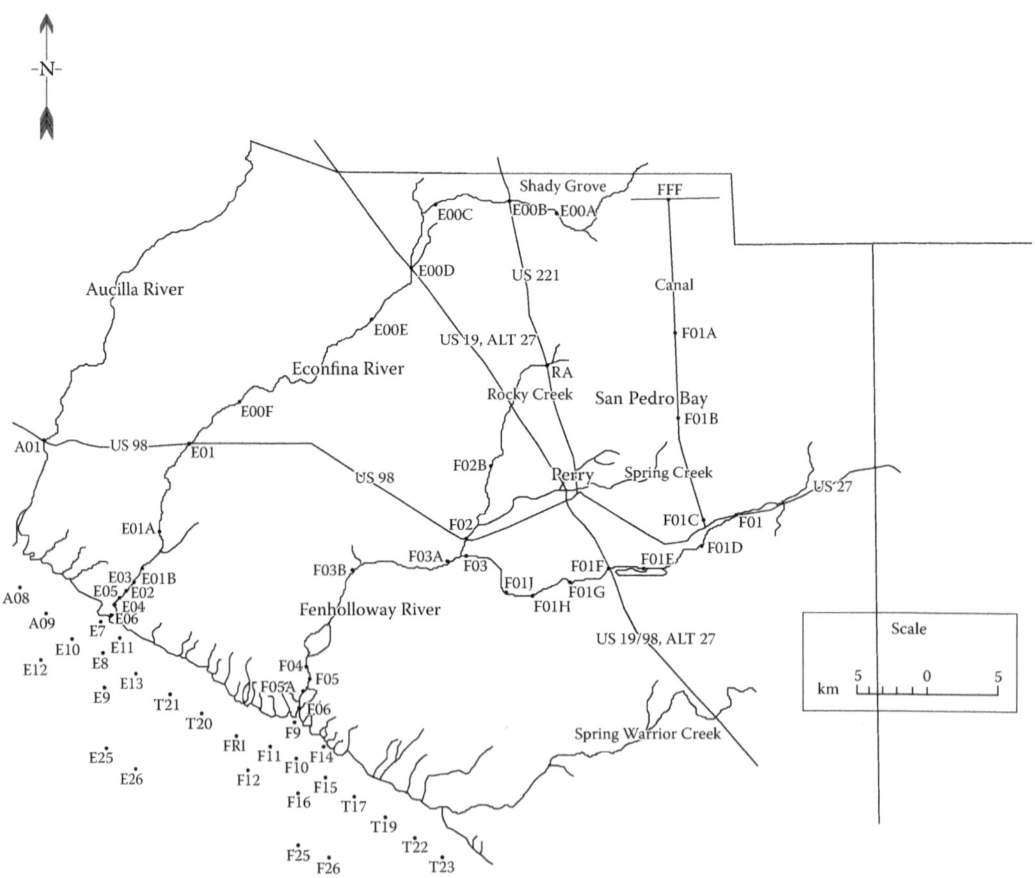

Figure 7.2 The Econfina and Fenholloway River–Estuaries in the Apalachee Bay showing sampling stations used in the long-term studies of this system from 1971 to 2004.

(Livingston, 2000). The inshore marine habitat of the Apalachee Bay serves as a physically stressed but highly productive nursery for larval and juvenile stages of offshore fishes and crustaceans, many of which form the basis of the sports species of the region.

Gradients of physical variables, together with high primary production mediated largely by seagrass beds, are the dominant features in terms of primary production and habitat distribution. Phytoplankton productivity is relatively low in the largely oligotrophic offshore parts of the Apalachee Bay. The highly complex but extremely productive food webs of the inshore marine ecosystems of this area are largely associated with the extensive development of seagrass beds.

Although the Apalachee Bay remains in relatively natural condition, the Fenholloway River–Estuary represents an area that has been adversely affected by pulp mill discharges since the 1950s. The Econfina River, on the other hand, remains one of the most pristine blackwater streams along the coast (Livingston, 2000). This stream provides an excellent reference area for studies in the Fenholloway system. Both the Econfina and Fenholloway drainages have similar dimensions. They share a common origin (San Pedro Swamp) and a common meteorological regime. River flow characteristics in the two drainages are comparable in rate and seasonal variation. However, the Econfina River follows regional rainfall conditions, whereas the Fenholloway River is not as affected by the pronounced changes of droughts and floods in the region due to the continuous input of pulp mill effluents to the river.

The shallow offshore habitats of Apalachee Bay along the NE Gulf Coast of Florida are dominated by seagrasses such as *T. testudinum*, *S. filiforme*, and *Halodule wrightii* (Iverson and Bittaker, 1986; Zieman and Zieman, 1989; Livingston, 2000, 2002, 2005). There is also a rich diversity of other forms of submerged aquatic vegetation (SAV) that includes various macroalgal types. The biomass-dominant seagrasses remain relatively undisturbed by human activity in the Apalachee Bay with the exception of one area off the Fenholloway drainage where the release of pulp mill effluents has affected the nearshore beds. Zimmerman and Livingston (1976a,b, 1979) found severely reduced SAV cover relative to similar areas in the Econfina reference system.

The area between the Econfina and Fenholloway drainages has been well studied. Different studies have addressed the interrelationships of the Apalachee Bay seagrass beds with the offshore biological organization (Livingston, 1984b). These analyses included examination of the role of seagrasses in primary production (Zimmerman and Livingston, 1976a,b, 1979) and the effects of pulp mill effluents on predator–prey responses (Main, 1983), physiological responses of pinfish (*L. rhomboides*) (Stoner and Livingston, 1978), avoidance reactions of fishes (Lewis, 1974; Lewis and Livingston, 1977; Livingston, 1982b), changes in organism distribution (Hooks et al., 1976; Heck, 1979; Stoner, 1979a,b, 1980c, Dugan and Livingston, 1982), ontogenetic (developmental) processes (Livingston, 1980a), trophic organization (Stoner, 1980a,b, Greening and Livingston, 1982; Livingston, 1982a, 1984b, Clements and Livingston, 1983, 1984; Leber, 1983; Stoner and Livingston, 1984; Luczkovich, 1986, 1988; Luczkovich et al., 2002, 2003), community composition (Livingston, 1975a, Lewis, 1982; Leber, 1983; Lewis and Stoner, 1983; Livingston, 1988a), and long-term changes of water-quality trends (Livingston, 1982b, 2002, 2005). These studies indicated that offshore gulf areas were characterized by complex biological associations that were influenced by both physical/chemical phenomena (such as pollution and natural background phenomena) and natural biological processes.

Relatively small changes in critical habitat features had measurable impacts on the biological organization of the Apalachee Bay system, but such changes were not necessarily predicated on simple or direct influences. Rather, the biological systems were shaped by diverse combinations of natural and anthropogenic stressors that varied due to seasonal and interannual (multiyear) patterns of rainfall/drought conditions, land runoff, and associated offshore habitat changes. In a combined field-descriptive/experimental and laboratory experimental study, Livingston et al. (1998b) showed that pulp mill effluents in the Fenholloway River were associated with increased loading of dissolved organic carbon (DOC), water color, and nutrients to offshore areas relative to the unpolluted (Econfina River) reference system. These loadings resulted in changes of water-quality factors and light transmission characteristics in the polluted system. Sediments in offshore Fenholloway areas were characterized by increased silt/clay fractions and altered particle size relative to reference sites.

Field analyses showed that the best predictors of SAV distribution in the Fenholloway offshore areas were photic depths, qualitative aspects of wavelength distributions (light transmission data), and water-quality factors such as water color, DOC, and chlorophyll *a* that were inversely related to seagrass biomass. Experimental analyses using mesocosms of dominant seagrasses (*T. testudinum*, *S. filiforme*, and *H. wrightii*) indicated that direct contact with pulp mill effluents resulted in adverse growth responses at concentrations as low as 1%–2%. Mesocosm experiments with light conditions that reproduced actual field conditions showed that light conditions in offshore areas affected by pulp mill effluents adversely affected growth indices of *S. filiforme* and *H. wrightii*. Field transfer experiments showed that growth indices of the three dominant seagrass species were adversely affected by water quality and sediment quality in the Fenholloway offshore receiving area (Livingston et al., 1998b).

Phytoplankton productivity in the Apalachee Bay is generally low compared to the large river–estuaries to the west (Livingston, 2000, 2002, 2005). Nutrient input in the Apalachee Bay is restricted to small stream loading, groundwater discharges, and runoff during storm events. Nutrient limitation experiments in the Econfina and Fenholloway systems indicated a complex set of nutrient–phytoplankton interactions (Livingston, 1993, unpublished data). Inshore areas of both systems were generally limited by nitrogen, whereas offshore areas were increasingly phosphorus limited with distance from shore. Relatively low TIN–TIP ratios in inshore areas during the experimental period also suggested nitrogen limitation. Farther offshore, higher TIN–TIP ratios showed increased phosphorus limitation. The general increase of the TIN–TIP ratios from inshore to offshore gulf areas off the respective river mouths provided evidence of the increasing importance of phosphorus to phytoplankton development in offshore areas of the gulf.

Estuarine phytoplankton populations in the Econfina system were more abundant than those in comparable areas of the Fenholloway system (Livingston, 2002, unpublished data). Estuarine and inshore parts of the Fenholloway system were low in phytoplankton species richness and diversity compared to the Econfina cognate stations. The highest numbers of phytoplankton were found in inshore parts of the Econfina and Fenholloway systems. Fenholloway offshore areas had higher numbers of phytoplankton than comparable Econfina areas. The suppression of phytoplankton species richness and diversity in the inshore Fenholloway system was most likely due to reduced light transmission, although other effects such as inhibition by high concentrations of ammonia could not be ruled out as an additional modifying factor.

Various studies (Hooks et al., 1976; Dugan and Livingston, 1982; Greening and Livingston, 1982) indicated that invertebrates taken in the Econfina offshore system were generally more abundant than those found in the Fenholloway system during the 1970s and that macrophyte abundance was directly associated with invertebrate numbers. Livingston (1975a) found that reduced fish species richness and diversity characterized the offshore Fenholloway system. These differences corresponded to previously described distributions of SAV and macroinvertebrates. The effects on fishes were also attributed to physiological impacts (respiration, growth, food conversion efficiency; Stoner, 1976) and/or behavioral responses such as avoidance of the mill effluents (Lewis, 1974; Lewis and Livingston, 1977). However, habitat changes due to the loss of seagrass beds, together with changes of trophic organization due to altered nutrient loading in the Fenholloway system, were the primary features that were associated with the loss of secondary production in areas associated with pulp mill effluent disposal.

Changes in the Fenholloway seagrass beds were associated with altered food web patterns in terms of species shifts and feeding alterations relative to reference areas (Livingston, 1975a, 1980a, 2002). Feeding habits of the dominant fish species were different in the offshore Fenholloway system than those in similar areas of the Econfina drainage. These differences were traced to basic changes of habitat due to altered seagrass beds in the Fenholloway system (Clements and Livingston, 1983, 1984; Livingston, 2002). Reduced SAV together with increased phytoplankton productivity due to nutrient loading from the paper mill altered the habitat conditions and fish trophic organization in the Fenholloway offshore area relative to the reference system. These effects were also associated with complex effects of sediment quality on population distribution and community organization.

Food webs in the Fenholloway system were directly influenced by shifts of primary producers and altered benthic habitats as seagrasses were turned into unvegetated or sparsely populated soft-sediment areas. In offshore Fenholloway areas, these food web shifts were also affected by increased phytoplankton production as dilution-enhanced light transmission in areas of increased nutrient concentrations. The benthic food web species were thus replaced in part by planktivorous fish species in the Fenholloway system due to habitat-related food web changes (Livingston, 2002). Improvement of water quality in 1974 due to the construction of treatment ponds by the pulp mill was associated with some recovery of SAV and with recovery of the fish trophic organization in outer areas of the Fenholloway system (Livingston, 1982a, 1984b, 2002). During the earlier period of untreated mill effluents, plankton-feeding fishes were dominant in areas affected by discharges, replacing grass bed species. Subsequent water-quality improvement due to the implementation of secondary treatment of the mill effluents (Livingston, 1975a) was accompanied by trophic shifts that followed habitat changes in the outer portions of the Fenholloway Estuary.

As part of a Fenholloway use attainability analysis in 1992–1993, our research group conducted an extensive study of nearshore gulf drainage systems. From this study, it was determined that approximately 24 km^2 (9 $miles^2$) of seagrasses offshore of the Fenholloway River had been eliminated when compared to the relatively natural Aucilla and Econfina nearshore areas (Livingston et al., 1998b). The mill effluents also adversely affected an additional 4 km^2 of seagrasses in adjoining areas. The study showed that light availability due to the higher color from the Fenholloway River caused by the discharge of treated wastewater from the Buckeye Florida Plant was the primary factor affecting seagrass growth. The study recommended that color reduction be based on annual averages found in what was considered to be a potential seagrass habitat about 1.5 miles offshore of the Fenholloway River (station F10; Figure 4.1). In order to accomplish a return of seagrasses in this area of the Fenholloway system, water-quality models developed by HydroQual, Inc. showed that a 50% reduction in the historical treated wastewater color from the Buckeye Florida Plant would be required for a return to a median color level of 36 platinum–cobalt units (PCUs), which would represent a 38% reduction of water color at station F10. These projected targets would be comparable to conditions in the Econfina system.

In addition to the primary variables affecting light transmission, the study also showed that nutrient loading and sediment characteristics in the Fenholloway nearshore area could affect seagrass growth. After a reduction of color in the treated wastewater by 50% in October 1998, it remained uncertain how nutrient loading (i.e., ammonia, orthophosphate) from the pulp mill would affect phytoplankton growth and possible bloom formation (Livingston, 2000, 2002). It was also necessary to understand whether sediment quality played a role in seagrass growth. The direct loading of nitrogen and phosphorus compounds to an oligotrophic system dominated by seagrasses was considered problematic in the sense that phytoplankton productivity changes are connected to the seagrass problem both in terms of altered light transmission and associated effects on water and sediment quality. There was a possibility that reduced color and increased light penetration in the Fenholloway Estuary could result in the production of plankton blooms similar to those observed in the Perdido Bay and Pensacola Bay systems (Livingston, 2000, 2002). Improvements in water color thus presented some concern in terms of the nutrient loading issue and possible nearshore phytoplankton response.

The response of the Fenholloway system to the color-reduction program was determined as part of the follow-up studies of 1999–2004 (Livingston, 2005). The target of 36 PCU at station F10 was evaluated relative to the response of the seagrass beds and associated fauna over the period 1999–2002. The study was carried out during a major drought over this period. The offshore sediments were high in sand-size particles with relatively low complements of silt and clay. The distribution of particle size was relatively uniform across stations in the Econfina and Fenholloway offshore areas. Percent organics were also relatively low in the offshore sediments. The most obvious differences between the sediments taken in the Econfina and Fenholloway systems were the uniformly higher concentrations of phosphorus in the Fenholloway coastal areas and higher concentrations of total organic nitrogen (TON) in parts of the Econfina system.

Rainfall and River Flows

Over the period from 1971 through 2004, the Apalachee Bay studies concentrated on the impacts of pulp mill effluents in the Fenholloway River and Estuary to that of the Econfina system. The longer-term changes in both systems over the study period were not examined in detail. However, there was a long-term trend of climatological changes in terms of increasing frequency and intensity of droughts in the Apalachee Bay region that started during the 1980s. These findings led to an examination of the potential effects of climatic variables on long-term responses of various biological associations in the Apalachee Bay by comparing the later results with those taken during the predrought conditions from 1971 to 1980.

Long-term rainfall patterns (1957–2012; Figure 7.3a) showed a general increase in the frequency and duration of drought periods from 1980 to 2012. The differenced rainfall data (Figure 7.3b) showed the earlier described pattern of increasing drought periods with decreasing highs of rainfall with time. The initial study period (1971–1980) was undertaken during a time that preceded the increased incidence of enhanced droughts.

The decade of the 1970s spanned a period of alternating high river flows with minor droughts that ended during the 1981–1982 low-rainfall period. There was a general trend of reduced rainfall over this initial study period. The second study (1992–1993) was carried out at the end of a prolonged and relatively severe drought (1986–1990) that ended with a major storm period. This study period was thus characterized by alternating drought and flood years. A limited monitoring period extended from 1993 to 1998; this period of study was followed by a more intensive field analysis from 1999 to 2004. It was during this period that there was another more severe drought that extended from 1999 to 2002. This was followed by a period of increased rainfall. The succeeding years included the severe droughts of 2006–2008 and 2010–2012.

This climatological pattern of the flows of the Econfina and Fenholloway Rivers tended to follow the long-term patterns of increasingly severe and lengthy droughts along the lines of the Chattahoochee River flow rates. Econfina River flows during the study periods followed the Perry, Florida, rainfall patterns. The Fenholloway flows, augmented by discharges of pulp mill effluents, appeared to follow the long-term rainfall patterns in only a general fashion. Peak flows in the Econfina River tended to be higher and lower than those in the Fenholloway, and the Econfina flows showed the effects of increased length and severity of the droughts that were only evident in the Fenholloway River in a general manner. The river flows during the lesser droughts of the 1970s were of shorter duration and were less intense than those during the later periods.

Overall, both the rainfall changes and the Econfina River flow patterns indicated a shift to longer and more extreme drought periods that started in the 1980s and continued to the present time. Due to inputs of the pulp mill, the Fenholloway River did not show drought effects to the same degrees as the Econfina.

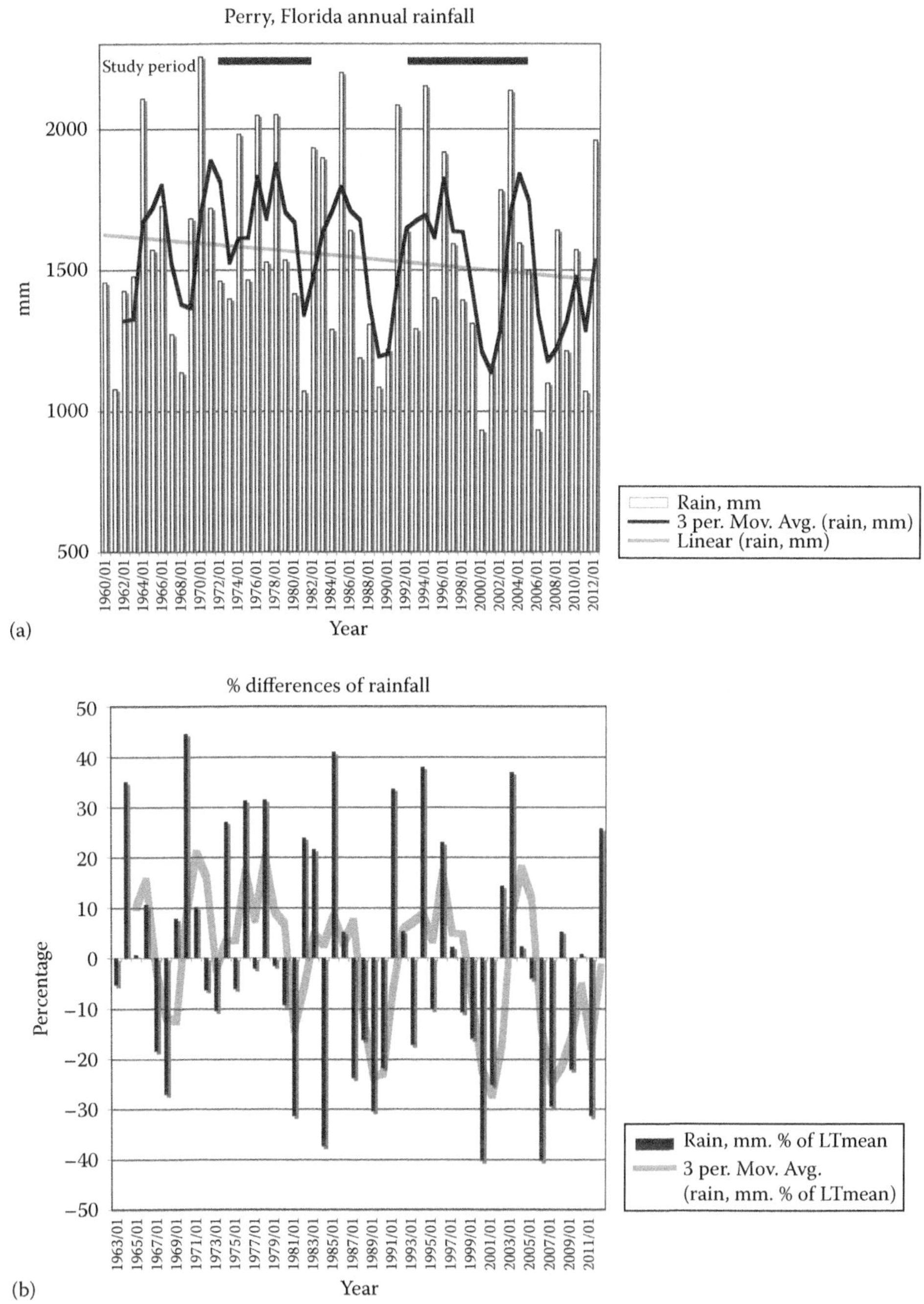

Figure 7.3 (a) Annual rainfall taken from 1957 through 2012. Data are also expressed as 3-year moving averages and as a linear regression. (b) Annual averages of regional rainfall presented as percentages of the long-term mean. Data are also shown as 3-month moving averages with illustration of periods of intensive studies.

Climatological Conditions and Nutrient Loading

Due to loading from the pulp mill, the Fenholloway River was a primary source of ammonia and orthophosphate to the Apalachee Bay system during most of the study period. Loading of both nutrients in the Fenholloway system was generally about two orders of magnitude higher than the reference Econfina system (Figure 7.4a and b). This loading of ammonia and orthophosphate maintained differences that were statistically significant ($p < 0.05$) over the sampling period. During the most recent drought, Econfina loading of ammonia and orthophosphate was low, whereas the continuous loading of these nutrients from the Fenholloway remained at relatively high levels. The pronounced reduction of ammonia and orthophosphate loading in the Econfina River extended from 1999 through 2002. There was a slight decrease in the Fenholloway loading over the last 5 years of sampling.

Trends of Water Quality

The regional rainfall and river flow conditions were associated with the salinity in the Econfina study area (Figure 7.5). In 1992, there was an increased salinity range from the drought to the floods of 1992–1993. During the extreme drought of 1999–2002, relatively high salinities were prolonged and reached the highest levels over the study period. High rainfall in 2003–2004 resulted in a return to low salinities. Once again, the range of variation of the rainfall was maximal during this period of the study. The data show that the more extreme post-1980 rainfall highs and lows produced a pronounced increase in the ranges of low and high salinities that were high during the 1992–1993 and 1999–2004 periods of intensive field research.

Surface temperatures were comparable between the Econfina and Fenholloway systems during the entire sampling period. In the Econfina offshore area (Figure 7.6), water temperatures were higher during the 1992–1993 study period than those during the 1970s. There was a major decrease of surface temperature during the succeeding rainfall period. The data indicate that average annual water temperatures in the inshore waters of the Apalachee Bay increased periodically during the later periods of study and the ranges of low to high temperatures increased during these times. Dissolved oxygen (DO) tended to decrease with increased temperature (Figure 7.6) and was thus affected by long-term trends of regional climatological conditions.

The concentration of DO in surface waters was comparable between the two systems during the first two sampling periods. The highest DO concentrations occurred in the Econfina inshore area (Figure 7.7) during the 1970s. There were general trends of lower DO in the Econfina offshore system as the water temperatures increased. The DO tended to decrease during or just after heavy rainfall. Salinity in the Econfina Estuary tended to be lower from the 1990s onward with greater ranges between low and high salinities that were particularly evident during the 2002–2003 transition from drought to heavy rainfall conditions.

Previous studies (Livingston et al., 1998b) indicated that water color and DOC were high in the Fenholloway Estuary and nearshore area (stations F06, F09, F10, F11, and F14) relative to the Econfina Estuary. Both factors tended to become equivalent between stations F10/E08 and F16/E09. Comparative water color levels were found somewhere around 3 km offshore. Water color was a key factor in the distribution of phytoplankton and submerged vegetation in the offshore sites of both estuaries. Long-term water color trends (Figure 7.8) indicated that this factor tended to decrease during drought conditions in the Econfina Estuary. The clearer water allowed increased light transmission characteristics in the unpolluted estuary as noted by Secchi depths. The effect of the droughts on water color levels in the Econfina offshore areas tended to increase with time. However, periods of heavy rainfall in the 1990s and 2002–2003 resulted in increased frequency of reduced Secchi depths relative to the 1970s.

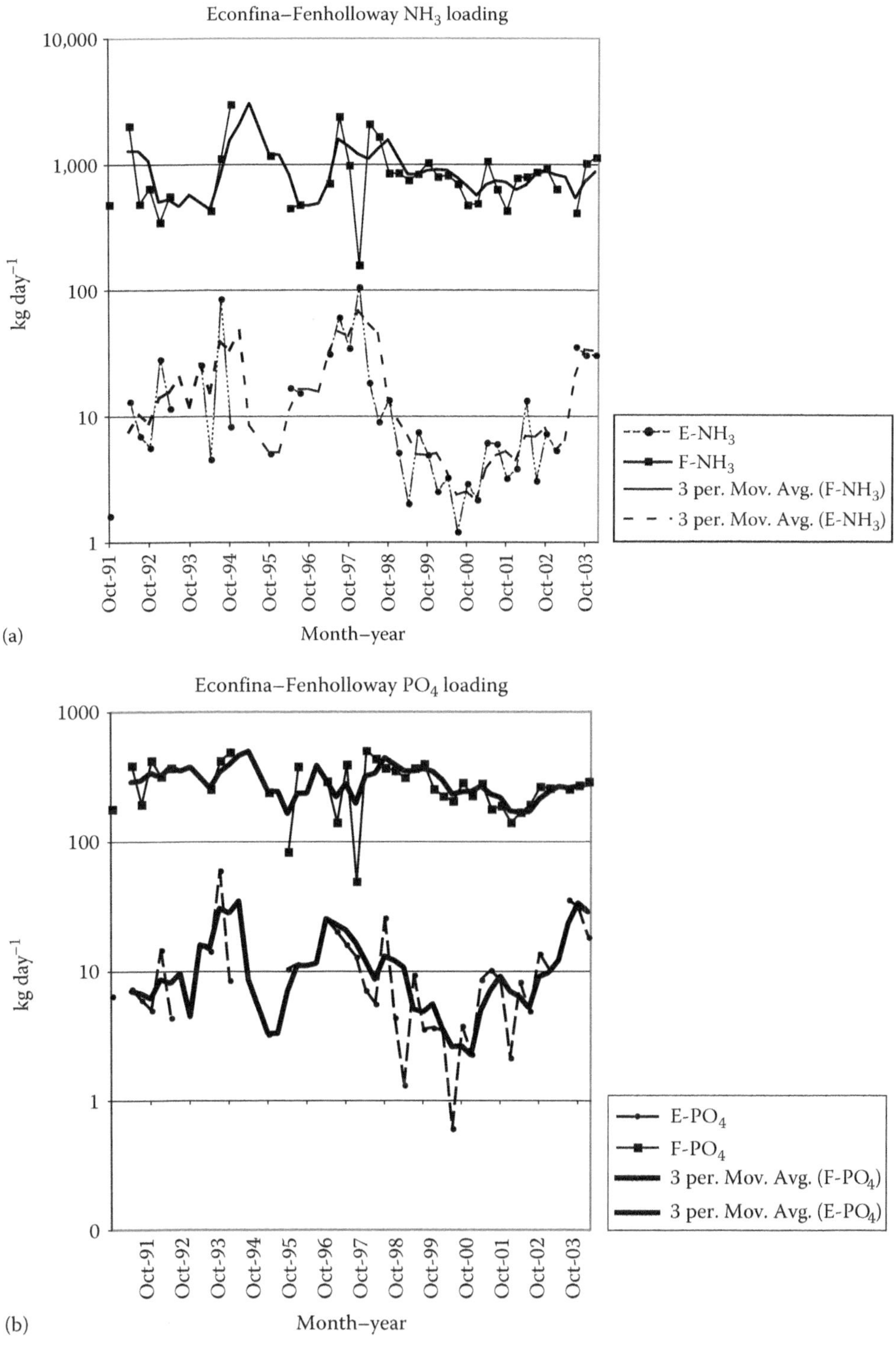

Figure 7.4 (a) Annual averages of ammonia loading to the Apalachee Bay in the Econfina and Fenholloway Rivers from October 1991 to January 2004. Three-month moving averages are shown. (b) Annual averages of orthophosphate loading to the Apalachee Bay in the Econfina and Fenholloway Rivers from October 1991 to January 2004. Three-month moving averages are shown.

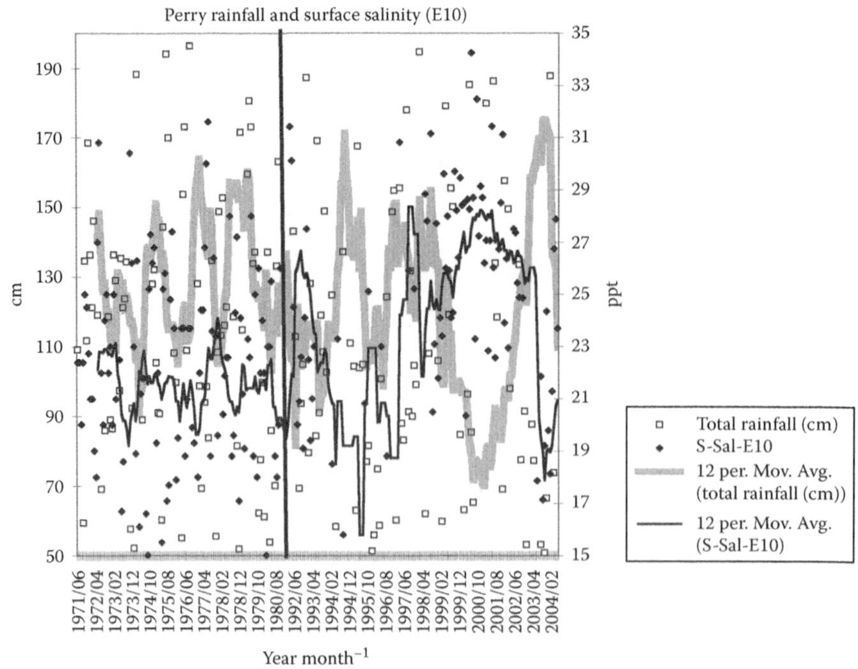

Figure 7.5 Perry rainfall and surface salinity at station E10 (Econfina Estuary) averaged by year over the long-term study period. Twelve-month moving averages are shown. Rain data were provided by the National Oceanic and Atmospheric Administration.

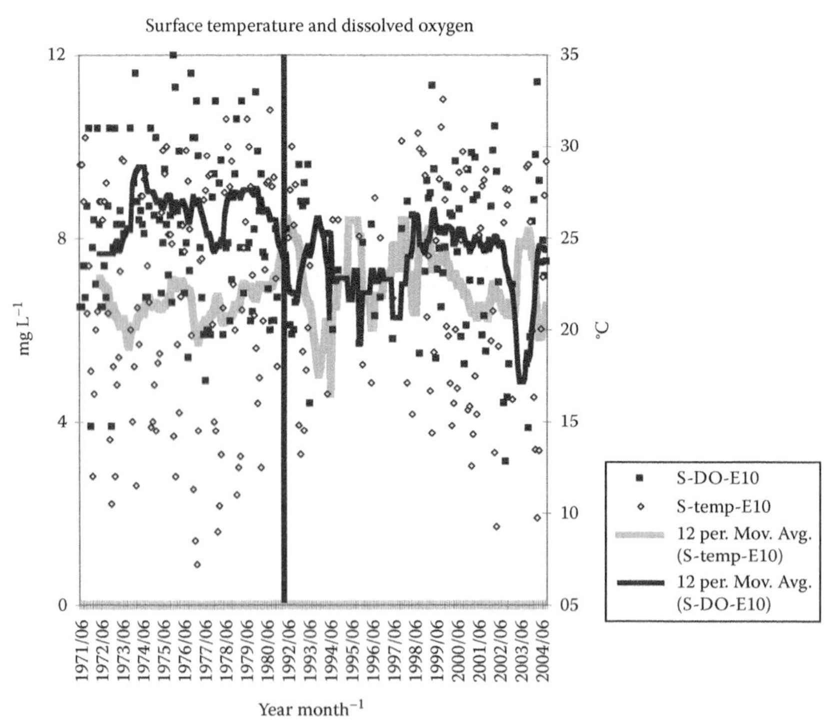

Figure 7.6 Surface temperature and DO at station E10 from 1971 through 2004. Data were averaged by year over the long-term study period. Twelve-month moving averages are shown.

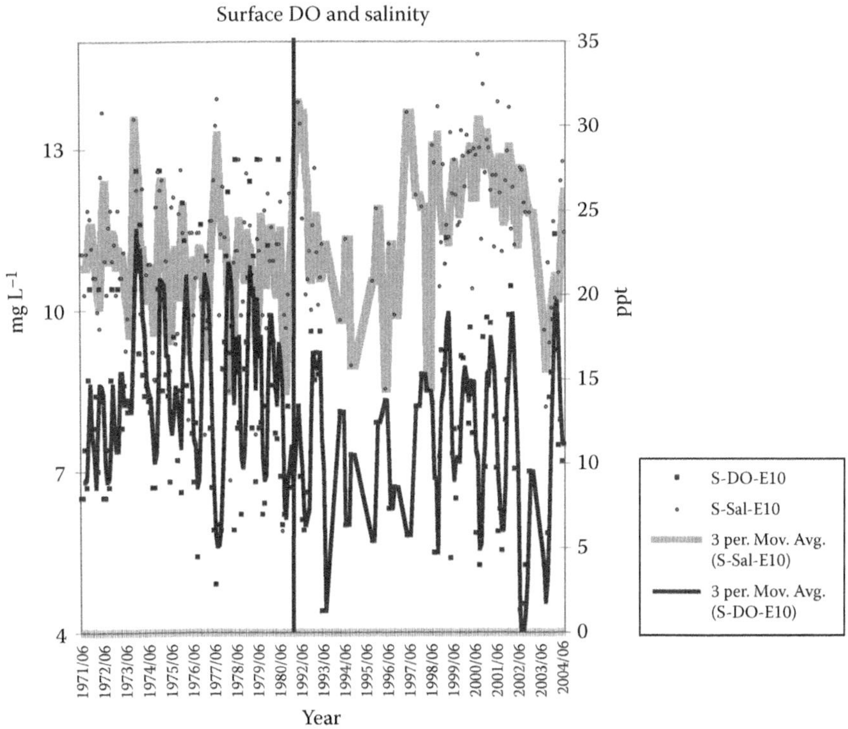

Figure 7.7 Surface DO and surface salinity at Econfina station E10. Data were averaged by year over the long-term study period (1971 through 2004). Three-month moving averages are shown.

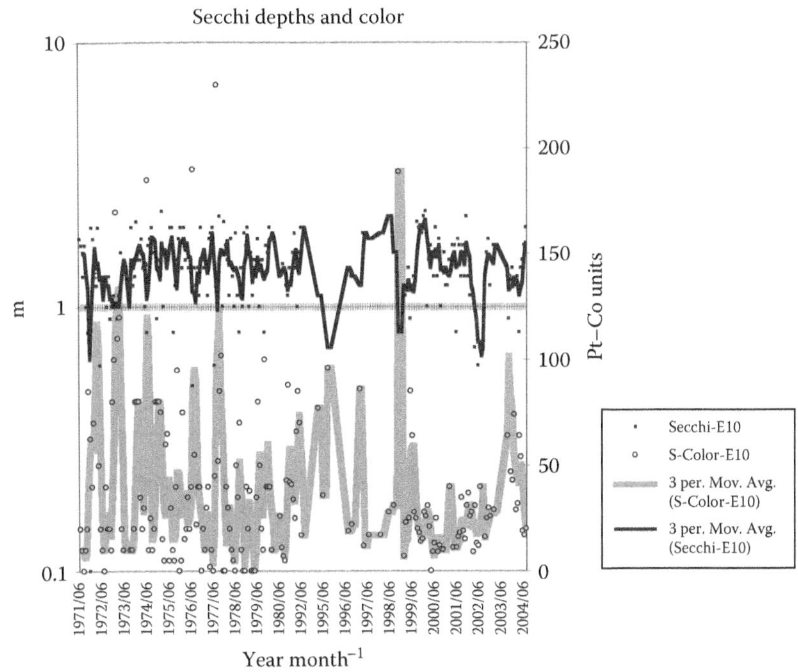

Figure 7.8 Water color and Secchi depths at station E10 averaged by year over the long-term study period (1971 through 2004). Three-month moving averages are shown.

Climatic Effects on Phytoplankton

An analysis was made of the phytoplankton communities in 1999–2003 (Figure 7.9, Table 7.1, Appendix C). In the Econfina system, phytoplankton numbers were somewhat higher in 1999 but generally decreased during subsequent drought years with a return to higher numbers with increased river flows and nutrient loading in 2003. The prolonged drought was thus associated with reduced

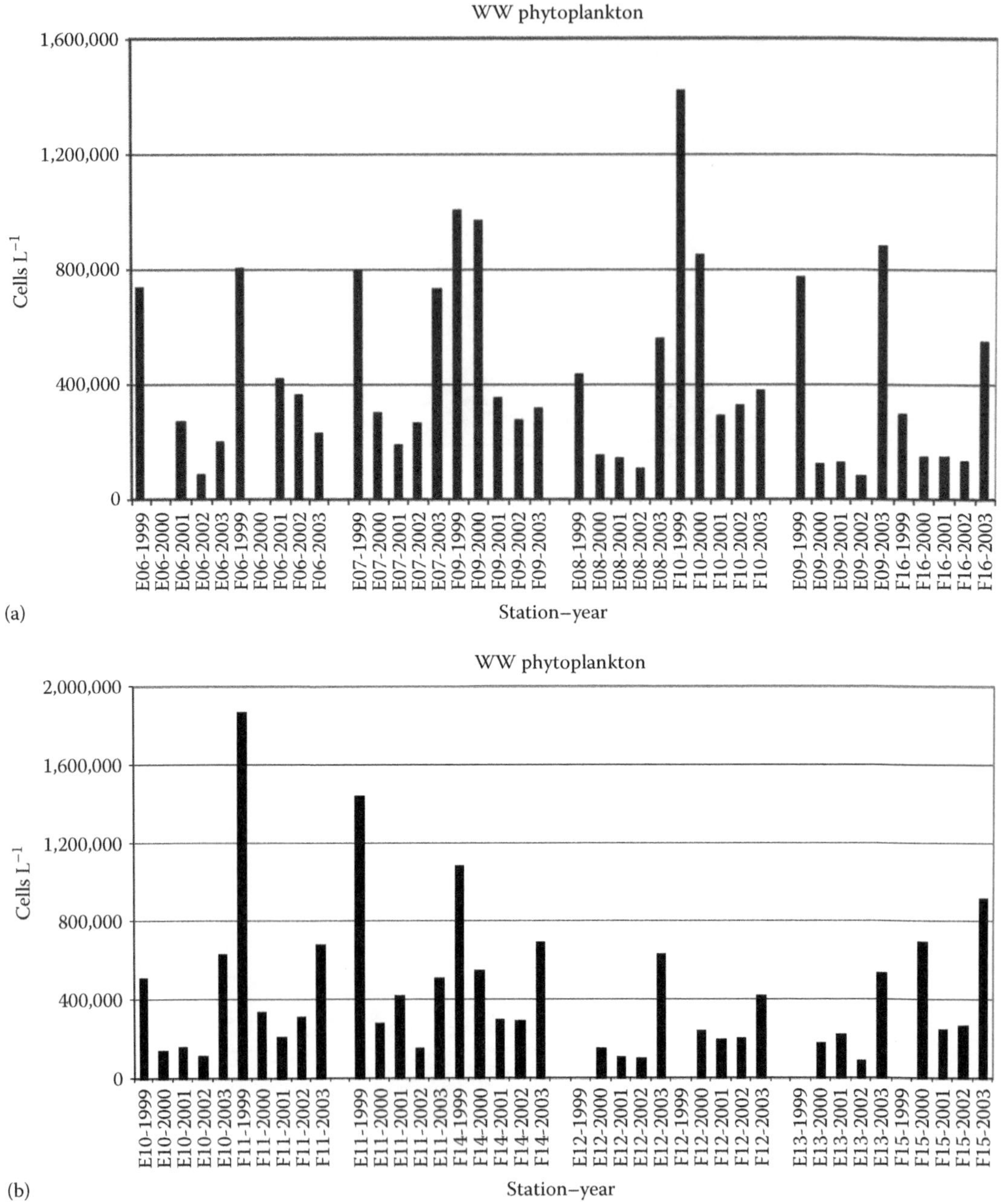

Figure 7.9 Comparison of annual averages of phytoplankton numbers of cells L^{-1} in the Econfina and Fenholloway systems from 1999 to 2003.

Table 7.1 Numerically Dominant Phytoplankton Species of Phytoplankton Taken with 25 μm Nets in the Econfina and Fenholloway Systems

Whole-Water Phytoplankton: No. of Cells L^{-1}				
Econfina (April 1992 to March 1993)				
Taxa	**E06**	**E07**	**E08**	**Total**
Unidentified nanoflagellates	53,169	52,856	75,818	362,212
Unidentified cryptophytes	21,340	54,642	63,058	344,464
Unidentified nannococcoids	20,147	17,059	34,011	130,602
Cyclotella sp. 17 (5–10 μm)	0	0	0	91,390
Thalassionema nitzschioides	3,191	4,950	32,104	62,450
Pyramimonas spp.	8,464	10,172	16,559	57,506
Navicula spp.	20,147	8,567	4,919	52,497
Unidentified pennate diatoms	13,459	14,531	4,904	50,635
Chaetoceros spp.	694	4,829	15,606	43,148
Cylindrotheca closterium	12,266	10,823	4,950	42,611
Whole-Water Phytoplankton: No. of Cells L^{-1}				
Fenholloway (April 1992 to March 1993)				
Taxa	**F06**	**F09**	**F10**	**Total**
Unidentified nanoflagellates	35,893	47,377	57,017	296,757
Skeletonema costatum	555	1,816	116,883	272,552
Unidentified cryptophytes	21,337	23,249	82,436	223,259
Unidentified nannococcoids	14,934	18,012	24,883	126,539
Navicula spp.	18,383	16,711	12,007	74,906
Pyramimonas spp.	1,748	4,541	11,526	57,568
Nitzschia section pseudo-nitzschia spp.	0	394	3,811	48,879
Cylindrotheca closterium	2,778	16,166	8,584	47,981
Nitzschia spp.	4,259	3,845	2,294	20,043
25 μm Net Phytoplankton: No. of Cells L^{-1}				
Econfina (April 1992 to March 1993)				
Taxon	**E06**	**E07**	**E09**	**Total**
Unidentified nanoflagellates	300	166	183	649
Cylindrotheca closterium	245	94	66	405

(Continued)

Table 7.1 (*Continued*) Numerically Dominant Phytoplankton Species of Phytoplankton Taken with 25 μm Nets in the Econfina and Fenholloway Systems

Navicula spp.	142	127	76	345			
Skeletonema costatum (narrow)	101	155	0	256			
Unidentified nannococcoids	145	32	59	235			
Asterionella japonica	4	163	68	234			
Amphora spp.	49	73	91	212			
Rhizosolenia stolterfothii	4	81	88	173			
Chaetoceros diversus	25	31	75	131			
Thalassionema nitzschioides	33	23	75	131			
25 μm Net Phytoplankton: No. of Cells L^{-1}							
Fenholloway (April 1992 to March 1993)							
Taxa	**F06**	**F09**	**F16**	**Total**			
Skeletonema costatum	453	10,388	14	10,854			
Unidentified nanoflagellates	265	236	183	684			
Cylindrotheca closterium	150	211	96	456			
Navicula spp.	176	115	93	384			
Nitzschia section pseudo-nitzschia spp.	0	0	257	257			
Chaetoceros diversus	3	15	206	224			
Thalassionema nitzschioides	10	17	148	174			
Chaetoceros compressus	21	0	134	155			
Unidentified nannococcoids	85	36	31	152			
Chaetoceros simplex	0	0	103	104			
Whole-Water Phytoplankton: No. of Cells L^{-1}							
Econfina (April 1999 to December 1999)							
Taxa	**E06**	**E07**	**E08**	**E09**	**E10**	**E11**	**Total**
Undetermined cryptophyte	320,614	184,092	63,585	81,119	96,228	299,027	1,044,665
Cyclotella choctawhatcheeana	2,072	42,718	17,428	23,707	141,212	386,243	613,380
Undetermined nanoflagellates	86,181	79,417	61,567	58,018	40,659	75,433	401,275
Johannesbaptistia pellucida	2,035	39,590	58,018	102,704	2,701	1,406	206,454
Asterionellopsis glacialis	37	296	2,590	138,760	74	93	141,850

(*Continued*)

Table 7.1 (*Continued*) Numerically Dominant Phytoplankton Species of Phytoplankton Taken with 25 μm Nets in the Econfina and Fenholloway Systems

Dactyliosolen fragilissimus	815	12,412	5,076	8,391	315	77,793	104,802
Chaetoceros laciniosus	2,186	70,665	20,343	815	0	334	94,343
Undetermined nannococcoids	2,629	5,924	22,501	7,812	46,650	3,367	88,883
Gymnodinium spp.	4,514	10,676	9,087	16,711	9,090	28,196	78,274
Cyclotella cf. *atomus*	5,665	6,106	10,476	8,936	15,695	23,757	70,635

Whole-Water Phytoplankton: No. of Cells L^{-1}

Fenholloway (April 1999 to December 1999)

Taxa	F06	F09	F10	F16	F11	F14	Total
Undetermined cryptophyte	358,064	270,651	138,850	70,457	229,703	195,946	1,263,671
Leptocylindrus danicus	18,755	206,469	493,799	10,028	15,985	1,554	746,590
Undetermined nanoflagellates	92,645	98,513	79,576	48,093	121,982	100,025	540,834
Cyclotella cf. *atomus*	59,715	54,184	41,313	352	126,789	80,402	362,755
Johannesbaptistia pellucida	2,206	6,697	39,886	3,293	279,721	20,805	352,608
Cerataulina pelagica	625	1,258	10,510	11,785	7,451	181,736	213,365
Undetermined nannococcoids	499	1,518	2,868	4,497	162,792	481	172,655
Cerataulina pelagica (veg.)	0	0	0	0	119,963	0	119,963
Dactyliosolen fragilissimus	1,915	3,145	23,348	1,647	74,009	10,361	114,425
Rhizosolenia setigera	17,402	7,552	66,382	13,988	4,289	4,367	113,980

25 μm Net Phytoplankton: No. of Cells L^{-1}

Econfina (April 1999 to October 1999)

Taxa	E06	E07	E08	E09	Total
Asterionellopsis glacialis	34	72	1,392	79,216	80,714
Johannesbaptistia pellucida	1,607	7,512	21,770	40,634	71,523
Chaetoceros laciniosus	413	24,921	17,888	556	43,778
Chaetoceros lauderi	3	9,374	1,041	58	10,476
Bacteriastrum hyalinum	33	2,799	2,699	108	5,639
Chaetoceros compressus	71	2,219	2,130	449	4,868
Chaetoceros cf. *subtilis*	411	2,224	0	0	2,635
Rhizosolenia setigera	42	928	1,480	158	2,607
Pseudonitzschia spp.	37	938	940	342	2,257
Peridinium quinquecorne	1,066	28	0	3	1,097

(*Continued*)

Table 7.1 (*Continued*) Numerically Dominant Phytoplankton Species of Phytoplankton Taken with 25 µm Nets in the Econfina and Fenholloway Systems

25 µm Net Phytoplankton: No. of Cells L^{-1}					
Fenholloway (April 1999 to October 1999)					
Taxa	**F06**	**F09**	**F10**	**F16**	**Total**
Leptocylindrus danicus	2,608	16,490	38,363	1,877	59,338
Rhizosolenia setigera	2,658	2,629	39,170	2,683	47,140
Falcula hyalina	16,081	15,966	9	4	32,060
Chaetoceros laciniosus	43	33	538	20,690	21,305
Chaetoceros lauderi	10	133	2,346	8,347	10,837
Chaetoceros compressus	54	11	8,745	573	9,384
Johannesbaptistia pellucida	0	28	1,604	5,925	7,557
Cerataulina pelagica	142	307	987	4,521	5,956
Thalassionema nitzschioides	90	173	2,073	72	2,409
Ceratium hircus	70	42	1,101	177	1,390
					212,452

Whole-Water Phytoplankton: No. of Cells L^{-1}					
Econfina (June 2000 to October 2000)					
Taxa	**E06**	**E07**	**E08**	**E09**	**E10**
Undetermined cryptophyte	123,543	1,038,628	254,579	178,489	275,058
Johannesbaptistia pellucida	0	38,961	85,415	99,734	91,576
Cyclotella cf. *atomus*	38,961	104,229	34,299	46,787	24,809
Undetermined nanoflagellates	24,975	119,215	44,789	43,124	40,960
Undetermined nannococcoids	0	24,309	67,100	22,645	112,555
Gymnodinium spp.	666	48,785	50,784	31,137	36,797
Cocconeis spp.	4,995	55,445	32,303	9,159	38,296
Pseudonitzschia spp.	666	12,987	73,260	12,988	24,144
Proboscia alata	0	999	7,993	62,438	1,499
Chaetoceros spp.	0	19,481	28,306	11,823	21,813
Taxa	**E11**	**E12**	**E13**	**Total**	**%**
Undetermined cryptophyte	682,984	262,071	297,037	3,112,388	38.6
Johannesbaptistia pellucida	3,497	114,219	233,600	667,001	9.6
Cyclotella cf. *atomus*	247,420	24,310	33,634	554,448	7.3
Undetermined nanoflagellates	89,078	42,458	57,777	462,375	6.4
Undetermined nannococcoids	33,633	139,861	55,279	455,381	6.4

(*Continued*)

Table 7.1 (*Continued*) Numerically Dominant Phytoplankton Species of Phytoplankton Taken with 25 μm Nets in the Econfina and Fenholloway Systems

Gymnodinium spp.	43,291	35,300	82,253	329,013	4.3
Cocconeis spp.	45,455	15,320	8,993	209,965	2.9
Pseudonitzschia spp.	2,166	56,112	11,656	193,978	2.7
Proboscia alata	1,166	500	79,088	153,681	2.4
Chaetoceros spp.	6,494	27,307	13,488	128,712	1.8

Whole-Water Phytoplankton: No. of Cells L^{-1}

Fe Fenholloway (June 2000 to October 2000)

Taxa	F06	F09	F10	F16	F11
Leptocylindrus minimus	0	2,137,860	2,601,396	360,306	102,731
Undetermined cryptophyte	78,921	594,072	175,492	384,283	170,830
Proboscia alata	102,231	364,302	549,950	282,384	170,164
Johannesbaptistia pellucida	0	215,118	165,501	190,976	201,798
Thalassiosira minima proschikinae complex	0	321,345	0	0	0
Chaetoceros spp.	333	79,254	101,566	71,263	141,859
Rhizosolenia setigera	1,332	42,291	68,765	50,117	52,615
Undetermined nanoflagellates	9,324	79,587	45,788	56,777	43,125
Dactyliosolen fragilissimus	666	666	99,567	666	1,333
Gymnodinium spp.	5,661	5,661	22,645	51,949	49,618

Taxa	F14	F12	F15	Total	%
Leptocylindrus minimus	234,932	1,288,544	0	6,725,769	37.4
Undetermined cryptophyte	590,909	175,159	138,030	2,307,696	12.8
Proboscia alata	232,767	458,208	41,292	2,201,298	12.3
Johannesbaptistia pellucida	55,112	199,634	164,670	1,192,808	6.6
Thalassiosira minima proschikinae complex	565,934	0	0	887,279	4.9
Chaetoceros spp.	42,291	80,254	7,659	524,478	2.9
Rhizosolenia setigera	29,804	202,964	1,332	449,220	2.5
Undetermined nanoflagellates	114,719	43,791	50,285	443,395	2.5
Dactyliosolen fragilissimus	102,731	103,563	5,495	314,687	1.8
Gymnodinium spp.	21,313	60,940	31,636	249,421	1.4

(*Continued*)

Table 7.1 (*Continued*) Numerically Dominant Phytoplankton Species of Phytoplankton Taken with 25 μm Nets in the Econfina and Fenholloway Systems

Whole-Water Phytoplankton: No. of Cells L^{-1}					
Econfina (June 2001 to October 2001)					
Taxa	**E06**	**E07**	**E08**	**E09**	**E10**
Undetermined cryptophytes	158,742	107,005	64,076	58,221	74,982
Undetermined nanoflagellates	18,848	13,931	12,544	10,074	11,766
Dactyliosolen fragilissimus	6,660	8,548	6,327	4,440	500
Gymnodinium spp.	4,996	11,017	29,500	32,634	31,775
Pyramimonas spp.	6,694	5,524	4,996	1,526	2,331
Navicula sp. 54 *small (F09 8/01)*	67	56	—	—	—
Thalassiosira minima proschikinae complex	—	333	—	—	—
Undetermined centric diatoms	26,341	2,499	1,831	1,249	2,027
Cyclotella cf. *atomus*	166	112	56	250	196
Cylindrotheca closterium	6,926	4,273	277	501	418
Taxa	**E11**	**E12**	**E13**	**Total**	
Undetermined cryptophytes	162,811	56,500	60,608	682,337	
Undetermined nanoflagellates	15,236	11,184	10,962	93,583	
Dactyliosolen fragilissimus	126,291	83	93,518	152,849	
Gymnodinium spp.	16,790	21,063	24,893	147,775	
Pyramimonas spp.	5,523	1,694	3,691	28,288	
Navicula sp. 54 *small (F09 8/01)*	—	—	—	123	
Thalassiosira minima proschikinae complex	—	—	—	333	
Undetermined centric diatoms	2,027	1,527	2,277	37,501	
Cyclotella cf. *atomus*	139	278	112	1,197	
Cylindrotheca closterium	4,247	556	167	17,198	

Whole-Water Phytoplankton: No. of Cells L^{-1}					
Fe Fenholloway (June 2001 to October 2001)					
Taxa	**F06**	**F09**	**F10**	**FII**	**F12**
Undetermined cryptophytes	135,809	118,115	135,420	100,844	81,170
Undetermined nanoflagellates	38,351	30,004	21,702	18,621	10,684

(*Continued*)

Table 7.1 (*Continued*) Numerically Dominant Phytoplankton Species of Phytoplankton Taken with 25 µm Nets in the Econfina and Fenholloway Systems

Dactyliosolen fragilissimus	—	67	7,659	3,968	7,964
Gymnodinium spp.	9,990	3,796	10,990	12,738	22,950
Pyramimonas spp.	24,309	16,085	10,268	5,829	3,887
Navicula sp. *54 small (F09 8/01)*	14,763	56,576	18,788	4,940	1,805
Thalassiosira minima proschikinae complex	63,270	27,072	1,554	—	389
Undetermined centric diatoms	5,384	9,523	4,107	3,081	2,136
Cyclotella cf. *atomus*	33,189	25,242	611	1,082	806
Cylindrotheca closterium	11,766	12,788	4,496	3,720	1,473
Johannesbaptistia pellucida	—	—	20,979	3,386	1,055

Taxa	**F14**	**F15**	**F16**	**Total**
Undetermined cryptophyte	159,574	106,811	66,906	904,649
Undetermined nanoflagellates	28,105	22,339	13,460	183,266
Dactyliosolen fragilissimus	7,592	7,937	889	36,076
Gymnodinium spp.	6,727	13,154	20,620	100,965
Pyramimonas spp.	25,475	16,790	3,470	106,113
Navicula sp. *54 small (F09 8/01)*	15,251	3,802	28	115,953
Thalassiosira minima proschikinae complex	333	7,076	56	99,750
Undetermined centric diatoms	4,497	2,137	1,083	31,948
Cyclotella cf. *atomus*	8,259	167	361	69,717
Cylindrotheca closterium	11,056	3,526	751	49,576
Johannesbaptistia pellucida	—	19,037	500	19,537

Whole-Water Phytoplankton: No. of Cells L^{-1}

Econfina (June 2002 to October 2002)

Taxa	**E06**	**E07**	**E09**	**E10**	**E11**
Undetermined cryptophyte	45,733	100,370	27,391	19,066	16,380
Cerataulina pelagic	—	—	—	—	—
Gymnodinium spp.	14,042	11,744	22,707	17,012	13,987
Thalassiosira cedarkeyensis	—	167	42	—	—
Undetermined nanoflagellates	9,713	14,259	4,670	4,394	3,105

(*Continued*)

Table 7.1 (*Continued*) Numerically Dominant Phytoplankton Species of Phytoplankton Taken with 25 μm Nets in the Econfina and Fenholloway Systems

Navicula spp.	1,166	605	480	293	480
Undetermined coccoid yellow-green algae	—	70,430	—	—	—
Rhizosolenia setigera	—	—	—	83	—
Pyramimonas spp.	611	1,500	272	209	83
Nitzschia spp.	722	1,187	626	1,084	584

Taxa	**E11**	**E12**	**E13**	**Total**
Undetermined cryptophyte	65,308	21,906	21,962	318,116
Cerataulina pelagica	—	—	—	0
Gymnodinium spp.	19,816	15,643	22,790	137,741
Thalassiosira cedarkeyensis	292	83	125	709
Undetermined nanoflagellates	11,476	3,688	4,918	56,223
Navicula spp.	876	542	792	5,234
Rhizosolenia setigera	—	—	—	70,430
Pyramimonas spp.	1,690	356	272	83
Nitzschia spp.	1,645	751	1,084	4,993

Whole-Water Phytoplankton: No. of Cells L^{-1}

Fe Fenholloway (June 2002 to October 2002)

Taxa	**F06**	**F09**	**F10**	**F11**	**F12**
Undetermined cryptophyte	62,504	56,303	111,063	86,177	63,437
Cerataulina pelagica	—	84	44,606	102,593	84,583
Gymnodinium spp.	1,168	999	39,543	21,616	33,301
Thalassiosira cedarkeyensis	123,975	81,729	3,835	20,761	2,289
Undetermined nanoflagellates	25,998	16,909	12,706	20,794	6,869
Navicula spp.	77,149	75,649	417	1,125	792
Undetermined coccoid yellow-green algae	—	—	—	—	—
Rhizosolenia setigera	167	1,665	13,208	5,162	2,123
Pyramimonas spp.	—	999	5,580	11,252	2,622
Nitzschia spp.	18,497	5,499	668	1,251	542

Taxa	**F14**	**F15**	**F16**	**Total**
Undetermined cryptophyte	84,244	56,169	60,441	580,338
Cerataulina pelagica	102,106	56,901	—	390,873
Gymnodinium spp.	15,157	40,670	22,172	174,626
Thalassiosira cedarkeyensis	28,178	1,541	42	262,350
Undetermined nanoflagellates	15,077	8,830	8,580	115,763

(*Continued*)

Table 7.1 (*Continued*) Numerically Dominant Phytoplankton Species of Phytoplankton Taken with 25 μm Nets in the Econfina and Fenholloway Systems

Navicula spp.	2,376	896	272	158,676
Undetermined coccoid yellow-green algae	—	—	—	0
Rhizosolenia setigera	6,496	12,780	229	41,830
Pyramimonas spp.	9,373	3,456	668	33,950
Nitzschia spp.	1,043	313	750	28,563

Whole-Water Phytoplankton: No. of Cells L^{-1}

Econfina/Fenholloway (August 2003)							
Taxa	**E09**	**E11**	**E12**	**E13**	**F11**	**F15**	**Total**
Pseudonitzschia pseudodelicatissima	433,190	39,996	53,932	121,099	695,925	894,443	2,238,585
Chattonella sp.	0	0	0	0	491,400	851,787	1,343,187
Bacteriastrum hyalinum	68,882	0	856,240	2,222	0	0	927,344
Johannesbaptistia pellucida	31,108	0	0	307,747	307,800	0	646,655
Undetermined cryptophyte	53,328	82,214	23,352	93,324	73,575	54,653	380,446
Cyclotella atomus var. nov.	0	222,200	0	49,995	14,175	0	286,370
Undetermined nanoflagellates	29,997	95,546	13,900	51,106	29,700	41,323	261,572
Thalassiosira cedarkeyensis	14,443	112,211	5,004	114,433	4,725	6,665	257,481
Thalassiosira spp.	0	26,664	2,224	0	5,400	59,985	94,273
Merismopedia spp.	0	0	0	0	0	93,310	93,310

phytoplankton productivity in the Econfina offshore area following a brief period of increased cell numbers during the initial year of the drought.

By 2000, Econfina phytoplankton numbers were down (Table 7.1); this trend continued during the drought. Species richness in the Econfina remained high in 2000. In the Fenholloway system in 2000, inshore numbers remained high. Species richness, diversity, and evenness tended to be low where such high numbers occurred (F09, F10, F14). These community changes reflected the incidence of phytoplankton blooms in the Fenholloway system during the drought years 1999 and 2000.

Peak whole-water phytoplankton numbers in the Fenholloway system were associated with phytoplankton blooms (number cells L^{-1} > 10^6) of *Leptocylindrus danicus* (1,831,582 cells L^{-1}, station F09, September 1999; 4,390,902 cells L^{-1}, station F10, September 1999), *Johannesbaptistia pellucida* (2,111,220 cells L^{-1}, station F11, September 1999), and *Cerataulina pelagica* (1,625,299 cells L^{-1}, station F14, October 1999). The diatom *L. danicus* has a worldwide distribution in a variety of tropical coastal environments and has adverse impacts on various fish species. *L. danicus* was a key pioneer bloom species in the Perdido Bay system during the initial blooms of 1994 (Livingston, 2007a,b). *J. pellucida* is a filamentous blue-green alga with a mucous sheath. It has a worldwide distribution and is not considered to be a noxious species (A. K. S. K. Prasad, pers. comm.). *C. pelagica* is a filamentous diatom that is a known bloom species that closely resembles *L. danicus* in body type and natural history (AKSK Prasad, pers. comm.). These were the first blooms noted in either of the Apalachee study areas.

In 2000, there were blooms during July in the Fenholloway Estuary and in the offshore Fenholloway system (station F15). The primary bloom species was *Leptocylindrus cordatum*. There was also high dominance of *J. pellucida*, *Rhizosolenia setigera*, and dinoflagellates (*Gymnodinium* spp.). These blooms were not as diverse as the previous year. The trends of both species of *Leptocylindrus* were somewhat different with high numbers of *L. danicus* in estuarine areas in 1999 and more generalized distributions in 2000. *Leptocylindrus minimus* only appeared in inshore Fenholloway waters in 2000. The species *Skeletonema costatum*, a primary dominant in 1992, virtually disappeared from the Fenholloway system during the later bloom periods. There was a more general distribution of the bloom species *Prorocentrum cordatum* in the Fenholloway in 2000.

In 2001, there were no blooms as cryptophytes and nanoflagellates predominated with bloom species such as *R. setigera* taken in the Fenholloway system in low numbers. Overall, there were qualitative differences of the phytoplankton biota between the Econfina and Fenholloway systems, but the bloom species that were present in the Fenholloway in 1999–2000 were taken in relatively low numbers. During summer–fall 2002, no blooms were noted in the Fenholloway system. Phytoplankton numbers were comparable to those in the previous year. Cryptophytes were once again the most abundant group in both systems. In 2003, the unusually high chlorophyll *a* concentrations in offshore areas of the Fenholloway drainage coexisted with the bloom species *Pseudo-nitzschia pseudodelicatissima* and *Chattonella* sp. *P. pseudodelicatissima* was present in both the Econfina and Fenholloway systems. The raphidophyte *Chattonella* sp., however, was noted only in the Fenholloway area. These species were present in bloom concentrations at intermediate offshore areas in the Fenholloway system in July and August 2003.

The data thus showed that during the early drought period of increased light penetration, there were increased phytoplankton numbers in both systems. However, with time during the drought, numerical abundance of phytoplankton went down. Increased dominance by the bloom species *L. danicus* noted at stations F09 and F10 in 1999 and the related species *L. minimus* in 2000 suggests the possibility that increased light penetration coupled with continued high ammonia and orthophosphate loading led to an overlap of deeper photic depths coupled with high nutrient gradients, a situation similar to that of the 1994 initiation of *L. danicus* blooms in the Perdido Bay under similar habitat conditions (drought + increased nutrient loading, Livingston, 2000, 2002).

Net phytoplankton data (Table 7.1) supported the observation that there was a basic shift of phytoplankton species composition and abundance in the Fenholloway system in 1999. Numbers, diversity, and evenness indices of the Fenholloway net phytoplankton in 1992 were relatively low compared to the Econfina. Species richness was generally lower in the Fenholloway during both sampling periods with particularly low numbers in the estuary. On the other hand, numbers of phytoplankton were higher in the Fenholloway than the Econfina in 1999, with comparable reductions of Shannon diversity and evenness.

Phytoplankton taken in August–November 2003 (increased rainfall and river flow) indicated that the dominant phytoplankton in the Fenholloway system were the bloom species *P. pseudodelicatissima* and *Chattonella* sp. *P. pseudodelicatissima* was present in both the Econfina and Fenholloway systems. The raphidophyte *Chattonella* sp., however, was noted only in the Fenholloway area. The relatively high rainfall and river flow conditions during summer 2003 was associated with major concentrations of these bloom species in the Fenholloway system. The annual trends of phytoplankton numbers and species richness indicated that estuarine areas (E06, E07; F06, F09) did not have any particular temporal pattern in terms of numbers of cells. However, the offshore stations in both systems were characterized by increased numbers of phytoplankton cells in 1999 (early drought) and 2003 (high rainfall).

These phytoplankton trends are consistent with the hypothesis that, during years of high rainfall that follow a prolonged drought, there are increased phytoplankton blooms in near offshore Fenholloway areas (anthropogenic loading of ammonia and orthophosphate) relative to cognate stations in the Econfina system. These blooms are generally associated with reduced species richness. During drought conditions, nutrient gradients overlap high color and reduced light transmission, thus precluding habitat conditions conducive to plankton blooms. High rainfall and river flows move the nutrients into water where color effects are diluted in the Fenholloway system and light transparency is enhanced. Such changes along with high nutrient loading thus lead to blooms. The discovery of the red tide dinoflagellate *Karenia brevis* at station F16 in September 2003 was the first time this species was noted in the study area. Phytoplankton trends were thus driven by rainfall and river flow trends.

The phytoplankton assemblages in estuarine areas of the Apalachee Bay are thus the result of complex interactions of nutrient loading and light penetration that are driven by rainfall/drought cycles. Drought conditions, after a brief period of increased phytoplankton productivity, are generally associated with reduced phytoplankton cells L^{-1} in the Econfina Estuary. These reductions of phytoplankton productivity in a natural estuary were associated with reduced nutrient loading from the river. During the initial year of drought, residual nutrients plus increased light penetration lead to increased phytoplankton production that quickly disappeared in the succeeding years of drought and reduced nutrient loading from land areas. Implicit in these findings is the fact that the increased intensity and incidence of droughts can be associated with overall reductions of phytoplankton production in the unpolluted parts of the oligotrophic Apalachee Bay system.

Submerged Aquatic Vegetation

The effects of pulp mill effluents on the Fenholloway offshore system have been well documented (Zimmerman, 1974; Zimmerman and Livingston, 1976a,b, 1979; Livingston, 1984b, Livingston et al., 1998b). Pulp mill effluents in the Fenholloway River were associated with increased loading of DOC, water color, and nutrients (orthophosphate, ammonia) in offshore Fenholloway areas relative to the unpolluted (Econfina River) reference system (Livingston et al., 1998b). This loading resulted in changes of water-quality factors and light transmission characteristics in the polluted system.

A statistical analysis was conducted on long-term trends of SAV in the study areas (Table 7.2). With the exception of offshore areas (E12, F12; E09, F16), SAV biomass was significantly lower in the Fenholloway system than the Econfina. This included periods from 1972–1973 through the intensive analyses of 1992–1993 to the most recent analyses (1999–2002).

Field data indicated that seagrass growth was highest during spring months with lesser growth during summer periods and losses of biomass during cooler months (Livingston et al., 1998b). An SAV quadrat analysis indicated that the dominant seagrasses in the study area (Table 7.3) were manatee grass (*S. filiforme*) and turtle grass (*T. testudinum*). The highest SAV biomass occurred in nearshore areas (stations E08, E10, E13) when compared to inshore sites (E07) and offshore areas (E09, E12). This was attributed to the effects of gradients of salinity, color, and nutrients (Livingston et al., 1998b). High water color and highly variable salinity inshore reduced SAV productivity. With distance offshore, reduced water color together with increased nutrients caused peak SAV development at intermediate distances from land areas. Farther offshore, SAV development was limited by reduced nutrients and depth (reduced light penetration). Seagrass distribution in the Apalachee Bay thus depended on complex gradients of salinity, nutrient loading, water depth, and light-attenuation features. The experimental evidence indicated likely enhancement of seagrass growth due to runoff from the natural rivers (Econfina) and coastal marshes, particularly during spring growth periods (Livingston et al., 1998b).

For the following series of figures, the data were grouped into averages across Econfina inner stations (E7, E8, E10, E11) that were called EIN or Econfina inner stations and Econfina outer stations (E9, E12, E13) that were called EOUT or outer stations. The grouped Econfina data were compared to cognate groups in the Fenholloway Estuary as defined by averaged Fenholloway inner stations (F9, F10, F11, F14) or FIN and the Fenholloway outer stations (F12, F15, F16) or FOUT.

Table 7.2 Long-Term Comparisons of SAV Indices Averaged over 12-Month Periods and Analyzed with t-Tests and Wilcoxon Means Tests for Differences (See Appendix B)

	SAV Comparison April 1972 to March 1973		SAV Comparison April 1974 to March 1975		SAV Comparison April 1992 to March 1993	
Factor Station Pair	Biomass (g m^{-2})	No. of Taxa	Biomass (g m^{-2})	No. of Taxa	Biomass (g m^{-2})	No. of Taxa
E07/F09	9.0/0.25	4.8/1.3	188.5/2.4	6.0/2.8	2.8/3.9	3.3/0.6
(River mouth)	WS ($p \leq 0.01$)	WS ($p \leq 0.01$)	WS ($p \leq 0.01$)	WS ($p \leq 0.01$)	WN	WN
E08/F10	1196/20	8.7/5.3	895/30.1	10/5.7	26.0/4.0	3.4/0.9
(Center, 1.5 km)	WS ($p \leq 0.01$)	WS ($p \leq 0.02$)	WS ($p \leq 0.01$)	WS ($p \leq 0.01$)	WS ($p \leq 0.02$)	WS ($p \leq 0.01$)
E09/F16	101/47.2	8.6/9.3	412/	12.0/	20.3/14.8	6.2/4.6
(Center, 5 km)	WN	WN	WN	WN	WN	WN
E10/F11	627/.17	9.5/2.7	641/5.8	10.2/4.0	7.1/0.41	5.7/0.6
(W shore)	WS ($p \leq 0.01$)	WS ($p \leq 0.01$)	WS ($p \leq 0.01$)	WS ($p \leq 0.01$)	WS ($p \leq 0.01$)	WS ($p \leq 0.01$)
E12/F12	1007/94.8	10.8/12.3	1007/227	12.0/10.5	57.6/14.0	8.0/6.3
(W, 3 km)	WS ($p \leq 0.01$)	WS ($p \leq 0.02$)	WS ($p \leq 0.01$)	WN	WS ($p \leq 0.01$)	WS ($p \leq 0.02$)
E11/F14	523/1.2	8.8/2.8	521/5.3	12.8/3.0	1.4/0.1	1.3/0.3
(E shore)	WS ($p \leq 0.01$)	WS ($p \leq 0.01$)	WS ($p \leq 0.01$)	WS ($p \leq 0.01$)	WS ($p \leq 0.01$)	WS ($p \leq 0.02$)
E13/F15	513/23.8	10.1/6.2	438.3/279	11.9/10.4	8.7/0.4	4.5/0.6
(E, 1.5 km)	WS ($p \leq 0.01$)	WS ($p \leq 0.01$)	WS ($p \leq 0.01$)	WN	WS ($p \leq 0.01$)	WS ($p \leq 0.01$)

Cognate Station Pair	SAV Comparison: April 1999 to March 2000		SAV Comparison: April 2000 to March 2001		SAV Comparison: April 2001 to March 2002	
	Biomass (g m^{-2})	No. of Taxa	Biomass (g m^{-2})	No. of Taxa	Biomass (g m^{-2})	No. of Taxa
E07/F09	29.0/1.1	5.0/2.6	20.8/5.3	6.1/5.9	35.5/7.5	7.8/5.2
(River mouth)	WS ($p \leq 0.01$)	WN	WS ($p \leq 0.02$)	WN	WS ($p \leq 0.01$)	WS ($p \leq 0.01$)
E08/F10	15.6/7.2	11.3/5.1	144/5.5	12/7.8	151/7.9	13.5/9.9
(Center, 1.5 km)	WS ($p \leq 0.01$)	WS ($p \leq 0.01$)	WS ($p \leq 0.01$)	WS ($p \leq 0.01$)	WS ($p \leq 0.01$)	WS ($p \leq 0.01$)
E09/F16	100.3/41.7	8.3/7.6	77.8/38.4	9.5/8.3	107/85.9	10.8/8.8
(Center, 5 km)	WS ($p \leq 0.01$)	WN	WS ($p \leq 0.01$)	WN	WN	WS ($p \leq 0.01$)
E10/F11	216/1.6	11.1/3.3	233/7.8	11.8/6.1	216/25.4	12.4/10.8
(W shore)	WS ($p \leq 0.01$)	WS ($p \leq 0.01$)	WS ($p \leq 0.01$)	WS ($p \leq 0.01$)	WS ($p \leq 0.01$)	WN
E12/F12	123/125	10.5/8.8	93/103	6.9/7.9	95.7/138	7.6/8.3
(W, 3 km)	WN	WN	WN	WN	WS ($p \leq 0.01$)	WN
E11/F14	35.1/7.6	9.1/4.2	35.6/10.5	8.6/7	23.9/7.6	8.8/6.3
(E shore)	WS ($p \leq 0.01$)	WS ($p \leq 0.01$)	WS ($p \leq 0.01$)	WS ($p \leq 0.01$)	WS ($p \leq 0.01$)	WS ($p \leq 0.02$)
E13/F15	212/19.5	12.5/8.1	201/18.8	13.6/13.4	178/23.2	13.6/13.4
(E, 1.5 km)	WS ($p \leq 0.01$)	WS ($p \leq 0.01$)	WS ($p \leq 0.01$)	WN	WS ($p \leq 0.01$)	WN

Note: ND, no data; WN, not sig. ($p = 0.05$); WS, sig. ($p = 0.05$); CV, RHO > critical value.

Table 7.3 Mean Biomass (g m^{-2}) of Dominant Seagrasses Taken in the Econfina System over the Long-Term Study Period

	Station						
SAV	**E07**	**E08**	**E09**	**E10**	**E11**	**E12**	**E13**
S. filiforme	0.5	1276.3	655.0	1023.8	0.4	258.5	1135.4
T. testudinum	0.3	42.1	285.4	126.7	0.1	1125.3	105.6
Total SAV	505	2654	1717	3927	611	1859	3480

Long-term changes of monthly SAV biomass in the Econfina and Fenholloway Estuaries (inner stations) are shown in Figure 7.10a. There were significant reductions of SAV biomass at the Fenholloway inner stations compared to the Econfina cognates throughout the study period. Fenholloway trends tended to go opposite those in the reference systems. This could have been a reaction to the increasing drought conditions and associated reductions of natural color in the Fenholloway Estuary that favored the SAV in the Fenholloway Estuary. There was a general decline of SAV biomass in the inner parts of the Econfina Estuary during the 1970s that was followed by extremely low biomass during the 1992–1993 study period. During the drought years of 1999–2002, there was a decline of SAV biomass in the Econfina inner areas that contrasted with the general increase of this factor in the Fenholloway areas.

Changes of SAV biomass in the outer stations of the two estuaries (Figure 7.10b) indicated different long-term trends of this variable. In the Econfina system, there was a general increase during the later part of the 1970s that was followed by major declines in 1992–1993 and less accentuated declines during the 1999–2002 drought. Once again, the SAV biomass in the Fenholloway outer areas was significantly lower than that in the reference area over the entire study period. The general SAV decline during the duration of the major drought could be a response of the Econfina Estuary to reduced nutrient flows and nutrient loading from land areas. Increased biomass in the Fenholloway during the 1970s could be attributed to increased light penetration due to the construction of holding ponds by the pulp mill during the early part of the decade.

Long-term changes of annual averages of the species richness of the SAV in the Econfina and Fenholloway systems are shown in Figure 7.11a and b. The lowest SAV species richness was evident in the inshore Fenholloway stations during all three sampling periods with the exception of increased plant species in the Fenholloway inshore and offshore areas during the latter part of the 1970s.

During the period following the 1992–1993 period, there was a reduction of SAV species richness in the Fenholloway inner station area. The index did not vary in the other Econfina areas from the levels noted during the previous study period. However, during the drought of 1999–2002, there was a significant increase of SAV species richness in inner and outer station areas in both systems. There were indications of such shifts toward the end of the 1992–1993 sampling period. These trends were noted in both the inner and outer parts of both estuaries.

This spectacular increase of SAV species richness during the latest sampling period was due to the addition of various macroalgal species. These algae were dominated by green algae (Chlorophyta), red algae (Rhodophyta), and brown algae (Phaeophyta) (Zimmerman and Livingston, 1979).

Long-term changes of the dominant seagrasses are shown in Figure 7.12a and b (turtle grass, *T. testudinum*) and Figure 7.13a and b (manatee grass, *S. filiforme*). Turtle grass was relatively abundant in the Econfina Estuary during the 1970s where it was almost completely absent in the Fenholloway Estuary. There was a declining trend to this species over this decade of sampling. During the two succeeding study periods, *Thalassia* was largely absent in both estuaries. These trends were comparable at both the inner and the outer stations of these systems. There was high variability of *S. filiforme* during the 1970s with low levels of biomass following drought periods. The highest concentrations of *Syringodium* were in the outer Econfina system with the lowest such concentrations in the inner Fenholloway area. This species was generally absent during the 1991–1992

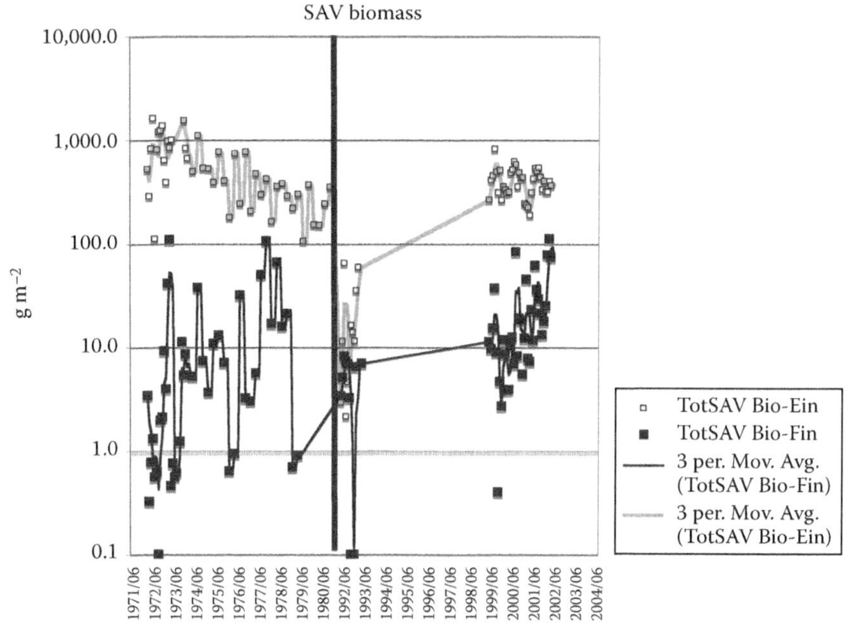

(a)

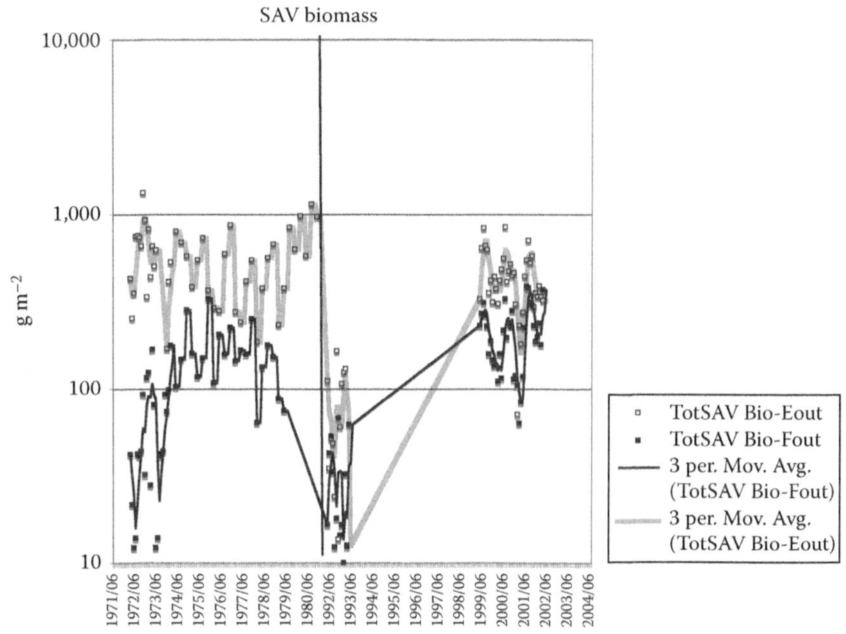

(b)

Figure 7.10 (a) Long-term comparison of SAV biomass (g m^{-2}) averaged over Econfina and Fenholloway inner stations (EIN, FIN). Analyses were run with 3-month moving averages. (b) Long-term comparison of SAV biomass (g m^{-2}) averaged over Econfina and Fenholloway outer stations (EOUT, FOUT). Analyses were run with 3-month moving averages.

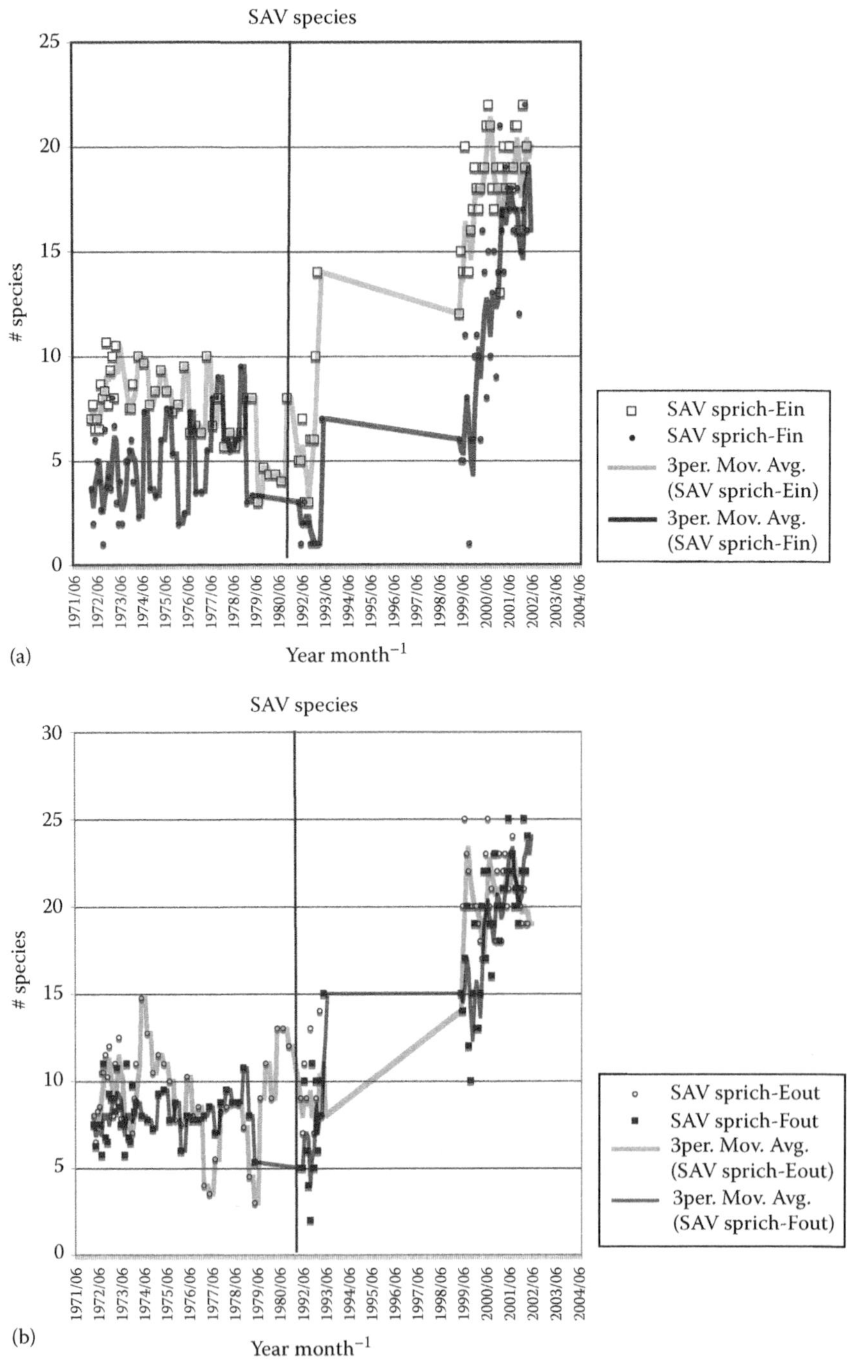

Figure 7.11 (a) Long-term comparison of SAV species richness averaged over Econfina and Fenholloway inner stations (EIN, FIN). Analyses were run with 3-month moving averages. (b) Long-term comparison of SAV species richness averaged over Econfina and Fenholloway outer stations (EOUT, FOUT). Analyses were run with 3-month moving averages.

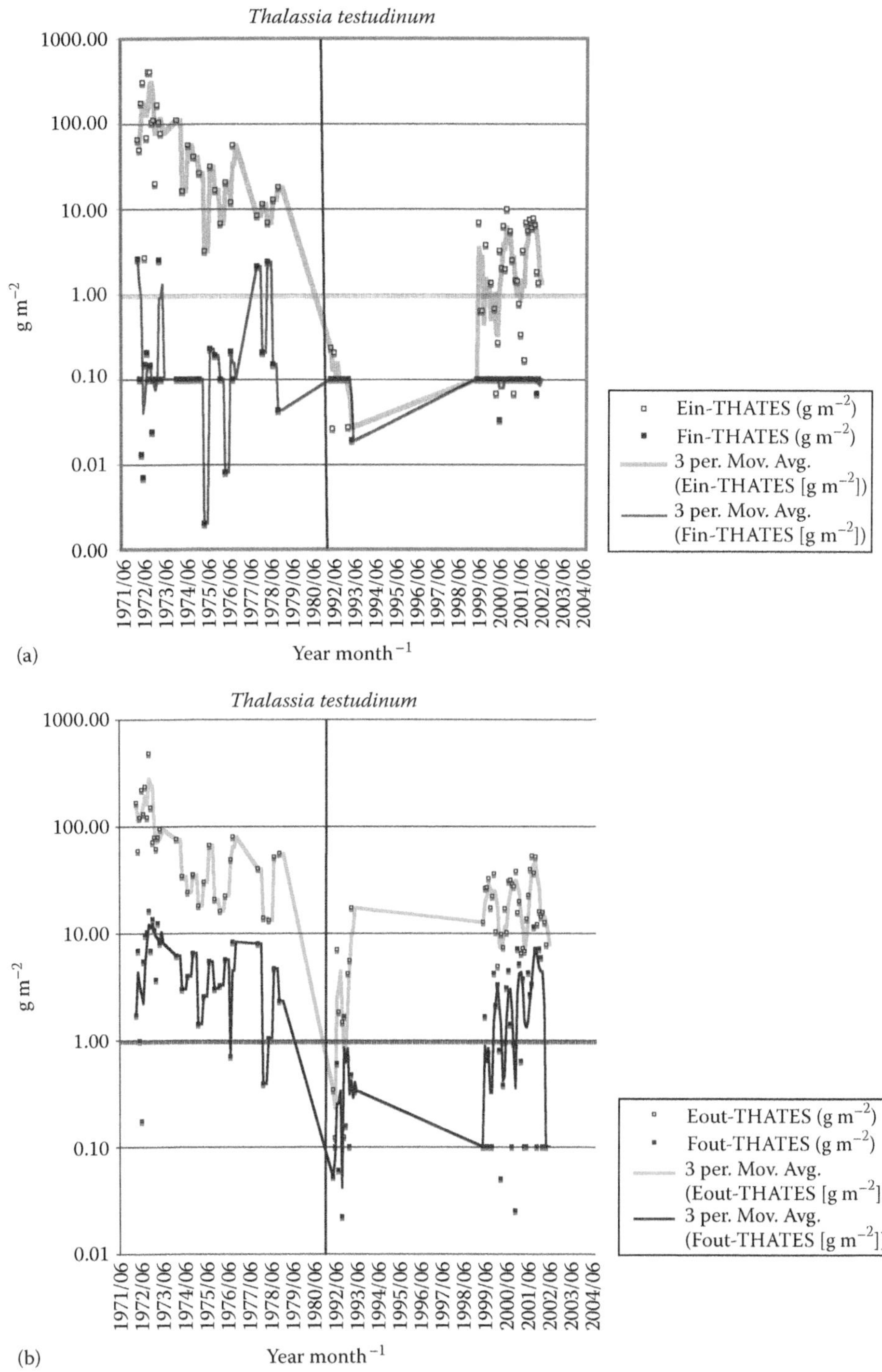

Figure 7.12 (a) Long-term comparisons of *T. testudinum* biomass (g m^{-2}) at inner stations of the Econfina and Fenholloway systems. Data analysis included the use of 3-month moving averages. (b) Long-term comparisons of *T. testudinum* biomass (g m^{-2}) at outer stations of the Econfina and Fenholloway systems. Data analysis included the use of 3-month moving averages.

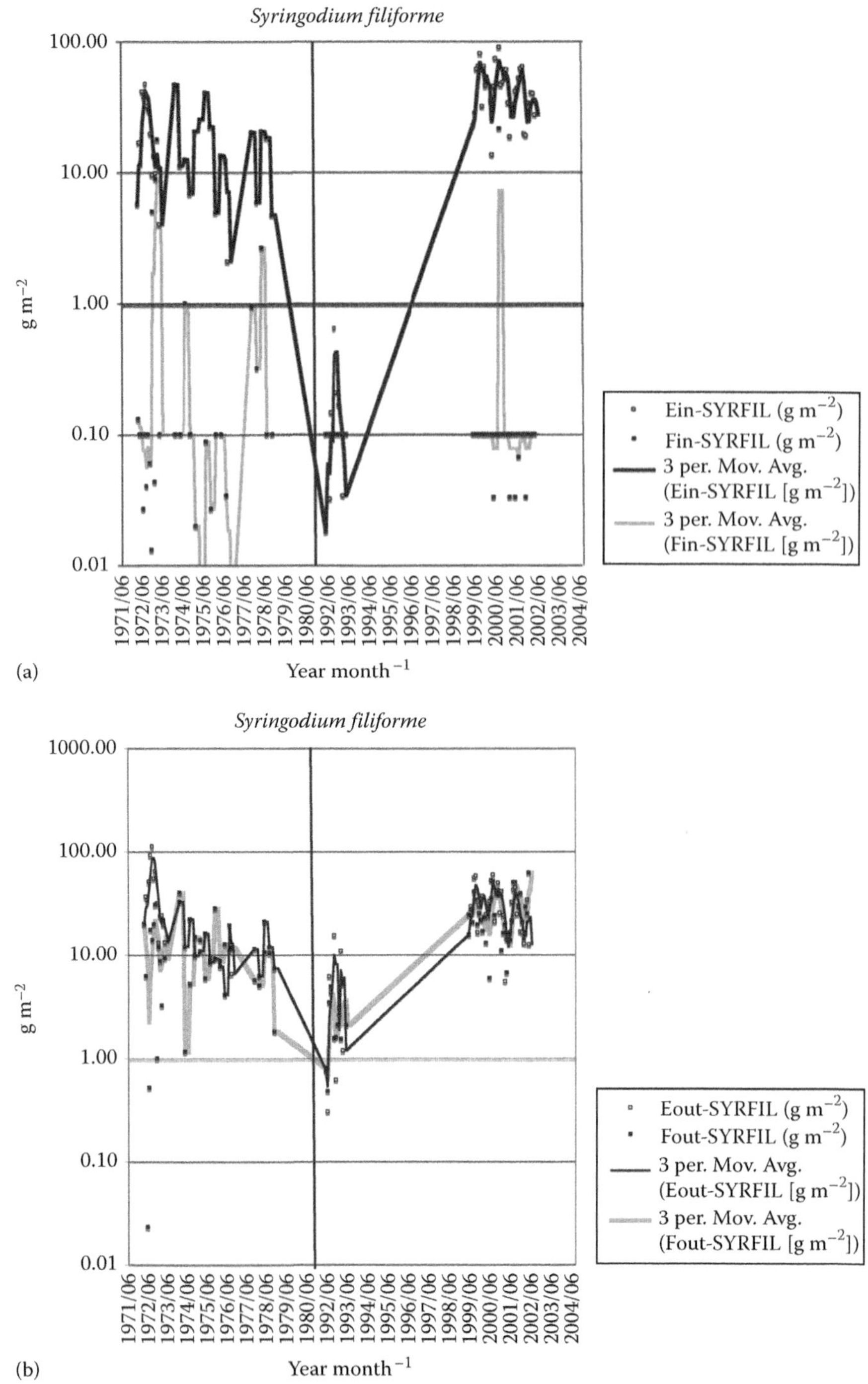

Figure 7.13 (a) Long-term comparisons of *S. filiforme* biomass (g m^{-2}) over inner stations with 3-month moving averages. (b) Long-term comparisons of *S. filiforme* biomass (g m^{-2}) over outer stations with 3-month moving averages.

sampling in all sampling areas. *Syringodium* recovered in the Econfina inner stations in 1999–2002 and at the outer stations of both systems during this period of the study. This recovery included relatively high biomass of this species at the inner stations.

A comparison of the macroalgal biomass in the inner and outer parts of the Econfina and Fenholloway Estuaries is given in Figure 7.14. The data indicate a declining trend after peaks during the early 1970s in both estuaries. These numbers reached low points during the 1992–1993 sampling period. However, during the last sampling period, there were significant increases in the macroalgal biomass in both estuaries with the highest such levels reached at inner stations. These increases coincide with the decreased biomass of *T. testudinum* in the study areas.

The long-term SAV data illustrate the extreme interannual variation of biomass in the study area with major decreases during periods of altered rainfall and river flows. The evidence showed that there was little recovery of SAV biomass in areas of the Fenholloway offshore that have been affected by the pulp mill effluents. These data represent a further indication that the nominal reductions of water color during the drought in the Fenholloway basin were not enough to bring back the benthic vegetation at a density comparable to that in the reference area. The increased species richness during the 1999–2002 drought was accompanied by species-specific shifts that included reductions of *T. testudinum* and increased biomass and species richness of the macroalgae. In any case, the SAV assemblages noted during the 1970s were followed by major shifts of the relative dominance of the seagrasses and macroalgae during the 1992–1993 and 1999–2002 study periods relative to the predrought period during the earliest sampling period.

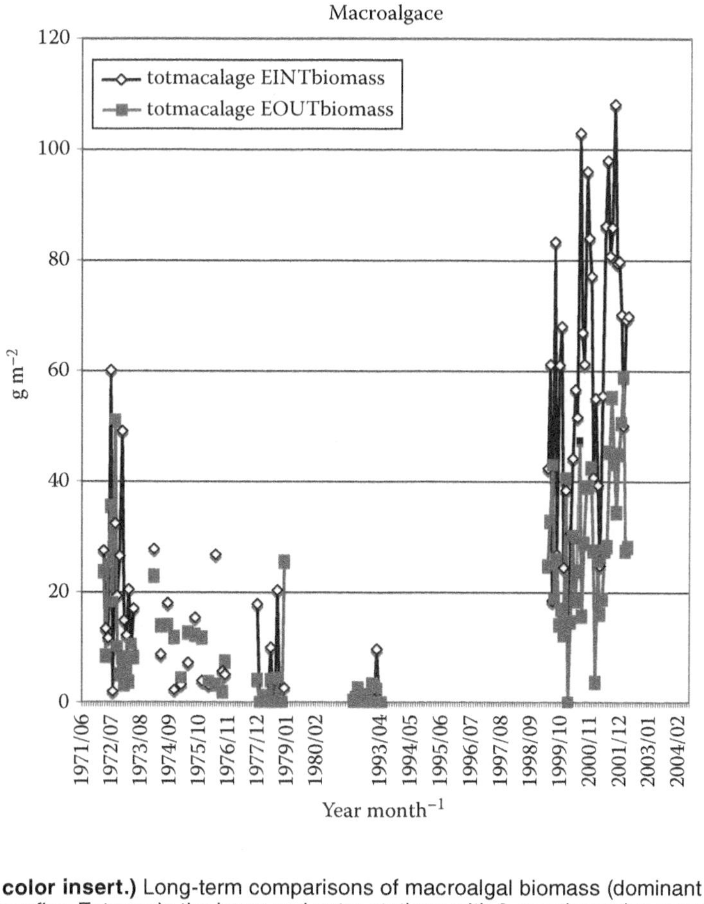

Figure 7.14 (See color insert.) Long-term comparisons of macroalgal biomass (dominant species; g m^{-2}) in the Econfina Estuary in the inner and outer stations with 3-month moving averages.

There were some other important changes in the SAV ecology of the Apalachee Bay area during the 1999–2002 study period. For the first time in almost 30 years of sampling, active flowering was noted in four dominant plant species (Table 7.4). The flowers and fruiting bodies were noted mainly in offshore stations in the Econfina and Fenholloway Estuaries during spring and summer months. This form of reproduction makes the expansion of seagrass beds in the Apalachee Bay more likely. The flowering of these species also coincided with substantial increases of water temperature and reduced water color during the study period. In addition, there was a major invasion of the Fenholloway system by *Halophila engelmannii*, a seagrass species that is usually found in deeper water areas of the gulf. The distribution of this species was most abundant at offshore stations to the east (F15, FT29, F16) with some increases at stations E11 and E13 in the Econfina Estuary.

This change in the reproductive mode of the seagrasses in the study area was a major development that could be attributed to the warming of the northern parts of the gulf.

Fishes

The analysis of averaged fish biomass (Figure 7.15a and b) indicated that this index in the Econfina system was relatively low following the slight drought during the early 1970s. In the inshore areas, the fish biomass was usually higher in the Econfina relative to the Fenholloway Estuary. There were reductions of fish biomass at inner station groups in both estuaries in 1992–1993. By 2004, there was little difference of fish biomass between the outer parts of the two areas of study. There was a general reduction of fish biomass in the outer parts of the Econfina Estuary during the 1999–2004 sampling. Overall, the biomass of fishes in the Econfina inshore area did not show much change relative to that of the 1970s. However, this index in the offshore Econfina declined relative to the fish biomass during the earlier sampling period, a trend that followed the distribution of the SAV biomass in the outer parts of the reference area.

By the end of the survey in 2004, the fish fauna in the inner and outer parts of both estuaries was in a severely deteriorated state, with relatively low biomass and reduced species richness relative to previous sampling periods. The reduced fish biomass was usually preceded by periods of drought that were followed by heavy rainfall and changes in habitat conditions that were consistent with the increased land runoff.

The abundance trends of the numerically dominant fish species gave further evidence of the long-term trends of fish populations in both parts of the Apalachee Bay. The distribution of pinfish *L. rhomboides* (Figure 7.16a) showed highest numbers (and high interannual variability) during the 1970s with the lowest numbers during or after floods and droughts at the beginning and end of this part of the study. The numbers of pinfish were generally lower in the Fenholloway Estuary than those of the reference area. In 1992–1993, there were severely restricted numbers of this dominant in both the Econfina and Fenholloway Estuaries that followed the trends of the SAV in both parts of the gulf. Pinfish numbers in the Econfina Estuary in 1999–2002 increased to levels comparable to those during the earliest sampling period but dropped during the 2004 sampling. The onset of storms appeared to adversely affect the numerical abundance of this species in both systems.

The planktivorous bay anchovies (*Anchoa mitchilli*; Figure 7.16b) were dominant in the inshore areas of the Fenholloway Estuary where seagrasses had been eliminated by pulp mill effluents. High nutrient loading in inshore Fenholloway areas had increased plankton food. Anchovy numbers were positively associated with rainfall and higher river flows in the region, and this species was somewhat reduced in the inner Fenholloway areas during the 2004 sampling period. This species was absent in the Econfina system during the 1999–2004 study period.

Perry (Florida) rainfall over the study period was differenced (following month – preceding month), and the results resembled those noted in the Chattahoochee River analysis (Figure 7.17). Both positive and negative deviations started to occur during the late 1970s. By the 1992–1993 study period, such deviations had reached a maximum extent with one such set occurring during

Table 7.4 Incidence of Flowering in Four Species of SAV Found in the Econfina–Fenholloway Systems during the Last Sampling Period (1999–2002)

Station	Date	Species	Dry wt g m^{-2}	Station	Date	Species	Dry wt g m^{-2}
F16	Jun-00	*Halophila engelmannii*	ND	F12	Apr-99	*Syringodium filiforme*	0.85
F15	Jun-00	*Halophila engelmannii*	2	F11	Jul-00	*Syringodium filiforme*	ND
F10	Jun-00	*Halophila engelmannii*	2.85	E13	Jul-00	*Syringodium filiforme*	ND
FT29	Jul-00	*Halophila engelmannii*	ND	E12	Jul-00	*Syringodium filiforme*	ND
E13	Jul-00	*Halophila engelmannii*	ND	E10	Jul-00	*Syringodium filiforme*	ND
E08	Jul-00	*Halophila engelmannii*	ND	E08	Jul-00	*Syringodium filiforme*	ND
FT29	May-01	*Halophila engelmannii*	0.25	E10	Aug-00	*Syringodium filiforme*	ND
F15	May-01	*Halophila engelmannii*	0.4	F12	May-01	*Syringodium filiforme*	0.45
F11	May-01	*Halophila engelmannii*	0.2	E13	May-01	*Syringodium filiforme*	1.6
FT29	Jun-01	*Halophila engelmannii*	0.25	E12	May-01	*Syringodium filiforme*	0.15
F16	Jun-01	*Halophila engelmannii*	0.15	E10	May-01	*Syringodium filiforme*	0.9
F15	Jun-01	*Halophila engelmannii*	0.25	E09	May-01	*Syringodium filiforme*	0.75
F11	Jun-01	*Halophila engelmannii*	0.05	E08	May-01	*Syringodium filiforme*	3.5
F10	Jun-01	*Halophila engelmannii*	0.1	F16	Jun-01	*Syringodium filiforme*	0.4
E08	Jun-01	*Halophila engelmannii*	0.15	E13	Jun-01	*Syringodium filiforme*	3.5
E07	Jun-01	*Halophila engelmannii*	0.1	E12	Jun-01	*Syringodium filiforme*	0.05
F11	Jul-01	*Halophila engelmannii*	0.1	E10	Jun-01	*Syringodium filiforme*	0.9
E13	Jul-01	*Halophila engelmannii*	0.05	E09	Jun-01	*Syringodium filiforme*	0.3
E10	Jul-01	*Halophila engelmannii*	0.05	E08	Jun-01	*Syringodium filiforme*	2.3
E08	Jul-01	*Halophila engelmannii*	0.05	E13	Jul-01	*Syringodium filiforme*	0.15
F15	Jan-02	*Halophila engelmannii*	0.5	E10	Jul-01	*Syringodium filiforme*	1.4
F15	Feb-02	*Halophila engelmannii*	0.1	E09	Jul-01	*Syringodium filiforme*	1.9
F11	Feb-02	*Halophila engelmannii*	0.2	E08	Jul-01	*Syringodium filiforme*	0.3
FT29	Mar-02	*Halophila engelmannii*	0.2	E08	Jan-02	*Syringodium filiforme*	0.1
F15	Mar-02	*Halophila engelmannii*	0.25	E13	Feb-02	*Syringodium filiforme*	0.2
F11	Mar-02	*Halophila engelmannii*	0.3	E10	Feb-02	*Syringodium filiforme*	0.2

(*Continued*)

Table 7.4 (*Continued*) Incidence of Flowering in Four Species of SAV Found in the Econfina–Fenholloway Systems during the Last Sampling Period (1999–2002)

Station	Date	Species	Dry wt g m^{-2}	Station	Date	Species	Dry wt g m^{-2}
F10	Mar-02	*Halophila engelmannii*	0.1	E09	Feb-02	*Syringodium filiforme*	0.9
E11	Mar-02	*Halophila engelmannii*	0.05	E08	Feb-02	*Syringodium filiforme*	0.9
E12	May-99	*Thalassia testudinum*	ND	F16	Mar-02	*Syringodium filiforme*	2.05
E09	Jun-01	*Thalassia testudinum*	0.55	F12	Mar-02	*Syringodium filiforme*	0.6
E12	Jun-00	*Thalassia testudinum*	ND	E09	Mar-02	*Syringodium filiforme*	0.1
E12	Jun-01	*Thalassia testudinum*	0.1	E08	Mar-02	*Syringodium filiforme*	0.6
E12	Jul-01	*Thalassia testudinum*	0.1	E11	Apr-01	*Ruppia maritima*	5.7
				E11	May-01	*Ruppia maritima*	2.05
				FT29	Jun-01	*Ruppia maritima*	0.2

the 1992 sampling and another occurring during the 2002–2003 period of field analysis. The period of increasing drought periods (1999–2002) was characterized by reduced deviations as might be expected. Since such increased frequency of both high- and low-rainfall peaks resulted in water-quality changes in nearshore waters of the Apalachee Bay, a closer analysis of the 1992–1993 and 1999–2004 study periods was made.

Perry rainfall differences during the 1990–1993 period (Figure 7.18) indicated a series of extreme rainfall incidences just prior to the 1992–1993 study period. There was another such incident in 1992. The SAV and fish biomass data (Figure 7.19) indicated reduced SAV throughout the sampling period. Fishes showed a peak biomass in the late 1992, but this index was generally low during this part of the study.

Rainfall and rain differences in 1999–2004 (Figure 7.20) were low during the drought from 1999–2002. With the break in the drought due to heavy rainfall in 2003, this later period was marked by increased differences in rainfall and river flow. The biological data for this period (Figure 7.21) marked decreasing SAV and fish biomass during the drought and a major decrease of fish biomass after the rainfall intervention of 2003. This period indicated the lowest level of fish biomass in the Econfina inshore waters over the entire study period.

Fish species richness in the outer stations of the Econfina and Fenholloway Estuaries (Figure 7.22) indicated comparable numbers of fish species in these areas. The analysis also indicated a major reduction of this index in both estuaries following the heavy rainfall in 2003. During this period, there was a loss of both C_2 and C_3 fish carnivores (Figure 7.23) with a severe reduction of the C_1 carnivores over the final year of sampling (2003–2004). Fish omnivores (Figure 7.24) at all levels of omnivory showed a similar collapse during the final year of sampling.

The biological response of Apalache Bay to the changing climate over the study period was complicated. The plants were driven by a complex of increasing light penetration and reduced nutrients, whereas the animals were driven by complex habitat and primary productivity trends that were mediated by freshwater runoff and associated nutrient loading. Interannual variability was extreme when related to such trends. However, the overall effects on the otherwise natural Econfina Estuary resulted in an overall decrease of SAV in the outer parts of the estuary and associated reductions of fish productivity in this area. The increased incidents of high rainfall and drought combined to result in a collapse of the fish organization during the last year of sampling.

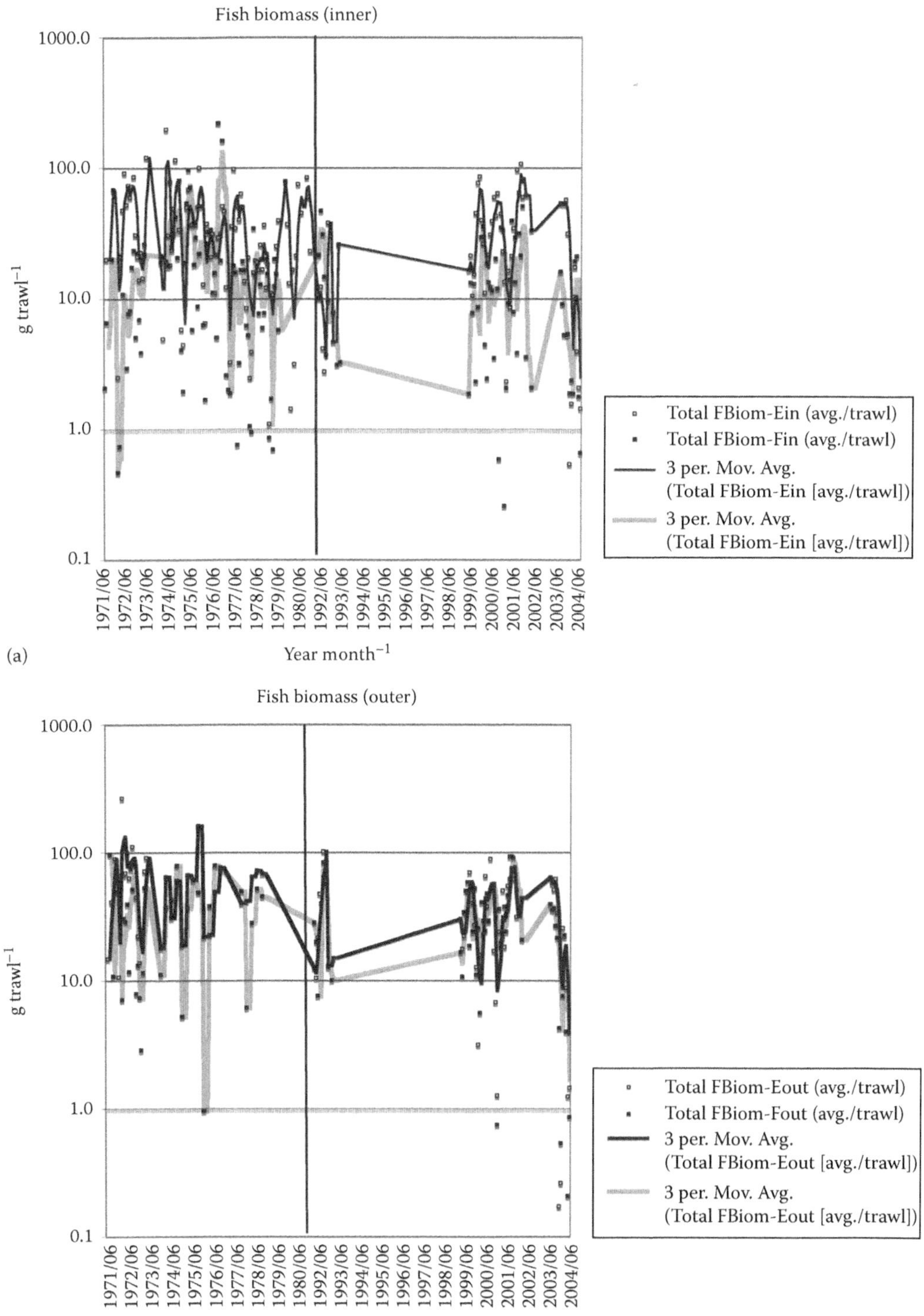

Figure 7.15 (a) Long-term comparisons of fish biomass per trawl tow on the inner stations of the Econfina and Fenholloway systems with 3-month moving averages. (b) Long-term comparisons of fish biomass per trawl tow on the outer stations of the Econfina and Fenholloway systems with 3-month moving averages.

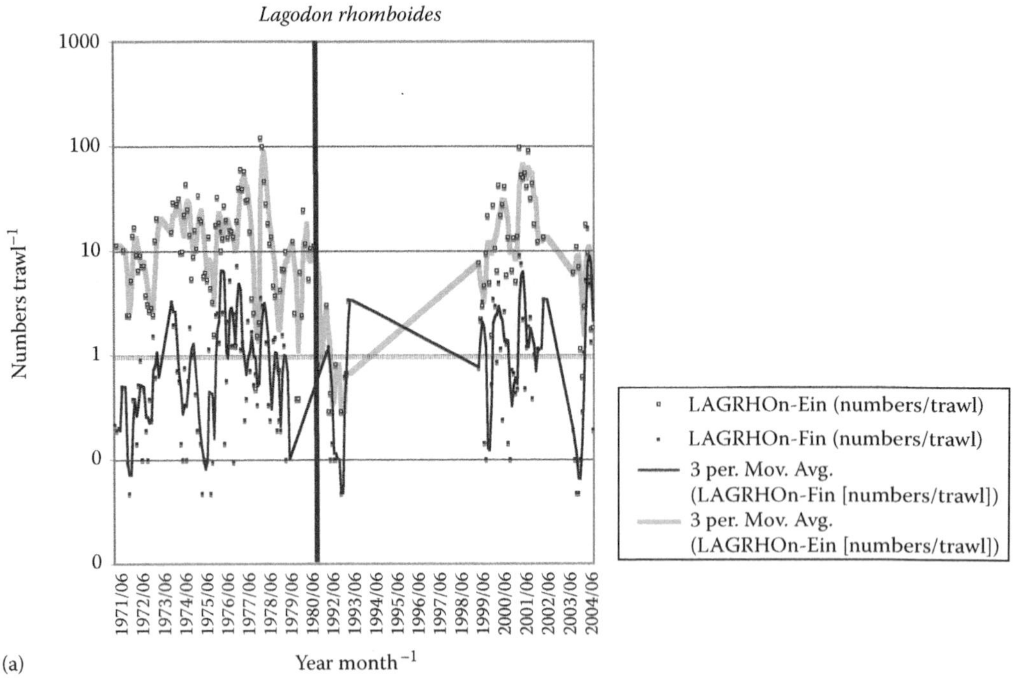

(a)

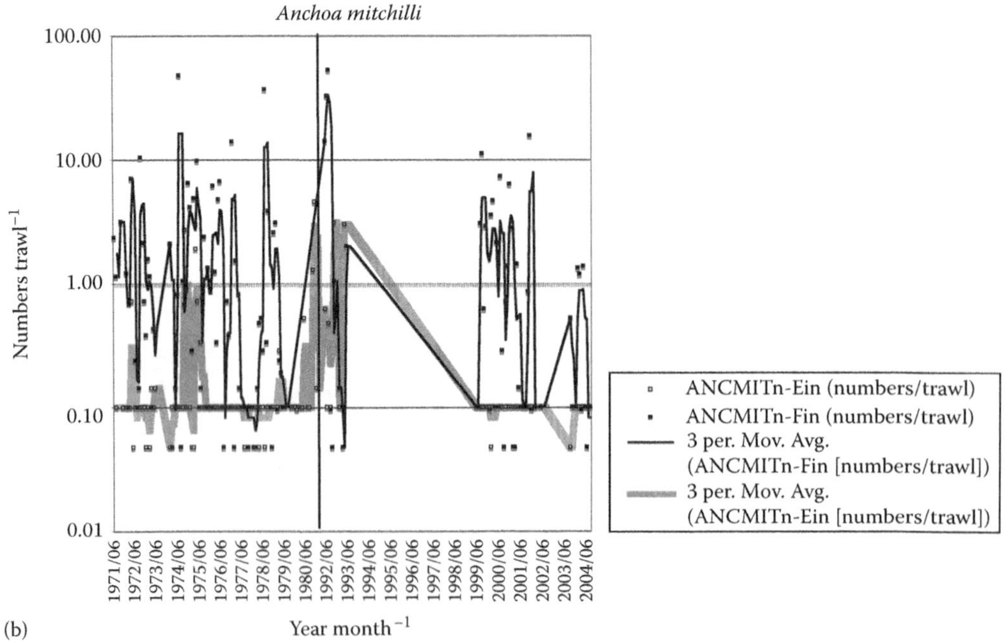

(b)

Figure 7.16 (a) Long-term comparisons of pinfish (*L. rhomboides*) numbers at the inner stations. Data analysis included 3-month moving averages of the pinfish numbers of individuals. (b) Long-term comparisons of bay anchovies (*A. mitchilli*) numbers at the inner stations. Data analysis included 3-month moving averages of the anchovy numbers of individuals.

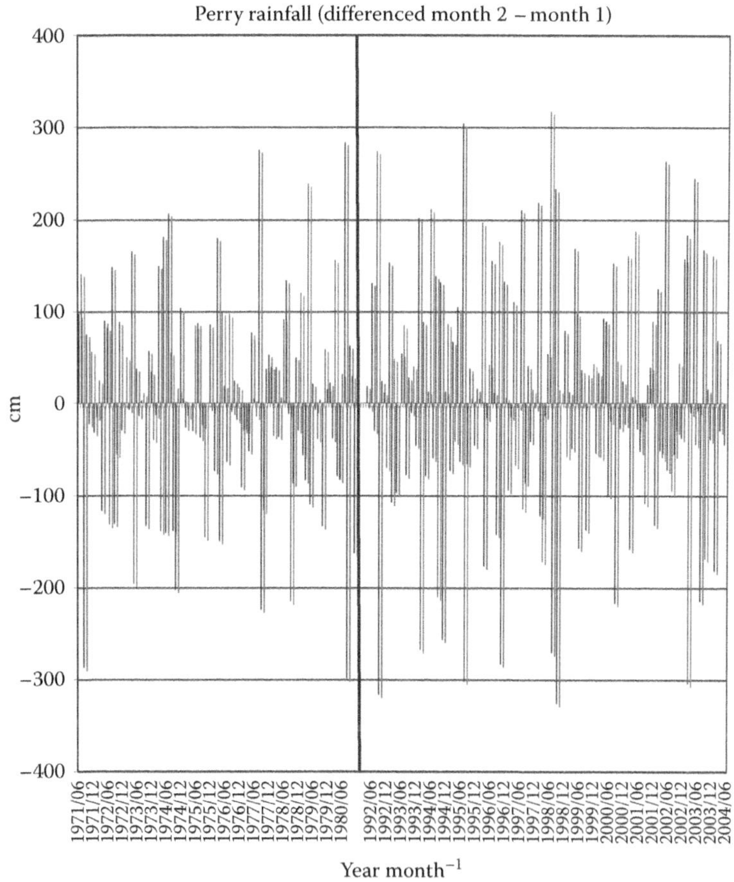

Figure 7.17 Perry monthly rainfall with differenced (month 2 – month 1) data from 1971 through 2004.

Invertebrates

There were differences in the inshore assemblages of epibenthic macroinvertebrates between the Econfina and Fenholloway Estuaries during the 1970s (Dugan and Livingston, 1982; Livingston, 1993a). During most periods, the numbers of the trawlable invertebrates were usually lower in the inshore parts of the Fenholloway coastal waters. These differences tended to decrease with distance from shore. There were particularly high numbers of blue crabs (*C. sapidus*) in the Fenholloway inshore waters. The high organic loading from the Fenholloway River was associated with various forms of infauna that served as food for the blue crabs. There were also relatively high numbers of penaeid shrimp taken in the inshore Fenholloway areas compared to the Econfina reference sites.

During the 1970s, the invertebrates in the study area (Appendix C, Table 7.5) were dominated by various crustacean species in the form of caridean shrimp. In both systems, the far offshore areas were quite different from the inshore and immediate offshore areas with dominance of species such as the ophiuroid echinoderm such as *Ophiothrix angulata*, the echinoids such as *Lytechinus variegatus* and *Diadema antillarum*, majid decapods such as *Mithrax pleuracanthus*, and porcellanids such as *Petrolisthes galathinus*. In areas where seagrass beds existed, the Fenholloway Estuary was characterized by higher numbers of individuals and species than the Econfina with higher dominance and lower species diversity and evenness than that in the Econfina Estuary.

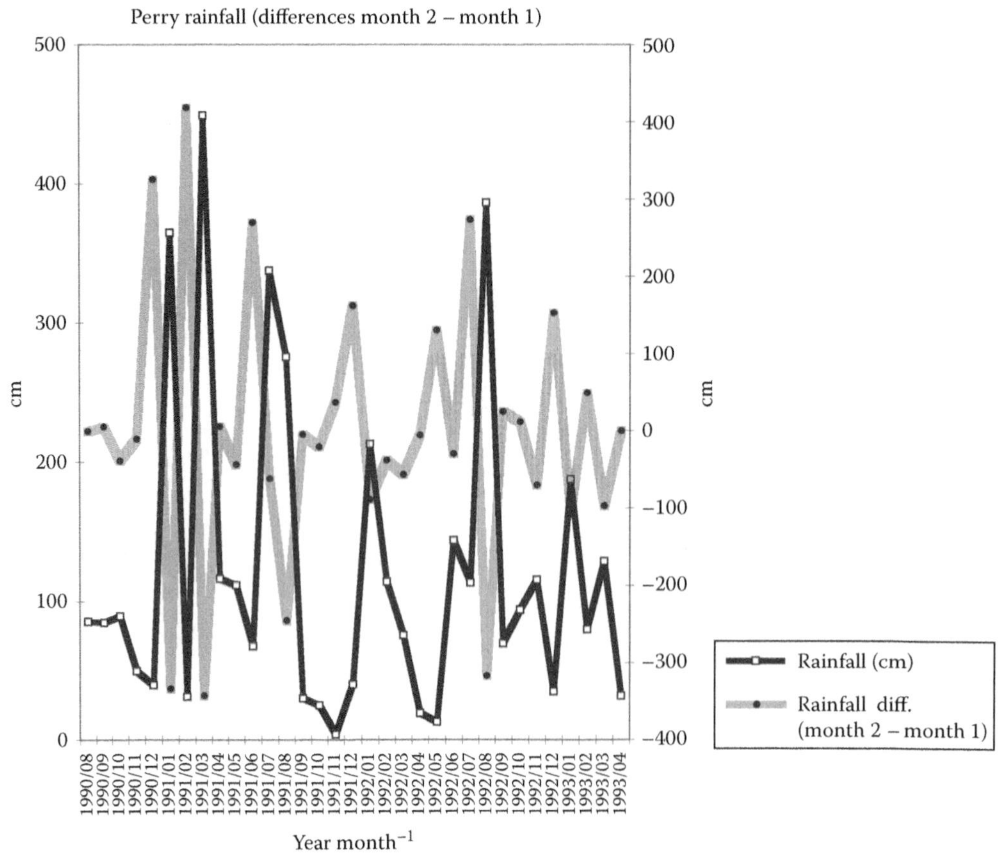

Figure 7.18 Perry rainfall and differenced data (month 2 – month 1) monthly from August 1990 through April 1993. Rain data were provided by the National Oceanic and Atmospheric Administration.

The primary habitats for the relatively small organisms that comprised the invertebrate biota in the Apalachee Bay were thus considered to be the SAV that provided cover and food for the primary feeding modes of detritivory and herbivory/omnivory.

Analyses of the invertebrate numbers averaged by inner and outer stations over annual periods are given in Figures 7.25 and 7.26. During the 1970s, invertebrate numbers in the Econfina Estuary went down during the final years of the decade. These decreases occurred during periods of increased fish biomass. By 1991–1992, there was a spread of such numbers between the two river systems, and such numbers tended to be considerably higher than those taken during the 1970s. The numbers tended to be higher in the Econfina Estuary. This pattern was repeated during the 1999–2002 drought with increasing invertebrate numbers in both systems over the study period. These trends of invertebrate numbers tended to be inversely related to the fish biomass in the study areas over the periods of observation. This relationship indicated a change in the predator–prey relationships with the reduction in fishes associated with increased numbers of their invertebrate prey.

Invertebrate species richness in inshore areas was higher in the Fenholloway Estuary during the 1970s, (Figures 7.27 and 7.28). These indices tended to decrease with time in both systems during the 1970s. Invertebrate species richness in inner and outer areas was much higher during the 1992–1993 and 1999–2002 surveys in both systems. These increases coincided with the reductions of fish biomass during these periods. This was a further indication that there was a possible rebound of invertebrates due to reduced predation pressure from the fishes during such periods. A complicating factor related to this hypothesis was that there was also a major increase in the SAV species

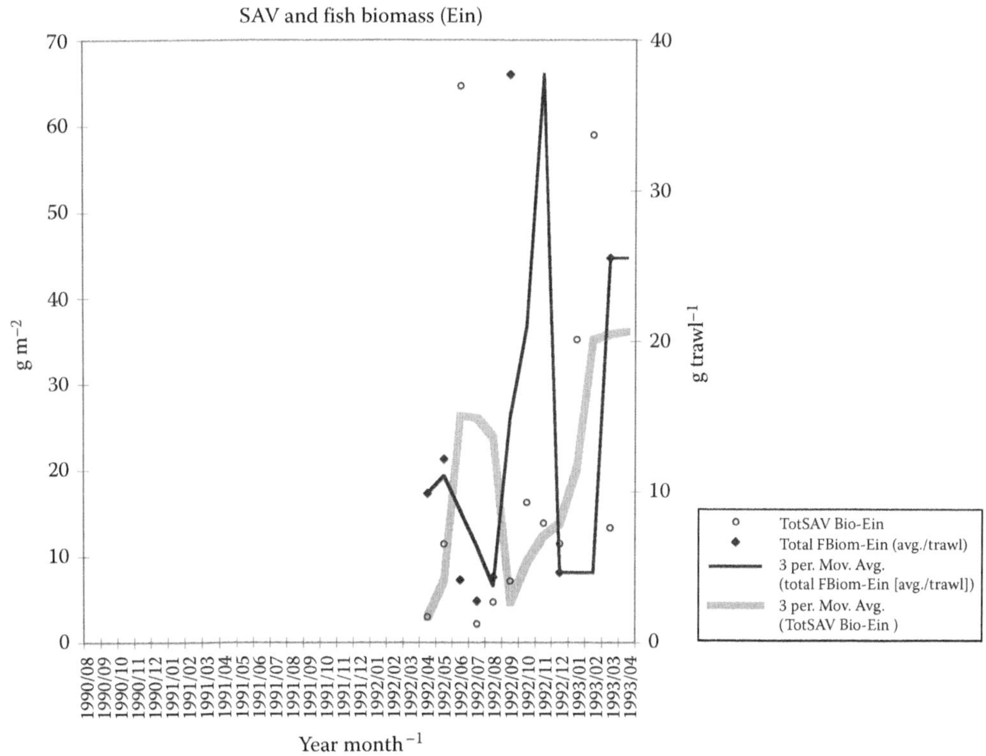

Figure 7.19 Total SAV and fish biomass taken in the inshore areas of the Econfina Estuary in 1992–1993. Data were analyzed with a 3-month moving average.

richness indices that could have contributed to the major increase of invertebrate abundance and species richness during the later periods of the long-term study.

The review of the invertebrate data indicated that the long-term changes of climatic conditions such as temperature and rainfall/river flow were associated with changes in the biological organization of the study areas with alterations of the benthic vegetation and associated populations of fishes and invertebrates that were complicated by biological interactions among these components. The combination of increasing drought conditions interspersed with periods of increasingly heavy rainfall was associated with changes in water-quality conditions in the study areas that in turn could be related to the major shifts of the trophic organization of the Econfina and Fenholloway Estuaries.

Conclusions

The Apalachee Bay is a large, relatively shallow estuarine system with a series of small rivers along the west coast of Florida. It is a high-salinity system with no major rivers. The bay is fringed with an array of extensive marsh developments along the landward fringe. The basis for the relatively high productivity in the bay is in the major formations of SAV that are dominated by several species of seagrass. The bay is one of the least polluted such bays in the conterminous United States. The only major industrial development is a pulp mill that discharges into the Fenholloway River. A long-term study (1970–2004) compared this river and its offshore habitats to a relatively natural blackwater stream (the Econfina River) and its unpolluted offshore area.

Phytoplankton productivity in the Apalachee Bay is generally low compared to the large river–estuaries to the west (Livingston et al., 1998a). Reductions of Fenholloway seagrass beds were caused by increased color and reduced light transmission. These changes were associated with altered food

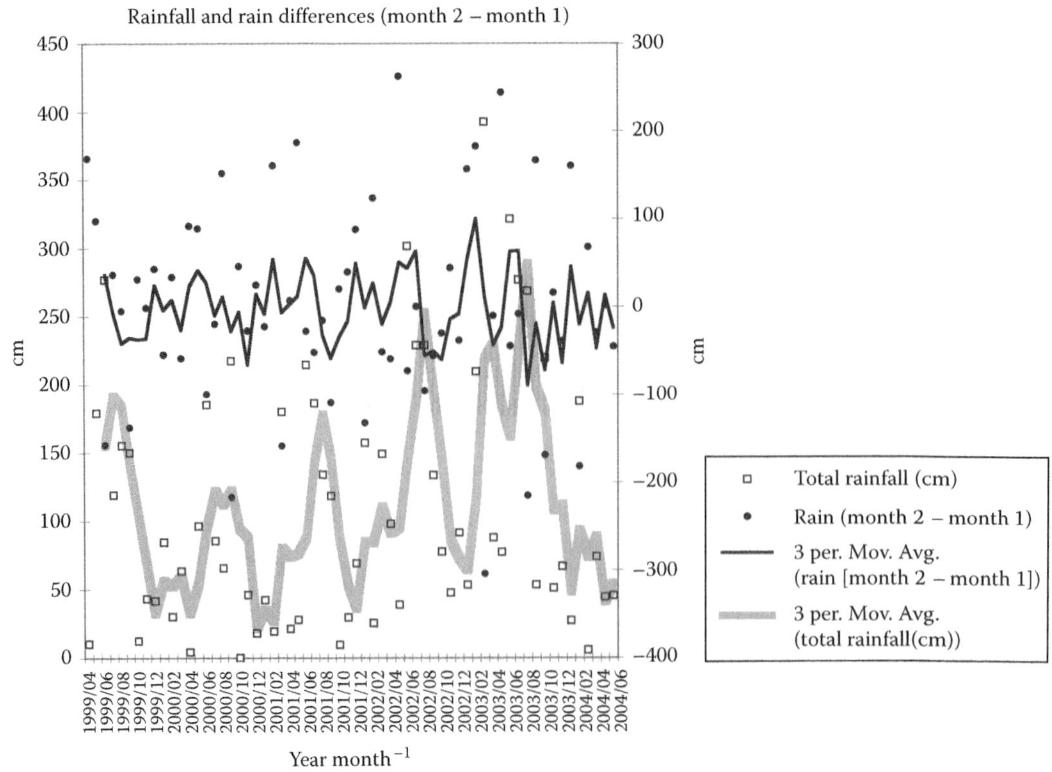

Figure 7.20 Perry rainfall and differenced data (month 2 – month 1) monthly from April 1999 through June 2004. Rain data were provided by the National Oceanic and Atmospheric Administration.

web patterns in terms of species shifts and feeding alterations relative to reference (Econfina) areas (Livingston, 1975, 1980; Livingston et al., 1998a). Feeding habits of the dominant fish species were different in the offshore Fenholloway Estuary than those found in similar areas of the Econfina Estuary. These differences were traced to basic changes in habitat due to the reduced SAV in the Fenholloway system (Clements and Livingston, 1983). In offshore Fenholloway areas, there was an increased phytoplankton production that was accompanied by increases in plankton-feeding fishes such as bay anchovies relative to the dominance of seagrass bed species such as pinfish in the Econfina Estuary.

Studies showed that there was a long-term trend of climatological changes in terms of increasing frequency and intensity of droughts in the region that started during the 1980s. The regional rainfall followed precipitation patterns of Georgia and major rivers along the NE Gulf Coast of Florida. Both the rainfall and the Econfina/Fenholloway River flow patterns indicated a shift to longer and more extreme drought periods interspersed with increased rainfall events that started in the 1980s and continued to the present time. Over the study period, there was an increased frequency of both droughts and rainfall peaks that led to changes in nutrient loading, salinity, DO, water color, and light penetration. Average annual water temperatures in the inshore waters of the Apalachee Bay increased over the long-term study period, and such increases were affected by the increased incidence of droughts.

There was a mixed, highly complex biological response to these long-term changes of climate and water quality. Following a relatively stable period of climate changes during the 1970s, there was a serious decline of SAV beds in both study areas with declines of a dominant seagrass species, *T. testudinum*. These declines followed a series of droughts and periodic floods that caused rapid changes in the inshore habitats of the Econfina and Fenholloway Estuaries. Offshore SAV beds during a major drought from 1999 to 2002 had biological changes with reduced turtle grass cover

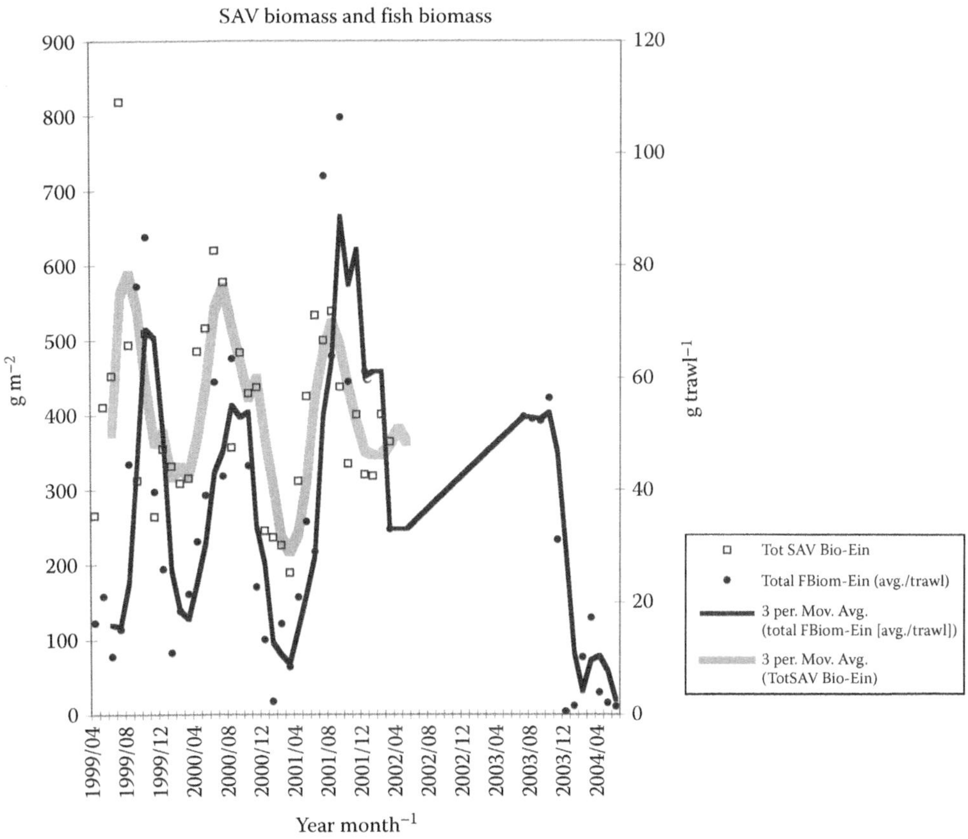

Figure 7.21 Total SAV and fish biomass taken in the inshore areas of the Econfina Estuary in 1999–2004. Data were analyzed with a 3-month moving average.

and increases in SAV species richness due to additions of various benthic macroalgal groups. There were declines in the fish populations during this period that followed enhanced rainfall conditions similar to that observed during the 1992–1993 declines of SAV and fish communities. Invertebrate numbers went up during these periods with major increases in numbers and species richness as a possible effect of reduced fish predation.

There were too few data to explain the reasons for the earlier noted trends in the Apalachee Bay biological organization that tracked the observed changes of temperature and rainfall. Increased water temperature was probably a major factor in the first observations of flowering in the major seagrass species during the later years of the study. This change could have been associated with an increased ability of the seagrasses to spread to adjoining areas.

The trophic organization of the fishes reflected both the changes caused by increased nutrient loading from the paper mill and the long-term responses of this community to habitat-quality alterations associated with the observed climate changes. Increased light penetration that accompanied the increased drought conditions was in direct conflict with reduced nutrient loading in the effects on the SAV coverage, density, and species composition. The decrease of the main seagrass (*T. testudinum*) combined with the increase of macroalgal biomass and species richness in unpolluted areas (the Econfina Estuary) denoted a significant change of the main source of habitat and food that appeared to be translated into various biological responses.

Zimmerman and Livingston (1979), in explaining the SAV trends in the Apalachee Bay, noted that "prolonged stress in an area would thus result in severely reduced biomass, changes in qualitative

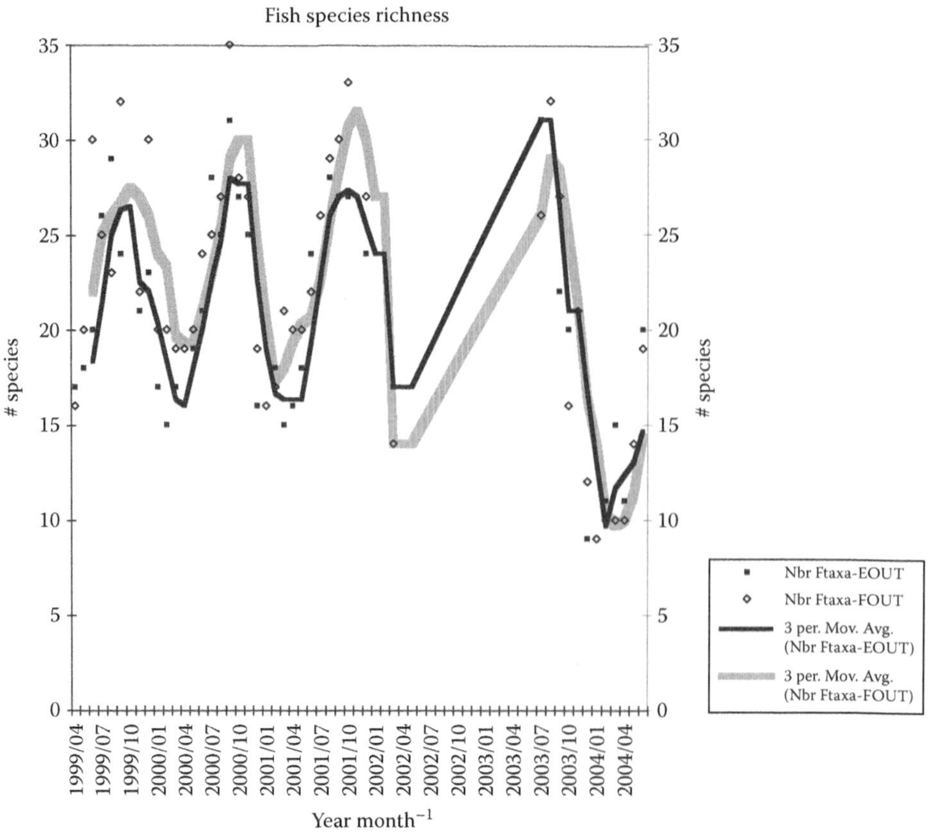

Figure 7.22 Fish species richness in the outer parts of the Econfina and Fenholloway Estuaries from April 1999 through June 2004. A 3-month moving average was used in the data analysis.

and quantitative relationships of the dominant species, and colonization by a series of relatively rare species capable of adapting to the available habitat." Although this observation was related to the SAV response to the pulp mill effluent discharge into the offshore Fenholloway Estuary, it could be applied to the response of the Econfina system to water-quality alterations resulting from the long-term climatological changes. Accordingly, decreased predation by fishes could account for the increase of invertebrate numbers and species as part of the altered trophic organization of the estuary.

Paine et al. (1998) reviewed the associations between compound perturbations and ecological surprises in a series of ecosystems. The authors considered that more serious ecological consequences could result from compounded perturbations with associated effects on the normative recovery time of the communities involved. Moe et al. (2013) and Stahl et al. (2013) developed this concept in greater detail.

As noted earlier, during years of high rainfall that followed prolonged droughts, there was an increase of phytoplankton blooms in near-offshore Fenholloway areas relative to cognate stations in the Econfina system. In this way, the increased rainfall exacerbated anthropogenic effects in the Fenholloway Estuary. The response of the trophic organization of a natural estuary (Econfina) relative to a polluted one (Fenholloway) demonstrated how compound anthropogenic impacts defined the responses to long-term climatological changes.

A conceptual diagram of such shifts is given in Figure 7.29. The reduction of plankton-feeding bay anchovies in the Fenholloway system during the storm-induced blooms (Figure 7.16b) indicated a different response of the Fenholloway food web patterns relative to that of the Econfina. The changes in the relationship of the seagrasses and macroalgae in the Econfina system could have

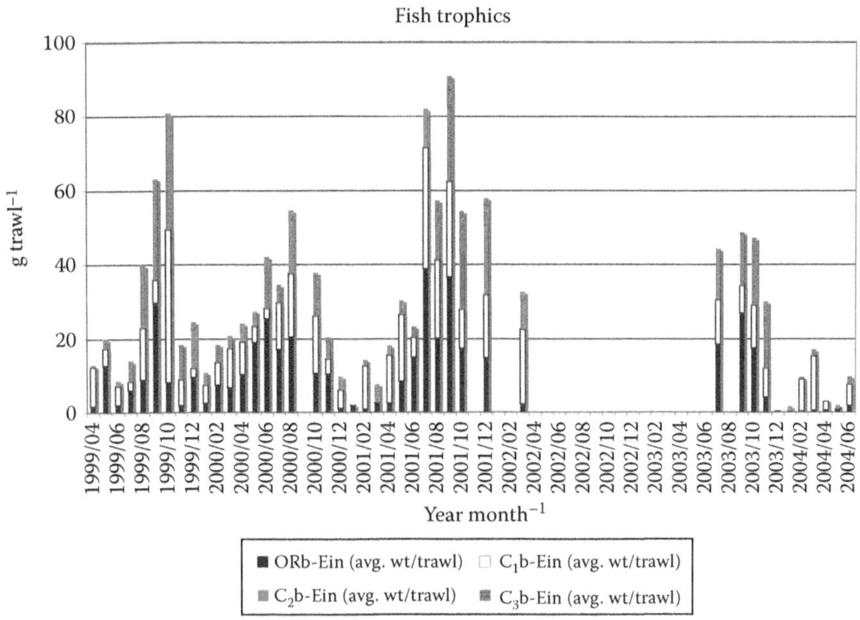

Figure 7.23 Fish trophic organization in the inner parts of the Econfina Estuary from April 1999 through June 2004. The data included biomass of C_1, C_2, and C_3 carnivores.

been due to complex biological interactions caused by the long-term climatological changes and reduced nutrient loading from terrestrial areas. Such interactions would be muted in the nutrient-enriched Fenholloway Estuary. By the same token, the interactions between fishes and their invertebrate prey could have been changed relative to the climatological alterations in ways not found in the primary Fenholloway food webs. The increase in macroalgal species relative to the seagrasses (especially turtle grass) could have been involved in the decreased fish populations and the increases in invertebrate numbers and species richness.

The combinations of multiple anthropogenic impacts on polluted areas of the Apalachee Bay resulted in different responses of the Fenholloway Estuary to the observed, long-term climate changes relative to those of the Econfina Reference Estuary. The Econfina offshore area responded to the increased drought occurrence (and reduced nutrient loading) with the observed changes of the SAV and the associated responses of the fish and invertebrate populations. Although it cannot be assumed that future changes in climate will follow the long-term trends noted since 1970, any prolongation of the trend of increased frequency and severity of droughts and storms could have an adverse effect on the overall biological productivity of the Apalachee Bay.

One indication of the processes involved in the observed changes could be provided by Holmquist (1997) who experimentally tested the effects of drift algal patches on underlying seagrass structure and fauna. Placement of large patches of macroalgae on seagrass plots was associated with degradation of the underlying seagrasses. The author noted the following:

"Abundance of mobile (invertebrate) fauna on experimental plots (increased drift algae) increased after 6 mo of algal cover relative to abundance on control plots.... There was also greater evenness in the canopy fauna of the algal plots which contrasted with the high level of dominance apparent on control plots.... Although the structurally complex mats formed by drifting algae provided short-term habitat enhancement for some (invertebrate) fauna, long-term effects on fauna are probably negative due to mat ephemerality and degradation of seagrass."

Reduced fish predation (fishes were not collected in the experiments) could have been involved in the noted increased invertebrate abundance.

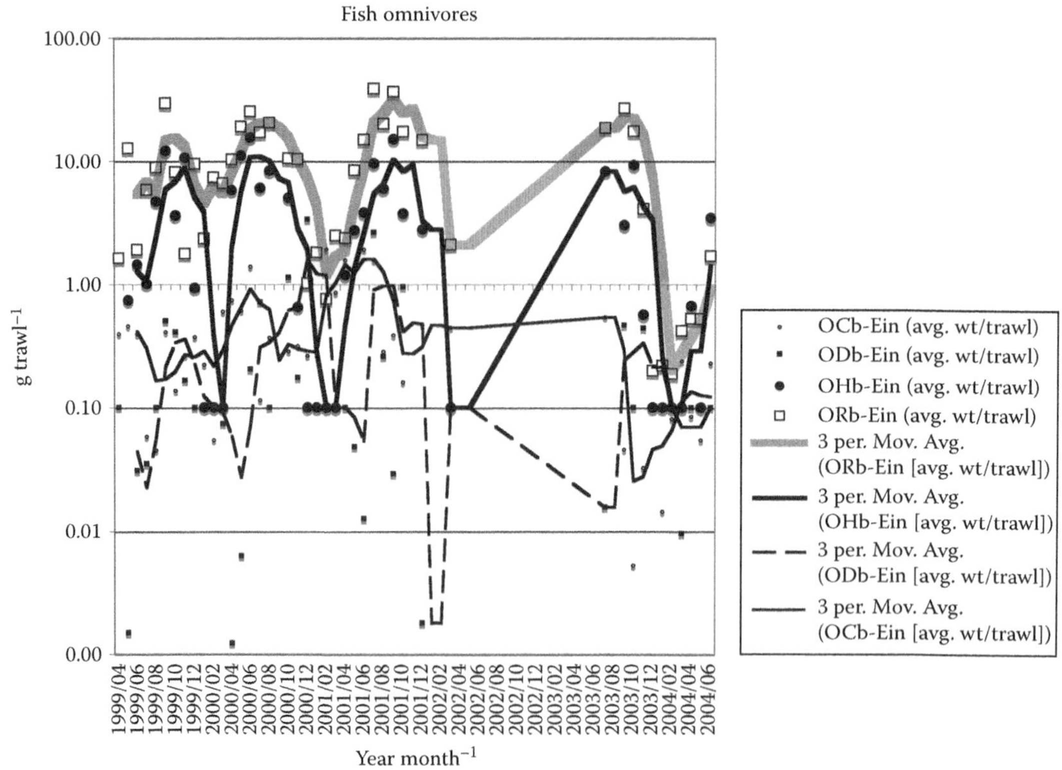

Figure 7.24 Fish omnivores (OD > 60% detritivorous; OH > 60% herbivorous; OC > 60% carnivorous; OR, detritus, plant, and animal feeders) in the inner parts of the Econfina Estuary from April 1999 to June 2004. Data are shown with 3-month moving averages.

If the replacement of natural populations of seagrasses by dense infestations of macroalgae has changed the habitat so that invertebrates are no longer available as food for the fish populations, resulting in the loss of fish productivity, and if such changes are widespread and continue to occur in the future due to alterations of the major climatological factors, this impact on the trophic organization of Apalachee Bay could have profound effects on the sports fisheries of the region.

APALACHICOLA ESTUARY

The Apalachicola River–Estuary is part of a major drainage area (the Chattahoochee–Flint–Apalachicola [CFA] basin) of about 48,500 km^2 (square kilometers). This ecosystem is located in western Georgia, SE Alabama, and northern Florida (Figure 7.30). There are 13 dams on the Chattahoochee River and 3 dams on the Flint River. The Apalachicola River, 21st in flow magnitude in the conterminous United States, is one of the few alluvial systems that remain free flowing in the country. The river flows 171 km from the confluence of the Chattahoochee and Flint Rivers (the Jim Woodruff Dam) to its terminus in the Apalachicola Estuary.

Background

Productivity along Florida's NE Gulf Coast is dominated by a series of large river–estuaries (Figure 3.1). Temperate, river-dominated estuaries are among the most productive and economically valuable natural resources in the world even though many such systems have

Table 7.5 Numerical Abundance of Dominant Invertebrate Species Taken in the Econfina and Fenholloway Coastal Systems over the Long-Term Study Period

Species	E07	E08	E09	E10	E11	E12	E13
Tozeuma carolinense		225	1621	540	1	4727	471
Costoanachis avara	243	1354	217	1310	242	244	1921
Hippolyte zostericola	14	211	1420	305	58	374	1015
Periclimenes longicaudata	10	511	1056	347	18	360	801
Neopanope texana	293	401	23	409	464	31	408
Pagurus sp.	64	271	217	521	119	217	540
Palaemon floridanus	319	160	122	176	232	251	404
Urocaris longicaudata	159	367	21	99	39	9	140
Neopanope packardii	53	79	172	122	98	42	73
Thor dobkini	18	279	28	108	23	14	108
Farfantepenaeus duorarum	29	53	36	150	25	117	54
Species	**F09**	**F10**	**F11**	**F12**	**F14**	**F15**	**F16**
Tozeuma carolinense		1	6	6553		1251	3120
Pagurus sp.	43	287	600	1168	218	411	30
Costoanachis avara	33	141	247	1707	183	350	89
Urocaris longicaudata	1	133	164	1198	16	497	422
Palaemon floridanus	33	145	165	1059	25	251	
Hippolyte zostericola	5	295	147	765	14	204	171
Neopanope texana	222	135	91	212	184	196	13
Ophiothrix angulata	1	34	3	176	2	19	226
Palaemonetes intermedius	262	2	54	1	23	15	
Neopanope packardii	1	14	10	228	11	24	35
Metoporhaphis calcarata	2	65	33	74	2	37	100

been seriously damaged by human activities. Estuarine primary production, based on loading of nutrients and organic compounds from associated alluvial rivers, is one of the most important functions in river-dominated estuaries (Howarth, 1988; Baird and Ulanowicz, 1989). Nutrient input from river sources comprises a major stimulus for autochthonous phytoplankton production in receiving estuaries. River-driven allochthonous particulate organic matter maintains detritivorous food webs in these estuaries. However, the relative importance of various sources of both organic carbon (dissolved and particulate) and inorganic nutrients can vary from system to system (Peterson and Howarth, 1987). These sources can be related to the specific tidal and hydrological attributes of a given estuary (Odum et al., 1979). Human sources of such compounds often have the exact opposite effect leading to hypereutrophication, phytoplankton blooms, deterioration of the estuarine food webs, and severe loss of secondary production (Livingston, 2000, 2002, 2005).

The association between alluvial freshwater input and estuarine productivity has been indirectly established in a number of estuaries (Cross and Williams, 1981). Deegan et al. (1986), using data from 64 estuaries in the Gulf of Mexico, found that freshwater input was highly correlated ($R^2 = 0.98$) with fishery harvest. Armstrong (1982) determined that nutrient budgets in Texas Gulf Estuaries were dominated by freshwater inflows and that shellfish and finfish production was a function of nutrient loading rates and average salinity. Funicelli (1984) found that upland carbon input was in some way associated with estuarine productivity. However, few studies actually evaluated the various facets of linkage of the freshwater river–wetlands and estuarine productivity (Livingston, 1981a). As a response to the projections of anthropogenic freshwater used by the state of Georgia

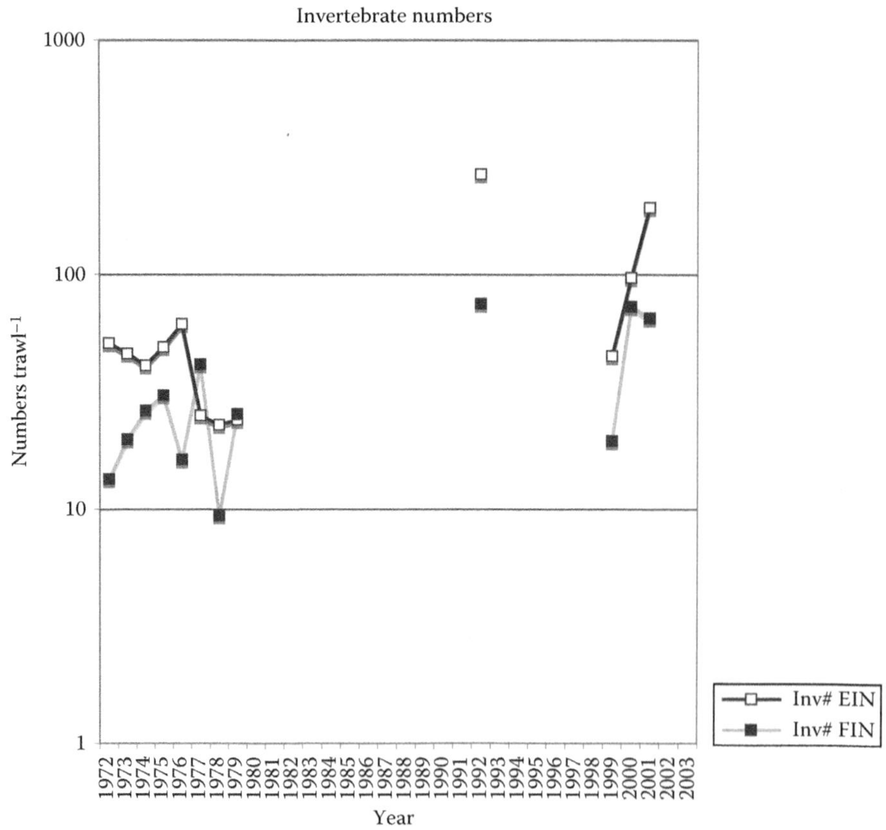

Figure 7.25 Invertebrate numbers in the inner parts of the Econfina and Fenholloway Estuaries from 1972 to 2004.

over the next 30–50 years (Livingston, 1988b), a long-term analytical program was initiated by our research group to analyze, quantitatively, the Apalachicola databases generated during the 1970s and 1980s and to determine how projected reduced flows of the tri-river system would affect the Apalachicola River and its receiving bay.

Mean flow rates of the Apalachicola River approximate 690 $m^3 s^{-1}$ (1958–1980) with annual high flows averaging 3000 $m^3 s^{-1}$ (Leitman et al., 1982, 1991). The forested floodplain, about 450 km^2, is the largest in Florida (Leitman et al., 1982) with forestry as the primary land use in the floodplain (Clewell, 1977). Other activities include minor agricultural and residential use, bee keeping, tupelo honey production, and sports/commercial fishing (Livingston, 1984a). The Apalachicola River is one of the last major free flowing, unpolluted alluvial systems in the conterminous United States. The importance of freshwater flows to the Apalachicola floodplain has been extensively studied (Cairns, 1981; Elder and Cairns, 1982; Mattraw and Elder, 1984; Light et al., 1998). Based on a long history of management efforts (Livingston, 2002), the unique characteristics of the river floodplain remain largely intact, a notable exception to the condition of most alluvial waterways in the United States today.

The Apalachicola floodplain gives rise to various freshwater fisheries. Some of these fisheries (e.g., striped bass, *Morone saxatilis*; sturgeon, *Acipenser oxyrhynchus*) have been destroyed or seriously impaired due to habitat destruction by channelization and damming along the Chattahoochee River (Livingston and Joyce, 1977; Livingston, 1984a). Dredging activities, mandated by the US Congress and continuing to the present time, have led to serious habitat damage along the

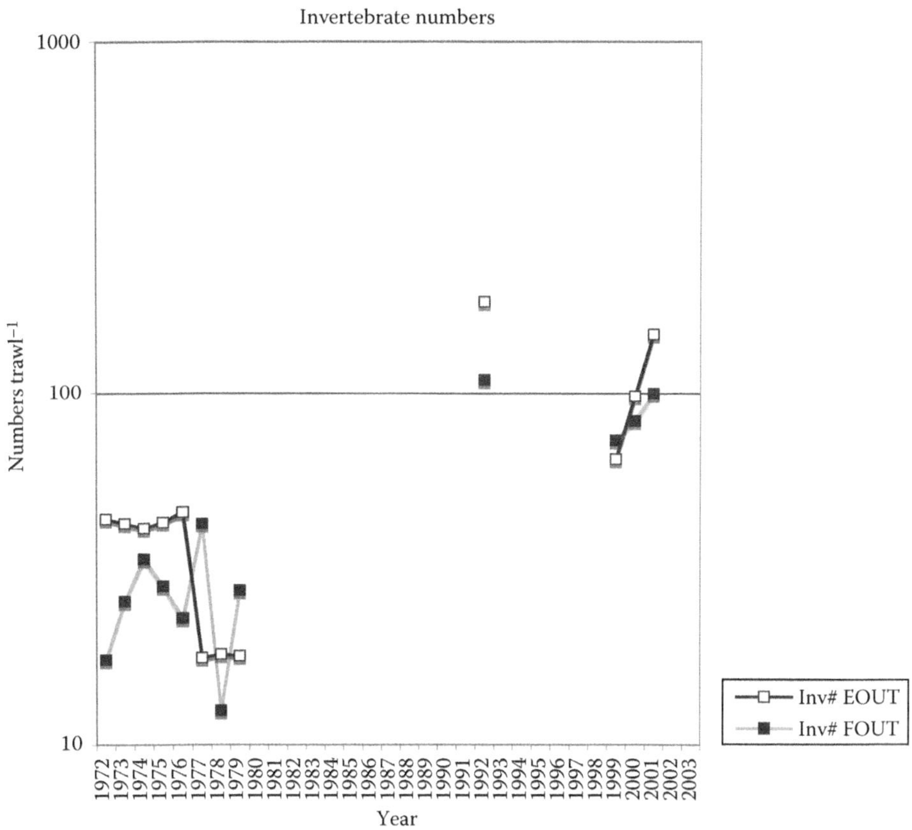

Figure 7.26 Invertebrate numbers in the outer parts of the Econfina and Fenholloway Estuaries from 1972 to 2004.

Apalachicola River with virtually no economic justification for such channelization (Leitman et al., 1991). Nevertheless, the Apalachicola River wetland system remains largely intact, and this river and its wetlands are one of the few such systems in the United States that is almost completely in public hands.

The Apalachicola River dominates the associated estuary (Figure 7.31) as a source of freshwater, nutrients, and organic matter; together with local rainfall, the river is closely associated with the salinity and coastal productivity of the region (Livingston, 1983d, 1984a, 2005, 2013; Livingston et al., 1997, 1999, 2000). The Apalachicola Estuary approximates 62,879 ha and is a shallow (mean depth 2.6 m) lagoon-and-barrier-island complex. The estuary is oriented along an east–west axis. Water movement is controlled by wind currents and tides as a function of the generally shallow depths (Livingston, 1984a, Livingston et al., 1999). Upland salt marshes grade into soft-sediment areas, oyster reefs, fringing grass beds, and a series of passes that are control points for the salinity structure of the system (Livingston, 1984a).

SAV contributes to the high level of estuarine production. These plants provide habitat and organic matter that forms the basis for important estuarine food webs. In the upper bay, there have been dense growths of SAV dominated by *Vallisneria americana*, *Ruppia maritima*, and *Potamogeton pusillus* (Livingston, 2008a). These SAV species are adapted to low light penetration and low, varying salinities. Recent salinity increases due to a prolonged drought have adversely affected the East Bay grass beds that represent an important habitat and source of productivity for the upper bay. Shoal grass (*H. wrightii*) is dominant in the highly productive seagrass areas off

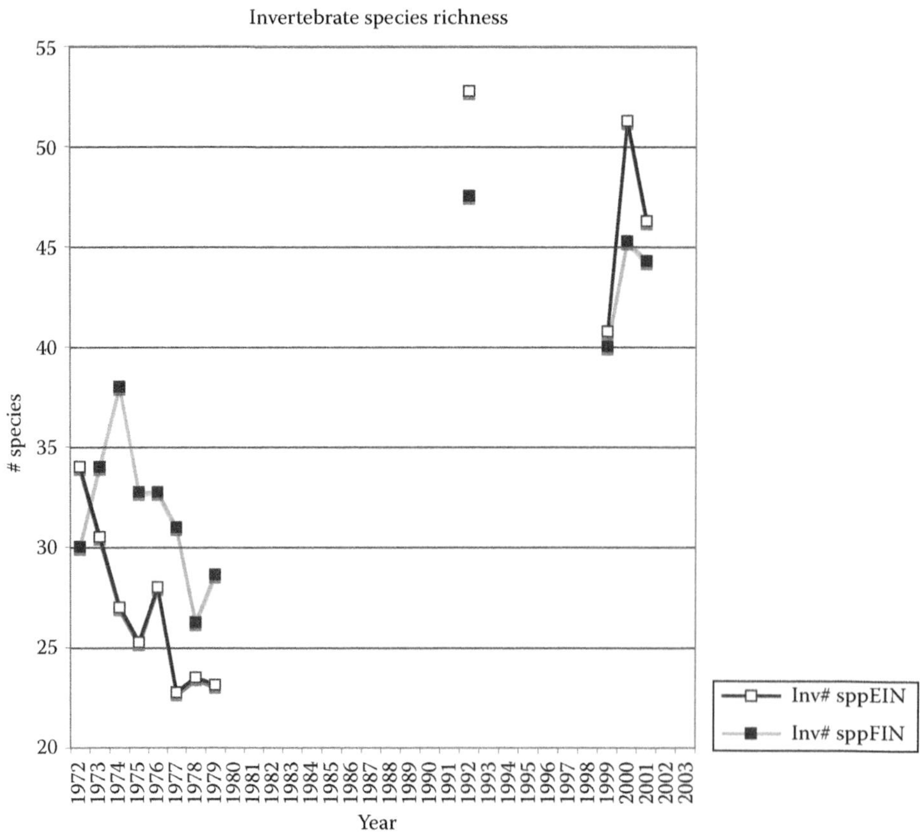

Figure 7.27 Invertebrate species richness in the inner parts of the Econfina and Fenholloway Estuaries from 1972 to 2004.

St. George Island. In areas of higher salinity (St. George Sound), turtle grass (*T. testudinum*) and manatee grass (*S. filiforme*) are dominant (Livingston, 2008).

Enhanced water color and turbidity and low salinities due to river input restrict the development of dominant seagrass species such as *T. testudinum* and *S. filiforme* in the upper estuary. These species are adapted to clear water and high salinities and are thus restricted to shallower areas in the outer regions of the Apalachicola Estuary such as along the St. James Island area from the city of Carrabelle to Alligator Harbor (Livingston, 2008). SAV thus complements the high phytoplankton productivity of river-dominated parts of the estuary, giving rise to an important sports fishing industry in the region. The intricate combination of freshwater input, nutrient loading, water-quality factors, high biological productivity, salinity distributions, and the diversity of natural habitats have contributed to a long history of high commercial and sports fishery values of the Apalachicola River–Bay system.

A sparse human population, together with a low level of industrial and municipal development, has been associated with continued high water and sediment quality in the region (Livingston, 1984a, 2000). Various studies have confirmed the wisdom of the early planning decisions and wetland purchases. Published results of the long-term bay research program included hydrology (Meeter et al., 1979), the effects of anthropogenic activities such as agriculture (Livingston et al., 1978a) and forestry (Livingston et al., 1976a; Duncan, 1977; Livingston and Duncan, 1979), and the importance of salinity to the community structure of estuarine organisms (Livingston, 1979).

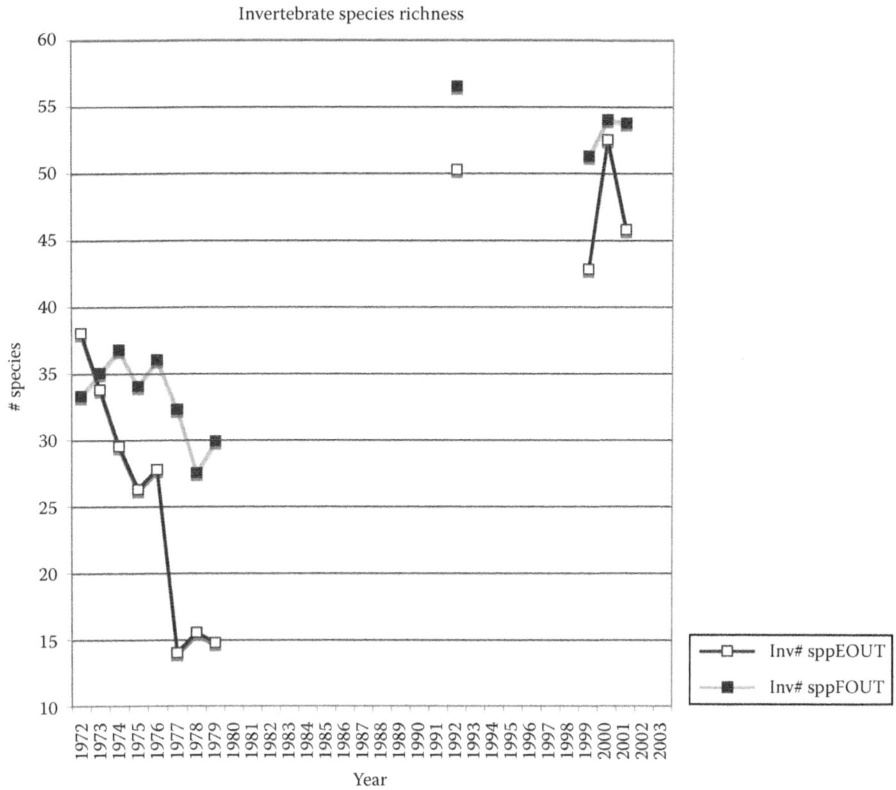

Figure 7.28 Invertebrate species richness in the outer parts of the Econfina and Fenholloway Estuaries from 1972 to 2004.

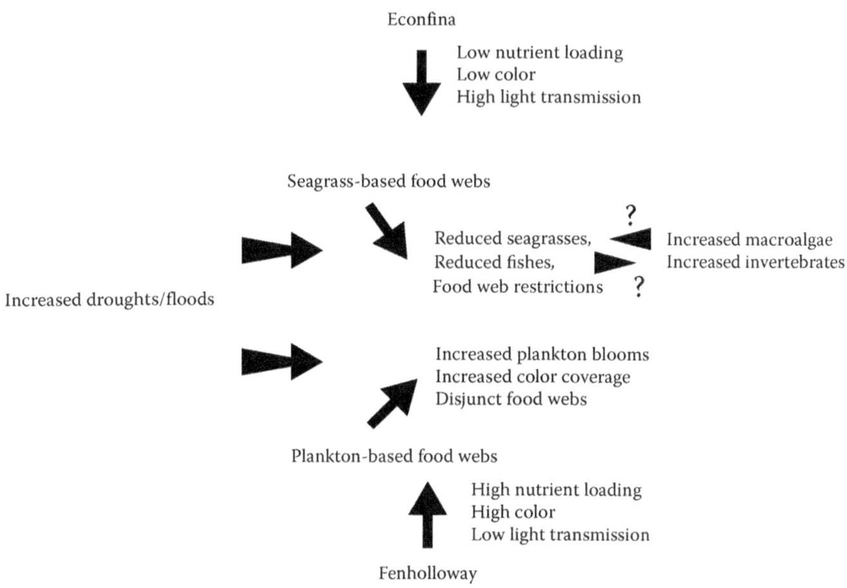

Figure 7.29 Schematic diagram of potential differences in the response of the unpolluted Econfina Estuary and the polluted Fenholloway Estuary to changes caused by long-term climatological changes in the Apalachee Bay.

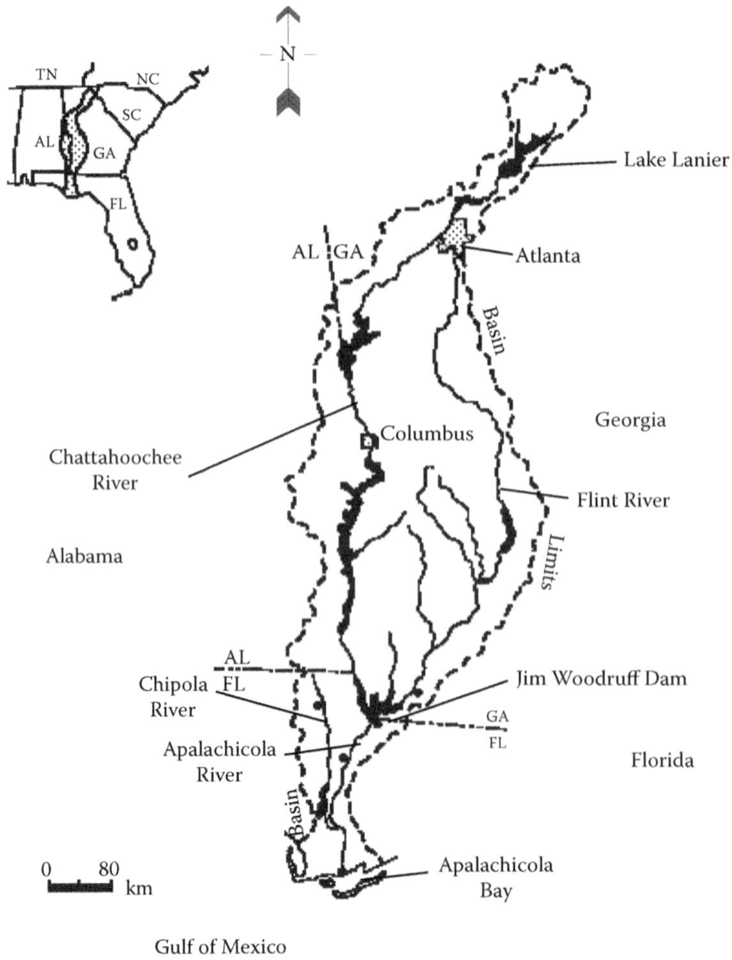

Figure 7.30 The ACF or tri-river drainage basin.

The basic distribution of the estuarine populations has been analyzed (Estabrook, 1973; Livingston et al., 1974, 1976a, 1977; Livingston, 1976b, 1977, 1981b, 1983c, Purcell, 1977; Edmiston, 1979; Sheridan, 1979; Sheridan and Livingston, 1979, 1983; McLane, 1980; Laughlin and Livingston, 1982; Mahoney, 1982; Mahoney and Livingston, 1982). Various studies have also been carried out concerning the trophic organization of the estuary (White et al., 1977, 1979a,b, Sheridan, 1978; Laughlin, 1979; Federle et al., 1983a,b; White, 1983; Livingston, 1984a, Livingston et al., 1997).

Nutrient loading to the estuary from the river is the highest of the major river systems along the Gulf Coast of Florida (Livingston, 2000) and remains relatively high without apparent widespread hypereutrophication in the bay. Seasonal river flooding provides the engine for mobilization and transfer of nutrients and detritus between the Apalachicola wetlands and associated river and bay areas. Food webs leading to high production of oysters, shrimp, blue crabs, and finfishes in the Apalachicola Bay are linked to river flows that also control salinity regimes and the nutrient/productivity dynamics of the estuary.

Nutrient loadings from the river, in the form of inorganic nitrogen and phosphorus compounds and various forms of particulate organic matter, colloidal conglomerates, and dissolved compounds,

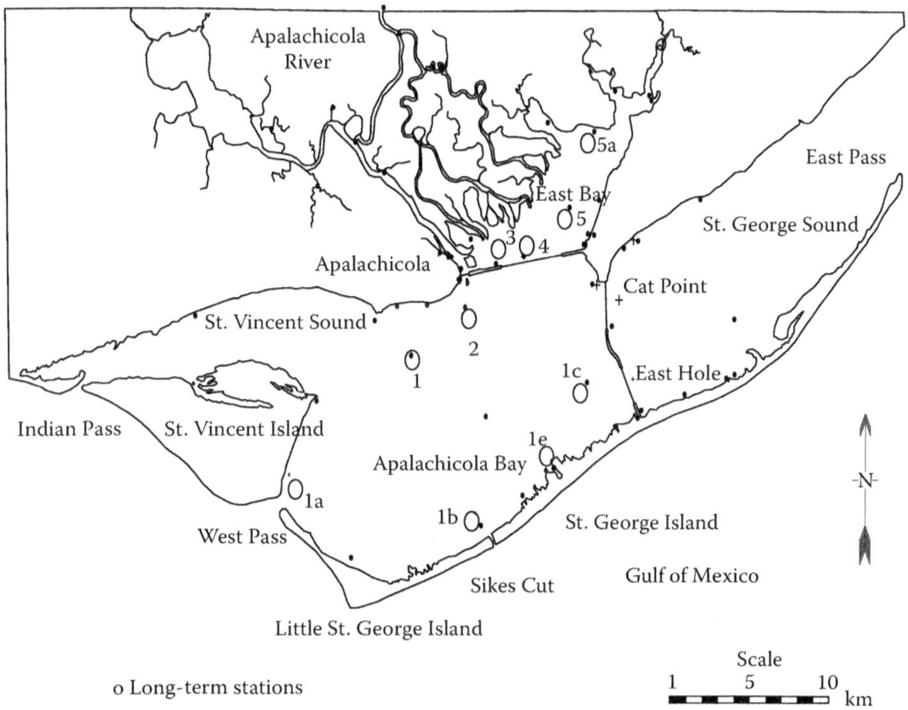

Figure 7.31 The Apalachicola Bay system showing sampling stations used in the long-term studies of this system from 1971 to 2012.

are loaded into the receiving bay. The inorganic nutrients (NH_3, NO_2, NO_3, PO_4) are taken up rapidly by phytoplankton (suspended microscopic plants), thus leading to relatively low concentrations of such nutrients in the bay. The resulting high phytoplankton productivity and particulate organic matter form the basis for highly productive food webs in the bay. This includes food for oysters and clams. Zooplanktons that feed on the phytoplankton support high numbers of bay anchovies (*A. mitchilli*) and Gulf menhaden (*Brevoortia patronus*). These species, in turn, represent primary food sources for various fishes and invertebrates in the estuary. Particulate organic matter from the bodies of plants and animals, together with particulate and DOC loaded from the river, accumulates in the sediments where infaunal microfauna, microflora, and macrofauna form important elements of detrital food webs of the bay.

Nutrients adhere to the particulate organic matter (detritus) that eventually falls into the sediments of the shallow Apalachicola Estuary. This nutrient-rich organic matter is colonized by microbial components that utilize the DOC from the river to form the basis of important detrital food webs (Livingston, 1984a). Infaunal, detritus-feeding macroinvertebrate assemblages that live in the sediments of the bay are dominated by various species of worms and crustaceans that include *Mediomastus ambiseta*, *Hobsonia florida*, and *Grandidierella bonnieroides*.

River–Bay Linkages

The linkage between the upland freshwater wetlands and the rich estuarine biota of associated estuaries has been the subject of considerable debate and research in the past (Livingston and Loucks, 1978; Livingston, 2000). Studies were made concerning the distribution of wetland vegetation in the Apalachicola floodplain (Leitman et al., 1982). It was determined that vegetation type was associated with water depth, duration of inundation and saturation, and water-level fluctuation. Stage range is reduced considerably downstream that indicates a dampening of the river flood stage by the expanding (downstream) wetlands. Litter fall in the Apalachicola floodplain (800 g m^{-2}) is higher than that noted in many tropical ecosystems and almost all warm temperate ecosystems. The litter fall of these systems is on the order of 386–600 g m^{-2} (Elder and Cairns, 1982). The annual deposition of litter fall in the bottomland hardwood forests of the Apalachicola River floodplain approximates 360,000 mt.

Seasonal flooding provides the mechanism for mobilization, decomposition, and transfer of the nutrients and detritus from the wetlands to associated aquatic areas (Cairns, 1981; Elder and Cairns, 1982) with a postulated, though unknown input, from groundwater sources. Studies (Livingston et al., 1974, 1976a) indicated that, in addition to providing particulate organics that fueled the bay system, river input determined nutrient loading to the estuary. Of the 214,000 mt of carbon, 21,400 mt of nitrogen, and 1650 mt of phosphorous that is delivered to the estuary over the period of a given year, over half is transferred during the winter–spring flood peaks (Mattraw and Elder, 1984). Reductions of the overall Apalachicola River flow rates due to climate changes and anthropogenic use of freshwater in the Chattahoochee and Flint Rivers were postulated as threats to the natural biota of this highly productive system (Light et al., 1998; Livingston, 1984a). In addition, such changes were projected to jeopardize millions of dollars of investments by the people of Florida in the various wetland purchases and management efforts over the past 30 years as the wetlands would disappear as a result of reduced flooding.

The aforementioned studies noted that the timed delivery of nutrients and dissolved/particulate organic matter was an important factor in the maintenance of the estuarine primary production (autochthonous and allochthonous). There were distinct linkages between the estuarine food webs and freshwater discharges (Livingston and Loucks, 1978; Livingston, 1981a, 1984a). The total particulate organic carbon delivered to the estuary followed seasonal and interannual fluctuations that were closely associated with river flow (Livingston, 1991a; $R^2 = 0.738$). The exact timing and degree of peak river flows relative to seasonal changes of wetland productivity were important determinants of short-term fluctuations and long-term trends of the input of allochthonous detritus to the estuary (Livingston, 1981a, 1984a). During summer and fall months, there was no direct correlation of river flow and detritus movement into the bay. By winter, there was a significant relationship between microdetrital loading and river flow peaks.

Nixon (1988a) showed that the Apalachicola Bay system ranked high in overall primary production compared to other such systems. Up to 50% of the phytoplankton productivity, which is the most important single source in overall magnitude of organic carbon to the estuary, was explained by the Apalachicola River flow (Myers 1977; Myers and Iverson 1977). During winter–spring periods of high river flow, there were major transfers of nutrients and organic matter to the estuary. Boynton et al. (1982) reported that the Apalachicola system had high phytoplankton productivity relative to other river-dominated estuaries, embayments, lagoons, and fjords around the world. Wind action in the shallow Apalachicola Bay system was associated with periodic peaks of phytoplankton production as inorganic nutrients, regenerated in the sediments, were mixed through turbulence into the euphotic zone (Livingston et al., 1974; Livingston, 1984a, Iverson et al., 1997). Iverson et al. (1997) noted that there had been no notable increase in chlorophyll *a* concentrations in the Apalachicola Bay over the past two decades despite increases in nitrogen loading due to increased basin deposition of this nutrient. They found that dissolved silicate did not limit phytoplankton production in the largely mesotrophic Apalachicola Bay.

In the Apalachicola Estuary, orthophosphate availability limited phytoplankton during both low- and high-salinity winter periods and during the summer at stations with low salinity (Iverson et al., 1997). Nitrogen, on the other hand, was limiting during summer periods of moderate to high salinity in the Apalachicola Estuary (Iverson et al., 1997). Light and temperature limitations were highest during winter–spring periods, thus limiting primary production during this time. High chlorophyll *a* levels during winter periods were attributed to low zooplankton grazing during the cooler months (Iverson et al., 1997). Nitrogen input to primary production was limited by the relatively high flushing rates in the Apalachicola Estuary. Flow rates affected the development of nutrient limitation in the Apalachicola system with this factor highest during low-flow summer periods.

Recent studies have documented the influence of the river on nutrient and organic carbon loading to the bay. Chanton and Lewis (1999) found that although there were inputs of large quantities of terrestrial organic matter, net heterotrophy in the Apalachicola Bay system was not dominant relative to net autotrophy during a 3-year period. Chanton and Lewis (2002), using $\delta^{13}C$ and $\delta^{34}S$ isotope data, noted clear distinctions between benthic and water column feeding types. They found that the estuary depended on river flows to provide floodplain detritus during high-flow periods and dissolved nutrients for estuarine primary productivity during low flows. Floodplain detritus was significant in the important East Bay nursery area, providing further proof that peak flows were important in washing floodplain detritus into the estuary.

The peak levels of macrodetrital accumulation that occurred during winter/spring periods of high river flow (Livingston, 1981a) were coincident with increased infaunal abundance (McLane, 1980). Four out of the five dominant infaunal species at river-dominated stations were detritus feeders. A mechanism for the direct connection of increased infaunal abundance was described by Livingston (1983a, 1984a), whereby microbial activity at the surface of the detritus (Federle et al., 1983a) led to microbial successions (Morrison et al., 1977) that then provided food for a variety of detritivorous organisms (White et al., 1979a,b; Livingston, 1984a). The transformation of nutrient-rich particulate organic matter from periodic river-based influxes of dissolved and particulate organic matter coincided with abundance peaks of the detritus-based (infaunal) food webs of the Apalachicola system (Livingston and Loucks, 1978) during periods of increased river flooding. Chanton and Lewis (2002) provided analytical support for these observations.

Mortazavi et al. (2000a) found that phytoplankton productivity in river-dominated parts of the Apalachicola Estuary was limited by phosphorus in winter (during periods of low salinity) and nitrogen during summer periods of high salinity. The dissolved organic nitrogen (DON) input was balanced by export from the estuary. Mortazavi et al. (2000b) determined temporal couplings of nutrient loading with primary production in the estuary. Around 75% of such productivity occurred from May through November, with main control due to grazing. Mortazavi et al. (2000c) gave detailed accounts of the nitrogen budgets of the bay. However, 36% of the dissolved organic phosphorus (DOP) was retained in the estuary where it was presumably to be utilized by microbes and primary producers (Mortazavi et al., 2000a). The data indicated that altered river flow, especially during low-flow periods, could adversely affect overall bay productivity.

Oczkowski et al. (2011) measured nitrogen (N) and carbon (C) stable isotopes ($\delta^{15}N$, $\delta^{13}C$) in macroalgae, surface water nitrate, surface sediments, and $\delta^{13}C$ of particulates. They found that oyster stable isotope values throughout Apalachicola Bay were complex in terms of derived food, but the diet of oysters was dominated by freshwater inputs, reflecting the variability and hydrodynamics of the riverine flows. The authors emphasized the importance of the river in linking the bay to the extensive Apalachicola floodplain. Wilson et al. (2010) found that biological production in Apalachicola Bay was highly dependent on riverine flows in two ways: (1) Economically important bivalves and crustaceans were supported by terrestrial organic matter supplied by river flooding, and (2) consumer fish species were dependent on in situ production that was reliant on nutrients supplied by the Apalachicola River. This study confirmed that consumers of the Apalachicola Estuary depended on river flow through estuarine production based on river-derived nutrients and organic detritus.

Population Distributions in the Bay

The ecology of the Apalachicola Estuary is closely associated with freshwater input from the Apalachicola River (Livingston, 1984a). The distribution of epibenthic organisms in Apalachicola Estuary (1972–1984) followed a specific spatial relationship to high river flows. Stations most affected by the river were inhabited by the bay anchovy (*A. mitchilli*), the spot (*L. xanthurus*), the Atlantic croaker (*Micropogonias undulatus*), the Gulf menhaden (*B. patronus*), the white shrimp (*Litopenaeus setiferus*), and the blue crab (*C. sapidus*). The outer bay stations were often dominated by species such as the silver perch (*Bairdiella chrysoura*), the pigfish (*Orthopristis chrysoptera*), the least squid (*Lolliguncula brevis*), the pink shrimp (*F. duorarum*), the brown shrimp (*Farfantepenaeus aztecus*), and other shrimp species such as the roughneck shrimp (*Trachypenaeus constrictus*). Sike's, an artificial opening to the gulf that is maintained by the US Army Corps of Engineers, was characterized by salinities that resembled the open gulf. This area was dominated by species such as the least squid, the bay anchovy, the sand sea trout (*Cynoscion arenarius*), the fringed flounder (*Etropus crossotus*), the iridescent swimming crab (*Portunus gibbesii*), and the sergestid shrimp (*Acetes americanus*).

River flow, as a signal determinant of bay habitat, was a controlling factor for the biological organization of the Apalachicola Estuary (Livingston, 1984a, 1991b,c). Field analyses of long-term changes of infaunal and epibenthic invertebrates and fishes were based on a series of studies executed monthly from 1972 to 1984 (Livingston, 1984a). Cross-correlation analyses demonstrated that numbers of species of fishes were positively associated with peak river flows (Livingston, 1991c). The responses of the dominant bay populations were complex due to species-specific responses to the river-directed habitat changes and responses of the food web to nutrient loading and phytoplankton production. These data analyses indicated that the various dominant Apalachicola Bay populations followed a broad spectrum of diverse phase interactions with river flow and associated changes in salinity.

The primary fish dominants, representing about 80% of total fish numbers taken over the sampling period (1972–1984), included bay anchovies, Atlantic croakers, sand sea trout, and spot (Livingston, 1997a). The numerically dominant invertebrates included the white shrimp and blue crabs that represented about 70% of the total numbers taken over the study period. Moving averages (12 months) of changes in river flow, salinity, and numbers/dry weight biomass of epibenthic fishes and invertebrates indicated long-term changes that were related in different ways to the river flow. Fish biomass tended to follow long-term trends of river flow conditions, whereas fish numbers peaked early during the drought of 1980–1981. Such peaks were due in large part to increases in the numbers of bottom-feeding spot in the bay. Anchovies, dominant during fall–winter periods, and Atlantic croakers were prevalent during winter–spring months and tended to follow long-term river flow trends.

Spot and croakers are in direct competition for food, especially during the young stages that are numerically abundant during the winter months of high river flow (Sheridan, 1979; Sheridan and Livingston, 1979). The long-term changes in population distribution of these species could be related to competition for food. The sand sea trout, a piscivorous fish that feeds primarily on anchovies (Sheridan and Livingston, 1979), reaches peak numbers during late spring and early summer. The long-term trends of invertebrate distribution indicate that invertebrate numbers were in some ways associated with river flow. This is based on very different relationships between white shrimp and blue crabs. The high invertebrate biomass early during the drought of 1980–1981 was due in large part to high numbers of blue crabs taken at that time. The high numbers of blue crabs could have been related to low shrimp numbers since blue crabs feed on shrimp (Laughlin, 1979; Livingston, 1984a).

A series of cross-correlation analyses indicated that river flow and rainfall follow a sine curve with peaks of river flow highly correlated with reductions in salinity (Meeter and Livingston, 1978;

Meeter et al., 1979). Apalachicola River flow is correlated with Georgia rainfall, indicating that the salinity of the Apalachicola Estuary is associated with precipitation trends in the tri-river upper basin. This positive association of important estuarine water-quality characteristics and upland rainfall patterns has certain implications for the management of the tri-river system. The relationship of the biological organization of the estuary with river flow is far more complex than the relatively straightforward correlations of the physical variables. Various species of fishes are associated with differing patterns of river flows. Overall fish numbers peak 1 month after river flow peaks, whereas invertebrate numbers are inversely related to peak river conditions with major increases during the summer months. Such data are understandable in that top fish dominants such as spot are prevalent in winter–spring months of river flooding, whereas the peak numbers of penaeid shrimp usually occur during summer and fall months.

Other top dominants such as anchovies reach numerical peaks 3 months before the Apalachicola River floods. Fish biomass shows a significant positive correlation with river flow at lags two and three, whereas invertebrate biomass shows a significant positive correlation with river flow peaks at lag four. These correlations do not prove any actual relationships of the various biological indicators with river flow conditions. Rather, the cross-correlation analysis indicates that the various estuarine biological components have a broad spectrum of diverse phase interactions with river flow and associated changes in salinity. The information presented here does give a qualified indication that river flow, as a habitat variable, is an important factor that is associated with the biological organization of the estuary (Livingston, 1984a).

Estuarine areas denoted by riverine (freshwater) input are associated with concentrations of the primary fishery species. These associations are different for the various ontogenetic feeding units. The dominant species in the Apalachicola Estuary form the basis for the high fishery potential of the estuary. The white shrimp, *L. setiferus*, is the numerically dominant penaeid species in the Apalachicola Estuary. The white shrimp represents the most commercially valuable population in the Apalachicola Estuary. This species is largely absent from the bay by mid- to late November through December. Young-of-the-year white shrimp concentrates in the East Bay during spring months (<25 mm; first trophic unit [TU 1]). With growth, the TU 2 of this species is in the East Bay during summer and fall months. The largest white shrimp (TU 3) is located mainly just west of the river mouth and in parts of the East Bay. The distribution of this species indicates to it remains in areas that receive freshwater inflows and the resulting food that accompanies such flows.

Pink shrimp (*F. duorarum*) and brown shrimp (*F. aztecus*) occur in lower numbers than the white shrimp. Pink shrimp are usually associated with higher salinities than white shrimp, and young-of-the-year pink shrimp also are most abundant during early fall in the East Bay. This appears to be related to food availability in areas receiving freshwater flow. Young brown shrimp are most abundant during late spring with the primary pattern of distribution just west of the river mouth and in the East Bay areas. Areas receiving direct freshwater runoff from the river are favored by both species.

The blue crabs (*C. sapidus*), another commercially important species in the Apalachicola Estuary, enter the bay as young-of-the-year (TU 1) during winter months and are largely concentrated in the East Bay and along the main river channel of the bay. Trophic stage 2 appears in the East Bay during February with secondary peaks during summer months. The largest blue crab TU 3 is found in highest numbers during summer months in the East Bay. In all three blue crab trophic stages, the East Bay nursery area appears to be the favorite habitat. Spatial–temporal blue crab distribution through time appears to be associated with freshwater inputs from the river.

The youngest bay anchovies (*A. mitchilli*) enter the bay during early summer and are located mainly in the East Bay. They eventually move to the river in the late fall. The second anchovy TU moves to the river area during fall months. Overall, this species is closely tied to freshwater flows from the Apalachicola River, and populations move from summer distributions in the East Bay to fall distributions in the Apalachicola River channel. The sand sea trout (*C. arenarius*), a piscivorous fish that feeds primarily on anchovies (Sheridan and Livingston, 1979), reaches peak numbers during late spring and early summer. The distribution of the first two TUs is located largely in the East Bay and around the Apalachicola River mouth from late spring to early fall. The larger sand sea trout are located mainly near the river channel of the bay. This distribution generally follows that of the bay anchovies.

Young-of-the-year spot (*L. xanthurus*) enter the bay during winter–early spring periods and are concentrated in the East Bay and areas near the river mouth. Older spot move to the outer parts of the Apalachicola Bay. This distribution is consistent with known distributions of infaunal macroinvertebrate distribution in space and time and is a trophic response to herbivorous and omnivorous species that respond directly to river inflows to the bay. Young-of-the-year Atlantic croakers (*M. undulatus*) enter the bay during winter–spring months and are located mainly in the East Bay and west of the river mouth. The larger forms move throughout the bay during summer months.

Oyster (*Crassostrea virginica*) distribution (Figure 7.32) is closely related to the spatial and temporal aspects of river flows into the estuary. The current distribution of oyster reefs in the Apalachicola Estuary does not differ appreciably from that at the turn of the twentieth century (Swift, 1896; Danglade, 1917). The Apalachicola Estuary is highly advantageous for oyster propagation and growth (Menzel and Nichy, 1958; Menzel et al., 1966; Livingston, 1984a). Growth rates of oysters in this region are among the most rapid recorded (Ingle and Dawson, 1952, 1953), and harvestable oysters can be taken 18 months after spat settlement due, in part, to relatively high year-round water temperatures (Ingle and Dawson, 1952) and the abundance of food in the estuary.

The commercial oyster industry of the Apalachicola Bay system depends on early sexual development, an extended growing period, and high growth rates (Hayes and Menzel, 1981). Oyster predation by various species such as the oyster drill, *Stramonita haemastoma*, and the

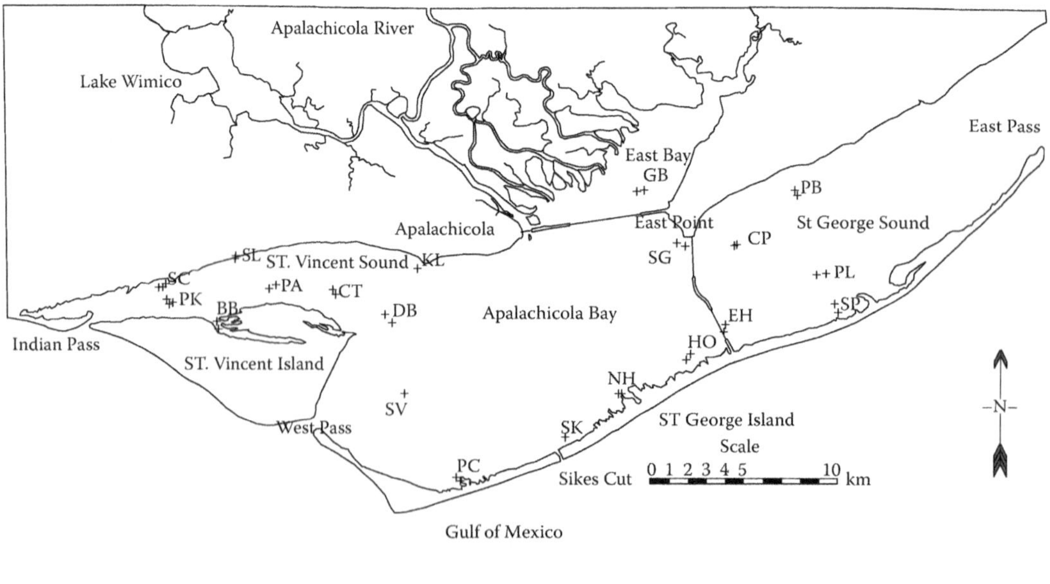

Figure 7.32 Map of the Apalachicola Estuary showing station locations and the distribution/extent of oyster reefs in the estuary. A permanent sampling station was located on each of the oyster bars shown. The following abbreviations were made: Scorpion (SC), Schoelles Lease (SL), Paradise (PA), Big Bayou (BB), Pickalene (PK), and Cabbage Top (CT) in the St. Vincent Sound; Kirvin's Lease (KL), Dry Bar (DB), St. Vincent (SV), Pilots Cove (PC), Sike's (SK), Nick's Hole (NH), Hotel (HO), and Sweet Goodson (SG) in the Apalachicola Bay; Gorrie Bridge (GB) in the East Bay; and Cat Point (CP), Platform (PL), East Hole (EH), Porter's Bar (PB), and Shell Point (SP) in the St. George Sound.

stone crab, *Menippe mercenaria* (Menzel et al., 1966), is accompanied by disease due to organisms such as the protozoan parasite *Perkinsus marinus* with the intensity of infection varying widely in space and time (Fisher et al., 1996; Oliver et al., 1998). The Apalachicola Estuary has accounted generally for 90% of Florida's commercial oyster fishery and 10% nationally (Whitfield and Beaumariage, 1977). Oyster populations evinced considerable resilience to storm-induced physical damage that was dependent on the timing and orientation of wind stress on the estuary (Livingston et al., 1999).

Rainfall and River Flows

The general increase of the frequency and severity of reduced river flows over the study period is shown in Figure 7.33. The effects of the brief drought of 1980–1981 and the later more extended drought of 1986–1988 were accompanied by analyses of the effects of reduced river flows on fish and invertebrate populations and on oyster production, respectively. Further experimental and field work and modeling efforts were continued with oysters during the 1990s. The combination of experimental analyses, field-descriptive data, and modeling was joined to the long-term determinations of the trophic organization of the Apalachicola Estuary to determine how the bay responded to long-term changes of climatological conditions.

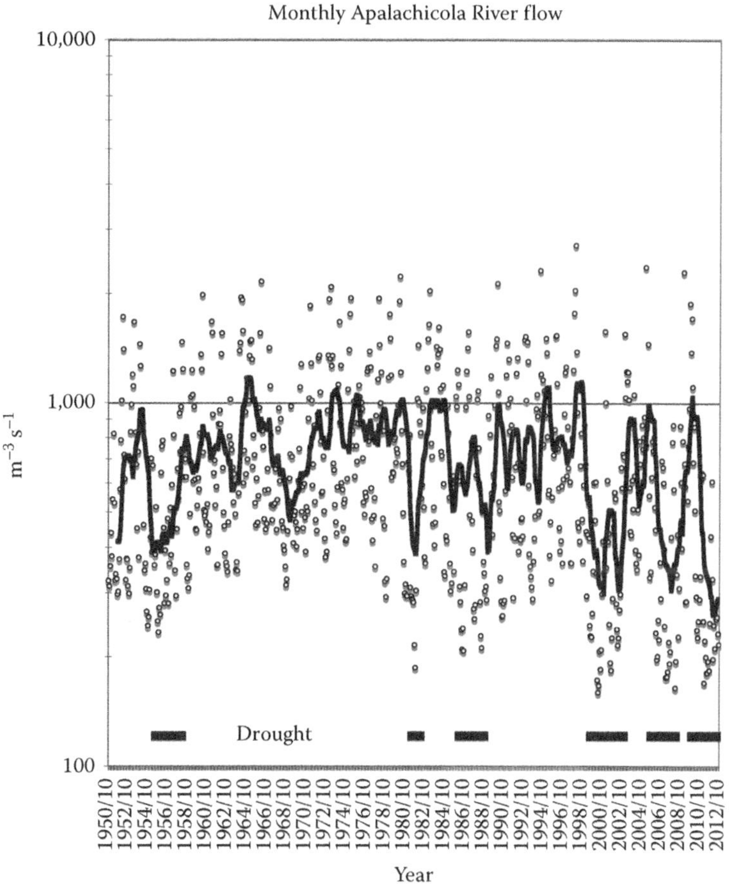

Figure 7.33 Daily Apalachicola River flows (Sumatra) from 1950 through 2012 with 3-month moving averages along with delineation of drought periods. River flow data are provided by the US Geological Survey Sumatra gauge no. 02359170.

The long-term trends and drought occurrences of the Apalachicola River closely followed those of the Chattahoochee River (Figure 7.33). From summer 1980 through the end of 1981, there was a short-term drought with substantially lower river flows during the winter–spring of 1981. Flow rates during the 20-month period prior to the winter of 1982 were consistently below the long-term (40-year) monthly means and were often less than 50% of what the East Bay usually received in the way of freshwater runoff from the river. The following 2.5 years were characterized by a general return to the prevailing patterns of Apalachicola River flow as noted during the period from 1978 through the winter of 1980. The river flows of 1986–1988 constituted the first long-term drought after the 1955–1956 drought and were followed by the major low-flow periods 1999–2002, 2006–2008, and 2010–2012.

The increased incidence and severity of reduced Apalachicola River flows followed that of the Chattahoochee River. A comparison of the incidence of Apalachicola River flows below 285 $m^3 s^{-1}$ is shown in Figure 7.34. The frequency and standard deviation of such observations increased with time over the study period. Once again, there was a straight-line trend of these observations showing the reduced time between the first years of each drought sequence. These data thus indicated a pattern of increasing frequency and severity of the droughts with time from the mid-1950s to 2012.

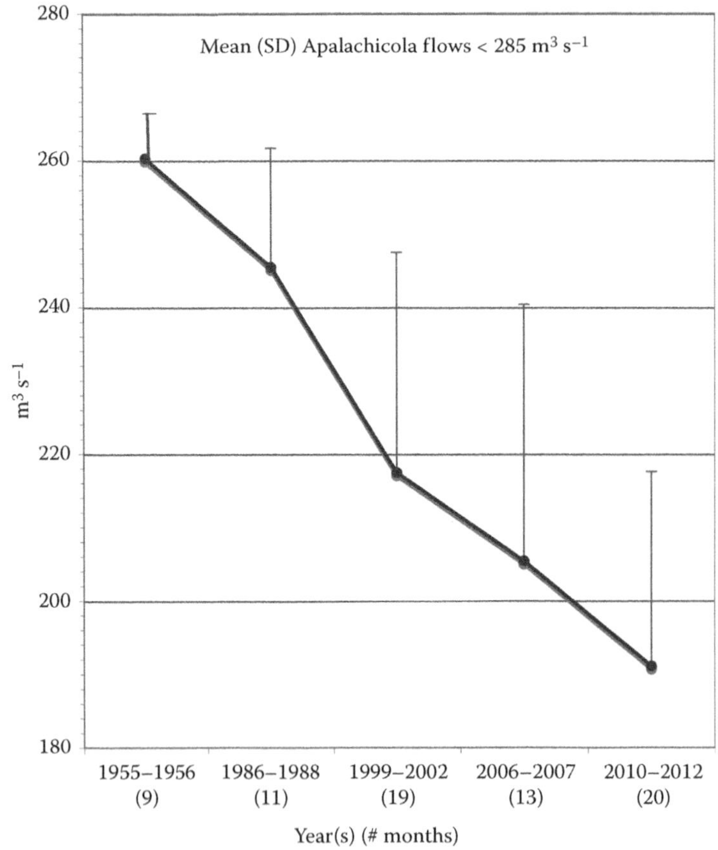

Figure 7.34 Mean (and standard) deviations of Apalachicola flow rates below 285 $m^3 s^{-1}$ over the period October 1959 through May 2012. Also shown are the numbers of months such low flows occurred.

Oysters

The distribution of oysters in the Apalachicola Estuary depends to a considerable degree on river flow entry points and current distributions along with the presence of an adequate substrate type. A detailed discussion of these relationships is given by Livingston et al. (2000). Average surface current velocities (Figure 7.35) were highest in the river, around the passes, and in the vicinity of the oyster reefs in eastern parts of the bay. High surface velocities were also apparent in the central Apalachicola Bay and at Sike's. Average bottom current velocities were highest at the passes and in the eastern region of the bay. Summer surface currents coming out of the East Bay meet the westward trending surface currents of the St. George Sound in the vicinity of the primary eastern oyster bars (Cat Point and East Hole; Figure 7.32).

The eastern oyster bars are thus aligned in a convergence of surface currents. The summer residual bottom currents flow in a westward direction from the St. George Sound with some incursion of sound water into the East Bay. In the Apalachicola Bay, the summer residual bottom currents show a northward direction from Sike's into the interior sections of the bay. The St. Vincent Bar appears to deflect these currents to the north, whereas the bar also appears as a barrier to the influence of

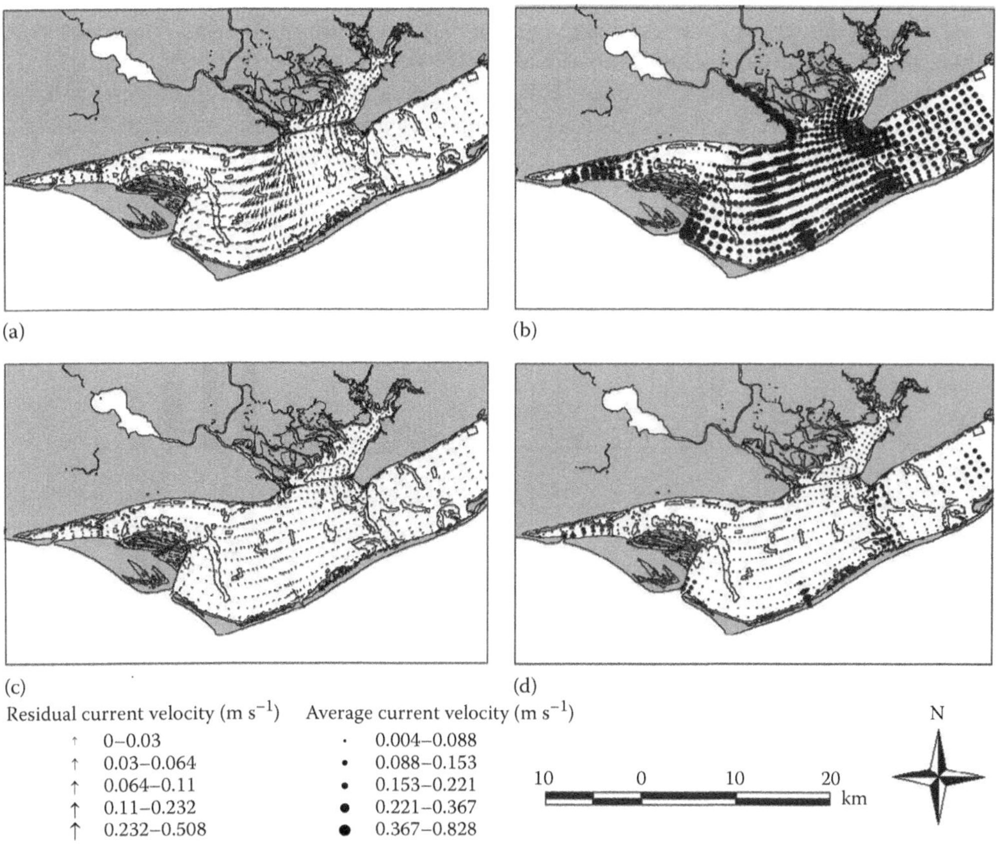

Figure 7.35 Surface/bottom current velocities (m s^{-1}) in the Apalachicola Bay system during summer (1985). (a) Residual surface currents, May–August 1985; (b) average surface current velocities (m s^{-1}), May–August 1985; (c) residual bottom currents, May–August 1985; (d) average bottom current velocities (m s^{-1}), May–August 1985. These data were provided by Glenn C. Woodsum (Department of Biological Science, Florida State University, Tallahassee, Florida 32306-1100 USA) and B. Galperin (Department of Marine Science, University of South Florida, St. Petersburg, Florida 33701).

the West Pass on the bottom current features of the bay. Small residual currents in the St. Vincent Sound move in an eastward direction during summer months.

There are major differences in the patterns of surface and bottom residual currents in the Apalachicola Estuary with oyster bars in the eastern parts of the bay situated in areas of surface convergences along with a strong influence of high-salinity water from the St. George Sound at the bottom. The convergences of the surface currents in the eastern parts of the bay are associated with high chlorophyll *a* (i.e., phytoplankton), a major food source for the oysters, while the high bottom salinities affected by periodic influxes of low-salinity water from the East Bay also favor high-quality habitat for the oysters as the salinity changes inhibit predation and the development of disease. Experiments indicate that oysters favor high salinities (Livingston, unpublished data). Sike's has a major influence on influxes of high salinities from the gulf into the outer part of the Apalachicola Bay. Residual currents in the St. Vincent Sound are minimal, trending to the west at the surface and to the east at the bottom.

Due to the convergence of surface and bottom currents and productivity trends in the estuary, oyster density (Figure 7.36) has been highest at the East Hole, Cat Point, and Platform Reefs (i.e., the eastern reefs) (Livingston et al., 1999, 2000). Oyster density was lowest in oyster reefs located in the St. Vincent Sound and was directly associated with average surface current velocity and bottom salinity. Oyster density was inversely associated with bottom temperature and surface color and maximum surface temperature. Individual new oyster growth was positively associated with surface salinity variation. Overall oyster production was concentrated on three eastern bars (Cat Point, East Hole, Platform) and was positively associated with surface water color and Secchi readings and average bottom current velocities. Thus, high oyster production in the bay occurs in areas subjected to a convergence of tannin-stained surface water from the East Bay (i.e., influenced by the Apalachicola River/Tate's Hell Swamp drainage) and high-velocity bottom-water currents moving westward from the St. George Sound. The convergence of nutrients leads to high phytoplankton

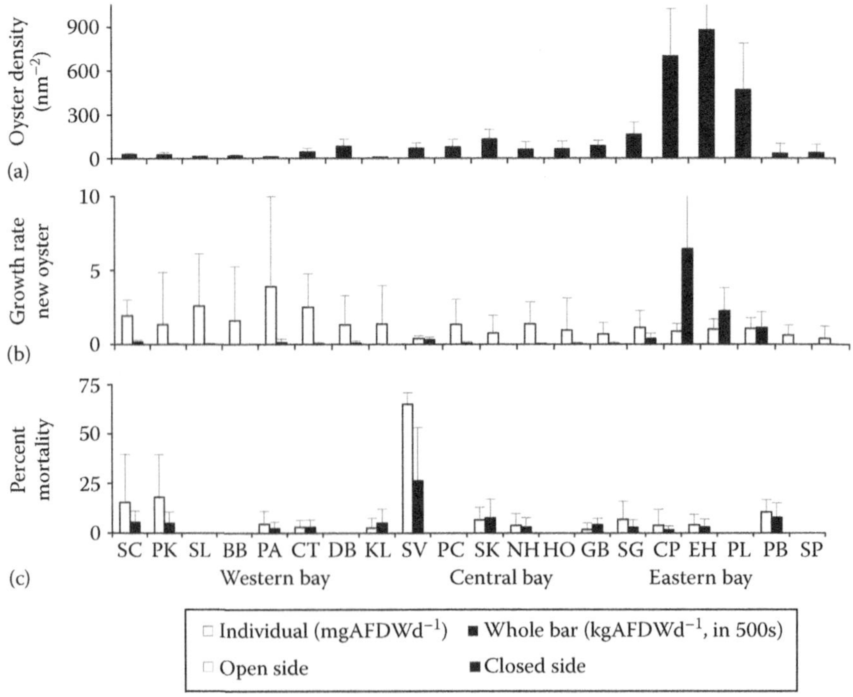

Figure 7.36 (a) Oyster density, (b) growth rates, and (c) percent mortality in the Apalachicola Estuary. Station locations are noted in Figure 7.31.

concentrations that serve as the main food source of the oysters in this part of the bay. Based on the distribution of oyster density, the primary oyster-growing areas exist in these eastern sections of the bay with maximum growth during periods of low water temperature and high salinity variation.

Overall, the surface and bottom current velocities in the Apalachicola Estuary are uniquely associated with the passes and create favorable habitats for the eastern oyster bars (Cat Point, East Hole, and Sweet Goodson). Variable bottom salinities discourage predation and disease of these oysters, a situation that enhances the productivity of these bars. The opening of Sike's, on the other hand, has allowed high-salinity gulf water and the entry of stenohaline predators into the central parts of the Apalachicola Bay, thus destroying what originally were some of the most productive oyster bars in the bay prior to the dredging of Sike's in the 1950s.

Livingston et al. (1999, 2000) outlined life-history descriptions of the Apalachicola oyster populations in the estuary. Hurricane Elena struck the Apalachicola system during fall 1985 and physically destroyed the major parts of the most productive oyster reefs in the bay. There was a decrease of about 95% of the oysters in the bay during this hurricane (Figure 7.37). Immediately after Elena, there was an adaptive response of the remaining oysters in the bay to the storm in the form of an unprecedented increase of spawning and spat fall success that resulted in a major resuscitation of oyster numbers throughout the bay. This resulted in the increase of numbers relative to the period before the hurricanes. A second hurricane (Katrina) hit the area a few months later with no lasting impacts on the bay. In this case, the oyster population of the bay responded to what amounted to a succession of major storms, overcoming a catastrophic reduction of oysters during the first storm. The recovery of the oyster proliferation after these storms indicated a form of reproductive resilience to what is a relatively common occurrence along the NE Gulf Coast.

The timing of major climatological events such as hurricanes relative to oyster populations is critical in terms of the actual impact of such storms on these bivalves. Had Hurricane Elena struck

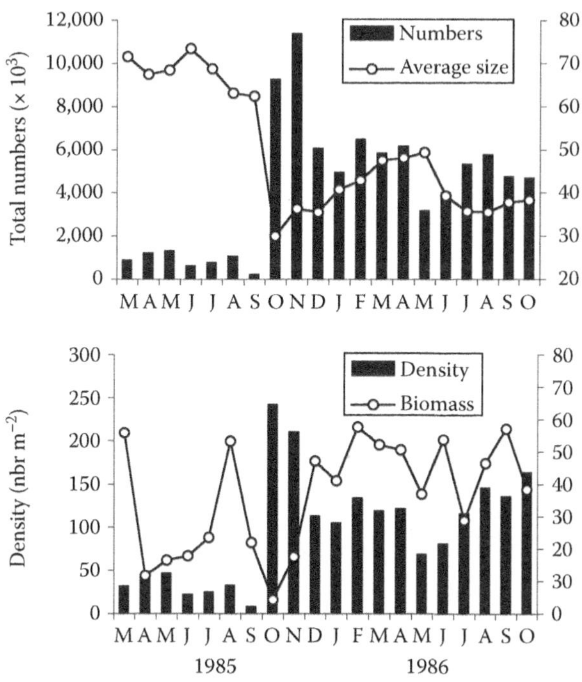

Figure 7.37 Oyster numbers, biomass (g m^{-2}), size (mm), and density averaged over all stations from March 1985 to October 1986. This included the period of Hurricanes Elena and Katrina in the Apalachicola system.

during the winter months, it is doubtful that there would have been a rapid recovery of this population. The hurricane appeared during a fall period and so acted as a stimulus for the rapid reproductive response of the surviving oysters during its natural spawning period. Conditions were such, in terms of adequate habitat, that the spat were able to survive in considerable numbers. This meant much to a fishery that is worth tens of millions of dollars to the region.

The comparison of the effects of a storm relative to the impacts of an extended drought is interesting here in terms of the resilience of the oyster population in the bay. There was a severe reduction of the size of the remaining oysters after the 1985 storms due to the loss of mature oysters and the subsequent recolonization that occurred during the months following the storm. However, during the drought of 1986–1988 (Figure 7.33), there were relatively low levels of oyster growth compared to the prestorm information (Figure 7.38). Oyster growth fell at all three of the most productive eastern stations during the drought months of 1986.

The recovery of the growth of the oysters lagged well behind that of numerical abundance, with the drought appearing to be an adverse influence on this aspect of the long-term population changes of this species. Oyster growth remained low during a 1990 sampling, indicating that the occurrence of the drought during 1986–1988 might have had a long-term, adverse effect on the growth of the oysters. Overall, the effects of the hurricane and subsequent drought on oyster ecology indicated the importance of the occurrence and timing of climatological factors in the life history of the Apalachicola oyster populations.

These oyster studies indicated that the combination of the storms and the incidence of drought conditions had a serious effect on the distribution and growth of the oysters at the major centers of bar production in the bay. Oysters are relatively well adapted to disturbances such as storms, even when the immediate effects include substantial damage to the existing population. Under such circumstances, the resilience of the oysters was enhanced by specific aspects of its life history that include rapid and massive spawning capabilities. The resurgence of the Apalachicola oyster

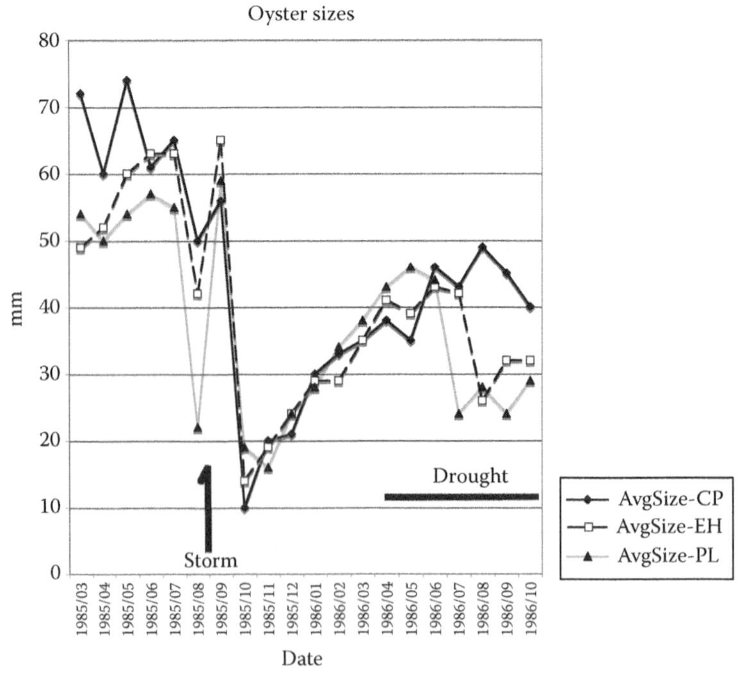

Figure 7.38 Average size of oysters (mm) at the Cat Point (CP), East Hole (EH), and Platform (PL) oyster bars by month from March 1985 to October 1986.

population numbers was aided by other by-products of the storm that included the absence of natural predators and the suspension of oystering in the bay almost a year after the storm. However, the occurrence of a prolonged drought after the storms appeared to have an adverse effect on the growth potential of the Apalachicola oyster populations. The results of these studies thus showed that the increased incidence of droughts over the past 30 years in the Apalachicola Estuary could have lasting effects on harvestable oyster product that could take years to return even if the river flows return to the pattern during the 1970s.

The issue of oyster mortality is important when evaluating the impact of drought occurrence on this species. Oyster mortality was highest at the St. Vincent Bar (SV) and areas associated with Sike's (SK), the dredged opening of the Apalachicola Bay to the gulf. These were the parts of the bay distant to river influence (with high salinity) but that were also in close proximity to the entry of oyster predators from the gulf. High-salinity gulf water moves through the passes and accumulates in deeper parts of the Apalachicola Bay. The most important oyster predator was the gastropod mollusk, *S. haemastoma*. Oyster mortality was low at the highly productive reefs in the eastern part of the bay (Cat Point, East Hole) due to variable salinities from the East Bay flows (Figure 7.36).

A series of studies featuring various statistical analyses and modeling efforts were carried out to determine how processes associated with prolonged droughts affected oyster productivity. Statistical analyses indicated that oyster mortality was positively associated with maximum bottom salinity and surface residual current velocity (Livingston et al., 2000). Mortality was inversely related to oyster density, bottom residual velocity, and bottom salinity. Oyster density was lowest in oyster reefs located in the St. Vincent Sound. Oyster bar growth (actual number of oysters in a given bar), and the density (number of oysters per unit area) were highest at the East Hole, Cat Point, and Platform Reefs (i.e., the eastern reefs). Bar growth, defined as oyster density times bar area, was directly associated with high surface water color and Secchi readings and average bottom current velocities.

We calculated a time-averaged model for summer oyster mortality by running a regression analysis with averaged predictors derived from a 3D hydrodynamic model and observed (experimental) mortality rates throughout the estuary. Based on the model, we determined that high salinity, relatively low-velocity current patterns, and the proximity of a given oyster bar to entry points of saline gulf water into the bay were important factors that contributed to increased oyster mortality (disease and predation) (Livingston et al., 2000). Mortality was a major determinant of oyster production in the Apalachicola Estuary with predation and disease as significant aspects of the loss of oysters. By influencing salinity levels and current patterns throughout the bay, the Apalachicola River was important in controlling such mortality.

Actual mortality data from basket experiments in the bay were plotted in Figure 7.39 so that the behavior of the model relative to field experimental mortality data could be observed. The distribution of mortality in 1985 (moderately low-river-flow year) was highest in areas directly affected by high salinity; such mortality was also near the entry points of oyster predators (St. Vincent Bar, Scorpion, Pickalene, Porter's Bar). Predation on the primary eastern oyster bars was relatively low. The projections of oyster mortality for 1986 (a drought year characterized by much lower river flow than 1985) were considerably higher, especially on the highly productive bars in the eastern sections of the bay. Experimental oyster mortality data taken in May 1986 (Figure 7.39) tended to confirm the model projections. In 1986, the projected predation on high-producing bars such as Cat Point, East Hole, Platform, and Sweet Goodson would have been extensive. These model projections were verified by the measured losses of oysters on the eastern bars during the significant drought in 1986.

We calculated a time-averaged model for summer oyster mortality by running a regression analysis with averaged predictors derived from a hydrodynamic model and observed (experimental) mortality rates throughout the estuary. Based on the model, we determined that high salinity, relatively low-velocity current patterns, and the proximity of a given oyster bar to entry points of saline gulf water into the bay were important factors that contributed to increased oyster mortality

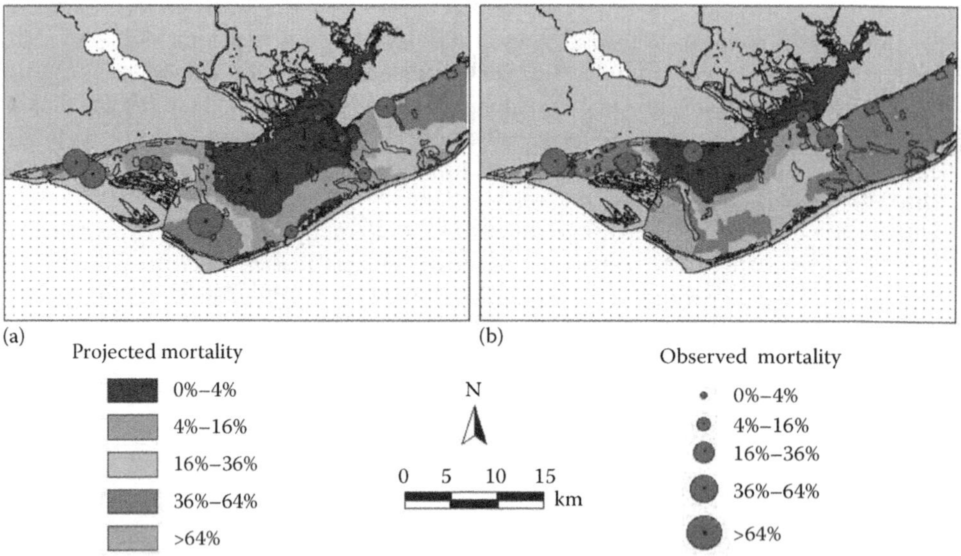

Figure 7.39 **(See color insert.)** Map of projected oyster mortality in the Apalachicola Bay system based on the statistical model for mortality in 1985 and the hydrodynamic model results for (a) 1985 and (b) 1986. Circles indicate observed mortality values from oyster predation experiments (1985, May–August average; 1986, May).

(disease and predation) (Livingston et al., 2000). Mortality was a major determinant of oyster production in the Apalachicola Estuary with predation as a significant aspect of such mortality. By influencing salinity levels and current patterns throughout the bay, the Apalachicola River was shown to be important as varying river flows eliminated such mortality.

In 1991, characterized by a river flow year intermediate between 1985 and 1986, oyster mortality was close to, but not directly impinging on the eastern bars. Model results indicated that reductions of river flow past this level would be accompanied by substantial reductions in oyster stocks.

The effect of river flow, as an indirect determinant of oyster mortality due to predation and disease through primary control of salinity regimes, was a major factor in the development of oysters in the Apalachicola Estuary. Model results indicated that reductions of river flow would be accompanied by substantial reductions in oyster stocks. Predation is an active factor in the determination of oyster production in the Apalachicola system. An example of its importance is the near-total demise of the St. Vincent oyster bar due to predation that followed the opening of Sike's in the mid-1950s (Menzel et al., 1957, 1966).

An analysis was carried out concerning the effects of an artificial pass (Sike's; Figure 7.40) on the salinity regime of the Apalachicola system and the potential impact of the pass on oyster mortality. A comparison was made between projected oyster mortalities with the cut open and closed. Modeling results indicated increased mortality due to salinity increases in the vicinity of Sike's that extended to localized oyster areas and the St. Vincent Bar. The model also indicated increased mortality in parts of the St. George and St. Vincent Sounds that included potential impacts on some of the productive oyster bars in these parts of the bay. Model results thus showed that the influence of Sike's on oyster survival extended beyond the lower regions of the Apalachicola Bay.

The very high oyster production rates in the Apalachicola Estuary depend on a combination of variables that are directly and indirectly associated with freshwater input as modified by wind, tidal factors, and the physiography of the bay as it relates to current distribution. Experiments and models predicted that river flow reductions, whether through naturally occurring droughts, through increased upstream anthropogenic (consumptive) water use, or a combination of the two, could have

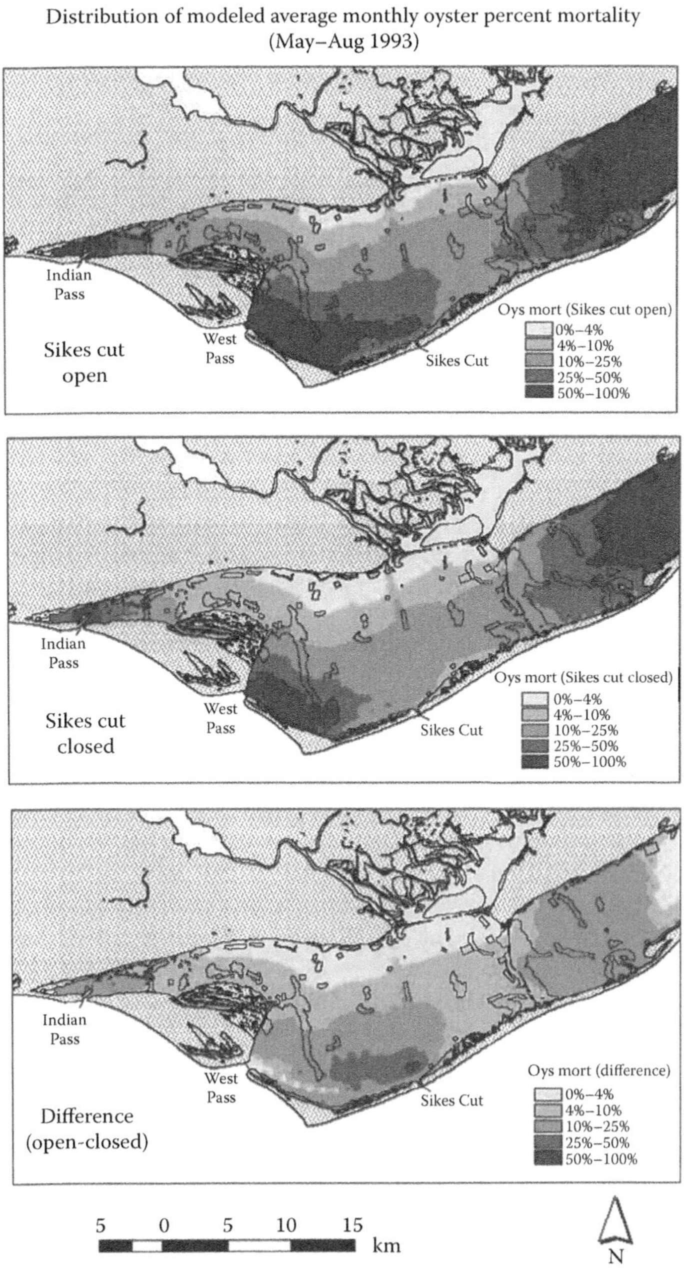

Figure 7.40 **(See color insert.)** Modeled oyster percent mortality data based on effects of an open Sike's relative to a closed Sike's. (Data are taken from an unpublished paper by R. J. Livingston, S. Leitman, G. C. Woodsum, H. Galperin, P. Homann, J. D. Christensen, M. E. Monaco, and R. L. Howell, IV.)

serious long-term, adverse consequences for oyster populations. Reductions of oyster production that are adapted to natural cycles of river flow were affected by increases in the frequency and severity of droughts in the Apalachicola drainage basin during the 1980s.

Long-term changes of the Apalachicola oyster population should be placed within the context of major habitat-controlling features such as Apalachicola River flow. Meeter et al. (1979) found that oyster landings from 1959 to 1977 were correlated negatively with river flow. The highest oyster landings actually coincided with drought conditions. The highest oyster harvests during this period occurred in 1980–1981, coinciding with a major drought. Wilber (1992), using oyster data from 1960 to 1984, found that river flows were correlated negatively with oyster catch per unit effort within the same year and positively with catches 2 and 3 years later.

Livingston et al. (1997) found that increases of the Apalachicola Bay nonoyster bivalve mollusk populations during droughts were based on changes in the trophic organization of the estuary. The relatively high oyster production during low-flow years was probably due to increased primary productivity as a function of altered physical conditions (i.e., increased light penetration) in the receiving estuary. The 2-year time lags between low-flow events and subsequent poor production during this period were probably associated with the reduction of nutrients and phytoplankton production as the drought continued without renewed nutrient loading from the river. Increased river flows contributed to increased growth rates and ultimately increased oyster production although such returns appeared to take years to result in a return of high oyster growth.

The responses to natural droughts (prior to the recent increase in low flows) in terms of oyster productivity were regular and occurred within long-term progressions of river flow fluctuations. However, the prolonged droughts over the past 30 years eventually induced serious reductions of oysters due to the effects of increased predation and disease. The oyster experiments and modeling that took place during the early stages of increased drought conditions during the 1980s indicated that oyster productivity would be seriously affected with an increase in the frequency and duration of droughts during succeeding decades. This prediction unfortunately came to pass during the droughts of 1999–2002, 2006–2008, and 2010–2012.

Fishes and Invertebrates

The predrought Apalachicola Estuary represents a fortuitous combination of natural physical and chemical characteristics that enhanced natural primary and secondary productivities. These conditions were centered around river flows into the bay that enabled populations of invertebrates and fishes to reach outstanding levels (Livingston, 1984a, 2000, 2002, 2005, 2008, 2013). This estuary harbored the principal nursery for blue crabs that ranged along the Florida Gulf Coast and was the center for white shrimp nurseries that fueled the offshore Big Bend shrimp fishery. It was the source of Florida's oyster production and was a renowned area for various sports fisheries.

The estuary was inhabited by euryhaline dominant species that are able to nurse in the extreme physical–chemical variations of the bay system. Extensive stomach content analyses were carried out with the dominant fish and invertebrate species. The long-term database was reorganized according to the ontogenetic feeding stages of the numerically dominant species. TUs were averaged by month over a 14-year study of the East Bay from 1972 to 1984. This study included a period of average drought–flood conditions of the river that were then succeeded by the drought of 1980–1981 that proved to be the beginning of the droughts that were to occur during the next 30+ years. The actual distribution of the key estuarine species was based on microhabitat organization that, in turn, depended on various conditions related to river flow processes. By transforming the data into the trophic organization of the bay, we were able to evaluate population distribution according to a basic, functional process (feeding) that defined the productivity of the nursing organisms.

Long-term studies of the Apalachicola Estuary indicated that there was a range of impacts on the Apalachicola ecosystem that could be associated with reduced river flows (Livingston, 1984a). These impacts were related to specific attributes of the upland wetlands and associated coastal areas. The relationships of flow rates, nutrient loading, and the resulting nutrient concentrations proved to be important in the initiation and sustenance of phytoplankton production that formed an important part of the detrital food webs in the river-dominated Apalachicola Estuary (Livingston, 2000, 2002, 2005, 2013). Nutrient loading into the Apalachicola coastal region remained relatively high without apparent hypereutrophication in the bay. This was due, in part, to the low concentrations of nutrients in the lower river and low residence time of these nutrients in the bay. The relationships of river flow rates, nutrient and organic carbon loading, current structure, the shallowness of the estuary, and the resulting nutrient concentrations in the bay were important determinants of the food web structure of the Apalachicola Estuary.

The most outstanding aspect of the bay salinity over the 1972–1984 study period was the sustained high salinities during the drought of 1980–1981. This reduction of Apalachicola River flow was associated with major changes in the physical/chemical characteristics of the East Bay (Figure 7.41). Water color (dissolved ions) was lowest and Secchi disk readings (light penetration) were highest during the low-flow period (Figure 7.41) with Secchi readings in 1980 being significantly higher ($p < 0.05$) than those of preceding years. Color levels during the drought years were significantly ($p < 0.05$) lower than those taken during the preceding and succeeding years. Light penetration, as affected by physical (color) and biological (turbidity) indicators, was an important part of the system response to reduced river flows. The relatively low turbidity levels during the second half of the drought indicated phytoplankton production was low relative to the spring and summer of 1981 when the turbidity (plankton numbers) was at an all time high. Secchi depths appeared to reflect the turbidity effects at this time when water color was at an all time low. With time, the indicators of such productivity went down accordingly as nutrients were used up without replacement from the river. All physical–chemical indices returned to previous levels during the period of increased river flows that followed the drought.

The period during the 1980–1981 drought was characterized by initially high positive values of DO (Livingston et al., 1997) that indicated high phytoplankton production during the beginning of the drought. There were negative oxygen anomalies during the last half of the drought (when Secchi depths were deepest), indicating that other biological activities occurred that could involve reduced primary production and/or increased secondary production. The high pH levels during the

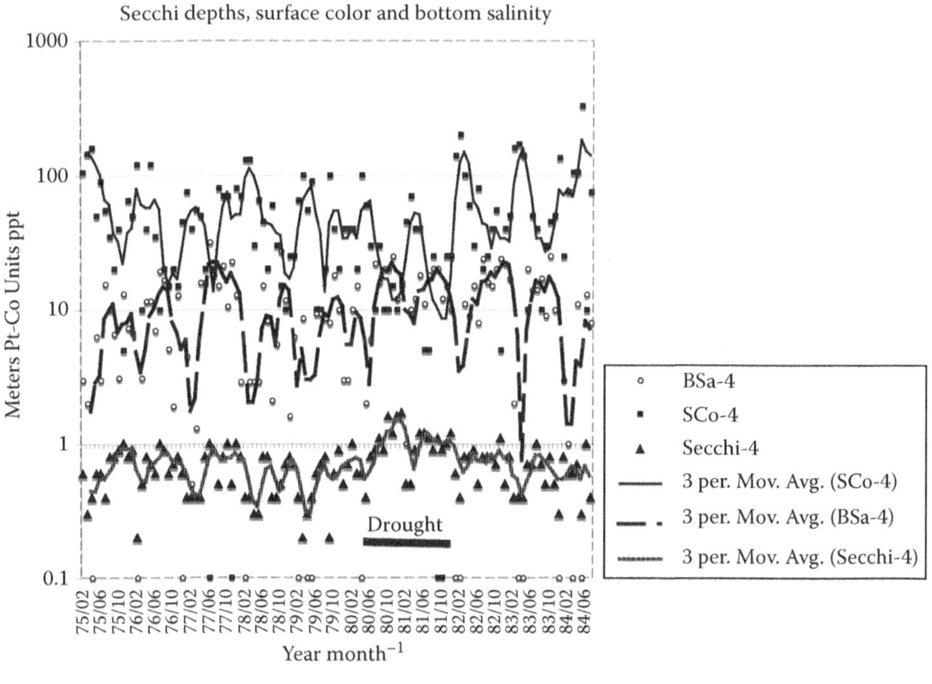

Figure 7.41 Secchi depths, surface water color, and bottom salinity at station 4 in the East Bay, monthly from February 1975 through June 1984.

initial months of the drought provided further proof of high primary production at this time. The extreme low pH deviations in 1983 tended to coincide with the tail end of the period of low-oxygen anomalies. These data support an association of high primary production during the early months of the drought that was followed by reduced plant activity during the period following the drought.

These data indicated a relatively uniform response to the drought conditions that were significantly different than those taken before and after the drought.

Statistical analyses of long-term trends of numerical infaunal abundance and trophic diversity (Shannon diversity calculated with the numbers of TUs) were determined. Infaunal abundance was highest during the beginning of the drought period with a sharp falloff of numbers by the spring of 1982 at which time the lowest numbers of the long-term data series occurred. Year-by-season trends were significantly different; the annual differences were also highly significant ($p < 0.05$; Tukey compromise, Scheffe's S). There were significant differences of these numbers in 1980–1981 relative to other years of record. The lowest numbers occurred following the drought (1982, 1983); such annual trends were significantly different than all other years of the survey with the exception of 1976–1977, a period marked by somewhat reduced river flows.

During summer 1981 (the latter part of the drought period), infaunal trophic diversity dropped to significantly ($p < 0.05$) lower levels than any other year; recovery of such diversity was not fully evident until the spring of 1983. A brief episode of low trophic diversity occurred during the reduced river flow of summer 1977 although such low diversity only lasted 3 months. Trophic diversity was significantly ($p < 0.05$) lower during warm periods than at other times of the year. The infaunal trophic diversity was thus an indicator of the interannual trends of low-river-flow conditions.

Sustained low numbers of epibenthic macroinvertebrate numbers ($p < 0.05$) occurred during the drought of 1980–1981. Trophic diversity showed extended low levels during the spring and summer of 1981 with recovery of this index evident by the spring of 1983. This pattern followed the changes in infaunal trophic diversity. There was evidence of a general decrease epibenthic invertebrate trophic diversity during the brief drought of the summer of 1977 although such reduction was not as pronounced as that during the 1981 period. These trends were significantly different only during the 1977 and 1981 periods of low diversity.

Fish numbers were significantly ($p < 0.05$) higher during winter–spring periods with particularly high peaks during the winter–spring of 1981. Fish trophic diversity was significantly ($p < 0.05$) lowest following the drought of 1981 with recovery of this factor evident by spring 1982. Such diversity was usually highest during summer–fall periods. Fish trophic diversity was also adversely affected by the drought, although the recovery followed a somewhat different pattern than that noted for infauna and epibenthic macroinvertebrates.

Species richness trends (Figure 7.42) indicated that infaunal numbers of species reached a peak during the drought, whereas the numbers of infaunal and invertebrate species reached low points during the end of the drought. Fish species richness remained low throughout the years following the end of the drought. These data indicated that adverse effects of the drought on important biological indicators lasted well beyond the period of low river flows.

A detailed analysis of the trophic data in the East Bay was given by Livingston (2002, 2013) and Livingston et al. (1997). The long-term trends of the trophic groupings (infauna, epibenthic invertebrates, fishes) in the bay are shown in Figures 7.43 and 7.44a,b. The mean monthly biomass of herbivores over the study period was 2.39 g m^{-2} with averages of 0.21 g m^{-2} for omnivores, 0.53 g m^{-2} for primary (C_1) carnivores, 0.04 g m^{-2} for secondary (C_2) carnivores, and 0.004 g m^{-2} for tertiary (C_3) carnivores. There was a marked increase in the herbivore biomass during the beginning of the drought. Herbivore biomass peaks coincided with low winter river flows, low water color, and increased Secchi depths. The more frequent periods of positive oxygen anomaly and high turbidity during the drought also coincided with the increased levels of herbivore biomass. There was a marked decrease of herbivore biomass that started during the second half of the drought. Herbivore biomass virtually collapsed during the following years from 1982 to 1984. Full recovery of the herbivores was

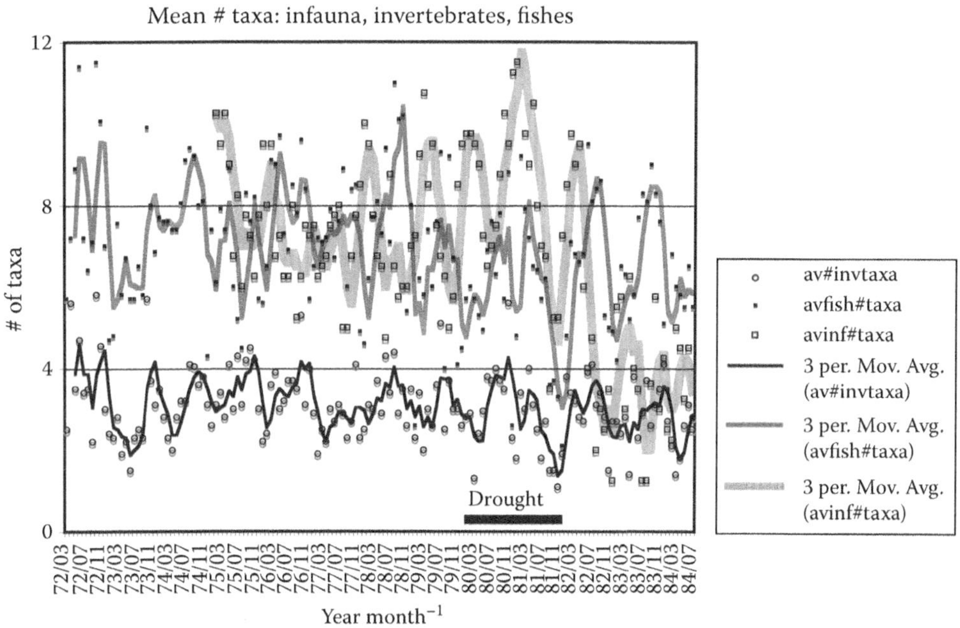

Figure 7.42 Infaunal, invertebrate, and fish species richness monthly from 1972 through 1984 in the East Bay system. Data were averaged across stations.

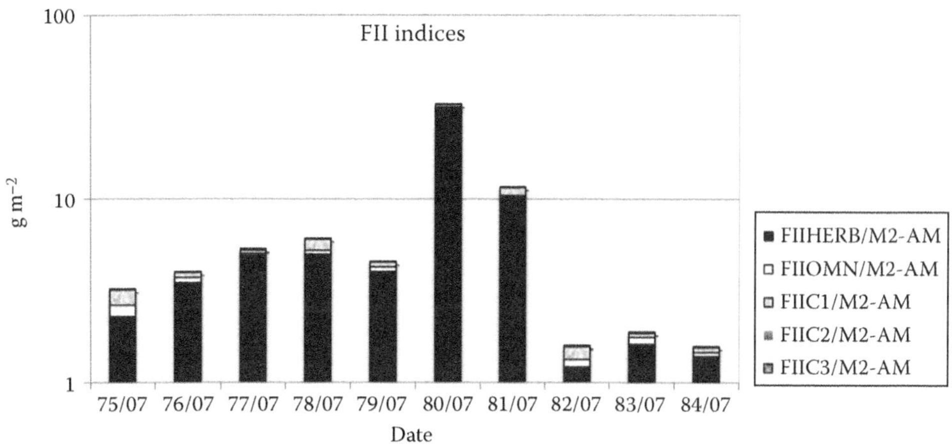

Figure 7.43 FII trophic levels (herbivores, omnivores, C_1 carnivores, C_2 carnivores, C_3 carnivores) taken in the East Bay (averaged across stations) annually from 1975 through 1984.

not complete by the end of the sampling period. These trends followed the changes in the indicators of primary productivity as noted by the various physical and water-quality changes during and after the 1980–1981 drought.

The drought-induced fish, infauna, and invertebrate (FII) biomass losses extended beyond the recovery of the Apalachicola River flows (Figure 7.44a). Omnivore biomass peaked during the period of increasing herbivore biomass in the drought period (Figure 7.44b). There was a general decrease of omnivore biomass during the 2-year postdrought period. Primary carnivore biomass increased during fall 1980 and continued at high levels through spring 1981. After the drought, primary carnivore biomass declined to levels below the predrought conditions. There was no recovery of the C_1 carnivores during the postdrought months of the study. The biomass of C_2 carnivores increased incrementally from the onset of the drought with peaks that generally coincided with those of the primary carnivores. There was no overt recovery of this group during the months following the drought.

These data indicate that, prior to the onset of the drought in 1980, there were relatively conservative temporal changes in the various trophic levels in the East Bay. Despite the relatively modest duration of the drought of 1980–1981, the reduced river flows during this period had profound effects on the trophic organization of the East Bay. The fact that the East Bay was the center of the highly productive nursery of the Apalachicola Estuary made such findings particularly important. This impact remained at various trophic levels during the 2.5-year period of postdrought river flow recovery.

Within a certain range of river flow variation, there did not appear to be significant deviations of the pattern with omnivores becoming dominant after the major floods of the first year of recorded data and the various other trophic levels showing no particular interannual patterns. The drought and reduced river flows led to changes in key water-quality variables such as increased salinity and light penetration. These changes coincided with changes in turbidity and DO that indicated increased immediate phytoplankton productivity. In turn, there was increased predominance of the herbivores during the drought that was followed in quick succession by major reductions of the various trophic levels during the postdrought years.

When viewed as total biomass m^{-2} y^{-1} as a function of river flow, there was a clear relationship between river flow and overall (invertebrate/fish) biomass in the East Bay. There were significant

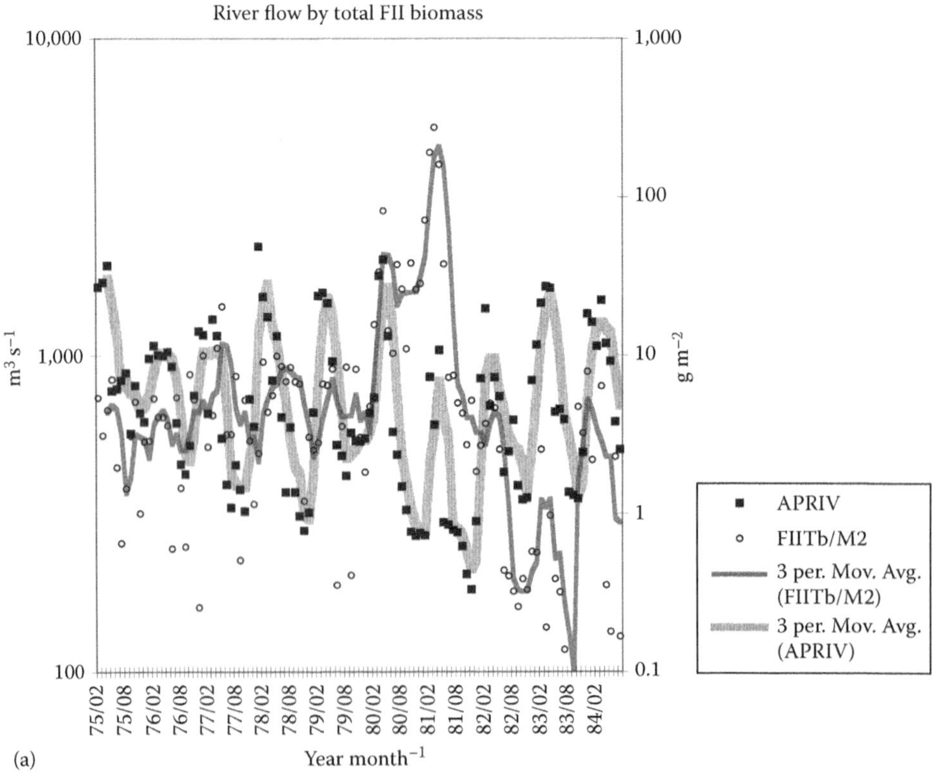

Figure 7.44 (a) Monthly Apalachicola River flow and total FII biomass averaged over East Bay stations and taken monthly from 1975 through 1984. River flow data are provided by the United States Geological Survey Sumatra gauge no. 02359170. *(Continued)*

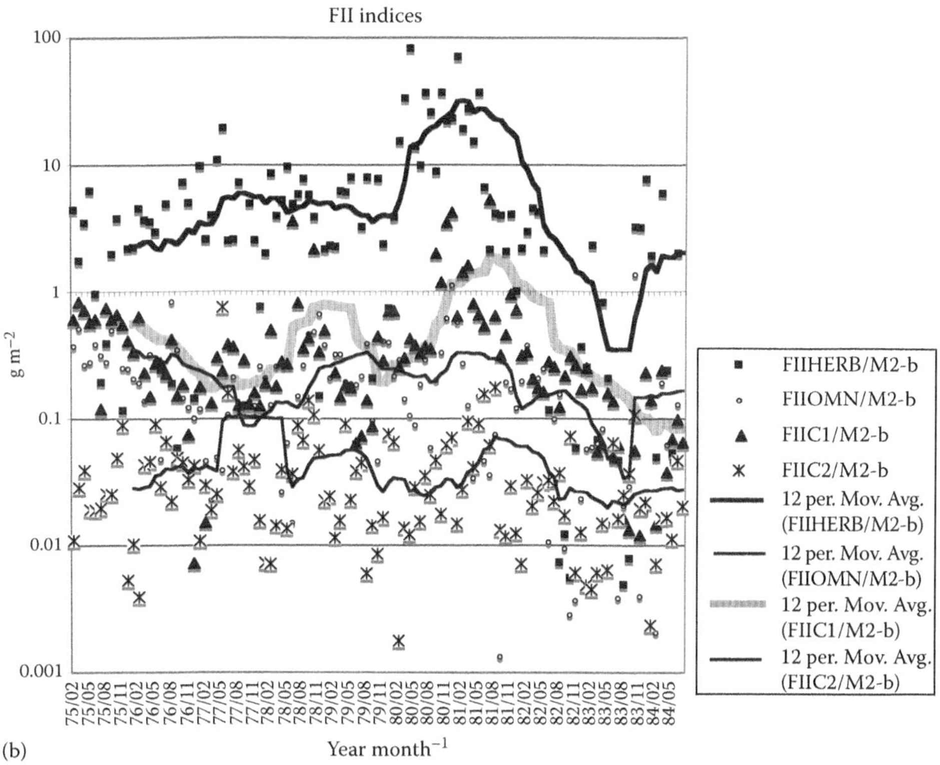

(b)

Figure 7.44 *(Continued)* (b) FII trophic levels (herbivores, omnivores, C_1 carnivores, C_2 carnivores) taken in the East Bay (averaged across stations) monthly from 1975 through 1984. Also shown are the 12-month moving averages of the data.

seasonal and interannual differences in the biomass changes. Late winter/spring biomass, coinciding with peak phytoplankton production in the bay, was higher ($p < 0.05$) than at any other season. During the first 5 years of sampling, river flow and total animal biomass remained within a relatively small level of interannual variance. Peak biomass years (1980–1981) coincided with major reductions in river flow, while significant decreases in biomass coincided with the recovery period following the drought. Herbivores increased significantly during the drought, after which nutrients eventually became limiting with coinciding downtrends of the various trophic levels of the bay along with severe reductions of animal biomass compared to the period prior to the drought.

The changes in the trophic organization relative to the 1980–1981 drought were quite extreme, considering that this drought was relatively minor compared to the succeeding droughts over the next 30+ years. These results indicate that based on the response of the bay trophic organization to a relatively minor drought, the changes in the estuarine food webs due to the succeeding more frequent and severe droughts could have been more extreme. This response could also be negatively affected by the cumulative impacts on bay resilience. The indications are that droughts are initially characterized by increased animal indices as productivity peaks due to the increased light penetration in the estuary. With time and reduced nutrient loading from the river, plant productivity wanes, and with it, the herbivore and omnivore (including the detritivores) biomass totals go down, followed by the various levels of carnivores. The relatively long period of the drought could be due to the time taken for recovery of the various processes associated with the nutrient and organic carbon-driven parts of the trophic organization of the estuary. It remains necessary then to evaluate the changes that take place when drought conditions occur with increased frequency and severity.

Long-Term Drought Effects on Bay Fisheries

There is considerable evidence that river-dominated estuaries are affected by key climatic factors that contribute to river flow characteristics. Paerl et al. (2009), working in the Neuse River–Estuary, found that multiannual, seasonal, and episodic hydrologic perturbations caused significant shifts in phytoplankton community structure. These oscillations, including droughts, can overwhelm anthropogenic nutrient inputs in terms of controlling algal blooms. Scavia et al. (2002) noted that estuarine productivity would change in response to alterations of the timing and amount of freshwater, nutrients, and sediments that are delivered to coastal estuaries although the impacts would vary from estuary to estuary. Changes in precipitation and river flow patterns could influence water residence time, nutrient delivery, vertical stratification, and control of phytoplankton growth rates (Howarth, 1988; Paerl et al., 2009). These studies point to the adverse impacts on fishery resources caused by increased incidence and severity of drought events.

During the severe drought of 1999–2002, there was a collapse of oystering in the highly productive eastern reefs from the Cat Point to East Hole. In a 2002 field assessment by the Florida Department of Environmental Protection (G. S. Gunter, pers. comm.), lowered oyster productivity in eastern bay reefs was accompanied by large numbers of predators that included oyster drills, crown conchs, scallops, and sea urchins. Hard and soft corals were noted on Porter's Bar during these field surveys (G. S. Gunter, pers. comm.). With a return of higher river flows in 2003, there was an increase of observed oysters on the Cat Point and East Hole bars with an accompanied reduction of oyster predators (G. S. Gunter, pers. comm.). During the 2002 collapse of the eastern oyster reefs, commercial oystering continued in the northern sections of western bars such as the Dry Bar and the upper St. Vincent Bar (G. S. Gunter, pers. comm.). The preceding observations during this drought represented field verification of model predictions made by Livingston et al. (2000). Trawl data (L. Edmiston and J. Wanat, pers. comm.) indicated that after initial increases of fish and invertebrate population numbers during the beginning of the 1999–2002 drought, there were subsequent reductions of such figures. These trends followed the model predictions noted by Livingston et al. (1997).

Grass beds in the upper bay, dominated by *V. americana*, were adversely affected by the drought of 2006–2008 (L. Edmiston, pers. comm.). At one time, SAV supported a productive food web in the East Bay (Purcell, 1977). Based on a survey of the SAV in the East Bay in June 2005, Hurricane Dennis took out all the SAV in both the lower river and the East Bay. The grass beds started to return the following year. However, these grasses all but disappeared again during the extended drought. The high salinities associated with the drought were thought to be responsible for the loss of the East Bay SAV due to the fact that such vegetation is adapted to low and varying salinity (L. Edmiston, J., pers. comm.).

Mr. Edmiston's trawl information (unpublished data) showed severe reductions of white shrimp in the East Bay and the Apalachicola Estuary during the drought of 2006–2008. Pink shrimp increased modestly in 2007 because of higher salinities, but the numbers of this species did not compensate for the loss of the white shrimp in the bay. Relatively few blue crabs were taken in the East Bay in 2007. Commercial fishing landings data taken in 2007 reflected decreases of white shrimp (down 90%), brown shrimp (down 55%), blue crabs (down 55%), and flounder (down 40%). The severely reduced landings of key commercial species represented an example of how low river flows had adverse effects on the Apalachicola Estuary that followed those noted by Livingston et al. (1997) during the 1980–1981 drought. The overall collapse of the various fisheries in the estuary could have been exacerbated by the repetitive nature of the drought incidences.

During the most recent, very severe drought of 2010–2012, the entire Apalachicola commercial fishery collapsed, causing a catastrophic economic condition in the Big Bend coastal region. Havens et al. (2013) noted that the 2012 oyster industry went through an "unprecedented" decline in oyster landings that was "most likely due to recruitment failure and high mortality of small oysters."

High levels of oyster parasitism were reported. This collapse included the destruction of the major oyster bars due to predation and disease and the loss of the lucrative white shrimp and blue crab industries. This drought was the worst of the line of recent low-flow conditions in the Apalachicola Estuary. The conditions were such that most of the commercial fishery operations of the estuary came to a halt. All of the predictions of how droughts would adversely affect the Apalachicola Estuary came true.

Prasad et al. (2010), working in the Chesapeake Bay, found that long-term climate indices, such as El Niño Southern Oscillation (ENSO) and North Atlantic Oscillation (NAO), exerted only moderate influences on the riverine discharge to the bay or over the ecosystem response in terms of chlorophyll *a* in the bay. Freshwater discharge was significantly correlated with both climate indices, and chlorophyll was only moderately related to NAO and weakly related to ENSO, suggesting increased complexity in factors that drive biological response relative to the hydrological response. Vinagre et al. (2011) found that increased drought frequency due to climate change lowered the connectivity of the estuarine fish food webs in the Tagus Estuary, causing habitat fragmentation and consequent loss of community complexity and resilience. The authors stated that the consequences of droughts should be taken into account in the determination of future river flows allowed by the upstream dams.

Livingston (2013) reviewed some of the meteorological information of the Apalachicola region. According to the Carnegie Institution for Science (Archer and Caldeira, 2008), there is evidence that the jet stream distributions in the northern hemisphere have changed with an increased northward movement in recent years. These changes fit the predictions of climate change models and have implications for the intensity of future storms, including hurricanes. Storm paths in North America may be likely to shift northward as a result of the jet stream changes. This change could lead to less rain in the already drought-stricken SE United States. There is no way to confirm the future long-term effects of these changes that are likely due to global climate trends.

The data presented here represent the ecological changes that occurred during the droughts over the past 30+ years without any predictions of future changes in the estuary. If the projections of reduced rainfall in the Apalachicola–Chattahoochee–Flint (ACF) River region are accurate, however, it would place even more emphasis on the need for an advanced water conservation effort in the tri-river basin. The increase of the frequency and intensity of droughts in the Apalachicola Estuary is a fact, and the potential for further reductions of rainfall and natural river flow should be included in calculations for water conservation efforts. Once again, no one can accurately predict the future with respect to the incidence of droughts and changes in water temperature and sea level. However, the current trends are clearly in favor of future exacerbation of the current effects of climate changes. Based on the ecological data, if such changes continue or become even more extreme in the future, the continued productivity of the Apalachicola River and Estuary would be in serious jeopardy.

The concepts of biological stability and resilience have been defined in various ways by researchers (see Harrison [1979] and Santos and Bloom [1980] for a review of the semantic problems encountered). Stability is defined as the ability of a given system, once perturbed, to return to its previous state. Resilience refers to the degree, manner, and pace of restoration of the initial system function and structure following a disturbance (Westman, 1978). Cairns and Dickson (1977) referred to various forms of a recovery index: proximity of recolonization sources, mobility of propagules, physical and chemical suitability of habitat for recolonization, toxicity of the disturbed habitat, and effectiveness of human management initiatives to facilitate rehabilitation. While the high degree of resilience of the Apalachicola oysters to a natural disturbance is evident, the same is not true with respect to the response of various biological indicators to an unprecedented change of climatic conditions as represented by the increased frequency and severity of droughts.

The recent droughts and the ecological impacts of such conditions should be viewed within the context of the resilience of the bay to repeated low river flows. The increasing severity of the biological impacts indicated reduced resilience of the bay with respect to the productivity of the primary populations. It is clear that the commercial resources of the Apalachicola Estuary are currently in danger of being lost for an as yet undetermined period. If the increased incidents of low-river-flow regimes during severe drought periods in recent years continue as a trend into the future, the bay productivity in terms of the key commercial species will inevitably give way to the replacement of these populations by less important species in terms of their fishery value.

The resilience of the estuary is being serially reduced by the increased severity and length of the reduced river flows during droughts. The extended time periods for recovery of the trophic organization of the bay are evidence that nonlinear responses of estuarine bay populations to reductions of river flows have already occurred. If such changes result in long-term habitat alterations, even the return of higher river flows may not induce a recovery of the bay commensurate with the observed recent losses. Recovery may not follow the lines of the previous trends of productivity if the resilience of the bay has been altered. The recent and current lack of ecological research in the Apalachicola drainage system does not bode well for a resolution of these research questions.

Under the natural range of river flow variation, the physically unstable estuary has been associated with a relatively stable biological system over interannual periods of time (Livingston et al., 1997). River flow changes within the specific limits of seasonal variation resulted in a generally stable biological progression through time. With consistent reductions of flow rates during a prolonged series of droughts, the clarification of a turbid system leads to rapid changes in the pattern of primary production, which, in turn, becomes associated with major changes in the trophic organization of the system. With the prolongation of the restricted river flows, the formerly physically controlled biological system gradually changes to a more biologically controlled system (Livingston, 2002). Eventually, permanent reductions of freshwater flows would then be associated with major changes of biological productivity. The river-dominated estuarine conditions could give way to a far less productive system in terms of sports and commercial fisheries.

Conclusions

The relatively natural Apalachicola River/Estuary, with its shallow depths and the general lack of widespread salinity stratification, has formed a highly productive river and bay system in terms of key sports and commercial populations. Nutrient-induced phytoplankton productivity has been an important component of estuarine food webs in the Apalachicola Estuary. The alluvial river also provides organic matter that fuels the key detrital food webs in the bay. Autochthonous and allochthonous organic carbon creates the resources for the trophic organization in the Apalachicola Estuary. The natural productivity of this ecosystem has been historically high relative to other such systems in the northern hemisphere.

Over the past 30+ years, there has been a series of increasingly severe droughts that have adversely affected the productivity of the bay. Recent analyses of bay sediments suggest a decrease in riverine organic matter and a change in sediment clay content supply and/or distribution in the estuary. This loss of detrital matter could adversely affect key estuarine food webs. A similar trend of reduced nutrients loaded into the bay from the river has been associated with the droughts that have, in turn, co-occurred with reduced secondary productivity and altered food webs in the estuary. These changes were directly associated with severe reductions of the dominant, commercially important populations of the Apalachicola Estuary (oysters, blue crabs, penaeid shrimp, sciaenid fishes). Increased salinities have allowed the entry of stenohaline predators from the open gulf to invade the estuary, thereby reducing the nursing function of the bay with specific impacts on oyster populations. Combined with the reduced nutrient and organic carbon loading of the river, the results, in terms of fishery potential, have been considerable. Commercial landings during and after the droughts have suffered major decreases.

The obvious changes in the regional climate conditions over the past three decades have thus led to the deterioration of one of the most productive estuaries in the country. These changes have been illustrated by the long-term ecological studies in the Apalachicola Basin.

PERDIDO ESTUARY

The Perdido Project was a court-ordered study that concerned the scientific determination of the impact of a pulp mill on the Perdido Estuary in NW Florida and eastern Alabama. The study was conducted over a 20-year period (October 1988–July 2007) (Table P.1).

Background

Seasonal and interannual attributes of the response to nutrient loading are complex. Jouenne et al. (2007) noted that the timing and intensity of plankton blooms vary at various levels of temporal events. The time-based factors include climatological events that result in progressions of phytoplankton community structure that, in turn, are intricately involved in the trophic organization of estuaries. Responses to nutrient loading and concentrations are modified by temperature, salinity, wind, tides, and sediment processes. Nutrient loading and responsive primary production represent processes that are central to our understanding of how highly productive estuarine and coastal systems function.

Hypoxic events are relatively common in coastal systems. Verity et al. (2006) related oxygen depletion in well-mixed estuaries in the SE United States to increased nutrient concentrations and increased decomposition of autotrophs below the pycnocline. The data indicated that hypoxia can occur from microbial respiration in well-mixed estuaries with strong vertical and horizontal mixing. Conley et al. (2007) reported recent (2002) hypoxic events in Danish coastal waters that were associated with high rainfall and nitrogen loading. The authors related such changes to those noted in the Chesapeake Estuary (Hagy et al., 2004) where bottom DO was "modulated" by nitrogen runoff from adjacent land areas. These trends of low DO were also associated with bottom-water transport, wind events, and surface water temperature. However, decreased nutrient loading was not associated with increased DO during recent times. Borsuk et al. (2004), on the other hand, found that riverine TN concentrations played only a minor role in determining chlorophyll *a* in the Neuse River–Estuary. River flow had a stronger effect on such productivity.

The relationship of plankton blooms and DO is complex and includes instances of both supersaturation and hypoxia. Hypoxia has been reported in most of the gulf estuaries (NOAA, 1997). Episodes of low DO during "jubilees" have occurred over relatively long periods in gulf coastal areas such as the Mobile Bay (Loesch, 1960, May, 1973). Lehrter (2008) noted that chlorophyll *a* was strongly related to freshwater flushing times. Estuarine phosphorus concentrations and photosynthetically active radiation were also good descriptors of chlorophyll *a* concentrations. Periodic hypoxia in some coastal systems can be the result of natural conditions (Seliger and Boggs, 1988).

Salinity stratification and the restriction of vertical mixing in estuaries have been associated with hypoxic conditions. Schroeder et al. (1990) found that river flow was the dominant control mechanism of salinity stratification in the Mobile Bay. Annual spring freshets flushed salt from the bay, while "the relative strengths of river discharge and wind stress changed the bay from highly stratified to nearly homogeneous and back on a variety of time scales ranging from daily to seasonal." Stratification events and DO levels were tightly coupled with variations in freshwater discharge and wind stress in the Pamlico Estuary (Stanley and Nixon, 1992). Stratification appeared in a matter of hours, with hypoxia forming if the water is mixed only every 6–12 days.

Phytoplankton communities are derived from multiple processes (Cloern and Dufford (2005). The magnitude of estuarine primary production and the periodicity of algal blooms can be directly related to variations in upper watershed rainfall and its subsequent regulation of downstream river flow (Mallin et al., 1993). Heil et al. (2007) noted that the quality of available nutrients drives the composition of phytoplankton assemblages. Paerl (2009) recently emphasized the importance of both phosphorus and nitrogen to the establishment of plankton blooms and the relevance of such dual limitation to development of effective management programs. Anthropogenic activities resulting in increased nutrient loading, restricted water circulation, altered phytoplankton production, and trophic anomalies also influence the frequency and severity of such episodes (Seliger and Boggs, 1988; Breitburg, 1990). Biological changes associated with nutrient enrichment of estuaries can lead to shifts from large to small phytoplankton species and from diatoms to dinoflagellates (Smayda, 1980); these changes can adversely affect shellfish populations. Small-celled diatoms such as *S. costatum* and *L. danicus* can give way to mixed communities of large-celled diatoms and dinoflagellates such as *Chaetoceros* spp. and *Prorocentrum* sp. Dinoflagellate blooms commonly occur in highly stratified conditions with predominance of *Gymnodinium* spp. and *Prorocentrum* spp. following the intermediate replacement of the smaller-celled diatoms (Smayda, 1980).

Blooms of raphidophytes of the genus *Heterosigma* are known to cause fish kills in New Zealand, Chile, and British Columbia (Chang et al., 1990). *Heterosigma akashiwo* is well known for ichthyotoxic blooms. Honjo (1993) found that *H. akashiwo* blooms were associated with river runoff, low bottom oxygen, and wind-induced turbulence of bottom sediments. Growth occurs from overwintering motile forms and/or germinating resting spores. This species has a high growth potential and can form a bloom in a short time, excreting allelopathic substances that can suppress diatom growth and may dramatically reduce cell numbers in other phytoplankton species. In addition to making vertical migrations, *H. akashiwo* can migrate to diverse spatial habitats in search of favorable growth conditions. Nutrient limitation stimulates migration tactics. There are relatively high nutrient requirements for *H. akashiwo.* Bactivorous feeding is stimulated by P limitation with Fe and Mn needed for bloom continuance. Blooms of this species have been associated with increased river flows in various regions. Temperature, nutrients, and competition appear to be limiting for bloom generation in *H. akashiwo.*

Phytoplankton blooms adversely affect invertebrate and fish populations (Shimada et al., 1983, 1996; Chang et al., 1990; Burkholder et al., 1992; Buskey and Stockwell, 1993; Buskey et al., 1997). Blooms also harm coastal fisheries (Smayda and Shimizu, 1993; Hallegraeff, 1995; Hallegraeff et al., 1995; Anderson and Garrison, 1997). Riegman (1998) reviewed the effects of altered nutrients on phytoplankton species composition although little is known about long-term changes of these coastal assemblages. Fryxell and Villac (1999) noted that data are rarely available for events leading up to blooms and even less is known about the effects of blooms on associated food webs. Sorokin et al. (1996), working in the Comacchio Lagoons (Ferrara, Italy), noted reduced zooplankton and losses of benthic organisms and fishes due to increased nutrient loading and dominance of "inedible cyanobacteria."

Roelke et al. (1997) developed models that, based on successions of phytoplankton populations in response to increased nutrient loading, provided guidelines for nutrient management. Interactions of nutrient loading, nutrient limitation, grazing, interspecific (phytoplankton) competition, and seasonal/interannual changes in plankton population distributions, all contributed to the trophic response of the Nueces Bay (Texas Gulf Coast) to sewage wastes. Bloom sequences can contribute to altered food webs although interactions of seasonal progressions with long-term changes related to drought–flood trends complicate interpretations of bloom impacts. Ryther (1954) noted the association of nitrogen fertilization of Long Island Sound with proliferation of chlorophytes and adverse effects on associated food webs. The impacts of plankton blooms are often cumulative and can occur in complex ways (Fryxell and Villac, 1999). A detailed review of the causes and effects of blooms on estuaries has been given by Livingston (2000, 2002).

The Perdido drainage basin (Figure 7.45) is composed of a series of freshwater drainages (the Perdido River and its tributaries, the Elevenmile Creek, the Bayou Marcus Creek). The primary

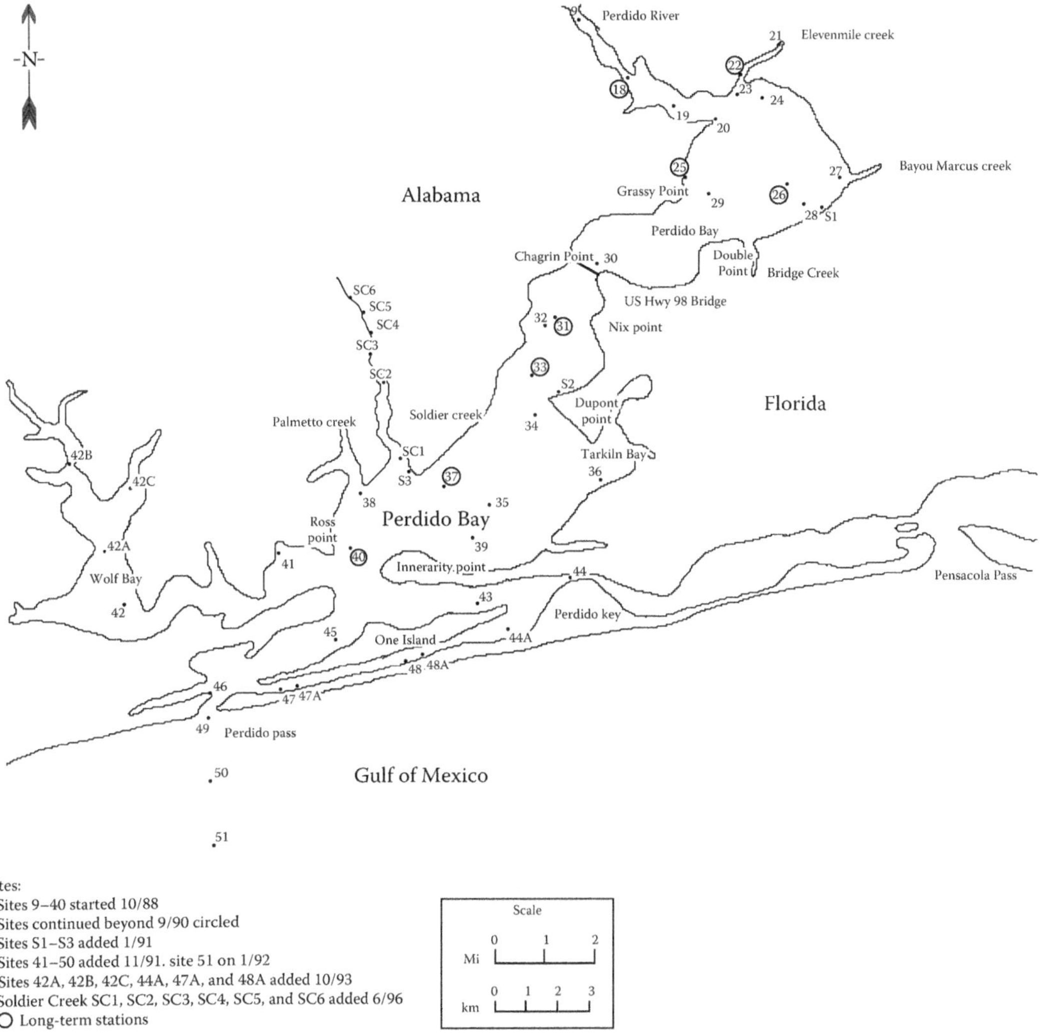

Figure 7.45 The Perdido Estuary showing long-term monitoring stations. The Florida Geographic Data Library (FGDL) provided geographic data for this figure.

source of freshwater to the estuary is the Perdido River that flows southward 96.5 km, draining an area of about 2937 km^2. The Elevenmile Creek is a small drainage basin that receives input from municipal and industrial point sources that include a kraft pulp mill and nonpoint runoff. The relatively small Bayou Marcus Creek and associated wetlands drain a residential area of western Pensacola with input from urban stormwater runoff. A nearby sewage treatment plant discharges through a marsh into the bay.

The Perdido Bay is a shallow (average depth, 2.2 m) inshore water body oriented along a NE–SW axis. The bay has four distinct geographic regions: the lower Perdido River, the upper Perdido Bay (north of the route 98 bridge), the lower Perdido Bay, and the Perdido Pass complex. As in the Choctawhatchee Estuary, there is a shelf that can extend up to 400 m in width around the periphery of the Perdido Estuary. This shelf usually does not exceed 1 m in depth. The upper bay is relatively shallow (average depth <2 m), while the lower parts of the estuary are somewhat deeper (average depth about 2–3 m). The deepest parts of the estuary are located at the mouth of the Perdido River and in the lower bay off the Ross Point. The Wolf Bay receives runoff from agricultural areas in

Alabama, and the lower bay receives stormwater runoff from municipal development. The Gulf Intracoastal Waterway (GIWW) runs through the lower end of the Perdido Bay about 5.6 km NE of the Perdido Pass. The US Army Corps of Engineers (USACOE) maintains the artificial Perdido Pass channel at about 4 m as part of the GIWW.

Prior to the opening of the Perdido Pass in the early 1900s, the Perdido Bay was a largely freshwater system, covered with freshwater plants (Brush, 1991). Access to the gulf was restricted by the shallow, shifting body of freshwater. At the time of an early survey (1767), the pass had a depth of ~2 m. During an outbreak of malaria in the early 1900s, locals with shovels opened the mouth of the bay to the gulf, making it the estuarine system it is today. This action resulted in the creation of the Ono Island/Old River complex. The history of the Perdido Bay (Livingston, 1992) is thus comparable to that of the Choctawhatchee Estuary (Livingston, 1986a,b) that was also opened at the mouth by another group of citizens with shovels in 1929.

In the early 1950s, the waters of the Elevenmile Creek turned black with concentrations of foam floating at the surface due to the discharge of pulp mill effluents. By 1968, more than 98,420 kL (26 million gallons) of effluent was flowing into the Elevenmile Creek–Perdido Estuary each day. The Florida Board of Health reported that the Elevenmile Creek was "grossly polluted" and that the Perdido Bay had been "greatly degraded within the 2.4-km (1.5-mile) radius of where Elevenmile Creek dumped into the bay." The report added that "during high temperatures and prevailing southeasterly winds, the chemical influence of Elevenmile Creek can be noted as far south as Lillian Highway Bridge at Highway 98."

In 1970, a "citizens' committee" was formed to fight pollution problems in Escambia and Santa Rosa Counties. This was the beginning of what was to be a series of citizens' groups that were to be organized to fight pollution of water bodies in the Perdido Basin. Following an inspection by federal authorities, and further protests from citizens in the area, a federal–state conference was held. In April 1970, the pulp mill was discharging 109,777 kL (29 million gal) per day and 33,384 kg (73,600 lb) per day of a 5-day biochemical oxygen demand. In 1971, the Alabama attorney general threatened to sue the paper company over its pollution of the Perdido Bay. In October 1972, the mill completed a new secondary effluent treatment system that resulted in marked improvements in the water quality of the Elevenmile Creek.

In 1984, a different company acquired the Pensacola Mill. Environmental issues concerning the mill in recent times have included DO, un-ionized ammonia, transparency (color), specific conductance, iron, zinc, and biological integrity of the receiving aquatic systems. Early scientific information concerning the Perdido Bay (Schropp et al., 1991) indicated that concentrations of DOC, Kjeldahl nitrogen, nitrate–nitrite, orthophosphate, particulate carbon, total nitrogen, total phosphorus, and total suspended solids were uniformly higher in the Elevenmile Creek than in the other tributary streams.

Since the turn of the century, there has been a steady increase in the human population in the Perdido Basin. By the 1980s, the Perdido Key was undergoing rapid residential and commercial development. Problems of sewage treatment, septic tank development, and urban stormwater runoff remain unresolved in various residential areas around the bay to the present day. Agricultural runoff, based largely in Alabama, contributes to nutrient loading in the Wolf Bay at the lower end of the basin. There was a long history of the impacts of the pulp mill that discharged into the Elevenmile Creek. A court order following a trial lodged against discharges by the pulp mill in Cantonment, Florida, led to the appointment of the author of this book to carry out a study of the Perdido system to determine the impacts of the pulp mill on the Perdido system. The results of the initial 3-year study were outlined by Livingston (1992). This study was followed by detailed, long-term analyses of the Perdido drainage system that were carried out from October 1991 to September 2007 (Livingston, 1992, 2000, 2002, 2005; Livingston et al., 1998a). The overall results of the 20-year study included intensive, bay-wide synoptic analyses at fixed stations (Figure 7.45) by a group of scientists and engineers (Appendix D).

During the initial 3-year study (1988–1991), the Elevenmile Creek was characterized by high conductivity (which eliminated various primary freshwater fishes and sensitive macroinvertebrates), high levels of free ammonia (which were associated by the U. S. Environmental Protection Agency [USEPA] criteria as prone to chronic toxicity), high concentrations of organic carbon (which was stimulatory to resistant chironomids and naidid oligochaetes that were found in high numbers in the upper creek), and periodically high water temperatures that could have had an additive effect to the toxicity factor resulting in the periodic occurrence of diseased fishes of the upper creek (Livingston, 1992).

The upper Elevenmile Creek was characterized by an almost complete lack of sensitive primary freshwater fishes (cyprinids, cyprinodonts, percids, and atherinids). On the other hand, secondary freshwater fishes (capable of withstanding high conductivity) such as the lepomis (*Lepomis macrochirus*, *Lepomis microlophus*, *Micropterus salmoides*) and various catfishes (*Ameiurus natalis*, *Ictalurus* sp.) were present in considerable numbers at the outfall station, along with estuarine types such as *Mugil cephalus* and fishes that were resistant to organic loading (*Dorosoma cepedianum*). These fishes were subject to periodic (summer) disease but were drawn by the rich source of macroinvertebrate food that was a direct response to the release of high levels of organic carbon. The trophic organization of the Elevenmile Creek was thus adversely altered as a direct result of the organic loading at the outfall site (Livingston, 2007a).

Pulp mill effluents represented a major source of nutrients to the bay, but there was considerable variation in the significance of this input depending primarily on the particular chemical species. The input of inorganic nitrogen in the form of ammonia together with inorganic phosphorus in the form of orthophosphate and the DOC and inorganic carbon (IC) components of carbon represented the most significant forms of input to the bay by mill discharges (Livingston, 1992). Chlorophyll *a* concentrations in the Perdido Bay showed distinct temporal and spatial variations over the 3-year period of the first part of the study. During most times and in most parts of the bay, chlorophyll *a* was relatively low in the Perdido Estuary compared to other reference bay systems such as the Apalachicola Estuary and Choctawhatchee Bay. However, the area of the Perdido Bay just off the Elevenmile Creek was affected by discharges of various nutrients that were associated with higher concentrations of chlorophyll *a* that were probably the result of phytoplankton blooms. These periodic blooms were probably the result of discharges of nutrients from the mill (Livingston, 2007a).

The results of a QUAL2E enhanced stream water-quality model (Brown and Barnwell, 1987) application to the upper Elevenmile Creek indicated that the pulp mill's wastewater discharge was the principal factor that caused low DO concentrations in the Elevenmile Creek, especially during warm periods. The principal factor producing low bottom-water DO levels (1–2 mg L^{-1}) in the Perdido Bay was salinity stratification. Sediment oxygen demand (SOD) in the bay was similar to that in other bays (without pulp mills) along the NE Gulf Coast. Modeling indicated that the wastewater discharge impact was calculated to decrease the bottom-water DO by less than 0.15 mg L^{-1}. In the surface layer, DO was usually relatively high (5–6 mg L^{-1}). The modeling data, combined with the descriptive field analyses and the field/laboratory experiments, indicated that hypoxia at depth in the Perdido Bay was the product of highly complex chemical interactions that were mediated in large part by the hydrographic characteristics of the estuary. The establishment of a strong halocline was a central feature of the development of hypoxia in the Perdido Bay (Livingston, 1992).

The results of incorporating a 2D transport developed from a Pritchard analysis and the calibration of the enhanced WASP3 water-quality model (Ambrose et al., 1986) indicated that local increases in phytoplankton productivity could be affected by nutrient loading from the mill, especially with regard to phosphorus releases during periods of low Perdido River flow (late summer–fall months). The WASP3 models confirmed that the lower bay was a major source for compounds that were identified as responsible for the consumption of DO at depth. Sediments in deeper parts of the Perdido Bay were characterized by liquid muds that were similar to those in other areas such as the Choctawhatchee Estuary. It was found that the entry of saline water from the gulf and the resulting stratification coupled with carbon, nitrogen, and phosphorus compounds in the estuary caused the observed hypoxic conditions at depth in the Perdido Bay. Periodic increases of phytoplankton production during increased nutrient loading indicated potential problems due to plankton blooms associated with mill discharges. This finding led to ongoing studies of the bay that were extended from October 1991 through fall 2007.

According to a US Fish and Wildlife Service Report (USFWS, 1990), SAV was largely concentrated in the lower bay. Historically, SAV has decreased by more than half from 1940–1941 to 1979. Dredging of GIWW in the 1930s and continuous pass enlargement and open-water spoil disposal of dredged sediments have been postulated as factors in the decline of SAV in the lower bay (Bortone, 1991). SAV development has been restricted to the Grassy Point on the west side of the upper bay with *V. americana* as the dominant species. Based on the success of previous grass bed transplant experiments, Davis et al. (1999) concluded that *V. americana* beds in the upper Perdido Bay were recruitment limited rather than constrained by water quality, toxic substances, light inhibition, or unsuitable substrate. Most natural growth of the *Vallisneria* beds did not extend deeper than 0.8–1.0 m in the bay (Livingston, 1992). However, following a major sewage spill with treatments using chlorine, the upper bay grass beds disappeared and did not recover by the end of the study.

Physical Structure of Perdido Estuary

The physical structure of the Perdido Estuary is closely associated with the effects of nutrient loading and water-quality distribution. According to Niedoroda (1989), there was a balance between river discharge and the intensity of stratification that varied over roughly annual periods. The circulation of the Perdido Bay was driven by a combination of the astronomical and meteorological tides as well as by the direct action of wind stress, river discharge, and density differences within the water mass of the bay. The area just south of the Lillian Bridge, where the incoming flood tide current meets the river current, had nearly zero flow (Niedoroda, 1989), which would explain the relatively high phytoplankton numbers taken here during periods of low nutrient loading. Under ordinary conditions, the slow river currents consistently flow down the bay throughout the tidal cycle. Only the speed of these currents changes. In the lower bay, the currents reverse between outgoing flows during ebb tide and incoming water during the rising tide.

The Perdido Estuary has a large freshwater input from the Perdido River compared to the volume of its tidal prism (Niedoroda, 1989, 1990, 1992, 1999). Freshwater tends to ride over a wedge of salt water along the bottom with minimal vertical mixing. Some mixing takes place so that the surface layer becomes progressively saltier as it slowly flows down the estuary. Since salt water is being entrained upward in spite of the stratification of the estuary, there is a continuous need for the salty bottom water to be replenished. This induces an up-estuarine flow in the bottom water of the stratified estuary. The general pattern is for the surface layer to slowly flow down the estuary, while the bottom layer slowly flows up the estuary. This two-layer flow typically represents a nearly steady component of the overall current structure in the estuary. The narrowness of the bay and its longitudinal (NE–SW) position restrict breakdown of the stratified water column due to wind mixing. Warm summer temperatures and the overall orientation of the bay with respect to the most frequent directions of strong winds contribute to the salinity stratification of the bay.

The changes in the hydrographic state of the Perdido Bay are driven by changes in river discharge (Niedoroda, 1989, 1990). Factors that affect this state include the distribution of water masses, mixing processes, development of strong density stratification, changes in flushing time, and differences in vertical diffusion of oxygen as a result of different turbulent mixing processes. The mouth of the Perdido River remains highly stratified except during high-flow events. Stratification in both the shallow northern and deeper southern portions of the upper bay is much more variable than the other parts of the estuary. Flushing times of the upper bay ranged from 0.5 to 15 days with whole upper bay lower layer flushing times ranging between 0.5 and 18 days. The upper layer of the middle bay has flushing times between 0.3 and 4.5 days. Corresponding values for the whole lower layer in the middle bay range between 0.3 and 6 days. Flushing times lengthened during prolonged drought periods presenting a factor in the increased bloom occurrence during such periods. Major discharge events upset the entire bay. The upper and often the middle bays are flushed out by freshwater inflows during such events. During the event itself, the flushing times are very short. The return of the bottom layer to the upper bay may take more than a month.

According to modeling studies (Hydroqual, 1992), the composition of the bottom layer water in the Perdido Bay is composed primarily by flows entering the bay at the Innerarity Point, ranging from nearly 100% at the mouth of the bay to 60% at the head of the bay. The remainder of the bottom layer water is mostly Perdido River water with 0.1%–2.0% pulp mill process water. Eighty percent of the water in the surface layer at the mouth of the bay is from the lower layer inflow at the Innerarity Point.

River Flows

Perdido River and Elevenmile Creek Flows (Figures 7.46 and 7.47) are characterized by considerable seasonal and interannual variabilities. Two-way ANOVA analyses, run by year and season using monthly averages within each season as replicates, showed significant ($p < 0.05$) differences between winter–early spring (high) and late summer/fall (low) flows of the Perdido River. There were two major droughts during the study period (1999–2002 and 2006–2007). This pattern followed that of the ACF rivers to a considerable degree. Major increased flow rates of the Perdido River occurred in 1990, 1998, 2003, and 2005. River flows during the prolonged drought periods were significantly ($p < 0.05$) different from those during the peak flows. The 1999–2002 dry period was the most severe such event in the north Florida region for the twentieth century (Livingston et al., 2000). The most recent drought tended to follow the lines of the previous low water period

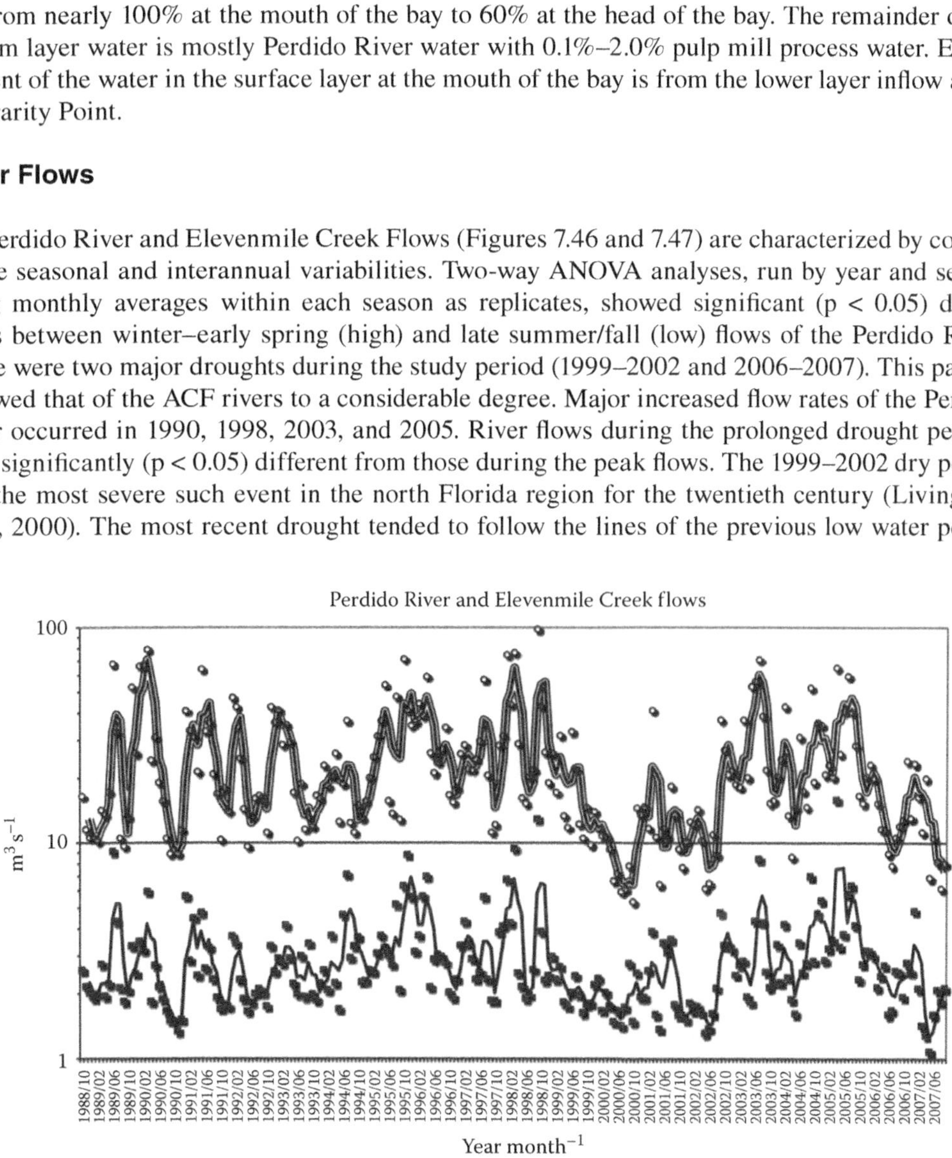

Figure 7.46 Monthly flows of the Perdido River and Elevenmile Creek from October 1988 through September 2007. Also shown are 3-month moving averages of the data.

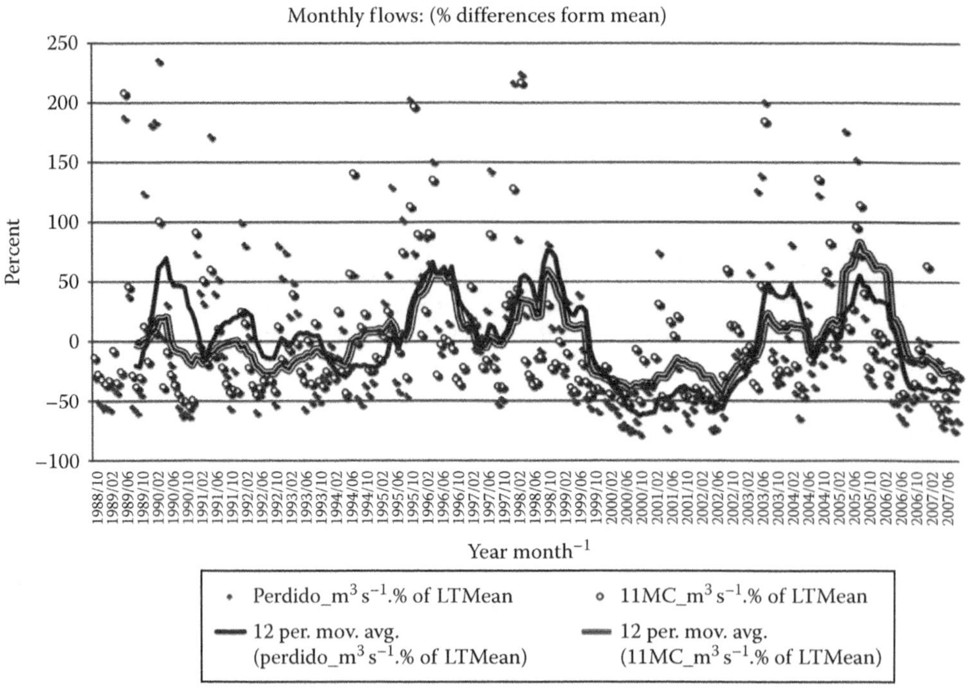

Figure 7.47 Monthly deviations of the total means of flows of the Perdido River and Elevenmile Creek from October 1988 to September 2007. Also shown are 12-month moving averages of the data.

in terms of severity. The long-term patterns of flow were similar in the Perdido River and the Elevenmile Creek. The creek had a relatively small flow rate compared to the river and tended to have extreme flash flooding during periods of heavy local rainfall.

Salinity Stratification and Dissolved Oxygen

Annual averages of surface and bottom salinity (Figure 7.48) and DO (Figure 7.49) indicate the associations between the stratification indices and bottom DO. With the exception of station P22, surface DO remained relatively high throughout the bay despite marked differences in salinity along the longitudinal gradient of the bay. The DO in the Elevenmile Creek was reduced by the breakdown of organic carbon compounds discharged from the kraft pulp mill mainly during warm months. The relatively shallow stations 23 and 25 were characterized by the least difference between surface and bottom salinities (i.e., low-salinity stratification). Bottom DO at these stations was relatively high relative to the bottom DO levels at stations P18 (high stratification due to depth and river flows) and P22 (high stratification and mill effluents). Low bottom-water DO was evident at deeper stations where salinity stratification was increased.

Salinity stratification increased along the longitudinal aspect of the bay, and bottom-water DO tended to follow these stratification trends with the possible exception of station P40 (the most seaward station) that showed an increased level of oxygenated bottom water relative to the other mid- to lower bay areas. This could be due to oxygenated water from the gulf that moved along the bottom toward the north.

A PCA–regression analysis of the bottom-water DO data was run against the primary (significant) variables (Figure 7.50). The data indicated that bottom DO was significantly and inversely related to depth, the stratification index (bottom salinity–surface salinity), and bottom salinities. There was also a negative relationship of the dependent variable with surface salinity and a positive association

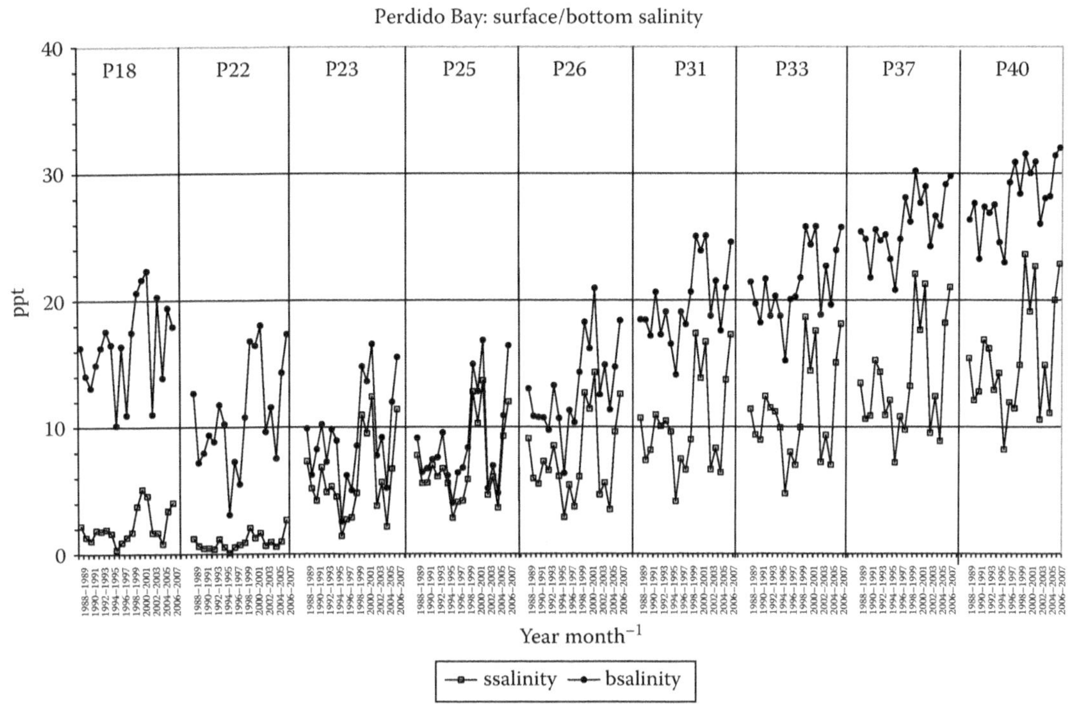

Figure 7.48 Station-specific annual averages of surface and bottom salinities (ppt) in the Perdido Bay from 1988 through 2007.

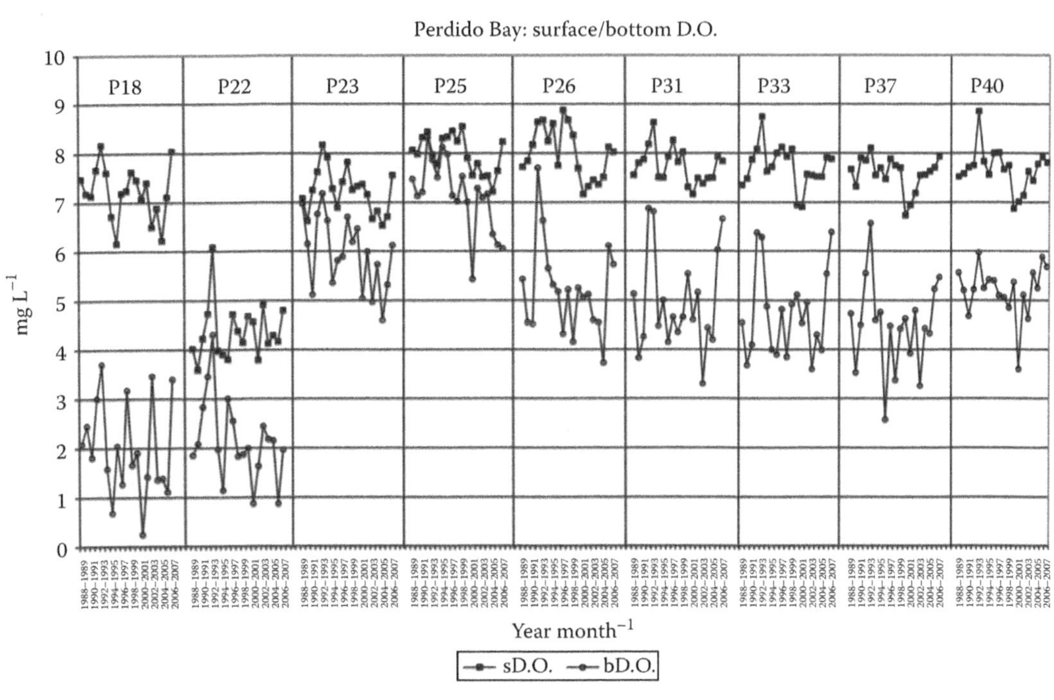

Figure 7.49 Station-specific annual averages of surface and bottom DO (mg L^{-1}) in the Perdido Bay from 1988 through 2007.

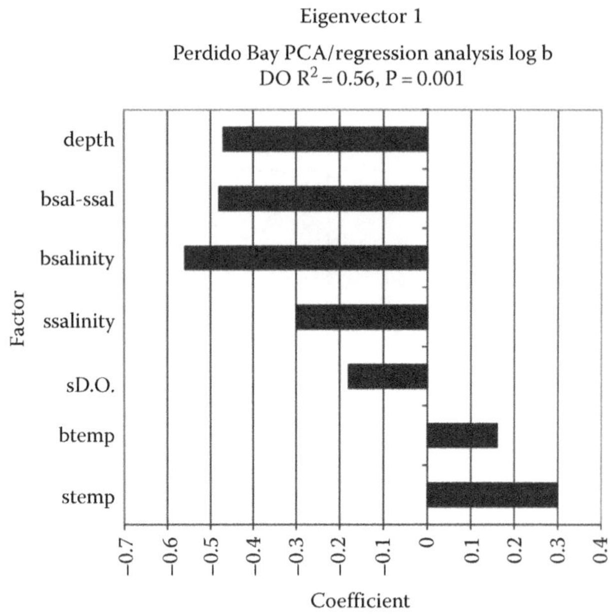

Figure 7.50 PCA/regression analysis of bottom DO (mg L^{-1}) in the Perdido Bay using annual means of regression variables.

with water temperature. Runs of bottom DO versus water-quality factors indicated significant negative relationships with various nutrient concentrations at depth although the R^2 value for these outputs was low. PCA–regressions run against phytoplankton indices were not statistically significant.

Reduced DO at depth tracked salinity stratification where the latter is driven by increased freshwater flow and a higher nutrient load. It can be hypothesized that lower DO will release more dissolved inorganic phosphorus (DIP) from sediments and $NO_2 + NO_3$ will be reduced to NH_3 and reach an equilibrium with overlying water.

When bottom-water DO was regressed against all variables at individual stations, the stratification index was significantly ($p < 0.05$) and negatively associated with this variable at stations 18, 22, 26, 31, 33, and 37, whereas bottom salinity was significantly ($p < 0.05$) and negatively associated with bottom-water DO only at the deepest stations (18, 22). Perdido River and Elevenmile Creek flows and ammonia loads were significantly ($p < 0.05$) and negatively associated with bottom DO concentrations at stations 31, 33, and 37. Water depth and phytoplankton indices at these stations were not significantly associated with bottom DO, and there were no significant ($p < 0.05$) associations of this variable with any of the independent variables at stations 23, 25, and 40.

Effects of Climate on Nutrient Loading

The trends of nutrient loading into the upper Perdido Bay (Figure 7.51a through c) indicated that the Elevenmile Creek comprised over 50% of the orthophosphate loads and over 72% of the NH_3 loads over the 20-year study period. Ammonia and orthophosphate loading in the Elevenmile Creek was due largely to pulp mill discharges. The Perdido River was responsible for most of the remaining loads of these nutrients. The Perdido River was the primary source of $NO_2 + NO_3$ providing around 63% of the total loading, whereas the Elevenmile Creek was only responsible for about 17% of the $NO_2 + NO_3$ delivered to the upper bay. The overland drainages (east and west stormwater runoff, Niedoroda, pers. comm.) and the Bayou Marcus Creek represented only a small part of the overall nutrient loading to the upper Perdido Bay.

Regressions of flow rates versus nutrient loading in the river and creek indicated significant ($p < 0.05$) associations of Perdido River flow for NH_3, PO_4, and $NO_2 + NO_3$. However, these nutrients were not significantly correlated with Elevenmile Creek flows, an indication of the almost continuous nutrient loading from the mill relative to interannual variations of Perdido River flows and nutrient loads. The NH_3 and PO_4 loading from the creek had considerably higher variances than

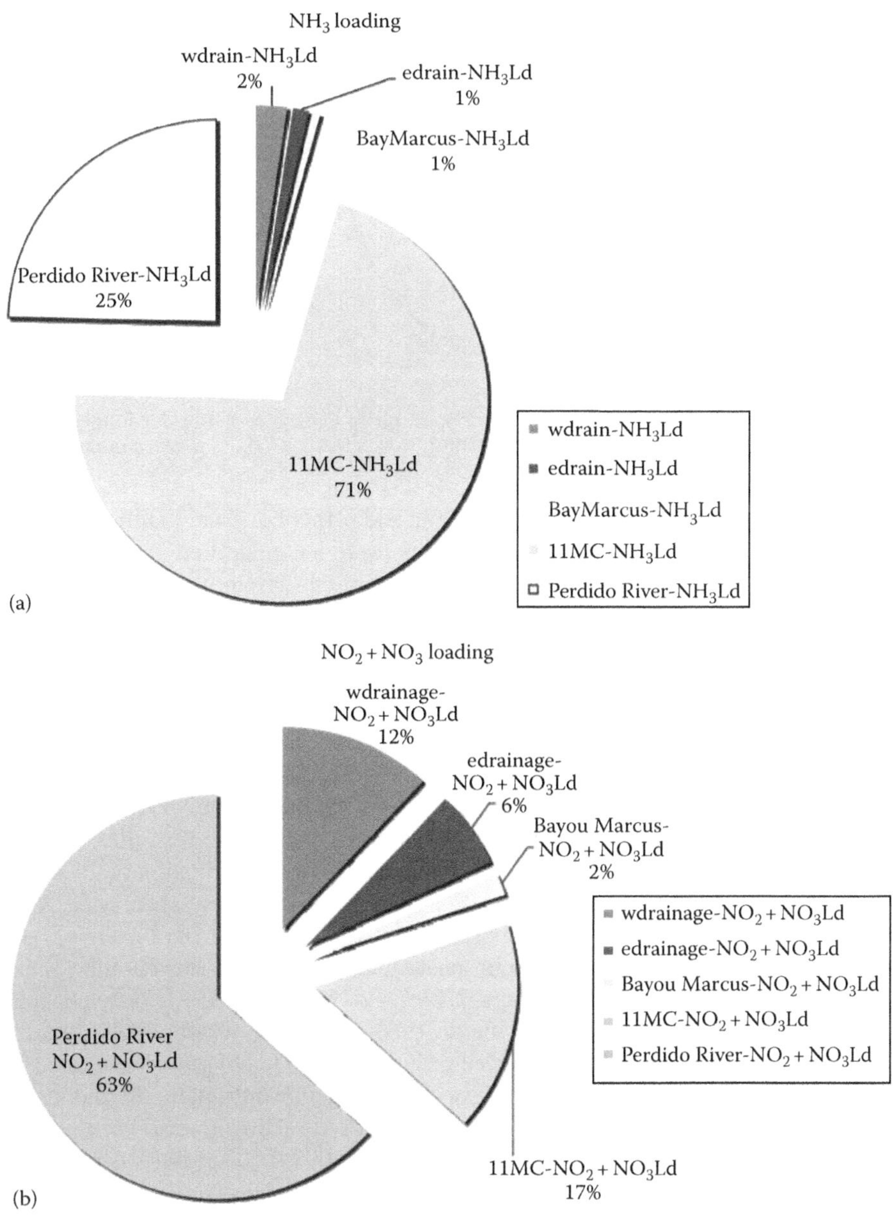

Figure 7.51 **(See color insert.)** (a) Ammonia loading to the upper Perdido Bay by system (Perdido River, Elevenmile Creek, Bayou Marcus Creek, western drainage area, eastern drainage area), 20-year means. (b) Nitrite + nitrate loading to the upper Perdido Bay by system (Perdido River, Elevenmile Creek, Bayou Marcus Creek, western drainage area, eastern drainage area), 20-year means. *(Continued)*

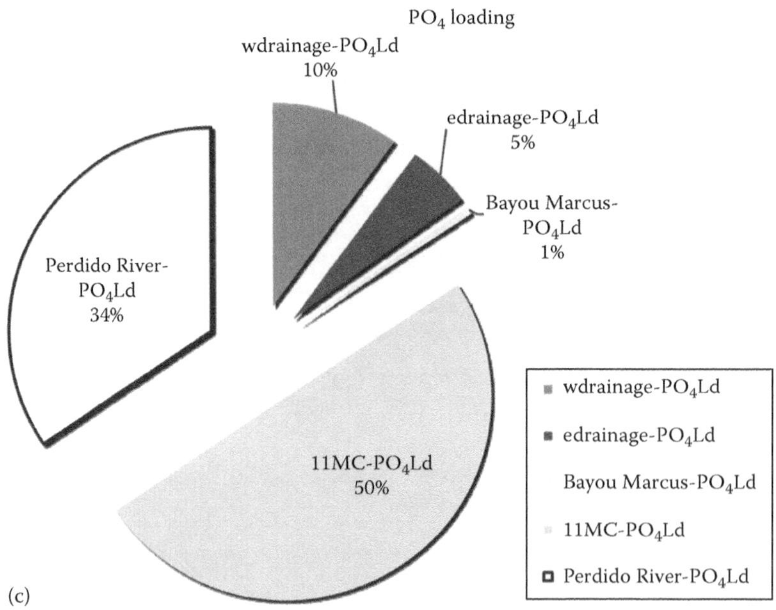

Figure 7.51 (*Continued*) **(See color insert.)** (c) Orthophosphate loading to the upper Perdido Bay by system (Perdido River, Elevenmile Creek, Bayou Marcus Creek, western drainage area, eastern drainage area), 20-year means.

those of the Perdido River. Single events of ammonia and orthophosphate loading from the sewage treatment plant near the Bayou Marcus Creek drainage were not quantified.

Monthly loading rates of ammonia in the Perdido River and Elevenmile Creek are represented in Figure 7.52a. During the first 3 years of sampling, Elevenmile Creek ammonia loading was comparable to that of the Perdido River system. There were periodic increases of ammonia loading in the creek starting in 1992–1993 with stepwise increases during warm months that peaked in 1998–1999. In 1995–1999, ammonia loading from the creek was significantly ($p < 0.05$) higher than that of the first 3 years. From 1999 to 2004, there were overall decreases of creek ammonia loading with periodic summer peaks. However, from 2004 to 2005, there was an increase of ammonia loading from the Elevenmile Creek (i.e., the mill) that was similar to that during the peak loading of 1997–1998. This was followed by reductions of such loadings during fall 2005–2007. There were general similarities of ammonia loading in the Perdido River with Perdido River flow trends. During the drought of 2006–2007, there were significantly ($p < 0.05$) lower ammonia loadings in the Perdido River, whereas the reductions of such loadings in the Elevenmile Creek were not significantly ($p < 0.05$) different from the base figures.

After initial peaks in 1988–1989, orthophosphate loading from the Elevenmile Creek was relatively low during the first few years (Figure 7.52b). From 1994 to 1997, orthophosphate loading was significantly ($p < 0.05$) higher than that in 1989–1991 (Livingston, 2000). Treatment with alum by the mill from fall 1997 to 1998 reduced orthophosphate loading to levels noted during the early sampling period. Orthophosphate loading decreased significantly in the Perdido River during the drought of 1999–2002. These decreases were pronounced relative to those in the creek. There were increased levels of creek orthophosphate loading in 2002–2003. After further reductions in 2004, orthophosphate loading from the Elevenmile Creek was again increased during spring–fall 2005 with peaks in May–June 2005 (Figure 7.52b). Such loading was also relatively high from the Perdido River system in April 2005. Orthophosphate loading was relatively low in both systems during the succeeding drought of late 2005–2007. These data indicated that orthophosphate loading tended to follow river flows in both systems although such loading in the Perdido River was more affected by the two major droughts relative to the creek orthophosphate loads.

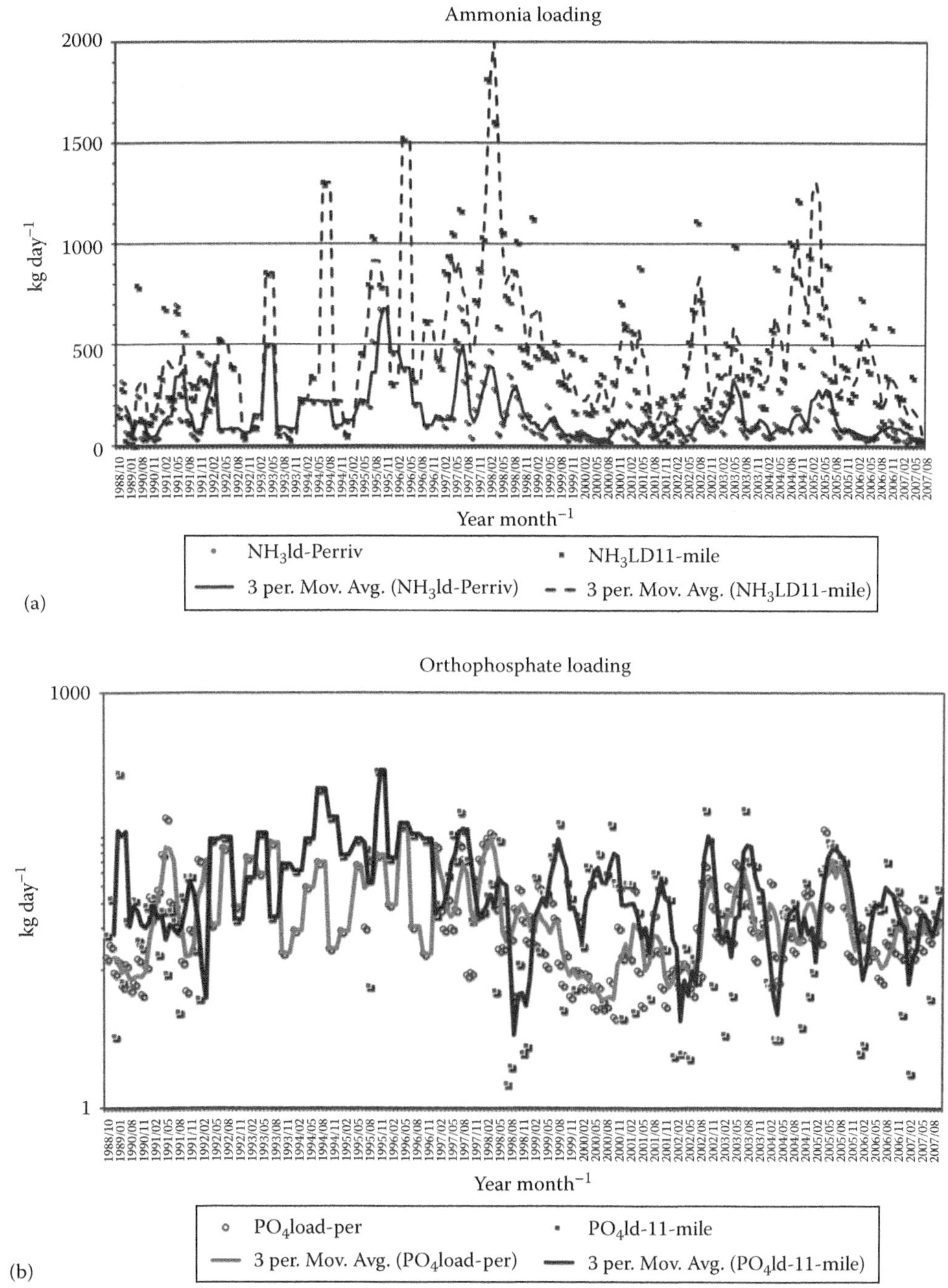

Figure 7.52 (a) Monthly averages of ammonia loading to the upper Perdido bay by the Perdido River and Elevenmile Creek from October 1988 to September 2007. (b) Monthly averages of orthophosphate loading to the upper Perdido bay by the Perdido River and Elevenmile Creek from October 1988 through September 2007. *(Continued)*

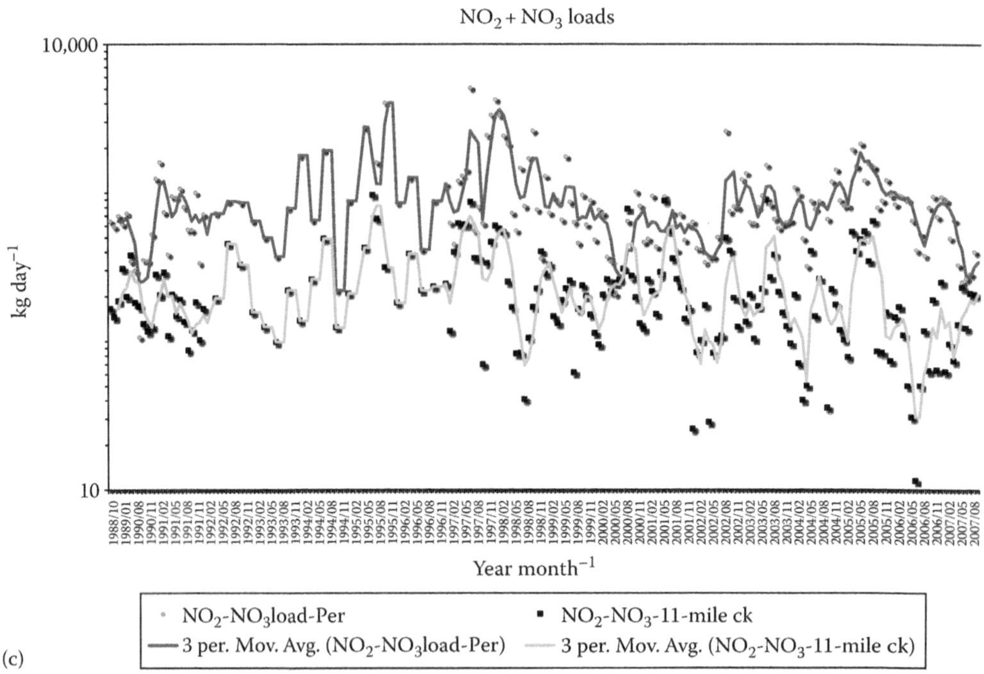

Figure 7.52 (*Continued*) (c) Monthly averages of nitrite + nitrate loading to the upper Perdido bay by the Perdido River and Elevenmile Creek from October 1988 through September 2007.

Nitrite–nitrate loading rates (Figure 7.52c) indicated that the Perdido River system was the primary source of this form of nitrogen to the Perdido Estuary. Once again, the extreme drought periods were typified by excessive reductions of such loading in the Perdido River relative to those in the Elevenmile Creek. Despite such reductions, the river loadings exceeded those of the creek throughout the 20-year study period.

A restoration program of proposed reductions of orthophosphate and ammonia loading by the pulp mill to the Elevenmile Creek was undertaken in 1999. For the first time in Florida, the Florida Department of Environmental Protection gave a permit to the paper mill that used the proposed orthophosphate and ammonia loading rates based on the long-term research in the Perdido Estuary. These loadings by the pulp mill were targeted for reductions that approximated observed nutrient loads during periods when the bay was free of blooms during the first 3 years of the study. A review of ammonia loading to the upper Perdido Bay and ammonia concentrations in the Elevenmile Creek showed that the level of ammonia loading by the mill should also insure that ammonia concentrations at the mouth of the Elevenmile Creek would remain below concentrations that could be toxic to phytoplankton species (Livingston, 2007a,b, unpublished data).

The problem of high NH_3 loading from the mill during periods of peak river flows remained after restoration efforts, whereas PO_4 loading was generally reduced by pulp mill actions. Overall trends of rainfall and drought conditions defined the loading of nutrients from the Perdido River with major reductions of these nutrients during the droughts that occurred in the second half of the study.

Nutrient Limitation Experiments

Flemer et al. (1998) reported on a series of nutrient limitation experiments carried out in the Perdido Bay. Phosphorus limited phytoplankton growth primarily during the cool season and nitrogen during the warm season. Frequency and magnitude of primary P limitation decreased with increasing salinity. Growth response to N enrichment usually was relatively weak compared to that of P and N + P. The N + P enrichment provided a joint or apparent "colimitation" with neither N nor P individually providing significant ($p < 0.05$) individual stimulatory responses on 5, 4, and 8 occasions, respectively, in the upper, middle, and lower bays. The N + P response of increased production occurred primarily during the warm months (June through October) in the upper bay. Of the four occurrences of N + P limitation at the middle bay station, only a single occurrence was detected during the cool period (February). At the lower bay station, N + P limitation occurred approximately in equal numbers during warm and cool seasons.

In the upper bay, dissolved inorganic nitrogen (DIN)–dissolved inorganic phosphorous (DIP) ratios decreased from a July maximum of 290 to a seasonal minimum of 20 by September, indicating generally potential P sufficiency. The lowest DIN–DIP ratios (e.g., 15–30) occurred in the mid- and lower bay from July through November, but these values still were within a N and P level of sufficiency. Ratios decreased below the threshold value of 10 only in October and November in the mid- and lower bay. From these observations, DIN potentially limited growth rates on an infrequent basis. Field PON–Chl g (wt/wt) ratios for N-saturated coastal phytoplankton crops usually ranged between 7 and 10 (Strickland, 1965). Ratios in the bay frequently approximated 3.0–7.0 suggesting some N limitation, especially in the upper and mid-bay regions. In nine of eleven available comparisons, PON–Chl *a* ratios were highest in the lower bay that indicated few occurrences of primary N limitation (i.e., an enhanced response with P addition; see Fisher et al. 1992 for full definition of terms), a response consistent with DIN concentrations in this region compared to mid-bay.

Primary P limitation approximated the river-dominated estuarine pattern by occurring more often during the cool season at lower salinities. This response pattern was somewhat out of phase with the low summer and late winter/spring DIP concentrations and declining DIN–DIP ratios detected during the late summer through fall at the upper bay station. These differences remained unexplained. Two cases of primary N limitation of biomass yield in the lower bay were less than expected based on the literature for river-dominated estuaries. The evidence suggests a source of DIN in the bay seaward of the lower bay station. For example, DIN concentrations often were higher in the lower bay than mid-bay, or concentrations frequently approximated each other, whereas salinities usually differed substantially between these stations. Macauley et al. (1995), working in the bay from January through December 1990, reported a similar pattern of DIN concentrations and salinity.

For algae, the role of each element is quite unique that results in stoichiometric requirements for growth for all phytoplankton species (Redfield et al., 1963; Hecky and Kilham, 1988). This relationship suggests that contemporaneous multiple nutrient limitation is not expected for unialgal cultures (Droop, 1974; Rhee, 1978) with a possible exception of some dinoflagellates (Sakshaug and Olsen, 1986). Abundance of dinoflagellates would not explain colimitation in our study as they were present but never numerically dominant. Nutrient colimitation (e.g., Fuhs et al., 1972; Gerhart and Likens, 1975; Norin, 1977; Caraco et al., 1987; Paasche and Erga, 1988; Suttle and Harrison, 1988; Pennock and Sharp, 1994; Louisiana estuaries, E. Turner, pers. comm.) or incipient colimitation (Powers et al., 1972; Graneli, 1978; Kivi et al., 1993) of phytoplankton growth rates or biomass yield has been reported from a variety of aquatic environments from oligotrophic to nutrient-enriched freshwater and marine systems.

Climatological Control of Plankton Blooms

Riegman (1998) related the occurrence of bloom species to macronutrient dynamics. He noted that diatom growth during blooms due to anthropogenic N + P loading can be inhibited by silica availability. This could be related to enhanced N–Si and P–Si supply ratios that favor nonsiliceous phytoplankton (Smayda, 1990). Shifts in N–P supply ratios can also cause changes in species composition. Nutrient enrichment with macronutrients can also be indirectly related to changes in the phytoplankton community composition through induced changes in predation, resource limitation, light requirements, and biological alterations of the sediments. Altered nutrient speciation can also influence these changes.

Most of the top 10 numerically dominant plankton species in the whole-water collections taken over the 20-year study period were considered to be bloom types (Table 7.6; Appendix C). The top numerical dominant, *Merismopedia tenuissima*, was generally restricted to the Elevenmile Creek and the immediate receiving area of the bay (station 23). The most toxic species, *H. akashiwo*, was the biomass dominant and was most prevalent at stations 23 and 26 in the direct drainage emanating from the Elevenmile Creek. *Cyclotella choctawhatcheeana* and *Chaetoceros throndsenii* were distributed throughout the bay, and *P. cordatum* was most abundant at stations P25 and P26. Dominant plankton populations were mainly located in the upper bay with the entry points of the primary nutrient loading of the upper bay. The key to understanding how nutrient loading was responsible for the stimulation of the blooms resided in both the spatial distribution of estuarine habitats and the temporal aspects of the occurrence of droughts and floods.

The various bloom species documented in the Perdido Bay system represent a taxonomically diverse group dominated by diatoms, raphidophytes, dinoflagellates, and blue-green algae (Table 7.6; Livingston, 2002). The numerically dominant Perdido Bay phytoplankton bloom species had highly variable structural components. The chain-forming diatom *L. danicus* has relatively large cells (29.0–47.3 μm in length; biomass of 404 pg Ash-free dry weight (AFDW) cell^{-1}). This species has numerous, well-developed chloroplasts. The diatom *C. choctawhatcheeana*, on the other hand, has relatively small cells (3.0–5.0 μm in length; 40 pg AFDW cell^{-1}) with few chloroplasts and little chlorophyll. The raphidophyte *H. akashiwo* has numerous chloroplasts. This species is relatively large (length, 8.8–12.5 μm; 207 pg AFDW cell^{-1}). The diatom *C. throndsenii* has very small chloroplasts and not much chlorophyll. This species is also very small (length, 5.4 μm; 19 pg AFDW cell^{-1}). The diatom *Synedropsis karsteteri* has only two chloroplasts, whereas *C. choctawhatcheeana* has more but not as many as *H. akashiwo* or *L. danicus*. *S. karsteteri* is somewhat larger (length, 20.0–23.1 μm; 26 pg AFDW cell^{-1}) than *C. choctawhatcheeana*. The dinoflagellate *P. cordatum* is photosynthetic with relatively small chloroplasts. This species is relatively large (length, 17.5–22.0 μm; 518 pg AFDW cell^{-1}). The raphidophyte *Chattonella subsalsa* is also relatively large (length, 23–43 μm; 1933 pg AFDW cell^{-1}).

Table 7.6 Average Numbers of Phytoplankton Cells of Numerically Dominant Species by Station Calculated over the 20-Year Study Period

Species	18	22	23	25	26	31	37	40
Merismopedia tenuissima	4,257	24,172,440	3,044,624	121,867	475,323	51,187	8,736	8,571
Cyclotella choctawhatcheeana	85,177	33,400	411,718	628,772	846,312	629,353	568,064	469,368
Chaetoceros throndsenii	13,765	633	77,853	186,093	251,806	164,456	613,180	198,879
Heterosigma akashiwo	143,936	15,805	415,519	169,731	354,966	101,609	60,807	25,421
Skeletonema costatum	7,641	784	6,023	11,594	10,522	26,234	50,196	58,750
Cyclotella cf. *atomus*	287	333	320	259	281	449	886	845
Prorocentrum cordatum	3,003	344	24,185	30,273	39,647	22,804	8,750	8,377
Urosolenia eriensis	35,305	2,519	24,925	97,908	93,441	74,089	44,169	26,197
Synedropsis karsteteri	1,206	367	78,795	41,438	30,867	18,286	195,321	3,492
Chaetoceros sp.	5,210	372	30,675	37,237	73,101	42,964	41,221	37,199

The colony-forming blue-green alga, *M. tenuissima*, is common in stagnant eutrophic freshwaters, mainly fertilized fish ponds. This species also occurs in brackish waters with some capacity to tolerate low salinities, particularly during the warm season of the year (Komarek and Anagnostidis, 1998). However, the small size of the cells (1–2 μm) makes it difficult to determine viability. This blue-green alga was reported primarily in the freshwater entry points to the upper Perdido Bay in the Elevenmile Creek. The fact that there is considerable variability among the various bloom species with respect to size, biomass, and chloroplast development could be related to the lack of a statistical relationship between the chlorophyll observations and the incidence and intensity of the blooms in the Perdido Bay.

The diatom *L. danicus* has a worldwide distribution in a variety of tropical coastal environments (Hargraves, 1990). It is often a major constituent of spring phytoplankton outbursts as well as fall growing periods (Marshall, 1988). This species has been associated with red tide events in Japan (Fukuyo et al., 1990). According to Fryxell and Villac (1999), *L. danicus* is believed to be harmful to caged sea trout (*Cynoscion regalis*), Atlantic salmon (*Salmo salar*), and smolt of coho salmon (*Oncorhynchus kisutch*) (Clement and Lembeye, 1993). Another diatom *C. throndsenii* by Marino, Montresor, and Zingone (Marino et al., 1987) has been associated with blooms stimulated by effluents from sewage and industrial outfalls (Marino et al., 1987). *C. throndsenii* was shown to be congeneric with *Chaetoceros ehrenberg* on the basis of fine structure of vegetative cell and resting spore morphology, and the resulting combination was *C. throndsenii* by Marino, Montresor, and Zingone (Marino et al., 1992).

The diatom *C. choctawhatcheeana* responds to a combination of environmental factors that range along inshore–offshore gradients of light and nutrients with salinity as a possible modifying factor (Livingston and Prasad, unpublished data). This species was first described from samples taken in the Choctawhatchee Bay system (Prasad et al., 1990) where short chains of *C. choctawhatcheeana* were found during each month of the year. Small blooms were observed at some stations in November and December with one such bloom appearing in April and persisting until June. Salinity ranges for *C. choctawhatcheeana* in the Choctawhatchee Bay were >20‰ at temperatures >25°C. Numbers decreased sharply as temperatures climbed to 30°C. Optimum salinities ranged from 15‰ to 20‰. This distribution indicated that *C. choctawhatcheeana* can survive and grow under varying salinity conditions. Carvalho et al. (1995) explained the noted differences in salinity preferences of this diatom as a product of possible competition that may have forced *C. choctawhatcheeana* into lower salinities. Cooper (1995a,b), working in the Chesapeake Bay, found that *C. choctawhatcheeana* had increased abundances that coincided with increased sedimentation, turbidity, and eutrophication of the estuary. Overall, this species has a high tolerance for wide ranges of salinity and is stimulated by increased nutrient loading. Cooper (1995a) suggested that *C. choctawhatcheeana* may be a good indicator of estuarine habitats that are characterized by various anthropogenic effects.

There are relatively high nutrient requirements for *H. akashiwo*. Bactivorous feeding is stimulated by P limitation with Fe and Mn needed for bloom continuance. Nutrient enrichment supports blooms of this species with dependence on river flows in various regions. Temperature, nutrients, and competition appear to be limiting for bloom generation in *H. akashiwo*. Smayda (1997c) indicated multifactorial control of blooms that, once released, have a "remarkably high degree of broad spectrum allelochemical allelopathic antagonisms" that are used against competing species and potential grazers.

The fish-killing mechanism of raphidophyte blooms is still poorly understood. Both physical clogging of gills by mucus excretion and gill damage by hemolytic substances such as polyunsaturated fatty acids may be involved (Shimada et al., 1983; Chang et al., 1990). There is accumulating evidence that the production of superoxide radicals represents the primary mechanism of fish mortality. Imai et al. (1997) reported details concerning life cycle and bloom dynamics of *Chattonella*, a genus with two known fish-killing species (*C. antiqua* and *C. marina*.). Nutrients and competitors (mainly diatoms) appear to affect the development of *Chattonella* populations. Anything interfering with diatom proliferation could give *Chattonella* an advantage. During a severe phytoplankton outbreak in the Seta Sea in 1972, a raphidophyte red tide killed 14 million cultured yellow tail fish. Effluent controls were then initiated to reduce the organic carbon loading and the discharge of phosphates from household detergents. Following a time lag of 4 years, the frequency of red tide events in the Seta Sea then decreased by about twofold to a more stable level (Hallegraeff, 1995). A similar pattern of long-term loading of coastal waters was evident for the North Sea in Europe (Smayda, 1990).

The dinoflagellate *P. cordatum* is a bloom species that is mostly estuarine and cosmopolitan in cold temperate to tropical waters. It is a toxic species associated with postulated shellfish poisoning and fish kills. Nakazima (1965) indicated poisonous effects on shellfish feeding on *Prorocentrum* sp. The toxic substance venerupin has been associated with *P. cordatum* although there is some question concerning this association. Hansen (1997) reported that *P. cordatum* can ingest certain cryptophytes (*Cryptomonas* sp.) and ciliates. Mixotrophy (being photosynthetic and phagotrophic) is widespread among dinoflagellates with a variety of feeding mechanisms found in this group. The mechanisms of prey selectivity are not well understood, but such selectivity is not by size alone.

The control of plankton organisms by phagotrophy is considered to be highly variable (Graneli and Carlsson, 1997). However, losses of plankton to pigmented flagellates must be considered according to these authors. Sellner et al. (1995) indicated that *P. cordatum* did not adversely affect oysters (*C. virginica*) in the Chesapeake Bay that effectively reduced these bloom-forming dinoflagellates. Lassus and Berthome (1988) reported that *P. cordatum* caused mortalities in old oysters. Woelke (1961) found that this species caused oyster (*Ostrea lurida* [new species, *Ostrea conchaphila*]) mortalities and cessation of feeding at high densities. Wikfors and Smolowitz (1993) found that increased abundance of *P. cordatum* could cause shell losses. This observation was based on feeding experiments where toxins from this species caused slow growth and mortality in shellfish. Okaichi and Imatomi (1979) reported that in 1942, shellfish poisoning by *P. cordatum* killed 114 out of 324 human victims of poisoned shellfish. The authors isolated the toxic substances believed to be operational in those deaths. Cardwell et al. (1979) found that *Gymnodinium splendens*, another dinoflagellate, was acutely toxic to oyster larval stages.

The details of plankton distribution in the Perdido Bay in space and time (the early years) have been given by Livingston (2002). An analysis was made of the net phytoplankton taken during the years of relatively low nutrient loading (1988–1991) and the period characterized by the initiation of consistently high-orthophosphate loading (1992–1994). In 1993–1994, there was a series of phytoplankton blooms dominated by diatom species. When compared to the phytoplankton taken in 1988–1989, there were increased phytoplankton numbers and corresponding reductions of phytoplankton species richness at all stations except in seagrass areas of the upper Perdido Bay. Numerical abundance of the net phytoplankton was significantly higher in 1993–1994 during all months. Species richness was significantly lower in 1994 in 7 out of the 9 months analyzed. There was an orderly progression of the relative abundance of various bloom species. During fall 1993, *Falcula hyalina* was dominant. That species was followed by *L. danicus* blooms in January–February 1994. In March 1994, *M. throndsenii* was dominant. There were extensive blooms of *C. choctawhatcheeana* in April 1994.

Analysis of the zooplankton indicated that, with the exception of February 1994, increased numbers and reduced species richness of the phytoplankton during the sustained increases of orthophosphate loading by the pulp mill were accompanied by increased zooplankton abundance. Peak zooplankton numbers at stations P23, P26, and P31 were associated with increased numbers of *C. choctawhatcheeana* and *M. throndsenii* during spring 1994. The zooplankton dominant, *Acartia tonsa*, tended to have comparable numbers between 1988–1989 and 1993–1994. Zooplankton species richness was higher during fall 1993 than at any other time. Although zooplankton species richness tended to go down in February and March 1994, there was no general reduction of this index during the 1993–1994 bloom period.

Increased zooplankton numbers were associated with high numbers of phytoplankton from January through March 1994. Significant increases of zooplankton numbers occurred in April and May 1994 in the Perdido Bay. Zooplankton species richness was significantly higher from September 1993 through January 1994 than that observed during the nonbloom period (1988–1989); however, relatively lower numbers of zooplankton species were noted during the succeeding bloom months that could have been associated with the high dominance of the leading species *A. tonsa*. The inverse relationship of zooplankton numbers and zooplankton species richness in terms of significant differences between the two study periods could be related to possible interspecific competition. Thus, the diatom blooms of 1993–1994 appeared to have little adverse effect on the zooplankton that, if anything, increased in numbers during the bloom period.

Overall, the plankton blooms of 1994 were significantly and negatively associated with phytoplankton species richness, infaunal numbers and species richness, and herbivorous bivalves. Zooplankton numbers, on the other hand, appeared to be positively associated with the blooms that were dominated by various diatom species. The data also suggest that during the initial phytoplankton blooms in the Perdido Bay, the seagrass habitat in the western sections of the upper bay was relatively unaffected relative to the unvegetated sedimentary environment. The resumption of increased Perdido River flow during early summer 1994 was associated with a cessation of phytoplankton bloom activity.

The trends of numbers and biomass over the study period were strongly affected by individual bloom species. Some species such as *P. cordatum*, *C. choctawhatcheeana*, *L. danicus*, and *M. throndsenii* were present in the bay before and after nutrient loading was increased in 1993–1994. Each of these species had specific seasons, or even months, when they were most likely to form blooms. The diatom *C. choctawhatcheeana* was a spring species with most blooms occurring in April. High concentrations of *P. cordatum* followed the presence of this species from the beginning of the study. This species usually increased during high-phosphorus-loading periods of the winters of 1993–1994, 1996–1997, and 1998–1999. The lack of blooms by this species during the low-orthophosphate loading period (1997–1998) further supports the hypothesis that high-orthophosphate loading during the winter stimulates *P. cordatum* blooms. Unlike the smaller-celled diatoms, these blooms accounted for relatively high percentages of the total phytoplankton biomass. Relatively high numbers of *P. cordatum* remained in the upper bay during the high-orthophosphate loading of spring 1999.

Phytoplankton blooms appeared to be affected by both the type and amount of the nutrients loaded to the bay as well as the seasonal period of such loading. There also appeared to be a sensitization of the bloom response (i.e., *P. cordatum*, *C. choctawhatcheeana*) over the period of exposure whereby these species required less nutrient stimulus to form blooms with time. The general increase of both phytoplankton numbers and biomass with time could reflect a generalized sensitization of certain bloom species after prolonged periods of high nutrient loading to the bay. The fresh- to brackish-water cyanobacterium, *M. tenuissima*, first appeared in the Perdido Estuary in 1996 and was usually found in high numbers during late summer–fall months and winter periods. It was during the 1996 sampling that extremely high concentrations of the raphidophyte *H. akashiwo* appeared in the Perdido Bay. This coincided with increased ammonia loading to the bay by the pulp mill so that the N–P percent differences were very high.

The blue-green alga *M. tenuissima* (Figure 7.53) was first noted in the Elevenmile Creek (station 22) in relatively low numbers after being absent from the bay during the early years of the study. The presence of *Merismopedia* appeared to be associated with the high levels of nutrient loading from the pulp mill. Although this species was found in bloom numbers in the Elevenmile Creek, it was not found in any appreciable numbers in the bay.

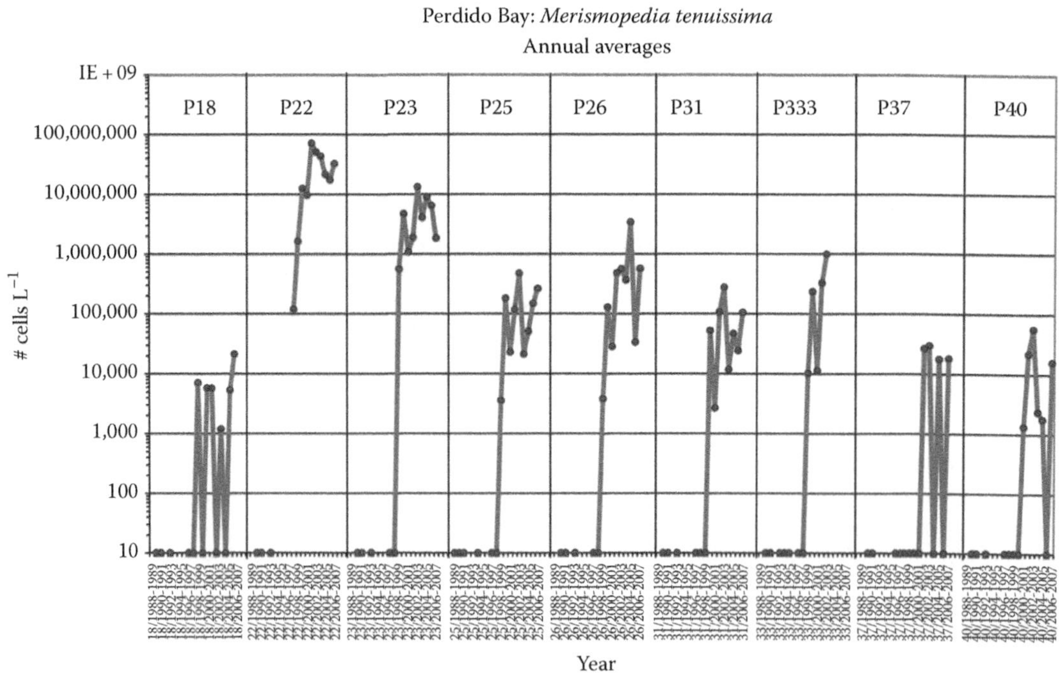

Figure 7.53 Annual averages of numbers of cells L^{-1} of *M. tenuissima* in the Perdido Bay from October 1988 through September 2007.

The raphidophyte *H. akashiwo* was first noted during the 1993–1994 bloom period (Figure 7.54). This species was present mainly during spring–summer periods. It was dominant at stations P23 and P26. During the period of peak nutrient loading to the Elevenmile Creek (1996–1998), there were major changes in the phytoplankton assemblages in the bay. The newly discovered raphidophyte *H. akashiwo* was now the dominant in the upper bay and was present in all parts of the bay. The *Heterosigma* cell numbers were generally lower in the middle and lower parts of the bay, but such numbers tended to increase during the latter years of the study. The diatom *C. choctawhatcheeana* was found to be a dominant in upper and middle parts of the bay. This species appeared to have an inverse relationship with *Heterosigma* cell numbers and tended to have relatively high interannual variability.

By the period of reduced nutrient loading (1999–2002), *M. tenuissima* was now the top dominant in the upper bay, and the total numbers of phytoplankton had increased due largely to expansion of the cryptophytes and nanoflagellates. In mid- and lower bay areas, *C. choctawhatcheeana* was still dominant as *H. akashiwo* was still present in all parts of the bay in reduced numbers. During the period of sewage nutrient loading into the bay (station P26) from the Bayou Marcus Creek area in 2002–2003, *H. akashiwo* was present in the most dense blooms in this part of the upper bay. At the same time, *M. tenuissima* was now a major dominant in the Elevenmile Creek drainage. There were reduced numbers of *H. akashiwo* in the Elevenmile drainage that accompanied the reduced nutrient loading in this area. This raphidophyte was also dominant in mid-bay areas and appeared to respond to nutrient loading trends in the upper bay from the sewage treatment plant. Cryptophytes in mid- and lower bay areas had increased to extremely high numbers, and the overall numbers of phytoplankton in the bay increased significantly due to the earlier noted changes in species representation bay-wide.

During the next high-nutrient-loading period from the pulp mill (2004–2005), *H. akashiwo* was dominant at stations P23 and P26, with greatly reduced numbers of *C. choctawhatcheeana* in these areas. Total numbers of phytoplankton were high throughout the bay. During the succeeding drought and low nutrient loading of 2006–2007, *M. tenuissima* was still dominant. There were now relatively

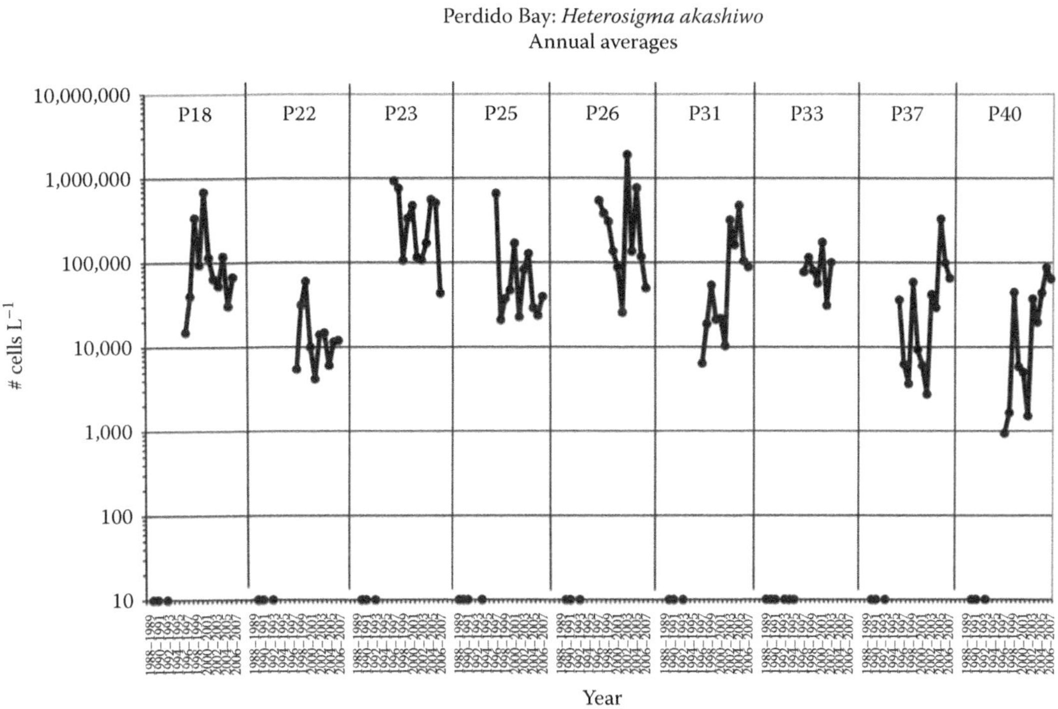

Figure 7.54 Annual averages of numbers of cells L^{-1} of *H. akashiwo* in the Perdido Bay from October 1988 through September 2007.

low numbers of *H. akashiwo* in the bay during this period. The diatom *C. choctawhatcheeana* was again dominant in mid- and lower bay areas.

Both nitrogen and phosphorus loadings from the Elevenmile Creek were significantly ($p < 0.05$) associated with *H. akashiwo* cell numbers although such numbers were not proportionate to the level of loading of these nutrients. There were increased cell numbers with time relative to the loading levels, showing a possible adaptive response of this species to nutrient loading. The number of *Heterosigma* blooms in the upper bay tended to be inversely related to the number of *Cyclotella* blooms, and, in the upper bay (stations 23 and 26), these species were inversely associated ($R^2 = -0.36$, $p < 0.05$). Annual averages of *Cyclotella* reached peaks with the downward trend of *Heterosigma* during the period of high but diminishing nutrient loading from the Elevenmile Creek.

The combination of high nutrient loading and bay nutrient concentrations, drought/flood trends, and biological interactions appeared to drive the long-term distribution of the dominant phytoplankton groups and the chief bloom species. A cross-correlation analysis and associated regressions were run with log numbers of the dominant plankton groups and species against logged independent variables (Table P.1), and a PCA–regression analysis was carried out with the significant factors. Numerical abundance of the blue-green alga *M. tenuissima* was negatively associated with ammonia and DIN loading in the Elevenmile Creek ($R^2 = -0.44$, $p < 0.05$; $R^2 = -0.46$, $p < 0.05$). There were also negative associations with surface and bottom ammonia concentrations ($R^2 = -0.52$, $p < 0.05$; R^2 -0.63, $p < 0.05$). The blue-green alga, *M. tenuissima*, was directly associated with bottom salinity ($R^2 = 0.38$, $p < 0.05$) and salinity stratification ($R^2 = 0.45$, $p < 0.05$). There were general increases in both factors during the drought periods of 1999–2002 and 2006–2007. This species was positively associated with phytoplankton numbers and biomass ($R^2 = 0.45$, $p < 0.05$; R^2 0.63, $p < 0.05$). These data substantiate the observations that connect the *Merismopedia* blooms with drought periods of reduced river and creek flows, high salinity, and increased salinity stratification.

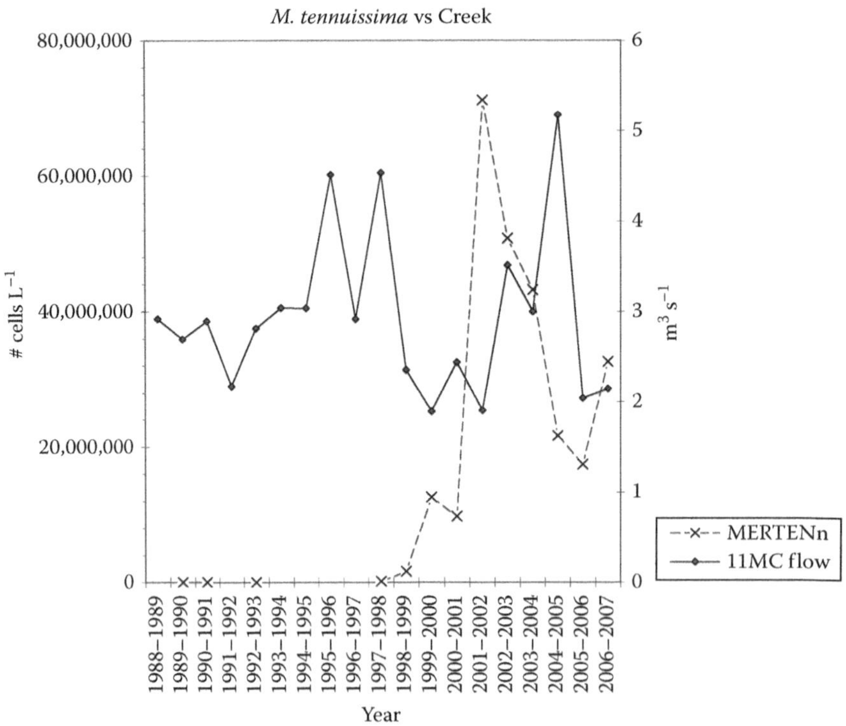

Figure 7.55 Annual averages of *M. tenuissima* cell numbers with Elevenmile Creek flows from 1988 through 2007.

The inverse association of log numbers of cells of the blue-green alga, *M. tenuissima*, with Elevenmile Creek flows is shown in Figure 7.55. With subsequent increases of Elevenmile Creek flows, there was a general decrease of cell numbers with a much reduced presence of *M. tenuissima* just after the floods of 2003 and 2005. There was then a partial recovery of this species during the ensuing drought period (2006–2007). This blue-green alga was restricted to the creek and tended to be dominant during cold months of the year. The occurrence of *M. tenuissima* was negatively ($p < 0.05$) associated with nutrient loading into the Elevenmile Creek. Biological indices were negative mainly at station P22 where the primary biomass of *M. tenuissima* was located.

Analysis of the PCA–regressions involving *M. tenuissima* cell numbers L^{-1} (Figure 7.56) confirmed previous analyses that indicated growth of this species during drought periods (high salinity) and low orthophosphate concentrations in the water.

The field and laboratory experimental evidence indicated that ammonia and orthophosphate loading to the Perdido Estuary was more powerful as a stimulus of plankton productivity than either nutrient by itself. The mean differences of N + P loading (Figure 7.57) indicated that this combination of nutrient loading was negatively associated with the two drought periods of 1999–2002 and 2006–2007. The levels of this index peaked during the increased flows of 2003 and 2005.

A comparison of monthly *H. akashiwo* cell numbers (averaged across stations 23 and 26) versus monthly rainfall (Figure 7.58) indicated that the peak *Heterosigma* blooms started in 1996 (pulp mill–increased nutrient loading) and continued during the following 2 years of increased rainfall and nutrient loads. These peaks were evident only when heavy rainfall occurred during the spring–early summer months when this temperature-sensitive bloom species was active. The blooms usually occurred during the month following the rainfall peaks. After reduced numbers during the 1999–2002 drought, there was a major *Heterosigma* bloom (sewage treatment plant spills) when

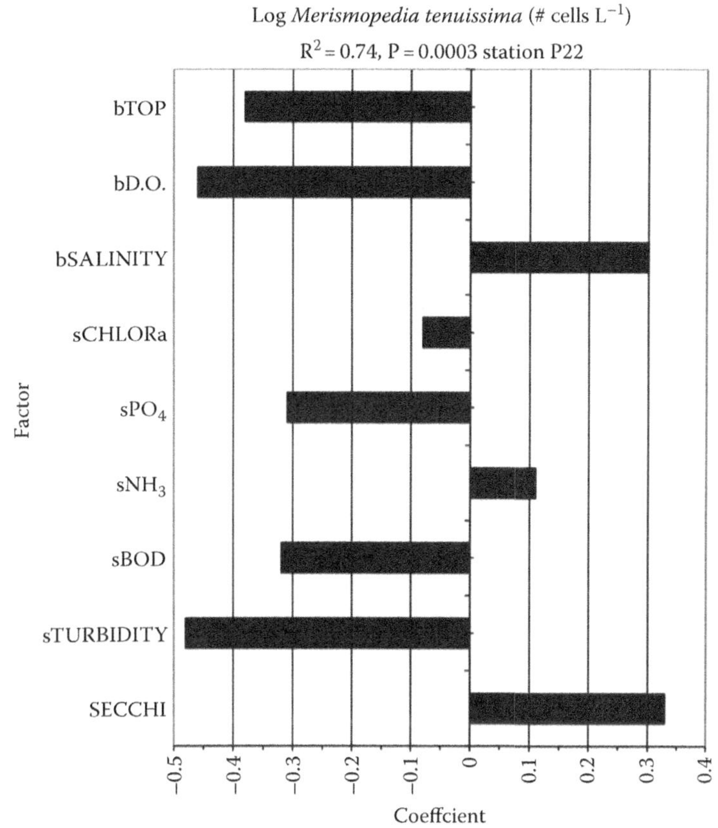

Figure 7.56 PCA/regression analysis of *M. tenuissima* cell numbers L^{-1} in the Perdido Bay using annual means of regression variables averaged from April to July over the 20-year study period.

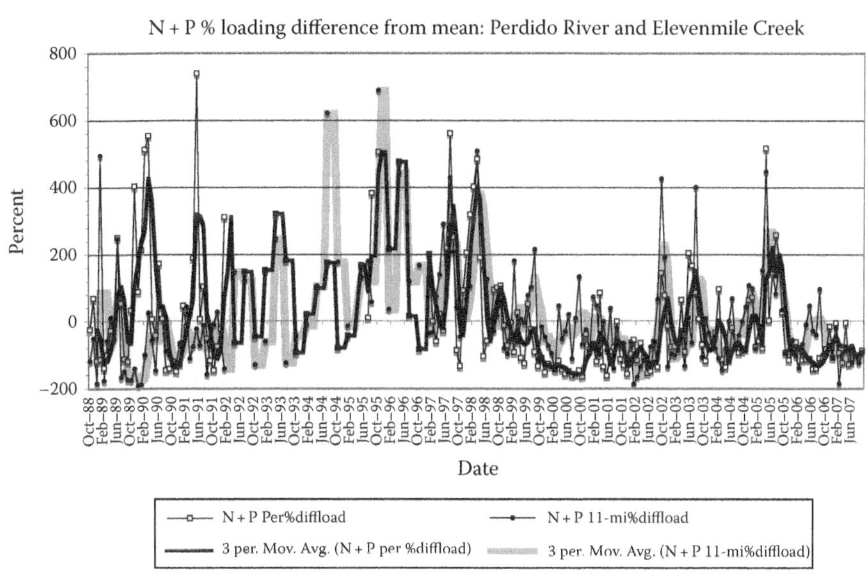

Figure 7.57 Percent differences from the mean of ammonia + orthophosphate loadings of the Perdido River and Elevenmile Creek taken monthly from 1988 through 2007.

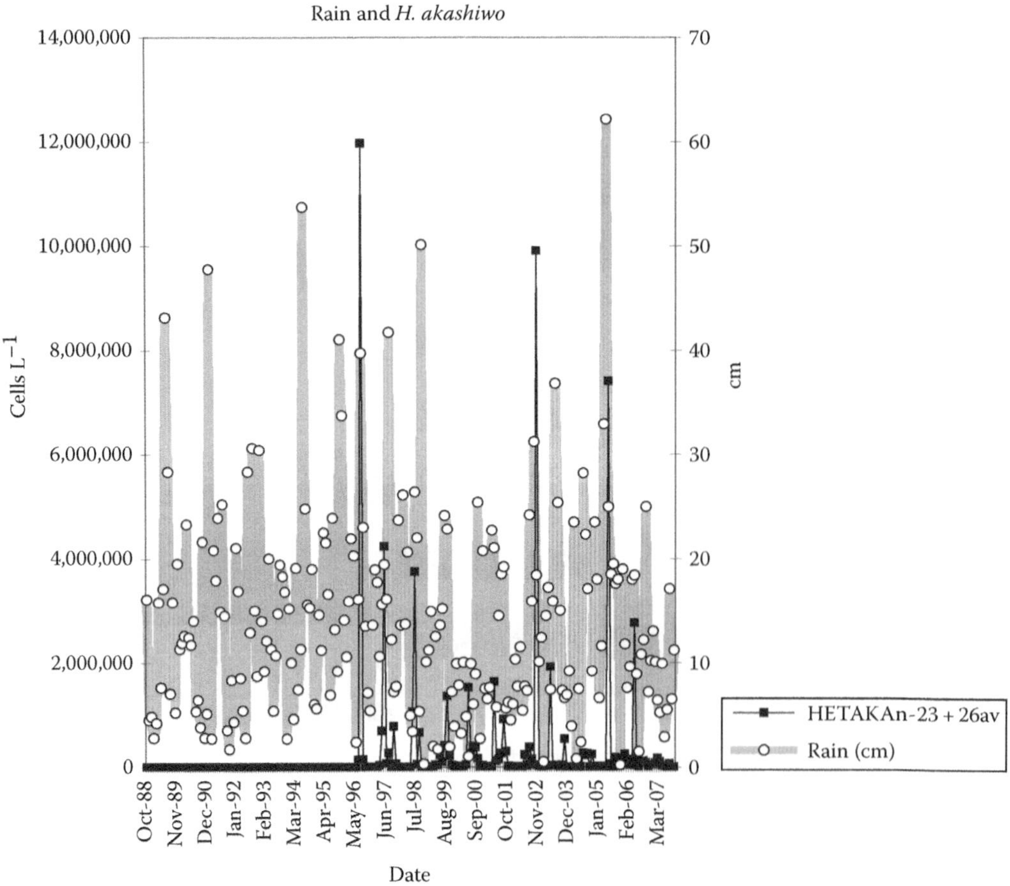

Figure 7.58 *H. akashiwo* numbers L^{-1} versus regional rainfall (cm) monthly from October 1988 through September 2007.

rainfall again peaked in 2002–2003. Another such bloom occurred during the major rainfall of 2005. Numbers of this bloom species then fell off during the drought of 2006–2007.

The bloom numbers of this raphidophyte were also associated with the N + P loading index (Figure 7.59). A statistical analysis of the data (Figure 7.60) indicated that *H. akashiwo* was significantly associated with flow rates of the creek, nutrient (N + P% differences from the grand mean) loading into the Elevenmile Creek, and nutrient concentrations in the bay along with the lowered salinities associated with the high stratification.

Overall, the numbers of phytoplankton increased after initiation of nutrient loading and remained high even after nutrient loading was reduced (Figure 7.61). The lowest numbers occurred at the mouth of the Perdido River (station P18), and the highest numbers occurred in the Elevenmile Creek (*M. tenuissima*) and at the mouth of the creek (station P23) where nutrient loading and high nutrient concentrations stimulated the major plankton blooms. Phytoplankton species richness (Figure 7.62) was generally higher at stations in the upper bay with interannual trends that tended to follow the occurrence of *Merismopedia* blooms in the creek and *Heterosigma* blooms in the bay. There was a general trend of increased phytoplankton species richness at mid-bay and lower bay stations over the 20-year study period. There were seasonal trends of the nutrient loading characteristics and the presence of dominant plankton species in the bay. Elevenmile Creek ammonia loading peaked during April, and orthophosphate loading peaked from May through September. When standardized

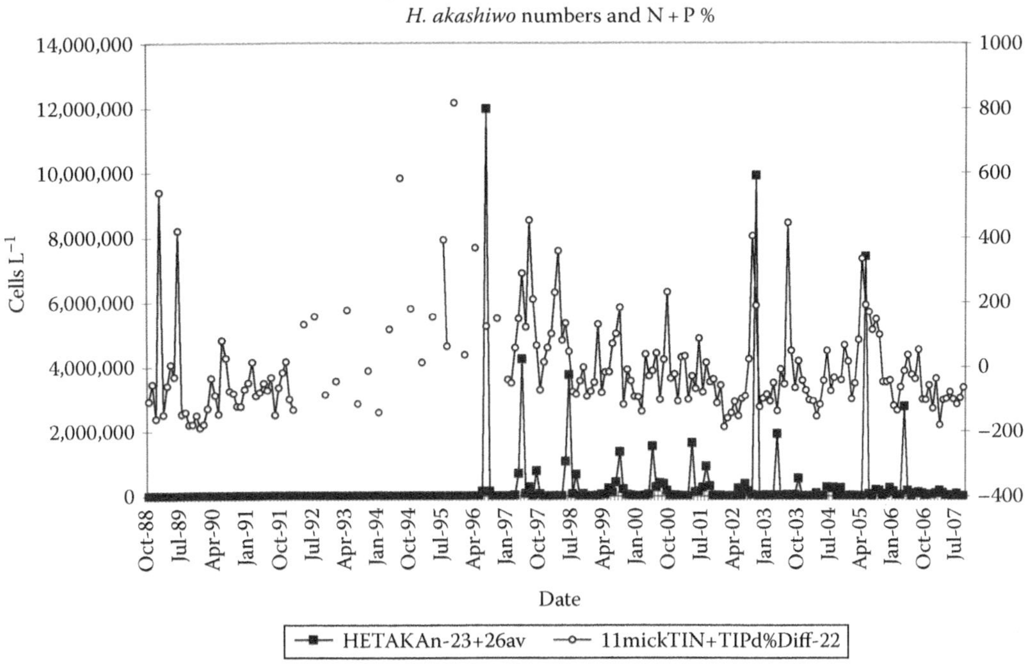

Figure 7.59 *H. akashiwo* numbers L^{-1} versus regional N + P loading as percentages from the overall mean (October 1988 to September 2007).

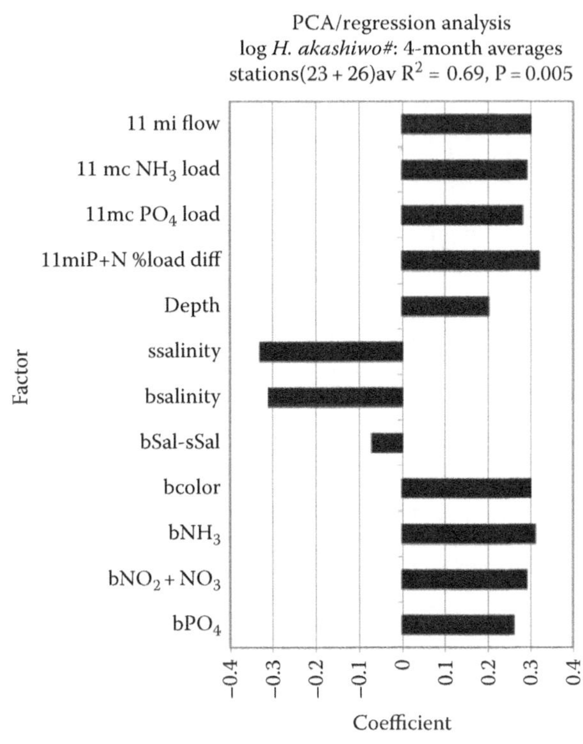

Figure 7.60 PCA/regression analysis of log *H. akashiwo* numbers (dependent variable) against major loading and nutrient water concentrations in the Perdido Bay. Data were taken from averages of stations 23 and 26 from April to July over the sampling period.

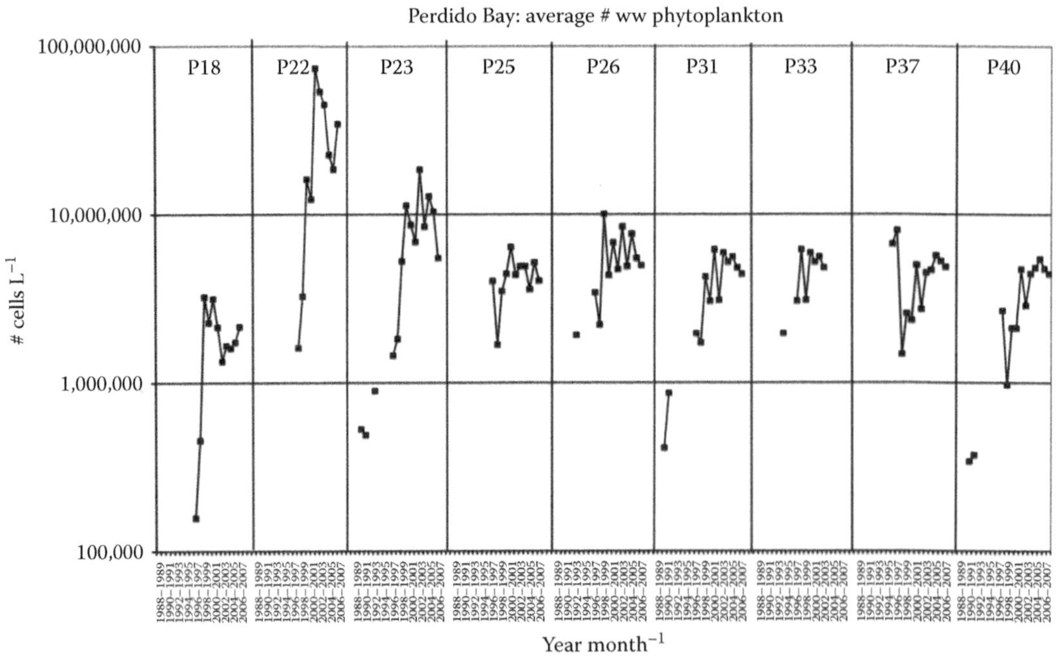

Figure 7.61 Annual averages of total numbers of cells L^{-1} of phytoplankton in the Perdido Bay from October 1988 through September 2007.

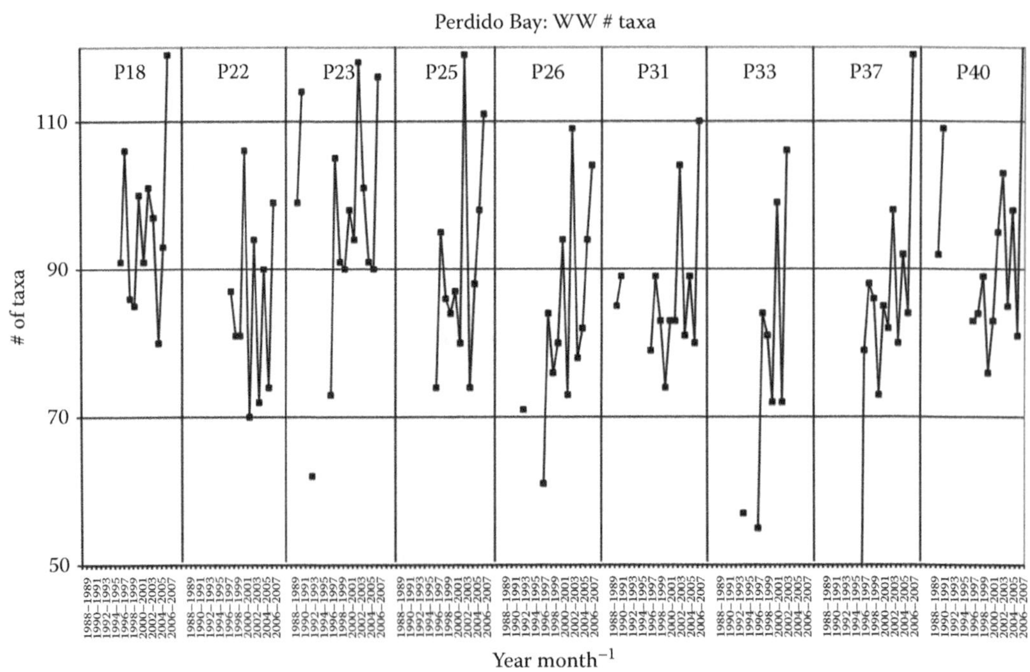

Figure 7.62 Annual averages of total numbers of species of phytoplankton in the Perdido Bay from October 1988 through September 2007.

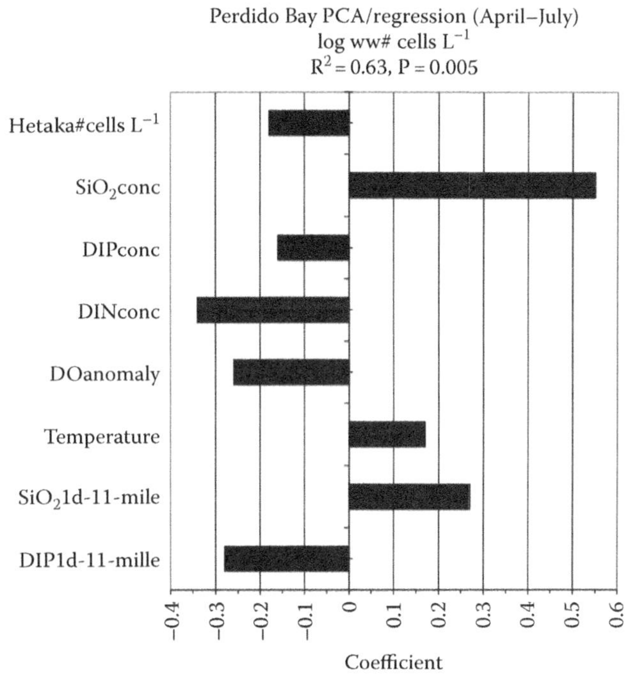

Figure 7.63 PCA/regression analysis of phytoplankton cell numbers L^{-1} in the Perdido Bay using annual means of regression variables averaged from April to July over the 20-year study period.

as percent loadings of these nutrients of the overall mean, there were peaks from April through August. Peak numbers of *H. akashiwo* occurred from April through July and *C. choctawhatcheeana* peaked during April and May. Bloom species occurrence was thus timed to both seasonal and interannual cycles of nutrient loading in the Elevenmile Creek.

The data were averaged from April to July by year, the period of the highest numbers of the bloom species, *Heterosigma*. The PCA–regression analysis of the log whole-water phytoplankton numbers of cells (Figure 7.63) indicated a strong positive relationship with SiO_2 concentrations and significant negative associations with DIP concentrations, DO anomalies, DIP loading, and *Heterosigma* cell numbers. Species richness of phytoplankton (Figure 7.64) was positively associated with water temperature and negatively associated with *Heterosigma* numbers and DIP concentrations. The strong negative response of both phytoplankton indicators to DIP water concentrations could be associated with the fact that inorganic phosphorus was limiting to *Heterosigma* blooms that, in turn, were limiting to phytoplankton productivity.

Water-quality data in the lower parts of the Perdido Estuary (Wolf Bay, Figure 7.45) indicated general increases of water color and turbidity since 1999 with periodic DO reductions (<4 mg L^{-1} at depth) (Livingston, 2007a,b). Both surface and bottom chlorophyll *a* concentrations increased steadily during this period. Concurrently, ammonia concentrations increased as the Wolf Bay received nutrient loading from upland (agricultural) sources (Livingston, 2007a,b). Phytoplankton collections from the Wolf Bay indicated that blooms of both *P. cordatum* and *H. akashiwo* started to occur in 2000, and these blooms continued through fall 2004. The plankton blooms showed seasonal periodicity similar to that described earlier (*P. cordatum* during cold months and *H. akashiwo* during warm months).

The Wolf Bay has been subjected to unregulated agricultural runoff from Alabama for years (Livingston et al., 1997; Livingston, 2000, 2002, 2005). A review of agricultural activities in Baldwin County, Alabama (A. Niedoroda, pers. comm.), indicated that the production of cotton in

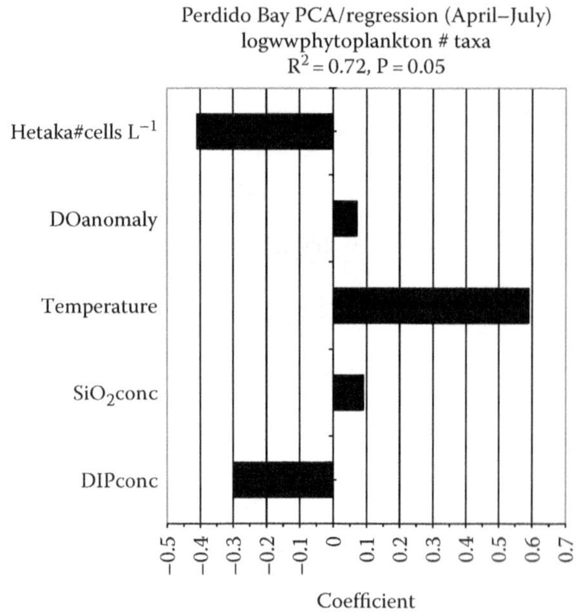

Figure 7.64 PCA/regression analysis of phytoplankton species richness in the Perdido Bay using annual means of regression variables averaged from April to July over the 20-year study period.

this area increased by a factor of 26 between 1988 and 1995 and that comparable shifts in the agricultural activities in this region have continued to the present time. Possible changes related to the altered use of fertilizers in this area of the bay due to shifts in agricultural activities are thus directly relevant to the observed deterioration of the Wolf Bay. The Wolf Bay drainage basin was inexplicitly removed from federal land-use plans.

The Wolf Bay system was sampled for water quality since October 1993. Whole-water phytoplankton samples were taken monthly in the Wolf Bay from August 1998 through 2007. Sediment quality in the Wolf Bay was similar to that in mid- to lower Perdido Bay (Livingston, 2000, 2007a,b). The relationship between rainfall and DO anomaly was complex and not linearly related ($p < 0.05$). From 1996 through 2001, there was a generally positive correspondence between these two factors. In 2002, there was no sign of any relationship, and from 2003 to 2007, there was usually an inverse relationship between these factors. After a period of no observable trends, peak summer water temperatures tended to increase from 2003 through 2007. During the latter months of this period, DO anomaly tended to peak just before the peak summer temperatures (spring–early summer). There was a general inverse relationship between water temperature and DO.

Nutrient trends in the Wolf Bay indicated increased ammonia from 2001 through 2004. During this period, orthophosphate concentrations were minimal. Peaks usually occurred during winter–spring periods. Orthophosphate peaks occurred in 1998–1999 and 2005–2007. These peak concentrations usually occurred during warm months of the year. Phytoplankton numbers of the dominant species in the Wolf Bay (Table 7.7) were dominated by several bloom species. Phytoplankton cell numbers in the Wolf Bay (Figure 7.65) tended to peak during spring–summer periods of 2000–2001, 2003–2003, 2005, and 2007. There was a general trend toward increased cell numbers of *H. akashiwo* over the study period. The Wolf Bay phytoplankton species richness tended to show a similar upward trend with time and was generally aligned with the peak numbers over the study period. The Wolf Bay phytoplankton cell numbers tended to peak during

Table 7.7 Dominant Phytoplankton Species Numbers Taken at Station P42 (Wolf Bay) over the Study Period

Species	Total
Skeletonema costatum	43,933,303
Chaetoceros spp.	41,713,896
Gymnodinium spp.	37,166,156
Prorocentrum compressum	32,107,543
Heterosigma akashiwo	28,441,832
Cryptomonas spp.	23,981,116
Pyramimonas spp.	16,045,208
Undetermined nannococcoids	14,617,698
Heterocapsa rotundata	10,047,511
Chrysochromulina spp.	8,764,616
Chlamydomonas spp.	8,175,427
Pseudoscourfieldia spp.	4,049,062
Heterocapsa pygmaea	3,114,995
Chaetoceros subtilis	2,434,921
Merismopedia tenuissima	2,090,144

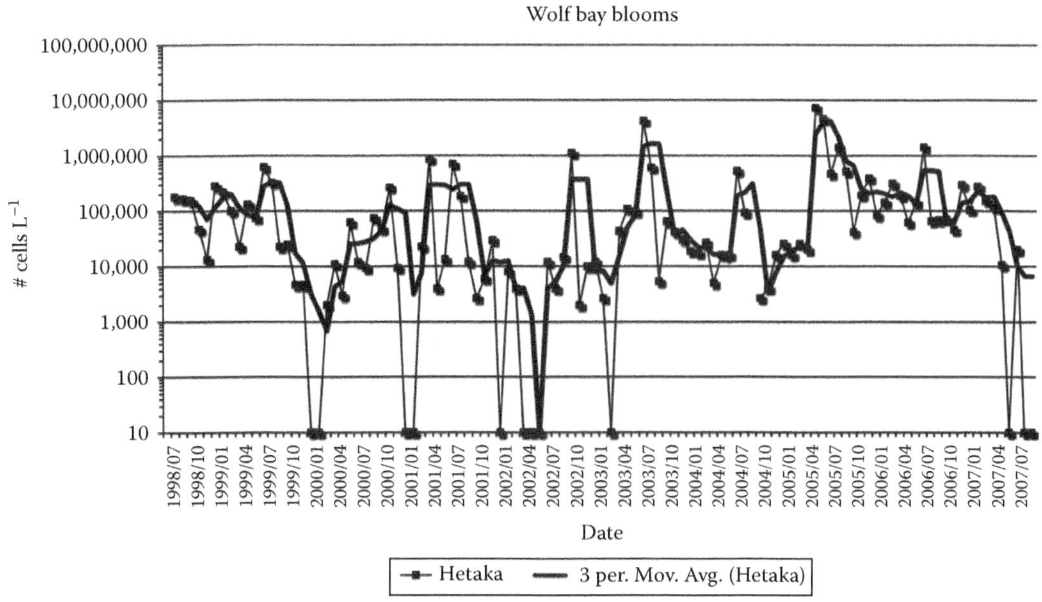

Figure 7.65 *H. akashiwo* cell numbers taken monthly at station P42 in the Wolf Bay from 1998 through 2007.

spring–summer periods of 2000–2001, 2003–2003, 2005, and 2007. There was a general trend toward increased cell numbers over the study period.

Phytoplankton collections from the Wolf Bay indicated that blooms of both *P. cordatum* and *H. akashiwo* started to occur in 2000–2001 and these blooms continued seasonally. The blooms showed seasonal periodicity similar to that described earlier (*P. cordatum* during cold months and *H. akashiwo* during warm months). *C. choctawhatcheeana*, a nontoxic diatom considered valuable to the bay food webs, was usually reduced during the *H. akashiwo* blooms. Both species occurred during spring periods. The bloom sequences in the Wolf Bay became more complex with increased numbers of bloom species and the increasing dominance of *H. akashiwo* during warm months. There was a decrease of this species during the drought of 2007 that followed trends in the Perdido Bay over the study period.

There was a general downward trend of infaunal macroinvertebrate numbers in Wolf Bay over the study period. Infaunal species richness also peaked during winter–spring periods from October 2003 to March 2005. However, from May 2005 through September 2007, there was a general collapse of the infaunal community in the Wolf Bay with virtually no species noted during warm months of the year. During the final year of sampling, the Wolf Bay infaunal macroinvertebrates remained in a severely deteriorated state. Regression analyses indicated that infaunal numbers were negatively associated with cell numbers of *P. cordatum* ($R^2 = 0.48$, $p < 0.05$) and *H. akashiwo* ($R^2 = 0.49$, $p < 0.05$). The polychaete dominants *M. ambiseta* and *Streblospio benedicti* decreased to insignificant numbers over the study period (Livingston, 2007a). Oligochaetes, which are indicators of reduced water/sediment quality, increased over the study period. These data indicate that the Wolf Bay deteriorated over the study period during a noted increase of plankton blooms of known toxic species.

Recently, there was a considerable increase in housing developments built along the western shores of the lower Perdido Bay system including the Wolf Bay. Livingston (1999, 2007a,b) noted salinity stratification and hypoxia in the lower Perdido Bay south of Ono Island, an area characterized by generally uncontrolled urban development. Agricultural and urban runoff (nonpoint sources of nutrients) differed from point sources (i.e., the pulp mill and sewage treatment plant) in that the pulp mill effluents were more or less continuous, whereas nonpoint sources followed rainfall patterns. The *Prorocentrum* blooms of winter 2003 followed increased rainfall during the previous fall, indicating that such reactions can be delayed by a matter of months. However, the trends were real and signified a basic difference between the mechanisms of bloom stimulation by continuous point sources and discontinuous nonpoint sources.

The occurrence of plankton blooms in different parts of the bay due to various sources of anthropogenic nutrient loading was associated with changes in factors that included water and sediment quality and animal associations (Table 7.8). By the end of the 20-year study period, major parts of the Perdido Estuary had been damaged by diverse anthropogenic activities. The biological components of deeper parts of the bay were adversely affected by salinity stratification due to the opening of the bay mouth early in the twentieth century. These changes represented cumulative effects over long-term periods. Anthropogenic nutrient loading from a pulp mill and a sewage treatment plant caused blooms in the estuary that were associated with severe reductions of infaunal and epibenthic macroinvertebrates and fishes. Agricultural runoff loaded nutrients in the Wolf Bay that resulted in bloom activity and the eradication of infaunal macroinvertebrates. There were other changes that were associated with climatological trends (severe droughts and enhanced rainfall) that were associated with stimulation of the two major bloom species of the estuary.

Long-term changes of the plankton production in the Perdido Estuary were thus influenced by complex combinations of nutrient loading, seasonal and interannual meteorological conditions, river and bay habitat factors, species-specific plankton trends, and the spatial–temporal status of the ecological components of the estuary. The complicated interactions of natural and anthropogenic nutrient loading and long-term changes of rainfall/drought patterns determined the temporal succession and severity of plankton blooms in the estuary. The primary blooms were associated with specific interactions between nutrient loading and climatological conditions. The blue-green bloom of *Merismopedia* occurred primarily in the Elevenmile Creek during drought periods. The raphidophyte (*Heterosigma*) blooms occurred in the upper bay areas of high nutrient loading during periods of high rainfall. The distribution of the blooms in the Perdido Estuary was also associated with the species-specific, seasonal cycles of plankton dominance as they related to the anthropogenic nutrient loadings.

Table 7.8 Status of the Perdido Estuary by September 2007

Factor	Water Quality	Sediment Quality	Bloom Occurrence	Status of Phytoplankton	Status of Infauna	Status of Invertebrates	Status of Fishes
Bay Area Elevenmile Creek	High nutrient concentrations	Reduced TP and TON	*Merismopedia tenuissima*	Damaged	Bloom associated reductions	Bloom associated reductions	Bloom associated reductions
Upper bay	Increased bottom NH_3	Increased TP and TON	*Heterosigma akashiwo*	Damaged	Bloom associated reductions	Lowest no. in 20 years*	Bloom associated reductions
Mid-bay		Increased TP and TON	*Heterosigma akashiwo*	Damaged	Bloom associated reductions	Lowest no. in 20 years*	Bloom associated reductions
Lower bay		Increased TP and TON	*Heterosigma akashiwo*	Damaged	Bloom associated reductions	Lowest no. in 20 years*	Bloom associated reductions
Wolf Bay	Low DO	Increased TP and TON	*Heterosigma akashiwo* *Prorocentrum cordatum*	Damaged	Bloom associated reductions	ND	ND

ND, no data.

Secondary Productivity and Trophic Organization

During the second half of the 20-year study, there were increased incidences of droughts (1999–2002, 2006–2008) relative to those encountered over the first decade of the study. The prolonged dry period from 1999 to 2002 marked a very severe event relative to records taken over the previous 100 years (Livingston, 2005). The most recent drought extended from fall 2005 to the end of the study in September 2007. This latter drought was almost as severe as the 1999–2002 event. These droughts followed long-term climatological trends in the general area of the NE Gulf of Mexico.

Infaunal macroinvertebrate numbers (Figure 7.66) were highest in the *Vallisneria* beds on the NW section of the Perdido Bay. Relatively low numbers were evident at the mouth of the Perdido River (P18, salinity stratification, low bottom DO), at the mouth of the Elevenmile Creek (P22, salinity stratification, low bottom-water DO, mill effluents), and at the mid- to lower bay stations (P26–P40, salinity stratification, low bottom-water DO). Station P23, the bay receiving area for mill effluents, was intermediate in terms of infaunal abundance. At all stations, there was a pronounced decrease in abundance over the 20-year sampling period.

The initial year of sampling was characterized by reduced numbers throughout the bay. This could have been due to the fact that, prior to 1998, there had been a moderate drought. During the final 2 years (after the loss of the *Vallisneria* beds), there was a sharp drop in infaunal numbers at station P25. There was also a general reduction of infaunal numbers during the two droughts (1999–2002; 2006–2007). Overall, however, the bay was characterized by a depauperate infaunal community that got generally worse with time over the sampling period.

Infaunal species richness (Figure 7.67) showed similar spatial and temporal trends to those of the numerical abundance of this group with the highest levels at stations P25 and P23 and the lowest levels at the various other stations with the possible exception of station P40 that showed somewhat

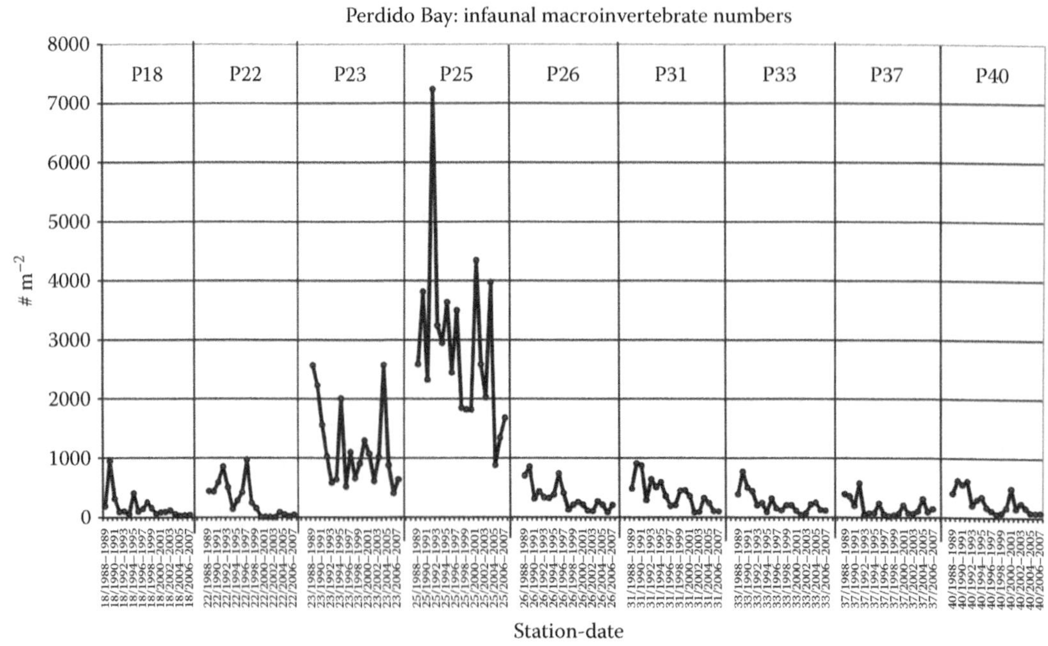

Figure 7.66 Annual averages of infaunal macroinvertebrate number m^{-2} in the Perdido Bay from October 1988 through September 2007.

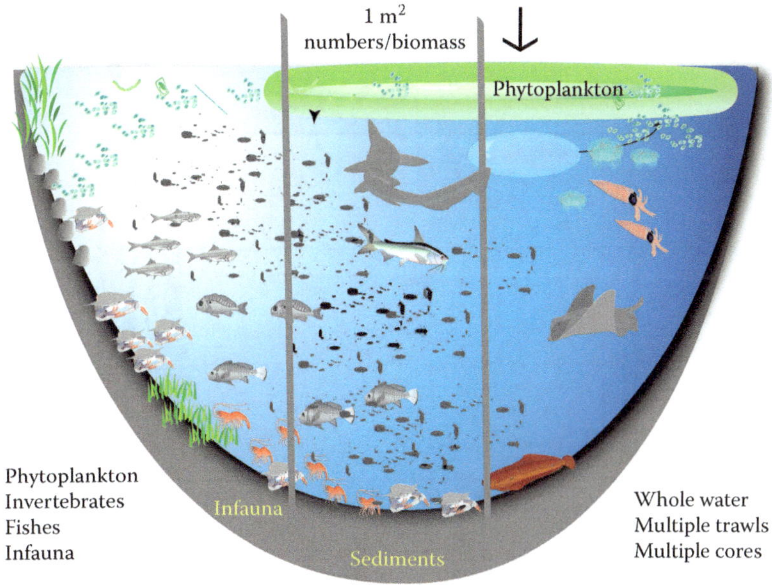

Figure 2.1 Field sampling for phytoplankton, infaunal and epibenthic invertebrates, and fishes for studies in the eastern Gulf of Mexico. Detailed sampling protocols are given in Appendix A.

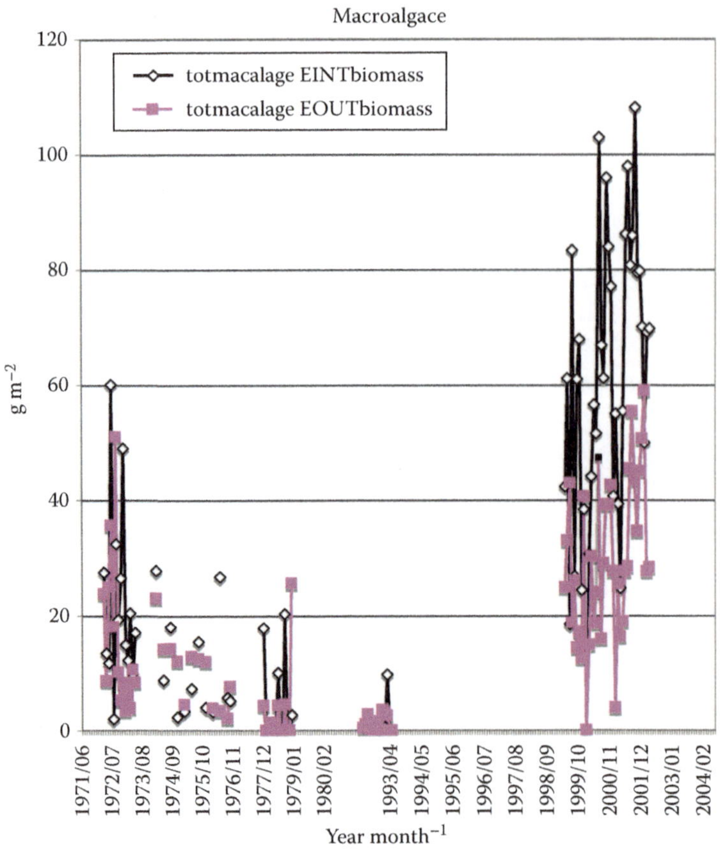

Figure 7.14 Long-term comparisons of macroalgal biomass (dominant species; g m^{-2}) in the Econfina Estuary in the inner and outer stations with 3-month moving averages.

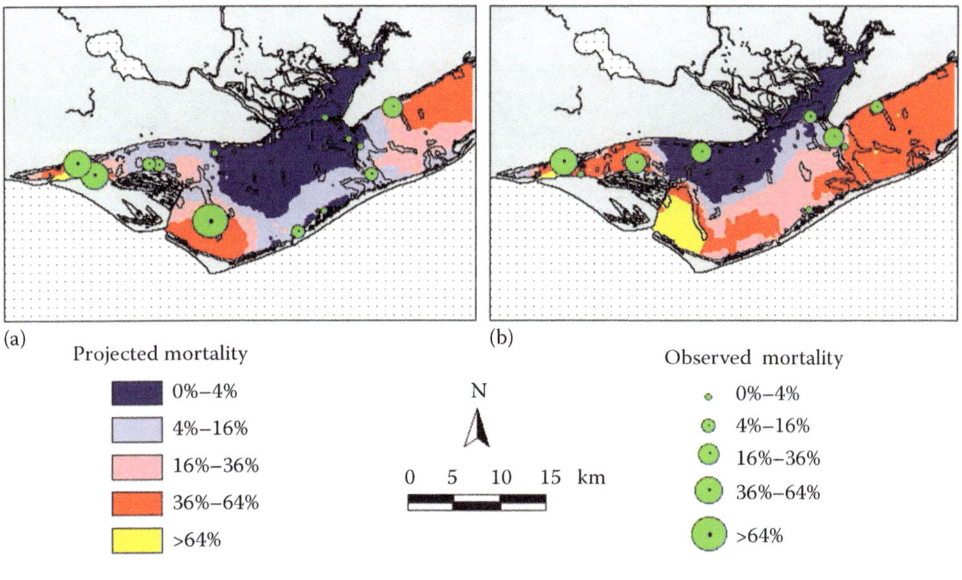

Figure 7.39 Map of projected oyster mortality in the Apalachicola Bay system based on the statistical model for mortality in 1985 and the hydrodynamic model results for (a) 1985 and (b) 1986. Circles indicate observed mortality values from oyster predation experiments (1985, May–August average; 1986, May).

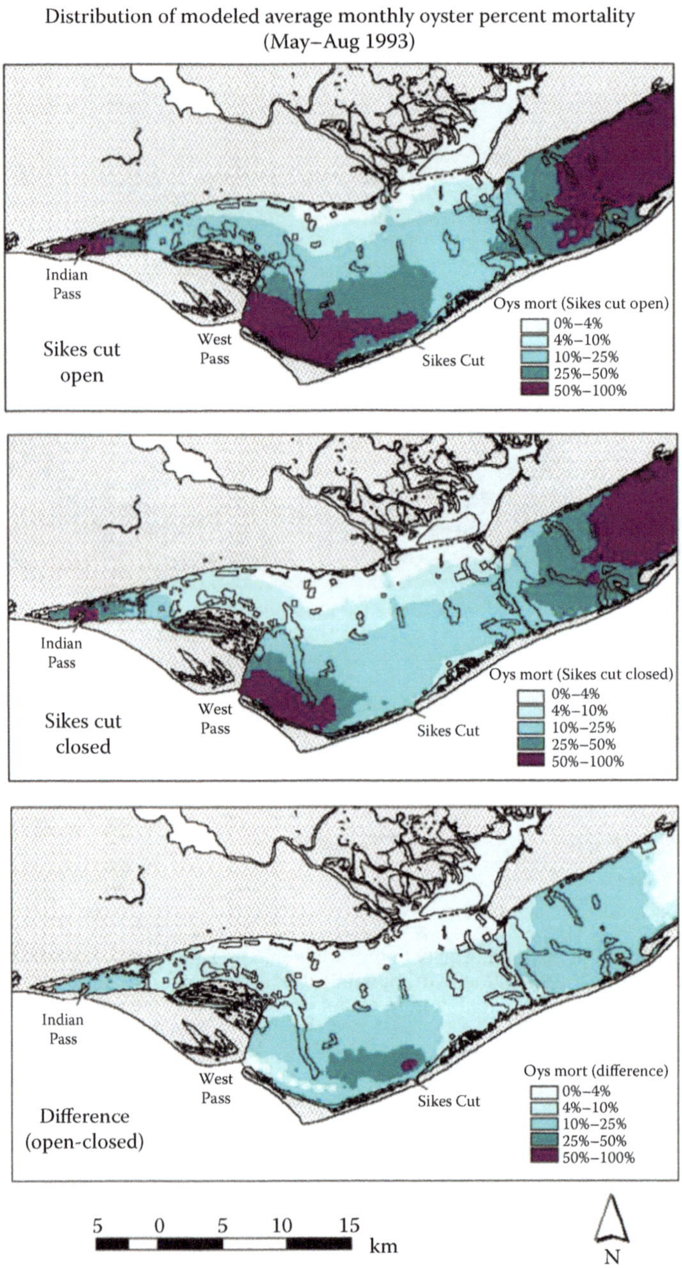

Figure 7.40 Modeled oyster percent mortality data based on effects of an open Sike's relative to a closed Sike's. (Data are taken from an unpublished paper by R. J. Livingston, S. Leitman, G. C. Woodsum, H. Galperin, P. Homann, J. D. Christensen, M. E. Monaco, and R. L. Howell, IV.)

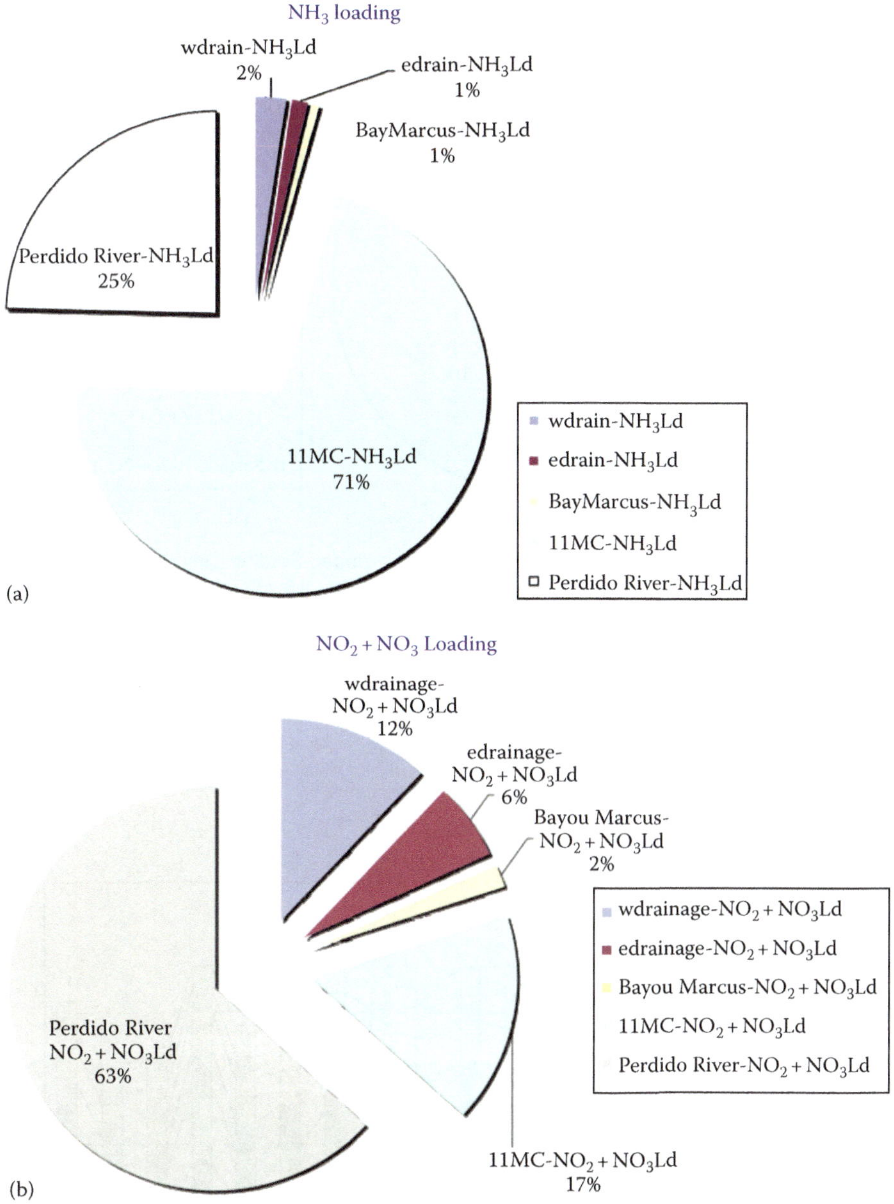

Figure 7.51 (a) Ammonia loading to the upper Perdido Bay by system (Perdido River, Elevenmile Creek, Bayou Marcus Creek, western drainage area, eastern drainage area), 20-year means. (b) Nitrite + nitrate loading to the upper Perdido Bay by system (Perdido River, Elevenmile Creek, Bayou Marcus Creek, western drainage area, eastern drainage area), 20-year means. *(Continued)*

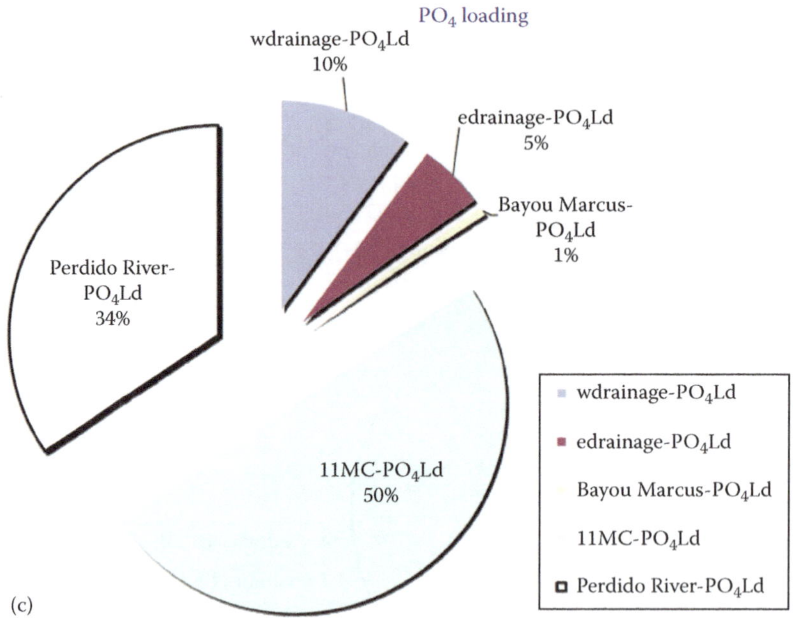

Figure 7.51 (*Continued*) (c) Orthophosphate loading to the upper Perdido Bay by system (Perdido River, Elevenmile Creek, Bayou Marcus Creek, western drainage area, eastern drainage area), 20-year means.

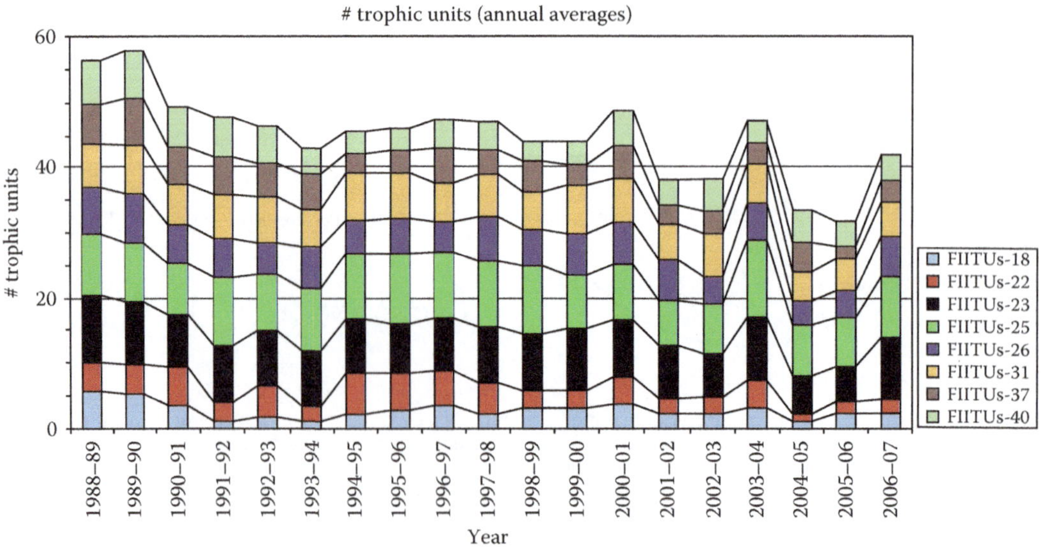

Figure 7.81 Numbers of TUs by station and averaged by year in the Perdido Estuary from 1988 through 2007.

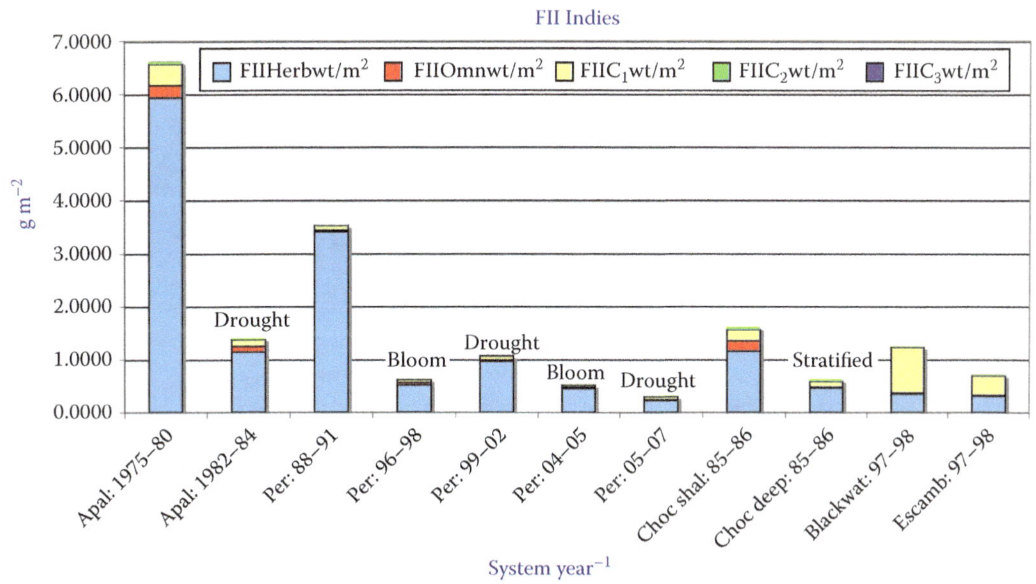

Figure 9.9 FII indices showing differences between periods of regular river flow rates and prolonged droughts, between periods of nonbloom and bloom outbreaks, and between deep (stratified) and shallow (unstratified) bay conditions. The Apalachicola data were averaged over regular (1975–1980) and postdrought (1982–1984) conditions. The Perdido data were organized by drought-free conditions with no plankton blooms (1988–1991), *Heterosigma* blooms (1996–98, 2004–2005), and drought (1999–2002, 2006–2007) conditions. The Choctawhatchee data were organized by shallow (unstratified) and deep (stratified) conditions.

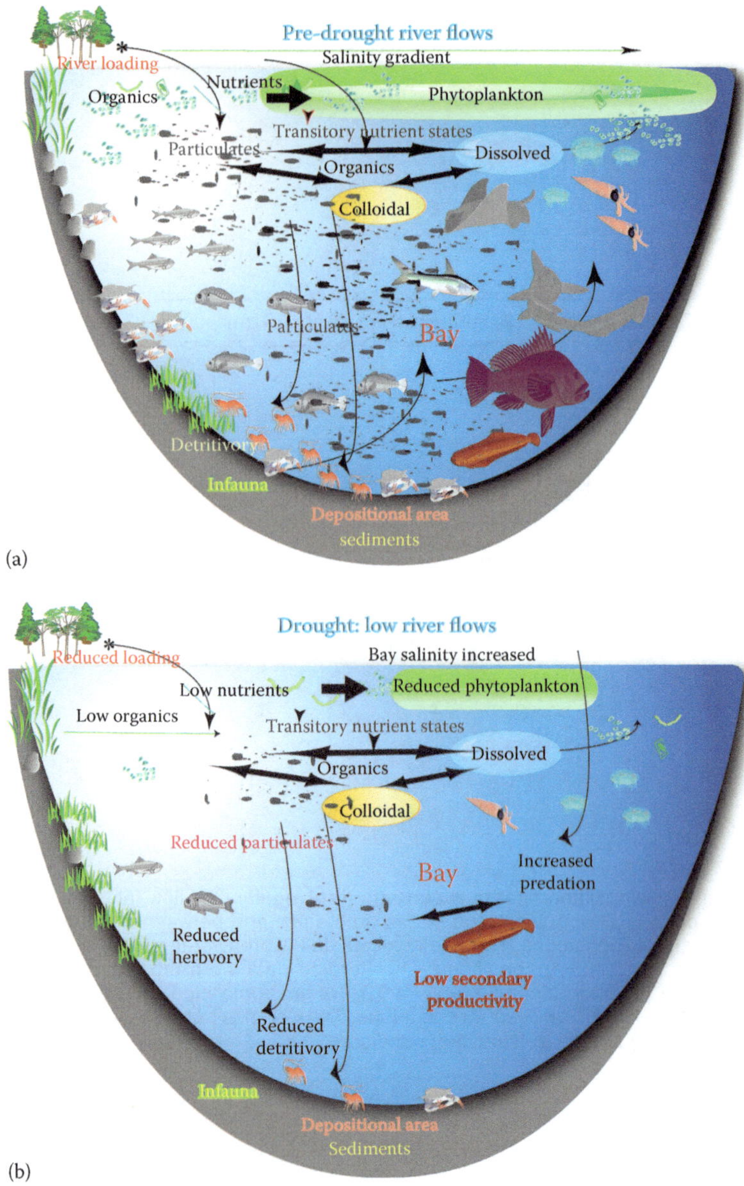

Figure 9.13 (a) Model of the connections of the food webs of the Apalachicola Estuary with river processes during natural fluctuations of flow rates. (b) Model of the connections of the food webs of the Apalachicola Estuary with river processes under the effects of prolonged droughts on flow rates.

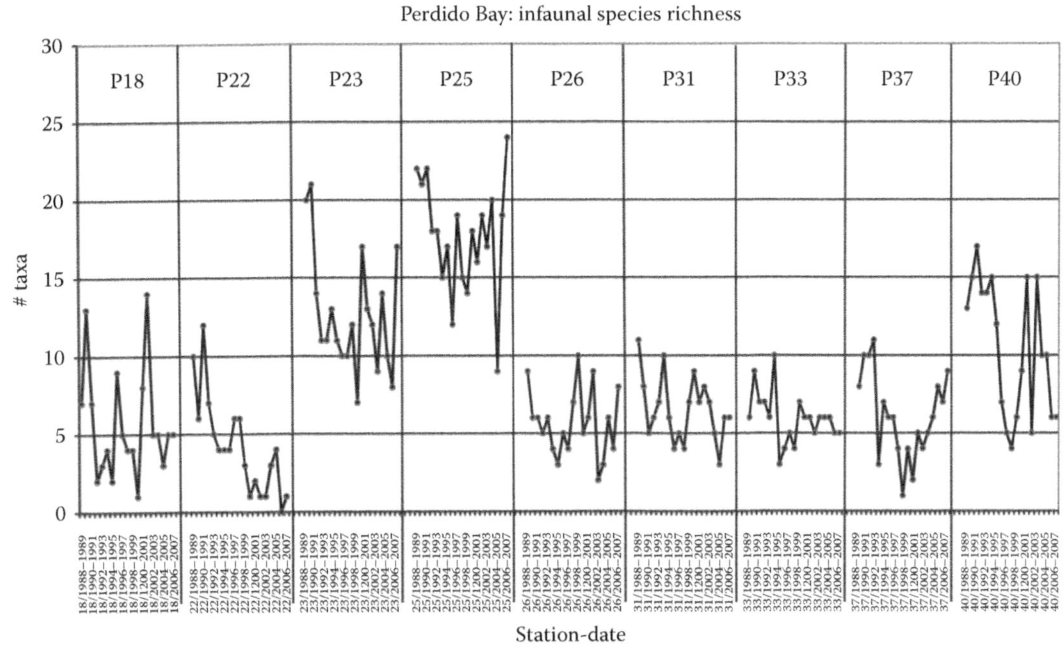

Figure 7.67 Annual averages of infaunal macroinvertebrate species richness in the Perdido Bay from October 1988 through September 2007.

increased species richness values relative to those in the upper and lower bays. At most stations, there was a relatively steep decline in infaunal species richness with time.

Numerical abundance of the dominant infaunal macroinvertebrates species (Appendix C, Table 7.9) indicated that the polychaete worms *M. ambiseta* and *S. benedicti* were the prime dominants at stations P25 and P23. The clam, *Rangia cuneata*, was prevalent at stations P23 and P25. Numbers of infauna were low in areas marked by high-salinity stratification (P18, P31, P22, P37, P40) and mill effluents (P22).

The numerical abundance of epibenthic macroinvertebrates (Figure 7.68) was generally low throughout the bay with particularly low numbers at the mouth of the river and at the stations in the lower bay. Invertebrate species richness (Figure 7.69) was also low throughout the bay with very steep reductions at stations P22, P33, P37, and P40 with time. These trends were not obvious in the upper bay stations P23 and P25.

The Perdido Bay invertebrates were dominated by brown shrimp (*F. aztecus*) and blue crabs (*C. sapidus*) (Appendix C, Table 7.10) that were located mainly in mid-bay areas. Overall numbers were generally low.

Fish numerical abundance (Figure 7.70) was somewhat different than that of the infaunal and epibenthic invertebrates with generally low numbers of fishes in the upper and lower bays and the highest numbers at mid-bay stations P26, P31, and P33. Once again, however, there was a very steep decline in fish numbers throughout most of the bay with time. By the end of the study, there were virtually no fish left in the entire bay. Fish species richness (Figure 7.71) was more evenly distributed throughout the bay, but there was again a serious decline in this index with time at all stations.

The primary fish dominants (Appendix C, Table 7.11) included spot (*L. xanthurus*) and bay anchovies (*A. mitchilli*) that were located primarily in upper and middle parts of the Perdido Bay.

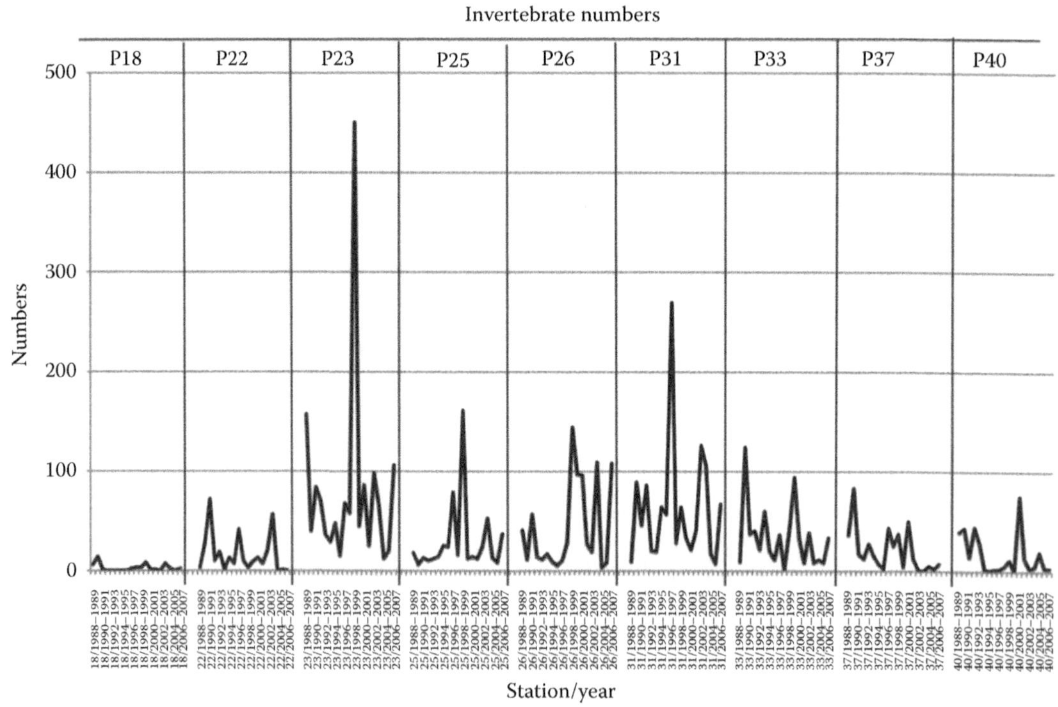

Figure 7.68 Annual averages of macroinvertebrate numbers in the Perdido Bay from October 1988 through September 2007.

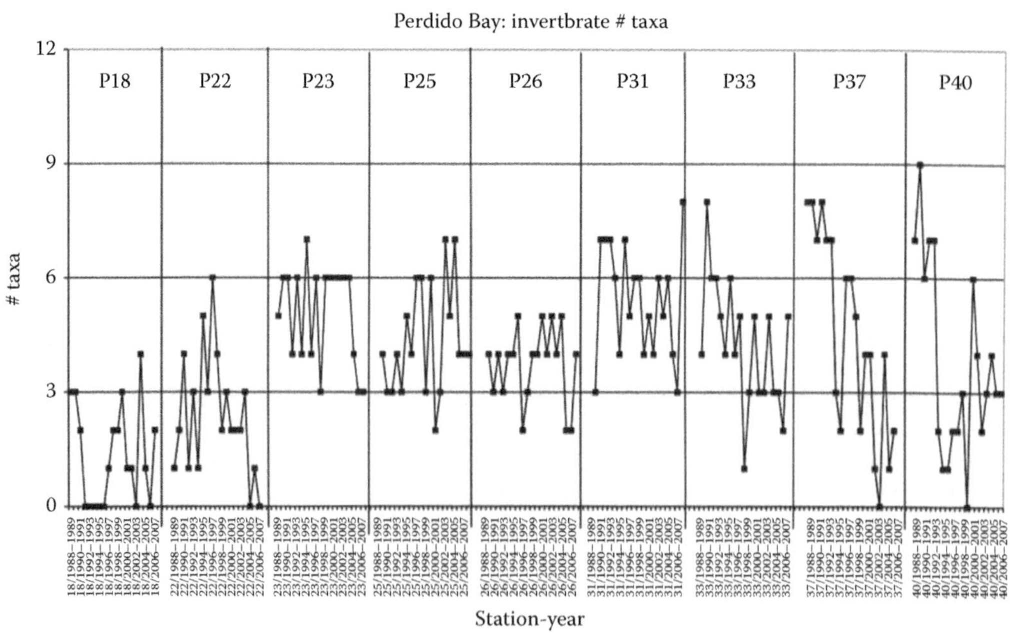

Figure 7.69 Annual averages of macroinvertebrate species richness in the Perdido Bay from October 1988 through September 2007.

Table 7.9 Numerical Abundance of Infaunal Macroinvertebrates, Averaged over the 20-Year Sampling Period by Station

Species	18	22	23	25	26	31	33	37	40
Mediomastus ambiseta	16.73	4.41	363.01	1423.86	131.81	206.77	82.79	20.59	34.29
Streblospio benedicti	118.08	162.58	283.33	411.01	128.18	119.85	109.81	76.81	57.51
Rangia cuneata	2.79	0.86	200.59	170.71	11.57	1.24	0.95	0.30	0.42
Nemertean sp.	2.98	0.67	18.17	51.32	6.89	7.65	5.61	1.82	11.63
Scottalana canadensis	—	0.19	5.55	4.21	0.10	0.57	0.10	0.30	0.21
Assiminea succinea	3.17	0.19	27.74	29.56	2.39	0.19	0.19	—	—
Hobsonia florida	4.62	20.49	21.81	35.51	2.01	0.19	0.10	0.30	—
Macoma mitchelli	0.19	0.19	35.58	35.97	11.00	2.01	2.09	0.50	1.66
Chironomus sp.	0.96	14.82	46.97	38.25	3.06	—	—	—	—
Leitoscoloplos fragilis	0.29	—	7.56	18.27	9.57	13.50	9.61	7.77	12.77

Table 7.10 Numerical Abundance of Epibenthic Macroinvertebrates, Averaged over the 20-Year Sampling Period by Station

Species	18	20	22	23	25	26	31	33	37	40
Farfantepenaeus aztecus	0.35	3.21	0.78	7.39	2.86	5.07	6.50	4.41	1.64	1.03
Farfantepenaeus sp.	0.09	1.88	1.28	7.07	2.19	2.09	5.29	7.59	1.04	0.15
Callinectes sapidus	0.22	4.67	1.76	5.30	1.29	1.72	0.77	0.63	0.28	0.24
Haminoea succinea	—	—	—	—	0.01	—	0.01	0.06	11.07	0.30
Farfantepenaeus duorarum	0.06	0.29	0.14	0.62	0.31	0.47	0.98	0.71	0.64	0.43
Acetes americanus	0.03	—	—	0.02	0.04	0.01	1.15	0.95	0.65	0.83
Litopenaeus setiferus	0.02	—	0.43	0.42	0.06	0.71	0.16	0.04	0.03	—
Callinectes similis	—	0.04	—	—	—	0.06	0.47	0.40	0.36	0.58
Lolliguncula brevis	—	—	—	—	0.01	0.01	0.24	0.24	1.05	0.64
Squilla empusa	—	—	—	—	—	—	0.02	0.09	0.34	0.25

The association of the blooms with nutrient loading should be viewed with respect to the combined N + P loading (Figure 7.72). The general trends of this loading were followed closely by *Heterosigma* biomass averages. Both indices followed closely the drought/flood trends with high rainfall/river flow incidences correlated with peaks of the DIP + DIN loading deviations and consequent *Heterosigma* blooms.

The response of the bay to all of this could be indicated by the long-term response of the dominant populations of the Perdido Estuary. The distribution of the various dominant populations of fishes and invertebrates indicated that many of these species were concentrated in the shallow upper bay stations (P23, P25, P26). Areas directly affected by the pulp mill (P22) and having strong salinity stratification due to increased depth (P18, P37, P40) were generally depauperate compared to those at the shallow, well-mixed stations. Depth was an important negative factor in the secondary production of the bay, probably due to the relationship with salinity stratification.

Mediomastus ambiseta

The polychaete, *M. ambiseta*, occurred primarily in the upper bay grass beds (station P25). There was a general decline in this polychaete worm over the study period (Figure 7.73a). There were

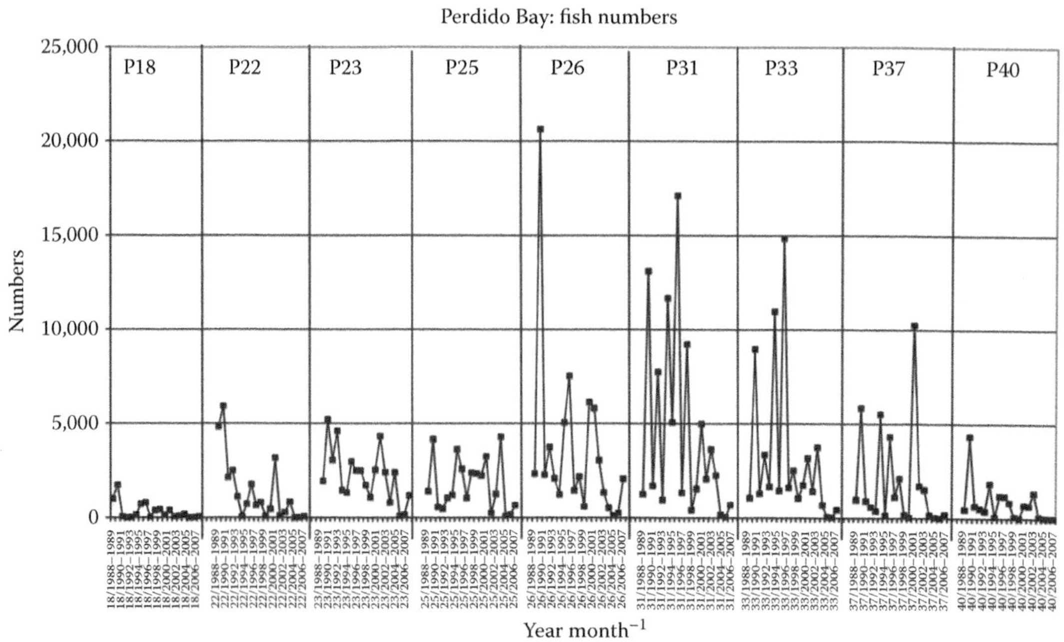

Figure 7.70 Annual averages of fish numbers in the Perdido Bay from October 1988 through September 2007.

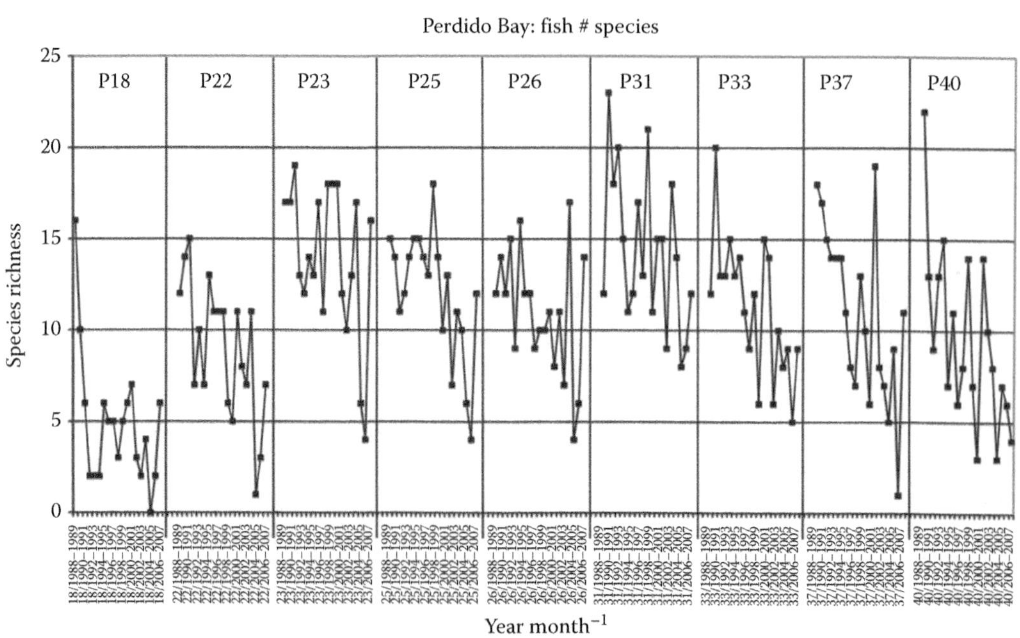

Figure 7.71 Annual averages of fish species richness in the Perdido Bay from October 1988 through September 2007.

Table 7.11 Numerical Abundance of Fishes, Averaged over the 20-Year Sampling Period by Station

Species	18	22	23	25	26	31	33	37	40
Leiostomus xanthurus	21.18	268.64	620.05	457.22	783.23	1068.10	758.04	302.93	80.70
Anchoa mitchilli	44.62	130.57	141.27	109.72	194.04	103.19	91.35	97.26	43.85
Brevoortia patronus	3.15	10.72	36.09	20.06	67.48	46.84	79.72	75.91	53.26
Micropogonias undulatus	3.42	28.89	35.14	10.79	27.72	60.88	48.47	10.51	2.72
Lagodon rhomboides	1.24	2.71	31.03	36.80	13.70	19.18	18.05	10.25	3.26
Arius felis	0.01	0.29	0.76	2.08	1.86	1.15	1.39	0.98	0.09
Eucinostomus argenteus	3.18	5.41	3.59	3.83	0.54	0.54	0.15	0.19	0.22
Sciaenidae sp.	0.35	0.99	0.36	1.32	0.52	0.30	0.89	4.34	0.41
Gobionellus boleosoma	0.04	0.32	0.84	0.98	0.58	1.14	0.45	0.40	0.81
Anchoa hepsetus	—	—	0.02	0.39	0.05	1.01	0.45	1.52	7.42
Anchoa nasuta	1.51	0.34	0.05	0.08	0.15	1.42	0.98	0.99	5.67
Trinectes maculatus	0.44	0.16	1.53	1.34	0.27	0.16	0.07	0.02	0.01

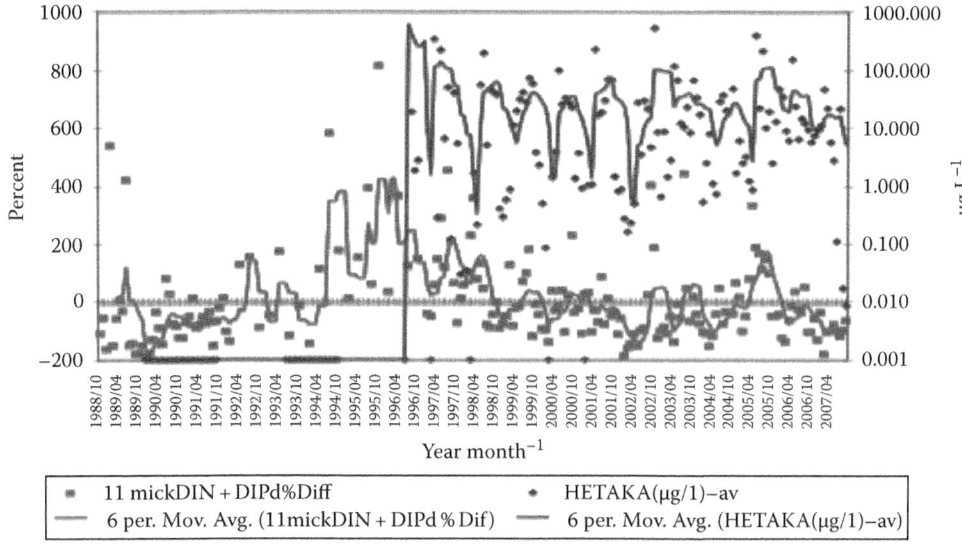

Figure 7.72 Elevenmile Creek DIP + DIN loading as percentages from the total mean plotted against annual averages of *H. akashiwo* biomass. Both indices are expressed as 6-month moving averages.

indications that blooms of *H. akashiwo* had an adverse impact on this species and that there was little recovery after the loss of the *Vallisneria* beds at station P25 following the 2002–2003 sewage spills and the increased pulp mill nutrient loads of 2005.

Streblospio benedicti

The polychaete, *S. benedicti*, was dominant at stations P23 and P25. The long-term changes of this species (Figure 7.73b) indicated adverse effects of both plankton blooms and the major droughts that led to the almost complete elimination of *Streblospio* from the Perdido Estuary by the end of the study.

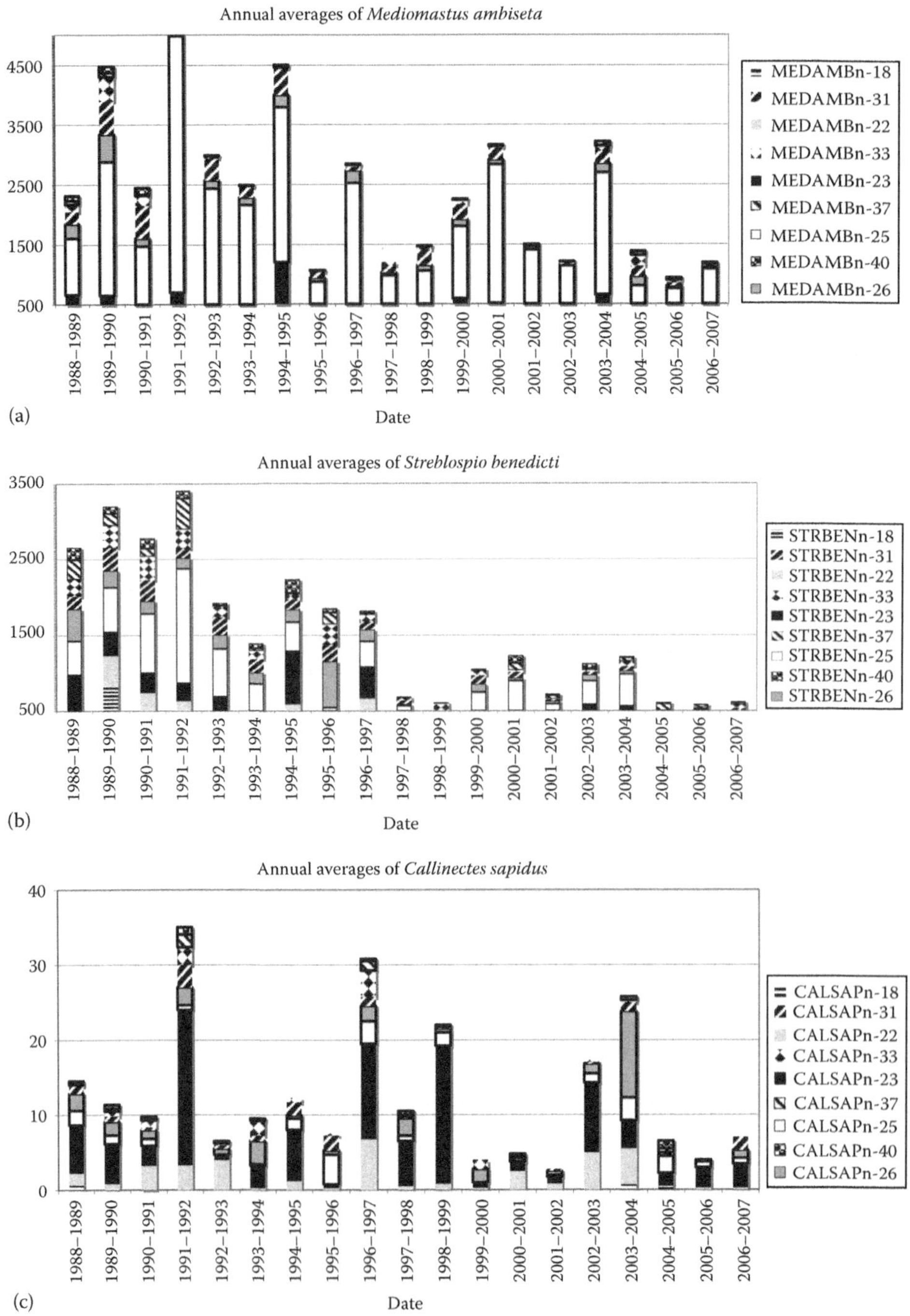

Figure 7.73 (a) Annual averages by station of *M. ambiseta* numbers m^{-2} in the Perdido Bay from October 1988 to September 2007. (b) Annual averages by station of *S. benedicti* numbers m^{-2} in the Perdido Bay from October 1988 to September 2007. (c) Annual averages by station of *C. sapidus* numbers in the Perdido Bay from October 1988 to September 2007. *(Continued)*

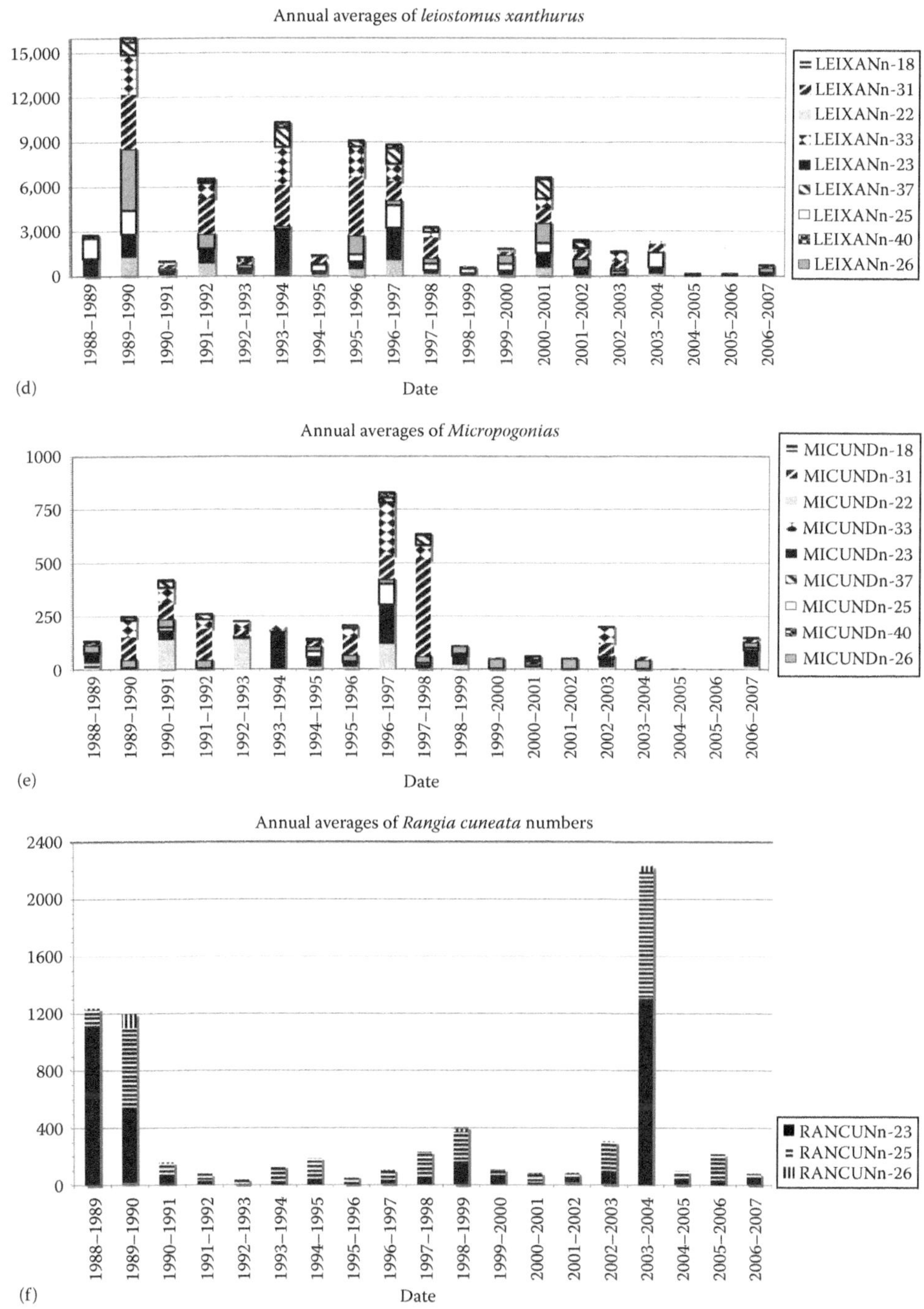

Figure 7.73 (*Continued*) (d) Annual averages by station of *L. xanthurus* numbers in the Perdido Bay from October 1988 to September 2007. (e) Annual averages by station of *M. undulatus* numbers in the Perdido Bay from October 1988 to September 2007. (f) Annual averages by station of *R. cuneata* numbers in the Perdido Bay from October 1988 to September 2007.

Callinectes sapidus

The blue crab, *C. sapidus*, was concentrated in the upper estuary (station P23). There were major numerical declines of this species that occurred after the initiation of plankton blooms and the prolonged droughts of 1999–2002 and 2006–2007 (Figure 7.73c). There were periods of recovery of blue crab numbers during years that occurred in between the blooms and reduced rainfall and river flows.

Leiostomus xanthurus

The spot, *L. xanthurus*, was concentrated in the upper and middle parts of the bay. This species went through a series of continuous decrease in numbers over the sampling period after periods of plankton blooms and drought periods (Figure 7.73d). At the end of the sampling period, there were virtually no spot left in the bay.

Micropogonias undulatus

Numerical abundance of the Atlantic croakers, *M. undulatus*, was concentrated in mid-bay areas. There were peaks in 1997–1999 with subsequent severe reductions of this species during the drought of 1999–2002 (Figure 7.73e). This species was also severely reduced by the end of the study.

Rangia cuneata

The common rangia clam, *R. cuneata*, was concentrated in the upper estuary at stations P23 and P25. *Rangia* was a biomass dominant in the Perdido Bay and was present mainly during the first 2 years of sampling and in 2003 and 2004, a period in between the major peak river flows of 2002–2003 and 2005 (Figure 7.73f). The second major increase in the clam numbers was composed mainly of young-of-the-year that were almost totally eliminated by increased river flows and accompanying *Heterosigma* blooms of 2005.

There were general declines of all of the dominant faunal populations of the Perdido Bay over the 20-year study period. Each species had different responses to the alternating flood and drought periods, but the overall patterns were similar with incremental declines ending in the low population numbers by the end of the study. The combination of disparate anthropogenic factors with the serial responses of the various dominant bloom plankton comprised the major elements that caused the impacts on the bay.

The long-term response of the trophic organization of a given coastal system to diverse anthropogenic activities depends on a series of interacting factors: the timing and qualitative/quantitative characteristics of anthropogenic nutrient loading, seasonal and long-term climatological conditions (drought and floods), the character and distribution of the habitats of the receiving system at the time of impacts, and the seasonal and interannual composition of existing phytoplankton assemblages and bloom species relative to the nutrient loading events (spring blooms during wet periods for *Heterosigma*; winter blooms during drought periods for *Merismopedia*). Food web components were thus subjected to the impacts of a complex combination of multiple factors within the context of the habitat distribution and the resilience of the system. The reactive nature of the plankton blooms was subject to climatic events that included a major hurricane and a series of prolonged droughts.

The 20-year study of the Perdido Bay system allowed a review of trophic relationships over a period of relatively significant climate change in the form of a direct hit by a major hurricane (Ivan) and two extended and severe droughts. The gradual long-term deterioration of the biological organization of the bay (Livingston 2007a,b) during this period should be analyzed relative to the trends of the interacting of key factors that proved to be statistically associated with the trophic processes of the bay. This would include salinity stratification, seasonal and interannual rainfall and river flow patterns, nutrient loading from the Perdido River and Elevenmile Creek, and associated plankton blooms in the Elevenmile Creek and Perdido Bay. The combined impacts of anthropogenic nutrient loading and extended droughts, together with an artificially stratified system (and associated hypoxia and sediment changes), contributed to the serial deterioration of estuarine populations and food web organization.

The reduced trophic organization in the Elevenmile Creek during the *Merismopedia* blooms was evident (Figure 7.74a). These trends were consistent with previous determinations of the origin and rise of the blue-green algae in the creek during the Perdido study. The FII TUs were highest at stations P25 and P23 where there was considerable interannual variability. The lowest such values occurred at stations in the Elevenmile Creek and at the mouth of the Perdido River. The general trends in these areas and at stations in mid- and the lower bay were downward over the 20-year sampling period. The FII herbivore distribution (Figure 7.74b) followed this pattern, whereas the C_1 carnivores (Figure 7.74c) underwent major decreases in 2003–2004 although similar trends were evident.

Data from the two upper bay stations (P23, P26) were averaged by year according to periods of *Heterosigma* seasonal abundance that peaked from April to July each year. Bloom biomass of this species was regressed against the primary factors (Table P.1). These data were used to construct PCA–regression analyses. With the exception of invertebrate herbivore and biomass and infaunal C_1 carnivore biomass, the *Heterosigma* blooms were negatively associated with the primary invertebrate and infaunal indices (Figure 7.75). Fish indices were also inversely related to the bloom numbers. With the exception of the FII omnivore index, all of the FII indices were negatively associated with *Heterosigma* biomass at stations 26 and 23. These analyses were consistent with

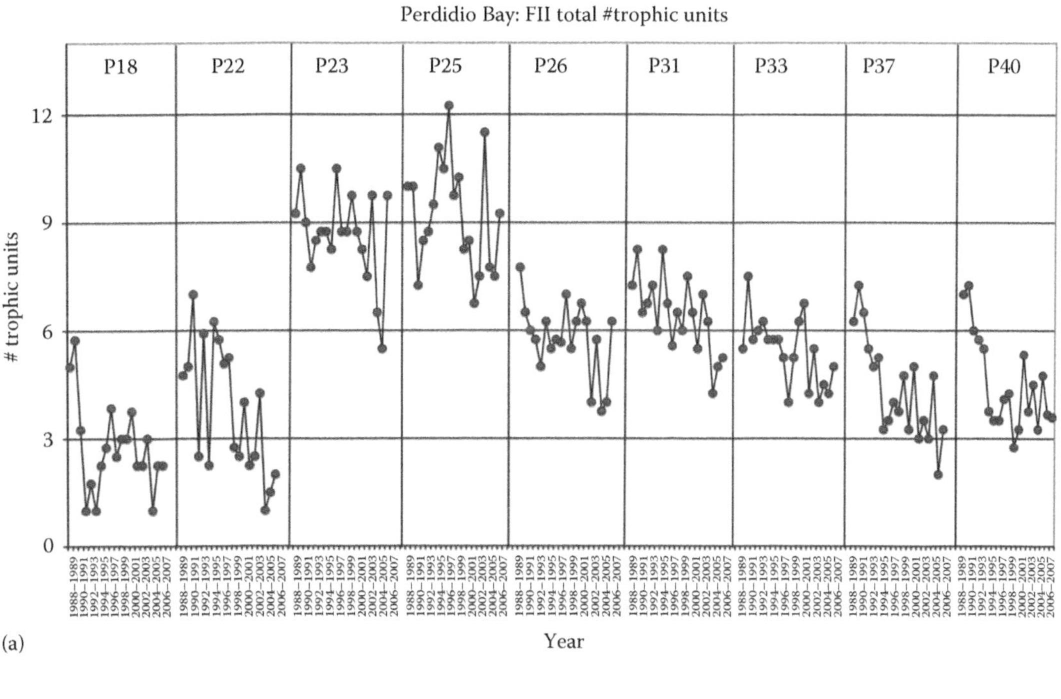

(a)

Figure 7.74 (a) Annual averages of FII total number of TUs in the Perdido Bay from October 1988 through September 2007. *(Continued)*

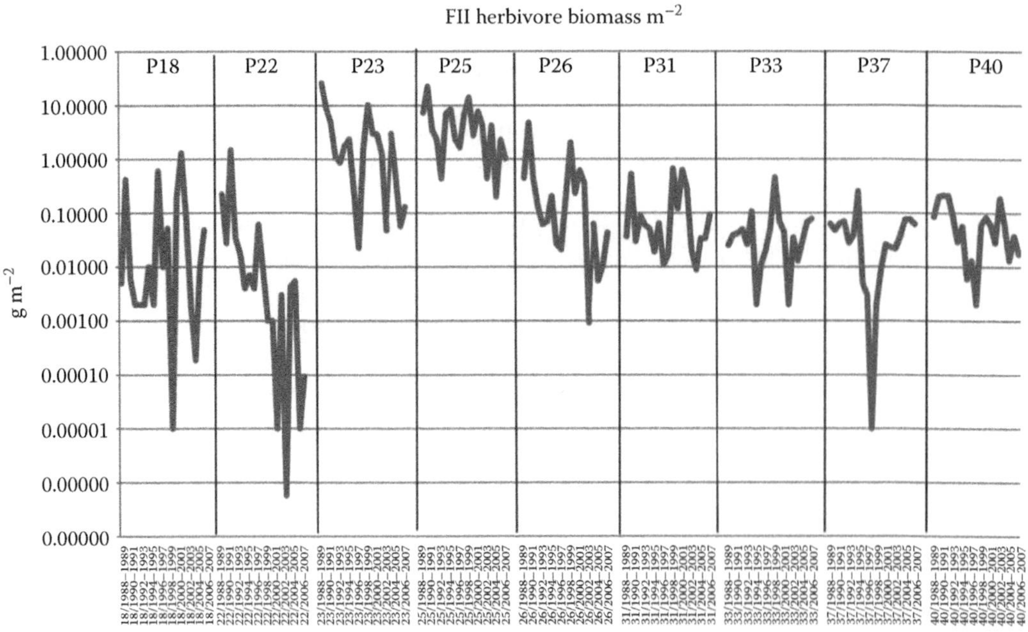

(b)

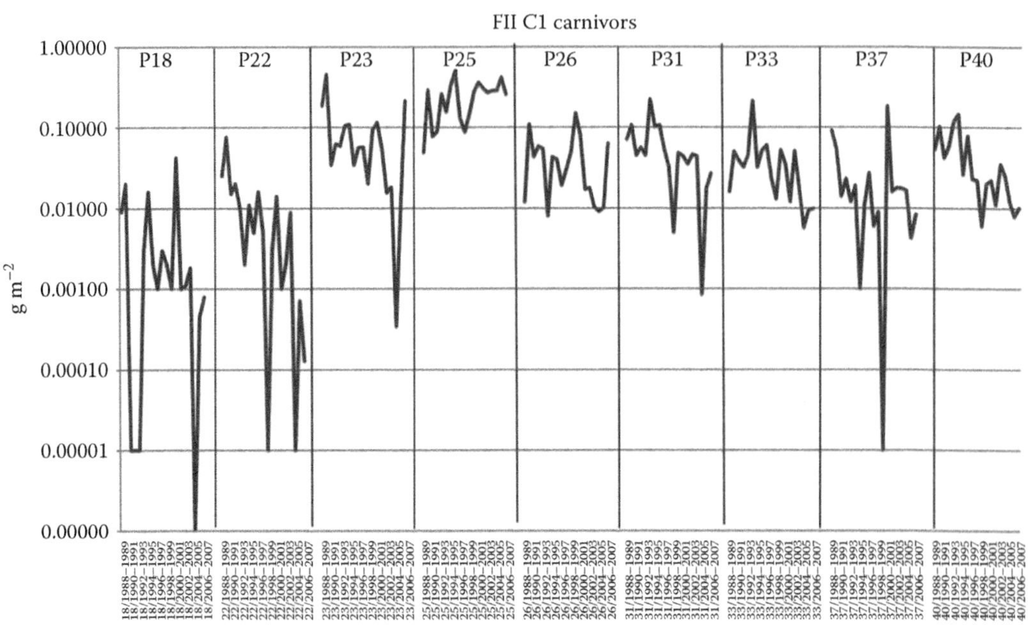

(c)

Figure 7.74 (*Continued*) (b) Annual averages of FII herbivores in the Perdido Bay from October 1988 through September 2007. (c) Annual averages of FII C_1 carnivores in the Perdido Bay from October 1988 through September 2007.

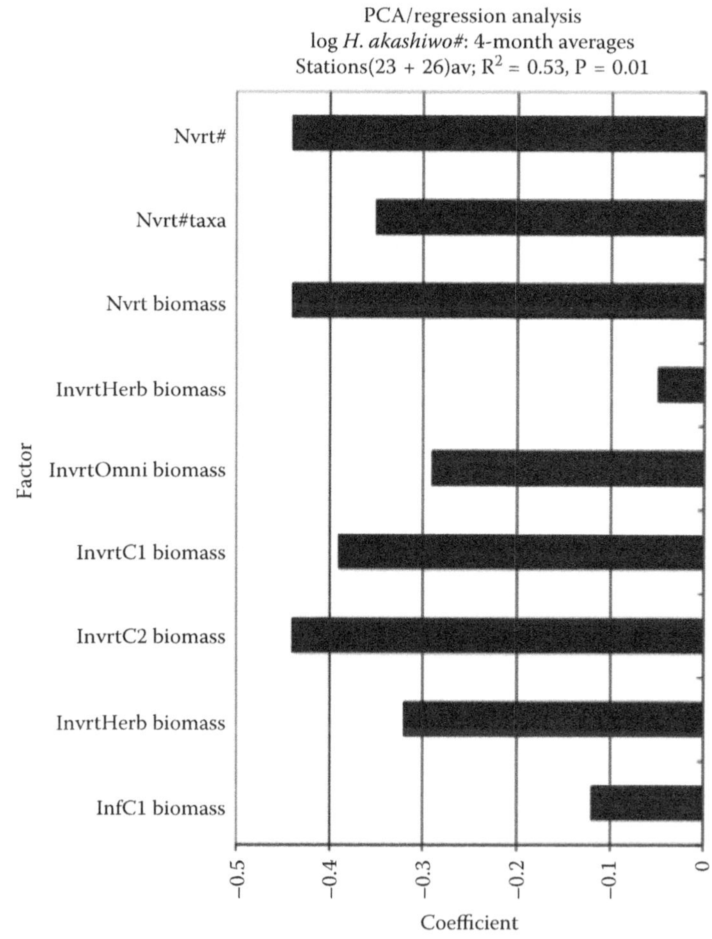

Figure 7.75 PCA/regression analyses of 4-month averages (April–July) of biomass of *H. akashiwo* from station averages of P23 and P26 versus infaunal and invertebrate indices (Table 7.11). Data were analyzed from October 1996 through September 2007.

the observation that the *Heterosigma* blooms in the Perdido Bay were closely associated with the temporal trends of the primary biological organization of the upper bay.

The most complete FII database was taken on stations 23, 31, and 40. Consequently, an analysis was made concerning the relationship of the main trophic indices with percent deviations from the grand mean of *H. akashiwo* cells L^{-1}. The raphidophyte had high numbers in 1995–1996 that continued during the following 2 years (Figure 7.76). Reduced nutrient loading during the succeeding years was followed by lowered bloom numbers, a situation that continued during the drought of 1999–2002. There was an increase in *Heterosigma* numbers during the increased nutrient loading in 2002–2003 and 2004–2005 that were followed by reduced blooms during the drought of 2006–2007.

Herbivore biomass was relatively high during the early years of sampling (Figure 7.77). This was followed by reduced biomass during the succeeding years characterized by plankton blooms. Increased herbivore activity occurred during the reduced *Heterosigma* blooms starting in 1998–1999 that was then followed by decreasing numbers of herbivores during the drought of 1999–2002. Very low levels of herbivore abundance occurred during the peak *Heterosigma* periods of 2002–2003 and 2004–2005. The drought years of 2006–2007 were characterized by low numbers of herbivores.

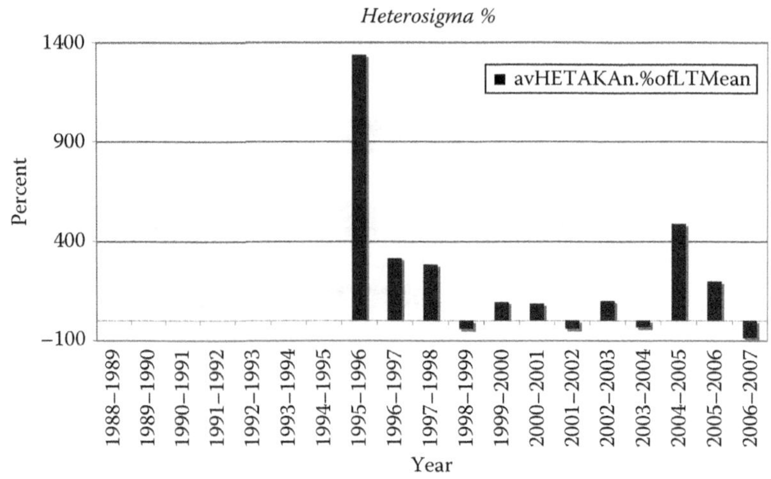

Figure 7.76 Percent deviations of mean numbers of *H. akashiwo* cells averaged from stations 23, 42, and 40 over long-term study period.

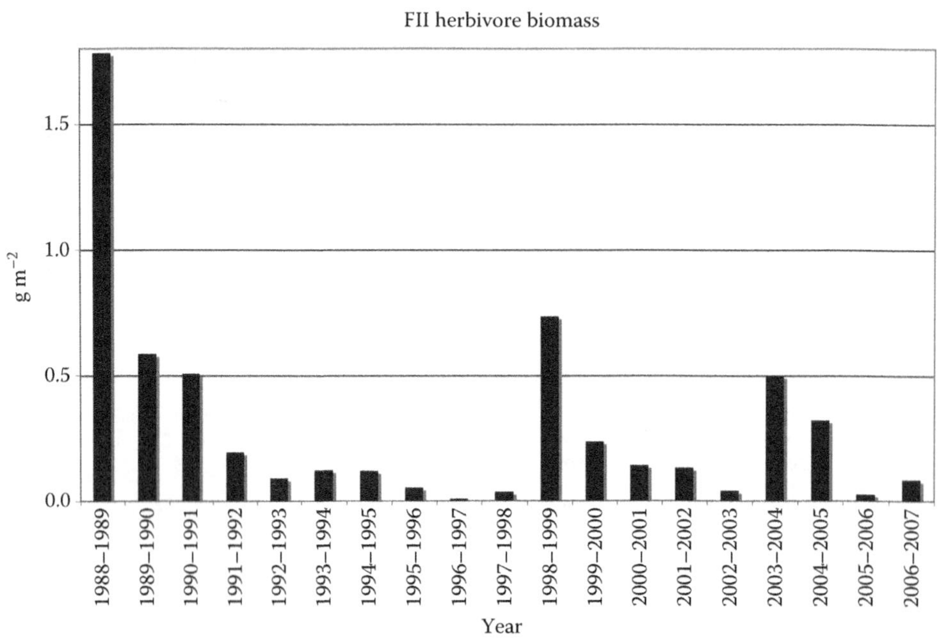

Figure 7.77 FII herbivore biomass averaged from stations 23, 31, and 40 over long-term study period.

Omnivore biomass (Figure 7.78) showed a different pattern with no particular association with the *Heterosigma* blooms. There was some indication of reduced numbers during the 1999–2002 drought although these trends were not substantial. It is obvious that the omnivore groups had a different response to the nutrient loading and plankton blooms as that of the herbivores.

The C_1 carnivores had biomass peaks during the first 3 years of the study (Figure 7.79). The period of *Heterosigma* dominance was characterized by serial reductions of the C_1 groups with some recovery during the early drought years. There was a pronounced reduction of this level of carnivory during the *Heterosigma* blooms of 2004–2005. Overall, there was a decrease of the C_1 carnivores over the 20-year period when compared to the first 3 years of the study.

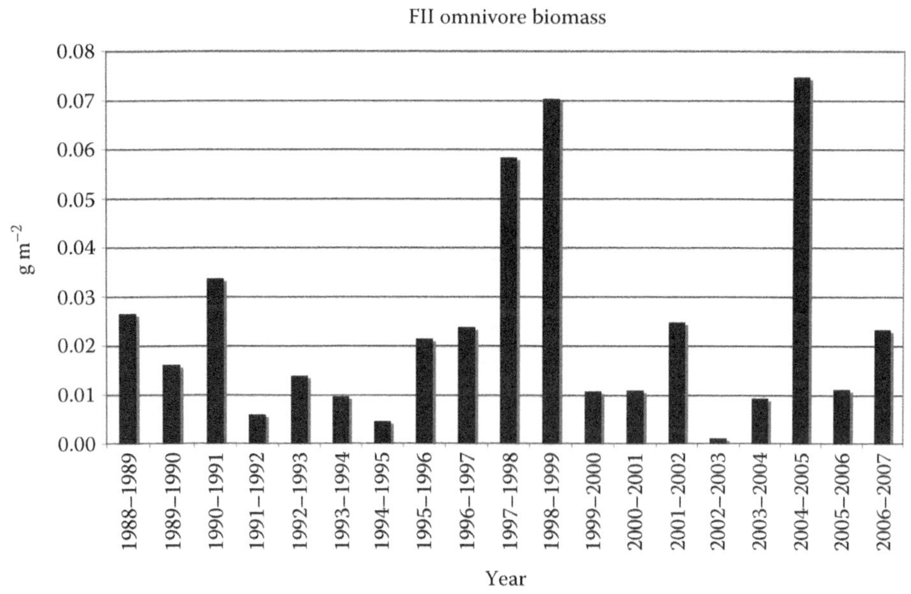

Figure 7.78 FII omnivore biomass averaged from stations 23, 31, and 40 over long-term study period.

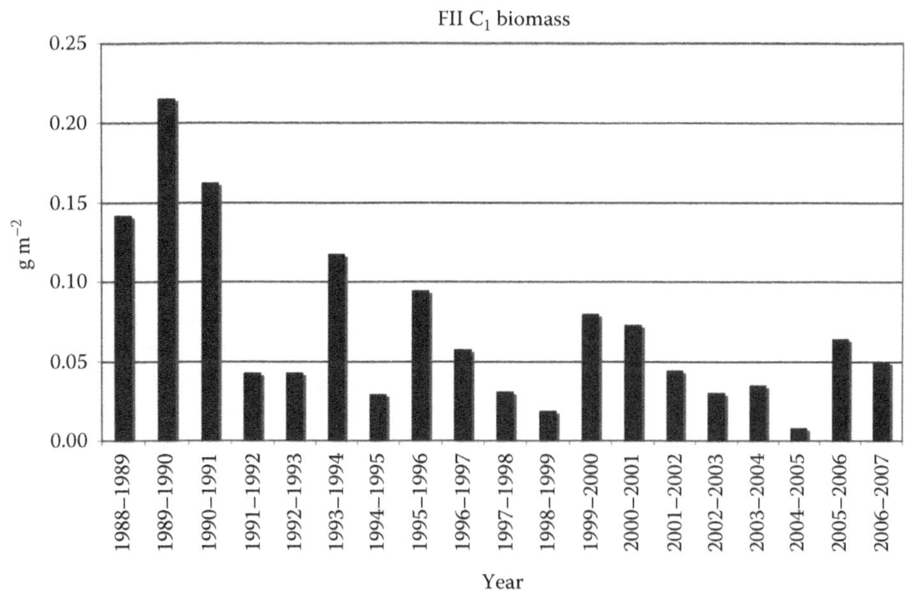

Figure 7.79 FII C_1 carnivore biomass averaged from stations 23, 31, and 40 over long-term study period.

A PCA–regression analysis (Figure 7.80) indicated no statistical association between omnivore biomass and *Heterosigma* cell numbers. However, there were strong negative correlations with various FII indices that included biomass of herbivore, C_1 carnivores, C_2 carnivores, total biomass, numbers of taxa, and TUs. This analysis presents further proof of the hypothesis that blooms of the raphidophyte, *H. akashiwo*, were statistically associated with reductions of the primary FII indicators of community structure and trophic organization.

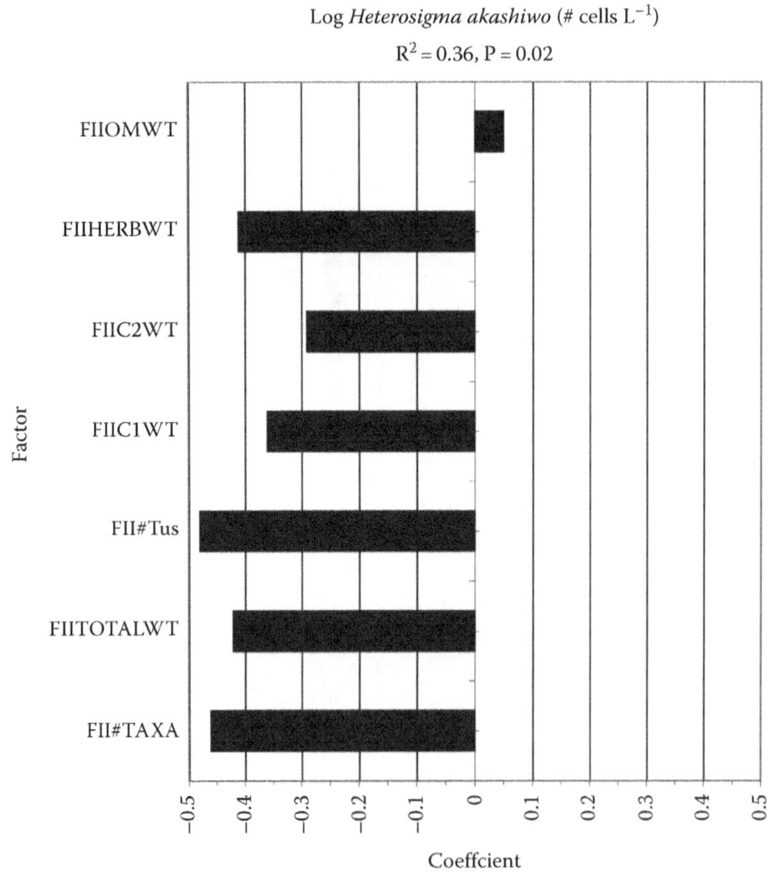

Figure 7.80 FII PCA–regression analysis of FII indices averaged from stations 23, 31, and 40 over long-term study period.

An analysis was carried out of the bay-wide distribution of trophic indices in the Perdido Estuary over the 20-year period of study. There was a reduction of the overall numbers of TUs from 1990–1991 to 1993–1994 followed by reduced TU diversity during the period of high but decreasing levels of *Heterosigma* biomass from 1994–1995 to 1999–2000 (Figure 7.81). The increased nutrient loading from the STP during fall 2002 and 2003 was associated with increased levels of *Heterosigma* biomass and associated reductions of the cumulative TUs. This was followed by reduced loading and associated declines of *Heterosigma* biomass. Increased nutrient loading during spring 2005 was associated with increased *Heterosigma* biomass and the lowest levels of cumulative TUs taken during the study. The drought of 2006–2007 was characterized by reduced nutrient loading and *Heterosigma* biomass and increased FII TUs.

The FII total numbers of species were highest during the first 3 years of the study (Figure 7.82). During this period, the species richness was relatively evenly represented at the various stations in the estuary. There were general declines of this index during the drought of 1999–2002. The lowest levels of the overall number of species in the bay were reached during the blooms of 2004–2005 and the beginning of the drought of 2005–2006. An overall downward trend of the FII species richness index was evident following the peaks during the first 3 years of the study.

The highest levels of herbivore biomass (Figure 7.83a) occurred at stations P23 and P25 during the first 3 years of sampling when there were moderate river flows and an absence of plankton blooms. The herbivores appeared to be particularly sensitive to the bloom periods of 1992–1994, 1995–1998,

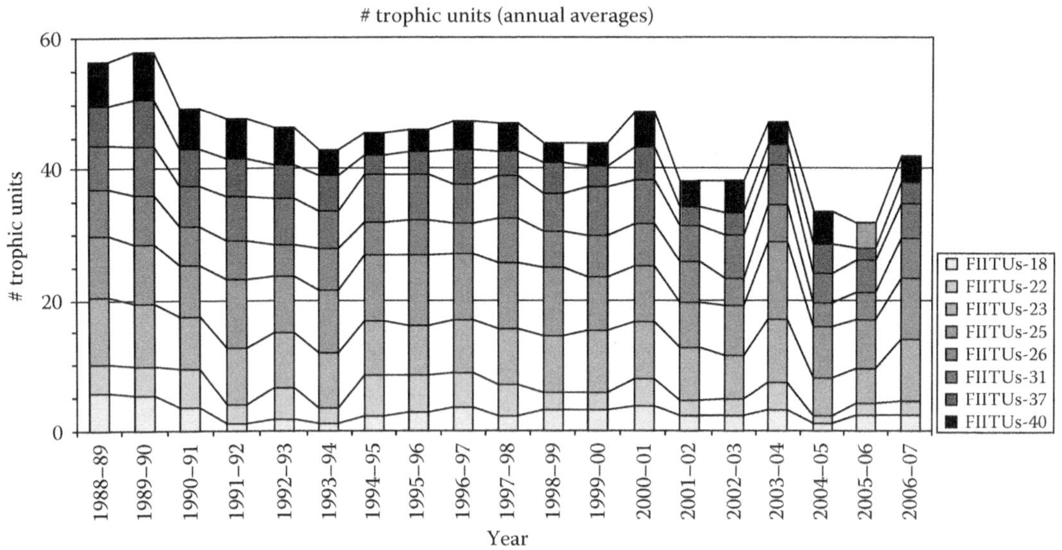

Figure 7.81 **(See color insert.)** Numbers of TUs by station and averaged by year in the Perdido Estuary from 1988 through 2007.

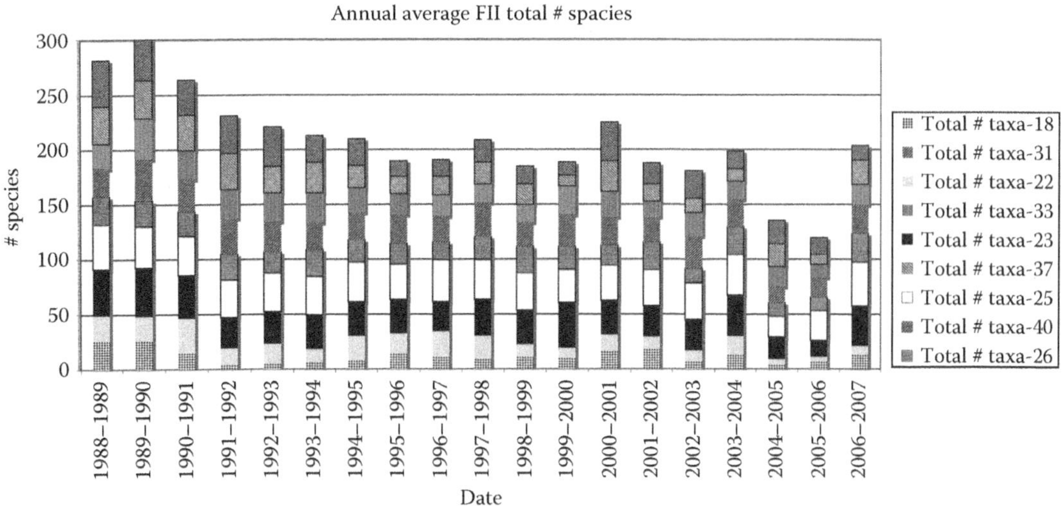

Figure 7.82 Annual averages by station of the FII total number of species in the Perdido Estuary from October 1988 to September 2007.

2003, and 2005. There were also reductions that occurred during the two major drought periods. The lowest levels of the herbivore presence in the bay occurred from 1995 to 1997 and from 2004 to 2007. During the final years of the study, the herbivores were largely absent from the Perdido Estuary.

Omnivore biomass (Figure 7.83b) showed a different long-term pattern with peak levels occurring at stations P23 and P25 during the period of enhanced *Heterosigma* blooms and sharply reduced herbivore biomass. This possible interaction of these two trophic levels in the Perdido Bay appeared to follow a pattern similar to that in the Apalachicola Estuary when the herbivore biomass declines during latter phases of the 1980–1981 drought were accompanied by increased numbers

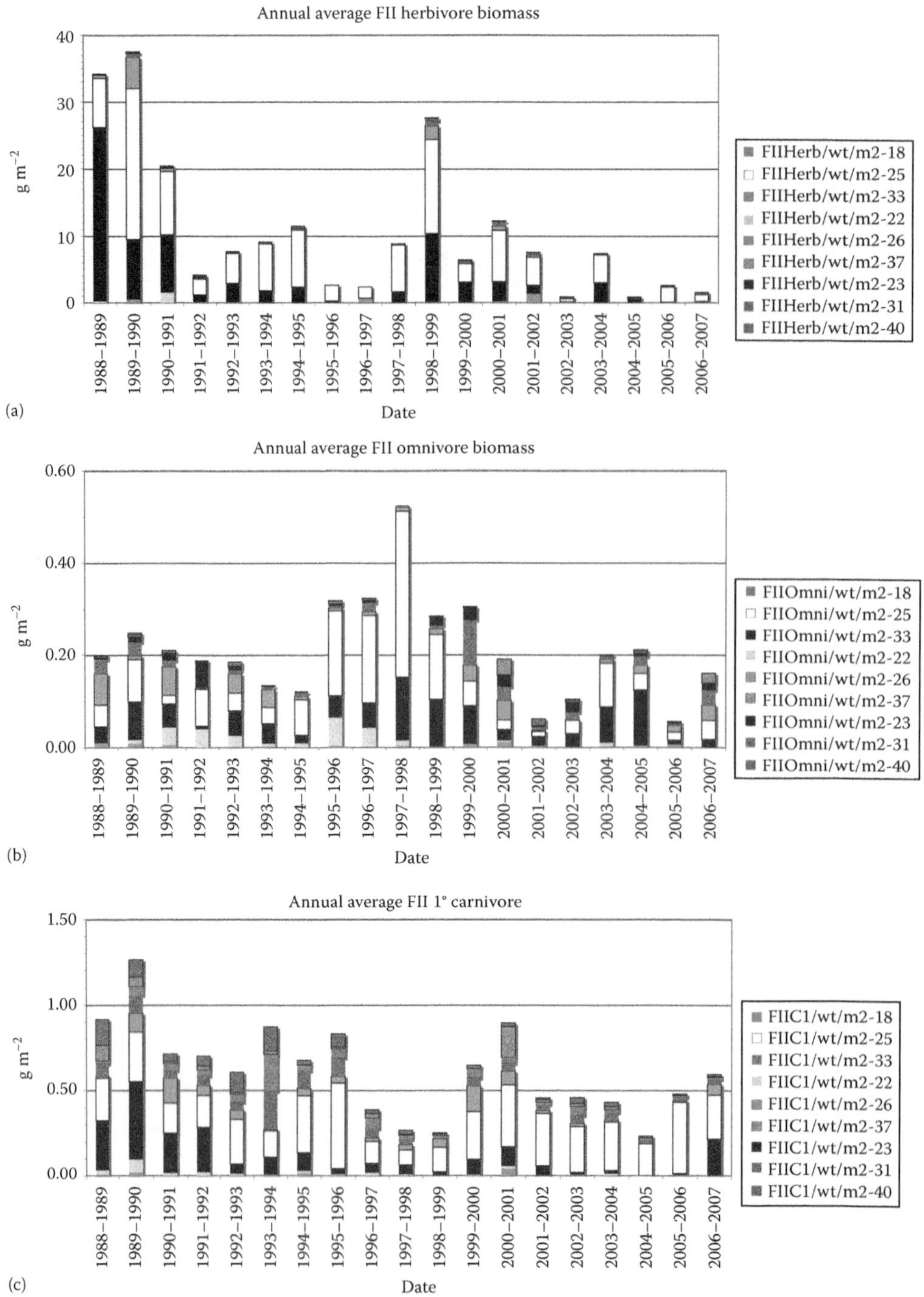

Figure 7.83 (a) Station-specific annual averages of FII herbivore biomass taken in the Perdido Estuary over the period from October 1988 to September 2007. (b) Station-specific annual averages of FII omnivore biomass taken in the Perdido Estuary over the period from October 1988 to September 2007. (c) Station-specific annual averages of FII C_1 carnivore biomass taken in the Perdido Estuary over the period from October 1988 to September 2007. *(Continued)*

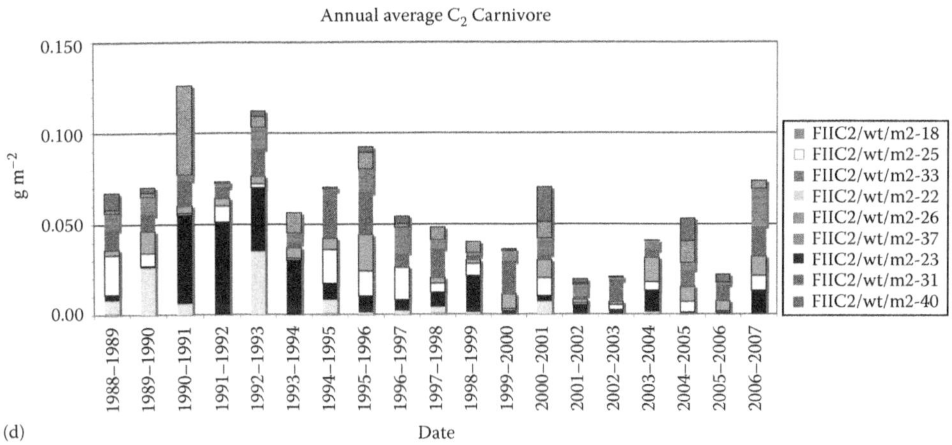

(d)

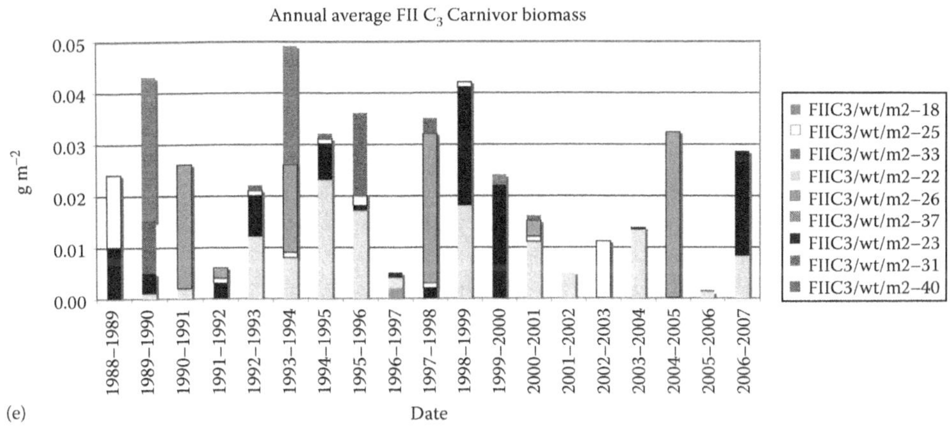

(e)

Figure 7.83 (*Continued*) (d) Station-specific annual averages of FII C_2 carnivore biomass taken in the Perdido Estuary over the period from October 1988 to September 2007. (e) Station-specific annual averages of FII C_3 carnivore biomass taken in the Perdido Estuary over the period from October 1988 to September 2007.

of omnivores. In the Perdido Estuary, the peak years of omnivore productivity occurred during the period of 1995–2000 during which time peak levels of raphidophyte blooms occurred.

The FII primary (C_1) carnivores (Figure 7.83c) were taken in highest biomass concentrations during the early years of sampling with major decreases occurring during periods of low herbivore presence from 1996 to 1998 and during later bloom periods. The primary carnivores at station P23 tended to decrease with time over the sampling period.

The FII C_2 carnivores (Figure 7.83d) and C_3 carnivores (Figure 7.83e) tended to decrease with time following a similar pattern of very low biomass figures that typified the herbivores and C_1 carnivores. The C_2 carnivores (Figure 7.83e) appeared to have a more erratic temporal distribution than the other levels of carnivory.

The FII species richness and trophic indices were relatively low at the mouth of the Perdido River (P18) and in the Elevenmile Creek (P22). The highest development of these indices was located at the mouth of the Elevenmile Creek (P23) and in the upper bay seagrass areas (P25) during the first 3 years of the study. Species richness was at an intermediate level at station P26. There was a general decrease of FII species richness down the bay (P31, P33, P37) with a slight increase at station P40. This pattern was followed by the TU index with the exception that there was no discernible increase at station P40.

The C_1 predators were most prevalent during the early years at stations P23 and P25. With increased nutrient loading, C_1 biomass at P23 decreased with time. By 2004–2006, C_1 predators were primarily restricted to station P25. The C_2 predators were distributed over a number of stations during the early years, with subsequent decreases following bloom periods. Low C_2 levels were reached during the blooms of 2001–2003 and 2005–2006. There was a general decrease of C_2 predator biomass with time. The C_3 predators were located mainly at stations P26 and P31 during the early years. Increased C_3 predators occurred at station P23 in 1998–2000. After that, there was a major decrease of this trophic level in most of the bay.

The *Heterosigma* blooms occurred primarily in the upper bay (P23, P26, P18) with lesser concentrations in mid- and lower bay areas. Blooms of this species originated at the mouth of the Elevenmile Creek (station P23) during periods of excessive nutrient loading from the pulp mill. The highest concentrations occurred at station P26 during increased nutrient loading in the Bayou Marcus marsh area (sewage treatment plant spills) in 2002–2003. These blooms affected the areas at lower bay stations. The blooms were not noted in the upper bay (P23) during this period but appeared to have had a major effect on both indices at station P26. Although *H. akashiwo* occurrence was not significantly associated with community indices in the lower bay, there were simultaneous reductions of FII species richness and trophic diversity during the blooms of spring 2005.

The annual FII biomass data were averaged across the various forms of impacts (blooms, droughts) and compared to averages taken during the period of regular river flows and no blooms (October 1988–September 1991) (Figure 7.84). The overall FII biomass was considerably higher during the first 3 years of sampling than the succeeding years. There was a trend during the earlier years of somewhat higher impacts caused by the *H. akashiwo* blooms compared to the droughts. However, during the last 5 years of sampling, the bloom and drought-induced impacts were comparable and were lower than the previous periods of such impacts. These trends indicate increasingly effective impacts of plankton blooms and droughts that were probably related to reductions of bay resiliency with time.

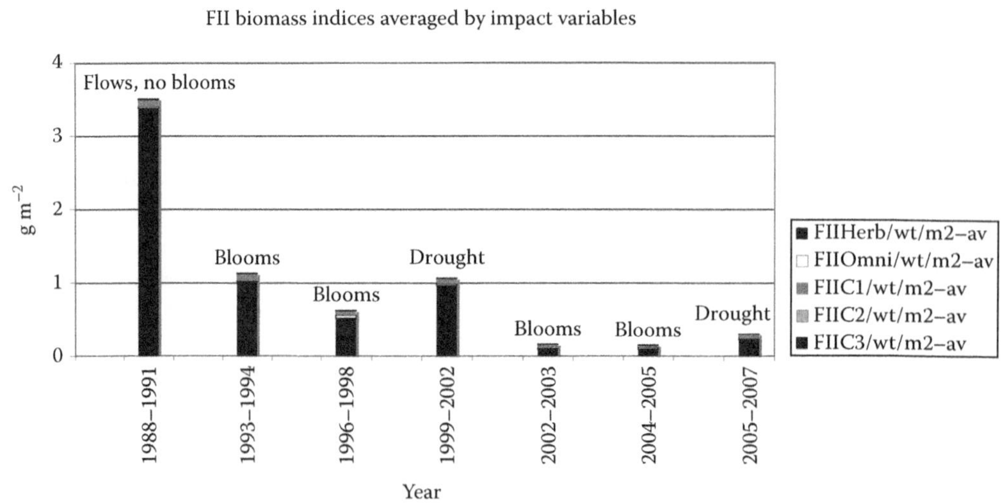

Figure 7.84 Annual averages of FII biomass taken in the Perdido Bay over the period from October 1988 to September 2007. Data were averaged across periods of river flows and no blooms (1988–1991), early bloom period (1993–1994), *H. akashiwo* blooms (1996–1998), the first main drought (1999–2002), *H. akashiwo* blooms (2002–2003), *H. akashiwo* blooms (2004–2005), and the final drought (2005–2007).

Conclusions

Each part of the Perdido Bay (river mouth, upper bay, lower bay) had specific relationships that were associated with the local habitat conditions. These factors were defined by water depth and proximity to areas receiving anthropogenic nutrient loading. In a stepwise regression, about 44% of the variability of DO distribution in the bay could be explained by depth, salinity, and temperature. DO distribution in the Perdido Bay was a function of temperature, depth, salinity concentration, and the level of stratification. With an increase in bottom salinity during warm periods of the year, there was an associated reduction of DO. The hypoxia at depth was statistically associated with the reduction of various biological indices. This loss of production was similar to the findings in other bay systems (Livingston, 1984a, 1989b; Livingston et al., 2000).

Temporal variation of phytoplankton assemblages has been ascribed to combinations of the timed nutrient loading events, nutrient limitation processes during such loading, light characteristics, and water temperature along with various biological and physicochemical modifying factors (Fabien et al., 2006). Salinity is a major limiting factor in coastal systems (Badylak and Philips, 2004). In river-dominated estuaries, river flow can control effects on down-estuary nitrogen delivery, residence time, salinity, and turbidity (Borsuk et al., 2004). In addition to the alignment of phytoplankton composition along salinity gradients, various habitat factors influenced the location of phytoplankton assemblages within distinct hydrological regions of the Perdido Estuary. Water depth, the pattern of prevailing water currents, and turnover rates affected the plankton distributions. Together with nutrient loading rates, nutrient concentration gradients, and the climatic conditions at the time of the loading events, these physiographic elements determined the source and distribution of plankton blooms in the estuary. The importance of interannual changes in the climate was evident in the long-term data.

There were species-specific seasonal successions of plankton blooms (Livingston, 2000). This proved to be an important factor relative to the long-term climatological changes over the study period. The dinoflagellate *P. cordatum* was usually dominant during winter months (January–February). The diatom *C. choctawhatcheeana* was usually dominant during spring months (April–May). The raphidophyte *H. akashiwo* prevailed from April through July, with blooms usually associated with rainfall/river flow peaks of nutrient loading. When *Heterosigma* co-occurred with *Cyclotella*, the diatom was repressed showing a marked interspecific interaction with the raphidophyte that affected bloom sequences.

The damaging blue-green algae (*M. tenuissima*) blooms occurred during various months of the year with particular high numbers during drought conditions. The diatom *Synedropsis* sp. was a summer bloom species, whereas the diatom *C. throndsenii* was a spring bloom. The diatom *Urosolenia eriensis* was a fall bloom species. The raphidophyte *C. subsalsa* bloomed during early summer, whereas populations of various dinoflagellates (*Gymnodinium* spp.) tended to peak during summer–fall months. There was thus a general seasonality in the occurrence of various bloom species in addition to a specific response, in most cases, to the timing as well as the type of nutrient being loaded.

In the Perdido Estuary, nutrient concentrations during drought conditions (i.e., relatively low nutrient loading rates) in the Elevenmile Creek stimulated the damaging blue-green algae (*M. tenuissima*) blooms, whereas the combination of high nutrient concentrations and high nutrient loading during episodic high-flow conditions led to sustained blooms of the toxic raphidophyte (*H. akashiwo*) populations in the receiving bay. The *Merismopedia* blooms were largely restricted to the Elevenmile Creek, whereas the *Heterosigma* blooms were started in the upper bay but were driven by currents to affect other parts of the system. These distributions of blooms were thus based on seasonal and interannual climatic trends and the proximity to anthropogenic nutrient loading.

Elevated nutrient concentrations from pulp mill and sewage treatment plant loading sparked the *Heterosigma* blooms in the immediate entry points of the bay. Nutrient loading from the Perdido River sustained the blooms as they moved down the bay. The dominant bloom species that had the most effect on the bay proved to be *M. tenuissima* in the Elevenmile Creek and *H. akashiwo* that bloomed mainly in the upper bay. Neither species was present during the early years of infrequent drought occurrence and low nutrient loading from the creek (pulp mill). In this way, the combination of seasonality, anthropogenic nutrient loading, and long-term climatic trends (and changes) led to the plankton blooms that did so much damage to the Perdido Bay.

Seasonal and interannual climatological conditions relative to the anthropogenic nutrient loading and concentration factors along with ecological characteristics of individual plankton species affected the initiation of blooms and the biological response of the estuarine food webs to such planktonic outbreaks. Drought conditions and high nutrient concentrations led to the *Merismopedia* blooms in the Elevenmile Creek. Peak river flows associated with storms were closely associated with the *Heterosigma* blooms in the upper bay. The raphidophyte blooms were dependent on orthophosphate loading that, together with the already high ammonia levels, were needed to sustain the outbursts of planktonic activity. Nutrient limitation experiments supported the influence of N + P loading in bloom initiation.

The timing of the nutrient loading depended on the pattern of long-term drought–flood conditions. *H. akashiwo* has a high growth potential and can form a bloom in a short time, excreting allelopathic substances that can suppress diatom growth and dramatically reduce cell numbers in other phytoplankton species. An understanding of the species composition of the plankton populations and the natural history of the bloom species of receiving areas of the bay proved to be fundamental to the determinations of bloom stimulation and impact. Interspecific plankton interactions affected the composition of plankton assemblages relative to bloom proliferation.

The diatom, *C. choctawhatcheeana*, a major dominant in the plankton assemblages in the Perdido Estuary, was inversely associated with the *Heterosigma* blooms. *Cyclotella* showed bay-wide distributions from blooms that originated in the upper bay and responded to a combination of environmental factors that ranged along inshore–offshore gradients of light and nutrients with salinity as a possible modifying factor. As a relatively important species in terms of acting as a food source for zooplankton, the effects on *Cyclotella* due to *Heterosigma* blooms could be a factor in the alteration of food webs in the bay.

During the first 3 years of the Perdido study, the bay had relatively high levels of primary and secondary productivities. There were no signs of hypereutrophication, and various parts of the upper bay had *Vallisneria* beds and productive shallow water habitats. Pulp mill effluents contaminated the Elevenmile Creek with high conductivity, high ammonia levels, and low DO that adversely affected the biological organization of the creek. The deeper areas of the Perdido Estuary (>2 m) were affected by water- and sediment-quality deterioration below a well-established halocline that adversely affected benthic biota. The upper (relatively shallow) bay was little affected by salinity stratification, a factor that could be related to the higher levels of secondary productivity in such areas.

During the first 3 years of study, there were signs of periodic high nutrient loading from the paper mill in the form of orthophosphate and ammonia. There were periods of increased phytoplankton productivity during high nutrient loading from the pulp mill. This finding led to an extension of the court-ordered study of the impacts of the paper mill on the Perdido Estuary. By the early 1990s, there was an outburst of plankton blooms that followed the increased loading of nutrients by the mill. This increase of orthophosphate and ammonia loading to the bay was accompanied by massive blooms of the raphidophyte *H. akashiwo* that, in turn, were associated with reduced biological productivity in affected parts of the bay. These blooms persisted during periods of high rainfall and river flow right to the end of the study in 2007.

Together with blooms caused by nutrient loading from a sewage treatment plant in the upper bay, the productivity at stations P23 ad P26 was reduced, and with the disappearance of the *Vallisneria* beds of the western sections of the upper bay (P25), biological productivity continued its downward trends during the remainder of the study. This included impacts on the marsh areas of the upper bay that were subject to *Heterosigma* blooms associated with nutrient loading from the pulp mill (Livingston, 2007b). The loading of orthophosphate, in the presence of high ammonia concentrations, stimulated the blooms during periods of high rainfall and river flows.

During the latter part of the study, major droughts (1999–2002, 2006–2007) were associated with reduced nutrient loading from the Perdido River. These periods were associated with the proliferation of blue-green algae blooms (*M. tenuissima*) in the Elevenmile Creek that caused with further reductions of biological productivity in the already polluted creek. Meanwhile, nutrient loading from agricultural land in the Wolf Bay drainage of the lower Perdido Estuary stimulated massive *Heterosigma* and *Prorocentrum* blooms that eliminated much of the benthic biota of this part of the estuary. Another factor that adversely affected estuarine productivity in deeper parts of the bay (P18, P31, P33, P37, P40) was the salinity stratification caused by dredging of a pass from the lower bay to the gulf. By the end of the study, virtually all parts of the Perdido system had been adversely affected by various forms of anthropocentric activities. The combination of climate changes in the form of droughts and floods, together with physical alterations and nutrient loading, nearly eliminated secondary production of the Perdido Estuary by the end of the study in 2007.

With the exception of the pulp mill, which responded to the study by reducing its nutrient loading and moving its effluent out of the Elevenmile Creek, not one of the other anthropogenic activities was subject to regulation by state or federal environmental regulatory agencies. The USEPA, which actually has a marine laboratory in the Pensacola Estuary, either was unaware of these problems or chose to ignore the scientific data. In fact, the Gulf Breeze EPA lab, together with the Florida Department of Environmental Protection, disputed the findings of the Perdido Bay study even though neither agency had much in the way of scientific information of its own. In short, what started as a court-ordered study of the impacts of the paper mill ended as a case analysis of the destruction of a complete estuarine system. To this day, the rapidly increasing urbanization around the Perdido Bay remains immune to any regulation by the Florida and Alabama state agencies or by the USEPA. Nutrient loading from the Alabama agricultural interests into the Wolf Bay was not acknowledged during the Perdido study.

The response of food web components to blooms should be placed within the context of habitat distribution relative to the point of loading, species-specific bloom composition, and temporal aspects (seasonal, interannual) of bloom occurrence. However, there are differential impacts within the overall trophic spectrum. The resilience of the trophic organization of the bay system relative to the repetitive nature of the plankton blooms was dependent on complex temporal changes (natural and anthropogenic) that occurred over long-term periods. Repetitive blooms had cumulative, adverse impacts on food web organization through time. These impacts were reviewed within the context of the occurrence of prolonged droughts that also had additive adverse effects on the bay populations. In this way, the size, location, and frequency of the species-specific blooms relative to nutrient loading changes and drought occurrences over time were implicated in reduced resilience and the almost complete elimination of secondary productivity in the Perdido Bay Estuary. Impacts on the trophic organization of the estuary were cumulative and resulted in different responses by the various trophic levels to the impacts of the blooms and increased drought occurrence.

Extended droughts that became increasingly frequent over the past 30+ years in the eastern Gulf of Mexico have been associated with reductions of nutrient loading in various river-dominated estuaries along the Gulf Coast. These droughts resulted in major reductions in the secondary productivity of the Perdido Estuary. The adaptive capacities of individual species represented a key factor in the structural breakdown of the estuarine food webs. The loss of resilience due to the sequential occurrence of plankton blooms and extended droughts was evident in the increased effectiveness of such impacts with time. The importance of seasonal and interannual variability of the various controlling habitat factors of the estuary should not be underestimated in the evaluation of the status of the Perdido Estuary.

A comparison of the effects of the various anthropogenic activities on the estuaries of the NE Gulf Coast is given in Figure 7.85. The Perdido and Apalachicola River–Estuaries are alike in that the estuaries receive flows from major rivers that deliver dissolved nutrients and dissolved/particulate organic matter that increases the overall productivity of these bays. The opening of these estuaries to the gulf increased the salinities of both systems. However, the impacts on the relatively shallow Apalachicola Bay were limited to the loss of some oyster beds, whereas the deeper Perdido Bay had extensive (bay-wide) salinity stratification, associated reductions of benthic DO, and the creation of liquid muds that combined to reduce benthic productivity in major parts of the estuary. Eventual anthropogenic nutrient loading from point and nonpoint sources in the Perdido Estuary caused damaging plankton blooms during low-flow (blue-green algae) and high-flow (raphidophytes) conditions. These blooms were associated with further reductions of secondary productivity; this represented another difference between the Perdido and Apalachicola Estuaries, the latter having little in the way of urban runoff and anthropogenic nutrient loading. Both estuaries were affected by droughts.

Apalachee Bay differed from the more western estuaries in that there were no major rivers on this estuary. A series of small streams emptied into the high-salinity bay that derived its major source of habitat and productivity from dense seagrass beds that covered major parts of the inshore waters. This difference with the plankton-dominated large river–estuaries resulted in major differences of the respective food web structures and the presence of major commercial fisheries (oysters, penaeid shrimp, blue crabs, fishes) in the large river–estuarine systems. The Econfina Estuary was a natural

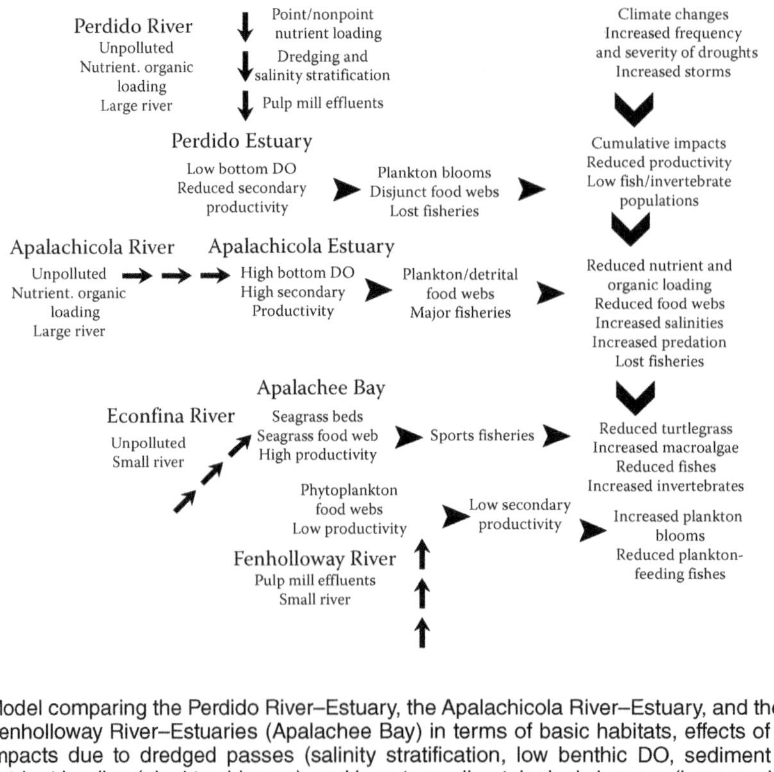

Figure 7.85 Model comparing the Perdido River–Estuary, the Apalachicola River–Estuary, and the Econfina and Fenholloway River–Estuaries (Apalachee Bay) in terms of basic habitats, effects of anthropogenic impacts due to dredged passes (salinity stratification, low benthic DO, sediment deterioration), nutrient loading (plankton blooms), and long-term climatological changes (increased frequency and severity of droughts and storm activity).

blackwater system dominated by seagrass-oriented food webs, whereas the polluted Fenholloway Estuary was characterized by the reduction of such seagrasses due to the impacts of pulp mill effluents. The offshore Fenholloway systems were dominated by plankton-related food webs of relatively low productivity in a way that superficially resembled the food webs of the large river–estuaries.

The impacts of long-term climatological changes in the study area included a series of droughts that become more frequent and severe with time. The impacts of such climatological trends on the subject estuaries were varied due to differences in the natural habitats, differing forms of anthropogenic activities (dredging, nutrient loading, pulp mill effluents), and the reaction of the different estuaries to changes in water quality and reduced nutrient loading during droughts. Both the Apalachicola and Perdido Estuaries underwent losses of secondary productivity due to reduced loading of nutrients and organic matter during droughts. The cumulative impacts of the various anthropogenic activities in the Perdido Estuary, with associated reductions of estuarine resilience, resulted in the lowest levels of secondary productivity noted in the various estuaries along the NE Gulf Coast.

The droughts had a devastating effect on the formerly very high productivity of the Apalachicola Estuary with the loss of major regional commercial fisheries. The impacts on the Apalachee Bay estuaries were more complex, and areas outside of the impacts of pulp mill effluents in the Fenholloway Estuary responded to the interplay of droughts and increased storms. In the unpolluted Econfina Estuary, there were changes in water quality and the loss of turtle grass along with increased biomass of macroalgae, reduced fish biomass, and increased invertebrate numbers. In the Fenholloway Estuary, there were changes in the planktonic food webs with increased plankton blooms and reduced numbers of plankton-feeding fishes such as bay anchovies during periods of storms that followed long-term droughts.

In sum, the climatological trends that were region-wide in the coastal study area had very different impacts on the estuaries along the NE Gulf Coast. The Apalachicola Estuary, adapted to very high nutrient loading from natural sources, underwent no hypereutrophication impacts. Indeed, the reduction of the loading of nutrients and organics led to the loss of the highly productive fisheries of this estuary, whereas increased anthropogenic nutrient loading in the Perdido and Fenholloway Estuaries led to the establishment of destructive plankton blooms. The effects of climatological changes thus depended on factors that were related to differences in natural habitat conditions and the relative impacts of anthropogenic nutrient loading and dredging in the individual estuaries.

PART IV

Impacts of Anthropogenic Nutrient Loading

CHAPTER 8

Estuarine Response to Urban Nutrient Loading

CHOCTAWHATCHEE ESTUARY

A study was carried out in the Choctawhatchee River–Bay system during the mid-1980s. The field methods used are outlined in Appendix A. Statistical methods (Appendix B) have been described by Livingston (1986a,b, 2000, 2002, 2005) and Livingston et al. (1997, 2002).

Background

The Choctawhatchee Estuary (Figure 8.1) is a drowned river plain surrounded by shallow shelf–slopes and inshore bayous (Livingston, 1986a,b) and is aligned along an east–west axis. Freshwater input is primarily provided by the Choctawhatchee River with secondary inflows from a series of bayous located primarily in northern sections of the bay. Saltwater input comes from the gulf through the dredged East Pass, a channel that was opened by a group of citizens during particularly heavy freshwater flooding in 1929. The "blowout" of East Pass was accompanied by increased salinity throughout a now stratified Choctawhatchee Bay. High salinities were associated by local observers with the loss of emergent (marsh, swamp) and submergent (seagrass) vegetation throughout the bay. Choctawhatchee Bay currently has one of the least well-developed fringing (emergent) vegetation zones of the various gulf estuaries in the region (Reyer et al., 1988). The artificial creation of the channel across Santa Rosa Island has also contributed to erosional and depositional patterns in areas of the Choctawhatchee Estuary.

The Choctawhatchee River, the third largest in Florida in terms of freshwater discharge, is formed by a series of tributaries in Alabama and Florida. Mean streamflow rates range between 156 and 198 m^3 s^{-1} with low flows approximating 96 m^3 s^{-1} and high flows of 736 m^3 s^{-1} (US Geological Survey, Tallahassee, Florida; unpublished data). Choctawhatchee Bay (Figure 8.1) has average depths of about 3 m (9.8 ft) in eastern sections and about 8 m (26.2 ft) in western sections. Over the past 35 years, there has been a major increase of urbanization in western parts of the bay (Livingston, 1986a,b, 2000). Eastern sections of the bay remain relatively undeveloped.

The Choctawhatchee River basin lies in a south temperate region characterized by mild winters and hot/humid summers. Land use within the Alabama portion of the Choctawhatchee basin is dominated by forestry (51.7%) and agriculture (cropland, 30.6%; pasture, 11.6%) (Alabama Water Improvement Commission, 1979). There is relatively little urban development in the region (3.2%) with the exception of extensive urban growth in western sections of Choctawhatchee Bay. The river drainage and eastern parts of the bay remains largely undeveloped with forestry and agriculture as the major land uses. There are high ambient values of nutrients in parts of the river system due to agricultural and urban runoff. There are numerous impoundments of this area. Upper Holmes Creek has water quality problems due to discharges from the Florida communities of Graceville (via Little Creek), Chipley (via Alligator Creek), Vernon (via Little Branch),

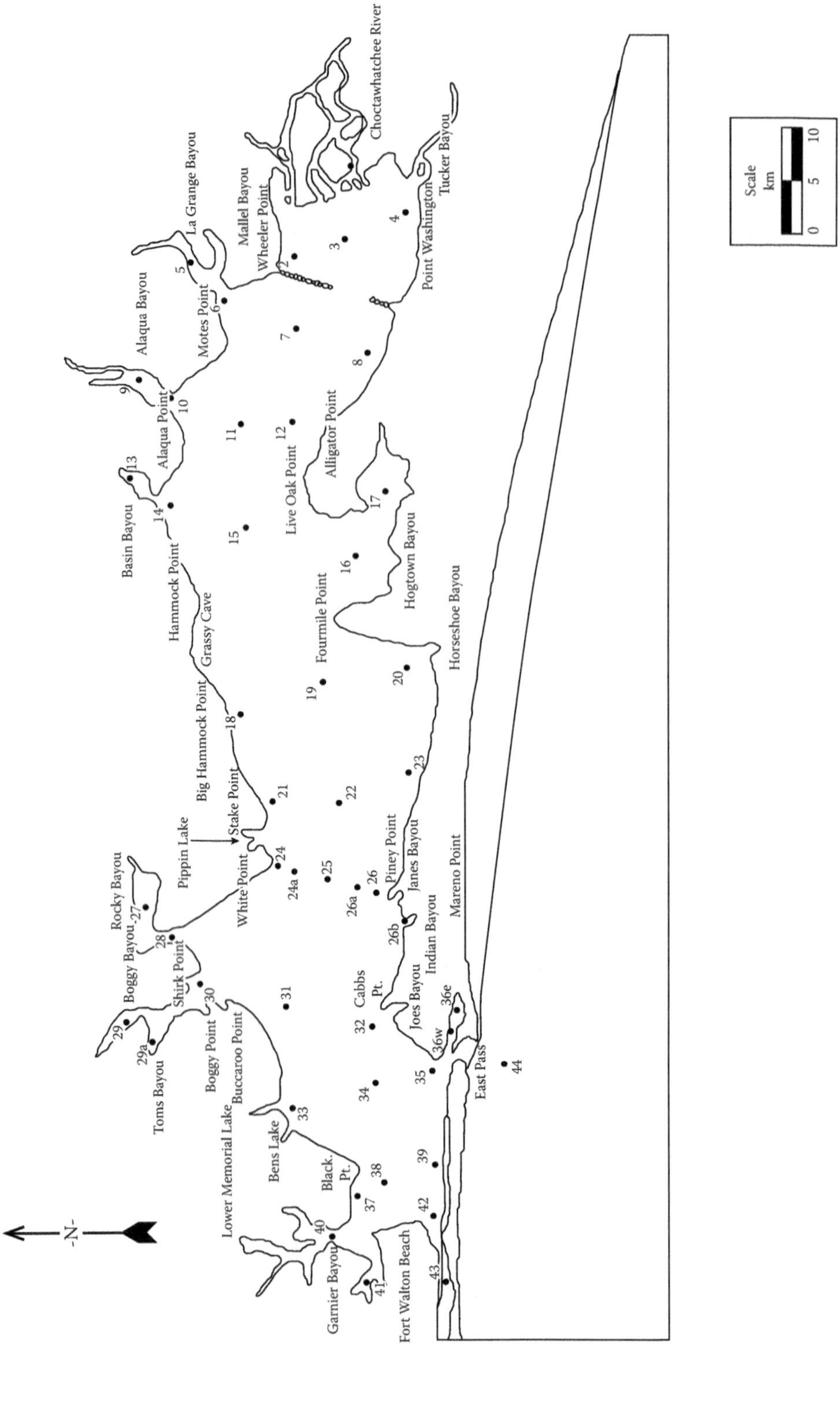

Figure 8.1 Choctawhatchee Estuary showing distribution of sampling stations.

and Bonifay (via Camp Branch). Such problems are due largely to sewage discharges. Holmes Creek also receives runoff from agricultural areas such as hog farms. West Sandy Creek also has degraded water quality due to sewage from DeFuniak Springs and Bruce Creek receives effluent from a chicken processing plant.

The Choctawhatchee Estuary represents a series of complex habitats that are strongly influenced by freshwater input (river, bayou drainages) and salinity stratification associated with the opening of the bay to the Gulf of Mexico. Various human activities associated with urbanization in the western Choctawhatchee basin (e.g., stormwater runoff and sewage spills) have contributed to the deterioration of natural aquatic resources in receiving bay areas. According to Livingston (1987a), discharges of sewage and stormwater associated with unrestricted land development in western sections of Choctawhatchee Bay were responsible for water quality deterioration in a series of bayou areas in the northern sections of the bay and in Destin Harbor (Old Pass Lagoon) to the south.

Several physiographic and geomorphological features contribute to observed ecological characteristics of the Choctawhatchee system (Figure 8.1). The bay has shallow shelf areas and relatively steep slopes that eventually level out to average depths of 3 m in eastern sections and 8 m in western sections. This arrangement amounts to a west-sloping drowned river plain surrounded by shallow shelf–slopes and inshore bayous. These features determine vertical and horizontal water quality and salinity distributions in Choctawhatchee Bay and associated ecological attributes of the estuary.

Aerial photographs from 1955 and 1985 indicated that there has been an overall loss of almost 20% of the submergent aquatic vegetation (seagrass beds dominated by shoalgrass, *Halodule wrightii*) in the western portion of the bay over this 30-year period. Loss of seagrass productivity was probably greater than that indicated by areal loss because the most dense seagrass beds were mainly affected. The deterioration of the seagrass beds was probably due to changes in the optical characteristics of the water column (i.e., possible changes in water color, turbidity, etc.). Most of the losses were in deeper (slope) regions, whereas increases in intermediate coverage occurred in shallower areas.

In recent times, there has been a major increase of urbanization within the Garnier Bayou complex, the Boggy Bayou basin, and the lower Rocky Bayou basin (Figure 8.1). High rates of development have also occurred in the White Point and Moreno Point (Destin) drainages, with more moderate increases in the Fort Walton Beach and Eglin Village areas. Relatively little increase in coastal basin usage was noted in the Walton County (eastern bay) regional drainage basins.

The Choctawhatchee Estuary was studied on a monthly basis from 1985 to 1986 during a period that preceded the drought of 1986–1988. The bay is characterized by a wide distribution of habitats that are defined by depth relationships and proximity to freshwater flows from uplands areas. Habitat delineation involved the identification of the following key areas: shallow-slope habitat (vegetated, unvegetated, oyster beds), deep regions of the central basin (unvegetated), bayous, river areas (delta), and a gulf pass (East Pass). Superimposed over such designations were temporal, horizontal, and vertical salinity distributions. Regional rainfall over the study period tended to peak during winter and summer months. Choctawhatchee River flow peaks during winter–early spring months, with low-flow levels during the summer and fall. Climatological conditions during fall 1985 were somewhat unusual because of the presence of three hurricanes in the northern Gulf of Mexico. The most important effects of these storms on the Choctawhatchee region included episodic increases in precipitation and some limited incursions of increased wind speeds.

Water temperature peaked from June to August, with low levels from December to February. The fall temperature transition was in November; the spring transition was in March. Salinity gradients in the bay followed river flow trends; lowest salinities occurred from December through April at the surface of the bay. Peak salinities were noted during summer–fall months. Turbidity and color were associated with river flow conditions; high color levels from December through March occurred in various bayous.

River Flow and Nutrient Loading

Choctawhatchee River flows (Table 5.1) were considerably less than those in the Apalachicola River with peaks in both rivers during winter–early spring months. The largely unpolluted Apalachicola system could be considered a reference area for nutrient loading. Ammonia loading (Table 5.1) in the Choctawhatchee system was comparable to that in the Apalachicola system but was lower during low-flow conditions. Nitrite–nitrate loading was also lower in the Choctawhatchee system. Nitrogen levels due to local runoff conditions were highest in western sections of the bay (Cinco, Garnier, lower Rocky Bayous, and Old Pass Lagoon). Orthophosphate loading in the Choctawhatchee River was generally higher than that in the Apalachicola system. Phosphorus levels were highest in the urbanized areas of Old Pass Lagoon, lower Rocky Bayou, and Boggy Bayou.

Overall, stormwater runoff and sewage disposal were thought to be associated with higher nutrient concentrations in both the river and the bay, which, in turn, could have affected sediment quality and water quality conditions (i.e., dissolved oxygen [DO]) in receiving parts of the bay. Direct associations were made between episodic surface water flows to the bay and increases in nutrients, particulate organic matter, chemical oxygen demand (COD), and low DO during summer periods.

Salinity Stratification and Dissolved Oxygen

Along the primary axis of Choctawhatchee Bay, midbay areas tended to be the deepest of those analyzed. There was a general east-west gradient of increasing depth. Regions along the periphery of the bay were generally less than 2 m deep. The deepest parts of the bay were positively correlated with high stratification.

DO was inversely correlated ($R^2 = -0.88$) with stratified conditions (Table 8.1 and Figure 8.2). There was a general decrease of such correlation with decreasing stratification. Areas above the halocline usually had averaged DO concentrations greater than 4.0 mg L^{-1}. A regression of

Table 8.1 Regressions of Bottom DO by Variables Taken at the Main Choctawhatchee Bay Stations with Data Averaged over the First 12 Months of Sampling (9/1985–8/1986)

Bottom DO Regressions					
Depth/Sediments		**Water**		**Biological**	
		Main stations			
Factor	R^2	Factor	R^2	Factor	R^2
Depth	−0.50	bSal	−0.34	Infaunal #	0.39
Mean GS	−0.59	bSal–sSal	−0.88	Inf # taxa	0.49
% organics	−0.26	sTemp	0.19	FII biomass	0.45
% sand	0.57			FII TUs	0.28
% silt	−0.53			FII # taxa	0.33
% clay	−0.55				
		Bayou stations			
Factor	R^2	Factor	R^2		
Depth	−0.76	bSal	−0.32		
Secchi	−0.80	bSal–sSal	−0.38		
		Scolor	0.37		
		Sturbidity	0.60		
		Schlora	0.41		

Note: Only significant ($P < 0.05$) relationships are listed.

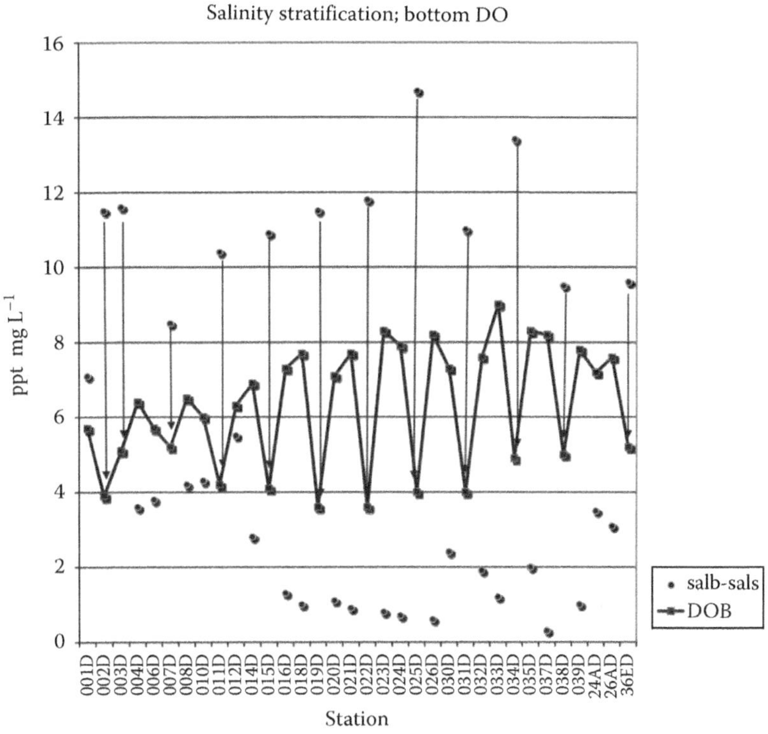

Figure 8.2 Salinity stratification and bottom DO based on 12-month averages from 1985 to 1986.

subhaloclinal DO with mean depths in the Choctawhatchee system showed a relatively high negative R^2 value (−0.50). The data thus indicated that there was a statistical association of depth, salinity stratification, and the extent of hypoxia in Choctawhatchee Bay.

Bottom DO was inversely related to silt/clay components and organics in sediments and was significantly (positively) associated with various infaunal and fish, infauna, and invertebrate (FII) indices (Table 8.1). Hypoxic conditions were noted in various western bayous and lagoons (lower Rocky, Boggy, Garnier, Old Pass Lagoon) at different times of the year (Figure 8.3). These areas coincided with high-density urban development. Garnier Bayou (station 40) was heavily developed, whereas Alaqua Bayou (station 9) was in a generally undeveloped state at the time of sampling. Bottom DO in Garnier Bayou was lower throughout the year. Data from the bayou stations indicated that high stratification in unpolluted upper Rocky Bayou was not associated with low bottom DO, whereas the various urbanized bayou areas (including lower Rocky Bayou: station 28) were highly stratified and were characterized by relatively low bottom DO.

Sediment and Water Quality Factors

Fine-grained, highly organic sediments were located in the deeper (central) parts of the bay and in the bayous (Figure 8.4). Shelf–slope habitats were characterized by sandy sediments. Sediment characteristics were determined largely by overland runoff and water depth. There was a definite alignment of high organic and silt concentrations in areas affected by pronounced salinity stratification. This effect was pronounced in the broad middle parts of the bay. High silt concentrations were statistically associated with low DO (Table 8.1). Liquid mud occurred in salinity-stratified parts of the bay.

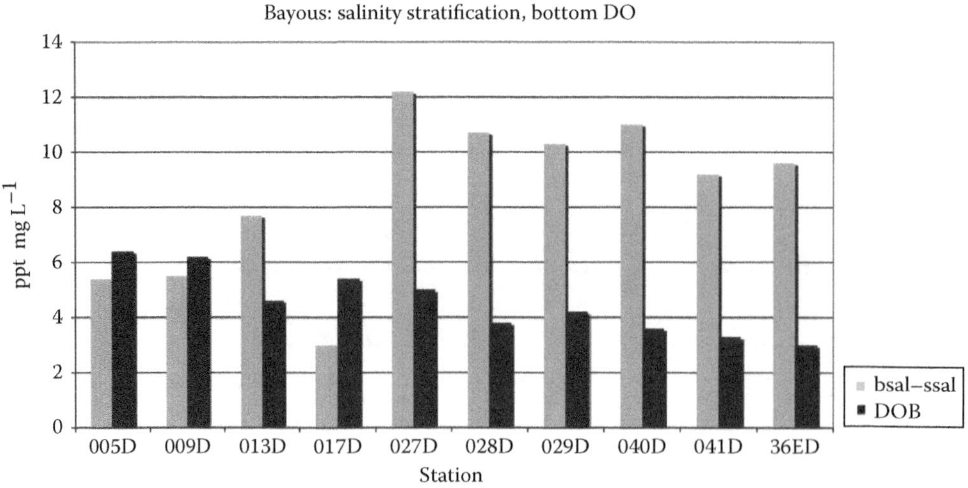

Figure 8.3 Twelve-month means of salinity stratification and bottom DO in bayous of the Choctawhatchee Estuary.

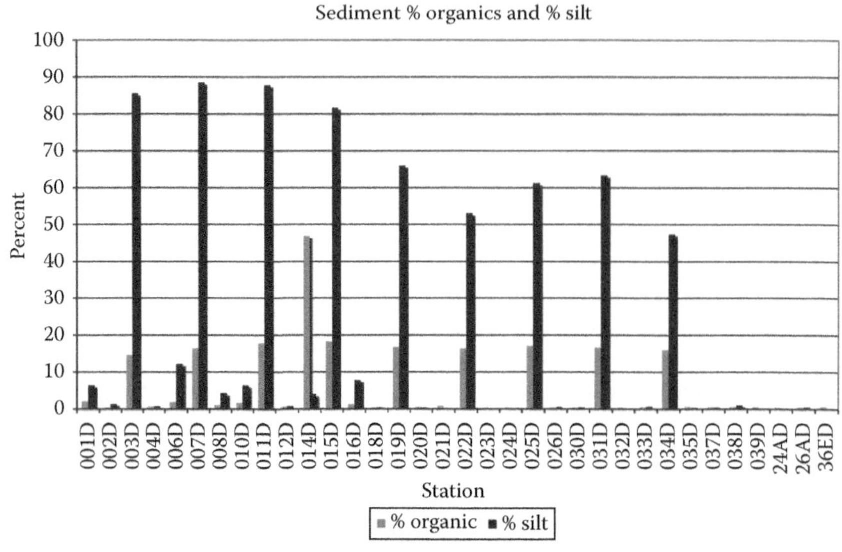

Figure 8.4 Sediment % organics and % silt of the Choctawhatchee Estuary.

Summer high COD in most of the deeper parts of the bay and areas bordered by high levels of urban development reinforced the likelihood that such areas were adversely affected by stormwater runoff; this is consistent with the periodic occurrence of hypoxia in such areas. Time-based, highly complex vertical changes in thermohaline stratification were noted in these areas of the bay. Low DO was most evident during summer months of uniformly high temperature and vertical water-column stratification. By August, virtually the entire bay was hypoxic to anoxic at depth.

Turbidity and water color in the upper bay were positively associated with river flow conditions; high color levels from December through March occurred in various bayous (Livingston, 1986a). Fecal coliforms were highest during summer months in the lower Rocky Bayou, Boggy Bayou,

Tom's Bayou, Cinco Bayou, and eastern Old Pass Lagoon. Summer high COD in most of the aforementioned areas reinforced the likelihood that such areas were adversely affected by urban runoff; this is consistent with the periodic occurrence of hypoxia in such areas. Nitrogen and phosphorus concentrations were higher in Old Pass Lagoon than in the upper bay. The nitrogen levels were generally highest in bayous in western sections of the bay (Cinco, Garnier, lower Rocky Bayous, and Old Pass Lagoon). Phosphorus levels were highest in Old Pass Lagoon, lower Rocky Bayou, and Boggy Bayou.

Direct associations could be made between episodic surface water flows from urbanized areas to the bay and increased nutrients, particulate organic matter, COD, and low DO during summer periods. Overall, stormwater runoff was thought to be associated with high nutrient loading that, in turn, affected sediment quality and water quality conditions (i.e., DO).

Plankton Distributions

Phytoplankton cell numbers (Figures 8.5 and 8.6) were lowest at the river mouth and were highest at stations C38 and C36e. Species richness was lowest in eastern parts of the bay and increased in western sections with peaks at stations C38 and C36e. Blooms (*Skeletonema costatum*, *Falcula hyalina*) were noted mainly in Old Pass Lagoon (C36e). The highest annual average phytoplankton numbers occurred in the urbanized Old Pass Lagoon and extreme western sections of the bay. This was further confirmation that areas affected by urban runoff were becoming culturally eutrophicated during this study.

The results of the PCA-regression analyses of key habitat factors in Choctawhatchee Bay and bayous with phytoplankton species richness are given in Figure 8.7. Phytoplankton species

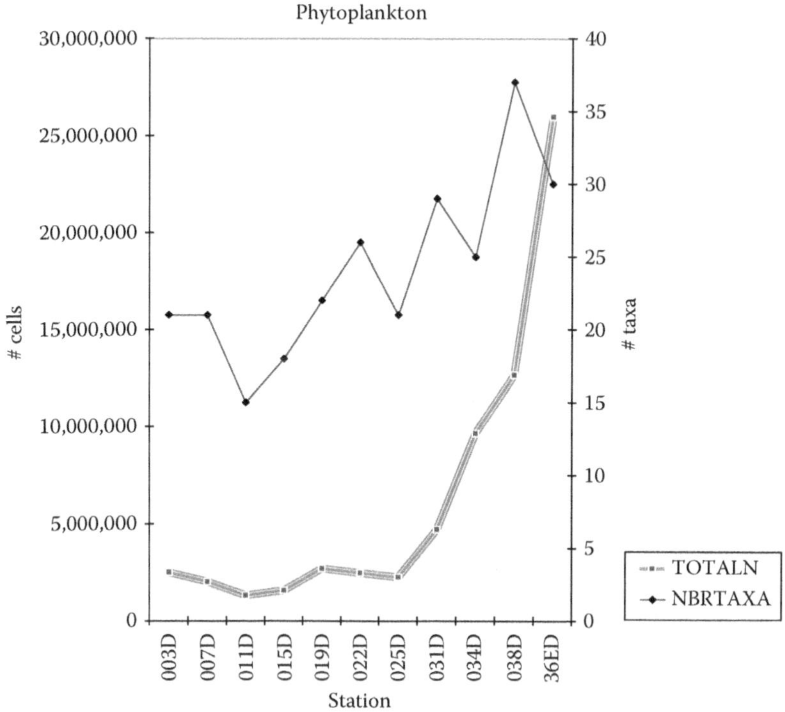

Figure 8.5 Phytoplankton numbers of cells and species richness in the Choctawhatchee Estuary.

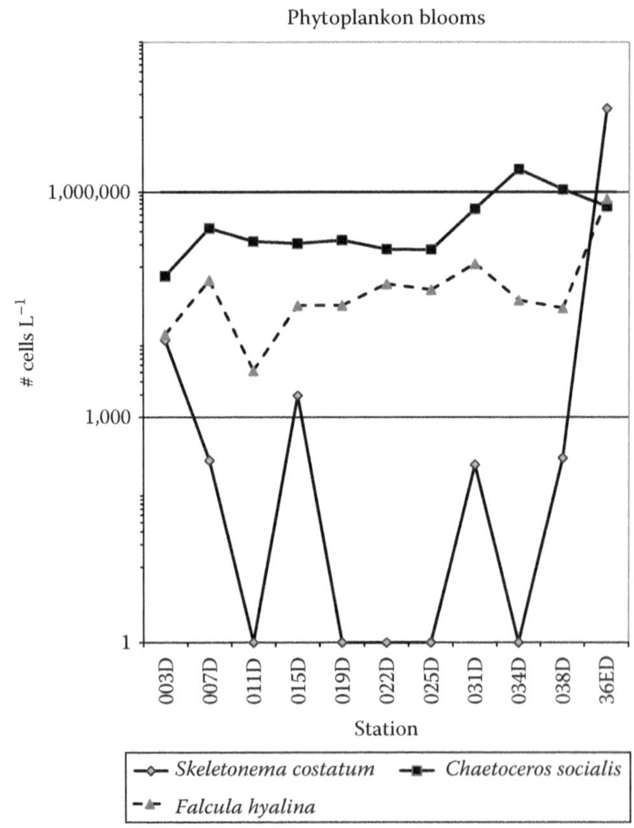

Figure 8.6 Phytoplankton bloom species numbers of cells in the Choctawhatchee Estuary.

richness was negatively associated with % silt in sediments, salinity stratification, and surface color, turbidity, and chlorophyll *a*. This index was positively associated with bottom DO and surface pH. These data indicate that phytoplankton species richness was inversely related to variables associated with urban runoff and increased incidence of blooms.

Results of the PCA-regression analysis of phytoplankton cell numbers (Figure 8.8) indicated that this factor was positively associated with pH, salinity stratification, and % silt content of sediments. These data show that phytoplankton blooms were positively associated with sediment quality factors and with salinity stratification. Phytoplankton cell numbers were negatively associated with reduced bottom DO. This ties together nutrient loading from urban areas, phytoplankton productivity (i.e., blooms), salinity stratification, and low bottom DO. The data that compare areas associated with urban development with undeveloped parts of the estuary indicate a relatively consistent pattern of urban impacts that included deterioration of water and sediment quality and altered phytoplankton composition with blooms and high numbers of cells that probably contributed to the reduced quality of the estuarine habitats.

Surface zooplankton numbers increased toward the west, with very high concentrations of such organisms along the Piney Point–White Point transect (Livingston, 1986a). Extremely low average annual numbers of zooplankton were found in the urbanized Old Pass Lagoon. The east-west gradient of zooplankton numbers also followed salinity trends in the bay. Peak numbers occurred in October, with secondary peaks in May and June (coinciding with low numbers of phytoplankton). Meroplankton distribution in the bay was directly opposite to that of the zooplankton, with increasing numbers from west to east. High meroplankton abundance was noted along the Piney Point–White

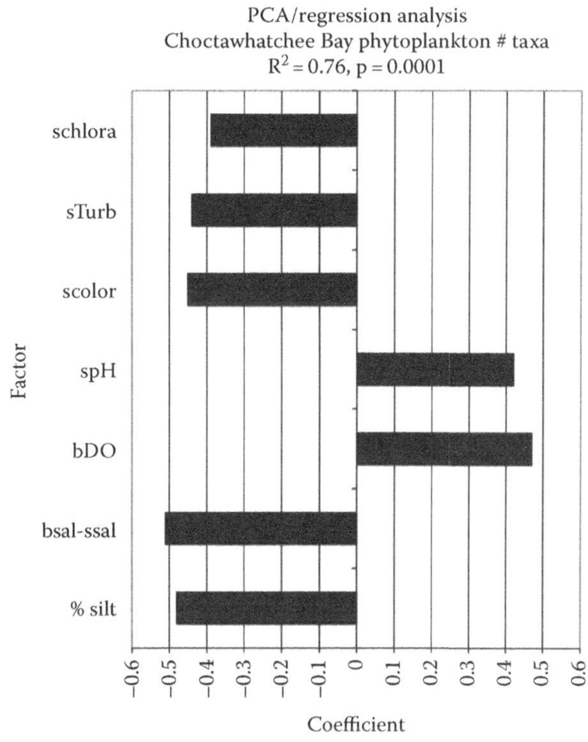

Figure 8.7 PCA-regression analysis of phytoplankton species richness run against key habitat variables in the Choctawhatchee Estuary over a 12-month period (1995–1996).

Point transect, an important region for zooplankton and meroplankton. Peak meroplankton abundance occurred during October to November and from April to June; lowest meroplankton numbers were noted in Old Pass Lagoon.

Annual average ichthyoplankton and fish egg numbers showed no east-to-west gradient. Low concentrations of fish eggs and larvae were found in the polluted Old Pass Lagoon. Peak numbers of fish eggs and larvae were taken in central and west-central surface areas from April through July. Overall, the animal plankton distributions showed considerable variation although some trends were clear: low numbers were found in polluted areas (Old Pass Lagoon) and peak numbers occurred during warm months.

The Association for Bayou Conservation, Inc. took a set of water samples in the study area on April 4, 2006. The samples were preserved in Lugol's solution and were allowed to settle for more than 48 h before they were concentrated to 50 mL volume. Several aliquots of concentrated samples were examined under Leica DMLB microscope, fitted with phase contrast and differential interference contrast optics in ca. 400× magnification. Dr. A.K.S.K. Prasad made species identifications and recorded detailed notes on the phytoplankton composition of Garnier Bayou.

No blooms were present, except for the Chula Vista surface sample, where bloom concentrations ($>10^6$ cells L^{-1}) of *Chaetoceros affinis* were observed. A potentially harmful diatom, *Pseudonitzschia* (possibly *Pseudodelicatissima*) sp., was present at almost all of the stations examined but not in bloom densities. According to the *Manual on Harmful Marine Microalgae* (Hallegraeff et al., 2003), increasing numbers of *Pseudonitzschia* species are being found to produce domoic acid, a toxin that causes amnesic shellfish poisoning (ASP). *Cyclotella choctawhatcheeana* and *Chaetoceros laevis* were also present. The overall assemblages in all of the stations were similar. Dinoflagellates, *Prorocentrum triestinum*, *P. micans*, and *P. cordatum* (syn. *P. minimum*), were present in very low numbers.

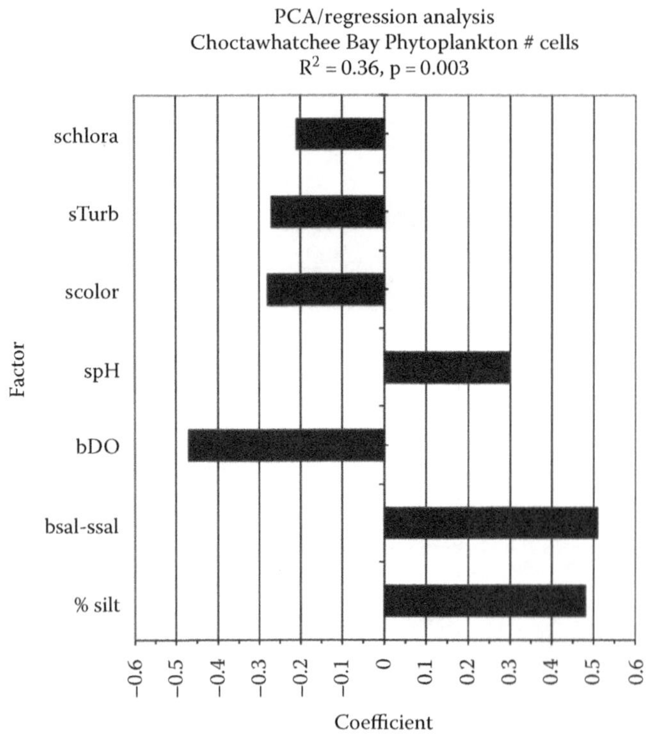

Figure 8.8 PCA-regression analysis of phytoplankton cell numbers run against key habitat variables in the Choctawhatchee Estuary over a 12-month period (1995–1996).

All three species are known to form red tide blooms. *Karenia brevis* was not observed. No harmful or harmless raphidophytes were encountered.

Anecdotal reports (K. Spencer, *The Destin Log*, 1999–2000) indicated that continued urban development in western sections of Choctawhatchee Bay had been associated with increased levels of urban runoff and sewage spills. These data also were consistent with signs of advanced hyper-eutrophication. This included hypoxia at depth, fish kills throughout the river–bay system, and reduced secondary production. These conditions were accompanied by novel extensive red tide (*K. brevis*) blooms both in the bay and in offshore areas. In addition to the usual signs of cultural eutrophication, there was a more ominous problem in the form of the deaths of sea turtles and "bottlenose dolphins" in the Choctawhatchee system and along the coast between Choctawhatchee Bay and St. Andrews Bay (K. Spencer, *Destin Log*, 1999–2000). About 121 dead dolphins were noted in this area between August 1999 and mid-January 2000. The deaths of some fish-eating terrestrial mammals such as raccoons and foxes in the region were also observed in various parts of the bay.

Infaunal Macroinvertebrates and Fishes

A comparison was made of key biological factors taken from annual averages taken over a 12-month period at stations throughout the bay. Infaunal invertebrate and fish numbers were generally lower at the deeper stations throughout most of the bay (Figure 8.9). Such numbers were reduced in the western bayou areas that were exposed to urban runoff. A PCA-regression analysis of the infaunal numbers (Figure 8.10) showed that this index was positively associated with surface chlorophyll *a* and bottom DO and negatively associated with surface temperature and sediment grain size.

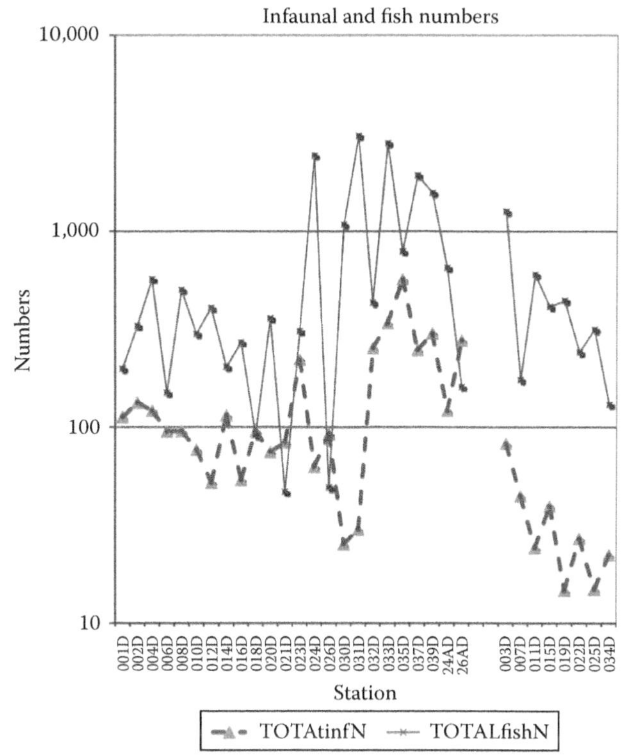

Figure 8.9 Infaunal and fish numbers the Choctawhatchee Estuary averaged over a 12-month period (1995–1996).

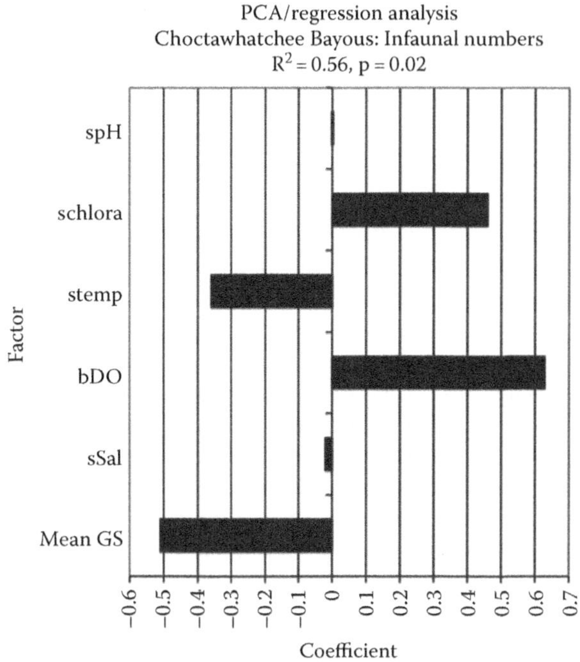

Figure 8.10 PCA-regression analysis of infaunal numbers run against key habitat variables in the Choctawhatchee Estuary averaged over a 12-month period (1995–1996).

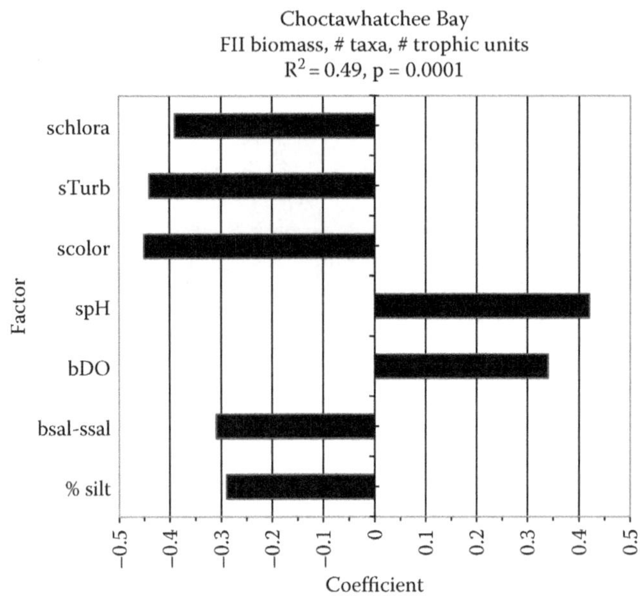

Figure 8.11 PCA-regression analysis of FII biomass, numbers of taxa, and trophic units run against key habitat variables in the Choctawhatchee Estuary averaged over a 12-month period (1995–1996).

FII biomass, numbers of taxa, and trophic units were positively associated with bottom DO and surface pH and were negatively associated with salinity stratification, sediment silt concentrations, and surface water color, turbidity, and chlorophyll *a* (Figure 8.11). These results are consistent with the observation that FII indices were negatively associated with water and sediment quality features at depth and in areas associated with increased urban runoff and salinity stratification.

Discussion

Reyes and Merino (1991) found that sewage releases and dredging adversely affected the DO in a lagoon in the Mexican Caribbean. Breitburg (1990) indicated that wind-driven salinity alterations were associated with severe hypoxia at depth. Stanley and Nixon (1992), working in the Pamlico River Estuary, found that hypoxia developed only under conditions of vertical water-column stratification and warm temperatures. Stratification and low DO events were associated with freshwater discharge and wind stress. Over the period of observation, there was no trend toward lower bottom water DO, and there were no demonstrated cause-and-effect relationships between nutrients, algal abundance, and bottom water DO levels. Proposed reductions of nitrogen were not expected to change the natural hypoxia and anoxia in the Pamlico system. Parker and O'Reilly (1991) reviewed long-term DO concentrations in Long Island Sound. They found that there was a spatial extension of the subpycnoclinal hypoxia eastward in terms of both severity and frequency of low oxygen events. Improvements in parts of the system were associated with upgrades in several sewage treatment plants in the drainage area. Welsh and Eller (1991) found that the ultimate control of the hypoxic conditions in Long Island Sound was salinity stratification with even a weak pycnocline effective in leading to low oxygen at depth.

The opening of Choctawhatchee Bay to high-salinity water from the gulf in the early twentieth century had a lasting adverse impact on the secondary productivity of the inshore bay system. Due to its physiographic characteristics and a relatively large freshwater (river) inflow compared to the volume of its tidal prism, the bay, formerly a relatively freshwater/brackish system, became a highly salinity-stratified estuary. The steep density gradient (pycnocline) was effective in suppressing vertical mixing. The relatively short fetch of the bay relative to prevailing wind currents enhanced stratification in deeper parts of the bay. Warm summer temperatures causing lower surface densities and the orientation of the bay with respect to the most frequent directions of strong winds both contributed to the high stratification. Downward diffusion of oxygen was arrested at the pycnocline and bottom waters become relatively depleted at the deeper, more salinity-stratified stations. DO profiles indicated reduced oxygen supply in the bottom layers of the bay.

Each part of the Choctawhatchee Bay (river–estuary, upper bay, lower bay) had specific relationships that were associated with the position of freshwater inputs into the bay and water depth. DO distribution in Choctawhatchee Bay was a function of temperature, depth, salinity concentration, and the level of stratification. With an increase of bottom salinity during warm periods of the year, there was an associated reduction of DO. The highly stable stratification conditions usually were directly related to hypoxic conditions. The hypoxia at depth was statistically associated with the reduction of various biological indices. This loss of biological productivity was similar to that found in other estuaries (Livingston, 1984a, 1989a; Livingston et al., 2000).

The relationship between nutrient and organic carbon loading from urbanized areas in the Choctawhatchee Estuary and hypoxia is complex and not as clearly delineated as that between salinity stratification and hypoxia. The direct connection between nutrient loading and phytoplankton production is often modified by other factors such as temperature, salinity, and physiographic features of the receiving system. Spatial/temporal trends of nutrient limitation of phytoplankton production also play a role in these relationships. In the case of the Choctawhatchee Estuary, there was a statistical relationship among the factors that included nutrient loading, phytoplankton density, salinity stratification, silt concentrations in the sediments, and reduced secondary productivity (infauna and fishes). In areas affected by anthropogenic nutrient loading and resultant plankton blooms, there were serious reductions of secondary productivity in the form of infauna and the FII trophic indices.

There is considerable (and legitimate) concern regarding the adverse impacts of anthropogenic nutrient loading into estuarine and coastal systems. However, little regard has been expressed concerning how habitat alterations and resulting changes in salinity stratification in such areas exacerbate the impacts of nutrient loading and bloom proliferations. Regardless of the nutrient and organic loading question, hypoxic conditions can be directly related to the physical components of depth and salinity stratification. Physical and stratigraphic components tend to establish the habitat conditions for hypoxia at depth, and anthropogenic loading of nutrients and organic carbon simply exacerbate conditions of low DO. Urban runoff has adverse effects on habitat variables such as sediment quality. Where such activities are accompanied by municipal development and discharges of organic carbon and nutrients, the DO regimes can be further (adversely) affected.

The observations that were made during the studies of the 1980s were followed by similar events during later periods. According to a FWC/FWRI report (Fish and Wildlife Research Institute, unpublished report, 2006),

> "Researchers attributed the aquatic animal mortalities in Choctawhatchee Bay to post-bloom brevetoxin exposure. Dead and dying species reported to the FWRI Fish Kill included Gulf sturgeon (*Acipenser oxyrinchus desotoi*), bay anchovies (*Anchoa mitchilli*), hickory shad (*Alosa mediocris*), and juvenile spot (*Leiostomus xanthurus*). Reports of dead invertebrates included lion's mane jellyfish (*Cyanea capillata*) and blue crabs (*Callinectes sapidus*). FWRI/FWC staff and volunteers collected samples of the reported species as well as live bivalves for health evaluation. In March 2006, FWC/FWRI staff sampled four dead and one live longnose gar (*Lepisosteus osseus*), one live mullet (*Mugil* sp.), and numerous dying juvenile spot fish from further east in Choctaw Beach and Basin Bayou (northern

shore of the central Bay). Samples collected on the southern shore of the Bay included bivalves and water from Sandestin and Alligator Point. Incidental bird mortalities were reported and tissue samples were collected and archived for future analysis. Since fall 2005, the National Oceanic and Atmospheric Administration (NOAA) with assistance from FWC/FWRI has been investigating a bottlenose dolphin (*Tursiops truncatus*) mortality event in the Florida Panhandle."

Through mid-April, no red tide was observed and only background levels of brevetoxin (toxin produced by *K. brevis*) were present in water samples from the area. Dying fish from the affected areas behaved as if they had been exposed to neurotoxic brevetoxins. High concentrations of brevetoxins found in the internal organs of fish (sturgeon, gar, and multiple samples of juvenile fish) indicate toxin exposure was postbloom. Most significantly, brevetoxin was detected in several food web components. Toxin levels were higher than would be expected since the last red tide in the area was in December 2005. Based on the results, brevetoxin is considered the primary cause of the fish kills.

These findings also indicate that a reservoir of brevetoxin is present in Choctawhatchee Bay. It has been well documented that brevetoxin has caused extensive marine animal mortalities in the Gulf of Mexico. While mortality events usually occur at the same time as red tide blooms, FWC/FWRI researchers have documented that effects can continue after the bloom has ended. Animals can be exposed to lethal doses of brevetoxins weeks to months after a bloom has dissipated. Toxins released from red tides often persist in the environment and circulate through the food web.

No other harmful algae bloom (HAB) species or toxins have been identified in connection with the Choctawhatchee Bay animal mortalities. Although background levels of a potentially toxic diatom (*Pseudonitzschia* sp.) were detected in some samples, no domoic acid (toxin produced by the diatom) was present in any biological or water samples tested. As part of the Choctawhatchee Bay mortality investigation, the Florida Department of Environmental Protection provided data on contaminant sampling from Garnier Bayou. These data did not provide any evidence to suspect any contaminant involvement in the ongoing fish kill. Reports of "pink water" in the vicinity of Garnier Bayou were not related to red tide. The "pink water" was likely caused by a high amount of purple bacteria in the water column. These bacteria can be attributed to local environmental conditions and are not considered responsible for the "Choctawhatchee Bay mortalities."

Critical, highly productive habitats include the river mouth (and associated freshwater wetlands), the remaining estuarine wetlands (emergent vegetation), the shallow submerged aquatic vegetation (*H. wrightii* beds), and the bayous. Although nutrient loading (orthophosphate) from the river has been relatively high compared to other systems, the primary problem of anthropogenic nutrient loading appeared to be centered in urban lagoons. For example, the situation in the Garnier Bayou, Tom's Bayou, Boggy Bayou, and Old Pass Lagoon combined the periodic loading of nutrients and organic matter and salinity stratification to cause water quality deterioration (hypoxia), high phytoplankton productivity, and associated sediment problems. These habitat conditions resulted in low numbers of herbivores (zooplankton, meroplankton), a depauperate infaunal invertebrate community, and relatively few epibenthic invertebrates and fishes.

Continued urbanization has led to massive plankton blooms and associated mortality of aquatic and terrestrial species. However, the lack of a scientific database has hampered the initiation of effective control of anthropogenic nutrient loading to the bay. The deterioration of major parts of the bay was also due to the effects of salinity stratification on water and sediment quality conditions. The combined impacts of urban runoff and salinity stratification caused an overall low level of secondary production in major parts of the Choctawhatchee Estuary.

The Choctawhatchee River–Bay study was a short-term effort, and longer-term trends remained undiagnosed. The effects of the extended droughts of the region have not been evaluated. No effort has been made by public authorities in Florida to carry out ecosystem-level analyses to evaluate the current status of the Choctawhatchee Estuary. There has been no effective management plan to correct known problems of nutrient loading to the bay, and the issue of salinity stratification of the estuary due to the opening of East Pass remains unaddressed. Recent events justified the warnings made in the original study (Livingston, 1986a).

PENSACOLA ESTUARY

Background

A detailed description of the primary habitats of the Pensacola Estuary has been provided by Thorpe et al. (1997). Other important reviews include those by Jones et al. (1992) (physical, sediments, and water) and Collard (1991a) (biological trends). Tides and water circulation in the system have been studied by Marmer (1954), Ellis (1969), Hopkins (1969), US Environmental Protection Agency (USEPA) (1971a,b), McNulty et al. (1972), Edwards (1976), and Ketchen and Staley (1979). Sediment studies have been carried out by Horvath (1968), USEPA (1971a), Glassen et al. (1977), the NWFWMD (1978), Young (1981), McAfee (1984), Science Applications International Corporation (1986), George (1988), and Seal et al. (1994). Water quality studies include those of Hopkins (1969), US Department of the Interior (1970), USEPA (1971a,b), Hannah et al. (1973), Glassen et al. (1977), Young (1981, 1985), Shuba (1981), Hand et al. (1996), and Livingston (1973, 1997d). Descriptions of submergent and emergent vegetation in the Pensacola Estuary have been provided by Hopkins (1973), Rogers and Bisterfield (1975), Stith et al. (1984), and Reidenauer and Shambaugh (1986). Coastal faunal studies have been performed by Cooley (1978), Little and Quick (1976), Young (1981), Butts and Ray (1986), Collard (1989), and Hudson and Wiggins (1996). The most quoted information source is Olinger et al. (1975) who conducted one of the only studies with interdisciplinary information (sediment conditions, water quality, circulation, and biology). Overall, much of the ecological data for the system are dated, with virtually no results from comprehensive (ecosystem-level) studies of the Pensacola Estuary available at this time.

The Pensacola Estuary is located in northwest Florida in the most heavily populated region in north Florida. The bay (Figure 8.12) is located in the Coastal Plain province along the NE Gulf of Mexico and is composed of a series of river basins (the Escambia, Blackwater, and Yellow). Parts of the ecosystem appear to be relatively unpolluted (Blackwater and Yellow River systems), whereas other areas, such as the Escambia Estuary, have had a history of degraded water quality due to point and nonpoint source discharges (Thorpe et al., 1997; Livingston, 1999). Some areas, such as the bayous in the city of Pensacola, have been severely affected by combinations of toxic substances and urban runoff. Other parts of the system, such as Blackwater Bay and East Bay, may be in better shape, although these areas are increasingly threatened by increased urban development and associated nonpoint source pollution.

The Pensacola Estuary includes five interconnected bay components: Escambia Bay, Pensacola Bay, Blackwater Bay, East Bay, and Santa Rosa Sound (Figure 8.12). The Escambia River extends from about 386 km (240 mi) from the northern end of Escambia Bay. It runs through Alabama as the Conecuh River. The drainage area of the Escambia River basin includes about 10,880 km^2 (4,200 mi^2), with approximately 90% of the basin in Alabama. The Blackwater River basin includes about 2230 km^2 (860 mi^2) of which about 81% is in Florida. The Yellow River basin extends 177 km (110 mi) from Blackwater Bay to a point northeast of Andalusia, Alabama. This basin includes an area of about 3535 km^2 (1365 mi^2), 64% of which is located in northwestern Florida.

Escambia Bay is located east of the city of Pensacola, with the Garcon Point peninsula to the east and the Escambia River delta to the northwest. The primary source of freshwater to the bay is the Escambia River; other sources include the Pace Mill Creek and Mulatto Bayou drainage basins in the upper bay and the Bayou Chico and Bayou Texar basins in the lower bay. Blackwater Bay receives freshwater inflow from the Blackwater River, with East Bay immediately downstream from the Blackwater Bay system. East Bay receives freshwater flows from the Blackwater River–Bay

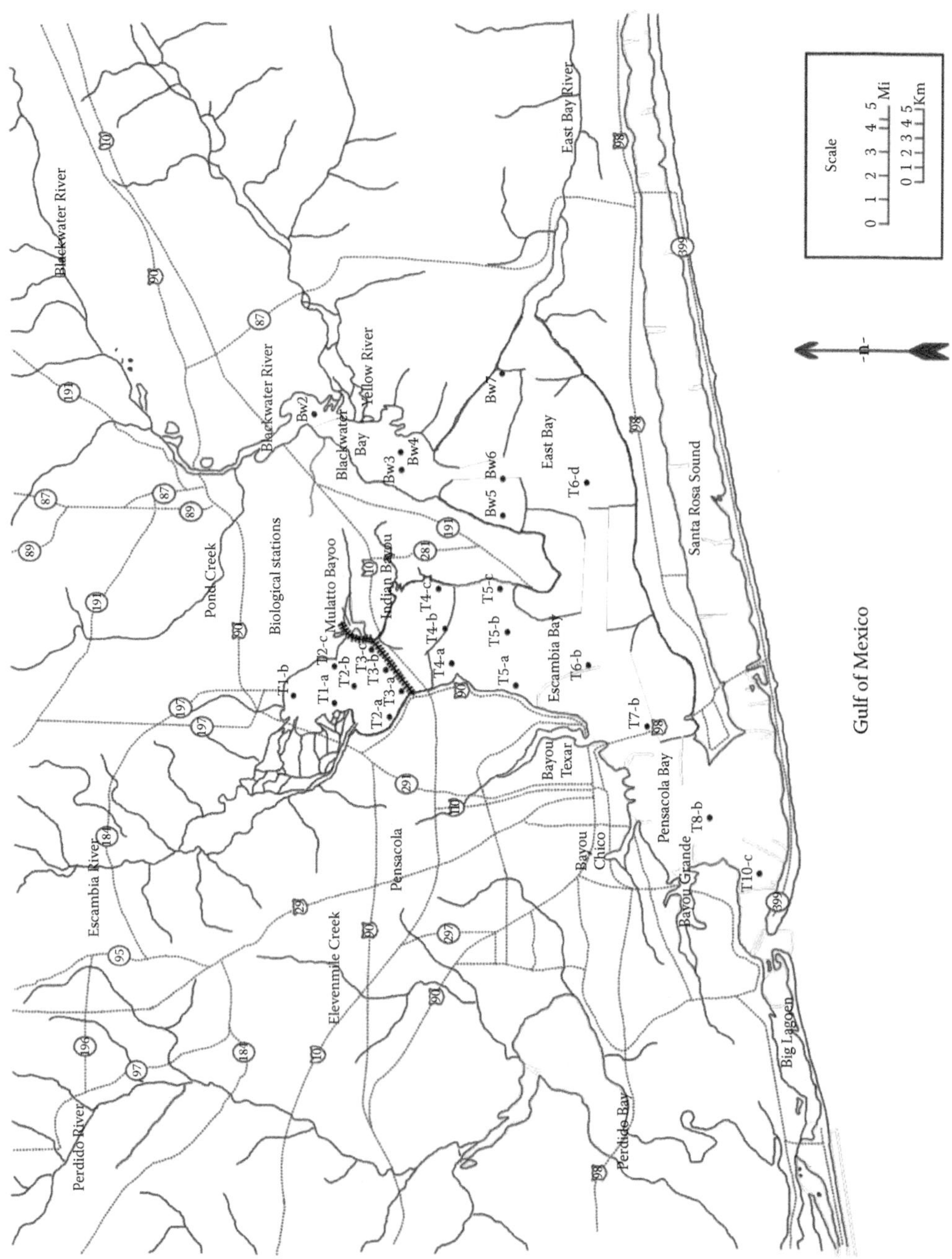

Figure 8.12 The Pensacola Estuary showing sampling stations.

system, the Yellow River, and the East Bay River. East Bay is bordered to the south by the Gulf Breeze peninsula. Pensacola Bay receives flows from Escambia Bay, East Bay, and Bayou Grande and is bordered to the north by the city of Pensacola and to the south by Santa Rosa Island. Pensacola Bay empties into the Gulf of Mexico through a pass at its southwestern terminus. Santa Rosa Sound is a lagoon between the mainland and Santa Rosa Island. This sound connects Pensacola Bay in the west with Choctawhatchee Bay to the east.

Circulation in Escambia Bay is strongly influenced by wind and tidal action as well as inflow from the Escambia River. There is a net southward flow of river water along the western shore, with water of higher salinity intruding along the eastern shore. This tends to produce a generally counterclockwise circulation pattern. Railroad and highway bridges may inhibit flushing and exchange between the upper and lower bay. Historically, Escambia Bay has had high levels of toxic agent contamination and nutrient loading. It has received substantial industrial and domestic wastewater discharges and is still affected by various point and nonpoint sources. Nonpoint source pollution is received from the city of Pensacola, unincorporated areas, and the river basin.

The Pensacola Estuary receives direct and indirect runoff from the city of Pensacola and unincorporated areas. On an outgoing tide, surface waters tend to move toward the pass from the more northerly and western portions of the bay (Reidenauer and Shambaugh, 1986). Olinger et al. (1975) found that circulation within the bay could be strongly influenced by surface winds, with effects not necessarily limited to the upper layers. Point source discharges to Pensacola Bay include the Main Street and NAS Pensacola wastewater treatment plants. Component bayous, formerly centers of productivity in the system, are now among the most polluted in the entire Pensacola system. In Bayou Texar, an area noted for fish kills related to nutrient overenrichment and hypoxia, nitrate has been reported to control primary productivity to a greater degree than phosphorus (Moshiri et al., 1987). Bayou Texar is also contaminated by toxic agents, with two USEPA-designated Superfund sites. These bayous act as sinks for sustained urban runoff and other nonpoint source pollution. Bayou Chico has also received substantial historic point source discharges. Incoming waters from these bayous tend to move along the bottom into the bay and then eastward along the southern part of the bay.

Long et al. (1997) carried out a study of the "magnitude and extent of sediment toxicity in four bays of the Florida Panhandle." The sediments of the Pensacola Bay system were compared to those of the Choctawhatchee, St. Andrews, and Apalachicola Bays. The greatest toxicity of all bays analyzed occurred in Bayou Chico in Escambia Bay, with the other developed bayous in the Pensacola area showing "relatively severe toxicity." Toxicity of the Pensacola Bay sediments was attributed to high-molecular-weight PAHs, zinc, DDD/DDT isomers, total DDT, and the highly toxic organochlorine pesticide known as dieldrin. There were elevated PAHs at the NOAA and NS&T Program stations in Pensacola Bay and Indian Bayou. In addition to the bayous and portions of Pensacola Bay, toxicity was also detected in upper and midportions of Escambia Bay. Toxicity was also noted in Blackwater Bay and East Bay. The NOAA study (Long et al., 1997) confirmed high concentrations of lead, mercury, and PCBs in the sediments of the urbanized bayous and in Pensacola Bay.

More recent studies by various government agencies have been conducted in the Pensacola Estuary including evaluations of Bayou Texar, Bayou Chico, and Bayou Grande. Analyses by the USEPA (Gulf Breeze, Florida, Laboratory) using annual samples of infaunal macroinvertebrates (Benthic Index) indicate relatively high indices in upper and midparts of Escambia Bay and in the Blackwater–East Bay areas (USEPA, 1997c). Reduced indices were noted in lower Escambia Bay and Pensacola Bay (USEPA, 1997c). Metals, polycyclic aromatic hydrocarbons, and pesticides were implicated in the deteriorated conditions in lower Escambia Bay and Pensacola Bay (associated with the city of Pensacola). Metals (AS, Ni, Cr) and pesticides were also found in areas associated with the Gulf Breeze Peninsula.

The classic response of an estuary to anthropogenic nutrient loading is extreme hypoxia and resulting disruption of the biological relationships of the receiving system. Escambia Bay, as part of the Pensacola Estuary, was subject to severe fish kills during the late 1960s and early 1970s due to such loading (Livingston, 1997d). At the time, Escambia Bay was subject to nutrient loading from various point and nonpoint sources that resulted in extreme fluctuations of DO (USEPA, 1971a). The Mulat–Mulatto Bayou (NE Escambia Bay) was a focal point for the extensive fish kills. This area had been subjected to extensive physiographic changes due to dredging associated with the construction of a nearby highway.

In 1965, the Florida Department of Transportation removed approximately 1,028,933 yd^3 (786,676 m^3) of sediment from Mulatto Bayou. A dredged channel opened the south end of the bayou to the bay, and deep borrow pits were created at the end of the channel. Further dredging was carried out in the bayou in 1970; at that time, a private firm dredged finger canals in the southeastern sector of the bayou. Dredge spoils obliterated surrounding wetlands. During that period, nutrients and toxic agents were released into Escambia Bay from a broad array of sources (Olinger et al. 1975), so that the aquatic conditions were altered. Dye studies in the Mulat–Mulatto Bayou (Livingston, unpublished data) indicated that the dredging created cul-de-sacs (i.e., the finger canals) that were isolated from direct tidal current exchanges. In addition, a natural cul-de-sac (i.e., Mulat Bayou) was seasonally isolated from tidal exchanges.

The most extensive fish kills in the scientifically documented literature occurred in Escambia Bay during the late 1960s and early 1970s (USEPA, 1971a; Livingston, 2005). Diurnal fluctuations of DO in the bayou over the period of study indicated considerable variation on a diurnal and seasonal basis. In Mulat Bayou during early July 1970, there was supersaturation during the day and hypoxic conditions at night. In August, this situation became more pronounced. By September, DO reached a low point during the early morning, at which time there was a massive fish kill. In October, conditions become more stable, and by the end of November, continuous high levels of DO were noted throughout the bayou, a situation that prevailed through the following spring (April). In the finger-fill canals, low DO at night during August was accompanied by a major fish kill. Subsequent increases were noted in October, with relatively high DO being maintained from November through the following April.

The seasonal pattern of fish kills represented a trend that occurred from June through October (Table 8.2). The fish kills in the Mulat–Mulatto Bayou, while directly associated with the DO regime, were actually part of a complex, interrelated series of events that was dependent on specific spatial–temporal controlling factors that reflected habitat conditions associated with the specific dredging events listed earlier. The extent of thermohaline stratification was also enhanced by summer rainfall, which peaked in August. Biochemical oxygen demand in the highly organic sediments (especially high in dredged holes) exacerbated the reduction of DO at depth. Vertical stratification in deeper parts of the bayou or in areas cut off from the main tidal and/or wind-driven circulation, together with nocturnally low oxygen in the water column, was thus associated with periodic hypoxia and anoxia at depth.

There was an exaggerated diurnal fluctuation of DO during summer–fall months of 1970 with supersaturation during the day and hypoxic conditions during nocturnal periods. Seasonal influxes of juvenile fish could have been a contributing factor to the low levels of DO at night. Overall, however, the timing and placement of the fish kills indicated that the combination of hypereutrophication and habitat alterations due to dredging were responsible for the fish kills. Dredging activities thus enhanced the effects of cultural eutrophication due to nutrient loading by sewage treatment plants, chemical factories, and urban runoff by inhibiting tidal and wind-produced exchanges of bay water. The fish kills were the final consequence of a distorted DO regime, which, in turn, was influenced by a sequential interaction of temperature, precipitation, altered water circulation patterns, nutrient-enhanced primary production, and a disjunct food web organization.

Submerged aquatic vegetation, formerly abundant in this system, disappeared during the 1970s (Collard, 1991a,b) and has not returned. Formerly productive oyster bars were also destroyed during

Table 8.2 Fishes Killed in Mulat–Mulatto Bayou (Pensacola Bay System) from 1970 to 1971

Location	Date	Visual Estimate
Mulat Bayou 7	June 21, 1970	250,000
Mulatto Bayou 7	June 29, 1970	750,000
Mulat Bayou 7	July 1, 1970	2,000
Mulat Bayou 7	July 5, 1970	Over 1,000,000
Mulat Bayou 7	July 1970	Over 10,000,000
Mulat Bayou 7	July 1970	10,000
Mulat Bayou 7	July 13, 1970	2,500
Mulat Bayou 7	July 1970	1,000,000
Mulatto Bayou	August 25, 1970	700
Mulatto Bayou	October 8, 1970	200,000
Mulatto Bayou 11a	August 12–13, 1971	Over 1,000,000
Mulatto Bayou 11a	August 22, 1971	Over 2,000,000
Mulat Bayou 7	September 10–12, 1971	Over 1,000,000

Note: Number of fishes estimated as killed are indicated.

this period and are no longer commercially viable although a small oyster fishery still exists. Virtually, all the fish camps around the bay went out of business during the early 1970s. Real estate values around the bay plunged due to stenches created by dead fishes and invertebrates. This event led to a public outcry and subsequent political action that reduced nutrient loading to levels that no longer led to the fish kills.

River Flows and Nutrient Loading

A comprehensive study was carried out in the Pensacola Estuary from 1997 to 1998 (Table P.1 Livingston, 1999). River flows of the major rivers (Table 5.1) indicated that, in terms of river flow, the Escambia River is the second largest of the rivers flowing into the eastern gulf. Peak flows occurred during winter–early spring months with the lowest during fall periods. Ammonia loading rates in the Escambia River were high relative to that in the Apalachicola River and ammonia concentrations were higher than those of the Apalachicola River (Table 8.3).

Sediment and Water Quality

Sediment quality is an important feature of the habitats of the Pensacola Estuary. Sediments in the deeper parts of Escambia Bay had the highest proportions of silt when compared to Blackwater

Table 8.3 River Flows, Nutrient Loads, and Concentrations in the Apalachicola and Escambia Rivers along the Eastern Gulf Coast

Means (21 Months)	Apalachicola River (94–95)	Escambia River (97–98)
Flow ($m^3\ s^{-1}$)	767.2	156.0
NH_3 load (kg day^{-1})	1115.2	949.6
$NO_2 \pm NO_3$ load (kg day^{-1})	22418.6	2257.5
PO_4 load (kg day^{-1})	436.2	185.5
NH_3 conc. (mg L^{-1})	0.040	0.058
$NO_2 \pm NO_3$ conc. (mg L^{-1})	0.354	0.196
PO_4 conc. (mg L^{-1})	0.018	0.012

Note: Data represent 21-month mean values.

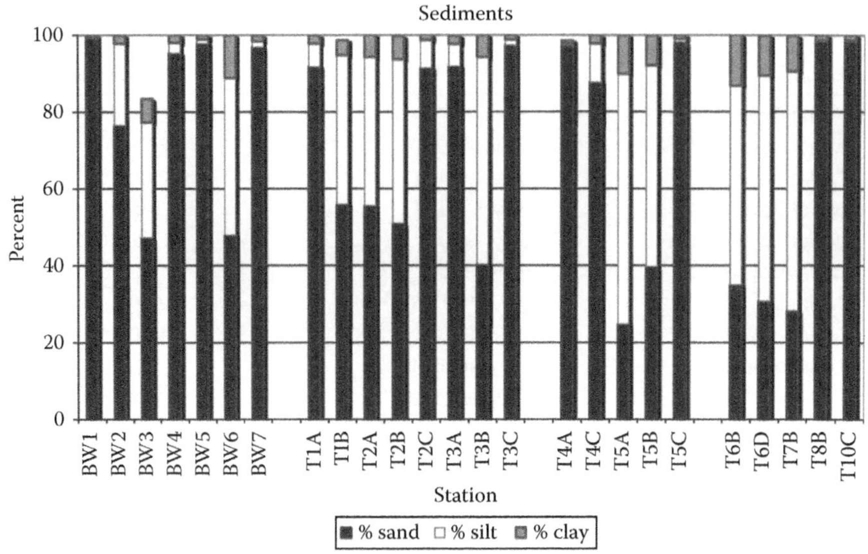

Figure 8.13 Sediment quality in the Pensacola Estuary.

and East Bay to the east and Pensacola Bay to the south (Figure 8.12). Within Escambia Bay, silt and clay concentrations tended to increase along the north-south gradient. This distribution indicated that depth and propinquity to the mouth of the Escambia River were leading determinants of the sediment quality in the receiving estuary.

Silt–clay fractions were comparable in the Blackwater and upper Escambia Bay systems. Both factors were high in lower Escambia Bay and Pensacola Bay (Figure 8.13). There was a general increase in the fines of sediments from upper to lower parts of the Escambia Bay–Pensacola Bay system. Percent organics also showed a general decline along the north-south axis of the Escambia Bay–Pensacola Bay system. Grain size was variable in all systems, but there was a tendency toward smaller grain size with distance from the mouth of the Escambia River. There was no discernible temporal trend of sediment oxygen demand (SOD) between spring and fall periods. There was a tendency for lower SOD rates in upper Escambia Bay relative to the Blackwater Bay and lower Escambia Bay. This is consistent with the grain size data.

Ammonia concentrations (Figure 8.14) tended to be relatively uniform in the estuary with the exception of higher concentrations at stations T1A and T2C. Nitrite/nitrate concentrations were highest in Escambia Bay and lowest in Pensacola Bay. Orthophosphate concentrations (Figure 8.15) were highest at surface stations in upper Escambia Bay and at the bottom of lower Escambia Bay. These bay nutrients were thus highest in areas receiving flows from the Escambia River.

The annual data for all variables were averaged over time (12 months, [5/1997–4/1998]) and stations (Blackwater Bay, BW3, BW4, BW5, BW6, BW7; upper Escambia Bay, T1A, T1B, T2A, T2B, T2C, T3A, T3B, T3C; lower Escambia Bay, T4A, T4B, T4C, T5A, T5B, T5C; Pensacola Bay, T6B, T6D, T7B, T8B, T10C). Both water color and turbidity were highest in upper Escambia Bay (Figure 8.16) as a result of the direct effects of the Escambia River flows into the estuary. Particulate and dissolved organic carbon (DOC) (Figure 8.17) was highest in the lower Escambia system. Secchi depths were lowest in upper Escambia Bay (Figure 8.18). There was an increase of this factor with distance from the Escambia River mouth. Blackwater Bay had deeper Secchi depths than Escambia Bay. Pensacola Bay was the deepest part of the system, whereas the other three parts of the system had comparable average depths.

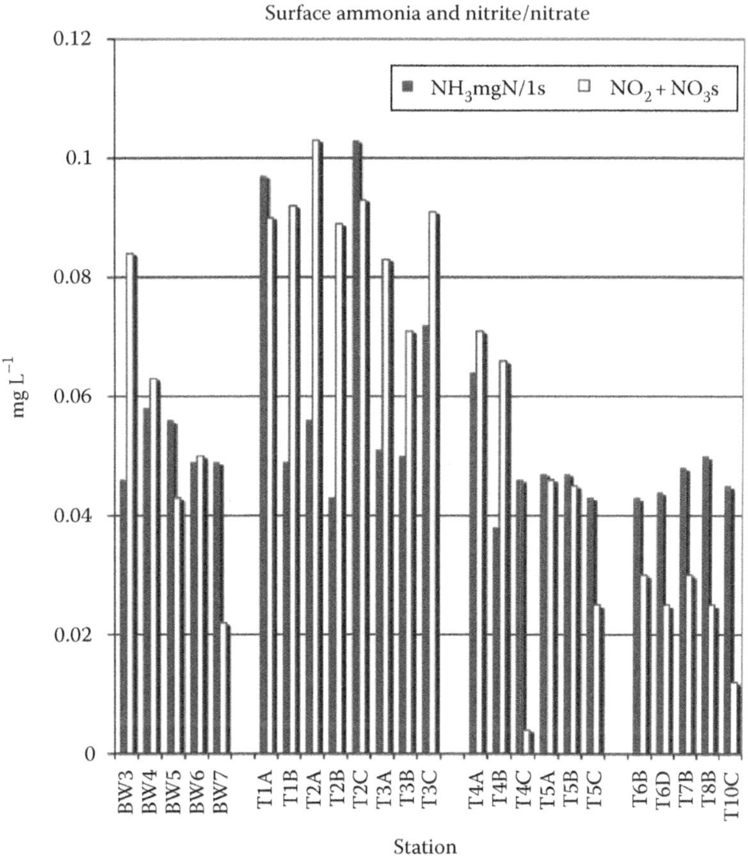

Figure 8.14 Surface concentrations of ammonia and nitrite/nitrate nitrogen in the Pensacola Estuary. Annual averages (1997–1998) are given.

Salinity Stratification and Dissolved Oxygen

High-salinity water from the gulf moves up the estuary and is concentrated in the deeper parts of Escambia Bay and East Bay. Overall, salinity was highly variable throughout the Pensacola system over both short- and long-term periods, with a family of cycles having periods related to diurnal and seasonal atmospheric controlling features, the physiography of the bay, and tidal/wind effects. High surface-salinity variation was noted in the upper and midportions of Escambia Bay: bottom salinity had the highest levels of variation in upper Escambia Bay and the lowest such levels in Pensacola Bay. In both bays, the influence of the river flows was apparent. The lowest salinities were noted during winter–spring periods of high river flow. The effects of the March 1998 peak flows of the Blackwater and Escambia Rivers were evident throughout the Pensacola Estuary. Peak salinities occurred during fall periods of relatively low river flows. North-south gradients of salinity were evident in the Blackwater and Escambia Estuaries. The respective rivers controlled the salinity regime in the receiving bay areas.

Salinity stratification was maximal in deeper areas of the lower and midportions of Escambia Bay (Figure 8.19). Temporal relationships between bottom DO concentrations and salinity stratification indices indicate that bottom DO in the Pensacola Estuary is controlled by salinity stratification in the upper and midparts of Escambia Bay and the Blackwater/East Bay system; such relationships do not hold in Pensacola Bay. Surface DO was highest in upper Escambia Bay. The lowest DO and oxygen

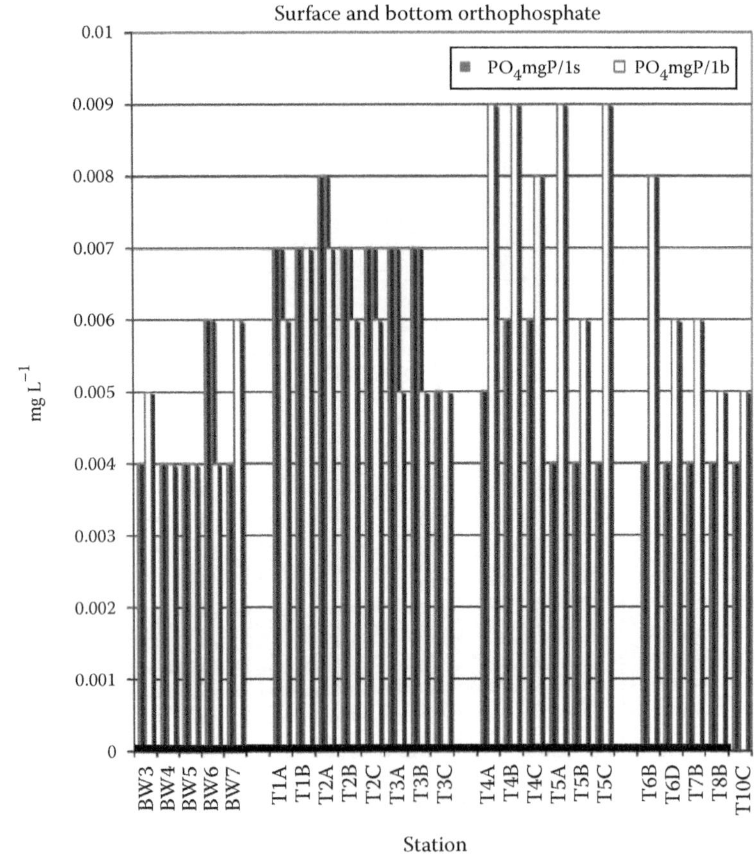

Figure 8.15 Surface and bottom concentrations of orthophosphate in the Pensacola Estuary. Annual averages (1997–1998) are given.

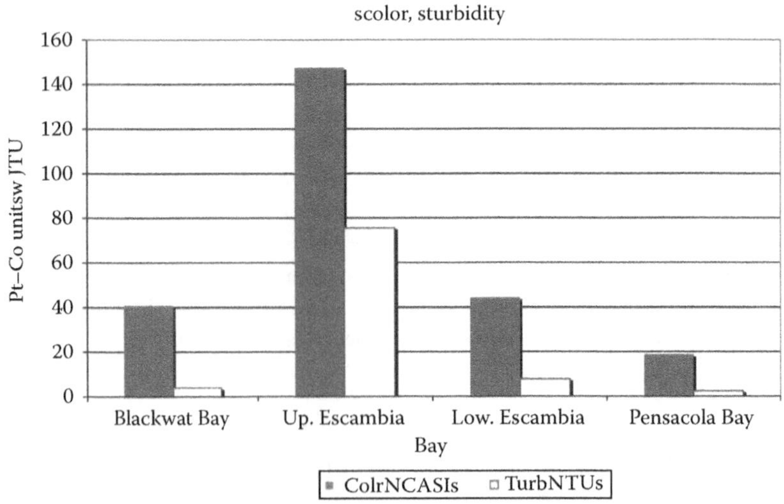

Figure 8.16 Surface water color and turbidity in the Pensacola Estuary. Annual averages (1997–1998) are given.

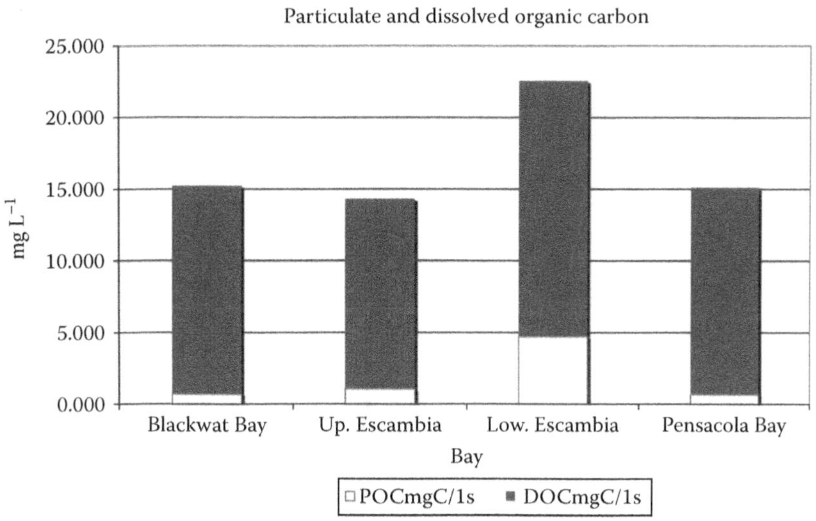

Figure 8.17 Surface particulate and DOC in the Pensacola Estuary. Annual averages (1997–1998) are given.

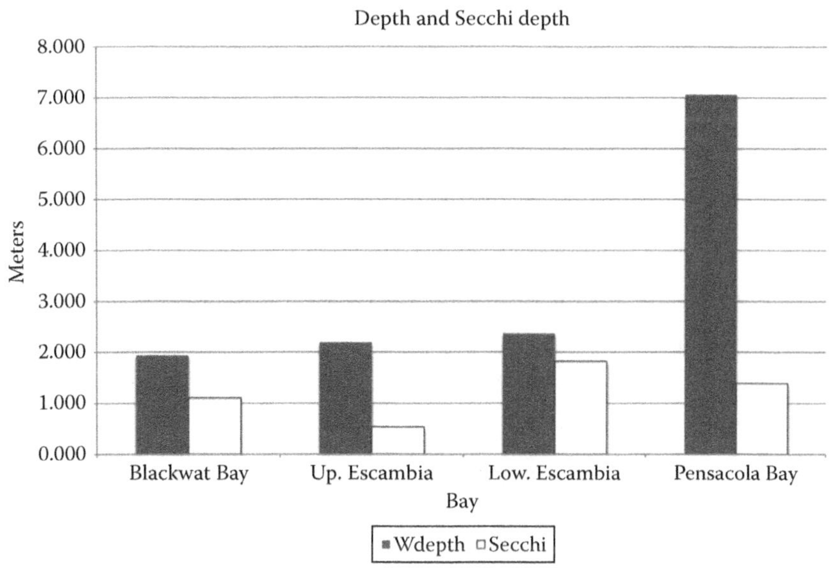

Figure 8.18 Water depth and Secchi depth in the Pensacola Estuary. Annual averages (1997–1998) are given.

anomalies occurred in areas of the highest salinity stratification indices. Regressions of DO with other physical/chemical variables indicate that surface DO was significantly (but weakly) associated with surface salinity, surface DOC, and surface TOC near the mouth of the Escambia River. Bottom DO was significantly (inversely) associated with bottom temperature and the salinity stratification index.

Chlorophyll *a*

The nutrient data indicate that loading from the alluvial Escambia River was a key factor in the general increases of inorganic nutrients in the receiving system (Escambia Bay) relative to the Blackwater Bay system. Surface chlorophyll *a* concentrations were highest in upper and

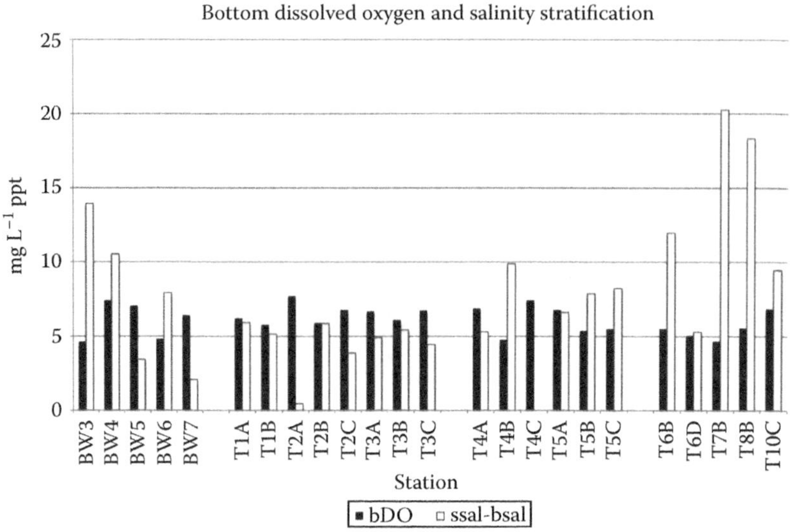

Figure 8.19 Annual averages of bottom DO and salinity stratification indices in the Pensacola Estuary (1997–1998).

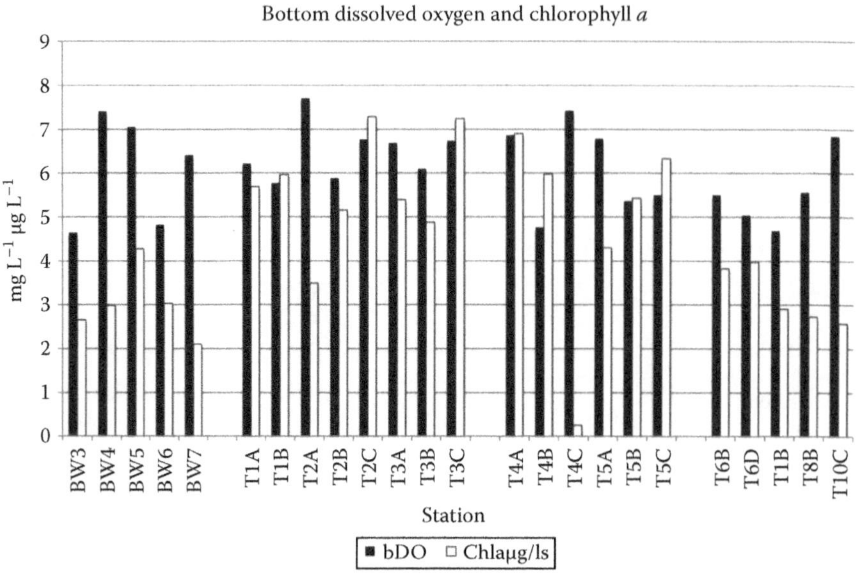

Figure 8.20 Annual averages of bottom DO and surface chlorophyll *a* in the Pensacola Estuary (1997–1998).

lower Escambia Bay relative to the Blackwater system and Pensacola Bay (Figure 8.20). This finding is congruent with the distribution of nutrient loading and nutrient concentrations in the bay. Chlorophyll *a* was related to bottom DO in the Pensacola system. In Escambia Bay, low surface chlorophyll *a* concentrations occurred relative to high bottom DO concentrations, and there was a significant (inverse; $P < 0.05$) relationship of these factors in upper Escambia Bay ($R^2 = 0.742$), lower Escambia Bay ($R^2 = 0.37$), and Pensacola Bay ($R^2 = 0.38$). These data indicate a possible link of phytoplankton activity to bottom DO in the bay with salinity stratification as a probable modifying factor for bottom DO in most parts of the bay.

Plankton Assemblages and Blooms

Species abundances by station are given in Table 8.4. Overall numbers of cells L^{-1} were highest in lower Escambia Bay due, largely, to the diatom *C. choctawhatcheeana*. Key subdominant species included the following: *P. cordatum*, *Heterosigma akashiwo*, and *Dactyliosolen fragilissimus*. The dinoflagellate *P. cordatum* and the raphidophyte *H. akashiwo* are well-known toxic bloom species that have been associated with damaged coastal systems. Both *Heterosigma* and *Prorocentrum* were concentrated in upper Escambia Bay. The lowest numbers were found at station T1A. Overall, there was no evidence of widespread plankton blooms in the Pensacola Estuary system over the sampling period.

Phytoplankton numbers were lowest during winter months with a peak during August 1997 (Figure 8.21). Species richness showed a different pattern, with the highest numbers of species observed during fall months (October–November 1997) and the following spring (March 1998). The lowest numbers of species occurred during July 1997. Stations T5B and T6B had the highest species richness. Numbers of individuals of the phytoplankton reached lows during December 1997, with subsequent increases in this index through the following spring.

This pattern of phytoplankton prevalence differed from the distribution of chlorophyll *a* and light/dark bottle productivity distributions (Livingston, 2002), where the highest values were observed in upper Escambia Bay. The discrepancy between chlorophyll *a* concentrations/primary productivity (highest in upper Escambia Bay) and the whole-water phytoplankton numbers (highest in lower Escambia Bay) could be due to competitive interactions with toxic plankton species, preferential feeding activities of zooplankton, and differences in the chlorophyll *a* concentrations among populations in upper Escambia Bay.

Table 8.4 Numerically Dominant Phytoplankton by Station Averaged over 12-Month Sampling Period in Escambia Bay

	T1A	T3B	T4B	T5B	T6B	Total
Cyclotella choctawhatcheeana	12,466	12,047	36,919	57,441	59,319	178,192
Prorocentrum cordatum	500	15,646	3,692	2,087	816	22,741
Heterosigma akashiwo	1,166	8,200	300	33	3,013	12,712
Cylindrotheca closterium	4,323	1,881	1,017	1,932	932	10,085
Dactyliosolen fragilissimus	278	82	117	1,483	4,543	6,503
Skeletonema costatum	0	166	533	1,449	3,599	5,747
Bacillaria paxillifer	111	84	1,116	2,577	933	4,821
Chaetoceros throndsenii	0	0	50	2,630	783	3,463
Caloneis sp.	998	0	250	1,811	99	3,158
Nitzschia reversa	1,139	229	167	1,437	150	3,122
Crucigenia apiculata	1,248	209	116	517	866	2,956
Amphidinium cf. vigrense	527	312	517	283	1,070	2,709
Chaetoceros spp.	332	187	500	633	699	2,351
Gymnodinium cf. splendens	0	0	0	0	2,230	2,230
Thalassionema nitzschioides	250	63	117	549	1,100	2,079
Cerataulina pelagica	0	0	67	1,717	166	1,950
Nitzschia sigmoidea	0	0	100	1,199	516	1,815
Chaetoceros compressus	722	416	134	233	0	1,505
Leptocylindrus danicus	28	0	150	566	750	1,494
Chaetoceros laciniosus	0	0	0	100	1,150	1,250
Mallomonas akrokomas	443	104	0	83	500	1,130
Paralia sulcata	83	42	84	334	482	1,025

Note: Bloom species in bold type.

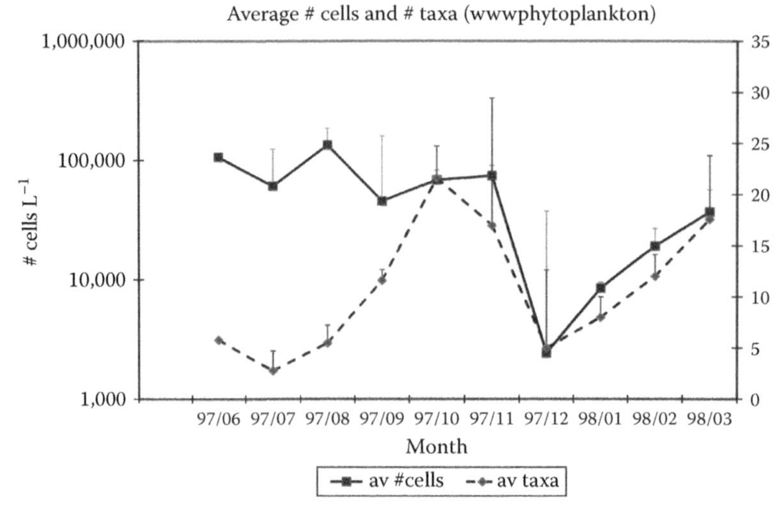

Figure 8.21 Average phytoplankton numbers of cells and species richness by month (6/97–5/98).

No blooms (numbers $>1 \times 10^6$ cells L^{-1}) of the various dominant and subdominant phytoplankton populations were noted during the sampling period in Escambia Bay. Dominant species of whole-water phytoplankton included *C. choctawhatcheeana* (a common bloom species in Perdido Bay), nanoflagellates, cryptophytes, *Cryptomonas* spp., *Nitzschia* spp., and *Navicula* spp. Among the numerical subdominants were *H. akashiwo* (one of the summer-bloom species in Perdido Bay), *Chaetoceros throndsenii* (Marino et al.) (one of the spring-bloom species in Perdido Bay), and *P. cordatum* (a winter-bloom species). The presence of these species in moderate numbers could be related to the history of the Escambia system (i.e., the result of previous hypereutrophic effects due to excessive nutrient loading in the past). It is also possible that these species responded to existing pollution loads going into Escambia/Pensacola Bays from point and nonpoint sources.

Animal Population Distribution and Trophic Organization

The spatial distribution of average infaunal biomass in the Pensacola Estuary is shown in Figure 8.22. The infaunal biomass was concentrated in the upper areas of the two bay systems: the highest such biomass was concentrated in the shallow shelf areas of the upper Escambia Bay and Blackwater Bay. There was a relatively low infaunal biomass along the eastern sections of upper Escambia Bay. In Blackwater Bay, there was relatively high biomass across the bay just below the mouth of the Yellow River. Infaunal species richness (Figure 8.22) was highest in the high-salinity sections of Pensacola Bay and lower Escambia Bay. Relatively high dominance was evident due to three species (*Mediomastus ambiseta*, *Lepidodactylus* sp., and *Streblospio benedicti*). The distribution of *Lepidodactylus* sp. was concentrated largely at station BW5, whereas the other dominants were more widely distributed in the upper parts of the Escambia Bay and Blackwater Bay.

Overall, the main concentrations of infauna in the Pensacola Estuary were in shallow areas that received direct runoff from the regional river systems. This distribution followed that of the nutrient loading and chlorophyll *a* concentrations of the respective systems with a seemingly direct response of increased infaunal biomass with increased nutrient loading, particularly orthophosphate. Other factors that could have influenced this distribution include the adverse impacts of pollution in eastern sections of upper Escambia Bay and lower Escambia Bay–Pensacola Bay (city of Pensacola) and salinity stratification patterns in sections of East Bay. Deeper areas in both bays had relatively low

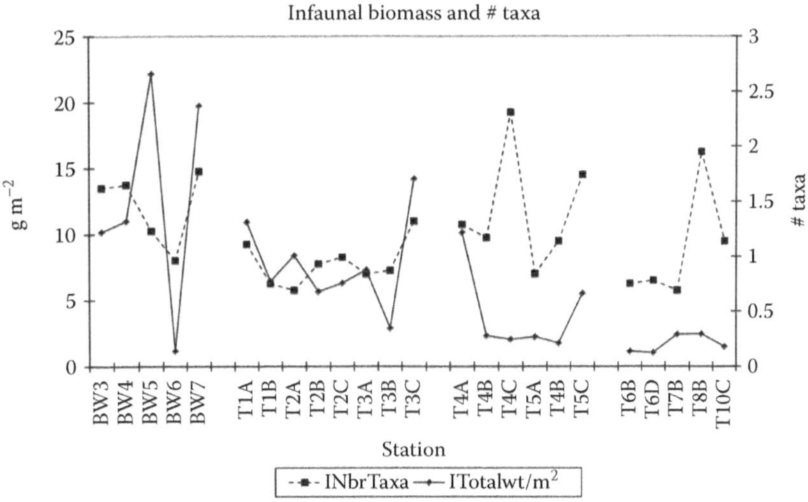

Figure 8.22 Infaunal biomass and number of species in the Pensacola Estuary averaged annually over the 1997–1998 study period.

infaunal numbers and low species richness that could also reflect sediment conditions in these areas and the distribution of bottom DO. Predation patterns could also affect these distributions.

The invertebrates in the Pensacola Estuary were dominated largely by brown shrimp (*Farfantepenaeus aztecus*), white shrimp (*Litopenaeus setiferus*), and blue crabs (*C. sapidus*) (Figure 8.23). The primary numbers of these commercially important species were located in upper Escambia Bay where the highest levels of primary production were located. The relatively low numbers of epibenthic invertebrates at stations in lower Escambia Bay (T5A, T5B, T6B) and Pensacola Bay (T7B, T8B, T10C) followed the low numbers of infaunal macroinvertebrates in this region of the bay: evidently, predation is not the reason for the low numbers of infauna in this part of the bay. It is possible that the low numbers of infauna are in some way the cause of the low numbers of their predators. The Blackwater/East Bay system was largely devoid of epibenthic macroinvertebrates.

It is likely that pollution in eastern sections of upper Escambia Bay and lower Escambia Bay–Pensacola Bay played a role in the distribution of epibenthic macroinvertebrates in the Pensacola Estuary (see discussion below). On the other hand, the increased primary productivity associated with discharges (and loaded nutrients) from the Escambia River resulted in the concentration of the commercially important invertebrates in the Pensacola Estuary in areas affected by the river. These commercially important species were not found to any extent in the largely unpolluted Blackwater Bay.

Planktivorous bay anchovies (*A. mitchilli*) were dominant in upper Escambia Bay (Figure 8.24). Two of the numerically most dominant fish populations (spot, *L. xanthurus*, and Atlantic croakers, *Micropogonias undulatus*) are benthic feeders, and both species were located largely in the central (deeper) areas of upper Escambia Bay. Relatively low numbers of fishes were collected in Pensacola Bay. There were also weaknesses of individual fish populations (*Cynoscion arenarius*, *Brevoortia patronus*, *A. mitchilli*, *M. undulatus*) in eastern sections of upper Escambia Bay. The fish data thus indicate possible adverse effects of point source(s) discharge(s) in this area of the bay. The concentrations of Atlantic bumpers (*Chloroscombrus chrysurus*) were highest in areas of the bay that appeared to be stressed by anthropogenic activity: eastern sections of upper Escambia Bay, western sections of lower Escambia Bay, and Pensacola Bay. This species could be an indicator of areas of the Pensacola Estuary that have been adversely affected by pollution sources in the Pensacola basin.

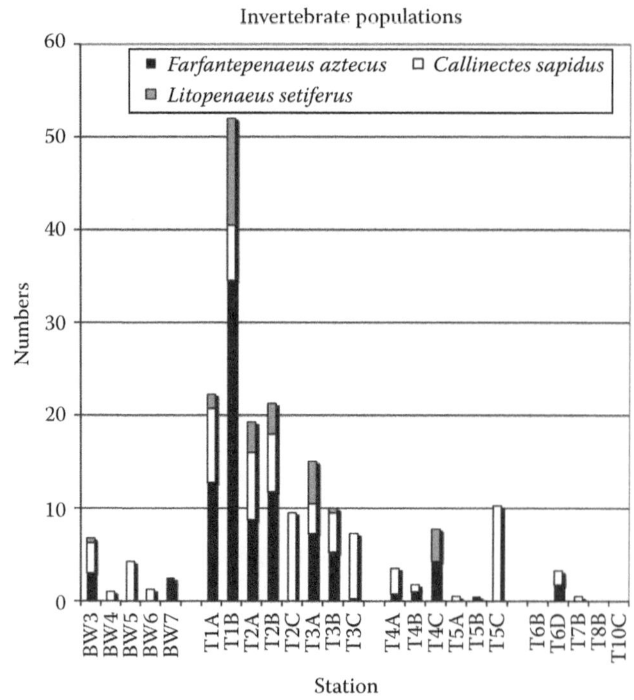

Figure 8.23 Dominant invertebrate populations in the Pensacola Bay system (averaged annually) over the 1997–1998 study period.

The upper parts of the Escambia and Blackwater Bays were represented by higher (total) species richness for infaunal and epibenthic macroinvertebrates and fishes. This includes the highest numbers of fish predators such as the sand sea trout, *C. arenarius*, and the primary planktivorous bay anchovy, *A. mitchilli*. The distribution of the closely related striped anchovy, *Anchoa hepsetus*, in East Bay could indicate a competitive interaction between the two plankton feeders. The other chief fish dominant, the gulf menhaden, *B. patronus*, was located in shallow areas of both Escambia and East Bays. This is a seasonally dominant planktivorous fish with relatively high numbers just above and below the I-10 bridge in Escambia Bay.

In terms of temporal distributions of dominant invertebrate and fish species, the brown shrimp were prevalent in upper Escambia Bay during summer–fall months. Blue crabs were more broadly distributed in space and time, with a presence in upper Escambia Bay during all four seasons. This species was a spring dominant in shallow western areas of East Bay and in eastern parts of lower Escambia Bay. *B. patronus* was present only during the spring in upper and midparts of Escambia Bay and in East Bay. *C. chrysurus* was only present during fall periods and was dominant in upper Escambia Bay and in western parts of lower Escambia Bay. Bay anchovies (*A. mitchilli*) were dominant in summer, fall, and winter periods in upper Escambia Bay, whereas spot (*L. xanthurus*) and Atlantic croakers (*M. undulatus*) were located in upper Escambia Bay during summer–fall months and in lower Escambia Bay during the spring 1998 river peaks. The sand sea trout (or white trout, *C. arenarius*) was dominant during summer–fall periods in upper Escambia Bay.

The field data from the Pensacola Estuary were used to establish a database based on the total ash-free dry biomass m^{-2} mo^{-1} as a function of the individual feeding units of the infauna, epibenthic macroinvertebrates and fishes (the so-called FII index), and related trophic transformations. Herbivores in the Pensacola Estuary were concentrated in upper

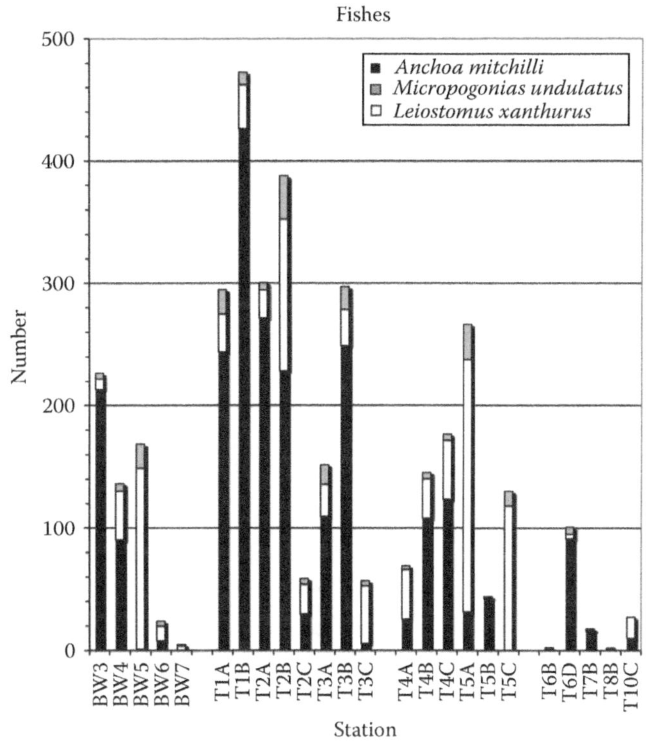

Figure 8.24 Dominant fish populations in the Pensacola Estuary (averaged annually) over the 1997–1998 study period.

Escambia Bay (Figure 8.25). Omnivores were most highly concentrated in lower Escambia Bay. The primary carnivore biomass was concentrated in shallow areas in upper Escambia Bay, whereas the biomasses of secondary and tertiary carnivores were highest in deeper parts of upper Escambia Bay (Figure 8.26).

Food web biomass was concentrated in the areas of highest chlorophyll *a* concentrations and primary productivity. In the Blackwater–East Bay system, the deeper areas (BW3, BW6) were characterized by low levels of biomass with dominance by herbivores. Shallower areas were characterized by higher biomass and representation mainly by primary carnivores. These distributions were not closely associated with either salinity stratification or DO anomalies. Chlorophyll *a* concentrations were generally low in the Blackwater–East Bay system with moderately low BOD concentrations at depth.

To summarize, during periods of high-salinity stratification, areas in upper Escambia Bay were associated with reduced DO at depth. Hypoxia was reduced along the north-south axis of Escambia Bay. Chlorophyll *a* concentrations were moderately high in upper Escambia Bay with a general decreasing trend along the north-south axis. Relatively low FII indices were noted at stations 2C and 3B in upper Escambia Bay and at stations 4B, 5A, and 5B in lower Escambia Bay. These areas tended to have reduced cumulative species richness indices, as well. The FII index levels were extremely low in Pensacola Bay as were the cumulative species richness indices. Overall, parts of Escambia Bay had the highest numbers of fish and invertebrate species (that included commercially important types) in the Pensacola Bay system. This was due to the nutrient loading from the alluvial Escambia River, showing the difference between such a system relative to a largely unpolluted Blackwater River and its associated bay.

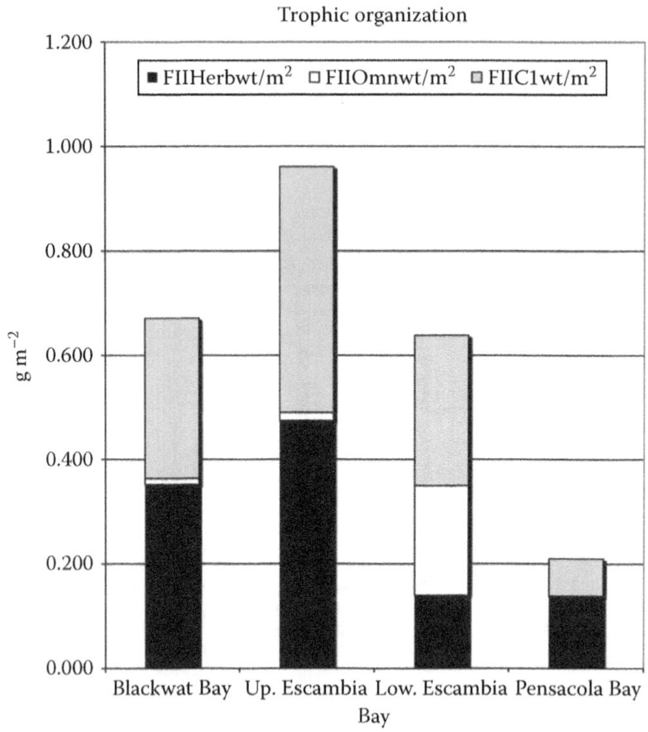

Figure 8.25 Trophic organization herbivores, omnivores, and C_1 carnivores in the Pensacola Estuary (averaged annually over months and stations) over the 1997–1998 study period.

Statistical Analyses

A regression analysis was made of the relationship of the chief biological features of Blackwater–East Bay, Escambia Bay, and Pensacola Bay to corresponding spatial/temporal physical/chemical habitat characteristics (independent variables: Table P.1). In Blackwater and Escambia Bays, infaunal biomass was positively associated with % silt in the sediments. The opposite was true in Pensacola Bay, with positive associations of this variable with % sand. Infaunal biomass was positively associated with bottom DO in Pensacola Bay. Invertebrate biomass in Escambia Bay was positively associated with bottom salinity and % silt and % clay; in Pensacola Bay, this factor was negatively associated with mean grain size and % sand and positively associated with % clay.

Infaunal species richness in Blackwater Bay was positively associated with % sand, whereas in Escambia Bay, this factor was positively associated with % silt and inversely associated with % sand. In Pensacola Bay, infaunal species richness was positively associated with % clay and inversely associated with % sand. Invertebrate species richness in Blackwater Bay was generally associated with high bottom DO, low chlorophyll *a*, and low salinity stratification. In Escambia Bay, invertebrate species richness was negatively associated with % sand and bottom DO. In Pensacola Bay, this index was positively associated with % sand, bottom DO, and depth. Fish species richness was positively associated with % sand and bottom DO in Blackwater Bay and Pensacola Bay. In Escambia Bay, however, this factor was positively associated with mean grain size and % silt and inversely associated with bottom DO. Thus, basic differences existed in the association of animal indices and habitat variables in Escambia Bay relative to the rest of the system.

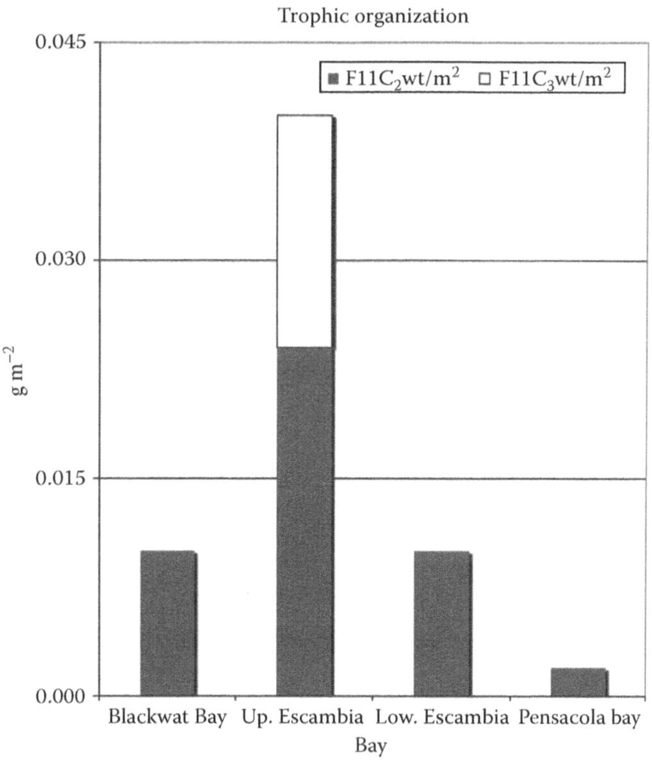

Figure 8.26 Trophic organization of C_2 carnivores and C_3 carnivores in the Pensacola Estuary (averaged annually over months and stations) over the 1997–1998 study period.

The FII herbivores in Blackwater–East Bay and Pensacola Bay were negatively associated with bottom salinity, sediment % organics, and % sand. The FII herbivores in Escambia Bay were positively associated with sediment % organics and negatively associated with bottom salinity. The C_1 component of Blackwater–East Bay was negatively associated with mean grain size and % organics in the sediments: in Pensacola Bay, the C_1 carnivores were positively associated with % sand and negatively associated with % clay. In Escambia Bay, this group was positively associated with % silt and surface/bottom BOD with negative correlations with % sand. The C_2 carnivores in Blackwater–East Bay were positively associated with bottom DO and negatively associated with chlorophyll *a* and mean grain size of the sediments. The C_2 carnivores in Pensacola Bay were positively associated with sediment % organics and % clay. These results are consistent with those described for the biomass and species richness data and illustrate a dichotomy of associations in the Escambia Bay system relative to the less productive Blackwater–East Bay and Pensacola Bay systems.

Based on the results of the regression analyses, a PCA-regression analysis was carried out using a series of biological factors (Table P.1: Figure 8.27) as the dependent variables. FII herbivores (Figure 8.27a) were positively associated with water color and turbidity and negatively associated with surface salinity (i.e., increased river flow). The FII herbivores were positively correlated with increased bottom total nitrogen and surface chlorophyll *a*. They were inversely correlated with FII omnivores. FII omnivores (Figure 8.27b) were negatively associated with surface SiO_2, $NO_2 + NO_3$, and chlorophyll *a*. Neither group was affected by salinity stratification or bottom DO. FII C_1 carnivores (Figure 8.27c) were generally similar to the herbivores (negatively associated with surface

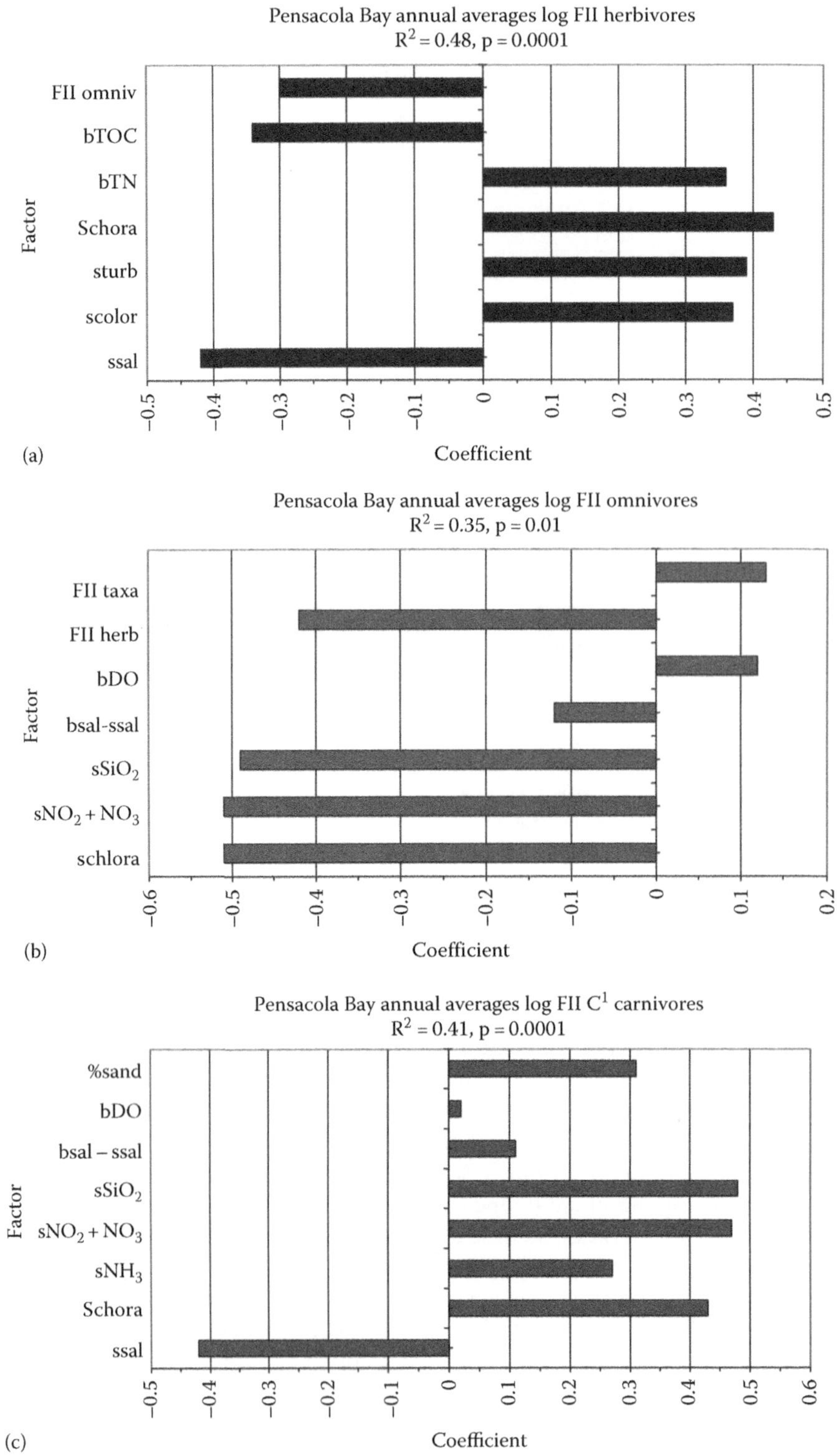

Figure 8.27 (a–f) PCA-regression analyses of variables taken in the Pensacola Estuary (all stations averaged) with data averaged annually (1997–1998). (a) FII herbivores, (b) FII omnivores, (c) C_1 carnivores.

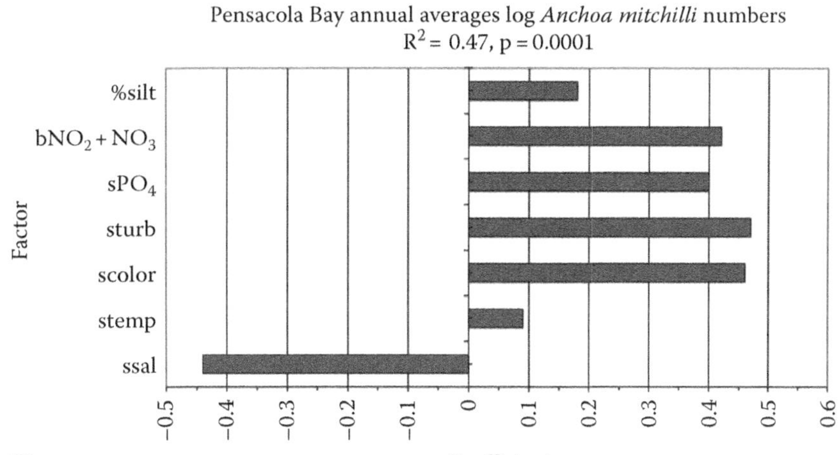

(d)

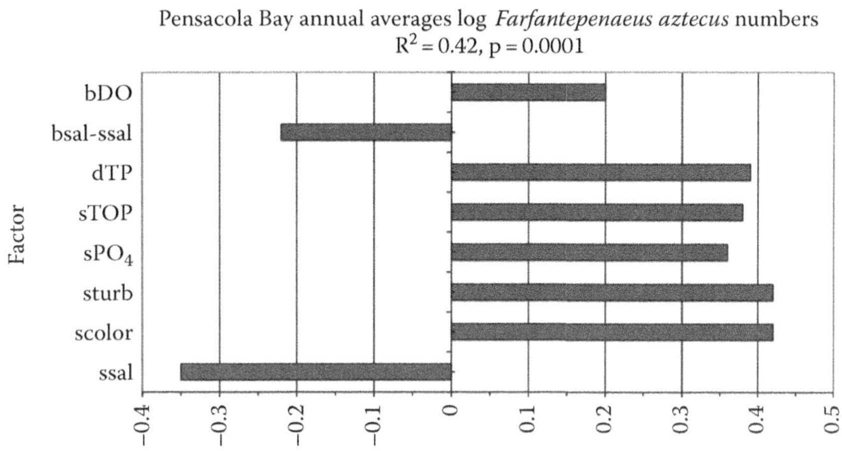

(e)

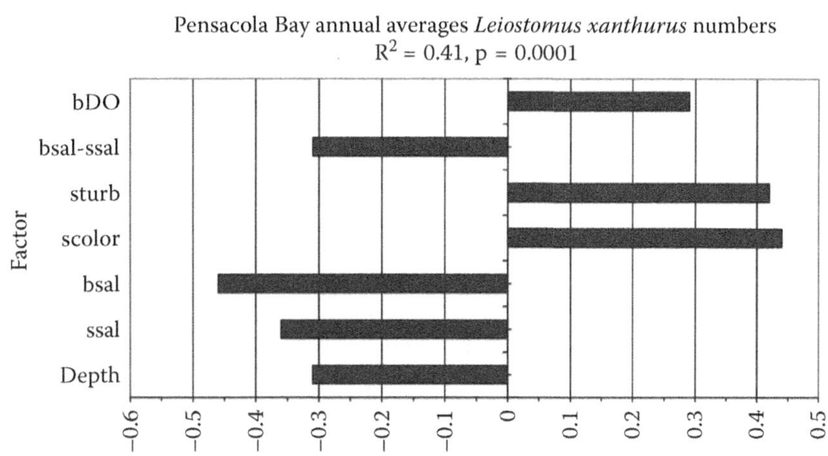

(f)

Figure 8.27 (*Continued*) (a–f) PCA-regression analyses of variables taken in the Pensacola Estuary (all stations averaged) with data averaged annually (1997–1998). (d) *A. mitchilli* numbers, (e) *F. aztecus* numbers, and (f) *L. xanthurus* numbers.

salinity and positively associated with river flow factors, nutrients, and chlorophyll *a*). Anchovies and spot and brown shrimp (Figure 8.27d through f) also followed this pattern. Spot numbers were the only factor that was negatively associated with salinity stratification and positively associated with bottom DO.

These data show the importance of Escambia River flow and nutrients to the biological organization of the Pensacola Estuary.

The results of this regression/PCA analysis indicate that organisms in the more productive areas of the estuary (i.e., upper Escambia Bay) have very different associations with sediment type and bottom DO than those in less productive areas of the system (i.e., Blackwater Bay and Pensacola Bay). Primary production, predator–prey interactions, and food web relationships may attain a higher priority than habitat type (sediments, DO) in upper Escambia Bay as a determinant of population distribution in space and time. The data are consistent with the hypothesis that nutrient loading into upper Escambia Bay, with associated increases in primary production and chlorophyll *a*, is the central determinant of secondary production in the Pensacola Estuary. The actual distribution of secondary production also reflects various impacts of pollution sources in this area.

The resulting trophic relationships represent primary determinants of the distribution of dominant populations at all levels of biological organization. While habitat factors, such as sediment type, DO, and salinity distributions, are important in determining the general distribution of organisms, these factors serve as secondary determinants of biological organization. Thus, the estuary-wide biological distributions can only be understood within the context of patterns of how the distribution of primary production in the system coincides with microhabitat distributions. The extremely low secondary production in eastern sections of upper Escambia Bay and in lower Escambia Bay–Pensacola Bay appears to serve as an indication of pollution coming from upper bay point sources and the city of Pensacola.

The relative importance of these interactions with respect to secondary production of the estuary can only be understood within the context of intersystem comparisons that will be treated below.

Discussion

The Pensacola Estuary has been subject to continuous anthropogenic stress for decades. Submerged aquatic vegetation, formerly abundant in this system, has been largely absent since the mid-1970s (Collard, 1991a,b). Formerly productive oyster bars are no longer commercially viable, although a small oyster fishery still exists. The most extensive fish kills in the scientifically documented literature occurred in Escambia Bay during the late 1960s and early 1970s (USEPA, 1971a). Although the massive fish kills during this period were ended by reduced nutrient loading to the bay from the various point sources, the destruction of habitat in the Pensacola Bay system due to nutrient loading from chemical plants and sewage treatment facilities had long-term consequences that continue to recent times.

Based on the qualitative and quantitative whole-water phytoplankton data, Escambia Bay does not appear to be in a hypereutrophic state. However, the presence of a number of bloom species as subdominants in lower Escambia Bay, which are known to cause problems when exposed to excessive nutrient loading, means that close attention must be paid to increased nutrient loading of Escambia Bay in the future. The presence of these species as subdominants could mean that Escambia Bay is predisposed to a hypereutrophic state, which is another reason why excessive nutrient loading in the future could cause bloom problems similar to those that have occurred in Perdido Bay.

The Pensacola Estuary is composed of a series of diverse habitats that are commonly found in river-dominated estuaries of the northern Gulf of Mexico. The Escambia River is an alluvial stream that is tannin-stained and turbid with low light penetration. River flow, wind and tides, the depth of the receiving basin, and the penetration of salinity from the gulf caused periodic water-column stratification within the Pensacola Estuary. Bottom DO was controlled by salinity stratification in upper and midparts of Escambia Bay and the Blackwater/East Bay system, whereas these relationships were not evident in Pensacola Bay.

River-associated nutrient loading was a prime determinant of overall primary and secondary production in Escambia Bay and Blackwater Bay. Nutrient concentrations were highest in upper Escambia Bay, with the Escambia River as the chief source of such compounds. Increased concentrations also occurred in mid- and lower Escambia Bay during periods of high rainfall. The highest mean concentrations of chlorophyll *a* occurred in upper Escambia Bay although such concentrations in the entire system were relatively low. There were no signs of hypereutrophication in the Pensacola Estuary over the sampling period (1997–1998). Whole-water phytoplankton data indicate that Escambia Bay is was in a modestly eutrophic state during the study. Dominant phytoplankton populations included various noxious bloom species found mainly in Escambia Bay. Urban stormwater runoff may be a factor in the maintenance of these populations.

The main concentrations of infauna (polychaete worms), epibenthic macroinvertebrates (brown shrimp, blue crabs), and fishes (sciaenids) were found in the relatively shallow parts of upper Escambia Bay, the center of primary and secondary production in the entire Pensacola Estuary due to nutrient loading from the Escambia River. The trophic organization emanating from the higher primary production of the upper bay was the main determinant of the distribution of dominant, commercially important populations. Habitat factors, such as sediment type, DO, and salinity distribution, while important in determining the general distribution of organisms in the estuary as a whole, had effects that could only be understood within the context of the distribution of primary production. Relative to other such areas in regional river-dominated estuaries, there was a relatively low biomass of infauna, epibenthic invertebrates, and fishes in the most productive parts of Escambia Bay.

The lower Escambia Bay–Pensacola Bay areas were among the most faunally depauperate of all estuaries sampled. This low level of secondary production could be the result of hypereutrophication due to habitat losses (oyster bars and seagrass beds) and changes in benthic areas due to anthropogenic nutrient loading in the late 1960s and early 1970s. Commercial populations, lost during the massive fish kills during this period, have not recovered according to the results of this study. Nonpoint pollution from the city of Pensacola and other urbanized areas of the Pensacola Bay basin, together with major pollution of the bayous (Texar, Chico, Grande) in this region, present serious problems in terms of continuing urban contamination. The Escambia Bay remains vulnerable to hypereutrophication from increased nutrient loading.

The results of the study on the Pensacola Estuary illustrate three basic trends. Nutrient loading from a major river is a major factor in the production of commercially valuable invertebrates and fishes. However, anthropogenic nutrient loading was responsible for the destruction of critical habitats and the debilitation of the major sports and commercial fisheries of the estuary. The effects of such loading during the 1960s and 1970s caused habitat debilitation that lasted for decades. Recent conditions are still subject to adverse effects due to urban runoff.

The limited time of study of this system seriously qualifies any conclusions regarding cause-and-effect relationships regarding the influence of long-term climate changes on the trophic organization of the Pensacola Estuary. The status of this estuary in terms of primary and secondary productivity can only be illustrated by a comparison of such factors with other, similar bay systems in the region. This comparison will be developed in the following.

The long history of pollution of the water in the Pensacola area was exemplified by a recent report issued by the Environmental Working Group (EWG; Top-Rated and Lowest-Rated Water Utilities—2009). In a comparison of drinking water quality in the cities across the United States, the group found that the worst such water quality was noted in Pensacola, Florida (Emerald Coast Water Utility). A news report stated that "Of the 101 chemicals tested for over 5 years, 45 were discovered. Of them, 21 were discovered in unhealthy amounts. The worst of these were radium-228, trichloroethylene, tetrachloroethylene, alpha particles, benzine, and lead. Pensacola's water was also found to contain cyanide and chloroform. The combination of these chemicals makes Pensacola's water supply America's most unhealthy." A report in *The Lancet Neurology Journal* (February 15, 2014) stated that toxic chemicals such as lead and polychlorinated biphenyls (PCBs) can cause a number of neurodevelopmental disabilities including attention-deficit hyperactivity disorder, autism, dyslexia, and other cognitive damage in children.

PART V

Comparative Analysis of Gulf Ecosystems

CHAPTER 9

Trophic Organization

INTERACTING PROCESSES

Background

Vinagre et al. (2011) noted that river flow variability influenced estuarine production, yet an understanding concerning flow effects on estuarine food webs is still scarce in the scientific literature. Using stable carbon and nitrogen isotopes in two main fish nursery areas of the Tagus Estuary, the authors showed that, during low river flow conditions, two distinct food webs were established in each nursery area. The food webs responded to small variations in the freshwater input. Winter floods disrupted localized trophic relationships established during low river flow periods. This led to the reestablishment of a wider food web. During wet years, this wide food web was maintained until spring when it underwent fragmentation into two localized and distinctive trophic organizations. Increased drought frequency caused by climate change was projected to lower the connectivity of the estuarine food webs, causing habitat fragmentation and consequent loss in complexity and resilience.

Loneragan (1999), working in the Logan River in southeast Queensland, Australia, noted that high river discharge had a strong positive effect on the production of coastal commercial and recreational fisheries and that the seasonal patterns of river flow were equally, if not more, important than the magnitude of the flows. Livingston (1984a) found that Apalachicola River peaks usually occurred during winter–spring periods of low evapotranspiration and that summer Florida rainfall had little effect on such flow rates due to the seasonal evapotranspiration cycling. The effects of climate change over regional rainfall amounts and patterns should thus be evaluated relative to altered seasonal patterns of river flow. Changes in seasonal river fluctuations are likely to have a significant effect on the production of associated coastal fisheries and, given the current pressures for water resource development, this is an important avenue for future research and evaluation.

Infaunal abundance and biomass have been associated with nutrient loading from river inflows (Montagna and Yoon, 1991). The influence of freshwater inflow is considered to be a function of interactive processes associated with water quality factors (e.g., salinity, dissolved oxygen), sediment interactions, nutrient enrichment, and biological processes such as column productivity, recruitment, and population variations due to stress (Montagna and Kalke, 1992). Although such useful productivity has been translated into economic as well as ecological benefits (Walsh, 1988), corresponding unprecedented human growth and development of near-shore estuarine and coastal systems has led to various forms of "stress" (Baird, 1996). Reduced or altered freshwater flow, osmoregulatory stress, extremes of temperature and dissolved oxygen, nutrient enrichment and hypereutrophication, fishing pressure, and toxic discharges have combined to reduce the fisheries production in many river-dominated estuaries in Florida (Livingston, 1990b) and elsewhere along the Gulf Coast (Livingston, 2000, 2002, 2005).

Evaluation of natural and human-related impacts is often based on population and community factors relative to temporal and spatial habitat trends. These trends in soft-sediment environments have been widely assessed (Thrush et al., 1989; Trueblood, 1991). Thrush et al. (1994) illustrated the difficulties of evaluating long-term changes because of spatial variation of soft-sediment macrofauna. Sampling scales relative to the distribution of subject organisms will affect the interpretation of the data (Livingston, 1987b). In many cases, sampling different populations and communities gives varying answers to the question of impact. A comparison of population dynamics with actual changes in a given habitat can lead to irrelevant conclusions since populations in estuarine/coastal areas vary considerably in space and time.

The use of individual populations as indicators of changes in state habitat variables is indirect at best. Factors such as spawning success and migratory behavior are involved in the ultimate expression of population distributions in coastal areas. Such influences cannot always be factored into the direct response of a given population to long-term trends of habitat change in a given estuary because of confounding issues related to the complex factors that control population changes. The individual population thus becomes relevant to process change in estuarine and coastal systems only as part of a more general trophic response to short- and long-term habitat alternations and productivity cycles. Hence, interannual responses of individual populations to habitat change and trophic organization represent species-specific forms of adaptation and the response of a given species to multiple variables may be masked by the actual position of such a population in the overall food web. The fact that such populations can be substituted trophically by other like populations of totally unrelated species in space and time in a given food web further complicates impact evaluation that is based on a taxonomically limited sampling effort. This contributes to the importance of trophic analyses in the evaluation of long-term trends.

Primary animal associations of shallow Gulf of Mexico estuaries are dominated by infaunal macroinvertebrates, epifaunal macroinvertebrates, and fishes (Sheridan, 1978, 1979; Laughlin, 1979; Sheridan and Livingston, 1979, 1983; Livingston, 1984a, 2000, 2002, 2005). Interactions of these groups are mediated by predator–prey relationships (Virnstein, 1977; Peterson, 1982; Whitlatch and Zajac, 1985; Livingston, 2002). The extensive literature concerning soft-sediment trophic organization tends to be based on issues that are often tangential to food web changes over extended periods of time. However, interactions between biotic and abiotic aspects of trophic dynamics in coastal systems remain poorly understood due to a lack of long-term, system-level studies. Few of the existing such studies take into account the spatial and temporal aspects of the physical environment or the historical determinants of trophic organization in coastal systems (Dayton et al., 1992). Thus, the primary interactions of river flow with estuarine productivity (with reference to changes in salinity, primary production, predator exclusion, and competition) and food web organization remain largely undetermined even though considerable data are available concerning short-term and simplified trophic relationships.

River-dominated estuaries along the Gulf of Mexico coast vary considerably in terms of geomorphology, habitat distribution, and dominant animal populations. Differences in the form of primary productivity can lead to major changes in secondary productivity. Sediment quality and distribution are fundamental to an understanding of the dynamics of relatively shallow coastal areas (Whitlach, 1977; Demas et al., 1996) with benthic communities closely associated with the physical properties of subaqueous soils (Rhoads, 1974). Sediment texture can vary considerably within a given estuary. As noted earlier, silty clays are often associated with deeper areas where there is less turbulence (Davies and DeMoss, 1982). Depth thus becomes a major limiting factor along with the related subject of salinity stratification, associated changes in sediment quality, and the occurrence of seasonal hypoxic to anoxic conditions. The importance of sediment quality and distribution is central to a comparative analysis of productivity in Gulf Coast estuaries.

Sediment Comparisons

Sediment analyses were averaged across stations and months for each of the seven study areas according to methods outlined by Livingston (1997b) (Figure 9.1). Sediments of the Econfina offshore area were composed mainly of sand with relatively small silt and clay fractions. In the offshore Fenholloway area, there was still a high sand content, but the silt and clay components were both higher than that of the Econfina due in large part to the discharge of highly organic pulp mill effluents into that coastal area. The relatively unpolluted Blackwater Estuary, another so-called Blackwater drainage, had a relatively large sand component in the sediment composition. There was a greater silt component and a lower clay component than that in the Fenholloway Estuary.

The four large river–estuaries had higher levels of silt than the blackwater estuaries. The influx of nutrients and particulate organic matter that was loaded into the offshore areas from the large rivers was responsible for the increased levels of silt. All four of the river-dominated estuaries also had larger plankton components than that in Apalachee Bay (as represented by the Econfina Estuary) and the Blackwater Estuary. These differences, due to the discharges of nutrients, from the rivers, had effects on the sediment composition of these estuaries and associated primary and secondary productivity. There were also increased levels of clay in the sediments of the river-dominated systems that reflected the addition of terrigenous components loaded into the receiving estuaries from river flooding.

Station-specific sediment composition of the upper parts of the large river systems (Figure 9.2) varied widely both within and among the subject study areas. The Perdido Estuary was typified by relatively high silt composition in deeper water at the mouth of the river that was a depositional

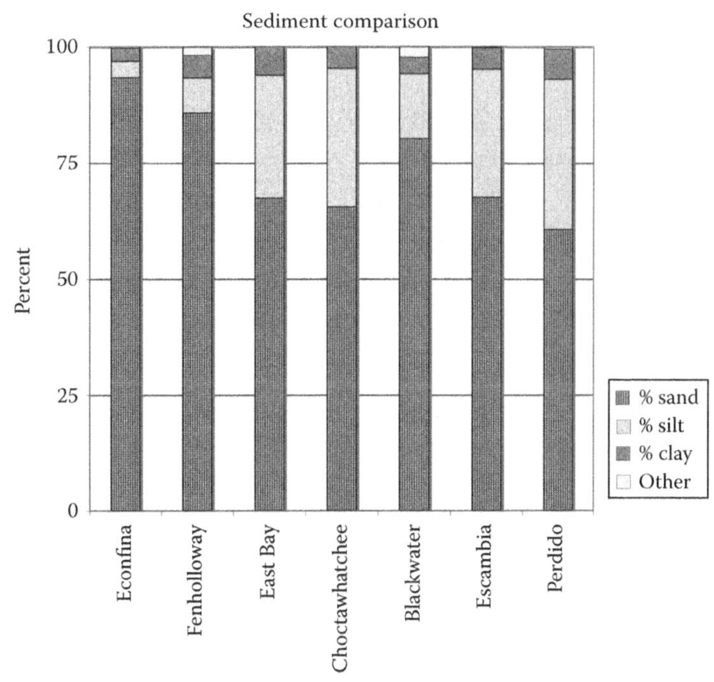

Figure 9.1 Sediment analyses of the major study sites in the long-term studies of the coastal estuaries of the NE Gulf of Mexico. Data were averaged over several sampling events for each system.

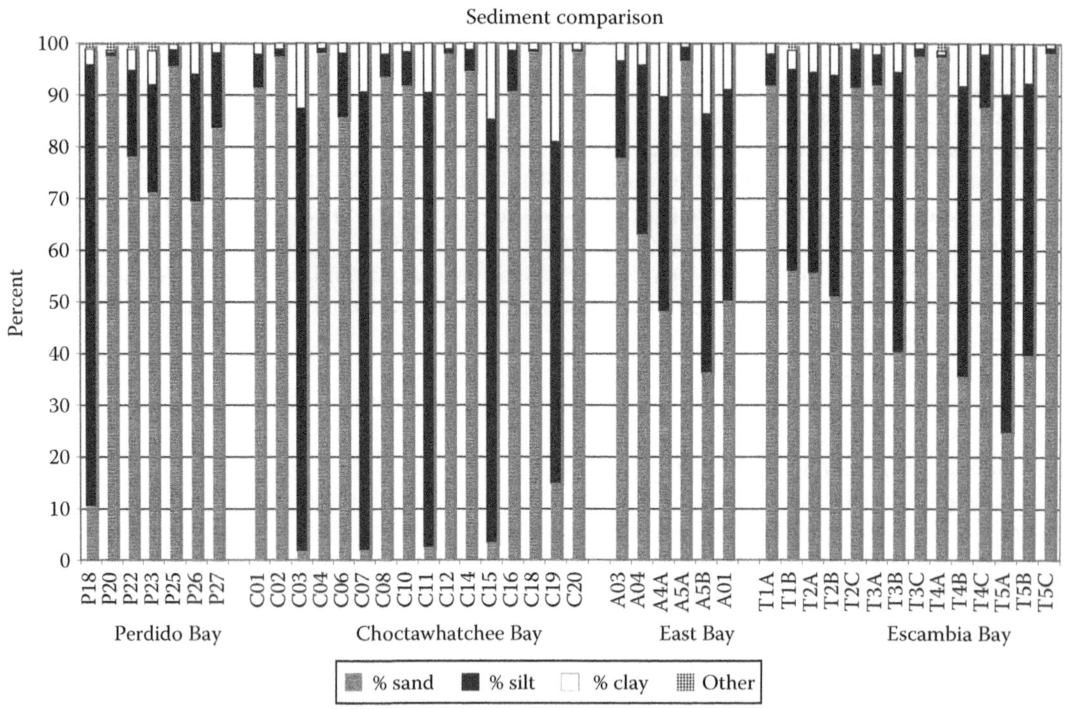

Figure 9.2 Station-specific sediment composition of the upper parts of the large river–estuaries of the NE Gulf of Mexico.

area for particulate organic matter loaded into the bay from the Perdido River. Both the Perdido and Choctawhatchee Estuaries had relatively sandy sediments in shallow habitats (<2 m) with relatively high silt and clay fractions in deeper parts that were salinity stratified. The relatively low levels of silt in most of the Choctawhatchee Estuary could be due to the large size of the estuary compared to the size of the river basin. By contrast, the East Bay area of the Apalachicola Estuary had relatively high silt–clay fractions that were generally comparable to those in the upper Escambia Estuary. The Apalachicola Estuary was subject to the highest riverine nutrient and organic carbon loading of the entire region, and this, together with the relatively small size of the immediate receiving area of East Bay, could account for this observation. The silt components of the Escambia Estuary were relatively high and increased with depth on the north–south axis of the bay.

A comparison of the percentage of organics of sediments of the four major river-dominated estuaries (Figure 9.3) indicated that sediment organic matter in the Escambia Estuary was much higher than any of the other subject bay systems. The relatively smaller size of the sediment particles in the Escambia Estuary (Figure 9.4) is consistent with the observed high organic components. There is no clear reason for these observations although the history of this part of the Pensacola Estuary at one time included some of the highest levels of plankton blooms in the entire Gulf of Mexico. These aspects of the sediment quality of the Escambia Estuary need further analysis for any definitive conclusions regarding ecological significance, but the quality of Escambia sediments could be related to the relatively low secondary productivity of this system. The deeper, more salinity-stratified parts of the Perdido and Choctawhatchee Estuaries had higher levels of organic matter than the shallower areas. The Choctawhatchee sediments had significantly higher organic concentrations than those in the Perdido Estuary that were comparable to those in the Apalachicola Estuary (East Bay) in the distribution of this sediment component. Mean grain size was generally smaller in the deeper stations of the various bay

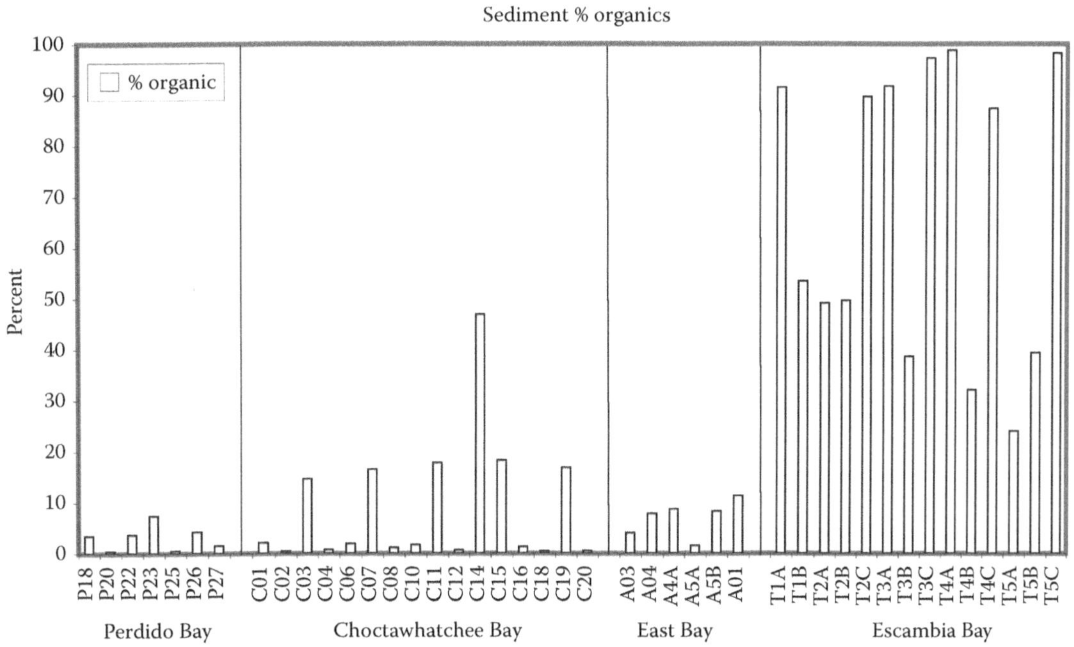

Figure 9.3 Station-specific percent organic sediment composition of the upper parts of the large river–estuaries of the NE Gulf of Mexico.

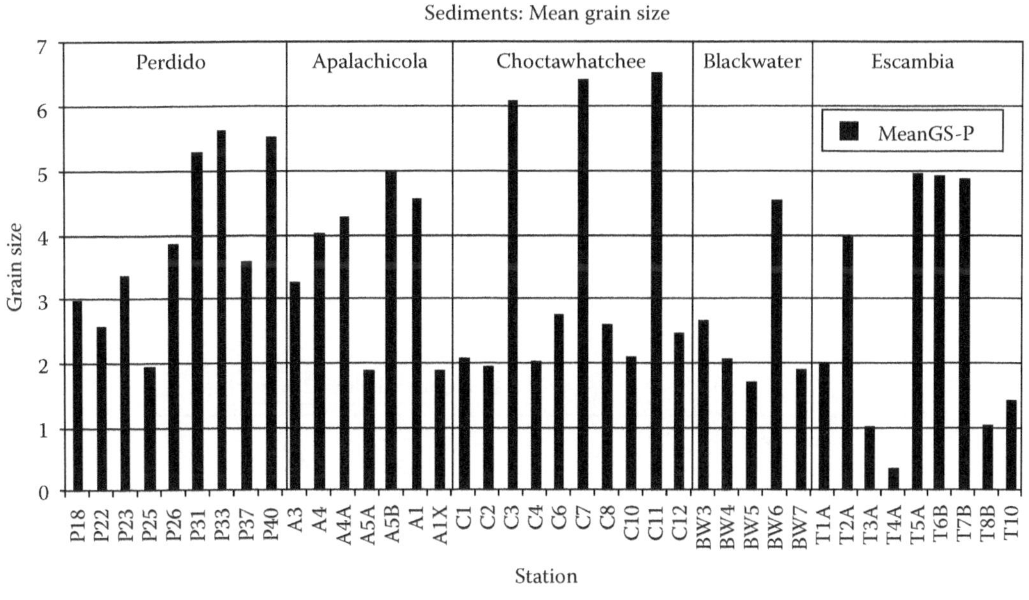

Figure 9.4 Station-specific sediment size (phi units) composition of the upper parts of the large river–estuaries of the NE Gulf of Mexico.

systems. The importance of depth and salinity stratification to sediment type was indicated by these data, but such distributions should be understood within the context of habitat types of the individual estuaries.

Salinity/Depth Relationships

Although salinity stratification was generally low, bottom salinities were highest in the offshore areas of the Econfina and Fenholloway Estuaries (Figure 9.5). Salinities in East Bay of the Apalachicola Estuary were low as were the associated stratification indices. Salinities in the Apalachicola Estuary were intermediate with relatively low stratification except in areas associated with the Sike's Cut area. In the Escambia Estuary, bottom salinities were low, with stratification levels comparable to those in the Apalachicola and Blackwater Estuaries. Mean salinities were comparable in the Perdido and Blackwater Estuaries. However, stratification in the Perdido Estuary was the highest of all the estuarie systems. In the Choctawhatchee Estuary, mean bottom salinities were relatively high and the stratification index was the second highest of all the study areas. It was comparable to that in the Blackwater drainage. The data show that salinity stratification was closely associated with the opening of the bays to the gulf through anthropogenic activities and that the resulting salinity stratification in both the Choctawhatchee and Perdido Estuaries was relatively stable even during storm events.

Station-specific stratification comparisons (Figure 9.6a) indicated that the East Bay of the Apalachicola Estuary was the least stratified of all the study areas, whereas the deeper stations in the Perdido and Choctawhatchee Estuaries were the most stratified. The Blackwater drainages and East Bay were the shallowest. The least stratified estuaries had the lowest levels of anoxia/hypoxia conditions of the gulf study areas (Figure 9.6b). The Choctawhatchee and Perdido Estuaries were deepest and had the highest anoxia/hypoxia percentages. The Apalachicola and upper Escambia Estuaries were intermediate in both indices.

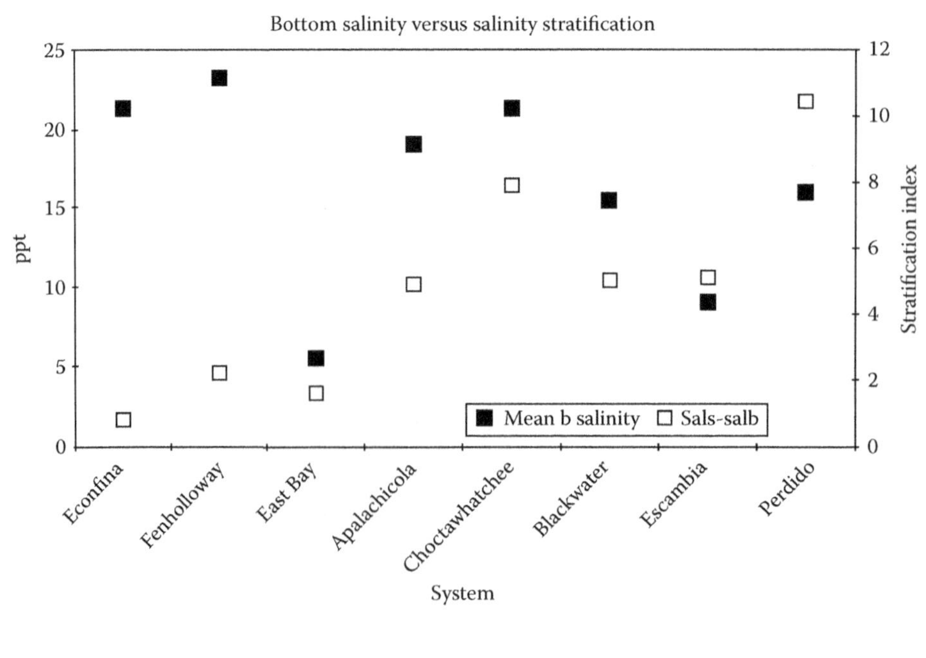

Figure 9.5 Mean bottom salinities and salinity stratification indices across stations in the coastal estuaries along the NE Gulf of Mexico.

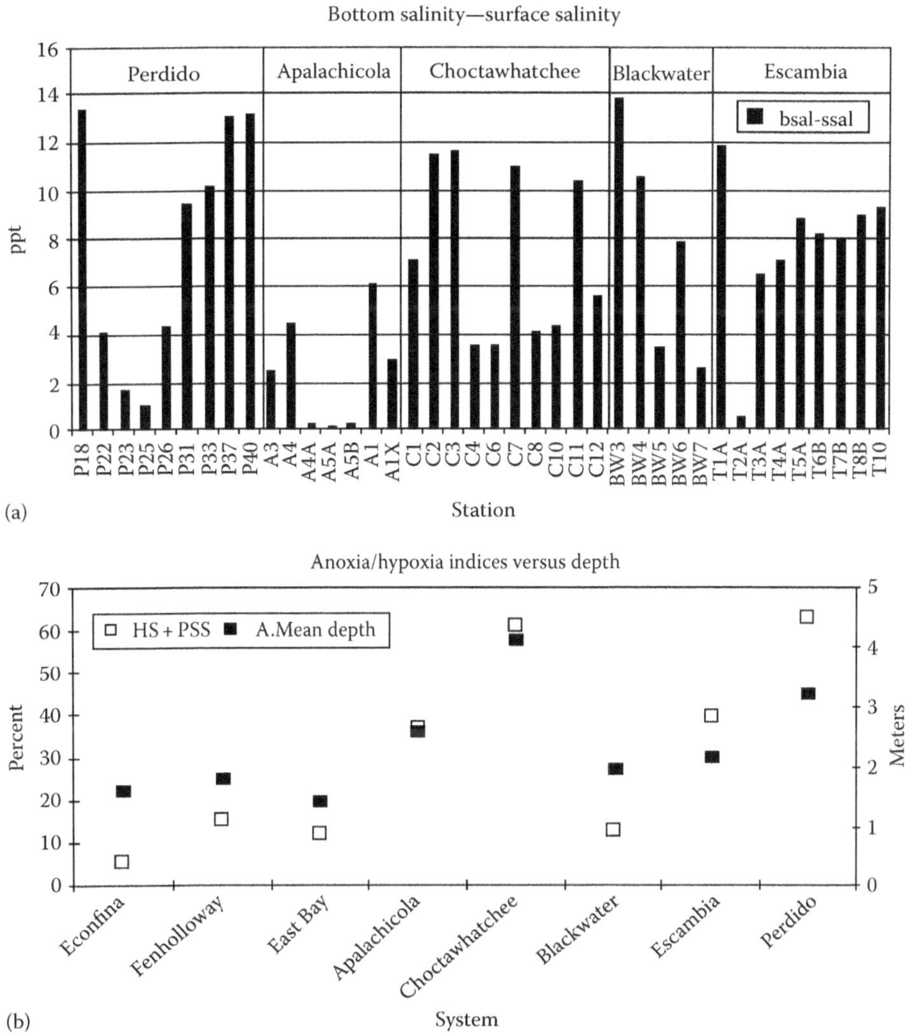

Figure 9.6 (a) Mean bottom salinities–mean surface salinities across stations in the coastal estuaries along the NE Gulf of Mexico. (b) Joint anoxia and hypoxia indices and mean depths across stations in the coastal estuaries along the NE Gulf of Mexico.

In most cases, dissolved oxygen above the halocline remained in excess of 4 mg L^{-1}. Dissolved oxygen concentrations in the shallow Apalachicola Estuary did not go below 4 mg L^{-1}. In the Choctawhatchee Estuary, there were seasonal episodes of bottom hypoxia in deeper areas. The most sustained periods of subhaloclinal hypoxia were noted in areas marked by high-salinity stratification. In the Perdido Estuary, hypoxia was most pronounced at depth at stations 18, 22, and 26 in the upper bay. Station 22, in Elevenmile Creek, had relatively low surface and bottom dissolved oxygen throughout most of the year; this was considered an effect of the loading of organic wastes by the pulp mill upstream. With the exception of Elevenmile Creek, the most sustained subhaloclinal hypoxia was noted at stations 31 and 37 in the lower bay. Overall, in both the Choctawhatchee and Perdido Estuaries, benthic hypoxia was associated with salinity stratification at depths exceeding 2 m. Station 40 in the Perdido Estuary was the main exception to this generalization where there was relatively little hypoxia in spite of the fact that the water was relatively deep and there was considerable salinity stratification with high concentrations of silt–clay fractions. The influx of gulf water at depth was responsible for this observation.

Diving surveys indicated that, in the Perdido and Choctawhatchee Estuaries, wherever there was a distinct halocline, the salt wedge was composed of clear, warm water during the warmer months of the year. The upper, less salty layer of water was more colored (tannins, lignins, fulvic acid complexes) and contained large amounts of particulate matter. This upper layer was usually characterized by water currents, whereas the water below the halocline was without detectable currents. The bottoms of the shallow periphery of the Perdido and Choctawhatchee Estuaries were characterized by sandy substrates. The shelf areas graded down a slope into the deeper parts of these bays that were characterized by liquid mud wherein the sediments were composed of suspended particles of a flocculent nature. The floc was composed of black particles about 2–3 mm in diameter. These particles were buoyant and disturbance set them in motion.

In the upper parts of Perdido Bay, the particles were larger and there was a distinct layer of sand beneath a shallow layer of silty flocculent material. At stations 31, 37, and 40, the sediments were very soft and dark, and no movement was noted in the liquid mud layer. Similar sediment conditions were noted in deeper parts of Choctawhatchee Bay. A salt wedge covered the deeper sections of the bay below the sandy shelf in a manner very much like that described in Perdido Bay. No such liquid muds were noted in the Apalachicola Estuary where sediments were continuously affected by turbulence in the relatively shallow water column. Depth was of considerable importance in the determinations of habitat conditions of these estuaries, and, together with salinity stratification and nutrient-loading characteristics, these factors determined the quality of the bottom water and associated sediments.

Comparison of FII Trophic Indices

The reorganization carried out to transform the original trophic analyses of infaunal and epibenthic invertebrates and fishes into a uniform database produced fish, infauna, and invertebrate (FII) indices that represented the five trophic levels. This transformation allowed between-system comparisons that avoided problems caused by differences based on the diverse populations that occur among the major river–estuaries along the NE Gulf Coast. These indices summarized the overall biomass of macroinvertebrates and fishes that occupied a square meter of water and sediments (10 cm in depth) at each station in every coastal system. The division into trophic units (TUs) and reorganization of cross-species TUs into a comprehensive food web database allowed a comprehensive determination of feeding interactions. Field data sampled over monthly to quarterly periods that extended anywhere from 2 to 20 years over time were then transformed into a simplified but representative trophic organization that could be related to long-term effects of various anthropogenic and natural drivers that included noticeable changes in the regional climate. The assumptions for the relevance of such transformations were based on the universality of the energy relationships based on the first and second laws of thermodynamics. The advantages of this approach included the avoidance of the diverse processes that caused species-specific population variations from system to system due to adaptive relationships and interspecific competition and predation.

A comparative analysis was made concerning the averaged FII data taken in the large river–estuaries along the NE Gulf Coast. The station-specific FII biomass data (Figure 9.7) averaged over the various sampling periods indicated that stations in East Bay of the Apalachicola Estuary had the highest overall biomass levels when compared with the other study areas. It should be noted that most of the data for the Apalachicola Estuary were taken during the 1970s, prior to the onset of the drought periods of the region that started in 1980–1981. This would make the Apalachicola FII database and the Apalachee Bay data of the 1970s important as background relative to the

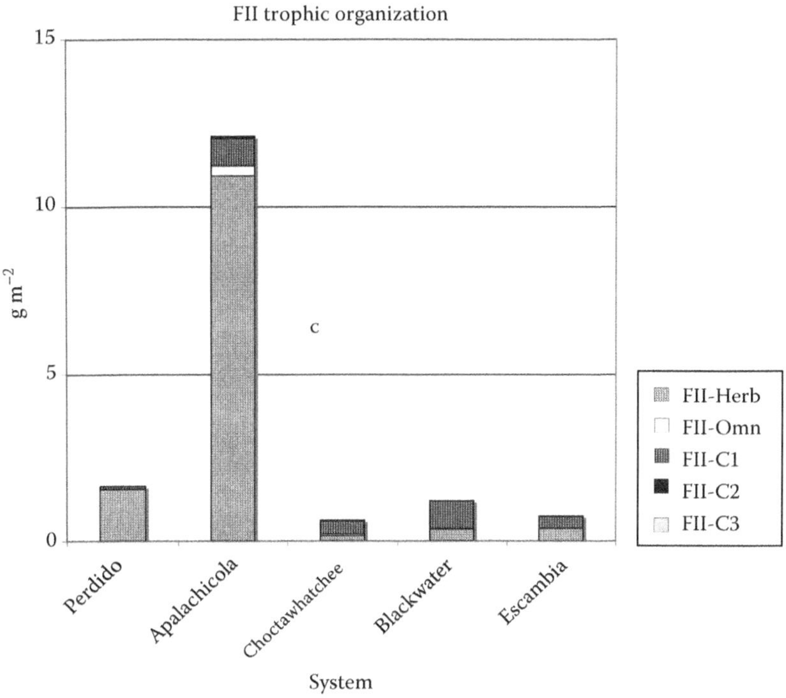

Figure 9.7 FII indices (biomass m^{-2}) averaged over stations and time for the major river–estuaries along the NE Gulf Coast.

subsequent climatic changes that took place over the following decades. The same could be said for the Perdido database where about half of the data were taken during a period prior to the onset of the major droughts 1999–2002 and 2006–2008.

The most obvious difference between the Apalachicola results and those of the other systems was the relatively high levels of herbivore biomass in the predrought Apalachicola system. The higher levels of omnivores in the Apalachicola system indicated a dependence of the lower levels of the estuarine food webs on detritivory relative to the other river–estuaries. These effects indicated a response of the trophic organization to basic differences in primary production and particulate organic carbon (POC) loading that influenced the composition and structure of the estuarine food webs. The predominance of the C_1 carnivores in the unbalanced food web organization of the less productive estuaries could be related to the changes taking place in the lower levels of the food webs. These differences could also be related to the disparate processes associated with the herbivore/omnivore groups dependent on available organic matter and the primary carnivores that were dependent on the lower levels of the food webs (Livingston et al., 1997).

Station-specific comparisons of the herbivore components of the different river-estuarine systems are shown in Figure 9.8a. In the Apalachicola Estuary, the highest levels of herbivores were found in upper parts of East Bay (receiving direct inputs from the river) and bay areas inshore of St. George Island (A1X) located in an area affected by both the river and the marshes that line the back-bay parts of the island. The lowest herbivore levels in the Apalachicola Estuary were in areas outside of the direct effects of river and land runoff (station A1). In the Perdido Estuary, stations P23 and P25 had the highest FII herbivore biomass levels that were somewhat comparable to those in the Apalachicola Estuary. These areas constituted the most productive parts of the Perdido Estuary prior to the increase of bloom activity and drought occurrence. These shallow, unstratified areas received direct inflows of river and creek freshwater runoff. The FII herbivore biomasses in the Choctawhatchee, Blackwater, and Escambia Estuaries were generally very low compared to that in the Apalachicola Estuary for reasons related to stratification effects, low Blackwater freshwater flows, and anthropogenic impacts related to nutrient loading in the respective bays. The combination of high river flows, shallow (unstratified) receiving areas, and the general lack of anthropogenic activities involving dredging and nutrient loading resulted in the ideal conditions for high herbivore productivity in the Apalachicola Estuary.

A comparison of the omnivore components (Figure 9.8b) indicated the much higher productivity of the lower end of the food web in the Apalachicola Estuary that had the highest levels of

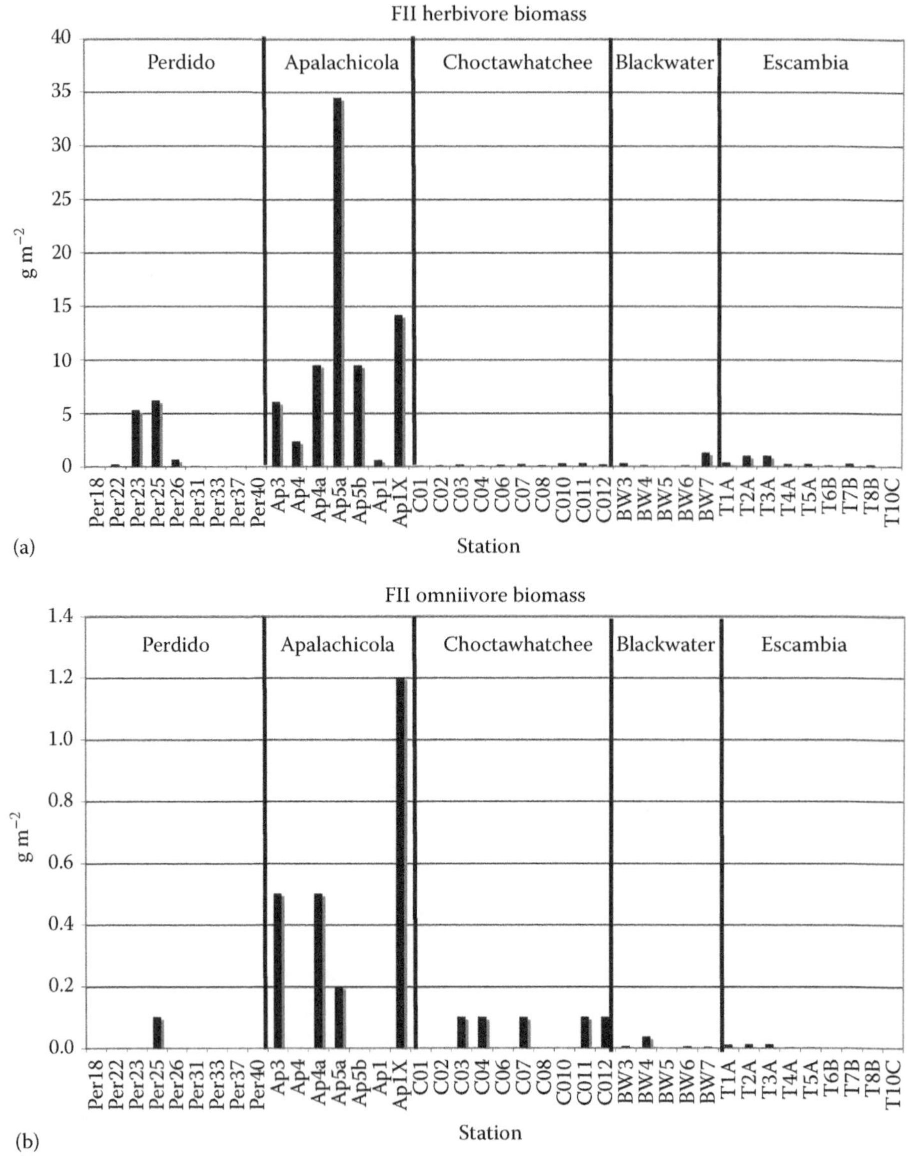

Figure 9.8 (a) FII herbivore biomass (g m^{-2}) by station averaged across dates for the river–estuaries in the NE Gulf of Mexico. (b) FII omnivore biomass (g m^{-2}) by station averaged across dates for the river–estuaries in the NE Gulf of Mexico. *(Continued)*

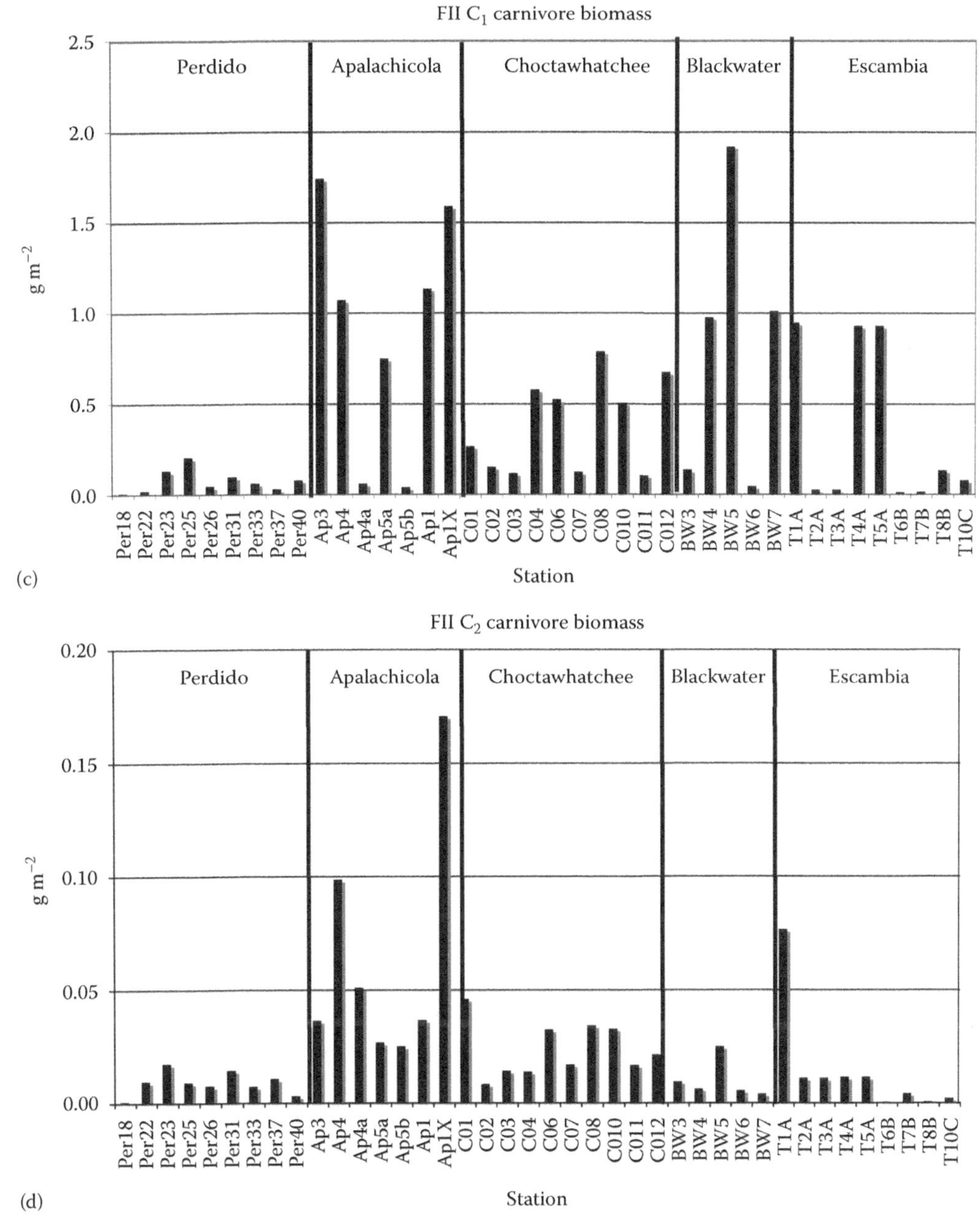

Figure 9.8 (*Continued*) (c) FII C_1 carnivore biomass (g m^{-2}) by station averaged across dates for the river–estuaries in the NE Gulf of Mexico. (d) FII C_2 carnivore biomass (g m^{-2}) by station averaged across dates for the river–estuaries in the NE Gulf of Mexico. (*Continued*)

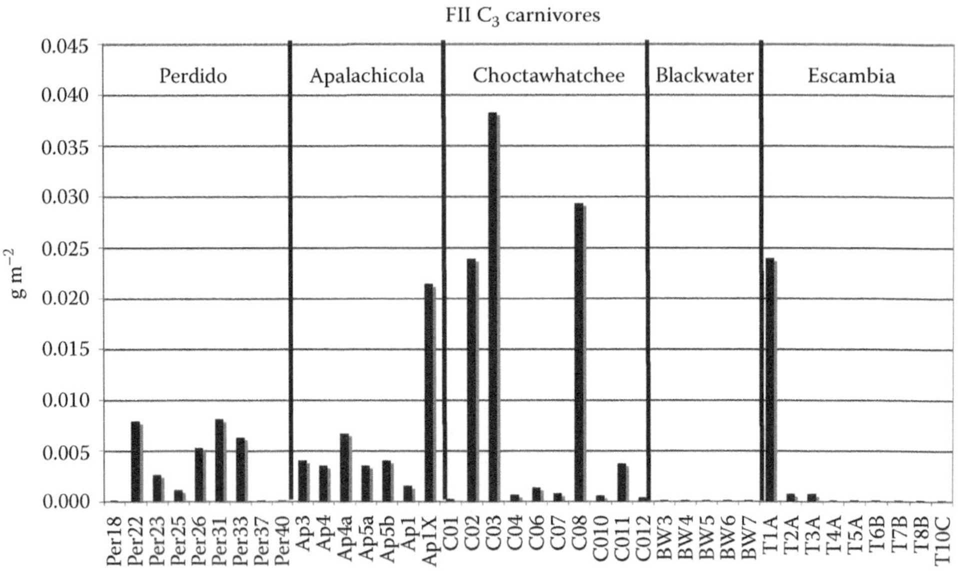

(e)

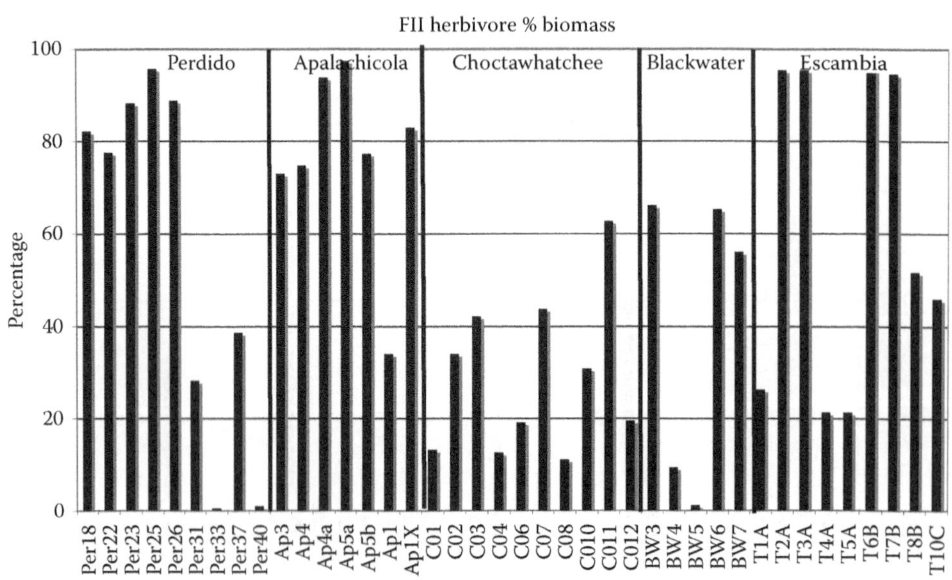

(f)

Figure 9.8 (*Continued*) (e) FII C_3 carnivore biomass (g m^{-2}) by station averaged across dates for the river–estuaries in the NE Gulf of Mexico. (f) FII herbivore percent biomass (g m^{-2}) by station averaged across dates for the river–estuaries in the NE Gulf of Mexico. (*Continued*)

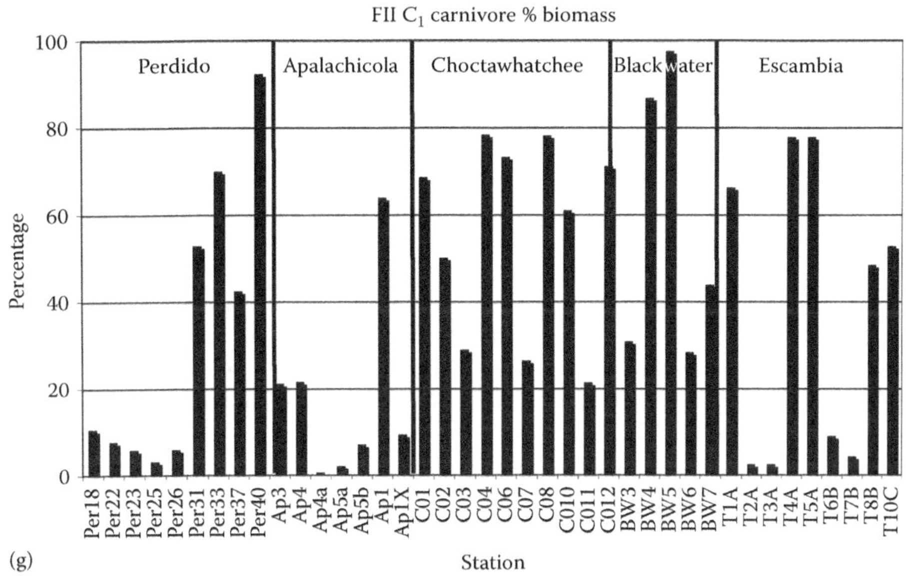

Figure 9.8 (*Continued*) (g) FII percent C_1 carnivore biomass (g m^{-2}) by station averaged across dates for the river–estuaries in the NE Gulf of Mexico. Data were taken from the overall sampling efforts in the respective coastal systems.

omnivory in those sections of East Bay that were affected by river flows (AP3), upper East Bay marshes (AP4a, AP5a), and runoff from the marshes off St. George Island (station A1X). There were low levels of omnivores at some Choctawhatchee Bay stations. The other bays had generally negligible concentrations of omnivores relative to the Apalachicola Estuary. The reasons for such high productivity of this level of the food webs in the Apalachicola Estuary resemble those given for the herbivore groups.

The C_1 carnivore data (Figure 9.8c) indicated again the higher levels of this part of the secondary productivity in the Apalachicola Estuary where four of the stations had higher biomass of C_1 carnivores than any of the other stations in the other estuaries with the exception of one station in Blackwater Bay. The highest levels of such carnivory were located in East Bay that received direct river flows (A3 and A4) and the offshore areas of Apalachicola Bay (AP1, AP1X) receiving runoff from island marshes. The C_1 carnivores were relatively high in several stations of the unpolluted Blackwater Estuary and at two stations in Escambia Bay. The C_1 biomass was particularly low in stratified parts of Perdido Bay and upper Choctawhatchee Bay and in most of Escambia Bay relative to that in the Apalachicola Estuary.

The dominance of the Apalachicola C_2 carnivores compared to the other estuaries was even more pronounced than that of the C_1 carnivores (Figure 9.8d). One station in upper Escambia Bay had C_2 numbers comparable to those in the Apalachicola Estuary. The overall biomass of the C_2 carnivores in Perdido, Choctawhatchee, Escambia, and Blackwater Estuaries was extremely low with no outstanding patterns concerning the general distribution of this level of trophic activity. Once again, the various factors that were associated with the high productivity of the lower levels of the Apalachicola food webs were directly associated with the predominance of the second level of carnivory in that estuary.

The relatively high levels of the C_3 carnivore biomass in upper parts of Choctawhatchee Bay (Figure 9.8e) could not be explained. The lack of sampling efficiency at this level (larger animals were better at avoiding the otter trawls than the smaller ones) could contribute to the relatively uneven distribution of this level of trophic organization in the subject study areas.

When viewed as the percent biomass of herbivores (Figure 9.8f), the highest levels of this index of the trophic organization occurred in the Apalachicola Estuary and in the unstratified (shallow) parts of the Perdido and Escambia Estuaries. The lowest concentrations of herbivores were noted in the Blackwater and Choctawhatchee Estuaries. This again was related to both depth and the absence of plant matter in the form of macrophytes and/or microphytes in these systems and the direct effect of loading of organic particulates in the larger river-dominated systems. When the percent biomass of C_1 carnivores (Figure 9.8g) was analyzed, in most cases, it was the reverse of the previous distribution of herbivores with the highest percentages of this trophic level noted in the stratified parts of the Perdido Estuary and in the Choctawhatchee and Blackwater Estuaries, two of the least productive estuaries in the study areas (Figure 9.7). The highest levels of overall secondary productivity were thus found in those areas with higher levels of herbivores. The Escambia Estuary had some areas of abundant herbivores without an accompanying higher level of secondary productivity. However, even here, the relatively abundant levels of commercially important species occurred in the Escambia Bay part of the Pensacola Estuary even though such concentrations were low compared to other gulf estuaries.

Factors such as water depth, light penetration, and salinity stratification appear as important variables that interact with river flows and the natural loading of nutrients and organic matter to define the trophic organization of a given river-dominated estuary. The trophic elements are dichotomous in that the herbivore/omnivore groups are directly linked to physical and chemical controlling factors that are associated implicitly with the autochthonous and allochthonous primary productivity of the estuary. The carnivores, on the other hand, are closely associated with biological factors—in this case, the lower trophic levels of a given food web. The different forms of carnivory are statistically associated with the herbivore populations rather than any given habitat feature of the system. This suggests that river flow and primary production are mainly associated with changes in the lower trophic levels and that the carnivores are associated primarily with other animal components. Changes in the nutrient and organic carbon loading due to droughts are thus a major influence on the herbivore/omnivore groups and changes at this level then reverberate up the food webs as the primary driving factor in the carnivore series. Habitat distribution, both as a result of natural and anthropogenic factors, is a major factor in the definition of the overall secondary productivity of a given system within the boundaries set by the productivity and species composition of the primary producers.

The data indicate that the Apalachicola Estuary was the most productive of all the areas of study. The higher flow rates of the Apalachicola River probably contributed to this situation although the size factor could not explain the relatively low levels of secondary productivity in the other river-dominated systems. The Blackwater Estuary had limited river input relative to the other systems. The Perdido and Choctawhatchee Bay Estuaries were adversely affected by widespread salinity stratification due to the artificial openings to high-salinity gulf waters. In addition, both systems were exposed to anthropogenic nutrient loading (industrial/sewage treatment facilities and urban/agricultural storm water runoff, respectively) that resulted in damaging plankton blooms. The Escambia Estuary had been subject to a past history of hypereutrophication from industrial and sewage loading and the continuing effects of toxic agents and urban storm water runoff.

In this way, each of the estuaries (other than the relatively natural Apalachicola) had a long history of anthropogenic impacts that could explain the relatively low levels of secondary productivity. There was also the matter of the increasing occurrence of severe droughts in the region that had major impacts on the FII trophic indices of the Perdido and Apalachicola Estuaries. In this way, by the end of the study, all of the study areas evinced severely damaged food webs with major losses of the commercial fisheries that had once been highly productive during previous periods of less anthropogenic activities.

The trophic data were organized by temporal periods that reflected different levels of microphyte blooms and drought events (Figure 9.9). These distributions reflect the changes that took place in the

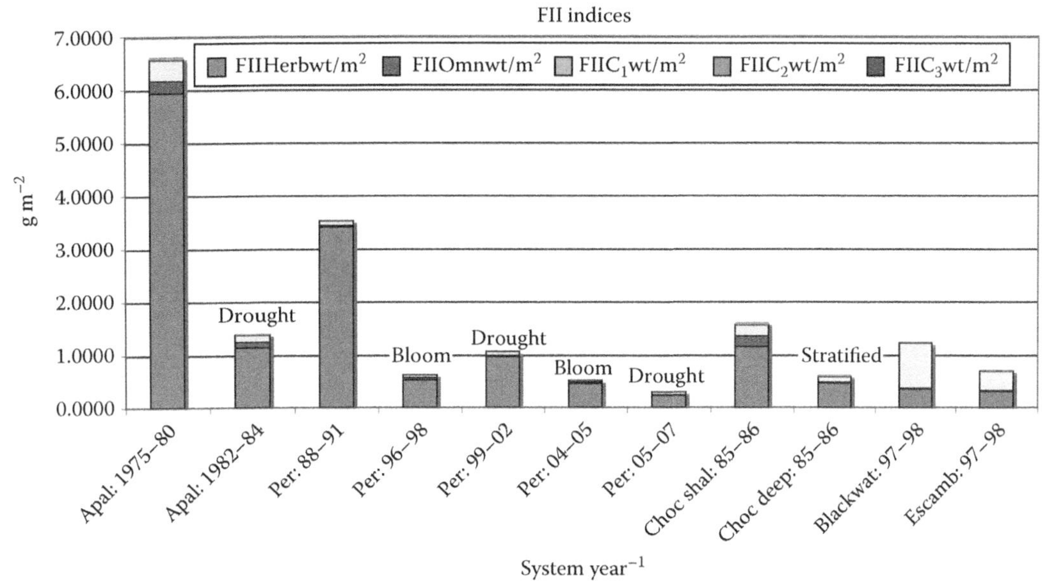

Figure 9.9 **(See color insert.)** FII indices showing differences between periods of regular river flow rates and prolonged droughts, between periods of nonbloom and bloom outbreaks, and between deep (stratified) and shallow (unstratified) bay conditions. The Apalachicola data were averaged over regular (1975–1980) and postdrought (1982–1984) conditions. The Perdido data were organized by drought-free conditions with no plankton blooms (1988–1991), *Heterosigma* blooms (1996–98, 2004–2005), and drought (1999–2002, 2006–2007) conditions. The Choctawhatchee data were organized by shallow (unstratified) and deep (stratified) conditions.

trophic organization of the different estuaries over time. The results of the Escambia, Blackwater, and Choctawhatchee Estuaries were based on relatively short-term field efforts and were thus not conducive to significant findings with regard to the temporal aspects of trophic response to various anthropogenic activities.

The adverse effects of depth (i.e., salinity stratification) were evident in the Choctawhatchee and Perdido Estuaries. The lack of a major drainage basin resulted in reduced nutrient loading and secondary productivity in the Blackwater Estuary. The relatively low productivity of the Escambia Estuary was probably related to habitat changes that occurred during the major blooms of the 1960s and 1970s and the continuous effects of urban runoff. The high organic content and small particle size of the sediments of this estuary could have been associated with the observed low secondary productivity. The resulting losses of seagrass beds and oyster bars and the alteration of sediment quality from the eutrophication events and ongoing urban runoff showed that, despite reduced nutrient loading and the lack of recent plankton blooms, the system had not fully recovered from habitat changes engendered during past periods of anthropogenic impacts. This suggests that the recovery of a damaged estuary does not necessarily proceed along a patterned reversal of previous conditions.

The effects of extended droughts were clearly established for the Apalachicola and Perdido Estuaries (Figure 9.9). The East Bay (Apalachicola) data concerning the loss of secondary production during postdrought conditions after the relatively modest drought of 1980–1981 were substantiated by the loss of most of the fisheries in the Apalachicola Estuary after the prolonged droughts during succeeding periods (1999–2002, 2006–2008, 2010–2012). The Perdido Estuary was adversely affected by the droughts and the plankton blooms along with the effects of salinity stratification associated with the opening of the bay to the high-salinity water of the Gulf of Mexico. The bloom effects were exacerbated due to nutrient loading from a pulp mill, a sewage treatment plant, and agricultural runoff. The result of a continuous series of these impacts led to the reduced resilience of Perdido Bay and cumulative effects that resulted in the lowest annual averages of secondary production of all of the systems studied in the NE Gulf.

Ecological efficiency is the rate at which energy is transferred from one trophic level to the next. Net production at one trophic level is generally only about 10% of the net production at the preceding trophic level (second law of thermodynamics). At any given time, because of the rapid transfer of energy through predator–prey relationships, there can be an anomaly called an inverted pyramid whereby the second law of thermodynamics does not seem to be applicable. This situation is often shown in estuaries with rapid turnover rates and can also occur with decomposers and parasites. Also, there are usually deviations of the 10% rule with efficiencies sometimes increasing up complex food webs due the more sophisticated forms of predation at higher levels of energy transfer.

An analysis was carried out concerning the differences in the ecological efficiency of biomass along the estuarine food webs of the subject estuaries (Table 9.1 and Figure 9.10). In the Apalachicola Estuary, during the early periods of regular changes of river flows prior to drought occurrence, there was a pattern that was within the variation of observed efficiency models with increasing efficiency up the food web. This pattern was somewhat disjunct within the period affected by droughts with somewhat increased C_2/C_1 transfer efficiencies. The Perdido Estuary from 1988 to 1991 had the closest efficiency distributions to the Apalachicola model. During periods of blooms (2004–2005) and droughts (2005–2007), these efficiencies became disjunct with no consistent patterns of energy flow through the food webs.

The Choctawhatchee food web efficiencies were strongly anomalous, particularly in the upper bay where C_1 carnivores were conspicuously higher than the herbivore biomass levels. This pattern was even more skewed in the Blackwater Estuary with a similar C_1/herbivore anomaly as that in the upper parts of the Choctawhatchee Estuary. This pattern was also evident in the analysis of the Escambia Estuary. The short period of analysis in these three estuaries relative to that in the Perdido and Apalachicola Estuaries could have had an effect on these data.

The changes that took place at the end of the Perdido and Apalachicola study periods due to bloom and/or drought effects suggest that these efficiencies went through basic alterations in trophic organization when compared with that of the original Apalachicola model.

The differences among the various coastal estuaries could have been influenced by variations of the physical and geographic distributions of habitat variables in the respective drainage basins. There are significant differences in the development of emergent and submergent vegetation in the subject coastal systems. Perdido Bay and the Pensacola Estuary have moderate concentrations

Table 9.1 Ecological Efficiencies of the Various Coastal Systems along the NE Gulf of Mexico

System	C_1/Herb	C_2/C_1	C_3/C_2
Apal:1975–1980	6.8	11.2	15.9
Apal: 1982–1984	11.9	18.5	11.3
Choc up: 1985–1986	241.3	5.8	40.4
Choc deep: 1985–1986	22.7	33.2	1.5
Choc shal: 1985–1986	18.1	24.9	5.3
Bwater: 1997–1998	225.7	1.2	1.0
Escam: 1997–1998	99.9	4.2	21.3
Per: 1988–1991	7.7	10.6	15.4
Per: 2004–2005	27.9	22.8	61.2
Per: 2005–2007	26.4	8.9	31.6

Note: Data are summed as means over the various study periods.

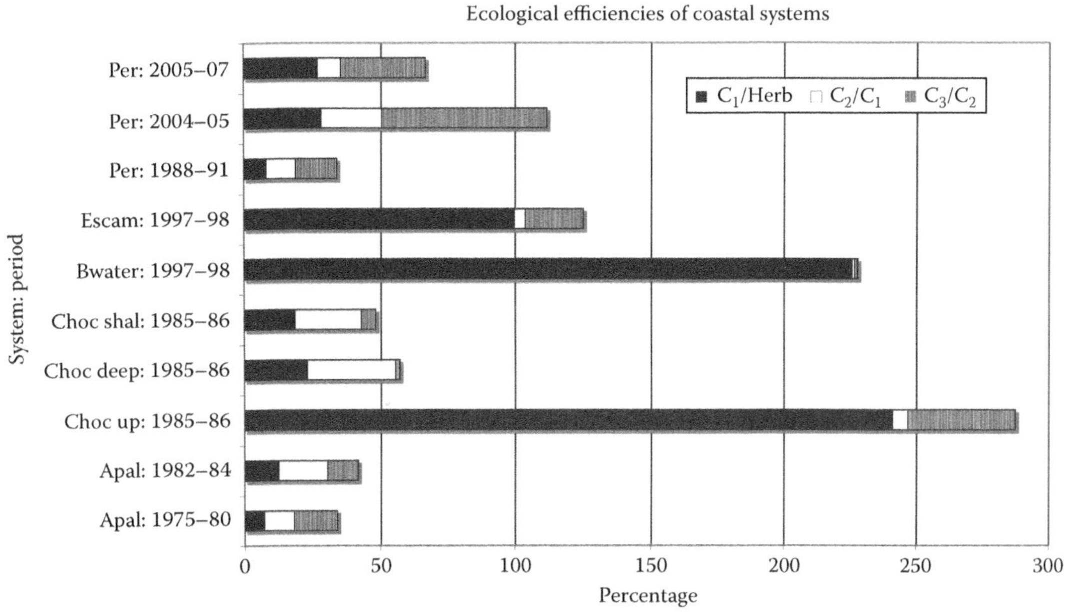

Figure 9.10 Trophic ratios (×100) of C_1 carnivore/herbivore, C_2 carnivore/C_1 carnivore, and C_3 carnivore/C_2 carnivore using data averaged across stations over the periods indicated. The Apalachicola data were averaged over regular (1975–1980) and postdrought (1982–1984) conditions. The Choctawhatchee data were organized by upper bay (up), deep (stratified), and shallow (unstratified) conditions. The Perdido data were organized by drought-free conditions with no plankton blooms (1988–1991), *Heterosigma* blooms (2004–2005), and drought (2006–2007) conditions.

of marshes, whereas the Choctawhatchee Estuary is practically devoid of emergent vegetation. This lack of marshes and swamps along with a relatively small river flow compared to the size of the receiving estuary could contribute to the relatively low rate of secondary productivity in the Choctawhatchee Estuary.

The Apalachicola Estuary has extensive and well-developed marsh systems that are located in a relatively large drainage basin with high rates of river flows and nutrient loading. The Econfina and Fenholloway Estuaries are distinguished from the major river–bay systems to the west by the relatively low flow/watershed and flow/open water ratios and by the relatively high development of marsh areas relative to the flow rates of the contributing rivers to Apalachee Bay. This bay has one of the highest concentrations of marshes and seagrass beds in the northern hemisphere. There were indications through seagrass experiments that the extensive SAV concentrations were stimulated by runoff from the upland marshes.

The relative shallowness of the bays and coastal areas of the inshore gulf regions and the differences in freshwater flows, human developments, and associated anthropogenic nutrient loading were contributing factors to the primary and secondary productivity of these areas. These factors were variously affected by drought conditions that were universal along this coast. There were also basic differences in the land–water interactions and resulting productivity between the large river systems (Perdido, Escambia, Choctawhatchee, Apalachicola) and the blackwater (Blackwater, Econfina, Fenholloway) drainages that were characterized by reduced basin sizes and lower flow rates and nutrient loadings. The trophic organization of the polluted Fenholloway Estuary resembled that of an unproductive Apalachicola system due to nutrient loading and the adverse impacts of pulp mill effluents on the submerged aquatic vegetation (SAV) beds.

The inshore marine habitats of the northwest Florida coast thus represent a diverse, series of coastal zones that once had relatively high natural primary and secondary productivity. Multiple sources (allochthonous, autochthonous) of nutrients together with more or less continuous disturbance by winds and tides provided an almost continuous supply of organic matter for associated food webs. Freshwater inflows, freshwater and saltwater wetlands, in situ phytoplankton productivity, and submerged aquatic vegetation all contributed to the coastal food webs. In the natural state, these areas were physically stressed but provided highly productive nursery areas for eurytopic developing stages of offshore fishes and crustaceans, many of which formed the basis of the sports and commercial fisheries of the region.

In a functional sense, these coastal estuaries encompass a biological continuum from upland, freshwater drainage basins to the offshore gulf that is characterized by certain common physical, chemical, and biological attributes. However, each system represents a unique combination of these habitats. Primary productivity differs widely both within a given system and among the various coastal areas. This unique identity can be characterized by the individual adaptive responses of indigenous species that contribute to the unique characteristics of the various food web associations and trophodynamic processes that characterized the individual estuaries.

Seasonal and interannual cycles of controlling factors such as rainfall and river flows together with intermittent storms and hurricanes provided the highly variable background for the area-specific characteristics of the productivity of these coastal systems. There was a trend of increased drought frequency and intensity in all of the coastal areas in the long-term studies. These estuaries reacted differently to the reductions of rainfall and river flow. The trophic responses to the droughts were strongly negative in the large river–estuaries that originally comprised the most important fisheries yields during periods of regular river flow trends. Apalachee Bay, on the other hand, had long-term changes in the SAV distributions and the fish and invertebrate populations. The trophic organization was changed over time with reductions of fish populations coinciding with increased species richness and biomass of invertebrates.

Overall, the comparative analyses of the trophic organization of the various river-dominated estuaries along the NE Gulf Coast allowed a comprehensive review of how the different estuaries responded to anthropogenic nutrient loading and the increasing frequency and severity of droughts in the region of study. The long-term trophic data showed how a series of once productive, river-dominated estuaries has been damaged by various anthropogenic activities. The FII trophic indices were richest in the predrought Apalachicola Estuary where increased drought conditions caused reductions of secondary productivity and disjunct trophic levels that ended in the loss of the major fisheries of the region. The Perdido Estuary responded to nutrient loading, serial plankton blooms, and droughts that eventually brought the once productive system to the lowest levels of secondary productivity in the study area.

Salinity stratification due to dredging caused reductions of benthic productivity in affected areas of the Apalachicola, Perdido, and Choctawhatchee Estuaries with the latter two ecosystems having debilitating plankton blooms due to nutrient loading that was associated with various human activities. The Escambia Estuary, subject to nutrient loading (and massive plankton blooms) from sewage disposal, urban runoff, and industrial wastes in the 1960s and 1970s, did not recover from the loss of habitat that caused reduced fisheries resources. This bay has been recently affected by urban storm water runoff.

The most important single factor that was associated with the observed changes in the various estuaries along the NE Gulf Coast was the short- and long-term changes of climatic conditions centered on the incidence and severity of droughts. The deterioration of the Perdido and Apalachicola Estuaries and changes in the trophic organization of Apalachee Bay over recent decades were directly and indirectly connected to long-term climatological changes that were typical of the entire region.

Grimm et al. (2013a,b) noted that there has been an intensification of the "flashiness" of hydrological cycles with increased droughts alternating with heavy rainfall incidence leading to altered food webs in streams. A recent review of the issue of climate changes relative to ecosystem processes (Sabo et al., 2010) is applicable to the changes noted along the NE Gulf Coast. Staudinger et al. (2013) pointed out that

> "Without better observational and empirical data regarding which environmental drivers and biological response influence shifts in biodiversity and alter ecological interactions, improvements to model performance is limited. ... however, time series of most observational data and empirical studies are insufficient to attribute and predict changes in species relationships and emergence of novel interactions or community assemblages."

Nelson et al. (2013) pointed out that sustainability of water resources will be exacerbated in many areas of the United States due to projected changes in the climate. Staudt et al. (2013) noted that climate changes will interact with other stressors to cause adverse changes in ecosystem processes. These authors indicated that there is an underlying lack of information concerning long-term changes in coastal systems due to a lack of comprehensive surveys of estuarine and coastal systems that extend over periods long enough to determine the effects of climate alterations in time. The paucity of such data reflects a long history of the lack of attention to climate as an important feature of how estuarine and coastal productivity is dependent on factors such as rainfall and river flow. The long-term avoidance of this issue by aquatic scientists has contributed to our lack of knowledge of how coastal ecosystems are being affected by climate change. The related lack of long-term ecosystem data in coastal areas has added to the uncertainty of the impact of the recent major oil spill in the Gulf of Mexico.

The data presented in this book indicate that climate change is already an important factor in the trophic organization and productivity of estuaries along the NE Gulf Coast. The impacts of such changes have already occurred and should not simply be relegated to some potentially future events. The fact that these changes took place with little awareness or recognition on the part of the scientific community or the state and federal environmental agencies is further evidence of the lack of attention to the impacts of long-term climate trends on the environmental well-being of coastal regions and associated resources.

Apalachicola Model

The Apalachicola River–Estuary during the 1970s was found to be one of the most productive such ecosystems in the world. Early on in the field studies of this estuary, it became evident that river input in terms of volume and timing of peaks was an integral part of the very high productivity in terms of sports and commercial fisheries. The high population numbers per m^2 in the river and the estuary reflected a set of conditions that favored high primary and secondary productivity. The combination of the lack of adverse anthropogenic activities and the natural attributes of the river–bay were somewhat unique when compared to the growing number of damaged rivers and estuaries worldwide. The onset of more frequent and severe drought conditions and extreme climatological conditions led to reductions in such productivity that reached peak levels by 2012.

The natural condition of the Apalachicola River–Estuary could best be defined by a conceptual model of how such a large system functioned. The extensive studies associated with the naturally high productivity of the river and the estuary provided the components for the model of how major river-dominated coastal systems function and why they are so productive in their natural states. The interactions between the river floodplain and the bay were well defined. The linkage between the upland freshwater wetlands and the rich estuarine biota of the Apalachicola Estuary was recognized and determined by a series of studies (Livingston, 1975b,c, 1980b, 1981a, 1983a,b,c,d, 1984a, 1985c, 1991a,b,c, 2013; Livingston and Loucks, 1978; Livingston et al., 1974, 1977; Livingston and Joyce, 1977). Studies were made concerning the distribution of wetland vegetation in the Apalachicola floodplain (Leitman et al., 1982). Litter fall in the Apalachicola floodplain (800 g m^{-2}) was found to be higher than that noted in many tropical systems and almost all warm temperate systems (Elder and Cairns, 1982). The annual deposition of litter fall in the bottomland hardwood forests of the Apalachicola River floodplain approximated 360,000 metric tons (mt) during the early studies. The largely natural interactions between the river and the freshwater wetlands were maintained with minimal impacts due to physical alterations such as dredging, damming, and river bank depositions.

Seasonal river flooding provided the mechanism for mobilization, decomposition, and transfer of the nutrients and detritus from the forested wetlands to adjacent aquatic areas (Cairns, 1981; Elder and Cairns, 1982). Of the 214,000 mt of carbon, 21,400 mt of nitrogen, and 1,650 mt of phosphorous that were delivered to the estuary during the early study years, over half was transferred during the winter–spring flood peaks (Mattraw and Elder, 1984). Flooding of the river was the key to river–wetland interactions. There were distinct links between the estuarine food webs and freshwater discharges (Livingston, 1981a, 2004a; Livingston and Loucks, 1978). The total particulate organic carbon delivered to the estuary followed seasonal and interannual fluctuations that were closely associated with river flow (Livingston, 1991a; $R^2 = 0.738$). Nutrient loading followed these trends.

Up to 50% of the phytoplankton productivity, which is the most important single source in overall magnitude of organic carbon in the estuary, was explained by Apalachicola River flow (Myers, 1977; Myers and Iverson, 1977). During winter–spring periods of high river flow, there were major transfers of nutrients and organic matter to the estuary. Boynton et al. (1982) reported that the Apalachicola Estuary had high phytoplankton productivity relative to other river-dominated estuaries. Wind action in the shallow Apalachicola Estuary was associated with periodic peaks of phytoplankton production as inorganic nutrients were regenerated in the sediments. Recent studies documented river influence on nutrient and organic carbon loading to the bay. Chanton and Lewis (2002), using $\delta^{13}C$ and $\delta^{34}S$ isotope data, noted clear distinctions between benthic and water column feeding types. They found that the estuary depended on river flows to provide floodplain detritus during high-flow periods and dissolved nutrients for estuarine primary productivity during low-flow periods.

Floodplain detritus was significant in the highly productive East Bay nursery. Winter–spring peaks of macrodetrital accumulation in the bay (Livingston, 1981a) were coincident with increased infaunal abundance (McLane, 1980). Four out of the five dominant infaunal species at river-dominated stations were detritus feeders. Studies of the loading of particulate organic matter (Figures 9.11 and 9.12) indicated that microdetritus and leaf matter loading followed the winter–spring river flow peaks as water floods into the upriver wetlands and drives the particulates from the wetlands down-river. Both sources of organic matter almost completely disappeared during the minor drought of 1980–1981. The lack of flooding during winter–spring months of available wetland organic matter was associated with this lack of loading to the estuary.

During peak flows of river flooding, leaf matter was deposited in areas of East Bay (A04) that were associated with river inflows where it then became a platform for microbial associations that formed the basis for the detrital food webs of the bay. A mechanism for the direct connection of increased infaunal abundance was described by Livingston (1984a) whereby microbial activity at the surface of the detritus (Federle et al., 1983a) led to successions (Morrison et al., 1977) that then provided food for a variety of detritivorous organisms (White et al., 1979a,b; Livingston, 1984a). The transformation of nutrient-rich particulate organic matter from periodic river-based influxes of dissolved and particulate organic matter coincided with abundance peaks of the detritus-based food webs of the Apalachicola system (Livingston and Loucks, 1979) during periods of increased river flooding. These connections between upriver flooding and estuarine productivity helped to explain the prolonged reductions of benthic productivity during drought periods.

Areas outside of the direct river inflows (station A05a in upper part of East Bay) had relatively low concentrations of terrigenous matter. The amounts of microparticulates and leaf matter were much reduced during the drought of 1980–1981 (Figures 9.11 and 9.12), a situation that corresponded to the reduced winter–spring flow rates. Bay studies (Livingston, 1976a, 1981a, 1983d, 1984a, 2000, 2002) corroborated the timing of these flow events with the delivery of nutrients and dissolved

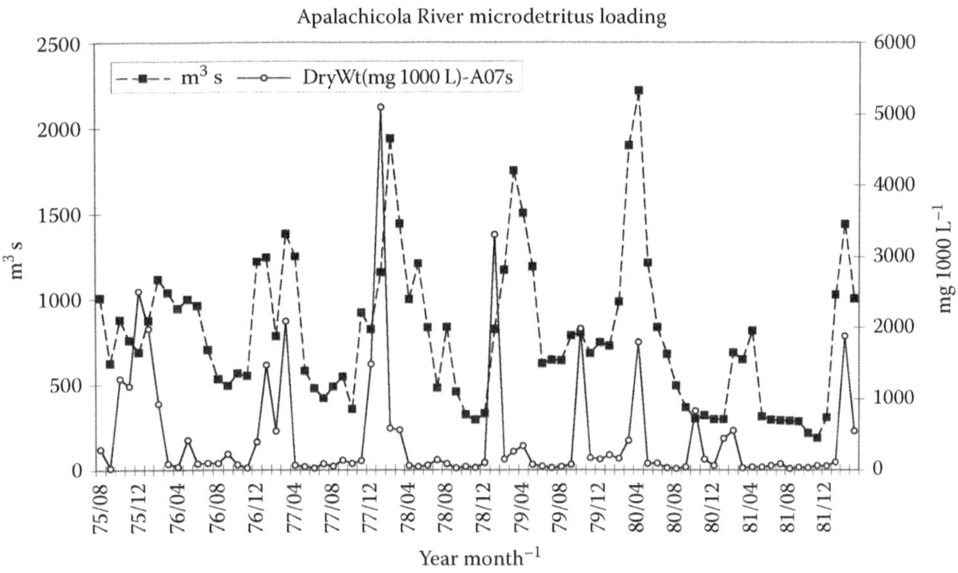

Figure 9.11 Microdetritus concentrations at station A07 in the Apalachicola River at the head of the bay from 1975 to 1982.

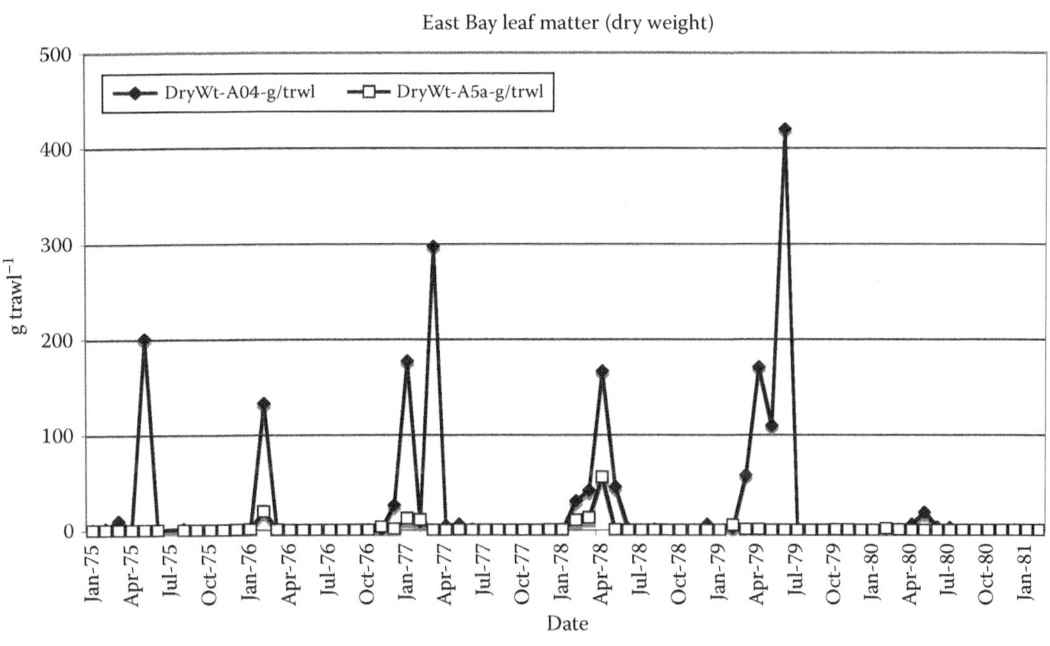

Figure 9.12 Leaf matter taken in otter trawls at station A04 and A05a in East Bay from 1975 to 1982.

and particulate organic matter as an important factor in the maintenance of the estuarine primary production (autochthonous and allochthonous). There were direct links between the estuarine food webs and freshwater discharges (Livingston, 1981a, 1983d, 1984a).

Particulate organic carbon delivered to the estuary thus followed seasonal and interannual fluctuations that were closely associated with river flow. During summer and fall months, there was no direct correlation of river flow and detritus movement into the bay. By winter, there was a significant relationship between microdetrital loading and river flow peaks (Livingston, 1991a; $R^2 = 0.738$). These data indicated several important points regarding the interaction between the Apalachicola River and its estuary. The movement of nutrients and organic matter was seasonal and occurred during the winter–spring floods at which time there was particulate organic matter in the adjoining wetlands. The occurrence of a winter–spring drought was accompanied by a reduction of river flooding and a consequent lessening of the loading of nutrients and particulate organic matter to the river and the bay.

These events were of particular importance to the phytoplankton productivity in the Apalachicola Estuary and to the detritivore food webs of the upper bay. The connections of both the estuarine oyster productivity and the FII food web development that were noted by various studies were linked both to the seasonal timing of the flood events and to the interannual progressions of droughts. These complex temporal interactions that were central to the high secondary productivity of the bay were crucial to an understanding of the overall trophic organization of the Apalachicola system. These data indicated that the timing of the peak river flow rates during winter–spring months of maximal available floodplain detritus was an important element in the transfer of organic matter from riverine wetlands to the river and the estuary and to the associated aquatic food webs.

One little understood problem that could impede the riverine interactions with the floodplain involves the decades-long efforts by the US Army Corps of Engineers to physically alter the river to provide easier passage for the barge traffic along the river. Despite the fact that this effort had one of the least favorable cost/benefit ratios in the country, the corps has proceeded to pursue an active program of destruction that has been ignored by just about everyone. This effort, which is ongoing, includes the deepening of the river, elimination of meanders, moving dredged sediments to adjoining wetlands, and blocking attachments of feeder streams to the main stem.

During the 1970s, the corps proposed a dam on the unimpeded Apalachicola River claiming that the lost of phosphorus behind the dam would not harm the river or its associated bay. This excuse was based on the unsubstantiated assumption that nitrogen was the limiting nutrient for the bay. However, data generated from the bay studies during this period indicated that both nitrogen and phosphorus were limiting bay productivity and that a dam on the river would reduce the loading of phosphorus to the bay thus reducing estuarine productivity. Such a dam would also impede the flow of particulate organic matter to the estuary. Threats of legal action and the overwhelming opposition to the dam by people along the river put a stop to the dam proposal. However, ongoing physical modifications to the river may be having adverse effects on the movement of nutrients and organic matter downstream. The current activities may have the effect of reducing river flooding and destroying the ecological well-being of the floodplain proper. The actual impact of these anthropogenic activities along the river has been ignored despite repeated efforts on the part of scientists to have a scientific evaluation of the situation.

Oczkowski et al. (2011) found that oyster stable isotope values throughout Apalachicola Bay were dominated by freshwater inputs that reflected the variability and hydrodynamics of the riverine inflows. Mortazavi et al. (2000a) found that phytoplankton productivity in river-dominated parts of the Apalachicola Estuary was limited by phosphorus in the winter (during periods of low salinity) and nitrogen during summer periods of high salinity. Mortazavi et al. (2000b) determined temporal couplings of nutrient loading with primary production in the estuary. Around 75% of such productivity occurred from May through November, with main control due to grazing. The data indicated that altered river flow, especially during low-flow periods, could adversely affect overall estuarine productivity.

Wilson et al. (2010) found that biological production in Apalachicola Bay was highly dependent on riverine flows in two ways: (1) Economically important bivalves and crustaceans were supported by terrestrial organic matter supplied by river flooding and (2) consumer fish species were dependent on *in situ* production that was reliant on nutrients supplied by the Apalachicola River.

This study confirmed that Apalachicola Bay consumers depended on river flow and its nutrients and river plain-derived detritus.

The key physical components (shallow, enclosed bay) with sediment quality and high primary productivity combined with the periodic peaks and low flows of the Apalachicola River to form a very high level of secondary productivity with a highly favorable environment for the bivalves and nurserying invertebrates and fishes that inhabited the bay (Figure 9.13a). Although various studies mentioned potential problems with continued freshwater flows due to upriver anthropogenic

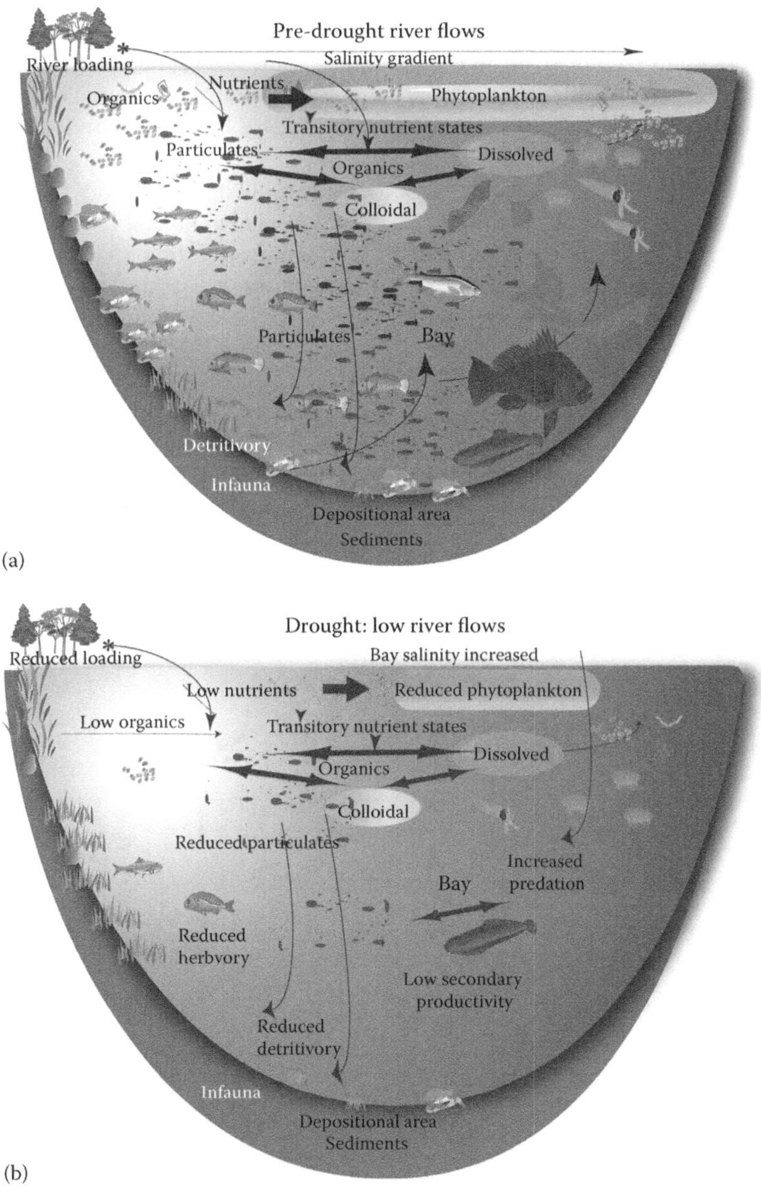

Figure 9.13 **(See color insert.)** (a) Model of the connections of the food webs of the Apalachicola Estuary with river processes during natural fluctuations of flow rates. (b) Model of the connections of the food webs of the Apalachicola Estuary with river processes under the effects of prolonged droughts on flow rates.

activities, the problems of extended droughts have become dominant in the restriction of river flows. The connection to the bay is trophic, and the cessation of upriver flooding has had dramatic adverse effects on the river floodplain and the Apalachicola Estuary (Figure 9.13b). These effects included increased salinities and predation from gulf predators and the reduction of the loading of dissolved nutrients and various forms of particulate and dissolved organic carbon that translated into the loss of phytoplankton productivity and detrital food webs.

Infaunal, detritus-feeding macroinvertebrate assemblages that live in the sediments of the bay are dominated by various species of worms and crustaceans that include the polychaetes *Mediomastus ambiseta*, *Hobsonia florida*, and *Streblospio benedicti* and the crustacean *Grandidierella bonnieroides* (Livingston, 1984a). The infauna form the food base for sciaenid fishes (Atlantic croaker [*Micropogonias undulatus*], spot [*Leiostomus xanthurus*], red drum [*Sciaenops ocellatus*], and sea trout [*Cynoscion* spp.]) that dominate the estuarine fish populations. Shallow depths and extremely high bottom productivity explain why the Apalachicola Estuary is a primary nursery along the Gulf Coast for blue crabs (*Callinectes sapidus*) and white shrimp (*Litopenaeus setiferus*). These species feed largely on the bottom and are dependent on the detrital food webs during their development in the bay. They also form the basis of highly lucrative sports and commercial fisheries in the region.

The gateway to the estuarine food webs is through the propagation and production of benthic herbivores and omnivores. The FII benthic herbivores, feeding on detritus and benthic microphytes had relatively regular seasonal and interannual numbers in East Bay during the periods of relatively regular river flows (Figure 9.14). After a slight increase during the early part of the drought, there was a general collapse of both groups of herbivores during the latter part of the drought that extended for 2 years after the cessation of low-flow conditions. This pattern was particularly pronounced for the detritivores and omnivores (Figure 9.15); these key levels of the food webs underwent prolonged major decreases during and after the 1980–1981 drought. It should be noted that this drought was relatively limited when compared to the succeeding periods of low flows from

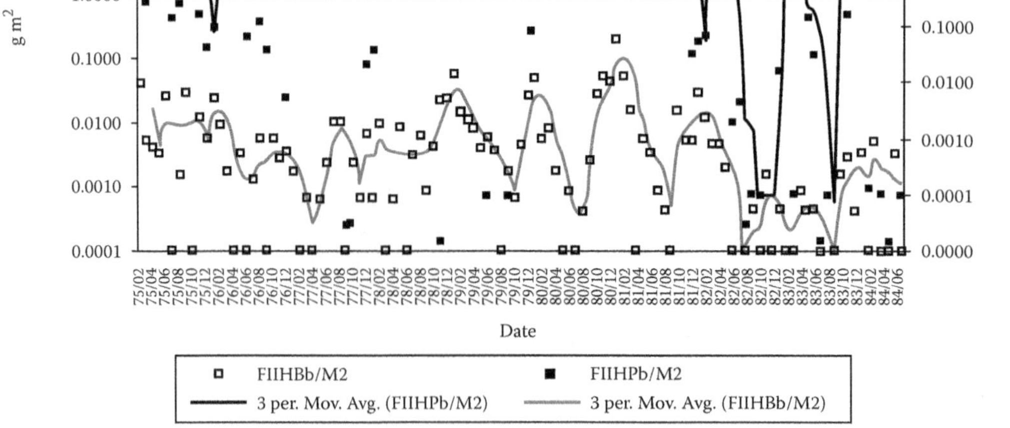

Figure 9.14 FII herbivores feeding on benthic plant matter (HBb) and benthic microphytes (HPb) monthly from 1975 to 1984. Three-month moving averages are also provided.

1986 to 1988, 1999 to 2002, 2006 to 2008, and 2010 to 2012. The extended effects of the short-term drought during succeeding years of increased river flows could have been due to the time needed for a recovery of the various stages of phytoplankton recovery and the reestablishment of detrital food webs. These effects included the delayed recovery of herbivores feeding on benthic plants and plankton and on the detritivores feeding on particulate organic matter (Figures 9.14 and 9.15).

These data indicated that for prolonged periods after a drought in the Apalachicola Estuary, there were long-term reductions of parts of the food webs that depend on particulate organic matter and benthic microflora. Animals that feed on the benthic herbivores and omnivores include penaeid shrimp, blue crabs, and the dominant sciaenid fishes in the upper bay. The increasingly severe losses of the commercial fisheries of the Apalachicola Estuary during subsequent droughts in the Chattahoochee–Flint–Apalachicola River Basin provided field evidence of the predicted connection between the droughts and the loss of key species in the estuarine system. The upper levels of the estuarine food webs (C_1, C_2, C_3 carnivores) were reduced due to the loss of their food supply.

The results of studies of the major estuaries along the NE Gulf Coast indicate that the primary feature of these estuaries is high primary productivity in the immediate receiving areas of freshwater runoff. The upper parts of these estuaries, which receive nutrient loading directly from the rivers, are often phosphorus limited in the NE Gulf during at least part of the year. Secondary production is highest where phytoplankton production and deposition of particulate organic matter are the greatest. Much of the secondary production in the river-dominated estuaries is associated with

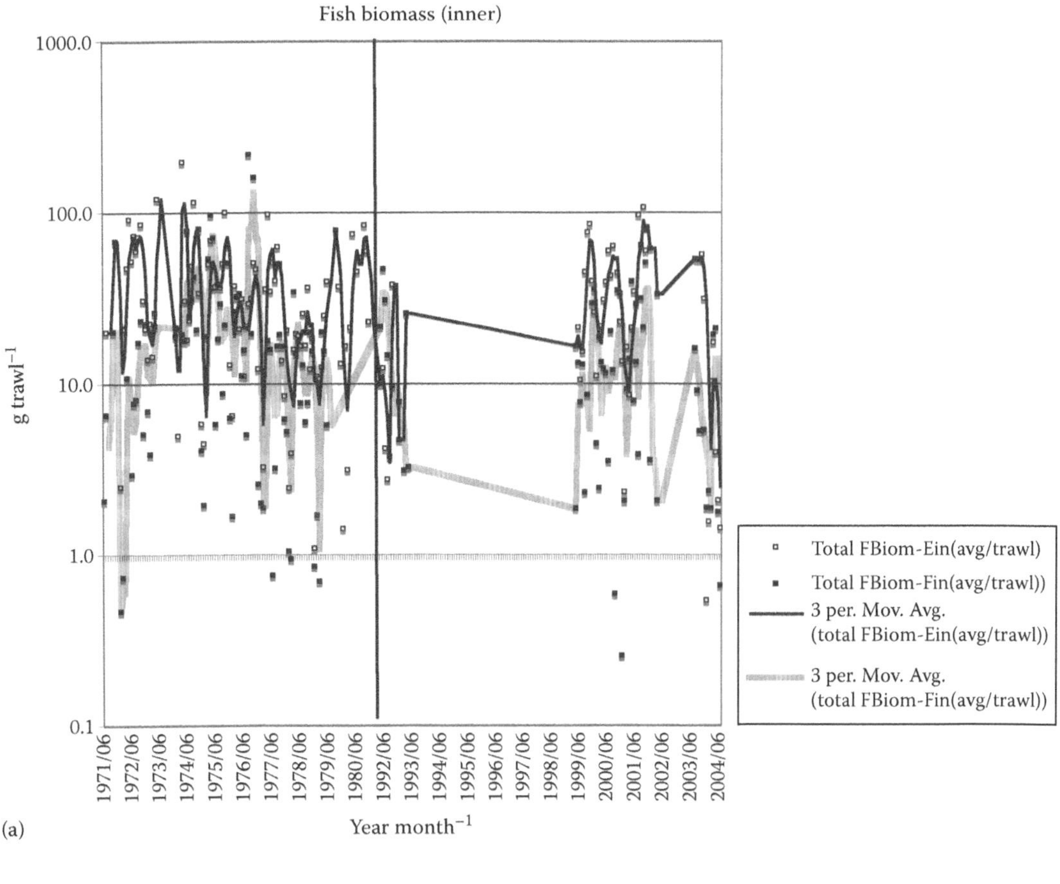

(a)

Figure 9.15 FII omnivores feeding on (a) detritus (ODb) and (b) benthic plant matter (OHb) monthly from 1975 to 1984. Three-month moving averages are also provided. *(Continued)*

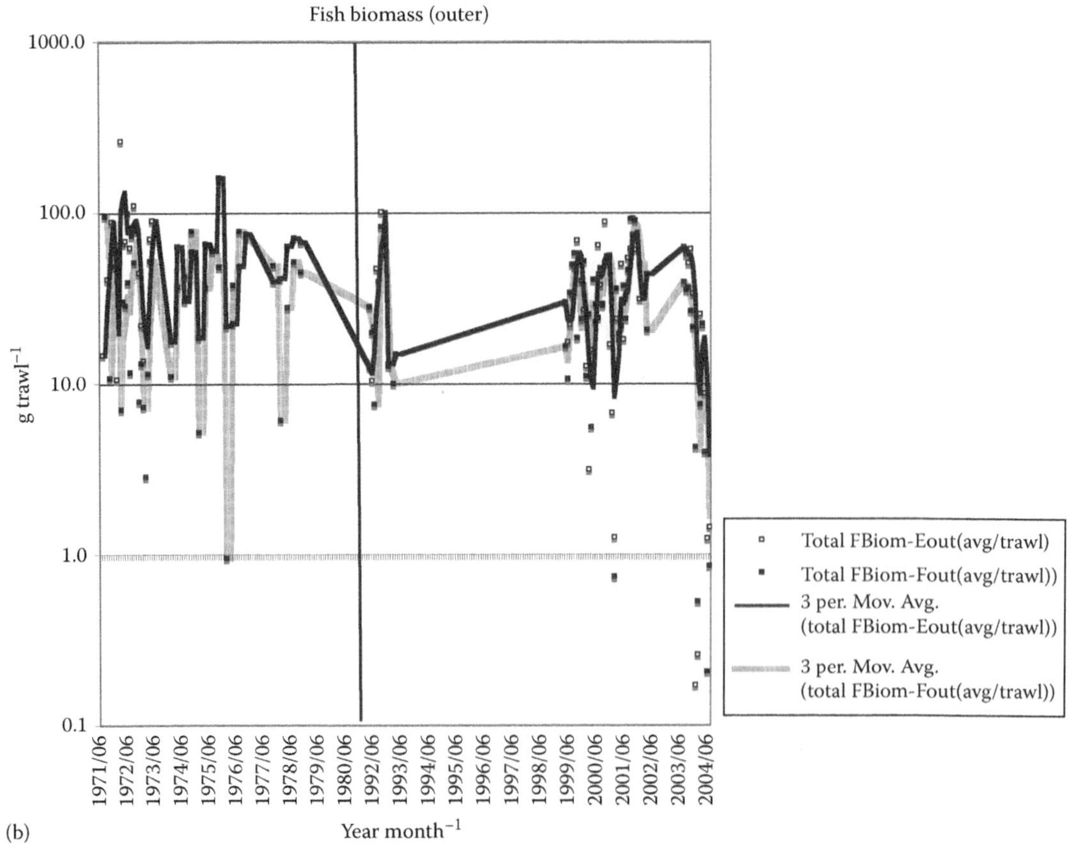

Figure 9.15 (*Continued*) FII omnivores feeding on (a) detritus (ODb) and (b) benthic plant matter (OHb) monthly from 1975 to 1984. Three-month moving averages are also provided.

benthic food webs. Sediment quality is thus an important factor of coastal secondary production in the NE Gulf as is the depth and current structure of individual receiving areas.

The relatively high productivity of the Apalachicola Estuary was thus due to several factors. The estuary was shallow, had low stratification features, and was in a generally natural state with no signs of cultural eutrophication due to anthropogenic nutrient loading. The naturally high levels of nutrient and organic carbon loading from the Apalachicola River enhanced the biomass of herbivores and detritivores at the base of the estuarine food webs. The carnivore groups were driven by this increase of the lower levels of the food web, a process that resulted in optimal ecological efficiency and high levels of secondary productivity that fueled regional fisheries. During the early years of sampling, these mechanisms were also operational in the Perdido Estuary with the result of a relatively high FII biomass and a balanced level of ecological efficiency in the shallow upper bay regions that received high nutrient and organic carbon loading from the Perdido River. The salinity stratification factor reduced the FII biomass in deeper parts of the estuary, but the overall FII productivity base in the upper bay still exceeded that of the other river-dominated estuaries that had a less developed food web base and corresponding disjunct ecological efficiencies due to combinations of adverse anthropogenic activities.

The integration of data from the different estuaries can be summarized in a basic model of climate impacts on major river-estuarine systems. Prolonged droughts cut off the loading of inorganic nutrients and organic matter in both the Apalachicola and Perdido Estuaries. Watercolor and turbidity in the estuaries decreased and the water became clearer. The remaining nutrients stimulated major increases in the phytoplankton and benthic microphytes. There was a resulting increase in herbivore and detritivore biomass. However, as the nutrients and organic matter were consumed without replacement, there were major reductions in the biomass of the herbivores and detritivores that were followed by major decreases of the carnivore trophic levels, a situation that led to reduced population numbers of dominant species. Disjunct ecological efficiencies followed the disproportionate loss of the lower parts of the food webs that led to altered trophic ratios.

A sudden increase in rainfall and river flow after the droughts in estuaries such as the Perdido, with enhanced anthropogenic nutrient loading, resulted in plankton blooms that further decreased the secondary productivity. These repeated impacts over time led to reduced resilience of the affected estuary. As was the case in Escambia Bay, this could lead to more or less permanent (decades or more) loss of habitats such as oyster bars, seagrass beds, and sediment quality. In this way, the interactions of natural habitat conditions, the working relationships associated with the different trophic levels of the food webs, and the various anthropogenic activities (dredging, nutrient loading, climate changes) that limited the natural attributes of the trophic organization of the various estuaries thus combined to account for the temporal patterns of the respective FII indices and the loss of useful productivity in the estuaries along the NE Gulf Coast.

The increasing severity of droughts in the study region has been associated with reduced primary and secondary productivity in the river-dominated coastal areas. These droughts have developed over the past 30+ years and have been associated with the loss of the major fisheries in the Apalachicola Estuary. The management of a river-dominated estuary should be based on protection and control of freshwater sources, nutrients, and organic matter with a minimization of physical alterations that often lead to increased salinity stratification and the associated loss of the nursery function of the estuary. These processes underlie the very basis of riverine and estuarine productivity.

It is clear that fisheries production, the nursery function, and the productivity of individual populations depend on species-specific responses to combinations of freshwater input, salinity changes, food web processes, and interspecific predation and competition. Because of the lack of science-based management of river–estuaries throughout the United States, these areas are quietly succumbing to the detrimental effects of a range of human activities (Livingston, 1984a, 2000, 2013). These effects can be avoided, but only if adequate research is directed to a scientific determination of the freshwater needs and nutrient-loading tolerances of coastal ecosystems.

The extensive studies of the Apalachicola Estuary that have been carried out since 1972 were used to develop an ecosystem-based management program that included the purchase of wetlands associated with both the river and the bay (Figure 9.16). The scientific data provided the objective basis for the development of the Apalachicola management program that has been documented by Livingston (1976b, 1977, 1980b, 1982b, 1983c, 1984a, 2000, 2002, 2013). During the early years of the study, there was excellent cooperation among local, state, and federal planning agencies and the political entities that cooperated with the major management efforts to protect the Apalachicola system.

Documentation of facts concerning the Apalachicola Basin (Livingston and Joyce, 1977) provided the initial details of the unique ecological status of the Apalachicola Estuary. The linkage between the upland freshwater wetlands and the Apalachicola Estuary via nutrient-loading analyses together with related bay studies led to purchases of river–wetlands by the Florida Department of Natural Resources as part of the Environmentally Endangered Land program (Chapter 259, Florida statutes). The initial purchase was 12,140 ha (30,000 acre) of hardwood wetlands in the lower Apalachicola in December 1976 (Pearce, 1977). This was to be the first of many wetland purchases in the Apalachicola region that have continued up to the recent time. Upland and coastal wetlands along the Apalachicola River and the East Bay area (Tate's Hell Swamp) have been recently purchased by state agencies.

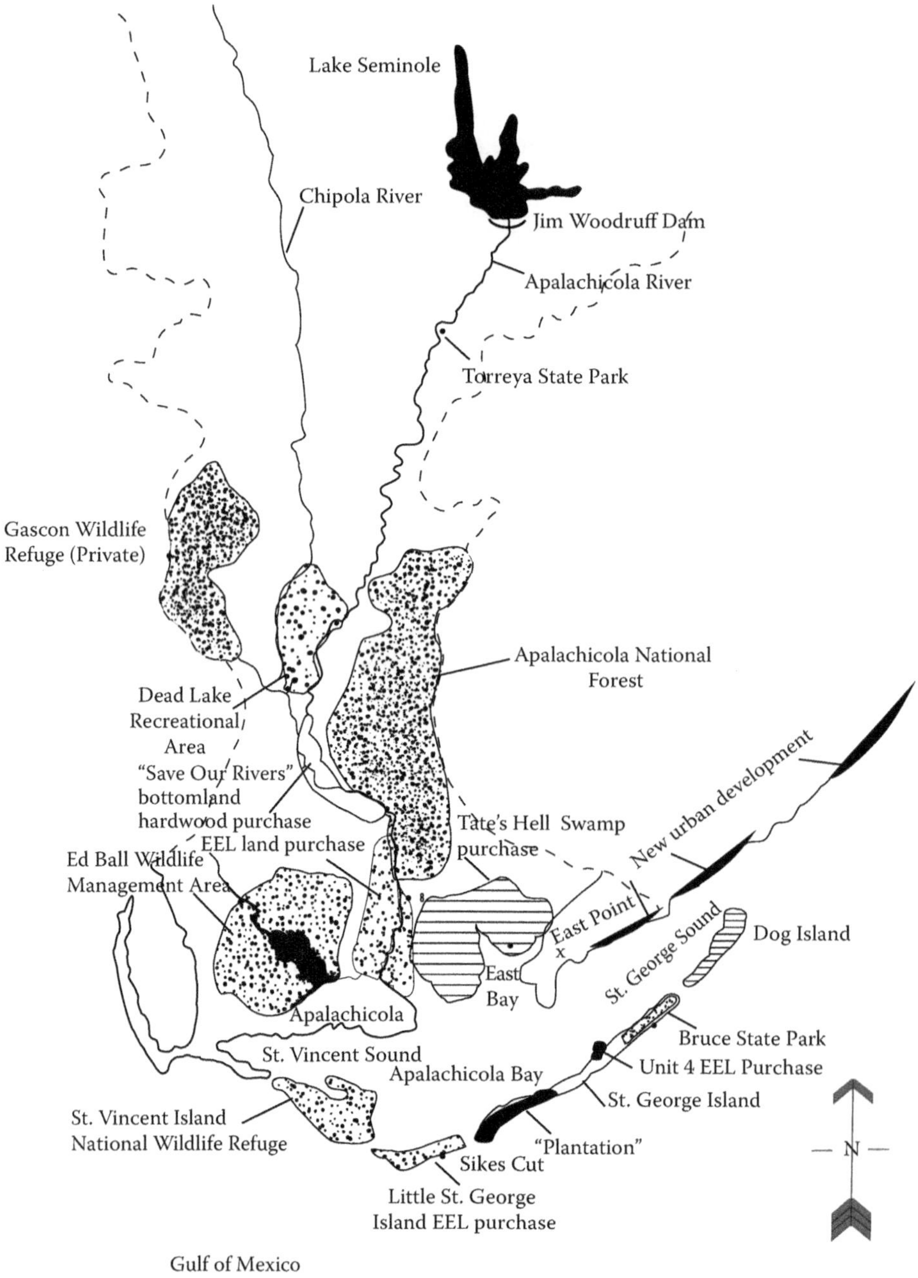

Figure 9.16 Local, state, and federal land purchases and management areas in the Apalachicola drainage basin.

The primary objectives of much of the early planning were related to maintaining natural flows to receiving areas and they included the following:

1. Purchases of environmentally critical lands in the Apalachicola drainage system that now include most of the river–bay wetland systems (Figure 9.8)
2. Designation of the Apalachicola system as an area of critical state concern (Florida Environmental Land and Water Act of 1972; Chapter 380, Florida statutes)

3. Creation of cooperative research efforts to determine the potential impact of activities such as ongoing forestry management programs, urban development, and pesticide treatment programs
4. Making provisions for aid to local governments in the development of comprehensive land use plans, a function that is vested primarily at the county commission level in Florida

Since the mid-1970s, the State of Florida has spent over $239 million through land acquisition to protect the wetland system of the Apalachicola River–Bay system (Livingston, 2013). Currently, major parts of the Apalachicola River–wetland system are held by public agencies for preservation and management (Figure 9.16). This is a unique situation in terms of protective measures of large river-estuarine systems in the United States. The State of Florida has devoted considerable time and expense to assure that the wetlands along the Apalachicola River–Bay system remain viable relative to the importance of such wetlands to the productivity of this drainage basin.

Thanks to the efforts of the same coalition of local and state personnel that instituted land planning in Franklin County in the early 1970s, the Apalachicola scientific database was used to establish the Apalachicola River–Bay Estuarine Sanctuary in 1979. This sanctuary, now designated a National Estuarine Research Reserve, included the purchase of about 78,000 ha (192,742 acre) in the bay system (Figure 9.16). The reserve remains one of the largest such systems in the country. The State of Florida purchased Little St. George Island in 1977 (Figure 9.16). Somewhat later, an area above the East Hole oyster beds (Unit 4 EEL) was purchased through the efforts of the Trust for Public Lands (Caroline Reusch, pers. comm.). St. Vincent Island was already a federal preserve administered by the US Department of the Interior. The east end of St. George Island is a state park. After considerable legal proceedings, most of the western section of St. George Island (the Plantation, Figure 9.16) was planned for maximum protection of island freshwater drainages, associated wetlands, and upland vegetation.

In a relatively short period, land purchases on the barrier islands were added to the purchases of the East Bay and Apalachicola River wetlands to complete a ring of publicly owned lands around the most environmentally sensitive areas of the Apalachicola River–Estuary. When combined with state regulations (aquatic preserve, area of critical state concern) and a Franklin County management program designed to protect the bay from urban runoff, most of the land–water interfaces were thus protected from the effects of human activities.

These progressive actions to protect the productivity of the Apalachicola ecosystem were carried out mainly during the 1970s and 1980s. During the late 1980s, the Apalachicola research funding from federal, state, and local sources was discontinued at the request of local and regional political operatives associated with land developers and associated financial concerns. The news media eliminated their former use of scientific information in their environmental coverage; skilled environmental reporters were no longer required, and reportage of ecological concerns was replaced by superficial coverage of ecological events by interviews with uninformed environmental interests and bureaucrats who would lose their jobs if they told the truth about what was happening to the natural resources of the region. Environmental issues were either dumbed down or were simply ignored by the news media, the prevailing political and economic factions, and the public at large, a situation that prevails to the present time.

PART VI

Information Dissemination

CHAPTER 10

Omission and Misrepresentation by Regional News Media

During the years following the end of World War II, the United States became a leader in the advancement of both scientific and economic matters. American government at all levels was eventually viewed as the enabler of material progress. However, despite these advances, there was a general lack of scientific information regarding what later became known as ecology, the study of the natural environment. Toward the latter decades of the twentieth century, there was an awakening in the United States to the value of natural ecological systems. The publication of Rachel Carson's *Silent Spring* in 1962 has been credited with the beginnings of the environmental movement in this country. Events such as the 1969 Santa Barbara oil spill added fuel to the fire that culminated in the passing of the National Environmental Policy Act (NEPA) of 1969 (NEPA was actually passed in January 1970). The primary purpose of the act was to ensure that environmental factors were considered in the decision-making processes of federal agencies.

The US Environmental Protection Agency (USEPA) was created on December 3, 1970, as the agency of the federal government charged with protecting human health and the environment. Personnel of the USEPA promulgated and enforced regulations based on laws passed by Congress. During the 1970s, a series of laws was instituted concerning various environmentally important issues with the EPA as the lead agency for compliance with and enforcement of such laws. The largely inert Federal Water Pollution Control Act of 1948 was reorganized and significantly expanded in the 1972 Clean Water Act making it unlawful to discharge pollutants from point sources into navigable waters unless a permit was obtained through the EPA's National Pollutant Discharge Elimination System (NPDES). The Toxic Substances Control Act (TSCA) was passed in 1976 to provide the EPA with the power to require reporting, record keeping, and testing requirements for toxic waste discharges.

The various states organized agencies with laws and issues of compliance and regulation similar to those of the federal agencies. In essence, there was a revolution of the manner in which environmental matters were addressed with a comprehensive legal structure as the backdrop for enforcement actions in this country. The high tide of environmental concern was reached during the 1970s and early 1980s. During that period, there was a growing concern regarding the impacts of various human activities on the natural environment. The science of aquatic ecology was born, and this field grew substantially during the early years of increased interest in environmental matters. In time, however, these concerns were tempered by an increasing tilt toward the economic ramifications of the environmental movement.

During the presidential reign of Ronald Reagan (1981–1989), there was a gradual erosion of public interest in ecological matters that was led by state and national political interests. There was a certain level of environmental disdain that leaked from the political leadership into the body politic:

> The American Petroleum Institute filed suit against the EPA [and] charged that the agency was suppressing a scientific study for fear it might be misinterpreted… The suppressed study reveals that 80 percent of air pollution comes not from chimneys and auto exhaust pipes, but from plants and trees.
>
> **Presidential candidate Ronald Reagan (1979)**

> I have flown twice over Mt. St. Helens out on our west coast. I'm not a scientist and I don't know the figures, but I have a suspicion that that one little mountain has probably released more sulfur dioxide into the atmosphere of the world than has been released in the last ten years of automobile driving or things of that kind that people are so concerned about.
>
> **Ronald Reagan (1980)**

> Trees cause more pollution than automobiles do.
>
> **Ronald Reagan (1981)**

> Facts are stupid things.
>
> **Ronald Reagan (1988)**

By the middle to late 1990s, public concerns for environmental issues were in retreat, although the various state and environmental federal laws were still operable. Environmental law grew more sophisticated during this period, and, together with important advances in ecological science, there were still advances in actions leading to improvements in environmental conditions.

During the first decade of the new millennium, there was a revival of more conservative political actions evident in state and federal environmental actions. Environmental laws came under increasing political pressure. Issues such as climate change were either not addressed or openly negated as the general public became less and less concerned with ecological matters. This apathy was matched by political inaction and withdrawal from ecological concerns at all levels of government. There were significant reductions of funding for environmental research, and laws were changed to make it more difficult to bring legal actions against major polluters. States such as Florida simply eliminated the legal regulatory framework with the burden of proof of impact shifted from the polluters to the state environmental agencies that had also been reduced or even eliminated by the Florida Legislature.

The news media were part of this shift against progressive environmental actions. During the early years of the environmental awakening, there had been numbers of well-informed writers who covered ecological stories in an objective and informed matter. This was soon to change. The American news media eventually followed (or led) the trend of ignoring scientific facts as they related to important environmental issues such as climate change. There was a general movement among those who influenced public opinion to add enthusiastically to the growing body of misinformation concerning environmental matters. Polls indicated that a majority of citizens did not take "global warming" seriously, and if it was occurring, most did not believe that human activities were to blame. Although these figures tended to improve somewhat with time, there was very limited political action by state and federal lawmakers with regard to what came to be known as climate change.

In north Florida, the two primary sources of regional news (the *Pensacola News Journal* and the *Tallahassee Democrat*) were located in the areas of highest population density. There was a gradual erosion of informed responses to environmental issues by these papers that paralleled the national trends. In recent decades, neither newspaper employed a full-time environmental writer. Indeed, there was virtually no attention paid to scientific findings concerning local and regional ecological matters.

When an environmental issue came up, the press played up the controversial issues by pitting unknowledgeable environmentalists against politicians, bureaucrats, and apologists for the promulgation of so-called economic progress. Scientific facts got short shrift, especially when it came to issues that affected the economic interests of "friends" of the news media.

In the Pensacola area, monopolistic control of the news by the daily *Pensacola News Journal* was in operation. It is no coincidence that one of the most polluted estuaries in the country was located in Pensacola. It was here where a federal action based on scientific information that showed a chemical company was actively polluting Escambia Bay was nullified by the Florida Legislature with an action that exempted the company from federal water-quality regulations. This was where political corruption was part of the ethos of the region and scientific inquiry was regarded with a skepticism that bordered on contempt. This included actions where a research boat that carried a black graduate student was shot at by citizens with shotguns and where the same student was denied service at a local restaurant.

There had been a long-standing controversy regarding the impacts of a pulp mill on Perdido Bay. During the early phases of mill activities, there had been major impacts on the bay due to the release of a tannin-stained, nutrient-/organic-enriched effluent. A major lawsuit was drawn against the pulp mill in a state court that pitted local environmentalist against the pulp mill. Faced with a lack of objective information, the court ordered the initiation of a scientific study of the Perdido Estuary. The Livingston research group was appointed to carry out the study that lasted from 1988 to 1991 with extensions that lasted through fall of 2007 (Livingston, 1992, 2007a,b).

The Perdido controversy was covered by the *Pensacola News Journal*. The study was carried out by a team of distinguished scientists, engineers, modelers, and computer experts drawn from across the United States (Appendix D). The results indicated that the pulp mill adversely affected Elevenmile Creek that received the initial loadings from the mill. Water-quality problems included high conductivity (total dissolved solids), high inorganic nutrients in the form of nitrogen (NH_3) and phosphorus (PO_4), and high dissolved organic carbon (DOC). The effluents caused hypoxia in various parts of the creek. It turned out that these parts of the mill effluent could not meet state and federal water-quality standards, which meant that the mill would eventually need to move its effluent out of the creek if it wanted to stay in business.

Results of the Perdido study during the first 3-year phase indicated that pulp mill discharges did not significantly affect sediment quality in Perdido Estuary (Livingston, 1992). No significant contamination from chlorinated compounds or metals occurred in sediments taken from Perdido Bay stations (Livingston, 1992). The concentrations of chlorinated compounds in organisms taken from Perdido Bay were similar to those in animals taken from unpolluted Apalachicola Bay. Model results indicated that the mill's wastewater discharges lowered the dissolved oxygen in the bay by relatively small amounts when compared to the other sources of nutrients and organic matter. The chief cause of lower dissolved oxygen at depth in the bay was found to be due to salinity stratification as a result of a dredged opening (the Perdido Pass) at the gulf end of the bay. Isotope analyses confirmed that dissolved organic matter from Elevenmile Creek did not significantly enrich bacterial activity in the upper Perdido Estuary. Whereas low dissolved oxygen conditions in Elevenmile Creek were due to mill discharges, low dissolved oxygen conditions at depth in Perdido Bay were thus not due to the release of mill effluents.

Results from the initial studies of Perdido Bay indicated that periodically increased nutrient loading from the mill, in the form of inorganic phosphorus and nitrogen, temporarily increased phytoplankton production in the upper bay, but not to the point of initiating destructive plankton blooms. Based on these data, it was recommended that nutrient loading from the mill should be studied more intensively. This concern for the impacts of nutrient loading and plankton blooms, which eventually proved to have major adverse impacts on the bay due to the promulgation of damaging plankton blooms, led to the long-term (20-year), court-ordered analyses of the Perdido River–estuary.

The data taken during the first 3 years indicated that, with the exception of estuarine portions of Elevenmile Creek, organic carbon levels in Perdido Bay were not higher than those in the

Apalachicola and Choctawhatchee Estuaries. Comparisons showed that there was no evidence that the primary productivity and organic carbon levels in the estuary were significantly higher than those in other (unpolluted) drainage systems elsewhere in the northeastern Gulf of Mexico. The biological results during the early study years indicated that Elevenmile Creek suffered serious losses of biological productivity due to pulp mill effluents, whereas the bay was not affected by such loading. Secondary productivity was adversely affected by salinity stratification and the resulting hypoxia and generation of liquid muds in affected areas. However, the secondary productivity of the shallow upper bay was comparable to that in the Apalachicola system, and the trophic organization of Perdido Bay during the first 3 years of sampling was similar to that of the predrought Apalachicola Estuary.

When the results of the first 3 years of the Perdido study came out, there was great consternation and "disappointment" on the part of the "environmentalists" whose fondest dreams were not verified by the facts. These views were prominently displayed by the print news media.

> POLLUTION FINDINGS BLASTED: Environmentalists in Alabama and Florida angrily denounced the findings of a study absolving a paper mill of blame for Perdido Bay pollution… "He is absolutely out of his mind" said Joy Morrill… "Any kind of development does have an impact, but the lower part of this bay is in far better shape than the upper part of the bay." Jackie Lane…also criticized Livingston's conclusions. Like Morrill, she had not seen a copy of his report.
>
> **The Associated Press**
> *August 8, 1992*

> HEALTH OF PERDIDO BAY STILL AT ISSUE: Dr. Ishphording (University of South Alabama) concluded from his mish-mash of studies that Champion was having no impact…he had to ignore the 6,000 pounds of sludge that come down Eleven-Mile Creek every day. And what about the dioxin?
>
> **Letter to the editor, "environmentalist"**
> *The Islander Gulf Shores, Alabama, April 21, 1992*

> TEST SIGNIFICANCE, METHODS AT ISSUE: Champion's effluent has "hormone-like" effects on many animals and causes reduced growth and reproduction. This hormone effect may very well be due to dioxin.
>
> **Letter to the editor, "environmentalist"**
> *The Islander Gulf Shores, Alabama, April 29, 1992*

> FISH FROM ELEVEN MILE CREEK SAID SAFE TO EAT: Fish from Eleven Mile Creek are safe to eat, the Florida Department of Health and Rehabilitative Services announced last week…Recent tests of fish caught in the creek have shown that dioxin has dropped to below the detectable level of one part per trillion, said Dr. Charles S. Mahan, HRS's deputy health secretary. "consuming fish from the creek no longer poses a heal threat" he said.
>
> **Michael Hardy**
> *The Mobile Register, Mobile, Alabama, September 5, 1992*

> WFSU-TV: The Coalition cited a report from the U. S. Fish and Wildlife Service which shows turtles and sediment in Perdido Bay and Elevenmile Creek in Escambia County are heavily polluted with dioxin.
>
> **Florida Public Radio Report**
> *February 11, 1994*

There were, according to the environmentalist version, mill-created toxic agents in the bay. According to these interests, the paper mill was responsible for seagrass losses throughout Perdido

Bay even though there was no scientific evidence of a light transmission problem in the upper bay, and independent studies showed that the primary losses of seagrasses were in the lower bay, far from the effects of the paper mill. According to the local environmentalist gospel, the dissolved oxygen problem in the bay was because of the mill despite the fact that the evidence indicated that salinity stratification of the bay accounted for most of the oxygen loss at depth as it had in Choctawhatchee Bay and other gulf estuaries. The agricultural and urban runoff problems in the bay were ignored by the local enthusiasts and, of course, by the news media.

The *Pensacola News Journal* became the chief outlet for sensationalized versions of the state of Perdido Bay with almost no attention to the scientific findings. Almost everyone had a good time. When, in 1997, the reports concerning the nutrient imbalances and phytoplankton blooms in Perdido Bay (1994–1997) were filed with the Florida Department of Environmental Protection, there was no response from anyone. Nutrient loading was not as intoxicating as toxic substances when it came to press reports. Statements by leading local authorities noted that the pulp mill was trying to get out of trouble by drawing attention to the nutrient question. Not only did the environmental groups misinterpret or simply ignore these reports, but when something of significance to the bay did occur, they missed it. They deliberately ignored the other problems in Wolf Bay and lower Perdido Bay, areas distant from the pulp mill. By this time, the effects of nutrient loading by the mill and other sources were evident in the increased level of plankton blooms and the impacts of such blooms on the biological organization of the estuary. These results were publicly disclosed in a series of reports.

Adverse effects by urban and agricultural interests were officially ignored by the *Pensacola News Journal*. In fact, a Perdido Basin land use study by the US Department of the Interior (Patrick, 1991) stated that "according to agreements with USEPA, the Wolf Bay drainage basin and Gulf Shores are not included in the Perdido Bay study." A review of agricultural activities in Baldwin County, Alabama (A. Niedoroda, pers. comm.), indicated that the production of cotton in this area increased by a factor of 26 between 1988 and 1995 and that comparable shifts in the agricultural activities in this region have continued to the present time. Nutrient-laden drainage from these agricultural areas went directly into the Wolf Bay part of the Perdido Estuary. The dredging of barrier islands continued all along the gulf coast. All of this, along with the active urbanization of lower Perdido Bay, was studiously ignored by all, including the news media.

Presently, nonpoint pollution due to agricultural and urban runoff remains largely unregulated by state and federal agencies in Florida and throughout the country as a result of political pressure from elected officials at all levels of government. Control of septic tanks is opposed by a public not willing to pay for the safe disposal of its wastes. Even though there was ample evidence that the pulp mill and other point and nonpoint sources of nutrients were adversely affecting Perdido Bay, the damaging plankton blooms were largely ignored by the environmental interests and the *Pensacola News Journal* in favor of lurid tales of toxic wastes loaded into the bay by the pulp mill.

> STUDY DOCUMENTS ENVIRONMENTAL DISASTER ON PERDIDO BAY: In late 1994, the paper mill converted to 100% chlorine dioxide bleaching. In November 1995, we measured Eleven Mile Creek to see if we could find evidence of chlorine dioxide. It was there...it is possible that chlorine dioxide, which is very close to carbon dioxide in chemical structure, could interfere with the photosynthetic process in plants... It is possible that chlorine dioxide is continuously released along the creek... Chlorate is the product from which chlorine dioxide is generated at the paper mill. We found it as well...With all the possibility of damaging effects from both chlorine dioxide and chlorate, no wonder Livingston found an environmental disaster in Perdido Bay. But of course, he blames this on too much phosphate. Why? Because phosphate can easily be controlled...Livingston has chosen the "politically correct" answer and solution to Perdido Bay's problems.
>
> ***Tidings* (newsletter of the Friends of Perdido Bay)**
> *13: No. 4, August 2000*

> NOTHING NEW: ...since the paper mill in Cantonment installed chlorine dioxide bleaching, much of the vegetation in Perdido Bay has disappeared— grass beds, aquatic vegetation, and even the scum algae.
>
> **Letter to the editor, "environmentalist"**
> *Pensacola News Journal, June 11, 2004*

The data on the nutrient loading and the impacts of plankton blooms on the bay were thus written off as a plot to revert to an "easily controlled" factor such as orthophosphate. The organized environmental groups joined local organizations in their denunciation of the pulp mill as the sole polluter in the region. The constant drumbeat of reports and letters to the editor to the *Pensacola News Herald* continued with an emphasis on the ways in which the paper mill affects Perdido Bay:

Oxygen-consuming materials (BOD) (1/2004)
High levels of dioxin and heavy metals (2/2005)
Dioxin (2/2005)
Low dissolved oxygen and high bacteria (8/2007)
Dioxins and heavy metals (1/2007)
Heavy metals, polychlorinated biphenyls (PCBs), dioxin, and 8,000 pounds of sludge (6/2007)
Organic material (equivalent of 38,000 septic tanks) (10/2008)
Barium and manganese (8/2009)
Environmental laws and agencies are shams (3/2009)

The *Pensacola News Journal* responded to reports during 2005:

> Over the years, residents near Perdido Bay have lamented its decline.. Dolphins pretty much have left the bay.. Mullet don't jump in the numbers they used to.. And a dark, slimy muck covers the bay's bottom, they say.. When Hurricane Ivan struck Sept. 16, the storm surge left a "slick, almost black layer of mud" that contained levels of dioxin....
>
> A second set of sediment samples from Perdido Bay has tested above Florida residential soil cleanup standards for dioxin and arsenic, according to data released by Friends of Perdido Bay.. The test results for the two toxic chemicals mirrored those of previously released results, lending credence to them, said Jackie Lane of the Friends group. The results were announced at a news conference.
>
> Mullet fished from throughout the Pensacola Bay system tested above federal-health safety levels for polychlorinated biphenyls, or PCBs.. The highest levels were found near a proposed $800,000 habitat-restoration project, according to researchers at the University of West Florida.. "The mullet was a big surprise for us, not just in PCB concentration but because of the amount consumed."

Despite this drumbeat of false information concerning the Perdido Estuary, the real problems due to toxic wastes were known by scientists to exist elsewhere. Collard (1989, 1991a,b) performed extensive reviews of the scientific literature concerning the Pensacola Estuary, noting a history of decreased biological integrity since the mid-1950s. Upper Escambia Bay, with reduced circulation due to both natural and anthropogenic conditions, was the subject of the most debilitating habitat destruction and fish kills; the nonpoint source pollution from the City of Pensacola and the Naval Air Station, together with major pollution of the bayous (Texar, Chico, Grande) in this region, presented a serious problem in terms of continuing, multisource contamination by toxic agents that was difficult to evaluate and even more difficult to mitigate. The Northwest Florida Water Management District (NWFWMD, 1993) conducted a study of nonpoint source loading in the Pensacola Bay system. The findings included the fact that Mulatto Bayou and the main residential areas of the Pensacola Bay system had the highest normalized loadings of the entire system.

Brim (1993) did a study of the toxic agents in Perdido Bay sediments. He found that the Perdido Estuary was generally free of toxic compounds (pesticides, metals, polynucleated aromatic hydrocarbons [PAHs], dioxin). However, field toxicity tests with bay water and sediments indicated that upper and lower bay sediments reduced feeding activity in some estuarine organisms. This was taken as a measure of reduced habitat quality. Seal et al. (1994) conducted statewide sediment tests for metals and organic contaminants. They found that urban stormwater runoff was the major cause of contamination of sediments in Florida's coastal areas. The highest concentrations of toxic agents were found in coastal sediments at sites near Tampa, Pensacola, Miami, and Jacksonville. The Pensacola Estuary was one of the most heavily impacted bays in the state in terms of contamination from highly toxic and long-lasting substances. Elevated levels of metal contamination were found in Bayou Grande (Cd, Pb, Zn) and Bayou Chico (Cr, Zn). Bayou Chico was also contaminated with PAHs and PCBs. In the Pensacola Estuary, PAHs and PCBs were found in sediments close to shore and in central parts of the bay. Phenolic compounds were found at one site near Pensacola Harbor.

Long et al. (1997) carried out a study of the "magnitude and extent of sediment toxicity in four bays of the Florida Panhandle." The sediments of the Pensacola Estuary were compared to those of the Choctawhatchee, St. Andrews, and Apalachicola Bays. The greatest toxicity of all bays analyzed occurred in Bayou Chico (Escambia Bay), with the other developed bayous in the Pensacola area showing "relatively severe toxicity." Toxicity of the Pensacola Bay sediments was attributed to high-molecular-weight PAHs, zinc, DDD/DDT isomers, total DDT, and the highly toxic organochlorine pesticide known as dieldrin. There were elevated PAHs at the National Oceanic and Atmospheric Administration's National Status and Trends Program for marine environmental quality stations in Pensacola Bay and Indian Bayou. In addition to the bayous and parts of Pensacola Bay, toxicity was also detected in upper and midportions of Escambia Bay. Toxicity was also noted in Blackwater Bay. The NOAA study confirmed high concentrations of lead, mercury, and PCBs in the sediments of the urbanized bayous and in Pensacola Bay.

Karouna-Renier et al. (2005) carried out an environmental health study of northwest Florida bays with a screening level assessment of contaminants in blue crabs (*Callinectes sapidus*) and oysters (*Crassostrea virginica*) collected in bays and bayous in the Pensacola area. Tissue samples were analyzed for mercury, arsenic, cadmium, chromium, copper, lead, nickel, selenium, tin, zinc, 17 dioxin/furan compounds, and 12 dioxin-like PCB congeners. The locations that exceeded SVs and had the highest carcinogenic or noncarcinogenic health risks were generally located in urbanized water bodies (Bayou Texar, Bayou Grande, and Bayou Chico of Escambia Bay) or downstream of known contaminated areas (Western Escambia Bay). Oysters collected from commercial oyster beds in Escambia Bay and East Bay (south of Blackwater Bay) and crabs collected from East, Blackwater, and Perdido Bays generally had the lowest levels of contaminants. Despite accounting for only 15% of the total tissue, inclusion of hepatopancreas in a crab meal increased contamination to levels above many SVs, and therefore, direct or indirect consumption of hepatopancreas from crabs in the Pensacola Bay system was discouraged.

During the early 2003, the Perdido report (Livingston, 2007b) outlined data showing that there were storm-related problems associated with effluent spills into upper Perdido Bay by a sewage treatment plant (STP). The data were given to personnel in charge of the STP. The facility was run by the Escambia County Utilities Authority (ECUA). ECUA officials responded with a memo that accused the report of misrepresentation of the facts, among other things. The *Pensacola News Herald* then wrote a series of articles that avoided the scientific results that implicated ECUA with plankton blooms while repeating the false accusations that various "environmentalists" had issued over the years that attacked the Perdido Bay research. Another story outlined how well the STP was designed and how it was impossible for such a facility to pollute the bay. A third article outlined the unflattering response of ECUA to Livingston's original Perdido report concerning the sewage spill.

The *Pensacola News Herald* choreographed its reportage to protect ECUA officials from any responsibility in the pollution of Perdido Bay. By concentrating on a preliminary report and the aggressive response by ECUA, the actual facts were largely ignored in favor of creating a confrontation between ECUA and the paper mill that was the actual intent of the series of newspaper articles. Even though mill personnel had nothing to do with the scientific findings, the *Pensacola New Herald* made it seem as if the data concerning the ECUA discharges were a trumped up diversion to the problems associated with pulp mill effluents on the Perdido Estuary. All of this was complicated by the fact that reporting of orthophosphate concentrations and nutrient loading by STPs was not required by the Florida Department of Environmental Protection.

The organized environmental groups joined the local environmentalists in the continuation of a long series of outrageous statements and reports that were published in the *Pensacola News Journal*. These people only cared about a narrow interest (their own) and were not really interested in the overall environmental problems in the Perdido system. The actual facts were all but ignored by everyone in their frenzied attempt to make the Perdido Estuary the center of the toxics universe. The real problems such as plankton blooms, dredging, and increased drought occurrence and severity that led to the almost total loss of secondary production were totally ignored by the Pensacola news media and the so-called environmentalists. Meanwhile, the results of the Perdido study led to the development of a nutrient restoration program by the pulp mill.

Determination of the water-quality-based effluent limits (WQBELs) by the Florida Department of Environmental Protection was made with the assumption that nutrient inputs other than the pulp mill into upper Perdido Bay would remain the same. The period 1988–1991 saw relatively natural conditions, including a balanced food web in upper Perdido Bay (Livingston, 1992). Therefore, the biological basis for the WQBEL recommendations was that the nutrient loads into upper Perdido Bay during the 1990 low-flow period could be safely discharged in the future and that these loads would be protective of the biological integrity of Perdido Bay. Effluent limits were instituted for ammonia nitrogen, orthophosphate, and total phosphorous. The Florida Department of Environmental Protection used the project data and did its own analyses, to establish the WQBELs for total ammonia, orthophosphate, nitrate–nitrite, total phosphorus, BOD, color, and soluble inorganic nitrogen. Final ammonia-N discharges from the mill would be limited to a maximum of 361 kg day^{-1}, based on a quarterly average as were the targets for orthophosphate.

A program was developed to divert the mill effluent into a salt marsh system at the head of the bay that would get the effluent out of Elevenmile Creek and reduce the nutrients to levels that would not have an adverse effect on the bay. This plan was attacked by a lawsuit by the environmentalists in the Florida Division of Administrative Hearings, Judge Bram D. E. Canter presiding. There were the usual sensationalized stories in the regional news media concerning the legal proceedings. Once again, the now well-understood scientific facts concerning the demise of the Perdido Estuary that were outlined in a series of public reports were ignored by all.

In the *Lane vs. International Paper* case, Judge Canter eventually found for the pulp mill (Appendix E), He stated the following:

> In December 1989, DER and Champion entered into Consent Order No. 87-1398 ("the 1989 Consent Order")...Champion was also required to submit a plan of study and, 30 months after DER's approval of the plan of study, to submit a study report on the impacts of the mill's effluent on DO in Elevenmile Creek and Perdido Bay and recommended measures for reducing or eliminating adverse impacts. The study report was also supposed to address the other water quality violations caused by Champion....A comprehensive study of the Perdido Bay system was undertaken by a team of 24 scientists lead by Dr. Robert Livingston, an aquatic ecologist and professor at Florida State University. The initial three-year study by Dr. Livingston's team of scientists was followed by a series of related scientific studies, which will be referred to collectively in this Recommended Order as the Livingston studies.

> In September 2002, while Champion's 1994 permit renewal application was still pending at DEP, IP submitted a revised permit renewal application to upgrade the WWTP and relocate its discharge. The WWTP upgrades consist of converting to a modified activated sludge treatment process, increasing aeration, constructing storm surge ponds, and adding a process for pH adjustment. The new WWTP would have an average daily effluent discharge of 23.8 million gallons per day (mgd). IP proposes to convey the treated effluent by pipeline 10.7 miles to a 1,464-acre wetland tract owned by IP, where the effluent would be distributed over the wetlands as it flows to lower Elevenmile Creek and upper Perdido Bay.

Dr. Livingston developed an extensive biological and chemical history of Perdido Bay and then evaluated the nutrient loadings from Elevenmile Creek over a 12-year period to correlate mill loadings with the biological health of the Bay. Because Dr. Livingston determined that the nutrient loadings from the mill that occurred in 1988 and 1989 did not adversely impact the food web of Perdido Bay, he recommended effluent limits for ammonia nitrogen, orthophosphate, and total phosphorous that were correlated with mill loadings of these nutrients in those years. The Department used Dr. Livingston's data, and did its own analyses, to establish WQBELs for orthophosphate for drought conditions and for nitrate-nitrite. WQBELs were ultimately developed for total ammonia, orthophosphate, nitrate-nitrite, total phosphorus, BOD, color, and soluble inorganic nitrogen.

Much of the water quality and biological data presented by Petitioners were limited in terms of the numbers of samples taken, the extent of the area sampled, and the time period covered by the sampling. Much of the expert testimony presented by Petitioners was based on limited data, few field investigations, and the review of some, but not all relevant permit documents.

The evidence is persuasive that the salinity stratification in Perdido Bay is a major cause of low DO in the Bay.[9] However, the stratification does not explain all of the observed changes in water quality, biological productivity, and recreational values. The stratification does not account for the markedly better conditions in the Bay that existed before the Cantonment paper mill began operations. The Livingston studies confirmed that when nutrient loadings from the mill were high, they caused toxic algae blooms and reduced biological productivity in Perdido Bay. As recently as 2005, there were major toxic blooms of Heterosigma in Tee and Wicker Lakes caused by increased nutrient loading from the mill.

The Livingston studies represent perhaps the most complete scientific evaluation ever made of a coastal ecosystem… (and) confirmed that when nutrient loadings from the mill were high, they caused toxic algae blooms and reduced biological productivity in Perdido Bay…WQBELs were ultimately developed for total ammonia, orthophosphate, nitrate-nitrite, total phosphorus, BOD, color, and soluble inorganic nitrogen….

Petitioners failed to prove that any new data in the December 2007 report of the Livingston team demonstrate that the proposed WQBELS are inadequate to prevent water quality violations in Perdido Bay.

However, with regard to many of their factual disputes, Petitioners' evidence lacked sufficient detail regarding the dates of observations, the locations of observations, and in other respects, to distinguish the relative contribution of the mill effluent from other factors that contributed to the adverse impacts in the Bay, such as salinity stratification, natural nutrient loading from the Perdido River and other tributaries, and anthropogenic sources of pollution other than the paper mill.

Dr. Livingston responded to the BOD and carbon issues that "these Petitioners raised over the years" by investigating them as part of the Livingston studies. He found no relationship between loading and DO. Dr. Livingston concluded that the mill was not having much effect on SOD.…Dr. Lane, who is a marine biologist, believes a major reason for low DO in Perdido Bay is "organic carbonaceous BOD." However, Dr. Lane presented no evidence other than statements of the theoretical process by which carbon from the mill would cause low DO in the Bay. She presented no scientific data from Perdido Bay to prove her theory.

The more persuasive evidence presented at the final hearing strongly indicates that the proposed Department authorizations would likely effectuate a significant improvement to the Perdido Bay system over the current discharge to Elevenmile Creek.

Deafening news media silence. Case closed.

During the 1960s, there were three ecosystems in eastern parts of north Florida that had attained international fame as representative of the most productive lakes (Lake Jackson), springs (Wakulla Springs), and estuaries (Apalachicola Estuary) in the southeastern United States. Wakulla Springs is the world's deepest and longest aquatic cave system with peak flows of over 14,000 gallons (53,000 L) per second. The main spring is located 22.5 km (fourteen miles) south of Tallahassee, Florida. After reaching the surface, the water enters the Wakulla River to the south, flowing to the Gulf of Mexico. The surface water manifestation is fed continuously by upwelling water naturally rich in nutrients forming one of the most productive aquatic systems in the world. The clear waters during these early years used to support massive arrays of bottom plants that once formed the habitat and base for extensive aquatic food webs. The cold, clear water had formed a diverse community that included various animals represented by many species of invertebrates, fishes, birds, and mammals. In recent decades, all of this has changed due to nutrient loading from urban centers to the north.

Lake Jackson is one of a number of solution lakes, many of which are located in Leon County, Florida (Livingston, 2007c). Dissolution of subsurface lime rock leads to the formation of what is known as a karst topography, a spongelike geological system of carbonate rock that forms a honeycomb of extensive interlocking caves and springs that riddle the north Florida area. Solution lakes are formed due to the erosion of the limestone as part of a low topography, a humid climate with ample rainfall, and the karst (i.e., infiltrated limestone) environment. At one time, Lake Jackson represented the almost perfect combination of natural habitats that allowed the growth of massive grass beds and bottom plants. These assemblages formed the basis of food webs that led to optimal growth of fishes such as largemouth bass (*Micropterus salmoides*). Lake Jackson became renowned as one of the greatest bass-fishing lakes in the country. That is no longer true.

By the end of the 1990s, there had been extensive urban development in the subbasins of the lakes and sinkholes of Leon County. Nutrient loading from unregulated stormwater runoff, septic tanks, and package sewage treatment systems led to the loading of inordinate amounts of nitrogen and phosphorus into the lakes of the region. This resulted in increased growth of introduced bottom plants such as *Hydrilla verticillata* and associated toxic plankton blooms. These effects of high nutrient loading from urban areas extended downstream into the Wakulla Springs. Within a few years, both Lake Jackson and Wakulla Springs had been adversely affected by the anthropogenic nutrient loading from upstream urban areas that produced plankton blooms and serious water-quality deterioration. These changes, in turn, caused alterations of the benthic plants that then led to altered food webs resulting in the massive loss of natural productivity. The benthic macrophyte beds were turned into sterile mud flats and *H. verticillata* masses that stretched to the surface. Wakulla Springs lost its populations of plants and animals, and Lake Jackson was no longer a world-class bass-fishing mecca.

During earlier years of long-term scientific analyses (1970–1997) of Lake Jackson, the results of the research were reported by a series of very able reporters for the *Tallahassee Democrat*. During the late 1990s, however, *The Democrat*, the regional news monopoly in eastern parts of north Florida, was lobbied by influential developers to stop publication of scientific findings concerning local lake problems. It seems that the publicity of the destruction of the area lakes had affected urban development in the region. Accordingly, the paper fired news personnel (reporters and editors) who were considered environmentally sensitive. A new regime was instituted in the regional news media whereby any articles concerning issues related to development interests in the region were suppressed or replaced by false or misleading stories, a situation that exists to this day. Even letters to the editor were not printed when complaints were made concerning the firings and intimidation of state environmental personnel by county elected officials.

The paper adopted a new approach to environmental reporting by ignoring the scientific data in favor of quoting the myths and excuses of "environmentalists" and bureaucrats. Sensationalism took the place of facts. Controversy flavored by the scientifically unintelligible rantings of both sides of a given argument replaced the results of decades of scientific studies concerning the lakes of the region. The new policies of the paper spread to the other forms of news media whereby the adverse effects of cultural eutrophication due to stormwater runoff, septic tank developments, and sewage spills were ignored. Political diatribes by bureaucrats replaced statements from knowledgeable scientists. "Environmentalists" whose main concern was the construction of bike paths and nature walkways, were favored. The loss of Lake Jackson as a regional ecological wonder was simply ignored as runoff continued to pour into the lake from roads and septic tanks.

The lake controversies were not the only form of omission and misinformation on the part of the press. One illustration of the newspaper's coverage of environmental events involved the Deepwater

Horizon (BP) oil spill of spring/summer of 2010. The so-called Macondo blowout of a gulf oil rig off the Louisiana coast was estimated to be the greatest marine spill in history (around 210 million US gallons [795 million liters] released over 87 days). This was greater than the other major Gulf of Mexico blowout, the Ixtoc spill of 1979–1980 (around 130 million US gallons [492 million liters]). In the previous fall, the *Tallahassee Democrat* reported extensively on a symposium sponsored by the Florida State University Institute for Energy Systems, Economics, and Sustainability that concerned potential impacts of oil drilling on marine systems. The symposium represented a response to a proposal by Republican leaders in the Florida Legislature to lift Florida's two-decade ban on oil drilling in Florida's offshore waters.

The newspaper's headline read "FSU drilling symposium experts say risk is low." According to the newspaper,

> Offshore drilling would pose relatively little risk to Florida's environment, according to academics and industry experts who participated Monday in a symposium sponsored by Florida State University. … Blowouts and spills are "rare by design," said Norman Guinasso, an adjunct professor in the Department of Oceanography at Texas A&M University and one of the developers of the Texas Automated Buoy System, which monitors water conditions in the western Gulf of Mexico. Guinasso described the 1990 explosion of the super tanker Mega Borg, which spilled 5.1 million gallons of light Angolan crude south of Galveston. Half of the crude burned off and a quarter of it evaporated, leaving about 1.3 million gallons in the water. Predictions that the spill would mar Corpus Christi didn't pan out, but tar balls were reported as far away as Louisiana. In 1995, another tanker collision spilled 2,000 barrels of oil south of Galveston. Guinasso said studies suggested the spills had little impact.

The timing and conclusions drawn here were somewhat less than prophetic. Soon after the public display of the merits of offshore oil drilling, the Macondo blowout occurred. A follow-up *Tallahassee Democrat* article headlined "Research Aims to Understand Post-Spill Gulf" (April 22, 2013) had the following quotes by various marine scientists involved in the gulf research:

> For the scientists who study the Gulf, their quest to understand the effects of the last disaster has underscored how little they know about the home …..The lack of knowledge of what existed before the spill has complicated their effort to trace its effects. (Marine scientist)
>
> The oil spill shown a light on the Gulf that said "My God…There is the supplier and nobody was paying any attention to how important it is." (Marine scientist)
>
> He (marine scientist) said that most marine biologists are lucky to venture out to sea once a year. In the wake of the spill, members of a consortium studying the deep waters of the Gulf have been able to take new expeditions every few months….
>
> The lack of investment of in basic scientific research in the Gulf has been a severe impediment. (Marine scientist)

Although the most intensive research efforts of the NE gulf have been in shallower inshore waters, the knowledge of the long-term changes in these systems was totally ignored even though such findings would have been of importance in the design of studies for the deeper parts of the gulf that were largely bereft of any kind of comprehensive or long-term scientific analysis. None of the extensive coastal research along the Gulf of Mexico was reported by the *Tallahassee Democrat.*

For decades, Florida, Alabama, and Georgia have been at odds concerning the right to use water in the tri-river (Apalachicola/Chattahoochee/Flint) basin. Georgia asserted its right to take water from the Chattahoochee River to support the rapid growth of metropolitan Atlanta. The issue was contested by Florida and Alabama in series of court cases over water rights in the tri-river system. This case was recently concluded with a court decision that favored the Georgia concerns. There were other issues that remained unresolved and largely overlooked such as the removal of water from the Chattahoochee and Flint Rivers by agricultural interests in Georgia.

With the increased frequency and severity of droughts in recent years and the loss of the fisheries in the Apalachicola Estuary, environmentalists and politicians in Florida seized on the urban withdrawal issue in Georgia as the primary cause of the catastrophic events along the northern Florida Gulf Coast. Their cause was taken up with great enthusiasm by the *Tallahassee Democrat* that proceeded to sensationalize this very narrow issue without once referring to actual scientific information. There were award-winning articles that concentrated on just about everything except the long string of droughts that had occurred over the past three decades that were at the heart of the loss of the major fisheries in the Apalachicola Estuary. Once again, controversy (Florida vs. Georgia) replaced the results of decades of research in the Apalachicola River–estuary.

According to a front page report of the *Tallahassee Democrat* (July 23, 2013),

> The issue of Atlanta's withdrawal of water from Lake Lanier and another nearby reservoir has been the subject of a decades-long dispute between Georgia and the two states— Florida and Alabama— that rely on the water supply for their own needs. At the center of the fight is the US Corps of Engineers, which manages the flow of water from federal reservoirs like Lake Lanier. Since the 1950s, the daily amount of water that metropolitan Atlanta has been allowed to withdraw from the lake under the Corps' watch has increased from 10 million gallons to 170 million, said Sen. Jeff Sessions, the Alabama Republican who presided over the hearing before the Senate Environment and Public Works Committee. As a result, that's put a severe strain on the Apalachicola-Chattahoochee-Flint River System, he said.

The gist of the preceding statement was then repeated in a series of articles by the *Tallahassee Democrat* with respect to the Apalachicola water crisis. A favored trick of the print media, the transformation of a myth into scientific fact through repetition, was put into play. The paper, and derivative articles in the *The Associated Press* and *The New York Times*, played up the public controversy between Georgia and Florida with regard to who gets the water flowing within the tri-river system. This illustrated another failing of the news media that pumped up controversial disputes to sell news while ignoring the boring factual details that formed the basis of the problem. The "water war" dispute was centered almost completely on the withdrawal of water from the Chattahoochee River for the city of Atlanta. There was, however, no attempt to identify the source or the validity of Senator Sessions' reference to the 170 million gallons per day (mgd) withdrawn from the Chattahoochee River for Atlanta's water needs.

What would be the ecological significance of the withdrawal of 170 mgd on Apalachicola River flows? Any calculation of such a withdrawal should include the significance of a return of a portion of the removed water to the river in the form of urban runoff (F. G. Lewis, pers. comm.). The withdrawals from the river would occur at the head of the Chattahoochee River; the relatively small drainage area involved would mitigate the significance of the volume of water removed in this part of the tri-river system. The fact that other withdrawals, such as those due to agricultural activities along the Chattahoochee and Flint Rivers, could be even more damaging to downriver interests was not even mentioned. Testimonials from politicians, bureaucrats, and "environmentalists" came forth in a torrent of outraged citizens, many of whom had virtually no knowledge of scientific basis of the interaction of stream flows and estuarine productivity. Not to be outdone, public officials in Georgia joined the battle for freshwater in the tri-river basin.

According to a letter from Georgia Governor Nathan Deal to the assistant secretary of the Army for Civil Works (January 11, 2012),

> As reflected in Mr. Turner's affidavit, based on current demographic information and, as a consequence of improved water conservation, Georgia now believes that 705 mgd will be sufficient to meet Georgia's water needs from Lake Lanier and the Chattahoochee River to approximately the year 2040.

According to the affidavit of Mr. Judson H. Turner, Director of the Georgia Environmental Protection Division of the Georgia Department of Natural Resources (January 10, 2013),

> The foregoing information affirms and updates Georgia's 2000 request that the Corps operate Lake Lanier to meet water supply needs of 705 mgd annual gross average withdrawal including 297 mgd annual gross withdrawal from Lake Lanier and 408 mgd annual gross withdrawal from the Chattahoochee River between the Buford Dam and the confluence of the Chattahoochee River and Peachtree Creek.

According to a subsequent article in the *Tallahassee Democrat*, the current withdrawal levels approximated about half of the amount recently requested by the Georgia governor, about 350 mgd.

A minimal flow of 5000 cubic feet per second (cfs) is guaranteed by the US Army Corps of Engineers. The operation of the various dams along the Georgia rivers was also a major concern of the Florida interests that claimed that the current policies of the COE favored Georgia. In any case, the news media then proceeded to fan the flames of controversy without any mention of the long-term climatological changes that had been the primary cause of the droughts that had led to the demise of productivity in the Apalachicola Estuary.

As part of a nationwide effort, the infamous climate issue had been made toxic by Florida's reactionary political regime that claimed there is no such thing as "global warming" occurring in the world and that, strangely enough, such global changes, if they do occur, have not been affected by human activities. The general public, along with state news media, have gone along with this mythology in what can only be described as a dedicated attempt to substitute abysmal levels of ignorance for scientific facts. The official policy of the state of Florida was thus observed in silence and obfuscation of scientific facts.

A recent article in the *Tallahassee Democrat* ("Apalachicola River water woes take center stage. US Senate hearing offers 'unprecedented opportunity,'" August 11, 2013) covered a meeting of the various Florida interests with the state's two US senators and a regional House member. These worthies debated the next step in the Apalachicola situation. Environmentalists, politicians of various types, state agency bureaucrats, and local fishermen gave their views that were quoted without ever once mentioning the chief cause of the droughts. The intensive coverage actually replaced football as headline news for a couple of days.

The main emphasis was the need to stop the withdrawals of water for the city of Atlanta. According to the paper,

> We hope to bring a large crowd out so the senators and Rep. Southerland can see for themselves how important this issue is to our whole region. To save the bay, Congress has to act. (Local environmentalist)
>
> The report (Florida Fish and Wildlife Conservation Commission) puts the cause of the oyster disaster squarely on the lack of freshwater flowing into the Apalachicola River and its estuary. The situation is a result of upstream consumption and Corps water management actions, exacerbated by a two-year drought that reduced water flow to the Apalachicola to its lowest level in recorded history.
>
> Recovery is possible, (a regional bureaucrat) said, but added: Even if the oysters in that oyster fishery start growing, that doesn't change the fact that we aren't going to get more water from the river because of the water management.
>
> We have asked (the Corps) for 20 years, You need to consider the needs of the Apalachicola River and Bay, and they say We are not authorized, (a local environmentalist) said, Right now there is not a level playing field. We are not even in the game. Congress has got to act to level the playing field.
>
> Sen. Nelson, however, said he is not optimistic Congress can come to an agreement on the issue either.

Various state officials took part in the hearing. NWFWMD Executive Director Jonathan Steverson said:

> The good Lord giveth and the court and Georgia taketh away.

The next day, the *Tallahassee Democrat* issued forth with another article ("Alabama quiet in new water war outbreak"). Florida Gov. Rick Scott, a new defender of the river, had issued his declaration of war against Georgia concerning the removal of water from the Chattahoochee River for Atlanta:

> This is a bold, historic legal action for our state, Scott said. But this is our only way forward after 20 years of failed negotiations with Georgia. We must fight for the people of this region. The economic future of Apalachicola Bay and Northwest Florida is at stake.

This was followed up by yet another *Tallahassee Democrat* article on August 14, 2013, titled "Scott says he'll sue Georgia for water flow to Apalachicola" in which the governor indicated "his intention to return the state's water war with Georgia into a live battle in court again, promising to seek an injunction from the US Supreme Court for more water to flow into Florida." This time, there were statements from Florida's two US senators and the House Representative Steve Southerland. Senator Marco Rubio declared that while money can help the problem, "we can't stop there." He said a real solution would have the three states equally divide water resources after looking at the entire volume available. In the ensuing debate, there wasn't even a glint of recognition of what was going on with the climate.

On August 16, 2013, the editorial of the *Tallahassee Democrat* ("More Water") fanned the flames of a "water war" with the following statement:

> But we side with Florida Agriculture Commissioner Adam Putnam who said Florida leaders should use "every arrow in their quiver" to save the river.

The editorial left out the most important arrow of all, namely, the long-term changes in the climatological conditions of the region that were the real cause of the current problems.

Even if the Florida interests prevailed in court and Georgia was forced to reduce its withdrawals of water for the Atlanta interests, what would be the outcome of such a decision? A review was made concerning the actual impact of the publicly stated withdrawals of river water for Atlanta. An analysis of the various scenarios regarding the Chattahoochee water issue that were trumpeted so widely is given in Table 10.1. These data indicate that the requested withdrawal figure of 705 mgd from the Chattahoochee River by Georgia would exacerbate the problems associated with low river flows during drought conditions. This would indeed have harmful effects downstream should such a withdrawal be allowed. However, these effects would not be significant during mean and high river flow conditions due to the relatively high flows of the Chattahoochee system.

However, the more modest withdrawals that were so roundly denounced were not shown to be as detrimental, especially when the expected returns of water to the river were taken into consideration. The so-called current withdrawals (350 mgd) showed a lesser impact with relatively low reductions of flow rates during drought conditions if the returns were factored into the calculations. This was even more true if the reported level of 170 mgd withdrawals (99.8 mgd with returns) was considered. This analysis showed that while the projected withdrawals from the current request from Georgia would exacerbate the impacts of severe droughts, the overall impacts of the Atlanta issue were not placed within a reasonable perspective. The stated range of such withdrawals would only exert minimal impacts during drought conditions and even less impacts during average and flood conditions of the river.

There certainly should be an overall plan for a more equitable division of the water resource that would include all freshwater usage along the tri-river system including a revision of the dam management by the Corps. However, there should be a scientific evaluation of all of the factors concerning river flows in this ecosystem that should include the recent (30+ year) trends of drought in the region. This evaluation should include agricultural withdrawls. The emphasis on

Table 10.1 Percentage Effects of Different Estimates of Water Withdrawals by Georgia from the Chattahoochee River (mgd)

Withdrawal	mgd	% Min Flow (5000 cfs[1])	% Mean Flow (21,390 cfs[2])	% High Flow (55,800 cfs[3])
Withdrawal without return to Chattahoochee River				
Requested by Georgia	705	21.8	5.1	2.0
Current (news media)	350	10.8	2.5	1.0
Reported (news media)	170	5.4	1.3	0.5
Withdrawal with 57% return to Chattahoochee River				
Requested by Georgia	400.0	12.4	2.9	1.1
Current (news media)	199.5	6.2	1.4	0.6
Reported (news media)	99.8	3.1	0.7	0.3

Data are presented with and without estimates of return flows given at about 57% of the withdrawal levels. Data for flow rates and the return flow estimates were given by Dr. F. Graham Lewis (NWFWMD). Withdrawal rates were given by Georgia Governor Nathan Deal (705 mgd), Senator Jeff Sessions (170 mgd), and *The Tallahassee Democrat* (350 mgd), the latter two figures representing current withdrawal levels.

1 Minimum low guaranteed by the US Army Corps of Engineers.

2 Long-term average at Chattahoochee Gauge.

3 High flow average that is exceeded only 5% of the time.

the Atlanta situation was misplaced and did not represent the scientific facts concerning the loss of the fisheries of the Apalachicola Estuary. The actual facts were not as exciting as the sensationalized version of the news media but were nevertheless more applicable to the problems created by the low Apalachicola River flows.

Having established the myth of the Georgia withdrawals, the *Tallahassee Democrat* followed up its series of overheated river stories with a follow-up article by its favorite opinion writer, Bill Cotterell, in an story printed on September 9, 2013, titled "There are three E's" in "reelection": for (Governor) Scott, going green "is good politics." According to Mr. Cotterell,

> … Scott used August to mount a sort of green campaign around the state. On Aug. 13, he went over to Apalachicola and announced the state will sue Georgia over water use from the Chattahoochee River… for pumping too much water out of the river for Atlanta….A week later, Scott and the Legislature announced a $40 million commitment to improve water being discharged from Lake Okeechobee.

Shades of the War on Georgia were again used by the paper to burnish Governor Scott's environmental accomplishments. As for the Okeechobee reference, *The New York Times* published an article ("In South Florida, a polluted bubble ready to burst; A release from Lake Okeechobee devastates estuaries and prompts calls for Action") on the same day as the Cotterell declamation:

> But environmentalists say Mr. Scott and the legislature have slashed the budget of the South Florida Water Management District which oversees the state's water flow, and have put in place some inexperienced managers.

There was no mention in the regional news of a state government that has the reputation of one of the least environmentally oriented administrations in the history of Florida.

According to an AP article "Environmental laws not enforced under Scott" (August 29, 2013),

> Enforcement of the state's environmental laws has plummeted under Gov. Rick Scott and the private company attorney he picked to lead the Department of Environmental Protection…. the department essentially has become nonfunctional under Vinyard, who worked for a shipyard before Scott appointed him secretary.

Finally, on September 19, 2013, the *Tallahassee Democrat* came out with a little article titled "House lawmakers opt not to wade into Florida-Georgia water fight":

> Members of the House Transportation and Infrastructure Committee voted Thursday against a proposal to require congressional approval whenever the US Army Corps of Engineers plans to divert at least five percent of water from its natural flow in a multi-state watershed. Rep. Steve Southerland, R-Panama City, who drafted the proposal, implored his colleagues on the panel to back it during debate on a larger water resources bill. Southerland wants to keep Georgia from drawing increasing amounts of freshwater from federal reservoirs such as Lake Lanier without federal approval. The water supplies residents of the Atlanta metro area. Since the 1950s, the Corps has increased the daily amount of water the area may draw from the lake from 10 million gallons to 170 million gallons. "Over the last five decades, water flow down the Apalachicola River (into Apalachicola Bay) has decreased over 50 percent," Southerland said. "And if we do not step in, it will continue to go down. This is a proud region, a place where heritage matters. And it's a place on the verge of extinction…." In August, Florida Gov. Rick Scott said he will file a lawsuit to block Georgia from diverting more water, but that effort is considered a long shot.

There is no doubt that the future withdrawals of water for Atlanta, if conforming with recent plans for increased water usage, would exacerbate the impacts of droughts on the Apalachicola River–estuary. However, the primary issue should be the overall trend over the past 30+ years of increasing frequency and severity of droughts in the region. If this climate trend continues, the proven damage of the past few decades would continue to the detriment of productivity along the gulf coast. It is somewhat ironic that prominent climate change deniers such as Congressman Steve Southerland and Senator Marcus Rubio should emerge as defenders of a system that has already undergone recent destruction due to changes in regional rainfall and river flows. It is even more ironic that Governor Scott, who, with the cooperation of the Republican Florida Legislature, has gutted the state's environmental regulations, should declare a water war with Georgia. The current political leaders did away with the Energy and Climate Commission and buried the Florida Office of Energy in the Department of Agriculture.

The Florida Legislature is already on record concerning climate change. State Representative Alan Hayes noted that

> The global warming hoax that has been perpetuated by Al Gore and his followers has been debunked.

These opinions have practical implications as is shown in the history of the Apalachicola River flows. The way that this situation has been handled by the news media, regional and national, has been sensationally irresponsible. *The Tallahassee Democrat* has not only ignored the extensive scientific data on the Apalachicola situation but has continued a long-term misrepresentation of important environmental facts concerning the demise of the major aquatic habitats of the region.

In a major *Tallahassee Democrat* article ("Growing Oysters," June 2013), the paper presented the option of growing oysters by farming efforts in Alligator Harbor, a small bay east of Apalachicola Bay. Such farming activities had already been tried in Apalachicola Bay and had failed for various reasons. While farming could be developed as a small, family-derived operation, such an effort could not even begin to replace the massive oyster production in the Apalachicola Estuary without a return to previous river flow conditions. Another article in *The Democrat* (May 20, 2013) by a University of Florida bureaucrat touted the efforts of researchers concerning the so-called RESTORE Act funded in part by BP's anticipated Clean Water Act fines for the restoration of the gulf coast after the *Deepwater Horizon*/Macondo oil spill. His comments again ignored the scientific data:

> Closer to home, the commercial harvest of oysters from Apalachicola Bay has declined significantly and for unknown reasons….We could, for instance, learn a lot about how freshwater flows from the

Apalachicola River influence oyster abundance in the bay simply by restoring some oyster bars near the river mouth, where freshwater from the river creates lower salinity habitats and others, farther away from the river mouth, where salinity is higher.

After many years of study, Livingston et al. (2000) noted the following:

Larval and spat stages (of oysters) were not well defined by the statistical results. This finding is not unexpected considering the highly variable nature of these biological features. The fact that spat numbers were not associated with oyster density seems to suggest that higher densities on eastern bars may be more a function of enhanced survival as opposed to higher recruitment. The measured predictors indicated that river-induced habitat changes, along with oyster density, were closely associated with factors such as mortality and bar growth. However, the hydrodynamic model results showed that maximum salinity and salinity variation were important influences on the mortality trends. Current velocities contributed to conditions for maximal bar growth. This appears to be reflected in the location of the highly productive eastern bars at the convergence of currents coming from river-dominated East Bay and high salinity Gulf water coming from St. George Sound. Current convergences may concentrate planktonic food while allowing the relatively low and variable salinity conditions that minimize losses due to predation and disease. Bottom currents appeared also to be a central factor in overall bar growth. This suggests that the movement of the high salinity St. George Sound water over the eastern oyster bars is beneficial to the overall oyster productivity of this area. Data derived from the physical models have thus expanded the scope and detail of our understanding of those factors that affect oyster growth and mortality in the estuary and have provided useful data concerning very distinct patterns of oyster productivity in the Apalachicola system.

Once again, many of the so-called unknown factors listed in the inaccurate articles by the *Tallahassee Democrat* have not only been well researched but have been published in peer-reviewed journals for all to see. Ignoring such information will hardly be beneficial for future oyster restoration efforts.

News concerning environmental events has been dumbed down and milked for the sensational aspects of many issues. Because of the complexity of the scientific data, it is easy to ignore the actual facts. These actions by the news media have contributed to unprecedented levels of public ignorance and indifference to environmental problems. The news media have played up the role of environmental heroes in their kayaks even though these people do not take the time to educate themselves with regard to the available scientific information. The cult of personality enhances the entertainment value of the environment without expanding public knowledge of the real issues.

News media continue to publish diatribes by "climate deniers." The associate editor of the *Tallahassee Democrat* (October 18, 2013) gave the following rationalization of why his newspaper published both sides of everything "scientific" in the "global warming" controversy:

The latest report from the Intergovernmental Panel on climate change says climate scientists are at least 95 percent sure that humans have caused most of the climate change since the 1950s. To borrow a comparison used often, the fact that humans have an impact on climate change is about as settled among scientists as the fact that cigarettes kill. But there's a problem. Some people don't trust scientists. Some say science is just another religion, a belief system. Some say scientists toe the line to keep grant money coming in. Some say climate change is part of a conspiracy based on politics and money. Hey, you and I know the oil companies are the ones with the money— I'm just telling you what they say. In other words, to a not insignificant portion of our population, the facts of climate change are not settled. So, do we simply declare their views invalid and ban them from the opinion pages?....Surely, we have concerns. Deniers can create a sense of "false equivalence," the idea that their view is just as accepted and valid as the prevailing view. Popular Science recently shut down online comments, saying "trolls" could alter the way readers understood a story. But this is not a science magazine, it's an opinion page. Our goal is to offer a marketplace of ideas, not to choose winners and losers on the day's important issues.

This position was backed up in a November 2013 article by a citizen in *The Democrat*:

> Climate change is dangerous, but this is not a good reason to censor mistaken opinions… Censure not censor!….We need to support free inquiry, the marketplace of ideas, debate and the input of experts.

As an epilogue to the string of misleading stories on the Apalachicola disaster, the *Tallahassee Democrat* published a political cartoon by a Rick McGee of *The Augusta Chronicle* on September 24, 2013, that pictured a giant, tattooed oaf holding "The Book of Gore" and stating the following: "There will be stronger and more frequent hurricanes; Global temperatures will rise; The Arctic ice will be gone by 2013." Standing next to him was a scrawny Chicken Little holding a sign that proclaimed "The sky is falling" and saying "…and they call me crazy."

Of course, we should not omit the public response to these issues. The following letter to the editor was published on August 3, 2013, by the *Tallahassee Democrat*:

> We shouldn't waste more on global warming trends
>
> Disregarding the fact there has been no global warming the last 15 years or the truth of the East Anglia emails: Suppose we embraced it anyway. We build a Department of Global Warming with budgets higher than the useless Department of Education's annual $65 billion. Maniacal zealots pounce on businesses with carbon-credit demands. The IRS will enforce, punish and extort. Gore, Soros and Wall Street will build their $10 trillion carbon exchange for fortunes. Despite our $17 trillion debt and poor economy: Fanatics will crush industries and jobs because man is environmentally evil and must be ruled. Public servants don't exist at the federal level. They are elite rulers with no accountability. Give them the unlimited excuse— "the world is being destroyed by global warming"— and they will dictate with impunity. Government will have unlimited and perpetual war-type powers. Individual freedom will have run its course in human history. We simply cannot fall for this ruse to endless tyranny.

And finally, a recent article in the *Tallahassee Democrat* gave an even more erroneous twist to the ongoing problems in the Apalachicola Estuary (Freshwater's return promising for oyster recovery: Near-record rainfall bodes well for ailing oyster population. The *Tallahassee Democrat*, January 12, 2014):

> Areas that feed freshwater into the Apalachicola Bay have seen record rainfall numbers over the past year after a severe drought in 2012. Climatologists say the reduction of freshwater was the main cause of the declining oyster population…After wide-reaching droughts throughout the Apalachicola-Chattahoochee-Flint Basin between 2009 and 2012, record rainfall is now flooding the freshwater-depleted region, which could bode well for the ailing oyster population in the Gulf…Florida State Climatologist David Zierden said Florida, South Carolina and areas of Georgia north of the Jim Woodruff Dam in Chattahoochee have seen near-record rainfall in the last 6 months of 2012 and throughout 2013. "We've had an unusual amount of rainfall in the drainage basin," he said. "So it certainly will be beneficial to the recovery of the estuaries and the oyster fishery."
>
> Zierden said the rebound really started during December 2012 as North America came out of a La Nina weather pattern that cast dry, cooler air over the Southeast and continued through the past summer. Helen, Ga., at the headwaters of the basin, which drains down to the Apalachicola Bay, had more than 101 inches of rain last year, he said.

According to data showing the daily flows of the Apalachicola River from 2006 to 2012 (Figure 10.1), the actual drought periods in terms of river flows were from 2006 to 2008 and from 2010 to 2012 (not 2009). There was no recovery of river flow during "the last 6 months of 2012," and the "rebound" of rainfall in December 2012 was not followed by high river flows during what amounted to a month with some of the lowest such flow rates during the long-term drought period.

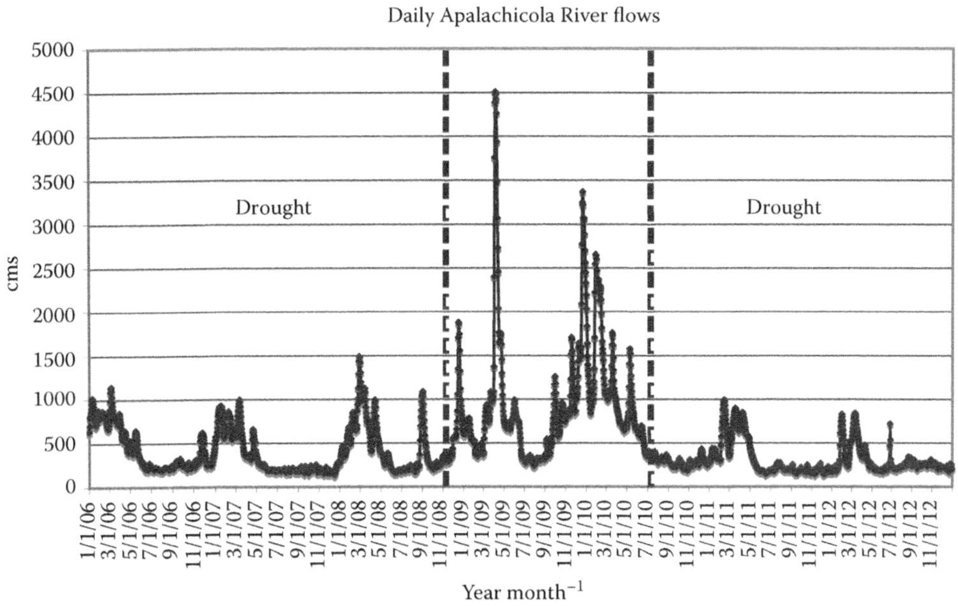

Figure 10.1 Daily Apalachicola River flows (Sumatra) from 2006 through 2012 along with delineation of drought periods. River flow data are provided by the US Geological Survey Sumatra gauge # 02359170.

The recovery of rainfall in 2013 could be associated with increased river flows, but even so, such a situation should be viewed as part of the overall (long-term) record of river flows. Information presented in the news feature represented a weather feature, not a climate change. There was an increase in such flows during 2009–2010 that occurred between two devastating droughts that resulted in major impacts on the Apalachicola drainage. There were periods of rain between the droughts: it is the trend of the increased frequency and severity of droughts that is troubling. Even with increased river flows during 2013, previous data with respect to the recovery of the oyster populations indicated that there would not necessarily be an immediate recovery of the usual growth levels of Apalachicola oysters. In addition, the concentration on oysters in the paper's series of Apalachicola articles is an oversimplification of the problems in the Apalachicola Estuary where losses of the penaeid shrimp are important. The shrimp industry represents a higher source of revenue than that of the oysters. The same lack of a mention of problems with the blue crab and finfish industries was also typical of the flawed stories in the *Tallahassee Democrat.*

As noted in the previous chapter, the integrated efforts by federal, state, and local agencies, together with long-term scientific investigations of the Apalachicola River and Estuary, were used to construct one of the most progressive resource management programs in the country. This effort was based on scientific facts and the translation of such information into realistic plans for the protection of a valuable natural resource. Anthropogenic nutrient loading was prevented and the entire river-estuarine wetlands in the Apalachicola basin were preserved. The results of the research presented in this book indicate that changes in climate in the form of sustained droughts over the past 30+ years are part of ongoing adverse impacts on estuarine productivity not only in the Apalachicola ecosystem but also in other river–estuaries along the NE gulf coast. The almost constant issuance of flawed information by regional news media along with the omission of important scientific data concerning the effects of recent long-term changes of climate represents a disservice to a public that remains misinformed and uninformed on significant environmental events.

It is true that no one can predict the future of the Apalachicola River flows. The research showing ecological impacts of the growing drought trend is not meant to be predictive. However, unless there is a recognition of what has happened over the past 30 years, the issue will continue to be ignored with no effort to address what could result in further losses of coastal productivity. Any recovery programs for the Apalachicola River and Bay should address the long-term trends of droughts that have resulted in losses of major parts of what was once a model of fisheries productivity. If it doesn't rain as it used to in the past, no one in the tri-river region will get enough water for their needs. To use this situation for political gain through overtly crude machinations and manipulation of the press is not only dishonest but is shortsighted in the continued maintenance of current public ignorance of one of the most important issues of our time. The noted changes of the climate go beyond the Apalachicola and include an entire region with effects extending along the NE gulf coast. The importance of the recent, extended trends of key climatological factors has been largely ignored by just about everyone, including the scientific community, but the ecological data in this book comprise evidence that climatological changes have already had an adverse impact on some of the most productive aquatic resources of the region.

PART VII

Closing

Conclusions

The results of a series of long-term (1970–2012), interdisciplinary studies of coastal systems along the northeastern (NE) Gulf Coast (Apalachee Bay, Apalachicola Estuary, Choctawhatchee Estuary, Escambia Bay, Blackwater Bay, Perdido Estuary) led to a relatively different approach from the usual field analyses of coastal areas. In this case, the reference to "long-term" involved decades (not months or years) in terms of project duration, and "interdisciplinary" meant an integration of the various physical, chemical, and biological field data that were taken in a carefully organized fashion. Sampling methods were quantified and standardized, and the biological data were analyzed to species. Based on extensive stomach content analyses and determinations of ontogenetic feeding units, the field data were then transformed into an overall trophic organization so that ecologically diverse river–estuaries could be compared.

Long-term, interdisciplinary ecological studies of major estuaries are not commonly undertaken. Such studies are expensive. Holding together a team of scientists, taxonomists, field-sampling crews, database specialists, and statistical/modeling personnel for decades represents a complex and difficult task. When funding is available, many such studies have often degenerated into a feeding frenzy of competing universities and research organizations where political concerns (not scientific considerations) have dominated the division of the work along territorial lines. There has been little attention over the years for the development of interdisciplinary field research, and the increasing emphasis on specialization has not produced investigators who are well versed in the needs and pitfalls of such research. Many scientists object to participating in cooperative field efforts with people outside their own specialties. The increased level of specialization in the environmental sciences has exacerbated this particular problem.

There are special considerations in the design of an interdisciplinary program that includes what to sample in the field and when and where to carry out the sampling. This requires a precise set of objectives for the study in question, a requirement that is seldom defined in a detailed fashion, especially with regard to the ultimate meshing of the different forms of field data into a cohesive base that can be analyzed with available statistical methods and modeling efforts. Studies show that success usually depends on one person who maintains control over the overall operations of the disparate parts of the research program. However, in most instances, a committee controls the work, the membership of which can be subject to change over the course of the study. This usually leads to a lack of a consistent focus of the study. Longevity of the principal investigators appears to be an important component in the development of a successful ecosystem effort since replacement of leading scientists in such studies is often accompanied by abrupt changes in program objectives and methods. Consistency here is not the "hobgoblin of little minds."

There are other problems with the initiation of such research. For example, there is the long-term delay in publication of the results that presents an obstacle for many principal investigators who labor under exigencies concerning the "publish or perish" phenomenon. The relatively monotonous sampling in the field involves more than just a so-called monitoring scheme. Merely taking endless data points (the "sorcerer's apprentice" approach) is not useful in the development of meaningful determinations of ecological variability and long-term trends of ecosystem changes. A continuous review of the data is necessary here to account for effectiveness of the field efforts. The initial sampling effort should include a comprehensive list of variables that can be reduced in time. Often, the reverse method of adding variables later is the rule. The ever-worsening lack of taxonomic experts needed for such studies presents another obstacle to the successful conclusion of long-term biological sampling efforts, especially when the work extends over periods of decades rather than years.

The main research effort of most investigators in coastal systems is usually carried out over a period of 1–3 years. Of course, there is the problem of environmental scientists who simply ignore the implications of how long-term trends of climatological factors affect the results of such short-term studies.

The fact that everything might change with the onset of the next drought or prolonged storm does not enter into the discussion. Among some circles, the taking of experimental field data is done in a way that just enough data are taken to test a favorite hypothesis at which time such collections are ended before the evidence becomes less conclusive. In any case, the question of whether or not a 2-year study is representative of ecosystem processes over longer periods rarely comes up.

An understanding of the interactions of short- and long-term changes of climatological factors is critical to impact evaluations and restoration efforts in coastal ecosystems. The scientific literature often reads like the efforts of thousands of blind specialists grasping ever-smaller parts of the elephant. Each year, the various ecological societies meet and literally thousands of researchers give an interpretation of their elephant based on their grasp of a small piece of the whole. Even the basic structure of such meetings by discipline precludes the advance of interdisciplinary efforts to a better understanding of how coastal and marine systems work. The reductionist mentality dominates, and the patch-quilt results have little to do with the tapestry of the long-term changes of some of the most complex ecosystems on earth.

In short, the lack of the successful development of long-term, interdisciplinary studies of coastal and marine systems is not difficult to understand considering past history. Unfortunately, the need for such studies is greater than ever before because of climate change. The following is a list of conclusions that can be gleaned from the long-term studies of riverine and estuarine systems in the NE Gulf of Mexico:

1. The long-term rainfall and river flow patterns of areas from Apalachee Bay to the Apalachicola Estuary along the NE Gulf Coast were similar with marked increases of the frequency and severity of droughts from 1980 to 2012 (1980–1981, 1986–1988, 1999–2002, 2006–2008, 2010–2012). Rainfall/river flow conditions in the Pensacola and Perdido Estuaries differed somewhat from estuaries to the east, with the drought trends extending from 1999 to 2012. There were increases in storms over the 30-year period that, together with the droughts, led to increased ranges of variability between low and high drainage events in time.
2. Apalachee Bay is one of the least polluted coastal ecosystems in the country. This high-salinity bay has no major rivers and is characterized by limited freshwater inflows from a series of small streams along the marsh-driven Florida west coast. The primary source of habitats and primary productivity is submerged aquatic vegetation (SAV) that is dominated by several species of seagrasses and a number of macroalgal species. Phytoplankton productivity in Apalachee Bay is low compared to the river-dominated estuaries to the west. The only point source of pollution on the bay is a pulp mill on the Fenholloway River. A comparison with an unpolluted blackwater system (Econfina River–Estuary) indicated adverse impacts of the pulp mill on the river, coastal marshes, and SAV in offshore areas. The loss of SAV was associated with long-term reductions of fish and invertebrate populations in the Fenholloway Estuary compared to those in the reference estuary.
3. Long-term changes of temperature and rainfall/river flow in Apalachee Bay included increased frequency and severity of droughts and associated periods of heavy rainfall that led to destabilization of water quality in offshore habitats. These trends were associated with complex changes of the biological organization of the offshore Econfina and Fenholloway estuaries over time. Increased water temperatures led to the initiation of sexual reproduction of seagrasses. Compared to the drought-free conditions during the 1970s, there were major increases of the species richness of macroalgae as the dominant turtle grass (*Thalassia testudinum*) biomass decreased by the end of the study (2005). Fish biomass decreased following the prolonged period of increased droughts and storms, whereas invertebrate abundance and species richness increased. These changes indicated complex biological responses in Apalachee Bay to the long-term climate changes. If the replacement of natural populations of seagrasses by dense infestations of macroalgae has changed the habitat so that invertebrates are no longer available as food for the fish populations, this could result in the loss of fish productivity. If such changes are widespread and continue to occur in the future due to alterations of the major climatological factors, these changes in the trophic organization of Apalachee Bay could have profound effects on the sports fisheries of the region. Previous experimental studies indicate impacts of macroalgae infestations on the long term productivity of seagrasses and associated birds. More research is needed to establish the extent of the observed changes.

4. The seasonally mediated high flows of the large rivers along Florida's west coast were associated with the high productivity of the receiving estuaries. Freshwater and estuarine species in the Apalachicola Estuary were adapted to a moderate range of natural interannual drought and flood flow cycles during the 1970s, a period that was relatively unaffected by major droughts. These flow rate fluctuations were consistent with those of previous decades with the exception of drought conditions during the mid-1950s. There were temporal successions of coastal species that were adapted to the seasonal and long-term changes of rainfall and river flow events. Conditions in the Apalachicola River–Estuary during this period represented a model of ideal conditions for maximal coastal development of estuarine populations and associated sports and commercial fisheries.
5. The high productivity of the Apalachicola Estuary depended on the seasonally directed loading of nutrients and organic matter from the river and the resulting autochthonous production of phytoplankton during the river flow cycles. Loading of particulate organic carbon fueled the dominant detrital food webs of the estuary. Varying salinities as a result of river flows reduced predation by stenohaline offshore gulf species on the euryhaline estuarine populations. Overall, the predrought Apalachicola Estuary had the highest rates of secondary production in the NE Gulf Coastal region. The resulting food webs provided for the base of important regional fisheries (oysters, penaeid shrimp, blue crabs, sciaenid fishes).
6. Oyster production in the Apalachicola Estuary was high and was characterized by dense populations of rapidly growing bivalves. Variable salinities reduced the impacts of predation and disease on the oysters. High oyster production in the estuary occurred in areas subjected to a convergence of tannin-stained, nutrient-enriched surface water from East Bay (i.e., influenced by the Apalachicola River) and high-velocity, high-salinity bottom water moving westward from St. George Sound. The convergence of nutrients due to the currents of the estuary led to high phytoplankton concentrations that served as a food source for the highly productive oyster beds with maximum growth occurring during periods of low water temperature and high salinity. The productivity of invertebrate and fishes was enhanced by different spatial and temporal components of the same factors that fueled the high level of oyster productivity.
7. A sustained effort was made to successfully translate the scientific results into a comprehensive system of resource planning and management for the Apalachicola ecosystem. This resulted in the development of one of the most comprehensive river–estuarine conservation plans in the country.
8. In Choctawhatchee Bay, the widespread occurrence of salinity stratification due to an artificial opening of the bay to the gulf was associated with reduced dissolved oxygen (DO) at depth, adverse impacts on sediment quality, and the loss of secondary productivity in large parts of the bay. The development of marshes in the bay was the lowest of all the estuaries in the area of study following the loss of such wetlands after the introduction of high-salinity water from the gulf. Anthropogenic nutrient loading from newly urbanized western parts of the bay led to plankton blooms that were associated with reduced water quality, massive fish kills, and the deaths of sea turtles, bottlenose dolphins, and land animals.
9. In the Choctawhatchee Estuary, biological indices such as fish, infauna, and invertebrate (FII) biomass, numbers of taxa, and trophic unit numbers were positively associated with bottom DO and surface pH. These indices were negatively associated with salinity stratification, sediment silt concentrations, and surface water color, turbidity, and chlorophyll *a*. Salinity stratification, combined with plankton blooms due to urban runoff in western sections of the estuary, resulted in an overall low level of secondary production compared to other estuaries in the NE Gulf.
10. Escambia Bay of the Pensacola Estuary was subject to continuous stress for decades due to runoff from urbanized areas and point source pollution from industrial activities and sewage treatment plants. Due to anthropogenic nutrient loading, the most extensive fish kills in the scientifically documented literature occurred in Escambia Bay during the late 1960s and early 1970s. The resulting hypereutrophication destroyed oyster reefs and seagrass beds and adversely affected sediment quality. The estuarine fisheries did not fully recover with time, even when the anthropogenic nutrient loading was reduced. Formerly productive oyster bars were no longer commercially viable, although a small oyster fishery still existed. The Pensacola Estuary remained largely bereft of seagrass beds. These trends indicate that recovery from major habitat changes due to urban nutrient loading and plankton blooms does not necessarily follow as a direct reversion back to

natural conditions with reduction of anthropogenic nutrient loading. The long-term trends in Escambia Bay indicated that recovery of estuaries after restoration efforts might not include a return to previous conditions of high secondary productivity.

11. Based on a recent study, Escambia Bay was not in a hypereutrophic state. However, the plankton community was dominated by a number of bloom species in the lower estuary. The main concentrations of infauna (polychaete worms), epibenthic macroinvertebrates (brown shrimp, blue crabs), and fishes (sciaenids) were found in the relatively shallow parts of upper Escambia Bay that continued to be the center of primary and secondary production in the Pensacola Estuary due to nutrient loading from the Escambia River. Nonpoint pollution from the City of Pensacola and other urbanized areas of the Pensacola Bay basin, together with major (toxic) pollution of the bayous in this region, presented serious problems in terms of multisource urban contamination. The trophic organization of the Pensacola Estuary was disrupted by the impacts described earlier, with disjunct relationships of ecological efficiency and altered food web processes that were associated with the losses of secondary productivity.
12. A 20-year study was carried out in the Perdido Estuary. A dredged pass to the gulf during the 1920s led to similar adverse impacts on habitat disruption and benthic productivity as those described for the Choctawhatchee Estuary due to resulting salinity stratification, low benthic DO, and reduced sediment quality. After an initial period of relatively natural conditions in the bay from 1988 to 1991, increased orthophosphate and ammonia loading from a pulp mill into Elevenmile Creek and the bay generated a series of plankton blooms dominated by diatoms, raphidophytes, dinoflagellates, and cyanobacteria. Blue-green algae blooms (*Merismopedia tenuissima*) in the receiving Elevenmile Creek were enhanced by drought conditions and increased nutrient concentrations that were statistically associated with reduced infaunal populations in the creek. Raphidophyte (*Heterosigma akashiwo*) blooms caused by the increased nutrient loading from the pulp mill and a sewage treatment plant in upper Perdido Bay during periods of storms and increased runoff were statistically associated with reduced invertebrate and fish populations and disruptions of bay food webs. Both the blue-green algal and raphidophyte blooms were driven by climatological conditions of drought and flood.
13. The climatological control of the outbreaks of plankton blooms in the Perdido Estuary was the result of complex interactions of nutrient loading conditions and the life histories of the plankton. The seasonal and interannual timing of climatalogical factors determined the response of the plankton populations. There were adverse effects on the overall secondary productivity of the Perdido Estuary due to recurring droughts and reduced nutrient loading from the Perdido River during the second half of the study. The trophic organization of the bay was disrupted by serial impacts due to plankton blooms and droughts. There were associated disjunct relationships of ecological efficiency and major losses of dominant populations of infauna, invertebrates, and fishes. There appeared to be a cumulative impact of the blooms and droughts on the bay through time with progressive loss of resilience of the bay to these climatologically controlled processes. In this way, the combination of various anthropogenic impacts on the estuary led to the almost complete elimination of secondary productivity in Perdido Estuary by the end of the study in 2007.
14. A comparative analysis of the various estuaries was carried out. Depth and salinity stratification due to dredged openings to the gulf were highest in the Choctawhatchee and Perdido Estuaries. Associated liquid mud formation and warm-season hypoxia at depth resulted in the reduced quality of benthic habitats in major parts of the Choctawhatchee and Perdido Estuaries. The dredged cuts had permanent, adverse effects on secondary production and the trophic organization of these estuaries. Salinity stratification was relatively low in the Apalachicola Estuary and Apalachee Bay and was intermediate in the Pensacola Estuary. Local effects of a dredged opening to the gulf in Apalachicola Bay were associated with the loss of productive oyster beds in the bay. The adverse effects of dredged passes to the gulf in various river-dominated estuaries were more or less permanent.
15. The impacts of climatological trends on the subject estuaries were varied due to differences in the natural habitats, differing forms of anthropogenic activities (dredging, nutrient loading, pulp mill effluents), and the reaction of the estuaries to changes in water quality and reduced nutrient loading during droughts and floods. Both the Apalachicola and Perdido Estuaries underwent losses of secondary productivity due to reduced loading of nutrients and organic matter during droughts. The droughts had a major adverse impact on the formerly very high productivity of the Apalachicola Estuary with the loss of major regional commercial fisheries. The cumulative impacts of the various

anthropogenic activities in the Perdido Estuary, with associated reductions of estuarine resilience over time, resulted in the lowest levels of secondary productivity noted in the various estuaries along the NE Gulf Coast. The impacts on the Apalachee Bay Estuaries were more complex, and areas outside of the impacts of pulp mill effluents in the Fenholloway Estuary responded to the interplay of droughts with increased storm activity with complex changes in biological organization. In the unpolluted Econfina Estuary, the droughts and increased storm activity led to changes in water quality and the loss of turtle grass along with increased biomass of macroalgae, reduced fish biomass, and increased invertebrate numbers. In the Fenholloway Estuary, there were changes in the planktonic food webs with increased plankton blooms and reduced numbers of plankton-feeding fishes such as bay anchovies. These changes reinforced the observation of differing responses of estuaries to changes in climate factors based on existing differences in habitat and trophic organization.

16. The Apalachicola River–Estuary during the 1970s represented an ideal ecological situation that maximized the production of oysters, blue crabs, penaeid shrimp, and sciaenid fishes leading to the highly productive fisheries in the study area. Combined with the reduced nutrient and organic carbon loading of the river and increased salinity during droughts, the food webs were disrupted leading to losses of various populations of oysters, penaeid shrimp, blue crabs, and fishes. The reduction of herbivore and omnivores due to reduced loading of nutrients and organic matter from the Apalachicola River was a key element in the trophic response to drought conditions. Detritivore food webs were adversely affected due to reductions of particulate organic matter. The C_1 and C_2 carnivore biomass levels followed the trend of reduced herbivores and omnivores that were directly affected by the altered estuarine conditions. The delayed recovery of secondary productivity after the end of a given drought was probably due to the processes involved in the rejuvenation of the phytoplankton and detrital parts of the trophic organization of the Apalachicola Estuary. These delayed recovery times included oyster growth rates and population recovery times of fishes and invertebrates. Increased drought frequency in the future could lead to prolonged recover times and estuary loss of overall frequency potential.
17. A comparative analysis of the trophic organization of the various river-dominated estuaries along the NE Gulf Coast allowed a comprehensive review of how the different estuaries responded to anthropogenic nutrient loading and the long-term changes of key climatic conditions in the study region. The highest levels of herbivores and omnivores occurred in the Apalachicola Estuary and in the unstratified (shallow) parts of the Perdido and Escambia Estuaries. The distribution of these components was related to the abundance of allochthonous and autochthonous plant matter in the large river–estuaries. The C_1 and C_2 carnivores had the highest biomass in the Apalachicola Estuary compared to the other estuaries. The highest levels of overall secondary productivity were found in those areas with the higher levels of herbivores and omnivores. The overall trophic organization of the respective estuaries was based on complex interactions of natural habitat conditions, anthropogenic physical alterations, and nutrient loading.
18. The Apalachicola Estuary was the most productive of all the areas of study. The higher flow rates of the Apalachicola River probably contributed to this situation, although the relative flow factor could not explain the low levels of secondary productivity in the other river-dominated systems. The Blackwater Estuary had limited river input relative to the other systems. The Perdido and Choctawhatchee Bay Estuaries were adversely affected by widespread salinity stratification due to the artificial openings to high-salinity gulf waters. In addition, both systems were exposed to anthropogenic nutrient loading (industrial/sewage treatment facilities and urban/agricultural stormwater runoff, respectively) that resulted in damaging plankton blooms. The Escambia Estuary had been subject to a past history of hypereutrophication from industrial and sewage loading and the effects of toxic agents and urban stormwater runoff. The increasing occurrence of severe droughts in the region had major impacts on the FII trophic indices of the Perdido and Apalachicola Estuaries and the biological organization of Apalachee Bay. By the end of the study, all of the large river–estuaries evinced damaged food webs and major losses of the commercial fisheries that had once been highly productive during previous periods of less anthropogenic impacts.
19. An analysis was carried out concerning the differences in the ecological efficiency of biomass along the estuarine food webs of the subject estuaries. In the Apalachicola Estuary during the early periods of regular changes of river flows prior to increased drought occurrence, there was a pattern of efficient

transfer of energy up the bay-wide food webs with somewhat increased C_2/C_1 transfer efficiencies in the upper parts of such webs. The Perdido Estuary from 1988 to 1991 had similar efficiency trends as those in the Apalachicola Estuary. During periods of blooms (2004–2005) and droughts, these efficiencies became disjunct with no consistent patterns. The Choctawhatchee food web efficiencies were strongly anomalous, particularly in the upper bay where C_1 carnivores were conspicuously higher than the herbivore biomass levels. This pattern was even more skewed in the Blackwater Estuary with a similar C_1/Herbivore anomaly as that in the upper parts of the Choctawhatchee Estuary. This pattern was also evident in the Escambia Estuary. The changes that took place at the end of the Perdido and Apalachicola study periods due to bloom and/or drought effects suggested energy efficiencies that went through basic alterations when compared with those of the original estuary models.

20. The relatively high productivity of the Apalachicola Estuary was due to several factors. The estuary was shallow, had low stratification features, and was in a generally natural state with no signs of cultural eutrophication due to anthropogenic nutrient loading. The naturally high levels of nutrient and organic carbon loading from the Apalachicola River enhanced the biomass of herbivores and detritivores at the base of the estuarine food webs. The carnivore groups were driven by this increase of the lower levels of the food web, a process that resulted in optimal ecological efficiencies, high levels of secondary productivity, and major regional fisheries. During the early years of sampling, these mechanisms were operational in the Perdido Estuary with the result of a relatively high FII biomass m^{-2} and a balanced ecological efficiency in the shallow upper bay regions that received high nutrient and organic carbon loading from the Perdido River. The stratification factor reduced the FII biomass in deeper parts of the Perdido Estuary, but the mean FII productivity still exceeded that of the other river-dominated estuaries that had less well-developed food web bases due to the adverse effects of anthropogenic activities.
21. The integration of data from the different estuaries can be summarized in a basic model of climate impacts on major river–estuarine systems. Prolonged droughts cut off the loading of inorganic nutrients and organic matter in both the Apalachicola and Perdido Estuaries. Water color and turbidity in the estuaries decreased and the water became clearer. These conditions, together with the remaining nutrients, stimulated major increases in the phytoplankton and benthic microphytes. There was a transient increase in herbivore and detritivore biomass. However, as the nutrients and organic matter were consumed without replacement, there were major reductions of the biomass of the herbivores and detritivores that were followed by decreases of the carnivores, a situation that led to reduced biomass of dominant species, and disjunct ecological efficiencies as the disproportionate loss of the lower parts of the food webs led to altered trophic ratios. A sudden increase in rainfall and river flow after the droughts in estuaries such as the Perdido, with enhanced anthropogenic nutrient loading, resulted in plankton blooms that further decreased the secondary productivity. These repeated impacts over time led to reduced resilience of the affected estuary. As was the case in Escambia Bay, this could lead to the more or less permanent (decades or more) loss of habitats such as oysters, seagrass beds, and sediment quality. In this way, the interactions of natural habitat conditions, complex food web processes, and anthropogenic activities (dredging, nutrient loading, climate changes) combined to alter the trophic organization of the various estuaries and reduce useful productivity.
22. The results of the research presented in this book indicate that changes in climate in the form of sustained droughts over the past 30+ years are part of a threat not only to the Apalachicola ecosystem but also to river–estuaries along the NE Gulf Coast. The almost constant issuance of flawed information by regional news media along with the omission of important scientific facts concerning the effects of long-term changes of climate on these coastal ecosystems represents a disservice to a public that remains uninformed and misinformed concerning significant environmental events.
23. Any recovery effort for the Apalachicola River and Bay should address the climatological trends that have resulted in losses of major proportion of what was once a model of fisheries productivity. Associated problems in the various estuaries along the coast should also be addressed. The importance of long-term changes in coastal ecosystems has been largely ignored by just about everyone, including the scientific community, but the data in this book represent an ecological view that such changes need to be understood in order to evaluate and address long-term trends of climate change. The data indicate that such changes are not only real, but have already happened, albeit at a prolonged pace, along the NE Gulf Coast.

Appendix A: Field/Laboratory Methods Used for the CARRMA Studies (1970–2012)

The research effort in the northern Gulf of Mexico is based on written, peer-reviewed protocols for all field and laboratory operations. Water quality and biological analyses have been taken by personnel from the Center for Aquatic Research and Resource Management (CARRMA, Florida State University, Tallahassee, FL) and Environmental Planning and Analysis, Inc. (Tallahassee, FL) with additional work carried out at a series of laboratories in various countries. Water quality methods and analyses and specific biological methods have been continuously certified through the Quality Assurance (QA) Section of the Florida Department of Environmental Protection (Comprehensive QAP #940128 and QAP#920101). Various methods that have been used to take field information and run laboratory analyses have been published in the reviewed literature. This includes the following: Flemer et al. (1998), Livingston (1980a, 1982a, 1984a,b, 2000, 2002, 2005), and Livingston et al. (1974, 1976a,b, 1997, 1998a,b, 1999, 2000, 2003).

SEDIMENT ANALYSES

Sediments were analyzed for particle-size distribution and organic composition according to methods described by Mahoney and Livingston (1982) as defined by Galehouse (1971).

Particle-Size Analysis

Sediment samples were taken with coring devices (Plexiglas: 7.6 cm d, 45 cm^2 cross section). Analyses were taken on 10 cm samples for regular percent organics and particle-size analysis:

(a) Unpreserved sediments were divided into coarse (>62 μm) and silt-clay (<62 μm) fractions.
(b) The coarse fraction was analyzed by wet sieving using 1/2 phi unit intervals.
(c) The silt-clay fraction was analyzed using a pipette method in 1/2 phi unit intervals from 4 to 6 phi and in 1 phi intervals for the finer fractions (6–10 phi).
(d) Statistical analyses were carried out using phi units and sediment designations as outlined in the Wentworth Scale.
(e) Statistical treatments of the granulometric results followed the method of moments for the calculation of mean grain size (in phi units, with 1 mm = 0.5 phi), skewness (a measure of nonnormality of distribution), and kurtosis (a measure of the spread of the distribution). Sorting coefficients represented the measure of grain size as follows: well sorted = 0.50; moderately well sorted = 0.71; moderately sorted = 1.00; and poorly sorted = 2.00.

Percent Organics

After placing the sediments in a drying oven set at 104°C for a minimum of 12 h, the sediments were ashed in a muffle furnace for 1 h or more at temperatures approximating 550°C. The percent organic fraction of the sediments was calculated as a percent of the dry weight.

Sediment analysis is a two-part process where the first consists of forming a data file and the second is the actual granulometric analysis. The first part may be accomplished by either running the EZseds program to enter and reformat raw data or by exporting previously entered sediment data from one of the Environmental Planning & Analysis, Inc. (EP&A) 4th Dimension Data Management System (DMS) sediment data files. EZseds is a FORTRAN program that prepares raw sediment granulometric data for analysis with the MacGranulo program. MacGranulo is a FORTRAN application that performs sediment grain-size analysis on the data in the input file. Results include summary statistics such as mean grain size (as determined by three different methods) and percent organics, particle-size distribution according to the Wentworth classification, and a frequency distribution plot.

Mercury Analyses

Sediment samples were collected at a series of sampling sites in the system. Three subsamples were taken at each established station. Sediment samples were placed on ice. The samples were shipped to EnviroTest Laboratories in Canada. Prior to shipping, all samples were logged into chain-of-custody forms and a sample logbook. Samples were examined to ensure that accurate labels had been affixed and that all lids were tight. Prior to closing each cooler containing samples, the chain-of-custody form was completed and placed in a ziplock bag that was then placed inside the cooler. The forms were placed in plastic bags. The originator retained a copy of each chain-of-custody form. As a general procedure, samples were taken in order from supposedly clean areas to possible polluted areas. No sampling was carried out in the rain.

Quality control samples (blanks) were labeled as the previous samples. After collection, identification, and preservation, the sample was maintained under chain-of-custody procedures discussed later. A blank consisted of an empty sampling container. For each group of samples (inorganic mercury, methyl mercury), one blank was used for each lot number of jars.

Samples of sediments were taken with corers (the top 2 cm was used for mercury analysis; the top 10 cm was used for sediment particle size and percent organics). A polyvinyl chloride (PVC) corer was used for the mercury samples. Chemical analyses were run on sediments taken from each of the sampling stations. Three replicate sediment samples were taken at each station for total mercury and methyl mercury analysis. Each of the three replicates was composed of three core subsamples that were cut into squares (all edges "shaved") to eliminate the outer parts of the core sample. The "shaving" of the sample was carried out with an acid (5%–10% nitric acid)-washed plastic knife. Samples were placed in acid-washed one pint, wide neck, Teflon-lined screw-top glass jars. Nitric acid (10% pesticide grade) was used to wash the jars. Samples were immediately preserved in the field with concentrated nitric acid to pH 2. Care was taken to mix the sediment and check for the proper pH. Each bottle was labeled as indicated previously. Acid preservation for each sample was checked with pH paper to make sure that the pH 2 target is maintained. Samples were then placed on ice and bubble wrap.

Sampling for toxic agents in sediments in other studies followed similar procedures. All sediment samples were taken with corers (the top 2 cm). A stainless steel corer was used for the organic toxicant samples and a PVC corer was used for the metal samples.

PHYSICOCHEMICAL SAMPLING AND ANALYSIS

Designated, fixed stations in a given system were sampled (surface and bottom) for water quality factors. Vertical collections of water temperature, salinity, conductivity, dissolved oxygen, depth, Secchi depth, and pH were taken at the surface (0.1 m) and at 0.5 m intervals below the surface to about 0.5 m above the bottom. Surface and bottom water samples were taken with Niskin bottles for further chemical analyses in the laboratory. All sampling was carried out on a synoptic basis with all samples in a given system taken within one tidal cycle.

A list of variables is given in the following:

1. Physical/Chemical Factors
   a. Temperature
   b. Salinity/conductivity
   c. Dissolved oxygen
   d. pH

   - e. Secchi depth
   - f. Depth
   - g. Light transmission
   - h. True color, National Council for Air and Stream Improvement (NCASI) colorimetric method
   - i. True color, spectroscopic method
   - j. Turbidity, nephelometric method
   - k. Total dissolved solids
   - l. Total suspended solids
   - m. Dissolved oxygen, YSI DO meter
   - n. Particulate organic carbon
   - o. Dissolved organic carbon
   - p. Total inorganic carbon (IC)
   - q. Total organic carbon
   - r. Total carbon
   - s. Chlorophyll *a*
   - t. Chlorophyll *b*
   - u. Chlorophyll *c*
2. Nutrients
   - a. Nitrogen nitrate and nitrite
   - b. Ammonia nitrogen
   - c. Organic nitrogen and organic phosphate
     - Total organic nitrogen and total organic phosphate
     - Dissolved organic nitrogen and dissolved organic phosphate
     - Particulate organic nitrogen (PON) and particulate organic phosphate (POP)
   - d. Total nitrogen
   - e. Total phosphorus
   - f. Orthophosphorus
   - g. Dissolved reactive silicate

Physical/chemical data and water samples for chemical analysis were taken according to methods defined by the US Environmental Protection Agency (EPA) (1983). Temperature and salinity were measured with YSI Model 33 S-C-T meters calibrated in the laboratory with commercial standards. Dissolved oxygen was taken with YSI Model 57 oxygen meters calibrated by the azide-modified Winkler technique. The pH was measured with an Orion Model 250A pH meter equipped with a calomel electrode. Aquatic light readings were taken with a Li-Cor LI-1800UW underwater spectroradiometer, which measures the spectral composition of photon flux density at 1–2 nm intervals from 300 to 850 nm.

A Beckman DU-64 spectrophotometer was used to analyze true color (APHA, 1989). Both ratio and nephelometric turbidity analyses were carried out. A ratio turbidimeter and a Hach model 2100A turbidimeter were used for turbidity analyses. Quantification of dissolved inorganic nitrogen (nitrite, nitrate) and dissolved inorganic phosphorus (orthophosphate) followed methods outlined by Parsons et al. (1984). Ammonium was measured with an ion electrode (US EPA, 1983, method 350.3) after raising the pH to 11. This method had some essential features (e.g., minimal interference from waters highly stained with humic materials and paper mill effluents) that required special attention to the results. Consequently, the level of detection was relatively high (e.g., 2.0 μm ammonium-N) but adequate for this study. PON (US EPA, 1983, method 351.4) and POP (US EPA, 1983, method 364.4) were collected on Gelman glass fiber filters combusted at 500°C, oxidized to inorganic fractions and organic N measured as ammonia and organic P measured as POP. Total dissolved solids (APHA [American Public Health Association] Method 209-B), total suspended solids (APHA Method 209-C), dissolved organic carbon (APHA Method 415.1), chlorophyll *a* (APHA Method 1002-G), and BOD (APHA Method 405.1) were carried out according to the methods outlined by the APHA (1989). Particulate organic carbon was analyzed according to Parsons et al. (1984) (method 3.1). Nutrient analyses were carried out according to APHA (1989).

QA

QA Objectives

#	Parameter	Method	Matrix	Precision (%RSD)	Accuracy (sd)	Completeness (%)
1	Ammonia (D)	*SM 4500-NH3-F*	L	10	1	100
2	B. O. D.	*EPA 405.1*	L	10	1	100
3	Chloride	*EPA 325.3*	L	10	1	100
4	Chlorophylls	*PML 4.1*	L	10	1	100
5	Conductivity	*EPA 120.1*	L	5	.2	100
6	Depth	*EPA*	L	5	.2	100
7	Diss. oxygen	*EPA 360.1*	L	5	.2	100
8	Effluent color	*NCASI 253*	L	12	5	100
9	Inorg. phos. (D)	*SM 4500-P-E*	L	10	1	100
10	Mercury	*SM 3112-B*	S/L	10	1	100
11	Nitrate (D)	*SM 4500-NO3-E*	L	10	1	100
12	Nitrite(D)	*SM 4500-NO2-B*	L	10	1	100
13	Nonvol. metals	*SM-3111C*	S/L	10	1	100
14	Org. carbon (D)	*EPA 415.1*	L	10	.5	100
15	Org. carbon (P)	*PML 3.1*	S	10	1	100
16	Org. N. (D)	*EPA 351.4*	L	10	1	100
17	Org. N. (P)	*EPA 351.4*	S	10	1	100
18	Org. phos. (D)	30–9	L	10	1	100
19	Org. phos. (P)	31	S	10	1	100
20	pH	*EPA 150.1*	L	10	.4	100
21	Salinity	*SM 2520B*	L	5	.2	100
22	Secchi	*EPA*	L	5	.2	100
23	Silicate	*PML1.7*	L	10	1	100
24	T. alkalinity	*EPA 310.1*	L	10	1	100
25	T. carbon (D)	*EPA 415.1*	L	10	1	100
26	T. nitrogen	1 + 11 + 12 + 28	L	10	1	100
27	T. org. carbon	14 + 15	L	10	1	100
28	T. org. nitrogen	16 + 17	L	10	1	100
29	T. org. phos.	19 + 20	L	10	1	100
30	T. phosphate (D)	*EPA 365.4*	L	10	1	100
31	T. phosphate (P)	*EPA 365.4*	S	10	1	100
32	T. phosphate	30 + 31	L	10	1	100
33	Temperature	*EPA 170*	L	5	.2	100
34	True color	*SM 2120C*	L	10	5	100
35	Turbidity	*EPA 180.1*	L	10	1	100

List of Abbreviations

EPA. *Methods for Chemical Analysis of Water and Wastes*. EPA-600/4-79-020.
SM. *Standard Methods for the Examination of Water and Wastewater*, 17th ed.
PML. Parsons, Maita, and Lalli. *A Manual of Chemical and Biological Methods for Seawater Analysis*, (Oxford: Pergamon Press, 1984).

NCASI. National Council for Air and Stream Improvement, *An Investigation of Improved Procedures for the measurement of Mill Effluent and Receiving Water Color*, Stream Improvement Technical Bulletin No. 253, December 1971.
BOD. Biochemical oxygen demand.
T. total, (D) dissolved, (P) particulate.

Sampling Procedures

The objective of sampling is to collect a portion of material small enough in volume to be transported conveniently and handled in the laboratory while accurately representing the material being sampled. This implies that the relative proportions or concentrations of all pertinent components will be the same in the samples as in the material being sampled and that the sample will be handled in such a way that no significant changes in composition will occur before the tests are made.

Sample bottles, usually half gallon plastic milk containers, must be rinsed with the water being sampled at least three times. Sample containers that are to be reused are washed with Liquinox™, rinsed at least three times with deionized water, dried, and then sealed to avoid any contamination. If phosphates are to be analyzed, the sample containers are acid washed with warm 10% HCl, and if trace metals are to be analyzed, the sample containers must be treated with 10% nitric acid.

Samples collected at a particular time and place can represent only the composition of the source at that time and place. Grab samples are collected with a Niskin bottle or pumped with a peristaltic pump at the bottom of the water column. Composite samples are taken with an integrated sampler. Avoid collecting detritus by taking the sample a few centimeters above the soil/water interface. Surface samples are collected by lowering an inverted sample container beneath the water/air interface and righting it. Avoid collecting any flotsam and jetsam by filling the sample container 5 cm beneath the surface. Avoid entrapping air in the filled sample container. The sample must be properly preserved until it is received at the laboratory. They are iced on the boat and maintained at 4°C thereafter.

When a source is known to vary with time, samples must be taken with appropriate frequency to monitor the extent of these variations. In such a situation, the location and the time of sample collection must be accurately duplicated. In open water, a GPS can assure site location to within a few meters; otherwise, landmarks must be judiciously chosen.

Samples are put on ice in the dark immediately to assure stability of constituents until they can be analyzed in the lab. Before delivery of the sample to the lab, a chain-of-custody form must be filled out detailing the volume of the sample, the location of the site, the date and time of sampling, the name of the samplers, the project and/or the parameters to be analyzed, the technique by which the sample was obtained, and the methods of preservation. Completed chain-of-custody forms must be received and signed by our laboratory personnel. Then they are filed at the laboratory where they are a record of sample history. All subsequent tests performed on these samples are recorded in the laboratory logbook and sample tracking forms.

Samples are to be delivered to the lab with all possible haste; if delivery time exceeds 6 h, correct preservation techniques must be observed. The unequivocal preservation of samples is fundamentally impossible. Regardless of the preservation technique, complete stability for every constituent can never be achieved. It is best to analyze samples as soon as possible after collection and then to judiciously determine the type of preservation to be utilized.

Measurement	Container	Preservative	Storage[a]
Alkalinity	P, G	Cool, 4°C	24 h/14 days
Ammonia	P, G	Cool, 4°C	24 h
Ammonia	P, G	Cool, 4°C, H_2SO_4 to pH < 2	7 days/28 days
BOD	P, G	Cool, 4°C	6 h/48 h
Chlorophyll	P, G	Cool, 4°C, dark	24 h/48 h
Color	P, G	Cool, 4°C	24 h/48 h
Conductance	P, G	Cool, 4°C	28 days
Filterable residue	P, G	Cool, 4°C	7 days
Kjeldahl nitrogen	P, G	Cool, 4°C, H_2SO_4 to pH <2	7 days/28 days
Mercury	P	Cool, 4°C, HNO_3 to pH <2	ASAP/5 weeks
Metals[b]	P	Cool, 4°C, HNO_3 to pH <2	6 months
Nitrate	P, G	Cool, 4°C	48 h
Nitrite	P, G	Cool, 4°C	48 h
Nonfilterable residue	P, G	Cool, 4°C	7 days
Oxygen, dissolved	P, G	None	0.5 h
Organic carbon	P, G	Cool, 4°C, H_2SO_4 to pH <2	7 days/28 days
Orthophosphate	G	Cool, 4°C, no acid	48 h
pH	P, G	None	2 h
Salinity	G	Wax seal	6 months
Silicate	P	Cool, 4°C, do not freeze	28 days
Solids	P, G	Cool, 4°C	7 days
Temperature	P, G	None	None
Total phosphate	G	Cool, 4°C, H_2SO_4 to pH <2	28 days
Total diss. phosphate	G	Cool, 4°C, H_2SO_4 to pH <2	24 h
Turbidity	P, G	Cool, 4°C, dark	24 h/48 h

[a] Recommended holding time, maximum holding time stated in the literature.
[b] Pb, Cu, Cr, Ni, Cd, Fe, Al.

Sample Custody

Sample custody is important from a legal standpoint. All work should be documented. Chain-of-custody forms cover the time the sample spends in the field after collection and before it gets to the laboratory. While in the laboratory, a logbook and sample tracking forms chart the various analysis performed on the sample. After all the experimentation is complete, the chemist performing the analysis initials the original data sheet. The original data sheets are handed to the QA officer by the laboratory director. The QA officer reviews the raw data and turns them in to the data analyst who files the original data sheets. The data are punched into our Macintosh computer system by the data analyst and his/her assistants. After the data are punched, they are computed and checked by the QA officer.

Field Sampling Operations

Equipment that goes in the field is prepared prior to each field trip. All field meters are calibrated in the laboratory by the QA officer prior to each field trip. Typically, field crews need a salinometer, a dissolved oxygen meter, a pH meter, and a thermometer. The salinometer we use in the field is a YSI SCT meter. It is calibrated against a standard saline solution before each field trip. The probe must be cleaned with an HCl/isopropanol solution. The temperature sensor is calibrated against a National Bureau of Standards (NBS) thermometer. These methods are detailed in the YSI operators manual (see SOP [standard operating procedure]). The DO meter we use in the field is a YSI DO meter. Before each field trip, it is calibrated in a YSI calibration chamber and the DO membrane is replaced. If response is sluggish, or if there is any difficulty in calibration, clean the silver electrode with a 14% ammonium hydroxide solution, and clean the gold electrode with an abrasive rubber eraser. The temperature sensor is calibrated against an NBS thermometer. These methods are detailed in the YSI operators manual (see SOP).

The pH meter we use in the field is a Corning 610 pH meter. Of all our field equipment, this is the most prone to water damage. If it gets very wet, the electronics will be damaged. The calomel electrodes are kept filled with a saturated KCl solution. If response gets sluggish, clean the semipermeable membrane in warm pH 10 buffer, and look for air bubbles in the ceramic junction. Before use, the meter is calibrated to three different pH buffers. The buffers are for pH 4, 7, and 10. These buffers are taken into the field, and the meter is calibrated prior to each reading in the appropriate pH range. These methods are detailed in the Corning Operators Manual (see SOP).

The meters, a graduated depth line, and the Secchi must be signed out to the field crew by the laboratory director, who makes sure all equipment is in full operating condition. The field crew labels and cleans all sample bottles after they have been cleared by the laboratory director. The date, location, and station identification number must be written on every sample bottle with an indelible marker. The field crew obtains field data sheets from the data analyst.

The field crew is responsible for custody of the samples and meters until they are released to a carrier or the laboratory. The chain-of-custody form initiated by the field crew states the location, time, date, method of sampling, number of duplicates, and method of preservation of each sample. The persons in charge of the sample until it reaches the laboratory must sign indicating the time when the sample came into their custody and when they relinquished the sample to someone else.

Laboratory Operations

The laboratory director is the sample custodian at the laboratory facility. He/she signs for incoming samples and obtains documents of shipment. Once the samples arrive at the laboratory, their progress through analysis is documented with sample tracking forms and in the laboratory log. All analysis, new reagents, standards, etc., are signed and dated in this book by the chemist.

The laboratory director is responsible for the analysis of the sample. Results of analysis are recorded on the serially numbered laboratory data sheets. These are a complete record of the results of all the analyses performed on the samples from a given field trip. The laboratory data sheets also record the standards that are run. The laboratory log details the observations and calibrations that may have occurred during the run. It provides an accurate record of the date and time of each run and of the preservation status of any given sample.

Calibration Procedures and Frequency

Spectrophotometers

(Beckman DU 64, Bausch & Lomb Spectronic 88, Turner Fluorometer Model 112)

A series of seven standards that bracket the range of concentrations anticipated for any given parameter are run with every analysis as specified in our SOP. A standard curve is generated from this data, and the slope and intercept are reported on the original data sheet. If the coefficient of variation is less than 0.90 on the standard regression, the spectrophotometer is serviced and the analysis is run again until a satisfactory calibration curve is obtained. All of these curves are kept on file by the QA officer. Each new one is compared with previous calibration curves for the same parameter. If there are significant differences between the curves, this is a sufficient cause for the analysis to be rerun.

All standards are made in our laboratory from ACS reagents according to EPA specifications. A stock solution is made first. The flask is dated and signed by the chemist who made it. A standard solution is then made from the concentrated stock solution. The flask is dated and signed by the chemist who made it. Standards run with the analysis are serial dilutions of the standard solution. They are made fresh for every analysis.

All stock and standard solutions are recorded in our laboratory logbook when they are made. Old standards are discarded as they go bad. All the standards are dated. We follow EPA guidelines in ascertaining their lifetime, and we check their age prior to use. We also do a visual inspection of the standard before use. The ultimate test of the standard is if it generates a good standard curve at the expected intensity; however, a bad standard curve can also be caused by malfunctioning equipment.

We can trace our standards by checking in our laboratory log for the current standard in use during any given analysis.

Mettler Balances

(Mettler H series and an Ohaus Triple Beam Balance)

Our Mettler balances are calibrated twice a year by certified Mettler technicians. If there are any problems with one of them, we have a backup triple beam balance.

Turbidimeters

(Hach model 2100A)

These are calibrated with every use with commercial Hach standards as specified in our SOP.

Colorimeters

(Hach DR-1)

These are calibrated with every use with commercial Hach standards as specified in our SOP.

DO Meters

(YSI model 51 B and a YSI 57)

These are air calibrated with every use. Winkler is calibrated quarterly as specified in our SOP.

SCT Meters

(2 YSI Model 33)

These are calibrated prior to each use with a standard saline solution as specified in our SOP.

Carbon Analyzer

Ionics model 555

This is calibrated prior to each use with organic carbon and IC standards. IC column is recharged with phosphoric acid bimonthly.

pH/mv Meters

(Corning pH meter #55A, and Hach pH meter #DR-1)

These are calibrated with buffered solutions obtained commercially at pH 4, 7, and 10. The mv scale used with an ammonia-sensitive electrode is calibrated with the seven-point standards used for the ammonia test as specified in our SOP.

Other Equipment

- **IEC Centra 7 precision centrifuge**: Lubricate and clean with alcohol.
- **Labconco Kjeldahl digestion rack**: Clean periodically with warm water.
- **General Electric vacuum pump**: Change oil quarterly and keep dry.
- **Precision scientific steam table**: Keep filled with water. Change the water quarterly. Clean top after use. Clean scale off of the heating element.
- **Wards refrigerator and freezer**: Defrost when necessary. Check temperature quarterly.
- **Labline Ambi-Hi-Lo incubator**: Check the temperature during use. Wipe out interior after use.
- **Thermolyne constant temperature bath and hot plates (2)**

Analytical Procedures

Only EPA-approved procedures are used in our laboratories. All our laboratory procedures are detailed in our SOP.

Data Reduction, Validation, and Reporting

All data are reported to our analyst. The analyst and the QA officer then preliminarily validate the data, which are then sent to the project manager for final validation. The criteria used to validate data integrity are adherence to our system of absolute standards, the conformation of duplicates, and the %RSD of spikes.

NUTRIENT LOADING MODELS

The U.S. Geological Survey (Tallahassee, FL; M. Franklin, pers. comm.) provided flow data for the Perdido River system, Bayou Marcus Creek, and Eleven Mile Creek. Nutrient sampling was carried out in various rivers leading into the bay. Monthly nutrient concentrations from these rivers were used in the models. Overland runoff from unmonitored areas was estimated using models devised by A. Niedoroda (pers. comm.). Loading models for the 11-year database were run on monthly, quarterly, and annual intervals.

The nutrient loading models were based on a ratio estimator developed by Dolan et al. (1981). This model corrects for the biases due to sparse temporal sampling. It uses auto- and cross-covariance values of flows and nutrients for correction of the loading calculations. Nutrient data for the models were usually taken monthly where river data were taken daily. Program modifications were made by A. Niedoroda and G. Han (pers. comm.) whereby (1) due to discontinued river monitoring of the Bayou Marcus Creek and the Blackwater River in September 1991, river flows for these areas were estimated from regressions (log/log, Eleven Mile Creek and Bayou Marcus Creek; Styx and Blackwater Rivers) and (2) stations 9 and 21 were sampled monthly from 3/93 to 10/94 and flows were then estimated by the following: St. 09, (2.52 × St. 05) + (1.01 × St. 04) + (1.10 × St. 07) and St. 21, (2.64 × [St. 13 − 32]) + 32. These data are considered questionable at best and should not be used for any serious conclusions.

Dr. Alan Niedoroda and Mr. Gregory Han of Woodward-Clyde designed the loading estimator program. Data were grouped according to methods described in previous reports. Model applications were carried out according to a formula given by Dolan et al. (1981):

$$\bar{\mu}_y = \mu_x \frac{m_y}{m_x} \left(\frac{1 + \frac{1}{n} \frac{S_{xy}}{m_x m_y}}{1 + \frac{1}{n} \frac{S_{x^2}}{m_{x^2}}} \right)$$

where

μ_y is the estimated load
μ_x is the mean daily flow for the year
m_y is the mean daily loading for the days on which concentrations were determined
m_x is the mean daily flow for the days on which concentrations were determined

$$S_{xy} = \frac{1}{(n-1)} \sum_{i=1}^{n} x_i y_i - n m_x m_y$$

$$S_{x^2} = \frac{1}{(n-1)} \sum_{i=1}^{n} x_{i^2} - nm_{x^2}$$

where

n is the number of days on which concentrations were determined

x_i is the individual measured flows

y_i is the daily loading for each day on which concentrations were determined

This formulation corrects for biases introduced by sparse (temporal) sampling by using auto- and cross-covariance values of the flows and nutrients to correct the loading calculations for variability in the river flow that influences the nutrient concentrations. This applies where flow data are usually taken regularly (usually daily), whereas nutrient concentrations are usually taken at monthly or quarterly intervals.

Designated, fixed stations in a given system were sampled (surface and bottom) for water quality factors. Vertical collections of water temperature, salinity, conductivity, dissolved oxygen, depth, Secchi depth, and pH were taken at the surface (0.1 m) and at 0.5 m intervals below the surface to about 0.5 m above the bottom. Surface and bottom water samples were taken with Niskin bottles for further chemical analyses in the laboratory. All sampling was carried out on a synoptic basis with all samples in a given system taken within one tidal cycle.

A Beckman DU-64 spectrophotometer was used to analyze true color (APHA, 1989). Both ratio and nephelometric turbidity analyses were carried out. A ratio turbidimeter and a Hach model 2100A turbidimeter were used for turbidity analyses. Quantification of dissolved inorganic nitrogen (nitrite, nitrate) and dissolved inorganic phosphorus (orthophosphate) followed methods outlined by Parsons et al. (1984). Ammonium was measured with an ion electrode (US EPA, 1983, method 350.3) after raising the pH to 11. This method had some essential features (e.g., minimal interference from waters highly stained with humic materials and paper mill effluents) that required special attention to the results. Consequently, the level of detection was relatively high (e.g., 2.0 μm ammonium-N) but adequate for this study. PON (US EPA, 1983, method 351.4) and POP (US EPA, 1983, method 364.4) were collected on Gelman glass fiber filters combusted at 500°C, oxidized to inorganic fractions and organic N measured as ammonia and organic P measured as POP. Total dissolved solids (APHA Method 209-B), total suspended solids (APHA Method 209-C), dissolved organic carbon (APHA Method 415.1), chlorophyll *a* (APHA Method 1002-G), and BOD (APHA Method 405.1) were carried out according to the methods outlined by the APHA (1989). Particulate organic carbon was analyzed according to Parsons et al. (1984) (method 3.1). Nutrient analyses were carried out according to APHA (1989).

LIGHT DETERMINATIONS

Light penetration depths were taken using standard Secchi disks. Field light transmission data were taken with a Li-Cor LI-1800UW underwater spectroradiometer. The underwater light field was characterized by incident radiant flux per unit surface area as quanta m^{-2} s^{-1}. Samples included 3–5 scans (replicates) taken at 1–2 nm intervals that were averaged for each reading. Flux measurements were taken for photosynthetically active radiation (400–700 nm; PAR) and individual wavelengths. For any series of collections, field samples were taken during relatively calm conditions between the hours of 1000 and 1400 h. Multiple air tight readings were taken to correct for short-term radiation variability during light measurements.

FIELD/LABORATORY METHODS

All sampling efforts were carried out in a similar fashion with respect to sample collection, transfer, and analysis. For all types of organisms, outside corroboration has been carried out by recognized taxonomists continuously. Laboratory specimen libraries have been established and continuously maintained. Chain-of-custody forms are filled out with each transfer with the usual procedure as follows:

1. Samples are delivered (chain of custody signed). A copy of the chain-of-custody form is given to the person transferring the samples; the original is kept with the samples. Samples are then stored in a secure area.
2. Samples are processed (if required) and date and time recorded (ID # kept with each sample).*
3. Organisms are picked (if required) and date and time recorded (ID # kept with the sample).
4. Preparation of specimens is made for identification (ID # kept with the sample).
5. Data are recorded in proper form for reporting to the customer.
6. Data, samples, and reference collection are transferred to the client (chain of custody signed) and a copy of the original requested.

Phytoplankton and Zooplankton

Net phytoplankton samples were taken with 2 25 μm nets (bongo configuration) in duplicate runs for periods of 1–2 min. Repetitive (3) 1 L whole water phytoplankton samples were taken at the surface. Phytoplankton samples were immediately fixed in Lugol's solution in its acid version (Lovegrove, 1960). Samples were analyzed by methods described by Prasad et al. (1990) and Prasad and Fryxell (1991).

Zooplankton were taken in various ways. Multiple tows were made with 202 μm nets run over varying time periods that depended on zooplankton density. Sample volumes were recorded so that a quantitative estimate could be made of zooplankton numbers. Another approach, used under certain conditions, involved pumping water (a volume that was determined in the field) through 202 μm nets. Zooplankton preservation was made using 10% formalin. All counts were made to species.

Periphyton

A set of slides were anchored at a given stations 2–2.5 cm below the surface. After 2–3 weeks, we randomly took five slides and placed them in labeled jars filled with distilled water. Jars were transported on ice to the lab where they were scraped on both sides and preserved with 5% formalin for a composite sample that will be stored in screw-cap vials. Each sample was then diluted with distilled water depending on the density of algae and thoroughly shaken, and three aliquots of 0.05 or 0.1 mL were placed on slide, covered, counted, and identified.

Benthic Macrophytes

Quantitative seagrass sampling was carried out using divers to take eight (randomly distributed) quadrat samples (1/4-m^2) according to methods described by Livingston et al. (1976a). Samples were identified to species. Dry weight determinations will be made for macrophytes (rooted and

* ID labels are kept on the outside of the sample container as well as inside the sample container with the organisms. This information includes project, sample location, and date and is made with indelible ink on a suitable material label outside and on a paper label inside (preferably white rag paper).

epiphytic) by species for above- and below-ground dry weight biomass. Samples were heated in the oven at about 105°C until there was no further weight loss (at least 12 h). Samples were then weighed to the nearest hundredth of a gram.

A map of the offshore seagrasses was developed from just west of the Aucilla River mouth to a point near the Spring Warrior Creek entrance (Fig. 1). A set of 166 natural color aerial photographs was taken (AeroMap U.S. Inc., pers. comm.) along four flight lines between 9:06 and 9:50 AM E.S.T. on November 15, 1992. Ground truthing was carried out as a series of diver transects at 320 m intervals from land to 5–7 km offshore: data included visual observations of seagrass species and density, habitat distribution, and bottom type. Seagrass density was estimated as % cover (bare, no observed macrophytes; very sparse, less than 10% coverage; sparse, 10%–40% coverage; moderate, 40%–70% coverage; dense, 70%–100% coverage). The aerial photography was transferred to a computerized seagrass basemap at a scale of 1:24,000. Overlay maps of the transect data were printed at the same scale. The center point of every photograph in the series was then positioned and interpreted using the diver transect data. The seagrass signature (density, distribution) was identified. Density classification was accomplished by visually comparing the photograph to examples of various density classes from an enlarged crown density scale. The final basemap was then photographically reduced in scale and scanned into a computer system. A detailed analysis was carried out concerning the inshore areas of the submerged aquatic vegetation (SAV) survey (within 3 km of shore). The survey region was divided into a series of 17 zones along the coast that were 2.5 km wide and 3 km offshore with the river mouths of the Aucilla, Econfina, and Fenholloway Rivers positioned in the center of individual zones. Absolute and relative areas of the different grass densities and other habitat classes, by zone, were then calculated and used for statistical analysis along with water/sediment quality analyses at stations within each zone.

Field Collections of Fishes and Invertebrates

All field biological collections were standardized according to statistical analyses carried out early in the program (Hooks et al., 1976; Livingston, 1976a; Greening and Livingston, 1982; Livingston et al., 1976; Meeter and Livingston, 1978; Stoner et al., 1983). Infaunal macroinvertebrates were taken using coring devices (7.6 cm diameter, 10 cm depth). The number of cores to be taken was determined using large initial samples (40 cores) and a species accumulation analysis based on rarefaction of cumulative biological indices (Livingston et al., 1976). Based on this analysis, multiple (10–12) core samples were usually taken randomly at each station with the assured representation of at least 80% of the species taken in the initial sampling. All infaunal samples were preserved in 10% buffered formalin in the field, sieved through 500 μm screens, and identified to species wherever possible.

Where roots, prevented coring, suction dredges were used as prescribed with multiple sweeps of specific time (number and time to be determined during first field trip). Animals were preserved in 10% buffered formalin and stored in 100% denatured ethanol.

Epibenthic fishes and invertebrates were collected with 5 m otter trawls (1.9 cm mesh wing and body, 0.6 cm mesh liner) towed at speeds of about 3.5–4 km h^{-1} for 2 min resulting in a sampling area of about 600 m^2 per tow. Repetitive samples (7) were taken at each site; sampling adequacy of such nets had been determined by Livingston (1976). All collections were made during the same lunar and tidal phase at the first quarter moon. Samples were taken within 2 h of high tide, with day and night trawls separated by one tidal cycle. Larger fishes (>300 mm SL) were taken with 100 m nylon trammel nets and stomachs were removed in the field and injected with a 10% buffered formalin solution.

All organisms were preserved in 10% buffered formalin, sorted and identified to species, counted, and measured (standard length for fishes, total length for penaeid shrimp, carapace width for crabs). Representative samples of fishes and invertebrates were dried and weighed and regressions were run so that data from the biological collections could be converted into dry and ash-free dry mass. The numbers of organisms were then converted to biomass/m^2. All biological data (infauna, epibenthic macroinvertebrates, fishes) were expressed as numbers m^{-2} mo^{-1} or biomass m^{-2} mo^{-1Z}.

Trophic Determinations

Gut-content analyses were carried out in the same manner throughout the 35-year field analysis period. Fishes and invertebrates were placed in 5, 10, or 20 mm size classes (depending on size), and food items were taken from the stomachs of up to 25 animals in a given size class with pooling by sampling date and station. Determinations similar to those described earlier were carried out, and it was determined that at least 15 animals were needed for a given size class × station × sampling period. Stomach contents were removed and preserved in 70% isopropanol and a dilute solution of rose bengal stain. In larger animals, we pumped the stomachs out and released the subject. The content analysis was carried out through gravimetric sieve fractionization (Car and Adams, 1972). Contents were washed through a series of six sieves (2.0–0.75 mm mesh) and frequency of occurrence of each food type was recorded for each sieve fraction. As the food items were comparable in size, the relative proportion of each food type was measured directly by counting. Dry weight and ash-free dry weight were determined for each food type by size class. General, mutually exclusive categories were used for the food types that included both plant and animal remains. Stomach contents by food type and by size class were then calculated and related back to station/time designations. Based on the long-term stomach content data for each size class, the fishes and invertebrates were then reorganized first into their trophic ontogenetic units using cluster analyses (Czekanowski [Bray–Curtis] or C-lambda similarity measure; flexible grouping cluster strategy [beta = −0.25]). The basis for this analysis is given by Sheridan (1978) and Livingston (1984).

All biological data (as biomass m^{-2} mo^{-1} of the infauna, epibenthic macroinvertebrates, and fishes) were transformed from species-specific data into a new data matrix based on trophic organization as a function of ontogenetic feeding stages of the species found in Perdido Bay over the multiyear sampling program. Ontogenetic feeding units were determined from a series of detailed stomach content analyses carried out with the various epibenthic invertebrates and fishes in the near-shore Gulf region (Sheridan, 1978, 1979; Laughlin, 1979; Sheridan and Livingston, 1979, 1983; Livingston, 1980, 1982, 1984b, unpublished data; Stoner and Livingston, 1980, 1984; Laughlin and Livingston, 1982; Stoner, 1982; Clements and Livingston, 1983, 1984; Leber, 1983, 1985). Based on the long-term stomach content data for each size class, the fishes and invertebrates were reorganized first into their trophic ontogenetic units using cluster analyses. Infaunal macroinvertebrates were also organized by feeding preference based on a review of the scientific literature (Livingston, unpublished data). The field data were then reordered into trophic levels so that temporal changes in the overall trophic organization of this system could be determined over the 8-year study period. We assumed that feeding habits (at this level of detail) did not change over the period of observation; this assumption is based on previous analyses of species-specific fish feeding habits that remained stable in a given area over a 6–7-year period (Livingston, 1980).

The long-term field data were used to establish a database based on the total ash-free dry biomass m^{-2} mo^{-1} as a function of the individual feeding units of the infauna, epibenthic macroinvertebrates, and fishes. Data from the various stations were used (g m^{-2}) for all statistical analyses. The data were summed across all taxonomic lines and translated into the various trophic levels that included herbivores (feeding on phytoplankton and benthic algae), omnivores (feeding on detritus and various combinations of plant and animal matter), primary carnivores (feeding on herbivores and detritivorous animals), secondary carnivores (feeding on primary carnivores and omnivores), and tertiary carnivores (feeding on primary and secondary carnivores and omnivores). All data are given as ash-free dry mass m^{-2} mo^{-1} or as percent ash-free dry mass m^{-2} mo^{-1}. In this way, the long-term database of the collections of infauna, epibenthic macroinvertebrates, and fishes was reorganized into a quantitative and detailed trophic matrix based not solely on species (Livingston, 1988) but on the complex ontogenetic feeding stages of the various organisms. A detailed review of this process and the resulting feeding categories is given by Livingston et al. (1977).

Oyster Studies

Oyster samples (multiple) were taken with full head tongs (16-tooth head; 4.5 m handles) at each station on a monthly basis in Apalachicola Bay from March 1985 through October 1986. The use of tongs was standardized with respect to opening widths and sampling effort. To quantify the sampling effort, a series of 30 standardized, random tong samples was taken at the Big Bayou and Cat Point reefs in February 1985. The cumulative size frequency distribution (in 10 mm increments) was determined and plotted for each sampling site. The number of samples necessary for a specified level of quantification was determined according to a method described by Livingston et al. (1976). The method allowed determination of the number of subsamples necessary to achieve specific levels of size class accumulation when compared to the results of 30 subsamples. Seven subsamples accounted for 80.8% (Big Bayou) and 87.5% (Cat Point) of the sampling variability for the total sample. Accordingly, this number of subsamples (located beyond the asymptote for size class accumulation) was considered to constitute a representative sample taken in each of the test regions. Numbers of oysters per tong were recorded and converted to numbers m^{-2}.

All oysters were measured to the nearest mm in the field according to the greatest distance from the dorsal umbo region to the ventral shell margin using linear calipers. A total of 140 oysters taken from four stations (n = 35 at each site) was used to determine the relationship of shell length and weight of oyster meat. Four separate length/weight equations were developed to account for known differential growth characteristics in different regions of the bay:

$$\ln(\text{AFDW}) = 2.505 \times \ln(\text{LEN}) - 10.980\left(\text{Cat Point, } r^2 = 0.83\right)$$

$$\ln(\text{AFDW}) = 2.303 \times \ln(\text{LEN}) - 10.306\left(\text{East Hole, } r^2 = 0.84\right)$$

$$\ln(\text{AFDW}) = 2.202 \times \ln(\text{LEN}) - 9.125\left(\text{Paradise, } r^2 = 0.90\right)$$

$$\ln(\text{AFDW}) = 2.465 \times \ln(\text{LEN}) - 10.190\left(\text{Scorpion, } r^2 = 0.86\right)$$

where
AFDW is the ash-free dry weight of oyster meat
LEN is oyster shell length in mm

F-tests ($\alpha = 0.05$) disclosed that equations from both bars in the eastern bay (Cat Point and East Hole) were significantly different from equations developed for bars in the western bay (Paradise and Scorpion). Although the differences were not significant within the respective regions, east and west, the site-specific equations were used for the transformations of the length data into ash-free dry weights. This allowed the most comprehensive and accurate use of the data for such transformations. All tong data (in terms of numerical abundance and ash-free dry weight) were calculated on a unit m^{-2} basis. These data were then transformed to estimates of total numbers and biomass on each bar (Table 1) based on the estimated size of the bars. Areas were established from computer simulations of oyster-producing areas (Livingston, unpublished data) based on historic records, interviews with oystermen and state environmental agencies, and our past and ongoing field studies (Livingston, 1984a).

The oyster tong data (oyster density and length frequency) were standardized by a quantitative comparison with data derived from a series of multiple (0.25 m^{-2}) quadrats taken on the same sampling sites (Cat Point Bar, East Hole, Paradise) over the same period of study by other researchers (Berrigan, 1990). Scattergrams of the respective density and length frequency databases were log-transformed to approximate the best fit for a normalized distribution. Statistical comparisons were made of the monthly data by station. Independence tests were used to examine each sample for autocorrelation at a number of lags. For each sample, autocorrelation was computed for the lesser of 24 or n/4 lags, where n is the number of observations. The program then computes the Q statistic of Ljung and Box (1978) for an overall test of the autocorrelations as a group, where the null hypothesis was that correlation at each lag was equal to zero. The parametric F-test for comparison of variances and t-test for comparison of means were used to test the hypothesis that the data came from normally distributed populations. Means testing was carried out with both parametric and nonparametric tests. For independent, random samples from normally distributed populations, the parametric t-test was used to compare the sample means. Where there was serial dependence of data from a given pair of stations, it was removed by differencing the observations and calculating and plotting the autocorrelations of the differences. If the differences were not serially correlated, we applied the Wilcoxon signed-rank test on the differences to compare the two sets of numbers ($p = 0.05$).

Spatfall accumulation was analyzed from a subset of 12 stations (Table 1). Spat baskets, constructed of plastic-coated wire (25 cm × 25 cm × 25 cm; 2.5 cm mesh), were filled with about 20 sun-bleached oyster shells and placed at each site. Bricks were placed at the bottom of each basket so that the oysters remained off the bottom to lessen problems with sedimentation. Samples were retrieved, and new sets of oyster shells were set out at 2- to 3-week intervals. One spat basket was used at each station. Seven shells were randomly chosen from each basket for analysis. Spat counts were made from the inner surface of each valve; this standardization was based on test results that indicated less variance of such adherence on inner surfaces than on outer surfaces.

Oyster data were grouped in two ways prior to analysis: (1) bay-wide totals and (2) eastern versus western reefs. In the latter grouping strategy, selected reefs from the eastern bay (Sweet Goodson, Cat Point, East Hole, Porter's Bar) were compared to selected reefs from the western bay (Paradise, Pickalene, Scorpion). These reefs were chosen because of their relative commercial importance to overall oyster production in the bay. Intervention models were used to analyze the effects of the two hurricanes on the monthly total oyster numbers (in thousands) and the monthly average shell length (in mm) of oysters in the bay (Box and Jenkins, 1978; Pankratz, 1991). Three major interventions occurred in the Apalachicola Bay during the study period: Hurricane Elena (combined with the cessation of commercial oystering after September 1985), Hurricane Kate, and the resumption of commercial oystering in May 1986. The models were used to test the possibility that the interventions were significantly associated with level changes in the monthly total oyster numbers $\{N(t)\}$ and the monthly average shell lengths $\{S(t)\}$.

Three indicator variables corresponding to the interventions are defined as

$$X_1(t) = \begin{cases} 0, t < \text{September 1985} \\ 1, t \geq \text{September 1985} \end{cases}$$

$$X_2(t) = \begin{cases} 0, t < \text{December 1985} \\ 1, t \geq \text{December 1985} \end{cases}$$

$$X_3(t) = \begin{cases} 0, t < \text{May 1986} \\ 1, t \geq \text{May 1986} \end{cases}$$

Using B as the backward shift operator such that $BX(t) = X(t - 1)$, the relationship between the monthly total oyster number $\{N(t)\}$ and the three interventions can be described by the following general intervention model:

$$N(t) = \beta_0 + \nu_1(B)X_1(t) + \nu_2(B)X_2(t) + \nu_3(B)X_3(t) + \xi(t), \quad (1)$$

where $\nu_1(B)$, $\nu_2(B)$ and $\nu_3(B)$ are polynomials with the typical form:

$$\nu(B) = \omega_0 - \omega_1 B - \cdots - \omega_R B^R$$

where $\{\omega_0, \omega_1, \ldots, \omega_h\}$ are parameters to be estimated.

In model (1), β_0 is a constant and $\xi(t)$ is a noise term that is often modeled as a stationary autoregressive moving average (ARMA; p, q) process where

$$\xi(t) - \phi_1\xi(t-1) - \cdots - \phi_p\xi(t-p) = \varepsilon(t) - \theta_1\varepsilon(t-1) - \cdots - \theta_q\varepsilon(t-q),$$

where

p is the order of the autoregression (AR) term

q is the order of the moving average (MA) term

The $\varepsilon(t)$'s are assumed to be independent and normally distributed with mean zero and variance σ^2.

Model (1) extends the traditional linear regression models in two directions. First, the monthly total oyster number series $\{N(t)\}$ may react to an intervention with a time lag. For example, the term $\upsilon_1(B)X_1(t) = \omega_0 X_1(t) - \omega_1 X_1(t-1) - \cdots - \omega_h X_1(t-h)$ in model (1) represents that the impact of the first intervention on N(t) is distributed across several time periods. Second, instead of assuming that the errors $\xi(t)$ are independently distributed, ARMA(p,q) models are used for $\xi(t)$, which incorporates possible serial correlations in the response series $\{N(t)\}$. Model (1) was fitted by using the *Linear Transfer Function Identification Method* proposed by Pankratz (1991, Chapter 5). The relationship between the monthly average shell length series $\{S(t)\}$ and the three interventions was modeled in a similar fashion.

In addition to the overall oyster data, intervention models were fitted to eastern and western sections of the bay using numbers m^{-2} and average shell length (in mm) of oysters from stations in these areas. The four series were denoted by EN(t) and WN(t) for eastern and western densities and by ES(t) and WS(t) for eastern and western shell lengths. Intervention models were fitted to the four series separately.

Modeling

Physical processes in Apalachicola Bay were simulated using a time dependent, 3D model of the Blumberg–Mellor family (Blumberg and Mellor, 1980, 1987). This model solves a system of coupled differential (prognostic) equations for free surface elevation, two components of the horizontal velocity, temperature, salinity, turbulence energy, and turbulence macroscale. The spatial integration is explicit in the horizontal and implicit in the vertical. For its forcing, the model admits a comprehensive database comprised of time-dependent temperature, salinity, surface heat and humidity fluxes and wind stress, tides, and residual signals, when available. In the horizontal, the model uses a body-fitted curvilinear orthogonal coordinate system that allows one to better adhere to the convoluted coastline. In the vertical, a sigma-coordinate system converts the free surface and seabed into coordinate surfaces and allows for better resolution of the surface and near-bottom boundary layers. The sophisticated second-order turbulence closure model of Mellor and Yamada (1982), as modified by Galperin et al. (1988), describes the vertical mixing.

The model was calibrated and verified using hydrographic data collected at 0.5 h intervals from 23 instruments located at sites throughout the bay during the 6-month period from June to November 1993 (Wu and Jones, 1992; Jones and Huang, 1996; Huang and Jones, 1997). Measured river inflows, surface wind stress, free surface elevations, temperature, and salinity signals at the open boundaries were utilized as time-dependent boundary conditions. The open boundaries were comprised of five single-point and multipoint grid locations through which Apalachicola Bay connects to the gulf directly. Freshwater runoff into the bay was drawn from four rivers: Apalachicola River and its tributaries, Whiskey George Creek, Cash Creek, and the Carrabelle River. Flow values measured by the U.S. Geological Survey's Sumatra gage were used to develop the input from the main stem of the Apalachicola River and its tributaries using a 1D hydrologic model (DYNHYD; Ambrose and Barnwell, 1989). Local rainfall data were used as input to hydrologic models (SWMM; Huber and Dickinson, 1988) for the Whiskey George Creek, Cash Creek, and Carrabelle River as well as the ungaged portion of the Apalachicola River downstream of the Sumatra gage. Wind forcing was obtained from a continuous record collected at a weather station at the St. George Island causeway. For modeling the 1985–1986 period used in the present study, real-time temperature and salinity signals at the open boundaries were unavailable; composite climatological forcing functions were developed based upon Fourier representation of the mean seasonal data in a strategy similar to that of Galperin and Mellor (1990a). Velocity estimates were not calibrated against field observations as currents are more sensitive to precise location. However, the salinity calibration efforts effected an indirect calibration of velocity since salinity is a direct result of velocity-induced advection and diffusivity transport processes.

For this study, model runs were made for entire years 1985 and 1986 in which hourly values were calculated for 715 cells in the Apalachicola Bay system. The model output included elevation and surface and bottom values for salinity, temperature, and components of current velocity (horizontal and vertical). Additional variables were derived from the raw model output and included averages, standard deviations, maximum and minimum values, and residual current velocities and direction. Such variables were computed for specific time periods as described in the following.

EXPERIMENTAL METHODOLOGY

Nutrient Limitation Experiments

We constructed 18-1, all-glass microcosms (width: 20, length: 21.5, and depth: 48 cm) using nontoxic sealant. A wooden rack assembly held microcosms at the surface of large temperature-controlled reservoirs. Bay ambient water temperatures were maintained within ±2°C through the use of heating and cooling coils placed in the thermal reservoir. Large volumes of bay water were gently pumped (Ruhl model 1500 rubber diaphragm pump) from approximately 0.5 m depth at each station into polyethylene containers by first passing through a set of nested Nitex plankton nets held in a standpipe system to equilibrate water pressure (i.e., a 202 μm net placed inside a 64 m net; 0.5 m × 2.0 m) and to remove larger zooplankton.

Sampling was conducted during late afternoon and samples were transported to an experimental "microcosm tank facility" near Tallahassee, FL, during early evening to reduce heating from daytime peak temperatures. Experiments began the next day. Because a continuous supply of estuarine phytoplankton was logistically infeasible, we used a modified static renewal approach with daily removal of 10% of the volume of each microcosm. A supply of the field sample for each station was maintained as renewal water. This water was filtered through a 25 μm Nitex® net and held at 4°C in the dark. Thus, early experimental artifacts such as production of allelopathic substances would be less likely in such a system compared to a static experimental design. Before use, all collecting vessels, culture tanks, and Tygon air lines were acid washed (10% HCl), rinsed twice with freshwater, rinsed once with deionized water, and air dried. Microcosms were sealed with nontoxic silicon sealant. Air was delivered to each microcosm to maintain mixing via Tygon tubing, plastic control valves, and "air stones" from an oil-free aquaculture air pump.

Water collected from a station the previous day was placed in a fiberglass mixing tank during early morning to minimize light shock and was gently mixed with a rubber paddle to ensure homogeneous filling requirements. Water was poured sequentially in small volumes into glass microcosms to ensure homogeneous filling requirements. Reference samples were taken from the mixing tank for chlorophyll *a*, dissolved inorganic phosphate (PO_4), ammonium (NH_3) nitrite (NO_2) and nitrate (NO_2), PON, and POP to characterize initial nutrient conditions and standing stock of phytoplankton. Chlorophyll *a* was collected on Gelman A/E glass fiber filters and extracted with 90% acetone buffered with MgCO and measured according to Parsons et al. (1984). Quantification of dissolved inorganic N (i.e., (NO_2 and NO_3) and P (PO_4) followed Parsons et al. (1984). Ammonium was measured as NH_3 with an ion electrode (US EPA 1983, method 350.3) after converting NH_4 to NH_3 by raising the pH to 11. This method has some essential features (e.g., minimal interference from waters highly stained with humic materials and paper mill effluents). However, the level of detection typically was relatively high (e.g., 2.0 μM NH_3–N) but adequate for this study. PON (US EPA, 1983, method 351.4) and POP (US EPA, 1983, method 364.4) were collected on Gelman glass fiber filters combusted at 500°C oxidized to inorganic fractions and organic N measured as NH_3 and organic P measured as PO_4.

Single samples were collected daily at approximately 0800 hours from each microcosm for chlorophyll analysis to measure changes in phytoplankton biomass. Nutrient enrichment experiments included the following treatments triplicated for each station: three control tanks, three P-enriched tanks at 10 μM PO_4 above ambient, three N-enriched tanks at 50 μm NH_3N above ambient NH_3, and three combined NH_3 + PO_4 (referred to as N + P) above ambient as described for single additions. We used NH_3 enrichments because of interest in potential increased riverine transport into the bay. Although aware of possible suppression of NO_3 uptake by NH_3, this process was judged to be relatively unimportant in bioassays lasting over 5–12 days (D'Elia et al., 1986) and at high NO_3 concentrations (Pennock, 1987). Dortch (1990) concluded from a comprehensive literature review that NH_3 suppression of NO_3 uptake under field conditions is quite variable and often undetectable. Nutrient additions were made at time 0 of the experiment. On day 2 and daily thereafter, 18 L of water was collected from each microcosm for chemical and biological analyses. This volume was replaced by an equal volume of renewal water adjusted to ambient temperature before introduction into the microcosms. Hydrographic variables (e.g., surface water temperature (°C) and salinity (8b) (YSI model 33 conductivity meter) and 30 cm all-white Secchi disk readings (m)) were collected generally within 1 week of samples for nutrient enrichment experiments.

A randomized block design was used with statistical significance of ($P < 0.05$) within and among stations. The 4 treatments in each row were randomly selected so that each of the 3 station-specific experiments had 12 treatments (randomized 4 treatments by 3 replicates for 3 stations). Treatments were oriented in a north/south direction, to minimize shading from changes in the annual sun angle, and to allow detection of a possible blocking effect. A repeated-measures, randomized analysis of variance (ANOVA) was run for each experiment. ANOVA assumptions were checked using residual boxplots, the Lilliefors test on the standard residuals, and Bartlett's test for equality of variances. A log transformation was necessary to achieve equality of variances. The two-way interaction P-values for treatment by day and for block by day was calculated. To determine how treatments differed by day, a randomized block analysis of variance was run for each station, month and day. As discussed previously, a log transformation was necessary to satisfy assumptions. Scheffe's S-test was used as a post hoc treatment.

Diver-Deployed (*In Situ*) Benthic Lander

Introduction

We have analyzed the relatively broad literature on this subject and the method chosen reflects our interpretation of results from the studies of others in addition to our own experience with SOD analyses in our South Carolina work. Replicate samples were used as an indication of experimental error. In situ tests have the most potential for error. The laboratory approach, while suffering from the problem of not being carried out in the field, has proven to be the most reliable in terms of variability of results. We followed techniques that have a record of being the most reliable on the basis of intersystem comparisons.

We worked with in situ samplers developed and constructed at Texas A&M University (Dr. G. Rowe). The diver-deployed benthic lander used in the SOD measurements was designed and built by a research team at Texas A&M University under the direction of Dr. Gilbert T. Rowe. A description of the instrument is given in a paper given by Gilbert T. Rowe and Gregory S. Boland (*Benthic Oxygen Demand and nutrient regeneration on the Continental Shelf of Louisiana and Texas*, Abstract for LUMCON Meeting, 1991).

Instrument Specifications

Duplicate samples are taken after the instrument is grounded in sediments at a given station. Dissolved oxygen probes (one for each of the duplicate units) are attached to YSI-dissolved oxygen meters, which are monitored at 2–5 min intervals at the surface. The battery-operated machine has dual stirrers that maintain a regular rate of stirring in each 7 L (v) chamber. Each chamber covers 0.09 m^2 of sediment. Syringe ports are available for removal of water within each chamber for chemical analysis. The lander is housed in a basic 0.19 cm aluminum frame to which is attached a battery housing with two power Sonic batteries. Ballast weights are used to keep the device in the sediments. Total negative weight is −28.23 kg and the total positive lift (largely from the 33 cm diameter instrument spheres) is 13 kg.

Operation of the Lander

Laboratory Preparations

1. The YSI meters should be prepared. This includes installation of probe membranes, redlining the instrument at salinity 0, adjustment of temperature to ambient for calibration, calibrating probes for ambient temperature, setting meter scales to desired dissolved oxygen range, and checking calibration with 5%–10% sodium sulfite solution (0 ppm oxygen). The temperature should be reset to the expected bottom temperature (based on previous experience). Salinity should also be set in a similar fashion. Glass and gasket should be cleaned with a solvent (toluene).
2. Make sure that the silicon oil fills the impeller housing.
3. Charge batteries to +13 V and place in housing.
4. Check o-ring on battery and use light treatment of silicon grease on the battery chamber o-rings Close chamber.
5. Check for chamber volume by measuring distance from the bottom of the rim to the flange (40 mm = 7.0 L).
6. Grease and clean the pins for attachment to the DO batteries (large pin is +). Attach battery wires and start/check stirring assemblies.
7. Secure device in frame and set/secure for travel.

Boat Operations (Deployment)

1. Turn on stirrers by connecting the battery leads. Calibrate the DO meters and place probes in housing.
2. Check syringes for the removal of water.
3. Slide apparatus off boat and let down gently with marker buoy attached to frame.
4. Divers should remove the chamber unit and secure the bungee cords. One diver on each side should lift the apparatus out and away from the frame. Then the divers should flip the chamber unit over to release all air bubbles. Make sure that stirrers are operating.
5. Look for flat area to place chambers and set chambers down gently on sediment, pressing down until the gray flange is sitting on the sediment surface. Again check the stirrers. DO chamber wires should be bundled together.
6. Take water samples (200 mL) with syringes. Filter samples through 0.2 μm filter.
7. Take DO and temperature readings at 5 min intervals, recording data on the sample sheet. Such sheets should be part of the overall data sheet (see BIFS field data sheet summary).
8. After a significant drop in DO has been recorded (1–3 h), divers should take a final water sample with the syringes (same as the earlier method). The chamber should be picked up and turned upside down to check for the impeller movement (sediment may interfere with such mechanism). Chamber unit should be placed in lander housing and attached with bungee cords. Dive lifts should be inflated and divers should escort apparatus to surface. The apparatus should be carefully placed on the boat and the battery leads detached.

Determination of Oxygen Consumption Rates

$$\text{Rate R} = \frac{(\text{Final O}_2\text{, mL L}^{-1} - \text{Initial O}_2\text{, mL L}^{-1})(\text{Volume of chamber} = 7\text{ L})}{\left(\text{Area of chamber, 0.091 m}^2\right)\left(\text{Total time, h} - \text{T}_{\text{final}} - \text{T}_{\text{inital}}\right)}$$

Conversion of mg to mL/L: mg = 0.7 mL

A.6.3 Laboratory (In Vivo) SOD Determinations

Four replicate samples of undisturbed core samples were taken with the SOD corer apparatus at each of the three stations. Sediment filled the corer to at least half the height of the PVC coring tube. Water filled the rest of the corer. Great care was taken so that there was minimal disturbance to the sample. Detachable cores were capped and placed in racks. The samples were sealed with PVC end caps and placed in an upright position in the rack in a Gott cooler. The seawater bath in the Gott cooler was maintained at the level of water in the cores to lessen pressure on the seals. The samples were taken immediately to the laboratory and equilibrated for 10 h with gentle aeration without disturbing the sediment. Laboratory water temperature was controlled at field levels.

We followed the batch method of analysis. The laboratory SOD measurements were made on undisturbed sediments in the original core tubes. Water circulation was maintained with an internal mixing device. Two replicates from each station set were poisoned with HCN; the remaining two samples were left without any additions or changes. Dissolved oxygen probes were pre- and postcalibrated using the micro–Winkler technique. Temperature and depletion of dissolved oxygen were monitored (without stirring) every 30 min for 4–6 h or until 1.5–3.0 mg L^{-1} DO depletion was obtained. If the initial DO reading in the cores was below 4.0 mg L^{-1}, the core water was aerated without sediment disturbance in order to have an initial DO reading of at least 4.0 mg L^{-1}.

DO depletion curves were constructed with the cumulative DO depletion data plotted against time. The slope was calculated from the linear portion of the DO depletion curve using the first degree regression function: the results are to be represented in units of g O_2 m^{-1} h^{-1}.

Primary Productivity (In Situ Light/Dark Bottle Study)

Primary productivity is the rate at which IC is converted to organic carbon via chlorophyll-bearing organisms such as phytoplankton. The plankton respiration term was an important part of evaluating the oxygen-uptake mechanism in estuaries during the 43-year study period. Therefore, we evaluated the phytoplankton net and gross photosynthesis using the classical light and dark bottle oxygen technique (Strickland, 1960, Standard Methods, 1989). Some unpublished data using the C_{14} uptake method were obtained by personnel of the Environmental Research Laboratory, Gulf Breeze for Perdido Bay. Data taken using this method were comparable to our own.

Presently, there is no available method of measuring primary productivity that is free of artifacts. Various authors have recently emphasized the importance of production processes occurring at strong physical gradients (Legendre and Demers, 1985), and an emerging area of emphasis has concentrated on ecohydrodynamics. The approach in this study examined bay primary productivity in terms of vertical stratification and its relationship to the euphotic zone, the longitudinal salinity gradient in the bays, and temporal patterns over the warm season when hypoxic conditions become prevalent. The scaling of measurements was made mostly at the meso- to macroscale; fine-grain scaling of observations was based on the present design. Integral water column rates of photosynthesis allowed an estimate of system production minus the relatively small contribution from SAV and a somewhat more important contribution of phytomicrobenthos.

The approach here is developed from that described in Standard Methods (1989). All procedures involving bottle cleaning, checking for supersaturation, the handling of water samples, and calculations are given in the previous citation. The oxygen method will be used whereby clear (light) and darkened (dark) bottles will be filled with water samples and suspended at regular depth intervals for appropriate intervals. The duration of exposure will depend on the rate of net photosynthesis and respiration with a need to avoid bubble formation in light bottles or hypoxia in dark bottles except at depths where hypoxia is present. We will start with sunrise to sunset deployment with possible adjustment to 0900–1500 hours or some equal distribution of time around noon (CST). Any given bubble development (>0.3 mm diameter) at a single depth will not result in a change in deployment schedule. Bubble development at different depths will be reason to change such deployment based on an initial screening by periodically checking the bottle strings.

Endpoints

Changes in the concentration of dissolved oxygen will be measured with a YSI model 58 DO meter equipped with a BOD probe. A magnetic stirrer and stirring bar will be required (a 12-volt system will be used). The specification of this DO unit is ±0.01 mg L^{-1} with an accuracy of ±0.03 mg L^{-1}. These specifications are minimally satisfactory for the warm season in Perdido Bay. More sensitive measurement equipment is not seen to be seaworthy in a small research vessel. If necessary, a more sensitive system was available at a shore-based system, which required transport of samples to this laboratory setup. The present method allowed a spatial coverage that was important to the project.

Methods

Two numbered light bottles and two dark bottles were used for each depth at each station (see Figure 1 for a schematic of the experimental setup). Nonbrass ("marine") wire was used to attach a strong (plastic or stainless steel) snap catch to the neck of each 300 mL BOD bottle. Wheaton "400" brand borosilicate (high purity) glass bottles were used. Plastic bottle caps were used to protect the stoppers. The bottles were to be acid cleaned (warm 10% HCl) and rinsed with distilled water prior to use. Just before the filling of each bottle, the container was rinsed with the water to be tested.

Phosphorus-containing cleaning agents were not used. The experimental setup was constructed so that the supporting lines or racks did not shade the suspended bottles.

Calculations

The increase in oxygen concentration in the light bottle during incubation is a measure of the net production (somewhat less than gross production due to respiration). The loss of oxygen in the dark bottle is an estimate of the total plankton respiration:

$$\text{Net photosynthesis} = \text{Light bottle DO} - \text{Initial DO}$$

$$\text{Respiration} = \text{Initial DO} - \text{Dark bottle DO}$$

$$\text{Gross photosynthesis} = \text{Light bottle DO} - \text{Dark bottle DO}$$

Results from duplicates were averaged. The calculations of the gross and net production for each incubation depth were plotted:

$$\text{mg carbonfixed/m}^2 = \text{mg oxygen released/l X} \times 12/32 \times 1000 \times \text{K}$$

where K = photosynthetic quotient ranging from 1 to 2, depending on the N supply. One mole of oxygen (32 g) is released for each mole of carbon (12 g) fixed. The productivity of a vertical column of water 1 m square is determined by plotting productivity for each exposure depth and graphically integrating the area under the curve.

COMPARATIVE ANALYSIS OF MAJOR COASTAL SYSTEMS

Study Area

This study is based on a series of long-term analyses of the subject river-dominated estuaries on the northern coast of the eastern Gulf of Mexico.

Field Collections

Sampling methods used for physicochemical and biological studies were identical in all three systems. Detailed descriptions of methods for the field collection of physical/chemical data are given earlier. River flow information was obtained from the U.S. Geological Survey in the respective river systems. Secchi depths were taken with standard Secchi disks. Samples for water chemistry analysis (surface and bottom) were taken with 1 L Kemmerer bottles at the fixed stations. Salinity was measured with temperature-compensated refractometers and salinity probes. Water color was measured with an APHA platinum-cobalt standard test (colorimeter, Pt-Co units). Turbidity was determined with a Hach model 2100-A turbidimeter. Dissolved oxygen and water temperature were taken with YSI-dissolved oxygen meters with calibration by modified Winkler methods. Dissolved oxygen anomaly was calculated from the field measurements as the difference between the measured dissolved oxygen and the oxygen solubility at a given temperature and salinity (Weiss, 1970). Sediment samples (top 2 cm) were taken with PVC corers (7.6 cm d) and were analyzed for particle-size distribution and organic composition according to methods described by Mahoney and Livingston (1982) as defined by Galehouse (1971).

Biological samples (infauna, epibenthic macroinvertebrates, fishes) were taken monthly at fixed stations in the respective systems. Infaunal macroinvertebrates were taken using coring devices (7.6 cm diameter, 10 cm depth). The number of cores taken (10–12) was determined using a species accumulation analysis based on rarefaction of cumulative biological indices (Livingston et al., 1976). All infaunal samples were preserved in 10% buffered formalin in the field, sieved through 500 μm screens, and identified to species wherever possible. Epibenthic fishes and invertebrates were collected with 5 m otter trawls (1.9 cm mesh wing and body, 0.6 cm mesh liner) towed at speeds of about 3.5–4 km h^{-1} for 2 min resulting in a sampling area of about 600 m^2 per tow. Repetitive samples (2 or 7) were taken at each site; sampling adequacy was determined by Livingston (1976a). All organisms were preserved in 10% buffered formalin, sorted and identified to species, counted, and measured (standard length for fishes, total length for penaeid shrimp, carapace width for crabs). Representative samples of fishes, invertebrates, and infauna were dried and weighed, and regressions were run so that data from the biological collections could be converted into dry and ash-free dry mass. The numbers of organisms were then converted to biomass m^{-2}.

A 2-year series of data in each system was chosen based on comparable river flow conditions; extremes of drought and flood conditions (Livingston et al., 1997) were avoided for comparative purposes.

Trophic Analyses

Methods of analysis of the feeding habits of invertebrates and fishes are described in detail by Livingston et al. (1997). An outline of the trophic relationships of most of the subject infaunal and epifaunal macroinvertebrates and fishes is given by Livingston (1997b) and Livingston et al. (1997). Ontogenetic feeding units were determined from a series of detailed stomach content analyses carried out with the various epibenthic invertebrates and fishes in the region (Sheridan, 1978, 1979; Laughlin 1979; Sheridan and Livingston, 1979, 1983; Livingston, 1980, 1982, 1984a, unpublished data; Stoner and Livingston, 1980, 1984; Laughlin and Livingston, 1982; Stoner 1982; Clements and Livingston 1983, 1984; Leber, 1983, 1985). Based on the long-term stomach content data for each size class, the fishes and invertebrates were reorganized into their trophic ontogenetic units using cluster analyses. Infaunal macroinvertebrates were also organized by feeding preference based on a review of the scientific literature (Livingston, unpublished data). The field data were then reordered into trophic levels so that monthly changes in the overall trophic organization of East Bay could be determined over the study period. The assumption that feeding habits (at this level of detail) did not change over the period of observation is based on previous analyses of species-specific fish feeding habits that remained stable in a given area over a 6–7-year period (Livingston, 1980). All biological data (as biomass m^{-2} mo^{-1} of the infauna, epibenthic macroinvertebrates, and fishes) were transformed from species-specific data into a new data matrix based on trophic organization as a function of ontogenetic feeding stages of the species found in the respective bays over the specified sampling periods.

Biomass data from various stations in Apalachicola Bay, Choctawhatchee Bay, and Perdido Bay were used (g m^{-2}) for all statistical analyses for the designated 2-year periods. The data were translated into the various trophic levels that included herbivores (feeding on phytoplankton and benthic algae), omnivores (feeding on detritus and various combinations of plant and animal matter), primary carnivores (feeding on herbivores and detritivorous animals), secondary carnivores (feeding on primary carnivores and omnivores), and tertiary carnivores (feeding on primary and secondary carnivores and omnivores) (Livingston et al., 1997). Data were given as ash-free dry mass m^{-2} mo^{-1} or as % ash-free dry mass m^{-2} mo^{-1}.

Statistical Analyses

Data Transformations and Sample Comparisons

Scattergrams of the long-term field data were examined and either logarithmic or square root transformations were made, where necessary, to approximate the best fit for a normalized distribution. These transformations were used in all statistical tests of significance. Statistical comparisons were made of the monthly field physical/chemical and biological data by station. Independence tests were used to examine each sample for autocorrelation at a number of lags. Autocorrelation was computed for the lesser of 24 or n/4 lags. The Q statistic of Ljung and Box (1978) was computed for an overall test of the autocorrelations as a group, where the null hypothesis was that correlation at each lag was equal to zero. The parametric F-test for comparison of variances and t-test for comparison of means were used to test the hypothesis that the data came from normally distributed populations. Means testing was carried out with both parametric and nonparametric tests. For independent, random samples from normally distributed populations, the parametric t-test was used to compare the sample means. Where there was serial dependence of data from a given pair of stations, it was removed by differencing the observations and calculating and plotting the autocorrelations of the differences. If the differences were not serially correlated, the Wilcoxon sign-ranked test was applied to the differences to compare the two sets of numbers.

Principal Components Analysis and Regressions

Principal component analysis (PCA) was determined using monthly data and the SAS™ statistical software package. A PCA was carried out as a preliminary review of the data using the river information, sediment data, and water quality variables. The PCA was used to reduce the physical–chemical variables into a smaller set of linear combinations that could account for most of the total variation of the original set. For the physical–chemical variables in this study, the series were stationary after appropriate transformations; thus, the sample correlation matrix of the variables was a good estimate of the population correlation matrix. Therefore, the standard PCA could be carried out based on the sample correlation matrix.

A matrix of the water and sediment quality data associated with the sampling stations in each system was prepared. The data were grouped over 2-year periods by month. Values for the dependent variables (herbivore biomass m^{-2}, omnivore biomass m^{-2}, primary carnivore biomass m^{-2}, secondary carnivore biomass m^{-2}, tertiary carnivore m^{-2}) were then paired with the water/sediment quality data (independent variables) taken for stations within each sector. Data from the lower trophic levels were added to matrices for the carnivore series. Unless otherwise defined, these statistics were run using SAS, Systat™, and SuperAnova™. Data analyses were run on the three independent data sets. Significant principal components were then used to run a series of regression models with the biological factors as dependent variables. Residuals were tested for independence using serial correlation (time series) analyses and the Wald–Wolfowitz (Wald and Wolfowitz, 1940) runs test. A chi-square test was run to evaluate normality. In most cases, the data were normally distributed whereas some data sets showed dependence.

OTHER METHODOLOGIES: BIOLOGICAL SOPS

Phytoplankton

Introduction

In addition to using the background (biological) data for hypothesis development in terms of experimental initiatives, various biological indices (which include phytoplankton, infaunal macroinvertebrates, epifaunal macroinvertebrates, and fishes) and the trophic equivalents (transformations based on extensive data concerning feeding processes and natural history phenomena) will be used to evaluate the present status of the Econfina and Fenholloway systems.

Our research group has carried out a series of studies concerning a series of drainage systems over the past 20 years, and we will use such data as part of the comparative analysis of the Econfina and Fenholloway systems. These studies have been carried out with submerged aquatic vegetation, infaunal and epifaunal macroinvertebrates, and fishes. Our phytoplankton and zooplankton collections are based on methods that are commonly used in the literature. We have quantified various other biological sampling techniques so that we can estimate relative effectiveness of such samplers and so that we can use a standard method for comparative purposes. In most cases, multiple pseudoreplicates (e.g., not true replicates in a statistical sense) were taken for the purpose of estimating sampler precision (accountability of variation within the capability of the given sampler). Quantification was based on an evaluation based on the rarefaction analysis of cumulative biological indices as a function of multiple subsamples. The asymptotic relationship of species accumulation curves was used to determine the sampling effort for a given sampler.

The results of these preliminary studies are given in a series of papers that have been published in peer-review journals (Livingston, 1976; Livingston et al., 1976b; Stoner et al., 1983). In this way, various methods that have been used to collect organisms in the Econfina/Fenholloway system will allow us to compare the results with those taken in other systems in a manner that is both consistent and standardized according to the demonstrated effectiveness of the various sampling techniques.

The following outline will provide the working order of the comparative analyses:

1. Determination of station comparability based on physical habitat conditions (e.g., salinity, surface, and bottom, in areas deeper than 1 m).
2. Determination, wherever possible, of comparable water years in terms of freshwater runoff in the respective systems.
3. A comparison of the age structure and size of important fish and invertebrate populations can provide insight concerning the ecological status of the Econfina and Fenholloway systems. We will focus on any indications of life history disruption or dysfunction when compared with systems that are known to be relatively free of human impact. This will include an analysis of habitat-associated trends in the Econfina and Fenholloway systems that could be associated with the possible absence of important groups of organisms. The evaluation of seasonally smoothed data will concentrate on the following indices of the above biological groups:
   a. Growth rates by population
   b. Numbers and biomass (total biota and top dominants)
   c. Species richness
   d. Species diversity and evenness
   e. Trophic units
   f. Guild associations

We will look for comparative trends that will include similarities and differences due to natural and anthropogenic stress.

Sample Collection

Phytoplankton were taken in various ways. In freshwater areas, phytoplankton samples will be taken in three 1 L samples, which were run through 15.2 cm (D) plankton nets (28 μm mesh). In the estuarine and open gulf areas, samples were taken with multiple (two nets, bongo configuration, duplicate runs) 64 μm nets that were run through the water for periods of 1–2 min. The intake hoses were moved up and down continuously. The water was run through 64 μm nets. In addition, whole water samples were taken with 1 L jugs at the surface and pumped in areas below the surface. During some samplings in the bay, three 1 L samples were collected for phytoplankton analysis at the surface and the bottom.

Phytoplankton samples are collected using the following:

1. Water bottles (for whole water sample)
2. Nets (25 μm and/or 64 μm micron mesh sizes)
3. Pumps

Water Bottles

Whole water samples may be collected using commercial bottles constructed from or coated with an inert material such as PVC to eliminate chemical contamination. Throndsen (1978) recommended the use of clear glass bottles only, since colored bottles such as brown glass bottles would make it difficult to ascertain the amount of fixing agent needed (if iodine is the preservative). Avoid plastic bottles as they will take up iodine from the solution. The volume of the sample is a second consideration. The minimum volume of the sample from which cells must be concentrated to give reasonable data varies from a few milliliter in rich coastal or estuarine environments to a liter or more in oligotrophic regions (Venrick, 1978).

Pump Sampling

Application of pumps for collecting plankton samples is known regularly albeit perhaps not commonly, and it is not unusual for samples for the study of phytoplankton population abundance and taxonomic composition to be taken from pumped water. The entire size spectrum of the phytoplankton can be sampled from the same source with a pump.

Nets

Plankton nets have been widely used as sampling devices in phytoplankton investigations because of their selective and now predictable filtering properties; plankton nets are not recommended for use in qualitative phytoplankton sampling. Methods for the evaluation of the volume of water filtered through a plankton net (e.g., use of flow meters), which have been developed primarily for quantitative zooplankton sampling, are of little or no value in phytoplankton investigations. Sampling with nets provides material for qualitative purposes. The advantage of nets is the ease with which large volumes of water can be filtered and organisms concentrated. The main disadvantage is the distorted species composition shown by net samples.

A plankton sample is obtained from a particular depth layer by towing the net horizontally while the weight holds the net at the selected depth. Phytoplankton may also be concentrated from a selected depth by pumping water from the depth and filtering the water through a net on board ship.

In waters poor in phytoplankton, the material collected in single haul (vertical) may be insufficient. A large catch from the same water layer is obtained from an oblique haul. The advantage of oblique hauls over vertical hauls is that more water is filtered from the same water layer. After the surface tow, phytoplankton material adhering to the net is washed down to the tail and collected in the bucket by spraying the outer or inner surface with filtered seawater while the net is hanging with the mouth upward. Freshwater may destroy the organisms and should, therefore, not be used. In order to reduce uncertainties about geographical distribution of species, the use of one particular net should be restricted to a limited geographical area. After washing, the net should be dried in the air and stored in a dark, cool place. Care should be taken to prevent nylon nets from exposure to sunlight when drying.

Handling and Preservation

The various phytoplankton samples will be preserved with 5% formalin and with Lugol's solution. The handling of samples after collection is a critical stage in any phytoplankton investigation. It is very important to minimize quantitative and qualitative changes in the phytoplankton composition before further treatment of the samples takes place. Phytoplankton in water samples will keep their viability for some time provided they are not subjected to rise in temperature or light intensity. In net hauls on samples from blooms, however, the viable period may be rather short since most of the photo-autotrophic, phagotrophic, and saprotrophic species are highly concentrated.

Water samples should be kept at sea temperature if the latter is below 10°C–15°C or cooled (2°C–10°C) if temperature is above 15°C and stored in total darkness (Throndsen, 1978). Net samples should not be too dense, and they should be kept in rather large amount of water (250–1000 mL). Some light may be necessary to avoid oxygen depletion in the sample. Temperature should be low, 2°C–5°C, even in summer because heterotrophic organisms, including bacteria, will multiply rapidly in such a dense sample.

Fixation

No single method can be expected to be suitable for the preservation of all types of phytoplankton. The fixing and preserving agent should be chosen with regard to the aim of the investigation. Even though a great number of fixatives and preserving agents have been described, only a few have been used extensively to give any experience worth drawing upon. However, some of the recent ones, particularly glutaraldehyde, have given promising results (Taylor, 1976a).

The two most widely used fixatives for phytoplankton are formaldehyde and Lugol's solution. The river net phytoplankton are fixed in neutral or slightly alkaline formaldehyde with hexamethylenetetramine, sodium borate, or sodium carbonate. The whole water samples from river and bay stations and the bay net samples are fixed in acidified Lugol's solution. The neutral formaldehyde fixation (when properly carried out) renders diatoms, thecate dinoflagellates, and coccolithophorids in identifiable conditions. However, formaldehyde fixation is qualitatively selective in that it distorts all shape of naked species and causes flagella to be thrown off in many flagellates. Cell content will bleach, making it very difficult to distinguish between pigmented and nonpigmented cells.

Preparation of the Neutral Formaldehyde Solution (Throndsen, 1978)

Prepare the solution by diluting p.a. (pro analysis) grade formalin (40% formaldehyde HCHO) to 20% with distilled water. Add 100 g hexamethylenetetramine to 1 L of the 20% solution. For water samples, add 100 mL of sample to 2 nL of the 20% neutral formaldehyde (final concentration of HCHO is −0.4%); shake the bottle immediately to facilitate an instantaneous fixation.

For net hauls, add fixing/preserving agent to make up about one-third of the volume if the sample is dense. Mix well. It is important that fixation should take place immediately after the collection of each sample (the water sample should be added directly to the fixing agent in order to prevent any adverse effects of light and temperature changes). This is particularly important for net hauls where the high cell density may otherwise cause rapid decay. The Lugol's solution (iodine dissolved in potassium iodide solution was used). For marine phytoplankton, it is often used in its acid version (Lovegrove, 1960). The weakly alkaline Lugol's solution (with acetate) will be better for coccolithophorids but inferior to the acid one (acetic acid).

Fixing and preserving agent were utilized as follows. One hundred grams of potassium iodide is dissolved in 1 L of distilled water; then 50 g iodine (crystalline) is dissolved and 100 mL of glacial acetic is added. As the solution is near saturation, any possible precipitate should be removed by decanting the solution before use. For water samples, add enough fixing agent to give the sample a weak brown color (0.4–0.8 nL to 200 nL sample: shake well). Use clear glass bottles only, since colored bottles such as brown glass bottles would make it difficult to ascertain the amount of fixing agent needed. Avoid plastic bottles as they will take up iodine from the solution. It is important that an adequate amount of fixing agent be added, as too much iodine may give a heavy staining; excessive staining can be cleared up by adding sodium thiosulfate, which reduces the iodine. The main advantage of Lugol's fixative is that a large number of flagellates will retain their flagella. Other organisms of the phytoplankton community will also fix reasonably well. The cells will stain brownish yellow and hence be more easy to observe during the counting procedure, though many organic and inorganic particles will be stained as well.

In a typical offshore area with dinoflagellate and diatom domination, an acid formaldehyde fixation may be applied; in a tropical area with coccolithophorids predominating, neutralized formaldehyde is appropriate; in inshore area with naked flagellates as an important part of the community, the acid iodine method may give the best result from a combined quantitative and qualitative view.

Sample Analysis

Identification of all algal samples will be to species using scanning electron microscopy (SEM) methods for ascertaining the species designations. The species encountered have been largely described in two atlases written for the Florida Department of Environmental Regulation (Prasad and Livingston, 1987). In all sampling efforts, the nets will be rinsed into numbered jars and preserved in 5% formalin or Lugol's solution. The volume of each sample will be measured in a graduated cylinder. The sample was then stirred with a magnetic stirrer for 1–2 min, and a 0.1 mL subsample was pipetted out. To limit clumping and cell damage, the stirrer was turned off between subsamplings.

Each subsample was placed in a Palmer–Maloney counting chamber and the numbers of cells were counted by the "strip" count method. In strip counting, the top and bottom of the grid will be the "count" and "no-count" boundaries, respectively, and plankters were counted as they move across the center vertical line. Dead cells or diatoms with broken frustules were counted. Empty centric and pennate diatoms were counted separately as "dead centric diatoms" or "dead pennate diatoms" for use in converting the diatom species proportional count to a count per milliliter. A preliminary analysis (10 samples) was conducted to determine how many subsamples were necessary for each count. Based on previous experience, three subsamples provided a number within 22% of the mean of the 10 net subsamples, and this number was used for all counts. All samples were processed to species-level identifications, if possible. After preparation, electron micrographs were taken on a Polaroid 4 × 5 land film type 55/positive–negative using JEOL-JEM-100CXII scanning and transmission electron microscope operating at an accelerating voltage of 20 kV. Light microscopy was also used with a Nikon biological microscope fitted with a phase condenser; photographs were taken on Kodak 35-mm Panatomic-x fine-grain black and white film using a Nikon camera. Phase-contrast illumination was used for diatom studies. Soft-bodied algae were photographed using wet mounts in distilled water. Methylene blue and India ink were used to determine the extent of sheath formation. The structure of the pyrenoid and enclosed starch caps, flagellar number, and insertion were studied with I_2 in KI solution.

A conversion will be made of all phytoplankton data (numbers per species per division). This will be carried out using seasonal collections of phytoplankton from various portions of the study area. Regressions will be determined from the experimental work (comparisons of the numbers vs. the ash-free dry weight biomass).

Identification Problems

Any plankton investigations involving cell counts usually involve identification of organisms as well. The goal of the investigation determines the appropriate taxonomic level for such identification. Cell numbers of the main groups—blue-green algae, diatoms, coccolithophorids, and other flagellates—offer some ecological information since to a certain degree spatial and seasonal distribution patterns do exist. Identification was carried out at the generic level, as many genera, particularly of dinoflagellates and coccolithophorids have their main distribution in particular geographic and climatic zones. Although every effort will be made to identify organisms to the lowest taxonomic level, which time and capability permit, it has to be realized that limitations are set by methods used for preservation, concentration, and microscopical (both light and electron microscope) examination.

Diatoms — Identification of larger planktonic diatoms at generic level usually has fewer problems since so many generics have a characteristic gross morphology. Identification at the species level, for instance, of *Chaetoceros* and most *Rhizosolenia* species can be made during routine analysis of water samples. Difficulties may be encountered if *Rhizosolenia fragilissima*, *R. delicatula*, *Leptocylindrus danicus*, and narrow specimens of *Cerataulina pelagica* are present in the same sample, which sometimes happens. *Lauderia annulata*, *Detonula pumila*, and *Bacterosira fragilis* could be easily confused with each other as well as with *Thalassiosira* spp. Only a few of the 100 *Thalassiosira* species described can be identified when observed in light microscope with uncleaned samples. Value structure is diagnostic for these genera and species and, therefore, effort should be made to observe them in value view. Pennate diatoms in ribbon-shaped colonies also must be examined in valve view to permit positive identification of genus as well as species.

The use of more sophisticated methods is mostly necessary for species identification and also for identification at higher taxonomic levels when dealing with this group. Species contributing to the greatest cell numbers or biomass are usually regarded as the important ones in a quantitative phytoplankton investigation. If the important species cannot be correctly identified in the preparation used for cell enumeration and isolation of single cells for further treatment is impracticable, a concentrate of any remaining water sample may be used for examination of the quantitatively important species.

Net hauls collected at the same stations as the water samples are very useful in such cases as additional material for species identification. Examination of net hauls prior to analysis of water samples is recommended in order to be acquainted with some of the species present.

Electronic microscopy has revolutionized the concepts of morphology and taxonomy of most of the phytoplankton groups. Methods in current use for the study of diatoms have been discussed in detail by Hasle (1978).

Raw Material in Water Mounts — This method of examination in LM gives sufficient information to identify following genera and most of their species: *Bacteriastrum*, *Chaetoceros*, *Corethron*, *Ditylum*, *Guinardia*, *Leptocylindrus*, *Lithodesmium*, *Rhizosolenia*, *Skeletonema*, *Stephanopyxis*, *Thalassionema*, and *Thalassiothrix*. These diatoms are readily recognized by the shape of the colonies, gross morphology of cell, special structures such as setae, processes, elevations, etc.

Raw Material in Permanent Slides — The resin hyrax or nephrax is well suited for mounting raw diatom material; this technique is described in detail by Hasle (1978). The whole, intact diatom frustule and colonies can be studied and also the nucleus can be stained as an indication that the cell was live when fixed. The method has all the advantages of water mount examination and, in addition, offers the possibility of studying the finer details of the diatom cell wall and the girdle.

Cleaned Material in Permanent Slides — The purpose of cleaning diatoms with chemicals is to remove the cell contents and to separate the single components of the siliceous frustule. Then, special structures such as valve process and septa of bands as well as areolation of valves and bands are easily available for examination. Whenever possible within the limitation of LM, cell dimensions, number of marginal processes, areolae, and striae should be measured by LM, either on water or permanent mounts or on photomicrographs made from these mounts.

One of the many methods in use for oxidation of the organic material of diatom cells is as follows (after Simonsen, 1974; Hasle, 1978):

1. Rinse the sample with distilled water by centrifugation or passive settling.
2. Add an equal amount of saturated KMn04. Agitate. Leave for 24 h.
3. Add an equal amount of conc. HCl (be careful, because of dangers) to the sample and KMnO4 mixture. The solution turns dark brown. Heat gently over an alcohol lamp until it becomes transparent and colorless or light yellow-green.
4. Rinse with distilled water until sample is acid-free.

Cleaned diatom material should be stored in distilled water to which are added a few drops of the mixture of formaldehyde and acetic acid to hinder growth of bacteria and fungi and the dissolution of silica. Glycerin may be added to present desiccation. Storage in 70% alcohol is preferable if the material is to be used for SEM later on (the stored material can then be used without rinsing).

Preparing Diatom Slides (after Hasle and Fryxell, 1970)

1. Clean coverslips with alcohol to remove any oil.
2. Lay cleaned coverslips on a labeled tray.
3. With a new, disposable pipette, place one to four drops of cleaned sample on each coverslip, depending upon the density of the sample. It is desirable to have specimens distributed evenly so that one can be viewed and photographed alone. Use the pipette for only one sample to avoid mixing.
4. Dry over gentle heat or leave overnight. Protect from dust.
5. Add two to four drops of resin with high refractive index such as hyrax or nephrax to give contrast to the silica of the frustules.
6. Dry over gentle heat or leave overnight until the resin becomes firm. Protect from dust.
7. Adjust hot plate to moderate heat and clean and mark microscope slides. Heat slides, leaving one end of the plate for easy handling.
8. Place one slide face down on a prepared coverslip. Turn it over quickly when the resin melts enough to stick, and replace the slide on the hot plate.
9. Heat until resin has spread under the entire coverslip. Do not boil. Gently tap the coverslip with a wood stick to remove bubbles.
10. Cool the slide. Trim excess resin with razor blade or use a solvent. Seal with nail polish. Affix permanent label.

SEM and Transmission Electron Microscopy (TEM)

SEM

1. Pipette out 2–4 drops of cleaned material on aluminum stubs and leave for 24 h.
2. Place the stubs with dried material in the sputter coater and coat them with gold/palladium.
3. Examine under scanning electron microscope at an accelerating voltage of 20 kV.
4. Use 4 × 5 Polaroid negative film for taking electron micrographs.

TEM — Cleaned material is pipetted out on the formvar-coated copper grids (200 mesh) and allow it to dry. Then, copper grids are directly examined under transmission electron microscope at 50 kV.

Dinoflagellates — The generalized procedure for the handling of this group is given in Taylor (1976, 1978).

Routine fixation and preservation of those species with cell walls consisting of heavily developed cellulose plates is readily accomplished with weak formaldehyde (1%–2%). However, forms with very thin plates, or no wall at all, are only poorly fixed by this method, many disintegrating rapidly on contact with the fixative. Lugol's iodine solution fixes a greater proportion of a natural sample, but the naked forms are often distorted. Nevertheless, Lugol's iodine seems to be the best choice for quantitative estimates. The flagella are shed very rapidly in this group, being thrown off in the presence of most fixatives and also in living cells under stress.

Fixation with osmium tetroxide (OsO_4), either in the form of exposure to the vapor of a 2% solution or by the addition of drops to the sample, is the most widely effective fixative. Glutaraldehyde ($CH_2(CH_2CHO)_2$) (2%–5% solution) is also fairly effective for naked species, if fixation is carried out in refrigerator. Although not as dangerous as osmium tetroxide, glutaraldehyde is unpleasant to smell and should not be inhaled for the long periods usually involved in light microscopic examination.

In studies relating to primary productivity, data referring to photosynthetic members should be distinguished from those referring to nonphotosynthetic species. Unfortunately, formalin eventually bleaches the pigments and it is difficult to determine if cells subjected to lengthy preservation were originally pigmented or not. The starch reserve of dinoflagellates stains bluish with iodine, but this is usually masked by the dark reddish-brown general staining of the photosynthetic forms that produce it, and it is therefore not very useful as an indication of group affiliation under routine conditions (Taylor, 1978).

Light Microscopy — Athecate forms are indentified principally by the shape of their cells, and the detailed observation of the girdle and sulcal furrows is often difficult. Various types of shallow focus light microscopy, such as Nomarski interference microscopy, are useful for such purposes, and regular bright field systems can be manipulated unconventionally to good effect, either by opening the iris diaphragm much more than usual or by offsetting the condenser laterally by a small amount.

Thecate dinoflagellates require the determination of the complete "plate pattern" for critical taxonomy, although in routine practice, this is often not possible and identifications are based on size, shape, and the pattern of some areas. Optical interference from the cell contents often obscures the thecal sutures. The cell contents may also be plasmolyzed by addition of sucrose or salt to see thecal details more clearly.

Dinoflagellate cells can be suitably positioned in water mounts by light coverslip pressure or by placing isolated cells in warmed glycerin jelly (Graham, 1942), the aspect being fixed once the jelly hardens.

Staining of Dinoflagellates — Staining the cellulose plates of dinoflagellates can be accomplished by several techniques. Combinations of iodine with oxidizing agents will stain cellulose bluish or red brown. However, iodine stain of the plates may not be very effective when applied to formalin-preserved sample (Taylor, 1978). If so, trypan blue (also known as diamine blue 3B) in a 0.2%–0.3% solution can be tried. This dye is also effective on live material and stains the theca a pale blue.

Thecal Dissociation (after Balech, 1959) — Another approach to the examination of dinoflagellate plates is to dissociate the theca in a controlled manner. Sodium hypochlorite (commercial "bleach") is effective for this purpose. A drop of a concentrated solution is run under the coverslip and allowed to act for a few minutes, at which time the plates will separate with light pressure on the coverslip.

Electron Microscopy — The SEM has proved to be a considerable aid in the examination of thecal surface features. Many thecate forms are washed with the distilled water and then air dried on coverslips, their preparation being extremely simple. Formalin-preserved samples frequently lack the outer amphiesmal membranes and provide a clearer view of the thecal plate surfaces when viewed with the SEM. Air dried materials on the coverslip are coated with 10–30 nm coating of gold/palladium.

Prymnesiophytes Including Coccolithophorids — The identification of prymnesiophytes is based almost entirely upon the structure and arrangement of scales (in some known as coccoliths), with only minor emphasis on other features such as the cell shape. Identification to genus and species level is hampered by the minuteness of the cells and their individual seals. As a matter of fact, prymnesiophytes can only be satisfactorily described through the combined use of light microscopy, SEM, and TEM. In order to obtain the most complete information for identification, it is recommended that the same specimen be studied under both light and electron microscopes whenever possible (Heimdal, 1978).

Light Microscopy — LM observations of water mounts give information on cell shape and on possible spines. Larger species can be identified using a dry objective (400×), while an immersion objective (1000×) is necessary to distinguish smaller forms measuring 5–10 less. The slides are generally examined with bright field illumination, the resolution being better in oblique light. Species with weakly calcified scales can be studied under phase contrast and dark field. Photomicrographs taken with such illumination should be accompanied by bright field photomicrographs for better comparisons and accurate measurements. The specimen being studied should be photographed with different levels of focus.

Electron Microscopy Techniques (after Heimdal, 1978) — TEM: Before examination in the electron microscope, salts and preservatives must be removed from the sample to prevent formation of crystals. If salts are not removed before the transfer of sample to grids for electron microscopy, the cells must be rinsed by overflowing them with neutral water. Another gentle way to remove salt crystals is to use bags made of dialysis tubing immersed in membrane filtered tap water at nearly neutral pH or buffered distilled water.

SEM: For examination under the SEM, the cleaned sample is mounted as for examination with the TEM or on a piece of coverslip. Then the grid or coverslip is cemented to a metal stub as described by Heimdal (1978). These are mounted on a specimen stub with colloidal silver and coated with a thin film of carbon and metal (gold/palladium). Thin coatings are desirable so that fine surface details will not be obscured and so that the specimen can be observed again under the TEM if required.

An accelerating beam voltage of 30 kV is generally used at high magnification (20,000×–50,000×) because significantly better resolution can be achieved.

Carbon replica must be prepared because the coccoliths are too opaque to be penetrated by the rather weak electron beam of the TEM. For replication, the grid should be placed with mounted material in a vacuum evaporator. The coccoliths are lightly shadowed by evaporating gold–palladium under vacuum at an angle of 30° or 45° to the horizontal and coated with carbon at an angle normal to the surface.

Nanoplanktonic organisms represent a major component of the phytoplanktonic biomass and productivity in oligotrophic marine water. These organisms involved in microbial food chain remain, however, poorly known probably because their identification requires time-consuming study of fine structural details using electron microscopy and also because their smallness and fragility make them largely unobservable when the traditional methods for plankton collection and light microscopic observation are utilized. The general problem related to the identification of nanoplanktonic flagellates is the lack of adequate fixing agents, and hence the study of living material is necessary for a reliable identification in the light microscope in some cases. For many species, the identification relies upon submicroscopical details to be revealed by the electron microscope only.

Examination of the material is preferably done in a compound light microscope equipped with phase contrast, differential interference contrast, and bright field optics. A good quality 100× objective is indispensable, and a calibrated ocular micrometer is a necessity observations on swimming behavior is preferable made at a medium magnification (200×–400×), whereas observation on cell morphology and anatomy should be made at 1000×–1200× as well. If necessary, cell motion may be slowed down by adding a drop of uranyl acetate solution (7 g of uranyl acetate + 100 mL of filtered Lea water).

Many "naked" nanoflagellates will be intact for only a few minutes under the microscope, also when a heat-absorbing filter is used (on the microscope lamp). Morphometric measurements and drawings of the cells in question are therefore important for further identification work. Another consequence of the short "microscopic life" of nanoflagellate naked cells is that it is very important to know what to look for.

Some of the characters that one should look for to identify the nanoflagellates are

I. Motility—number, equal or unequal nature, mode of use of flagella, presence or absence, and mode of use of haptonema
II. Pseudopodia, cell shape, cell symmetry, and asymmetry
III. Chloroplasts—its number, shape, and position
IV. Eyespot—whether intraplastidal on extraplastidal
V. Canal or gullet
VI. Special features such as ejectosomes periplast stripes, skeleton, and collar
VII. Cell wall, present or absent
VIII. Lorica—present or absent
IX. Storage products

After careful observation of the living material, following additional preparations stained and/or preserved in various ways may give further information that may help in identification:

1. Lugol staining/fixation will facilitate counting of flagella and measurement of their length; starch will be revealed. Lugol's solution (sensu Jane 1942) 6 g of KI, 4 g I_2, 100 mL of distilled water
2. Sudan Black—saturated solution in 70% C_2H_5OH (ethanol)
3. Sudan Blue—saturated solution in 50% ethanol
4. Uranyl acetate: 7 g uranyl acetate in 100 mL filtered seawater
5. Osmic acid (extremely dangerous): 200 mg OsO_4 in 10 mL distilled water
6. Brilliant cresyl blue (sensu Jane 1942)

The "naked" nanoflagellates often show a great variation in shape, and hence it is important to examine many specimens of the same species when a reliable identification is desired. Only occasionally will the natural concentration of cells be high enough to fulfill this requirement, and the concentration of the material has to be increased. Cultivation is a suitable method for increasing the concentration. Two methods are used: crude cultures and serial dilution cultures. Crude cultures are established by mixing seawater (inoculum) with a suitable seawater medium and leaving the culture in light. Dilution cultures are set up by stepwise dilution of the inoculum and adding subsamples of it to tubes containing seawater medium.

The satisfactory description of nanoplankton flagellates requires correlated light and electron microscopical observations. Three aspects of electron microscopy are important for the identification of flagellates: (a) shadow cast whole mounts observed in transmission electron microscope, (b) ultrathin sectioned material observed in TEM, and (c) SEM.

Shadow cast whole mounts permit satisfactory identification of species with distinctive extra cellular structures such as flagellar appendages for preparation of whole mounts; samples must be concentrated by gentle centrifugation usually in 10 mL tubes. Once a visible pellet of cells has been produced, samples are fixed for 5 min with 1% osmium tetroxide in 0.1 M cacodylate buffer at pH 7.0.

Subsequently, cells require a thorough washing in distilled water. To ensure clean preparations, the cells should be concentrated and resuspended in at least three separate washes of distilled water.

Drops of final concentrate may then be placed on formvar-coated copper grids and allowed to dry. Drops should also be placed on freshly cleaned slides or coverslips and these can subsequently be used for light microscopy or SEM. Shadow casting usually with chromium or an alloy of gold and palladium is carried out in a vacuum coating unit.

Some characters of importance were noted in the identification of flagellates (e.g., the number of microtubules in the haptonema, pyrenoid form can only be determined from material embedded and sectioned for electron microscopy. This involves fixation of cells in 5% glutaraldehyde followed by a wash in buffer solution and post osmication in 2% osmium tetroxide solution. After dehydration in an ethanol series, cells are embedded in an epoxy resin and sections cut with a glass or diamond knife. Full details of fixation and embedding procedures are given in Leadbeater (1972, 1978).

Whole mounts on slides, coverslips, or aluminum stubs may be uniformly coated with gold/palladium and used for SEM. Specimens can also be dried in a critical point drying apparatus so as to minimize the distortion caused by collapse of the cell during drying. Valuable information on flagellar insertion, arrangement of cell wall scales costal, can be obtained from SEM observations.

Estimating Cell Numbers

Phytoplankton abundance has primarily been measured and expressed as cell numbers based on enumeration of the phytoplankton in an aliquot of the sample. The use of numerical abundance as a measure of the phytoplankton standing stock is not without problems, however, partly attributable to the considerable interspecific and intraspecific differences in cell size characteristics of the phytoplankton. The significance of such size variability is that species that are relatively unimportant numerically may be very important in terms of biomass. Hence, abundance based on numerical census tends to overestimate small cells and underestimate the contribution of large cells (see Smayda, 1978). Biomass can be measured directly from chlorophyll, ATP, carbon, nitrogen, etc., or indirectly from the cell volume characteristics of the enumerated population. Biomass measurements do not estimate phytoplankton concentration but provide a measurement of some constituent common to the entire population that reflects numerical abundance.

Some species of phytoplankton are heterotrophic, notably dinoflagellates. Care must be taken not to ignore them as dead cells. Most pelagic species of diatoms form chains of attached cells. However, since the cell is the basic replicating unit, each cell in a colony or filament must be counted. The trichomes of blue-green algae such as *Oscillatoria* and *Trichodesmium* often collect into bundles. For such species, in which enumeration of the individual trichomes is difficult (Smayda, 1978), the number of colonies is enumerated. Unidentifiable cells should also be recorded. For this purpose, one may use either such broad designations as "flagellates" and "coccoids" or some appropriate description that may eventually permit taxonomic assignment.

Phytoplankton cells are counted with a standard (noninverted) light microscope furnished with a counting stage as described by Semina (1978). Before counting, the sample is agitated thoroughly by continuous shaking. Water with phytoplankton (0.05 or 0.1 mL) is taken with a pipette and transferred onto a glass slide, and the drop on the glass slide is covered by a cover glass and placed on the microscope counting stage. Cells are counted under with a magnification of 400×. Depending on the cell concentration, either the whole sample or a part of it is counted. If the sample was taken with a bottle, the cell number per liter is calculated after making an adjustment for the sample having been concentrated. For example, suppose the initial sample volume (Vi) was 1 L. Then the sample was concentrated to 10 mL (1/2) from which three subsamples of 0.05 mL each (V3) was taken. Assuming that no cells were lost during concentration, the number of cells per liter is obtained by multiplying the result of the count by N = V2/V3 = 666.6.

Counting Using the Inverted Microscope (Utermohl, 1931, 1958) — The microscope is inverted in the sense that the light source and condenser illuminate the chamber from above and the objectives view the specimens from below through a thin bottom plate or chamber. Chambers in 2, 10, or 50 mL can be used, depending upon the density of cells. This technique is reviewed in detail by Hasle (1978).

Counting Slides — The techniques are available for counting phytoplankton in the water sample. In the usual one, a concentrated preserved sample is examined in one or more counting slides such as Palmer–Maloney cell having chambers. In the second method, a preserved sample is concentrated by sedimentation in the separable plankton chamber designed for use with the inverted microscope.

The Palmer–Maloney Slide (Palmer and Maloney, 1954) Arthur H. Thomas Co., Philadelphia, PA.

Cover the chamber with a round or square 22 mm no. 1.5 coverslip, making secure contact with the glass or plastic chamber ring. Tilt the slide slightly (to right or left) and add 0–1 mL of sample via the lower of the two entry channels with a large-bore Pasteur pipette. A small piece of paraffin paper placed over each loading channel prevents evaporation and does not interfere with observation. It has a loading channel (slot) on each side. High dry objectives can be used. Cells as large as 150 nm will enter and be reasonably well distributed in this chamber.

GUIDE TO THE LITERATURE FOR IDENTIFICATION AND TECHNIQUES

There is no "plankton flora" of the world ocean. Nor does there exist a world review of any taxonomic group such as diatoms or dinoflagellates. The following selective list includes monographs, manuals, and other contributions that will be helpful in the phytoplankton identification, their preparation, and counting techniques.

Bushnell, V. C. (Ed.). 1972. Chemistry, primary productivity, and benthic algae of the Gulf of Mexico. Serial *Atlas of the Marine Environment*. Folio 22: 29 pp. + 6 plates. American Geographical Society, New York.

Hallegraeff, G. M. 1988. Plankton: A microscopic world. CSIRO Division of Fisheries, Hobart, Tasmania Australia. 112pp.

Hallegraeff, G. M. 2002. Aquaculturists' guide to harmful Australian microalgae. CISRO Division of Fisheries, Hobart, Tasmania, Australia. 111pp.

Prescott, G. W. 1962. *Algae of the Western Great Lakes Area*. W.C. Brown Co., Dubuque, Iowa. 977pp. [All sources I had gave printing date as 1962 and not 1971.]

Blue-Green Algae

Desikachary, T. V. 1959. *Cyanophyta*. Indian Council of Agricultural Research, New Delhi, India. 686pp.

Geitler, L. 1932. Cyanophyceae von Europa unter Beruksichtigung der anderen Kontinente. In: L. Rabenhorst (Ed.), *Kryptogamen-flora*. Akad. Verlag. Leipzig, Saxony, Germany. 14: 1–1196.

Diatoms

Barber, H. G. and E. Y. Haworth. 1981. A guide to the morphology of the diatom frustule. *Freshw. Biol. Assoc., Sci. Publ.* 44: 1–112pp.

Cupp, E. E. 1943. Marine plankton diatoms of the West Coast of North America. *Bull. Scripps. Inst. Oceanogr. Univ. Calif.* 5: 1–238.

Desikachary, T. V. 1988. Atlas of Diatoms. In: T. V. Desilachary (Ed.), *Marine Diatoms of the Indian Ocean Region*. Madras Science Foundation, Madras, India. Vol 5: plates 401–621.

Germain, H. 1981. *Flora des Diatomees*. Societe Nouvelle des Editions Boubee, Paris, France. 444pp.

Hagelstein, R. 1938. Scientific survey of Puerto Rico and the Virgin Islands. Botany of Puerto Rico and The Virgin Islands. Diatomaciae. *N.Y. Acad. Sci.* 8: 313–450.

Hendey, N. I. 1964. An introductory account to the smaller algae of British coastal waters. Part V: Bacillariophyceae (Diatoms). Fisheries Investigations, Series IV. Her Majesty's Stationery Office, London, England, U.K. Series 4: 317pp., 45 pls.

Hustedt, F. 1930. Die Kieselalgue. Part I. In: L. Rabenhorst (Ed.), *Krytogameu-Flora von Deutsehland, Osterreich und der Schweiz*. Akad. Verlag, Leipzig, Saxony, Germany. 7: 1–920.

Jensen, N. G. 1985. The Pennate Diatoms. A translation of F. Hustedt's (1959) "Die Kieselalgeu, 2. Teil" Koeltz Scientific Books, Koenigstein, Germany. 917pp.

Lebour, M. V. 1930. The planktonic diatoms of northern seas. *Ray Soc. Publ.* 116: 1–244.

Patrick, R. and C. W. Reimer. 1966. The diatoms of the United States. Vol. 1. *Monogr. Acad. Nat. Sci. Philad.* 13: 688pp., 64 pls.

Patrick, R. and C. W. Reimer. 1975. The diatoms of the United States. Vol. 2. Part I. *Monogr. Acad. Nat. Sci. Philad.* 13: 213pp., 28 pls.

Peragallo, H. and M. Peragallo. 1908. Diatomees marines de France et des districts maritimes voisins. Reprinted by A. Asher & Co. 1965. Amsterdam, the Netherlands. 491pp., 137 pls.

Rampi, L. and M. Bernhard. 1978. Key for the determination of Mediterraneau pelagic diatoms. CHEN. Dipartimento Radiazioni e Ricerche di Sicurezze e Protezione, Fiascherino, Roma, Italia. 71pp.

Round, F. E., R. M. Crawford, and D. G. Mann. 1990. *The Diatoms: Biology and Morphology of the Genera.* Cambridge University Press, Cambridge, Massachusetts. 747pp.

Saunders, R. P. and D. A. Glenn. 1969. Diatoms. Memoirs of the hourglass cruises. St. Petersburg, Florida. Fla. Dep. Nat. Resour. *Mar. Res. Lab.* Vol. 1: 1–119.

Simonsen, R. 1974. The diatom plankton of the Indian Ocean expedition of R/V "Meteor" 1964–1965. *Meteor Forsch. Ergebnisse, Riehe D.* 19: 107pp., 41 pls.

Dinoflagellates

Balech, E. 1959. Two new genera of dinoflagellates from California. *Biol. Bull.* 116: 195–203.

Dodge, J. D. 1975. The prorocentrales. *Bot. J. Linn. Soc.* 71(2): 103–125.

Kofoid, C.A. and O. Swezy. 1921. The free living unarmoured dinoflagellates. *Mem. Univ. Calif.* 5: 1–562.

Kofoid, C. A. and T. Skogsbert. 1928. The Dinoflagellata: The Dinophysoideae. *Mem. Mus. Comp. Zool. Harvard Col.* 51: 1–766.

Lebour, M. V. 1925. *The Dinoflagellates of the Northern Seas.* Plymouth: Marine Biological Association of the United Kingdom. 250pp.

Popovsky, J. and L. A. Pfiester. 1990. Süßwasserflora von Mitteleuropa. Vol. 6. Dinophyceae (Dinoflagellida). Gustav Fischer, Stuttgart, Germany. 272pp.

Schiller, J. 1931–1937. Dinoflagellatae (Peridineae). In: L. Rabenhorst (Ed.), *Krytogameu-Flora von Deutsehland, Osterreich und der Schweiz.* Leipzig Akad. Verlag, Saxony, Germany. Vol 10(3). Teil 1, pp. 1–617 (1931–33); Teil 2, pp. 1–590 (1935–1937).

Steidinger, K. A. and Williams. 1970. Dinoflagellates. Memoirs of the hourglass cruises. St. Petersburg, Florida. Fla. Dep. Nat. Resour. *Mar. Res. Lab.* Vol. 2(1): 1–251.

Subrahmanyan, R. 1968. The Dinophyceae of the Indian Seas. Part 1. Genus *Ceratium* Schrank. *Mar. Biol. Assn. India. Mem.* 2(2): 1–129.

Subrahmanyan, R. 1971. The Dinophyceae of the Indian Seas. Part 2. Family Peridiniaceae Schütt emend. Lindemann. *Mar. Biol. Assn. India. Mem.* 2(2): 1–334.

Taylor, F. J. R. 1976. Dinoflagellates from the International Indian Ocean Expedition. A report on material collected by the R. V. "Anton Bruun" 1965–1965. *Biblioteca Botanica* 132: 1–234, 46 pls.

Wood, E. J. F. 1954. Dinoflagellates in the Australian region. *Aust. J. Mar. Freshw. Res.* 5: 171–351.

Other Phytoflagellates (Including Chlorophyta, Euglenophyta, Chrysophyta, and Cryptophyta)

Butcher, R. W. 1959. An introductory account of the smaller algae of British coastal waters. Part I. Introduction and Chlorophyceae. Fisheries Investigations, Series IV. Her Majesty's Stationery Office, London, England, U.K. 74pp.

Butcher, R. W. 1961. An introductory account of the smaller Algae of British coastal waters. Part VIII. Euglenophyceae—Eugleninae. Fisheries Investigations Series IV. Her Majesty's Stationery Office, London, England, U.K. 17pp.

Butcher, R. W. 1967. An introductory account of smaller algae of British coastal waters. Part IV. Cryptophyceae. Fisheries Investigations, Series IV. Her Majesty's Stationery Office, London, England, U.K. 54pp.

Ettl, H. 1978. Xanthophyceae. In: H. Ettl, J. Gerloff, H. Henig, and D. Mollenhauer (Eds.), *Süßwasserflora von Mitteleuropa.* Fischer Verlag, Stuttgart, Germany. Vol. 3: 530pp.

Ettl, H. 1983. Chlorophyta I. Phytomonadina. In: H. Ettl, J. Gerloff, H. Heynig, and D. Mollenhauer (Eds.), *Süßwasserflora von Mitteleuropa*. Gustav Fischer Verlag, Stuttgart, Germany. 807pp.

Gaarder, K. F. and G. R. Hasle. 1971. Coccolithophorids of the Gulf of Mexico. *Bul. Mar. Sci. Gulf Carib.* 21: 519–544.

Iyengar, M. O. P. and T. V. Desikachary. 1981. *Volvocales*. Indian Council of Agricultural Research, New Delhi, India. 532pp.

Leedale, G. F. 1967. *Euglenoid Flagellates*. Englewood Cliffs, Prentice Hall, NJ. 242pp.

Schiller, J. 1930. Coccolithineae. In: L. Rabenhorst (Ed.), *Krytogameu-Flora von Deutsehland, Osterreich und der Schweiz*. Akad. Verlag, Leipzig, Saxony, Germany. 10: 89–273.

Starmach, K. 1985. Chrysophyceae and Haptophyceae. In: H. Ettl, J. Gerloff, H. Heynig, and A. Pascher (Eds.), *Süßwasserflora von Mittleuropa*. Gustave Fischer Verlag, Stuttgart, Germany. 515pp.

Steidinger, K. A. 1972. Dinoflagellate species reported from the Gulf of Mexico and adjacent coastal areas. In: V. C. Bushnell (Ed.), *Chemistry, Primary Productivity and Benthic Algae of the Gulf of Mexico*. Serial atlas of the marine environment. Folio 22. American Geographical Society, New York. pp. 14–15.

Throndsen, J. 1969. Flagellates of Norwegian coastal waters. *Nytt Mag. Bot.* 16: 161–216.

Preparation and Counting Procedures

Beers, J. R. 1978. Pump sampling. In: A. Sournia (Ed.), *Phytoplankton Manual*. UNESCO Press, Paris, France. Monogr. Oceanogr. Method. 6: 41–49.

Hasle, G. R. 1978. Using the inverted microscope. In: A Sournia (Ed.), *Phytoplankton Manual*. UNESCO Press, Paris, France. Monogr. Oceanogr. Method. 6: 191–196.

Hasle, G. R. 1978. Diatoms. In: A. Sournia (Ed.), *Phytoplankton Manual*. UNESCO Press, Paris, France. Monogr. Oceanogr. Method. pp. 136–142.

Hasle, G. R. and G. A. Fryxell 1970. Diatoms: Cleaning and mounting for light and electron microscopy. *Trans. Am. Micro. Soc.* 89(4): 469–474.

Heindal, B. R. 1978. Coccolithophoids. In: A. Sournia (Ed.), *Phytoplankton Manual*. UNESCO Press, Paris, France. Monogr. Oceanogr. Method. 6: 148–150.

Leadbeater, B. S. C. 1978. Other flagellates. In: A. Sournia (Ed.), *Phytoplankton Manual*. UNESCO Press, Paris, France. Monogr. Oceanogr. Method. 6: 151–153.

Lovegrove, T. 1960. An improved form of sedimentation apparatus for use with an inverted microscope. *J. Cons. Perm. Internatl. Explor. Mer.* 25: 279–284.

Palmer, C. M. and T. E. Maloney. 1954. A new counting slide for nannoplankton. *Am. Soc. Limnol. Oceanogr. Spec. Publ.* 21: 6pp.

Semina, H. J. 1978. Treatment of an aliquot sample. In: A. Sournia (Ed.), *Phytoplankton Manual*. UNESCO Press, Paris, France. Monogr. Oceanogr. Method. 6: 181.

Smayda, T. J. 1978. Estimating cell numbers. In: A. Sournia (Ed.), *Phytoplankton Manual*. UNESCO Press, Paris, France. Monogr. Oceanogr. Method. 6: 165–166.

Sournia, A. (Ed.). 1978. *Phytoplankton Manual*. UNESCO Press, Paris, France. Monogr. Oceanogr. Method. 6: 1–337.

Tangeu, K. 1978. Nets. In: A. Sournia (Ed.), *Phytoplankton Manual*. UNESCO Press, Paris, France. Monogr. Oceanogr. Method. 6: 50–58.

Taylor, F. J. R. 1976. Flagellates. In: H.F. Steedman (Ed.), *Zooplankton Fixation and Preservation*. UNESCO Press, Paris, France. 4: 259–264.

Taylor, F. J. R. 1978. Dinoflagellates. In: A. Sournia (Ed.), *Phytoplankton Manual*. UNESCO Press, Paris, France. Monogr. Oceanogr. Method. 6: 143–147.

Throndsen, J. 1978. Preservation and storage. In: S. Sournia (Ed.), *Phytoplankton Manual*. UNESCO Press, Paris, France. 6: 69–74.

Utermöhl, H. 1931. Neue Wege in der quantitativen Erfassung des Planktons. *Verh. Int. Ver. Theor. Angew. Limnol.* 5: 567–596.

Utermöhl, H. 1958. Zur Vervollkommung der qualitativeu phytoplankton methodik. *Mitt. Int. Ver. Meor. Angew. Limnol.* 9(1): 1–38.

Willen, T. 1975. Biological long-term investigations of Swedish lakes. *Verh. Inter. Ver. Angew. Limnol.* 19: 1117–1124.

Phytoplankton Literature

Edmondson, W. T. 1972. Nutrients and phytoplankton Lake Washington. In: G. Likens (Ed.), Nutrients and Eutrophication. Am. Soc. Limnol. Oceanogr. Spec. Symp. 1: 172–193.

Geissler, U. and R. Jahn. 1984. Infraspecific taxa of diatoms as indicators of water quality? In: M. Ricard (Ed.), *Proceedings of the 8th International Diatom Symposium*, Paris, France, Aug. 17–Sept. 1 1984. Koeltz Scientific Books, Koenigstein, Germany. pp. 766–772.

Maestrini, S. Y., D. J. Bonin, and M. R. Droop. 1984. Phytoplankton as indicators of sea water quality: Bioassay approaches and protocols. In: L. E. Shubert (Ed.), *Algae as Ecological Indicators*. Academic Press, New York. pp. 71–132.

Marshall, H. G. 1982. Meso-scale distribution patterns for diatoms over the northeastern continental shelf of the United States. In: D. G. Mann (Ed.), *Proceedings of the 7th International Diatom Symposium, Philadelphia, Pennsylvania*, Aug. 22–27, 1982. Koeltz Scientific Books, Koenigstein, Germany. pp. 393–400.

Patrick, R. 1984. Diatoms as indicators of changes in water quality. In: M. Ricard (Ed.), *Proceedings of the 8th International Diatom Symposium*, Paris, France, Aug. 17–Sept. 1, 1984. Koeltz Scientific Books, Koenigstein, Germany. pp. 759–765.

Potts, M. 1980. Blue-green algae (Cyanophyta) in marine coastal environments of the Sinai Peninsula; distribution, zonation, stratification and taxonomic diversity. *Phycologia* 19: 60–73.

Premila, V. E. and M. Umamaheswara Rao. 1977. Distribution and seasonal abundance of *Oscillatoria nigroviridis* Thwaites ex. Gomant in the waters of Visakhapatnam Harbour. *Indian J. Mar. Sci.* 6: 79–91.

Reddy, P. M. and V. Venkateswarlu. 1986. Ecology of algae in the paper mill effluents and their impact on the River Tungabhadra. *J. Environ. Biol.* 7: 215–223.

Schoeman, F. R. and E. Y. Haworth (Convenors). 1984. *Eighth International Diatom Symposium*, Paris, France, Aug. 17–Sept. 1, 1984.

Shubert, L. E. (Ed.). 1984. *Algae as Ecological Indicators*. Academic Press. New York. 434pp.

Squires, L. E. and N. A. Sinnu. 1982. Seasonal changes in the diatom flora in the estuary of the Damour River, Lebanon. In: D. G. Mann (Ed.), *Proceedings of the 7th International Diatom Symposium*, Philadelphia, PA, Aug. 22–27, 1982. Koeltz Scientific Books, Koenigstein, Germany. pp. 359–372.

Stoermer, E. F. 1984. Qualitative characteristics of phytoplankton assemblages. In: L.E. Shubert (Ed.), *Algae as Ecological Indicators*. Academic Press, New York. pp. 49–67.

Sullivan, M. J. 1984. Mathematical expression of diatom results: are these "pollution indices" valid and useful? In: M. Ricard (Ed.), *Proceedings of the 8th International Diatom Symposium*, Paris, France, Aug. 17–Sept. 1, 1984. Koeltz Scientific Books, Koenigstein, Germany. pp. 772–776.

Wilderman, C. C. 1984. Techniques and results of an investigation into the autecology of some major species of diatoms from the Severn River Estuary, Chesapeake Bay, Maryland, U.S.A. In: M. Ricard (Ed.), *Proceedings of the 8th International Diatom Symposium*, Paris, France, Aug. 17–Sept. 1, 1984. Koeltz Scientific Books, Koenigstein, Germany. pp. 631–643.

INFAUNAL MACROINVERTEBRATES

Collection

Multiple core samples (7.6 cm diam., 4.5 cm^2, 10 cm deep) were taken at each station. The number of cores was determined using large samples (40 cores) and a species accumulation analysis. Twelve cores per station have in the past given a representative sample of the benthic macroinvertebrates based on this analysis (75% of all species taken in the 40-core total sample). In cases where cores were not taken, we used suction dredges. Suction was provided by a 3.0-HP gasoline engine attached to a water pump. Samples were taken by inserting a 20.0 cm (d) PVC tube into the substrate and all of the material was removed to a depth of 10 cm, if possible. In addition, a 320 cm^2 sampling quadrat was placed 10 cm into the substrate and all of the material was taken with the suction device. Three samples per station were taken and placed in 50 μm bags.

As part of the monitoring program, we used leaf packs to determine the biota of the Econfina and Fenholloway Rivers. The leaf pack studies were based on a series of analyses carried out as described by Livingston (1985). Due to the results of the species accumulation tests with multiple samples, we found that asymptotes were noted somewhere between four and five leaf pack samples. Subsequent colonization tests using MacArthur–Wilson calculations indicated equilibria somewhere between one and two weeks of field exposure. Five leaf packs were placed at each station (stations determined at a later date). Each leaf pack consisted of 20 3.0 cm^2 Teflon "leaves," which were placed in an 8.0 mm mesh bag. The area of the Teflon "leaves" approximated 1000 cm^2. The leaf packs were retrieved about 4 weeks after placement. The leaf packs were tethered at each site by lines that were anchored to various overhangs. Leaf pack samples were washed through 500 μm sieves; the faunas were preserved in 10% buffered formalin, counted, and identified to species.

Handling and Preservation

All samples were preserved in 10% buffered formalin in the field. Samples were sieved through 500 μm sieves, US Standard #35, in the laboratory. Rose bengal stain was added to aid in sample picking. Picking and sorting will be carried out with continuous quality control including sample exchanges and blind counts by third parties on every fifth sample.

Sample Analysis

All macroinvertebrates were identified to species where possible with specific determinations verified by outside experts. Reference specimens of all species and a complete voucher collection are maintained in the laboratory.

FISHES AND EPIBENTHIC MACROINVERTEBRATES

Sample Collection

Fish and epibenthic macroinvertebrate populations will be monitored monthly in the respective streams using seines and drop nets. Estuarine samples will be taken using otter trawls and a variety of fish traps. Offshore (gulf) samples will be taken with otter trawls:

1. Otter trawl. A 5 m otter trawl (20 mm mesh wing and body; 6 mm mesh liner) will be used to collect fishes from the estuary. Repetitive (7) 2 min tows at speeds of –3 km h^{-1} are made at each station.
2. Seine (freshwater). A 16′ by 4′ minnow seine with 1/4″ mesh will be used in conjunction with multiple core samples (7.6 cm diam., 4.5 cm^2, 10 cm deep) that were taken at each station. The number of cores was determined using large samples (40 cores) and a species accumulation analysis. Twelve cores per station have in the past given a representative sample of the benthic macroinvertebrates based on this analysis (75% of all species taken in the 40-core total sample). In cases where cores were not taken, we used suction dredges. Suction was provided by a 3.0-HP gasoline engine attached to a water pump. Samples were taken by inserting a 20.0 cm (d) PVC tube into the substrate and all of the material was removed to a depth of 10 cm, if possible. In addition, a 320 cm^2 sampling quadrat was placed 10 cm into the substrate and all of the material was taken with the suction device. Three samples per station were taken and placed in 50 μm bags.

As part of the monitoring program, we used leaf packs to determine the biota of the Econfina and Fenholloway Rivers. The leaf pack studies were based on a series of analyses carried out as described by Livingston (1985). Due to the results of the species accumulation tests with multiple samples, we found that asymptotes were noted somewhere between four and five leaf pack samples. Subsequent colonization tests using MacArthur–Wilson calculations indicated equilibria somewhere between one and two weeks of field exposure. Five leaf packs were placed at each station (stations determined at a later date). Each leaf pack consisted of 20 3.0 cm^2 Teflon "leaves," which were placed in an 8.0 mm mesh bag. The area of the Teflon "leaves" approximated 1000 cm^2. The leaf packs were retrieved about 4 weeks after placement. The leaf packs were tethered at each site by lines that were anchored to various overhangs. Leaf pack samples were washed through 500 μm sieves; the faunas were preserved in 10% buffered formalin, counted, and identified to species.

Handling and Preservation

All samples were preserved in 10% buffered formalin in the field. Samples were sieved through 500 μm sieves, US Standard #35, in the laboratory. Rose bengal stain was added to aid in sample picking. Picking and sorting will be carried out with continuous quality control including sample exchanges and blind counts by third parties on every fifth sample.

Sample Analysis

All macroinvertebrates were identified to species where possible with specific determinations verified by outside experts. Reference specimens of all species and a complete voucher collection are maintained in the laboratory.

Beach seines were used to collect fishes from the river stations. Sweeps will be made along the shoreline until no new specimens are collected. Efforts are made to adequately sample major microhabitat types.

All fishes collected that are under 200 mm standard length will be preserved immediately in 10% buffered formalin and returned to the laboratory for identification, measurement, and trophic analysis. Larger fishes will be identified, measured, and released.

Handling and Preservation

Invertebrate samples were stored in glass quart jars with white plastic covers. After the cores were picked, samples were stored in 5 mL shell vials. Cores were preserved in 10% buffered (borax 200 g 2.5 L^{-1}) formalin and stained with 50 mL rose bengal (0.75 g rose bengal L^{-1} water). Pickled samples were stored in100% denatured ethanol. Cores were taken with a 7.5 cm (3 in.) PVC pull coring device. Cores were then placed in glass quart jars, preserved with formalin, and stained with rose bengal. Materials used included the following:

- Pull corers on gravity cores (no snatches or grabs), all 7.5 cm PVC heads
- Sieve 500 μm stainless steel
- Glass quart jars

The following materials were used in the storage of the fish collections:

- Glass quart jars with plastic lids.
- 10% buffered formation (Borax 200 g 2.5 L^{-1}).
- Stored in 50% denatured isopropyl.
- Small fish were placed in quart jars (less than 200 mm), and large fish were measured, identified, sexed (if possible), and released.

MACROINVERTEBRATE TAXONOMY AND ANALYSIS

a. Laboratory's macroinvertebrate Reference Collection catalogue listed in Appendix C
b. Equipment used to identify the organisms after they have been isolated from the sample
   - Wild M5A dissecting microscope
   - Dyonics fiber-optics light source
   - Nikon Alphaphot compound microscope

Sample Custody

1. Samples were delivered (**chain-of-custody form signed**). A copy of the chain-of-custody form was given to the person transferring the samples; the original was kept with the samples. Samples were then stored in a **secure area.**
2. Samples were sieved (if required) and date and time recorded (ID # kept with each sample).*
3. Organisms were picked (if required) and date and time recorded (ID # kept with the sample).
4. Preparation of specimens was made for identification (ID # kept with the sample).
5. Data were recorded in proper form for reporting to the customer.
6. Data, samples, and reference collection were transferred to the client with a signed chain-of-custody form and a copy of the original requested.

Analytical Procedures

Organisms found in freshwater systems often occur in large numbers; these include Chironomidae larvae and Oligochaete worms. Frequently, a compound microscope is needed for proper identification of these groups.

Slides must be made for the compound scope with the proper care being taken for needed character orientation.

Guidelines for determination of numbers mounted are as follows:

1–10	Mount all
11–50	Mount 5
51–150	Mount 10%
151+	Mount 15

Notes should be made on the slide label as to how many larvae were mounted and how many were in the group. For example, if 261 larvae were in the group and 15 were mounted, then notation should be 15/261. Information written on the slide includes project, sample location, and date and is made with indelible ink. The number of slides and the number of that particular slide should be noted on each slide (i.e., 1 of 3, 2 of 3, and 3 of 3).

Calibration Procedures and Frequency

Instruments Used in Identification

Compound microscope— -Nikon - Labophot-2 (1990) serial # 435089
lens types (Nikon)
CF-E Plan DL Phase Achromat 10×/0.25
CF-E Plan DL Phase Achromat 20×/0.40

* ID labels were kept on the outside of the sample container as well as inside the sample container with the organisms. This information included project, sample location, and date and was made with indelible ink on a suitable material label outside and on a paper label inside (preferably white rag paper).

CF-E Plan DL Phase Achromat 40×/0.65
CF-E Plan DL Phase Achromat 100×/1.25 oil
Dissecting scope— -Carl Zeiss - Sterio-4 microscope with diascopic base. serial # 42549
Illumination— -Cole-Palmer - Low noise illuminator— model # 9741-50.

Calibration of Reticle

A reticle is used in the compound scope and dissecting scope for measurements; this is calibrated with a micrometer at the magnification the measurements are made at.

Preventive Maintenance

Preventive maintenance on microscopes basically comes down to a few sensible procedures:

1. The microscopes should be kept in an air-conditioned environment. (An air cleaning, dust removing, air-conditioning unit is preferred, with frequent changing of filters a necessity.)
2. Dust covers should be kept covering the instruments when not in use. Stage should be left in highest position, lowest objective in place.
3. The objectives should be kept free of dust particles and other matter by using recommended procedures for removal. Frequent wiping should be avoided since this removes lens coatings.
4. No eating or other potentially harmful activities should be engaged in while using microscopes. When using the microscopes, follow proper procedures.
5. Light source should be turned on and kept on at low setting until instruments are no longer needed. The constant turning of the light source on and off will cut down bulb life.

Keys and Literature Used for Invertebrate Identification

Amphipoda

Bousfield, E. L. *Shallow-Water Gammaridean Amphipoda of New England* (1973) Comstock Publishing Associates, Ithaca, NY.

Annelida

Brinkhurst, R. O. *Guide to the Freshwater Aquatic Microdrile Oligochaetes of North America.* Department of Fisheries and Oceans, Institute of Ocean Sciences, Sidney, British Columbia.

Brinkhurst, R. O. and Jamieson. 1971. *Aquatic Oligochaeta of the World.* Oliver and Boyd, Edinburgh, 860pp. All genera to 1971, with probable type species and short diagnoses.

Donald J. Klemm. *A Guide to the Freshwater Annelida (Polychaeta, Naidid and Tubificid Oligochaeta and Hirudinea) of North America.* Edited by (1985) Kendall/Hunt Publishing Company, Dubuque, IA.

Coleoptera

Barr, C. B. and J. B. Chaplin. *The Aquatic Dryopoidea of Louisiana (Coleoptera: Psephenidae, Dryopidae, Elmidae)*, 1988, Department of Entomology, Louisiana Agricultural Experiment Station, Louisiana State University Agricultural Center, Baton Rouge, Louisiana. Vol. 26, Number 2.

Crustacea

Fitzpatrick, J. F. 1983. *How to Know the Freshwater Crustacea.*
The Pictured Key Series., Wm. C. Brown Company Publishers, Dubuque, IA.

Diptera

Ashe, P. 1983. A catalogue of chironomid genera and subgenera of the world including synonyms (Diptera: Chironomidae). *Ent. Scand. Suppl.* 17:1–68.

Borkent, A. 1984. The systematics and phylogeny of the Stenochironomus complex [Xestochironomus, Harrisius,and Stenochironomus] [Diptera; Chironomidae]. *Mem. Entomol. Soc. Canada* 128, 269pp.

Epler, J. H. 1983. Taxonomic revision of the Nearctic Dicrotendipes Kieffer, 1913 [Diptera: Chironomidae]. MS thesis, Florida State University, 283pp.

Epler, J. H. 1992. In Prep. *Identification Manual for the Chironomidae (Diptera) of Florida.*

Fittkau, E. J. and S. S. Roback. 1983. The larvae of Tanypodinae [Diptera: Chironomidae] of the Holartic region—Keys and diagnoses. *Entomol. Scand. Suppl.* 19: 33–110.

Hirvenoja, M. 1972. Revision der Gattung Cricotopus van der Wulp und ihrer Verwandten [Diptera; Chironomidae]. *Ann. Zool. Fenn.* 10: 1–363.

Hudson, P. L., D. R. Lenat, B. A. Caldwell, and D. Smith. 1990. Chironomidae of the southeastern United States: A checklist of species and notes on biology, distribution, and habitat. *Fish and Wildlife Research* 7: 1–46.

Johannsen, O. A. 1934–1937. *Aquatic Diptera.*[Reprinted 1969 by Entomological Reprint Specialists].

Manual of Nearctic Diptera, Volume 1, 1981. Biosystematics Research Institute, Ottawa, Canada.

Manual of Nearctic Diptera, Volume 2, 1987. Biosystematics Research Centre, Ottawa, Canada.

Roback, S. S. 1957. The immature tendipedids of the Philadelphia area. *Monogr. Acad. Nat. Sci. Philadelphia* 9: 1–152. 1976 The immature chironomids of the eastern United States I. Introduction and Tanypodiinae-Coelotanypodinae. *Proc. Acad. Nat. Sci. Philadelphia* 127[14]: 147–201. 1977 The immature chironomids of the eastern United States II. Tanypodinae- Tanypodini. *Proc. Acad. Nat. Sci. Philadelphia* 128[5]: 55–87. References [continued] 1978 The immature chironomids of the eastern United States III. Tanypodinae-Anatopyniini, Macropelopiini and Natarsiini. *Proc. Acad. Nat. Sci. Philadelphia* 129[11]: 151–202.

1980 The immature chironomids of the eastern United States. IV. Tanypodinae-Procladini. *Proc. Acad. Nat. Sci. Philadelphia* 132: 1–63.

1981 The immature chironomids of the eastern United States. V. Pentaneurini-Thienemannimyia group. *Proc. Acad. Nat. Sci. Philadelphia* 133: 73–128.

Saether, O. A. 1969. Some Nearctic Podnominae, Diamesinae and Orthocladiinae {Diptera: Chironomidae]. *Bull. Fish. Res. Bd. Canada* 170: 1–154. 1976. Revision of Hydrobaenus, Trissocladius, Zalutschia, Paratrissocladius and some related genera [Diptera: Chironomidae] *Bull. Fish. Res. Bd. Canada* 195: 1–287. 1977. Taxonomic studies on Chironomidae: Nanocladius, Pseudochironomus, and the Harnishia complex. *Bull. Fish. Res. Bd. Canada* 196: 1–143. 1983. The larvae of Prodiamesinae [Diptera: Chironomidae] of the Holarctic region—Keys and diagnoses. *Entomol. Scand. Suppl.* 19:141–147.

Saether, O. A. 1980. Glossary of chironomid morphology terminology (Diptera: Chironomidae). *Ent. Scand. Suppl.* 14: 1–51.

Simpson, K. W., R. W. Bode, and P. Albu. 1983. Keys for the genus Cricotopus adapted from "Revision der Gattung Cricotopus van der Wulp und ihrer Verwandten [Diptera: Chironomidae]" by M. Hirvenoja. *N. Y. St. Mus. Bull.* 450: 133pp.

Soponis, A. R. 1977. A revision of the Neararctic species of Orthocladius [Orthocladius] Van der Wulp [Diptera: Chironomidae]. *Mem. Entomol. Soc. Canada* 102: 1–1187.

Townes, H. K. 1984. The Neartic species of Tendipedini [Diptera: Tendipedidae]. *Am. Midl. Nat.* 34: 206.

Wiederholm, T. (Ed.). 1983. Chironomidae of the Holarctic region. Keys and diagnoses. Part 1. Larvae. *Ent. Scand. Suppl.* No. 19.

Wiederholm, T. (Ed.). 1983. Chironomidae of the Holarctic region. Keys and diagnoses. Part 2. Pupae. *Ent. Scand. Suppl.* 28: 1–482.

Wiederholm, T. (Ed.). 1983. Chironomidae of the Holarctic region. Keys and diagnoses. Part 3. Adult males. *Ent. Scand. Suppl.* 34: 1–524.

Ephemenoptera

Berner, L. and M. L. Pescador. 1988. *The Mayflies of Florida.* University Presses of Florida, Florida A & M University Press/Tallahassee and University of Florida Press/Gainesville.

Gastropoda

Harman, W. N. and C. O. Berg. 1971. *The Freshwater Snails of New York, With Illustrated Keys to the Genera and Species.*

Thompson, F. G. 1984. *Freshwater Snails of Florida, A Manual for Identification.* University Presses of Florida, University of Florida Press, Gainesville.

Hydracarina

Pluchino, E. S. *Guide to the Common Water Mite Genera of Florida.* State of Florida, Department of Environmental Regulation, Technical Series, Vol. 7, No. 1, May 1984.

Isopoda

Freshwater Isopods (Asellidae) of North America. U.S. Environmental Protection Agency. Office of Research and Development, Environmental Monitoring and Support Laboratory, Biological Methods Branch, Aquatic Biology Section, Cincinnati, Ohio. 45268.

Odonata

Daigle, J. J. (Dec. 1991). *Florida Damselflies (Zygoptera): A Species Key to the Aquatic Larval Stages.* State of Florida, Dept. of Environmental Regulation. Technical series, Volume 11, Number 1.

Dunkle, S. W. *Damselflies of Florida.* (1990) Scientific Publishers Nature Guide # 3. Gainesville, FL.

Dunkle, S. W. *Dragonflies of the Florida Peninsula, Bermuda and the Bahamas* (1989).

Plecoptera

Stewart, K. W., B. P. Stark, and J. A. Stanger. 1988. *Nymphs of North American Stonefly Genera (Plecoptera).* Entomological Society of America.

Trichoptera

Wiggins, G. B. 1977. *Larvae of the North American Caddisfly Genera (Trichoptera)*, 2nd edn. University of Toronto Press. Toronto Buffalo London.

Aquatic Insects and Invertebrates

Brigham, A. R., W. U. Brigham, and A. Gnilka. *Aquatic Insects and Oligochaetes of North and South Carolina.* (1982) Midwest Aquatic Enterprises, Mahomet, IL 61853.

An Introduction to the Aquatic Insects of North America. Edited by R. W. Merritt, K. W. Cummins. (1984). 2nd edition. Kendall/Hunt Publishing Company.

Marine Invertebrates

Abele, L. G. and W. Kim. *An Illustrated Guide to the Marine Decopod Crustaceans of Florida* (November 1986). Volume 8, Number 1, Part 1 & 2.

Heard, R. W. *Guide to Common Tidal Marsh Invertebrates of the Northeastern Gulf of Mexico* (1982). Mississippi/Alabama Sea Grant Consortium. MASGP-79-004.

Williams, A. B. *Shrimps, Lobsters and Crabs of the Eastern United States, Maine to Florida* (1984). Smithsonian Institution Press, Washington D.C.

Other Taxonomic References

Oligochaetes

Brinkhurst, R. O. 1986. *Guide to the Freshwater Aquatic Microdrile Oligochaetes of North America.* Can. Special Pub. Fisheries and Aquatic Sciences No. 84, Canadian Dept. Fisheries and Ocean, Ottawa, Canada, 259pp.

Klemm, D. J. (Ed.). 1985. *A Guide to the Freshwater Annelida Polychaeta, Naidid and Tubifield Oligochaeta and Hirudmea) of North America.* Kendall/Hunt, Dubuque IA, 198pp.

Crustacea

Fitzpatrick, J. F. 1983. *How to Know the Freshwater Crustacea.* Wm. C. Brown Co., Dubuque, IA, 227pp.

Holsinger, J. R. 1976. *The Freshwater Amphipod Crustaceans (Gammaridae) of North America.* U.S.E.P.A. Water. Poll. Control Res. Series 18050 ELD04/72, 89pp.

Williams, W. D. 1970. A Revision of North American Epigean Species of Asellus (Crustacea: Isopoda). *Smiths. Contri. Zool.* 49: 80.

Williams, W. D. 1976. *Freshwater Isopods (Asellidae) of North America.* U.S. E.P.A. Water Poll. Control. Res. Series 18050 EL D05/72, 45pp.

General

Pennak, R. W. 1978. *Freshwater Invertebrates of the United States*, 2nd ed. Wiley Interscience, John Wiley & Sons, New York, 803pp.

Molluscs

Burch, J. B. 1972. *Freshwater Sphaeriacean Clams (Mollusca: Pelecypoca) of North America.* U.S. E.P.A. Water Poll. Control. Res. Series Identification Manual No. 3, 31pp.

Harman, W. N. and C. O. Berg. 1971. *The Freshwater Snails of Central New York*. Search-Agriculture. 2: 68pp. Cornell Univ. Ag. Expt. Stn., Ithaca, NY.

Insecta

Berner, L. and M. L. Pescador. 1988. *The Mayflies of Florida*. Univ. of Fla. and Fl. A.&M. Univ., Tallahassee, FL. 415pp.

Borkent, A. 1984. *The Systematics and Phylogeny of the Stenochironomus Complex [Xestochironomus, Harrisius,and Stenochironomus][Diptera; Chironomidae]*. Mem. Entomol. Soc., Canada, 128. 269pp.

Brigham, A. R., W. U. Brigham, and A. Gnikla (Eds.). 1982. *Aquatic Insects and Oligochaetes of North and South Carolina*. Midwest Aquatic Enterprises, Mahomet, IL. 837pp.

Caldwell, B. A. *Description of the Immature Stages and Adult Female of* Unniella multivirga. Saether. [Diptera: Chironomidae] with comments on phylogeny.

Cranston, P. S., D. R. Oliver, and O. L. Sather, 1983. The larvae of Orthocladiinae [Diptera: Chironomidae] of the Holarctic Region—Keys and diagnoses. *Entomol. Scand. Suppl.* 19: 149–291.

Edmunds, G.F., S. Jensen, and L. Berner. 1979. *The Mayflies of North and Central America*. Univ. of Minnesota, Minneapolis, MN. 330pp.

Epler, J. H. 1983. Taxonomic revision of the Nearctic Dicrotendipes Kieffer, 1913 [Diptera: Chironomidae]. MS thesis, Florida State University. 283pp.

Fittkau, E.J. and S.S. Roback, 1983. The larvae of Tanypodinae [Diptera: Chironomidae] of the Holarctic region—Keys and diagnoses. *Entomol. Scand. Suppl.* 19: 33–110.

Johannsen, O. A. 1934–1937. *Aquatic Diptera*. [Reprinted in 1969 by Entomological Reprint Specialists].

Pinder, C. V. and F. Reiss. 1983. The larvae of Chironominae [Diptera: Chironomidae] of the Holarcticregion—Keys and diagnoses. *Entomol. Scand. Suppl.* 19: 293–435.

Roback, S. S. 1957. The immature tendipedids of the Philadelphia area. *Monogr. Acad. Nat. Sci. Philadelphia* 9: 1–152.

Soponis, A.R. 1977. A revision of the Nearctic species of Orthocladius [Orthocladius] Van der Wulp [Diptera: Chironomidae]. *Mem. Entomol. Soc. Canada* 102: 1–1187.

Tanypodinae—Coelotanypodinae. 1976. The immature chironomids of the eastern United States I. *Proc. Acad. Nat. Sci. Philadelphia* 127[14]: 147–201.

Tanypodinae—Tanypodini. 1977. The immature chironomids of the eastern United States II. *Proc. Acad. Nat. Sci. Philadelphia* 128[5]: 55–87.

Tanypodinae—Anatopyniini, Macropelopiini and Natarsiini. 1978. The immature chironomids of the eastern United States III. *Proc. Acad. Nat. Sci. Philadelphia* 129[11]: 151–202.

Tanypodinae-Procladini. 1980. The immature chironomids of the eastern United States IV. *Proc. Acad. Nat. Sci. Philadelphia* 132: 1–63.

Taxonomic studies on Chironomidae: Nanocladius, Pseudochironomus, and the Harnishia complex. 1977. *Bull. Fish. Res. Bd. Canada* 196: 1–143.

Townes, H. K. 1984. The Nearctic species of Tendipedini [Diptera: Tendipedidae]. *Am. Midl. Nat.* 34: 206.

Wiggens, G. 1978. *Larvae of the North American Caddisfly Genera (Tricoptera)*. Univ. of Toronto Press, Toronto-Buffalo-London.

Polychaetes

Banse, K. and K. D. Hobson. 1974. Benthic errantiate polychaetes of British Columbia and Washington. *Bulletin of the Fisheries Research Board of Canada* 185: 111.

Blake, J. A. 1971. Revision of the genus Polydora from the East Coast of North America (Polychaeta: Spranidae). *Smiths. Contri. Zool.* r75, pp. 32.

Day, J. H. 1967. *A Monograph on the Polychaeta of Southern Africa. British Museum (Natural History)*. London, Publication No. 656, 2 volumes.

Day, J. H. 1973. New Polychaeta from Beaufort, with a key to all species recorded from North Carolina. NOAA Tech. Rep. NMFS CIRC-375, 139pp.

Fauchald, K. 1977. *The Polychaete Worms: Definitions and Keys to the Orders, Families, and Genera*. Natural History Museum of Los Angeles County, Science Series 28, 188pp.

Gardiner, S. L. 1975. Errant polychaete annelids from North Carolina. *J. Elisha Mitchell Soc* 91: 77–220.

Gardnier, S. L. 1977. New records of polychaete annelids from North Carolina with description of a new species of Sphaerosyllis (Syllidea). *J. Elisha Mitchell Sci. Soc.* 93: 159–172.

Light, W. J. 1978. *Spionidae Polychaeta Annelida. Invertebrates of the San Francisco Bay Estuary System.* Boxwood Press, Pacific Grove California, 211pp.

Maciolek, N. J. 1984. New records and species of *Marenzelleria* Mesnil and *Scolelepides* Ehlers (Polychaeta: Spionidae) from Northeastern North America. *Proc 1st Int. Polychaete Conf. Sydney*, P. A. Hutchings (Ed.), Linn. Soc. NSW. pp. 48–62.

Milkkelsen, P. S. and R. W. Virnstein. 1982. An illustrated glossary of polychaete terms. Technical Report No. 46, Harbor Branch Foundation, Inc., Fort Pierce, FL.

Perkins, T. H. 1979. Lumbrineridae, Arabellidae, and Porvilleidae (Polychaeta), principally from Florida, with descriptions of six new species. *Proc. Biol. Soc. Wash.* 92: 415–465.

Perkins, T. H. 1980. Review of species previously referred to Ceratonereis, Mirabilis, and descriptions of new species of ceratonereis, Nephtys, and Goniada (Polychaeta). *Proc. Biol. Soc. Wash.* 93: 1–49.

Perkins, T. H. 1980. Syllidae (Polychaeta), principally from Florida, with descriptions of a new genus and twenty-one new species. *Proc. Biol. Soc. Wash.* 93: 1080–1172.

Perkins, T. H. 1984. Revision of *Demonax* Kinberg, *Hypsicomus* Grube, and *Notaulax* Tauber, with a review of *Megalomma* Johansson from Florida (Polychaeta: Sabellidae). *Proc. Biol. Soc. Wash.* 97: 285–368.

Perkins, T. H. 1984. New species of Phyllodocidae and Hesionidae (Polychaeta), principally from Florida. *Proc. Biol. Soc. Wash.* 97: 555–582.

Pettibone, M. 1963. Marine polychaete worms of the New England region. I. Families Aproditidae through Trochochaetidae. *Bull. U. S. Nat. Mus.* 227: pp 356.

Pettibone, M. H. 1971. Revision of some species referred to Leptonereis, Nicon, and Laeonereis (Polychaeta: Nereidiae). *Smiths. Contri. Zool.* 104: 53.

Uebelacker, J. M. and P. G. Johnson (Eds.). 1984. *Taxonomic Guide to the Polychaetes of the Northern Gulf of Mexico*. Final Report to the Minerals Management Service, Contract 14-12-001-29091. Barry A. Vittor and Associates, Inc. Mobile, Alabama. 7 volumes.

Wainwright, S. C. and T. H. Perkins. 1982. *Gymnodorvillea floridana*, a new genus and species of Dorvilleidae (Polychaeta) from southeastern Florida. *Proc. Biol. Soc. Wash.* 95: 694–701.

Wern, J. O. 1985. First record of the spionid polychaete, *Boccardiella ligerica* (Ferronniere, 1898) from the Gulf of Mexico. *Contr. Mar. Sci.* 28: 123–128.

Oligochaeta

Brinkhurst, R. O. and H. R. Baker. 1979. A review of the marine Tubificidae (Oligochaeta) of North America. *Can. J. Zool.* 57: 1553–1569.

Brinkhurst, R. O. and B. G. M. Jamieson. 1971. *Aquatic Oligochaeta of the World.* Univ. Toronto Press, Toronto, 860pp.

Amphipods

Barnard, J. L. 1969. The families and genera of marine Gammaridean amphipoda. *Bull. U. S. NAt. Mus.* 271: pp. 535.

Bousfiled, E. L. 1973. *Shallow-Water Gammaridean Amphipoda of New England.* Comstock Publishing Associates, Cornell University Press, Ithaca, NY, 312pp.

Bynum, H. H. and R. S. Fox. 1977. New and noteworthy amphipod crustaceans from North Carolina, USA. *Ches. Sci.* 18: 1–33.

Fox, R. S. and K. H. Bynum. 1975. The amphipod crustaceans of North Carolina estuarine waters. *Ches. Sci.* 16: 223–237.

Myers, A. A. 1981. *Amphipod Crustacea I. family Aoridae. Memoirs of the Hourglass Cruises.* Vol. V. Part V., 75pp.

Shoemaker, C. R. 1947. Further notes on the amphipod genus Corophium from the east coast of North America. *J. Wash. Acad. Sci.* 37: 47–63.

Isopods

Menzies, R. J. and D. Frankenberg. 1966. *Handbook on Common Marine Isopods Crustacea of Georgia.* Univ. of Georgia Press, Athens, GA, 93pp.

Schultz, G. A. 1969. *How to Know the Marine Isopod Crustaceans.* Wm. C. Brown Co, Dubuque, IA, 359pp.

Molluscs

Abbott, R. C. 1974. *American Seashells*, 2nd ed. Van Nostrand Reinhold, New York, 663pp.

Crustaceans

Chance, F. A. 1972. *The Shrimps of the Smithsonian-Bredin Caribbean Expeditions with a Summary of the West Indies Shallow-Water Species (Crustacea: Decapoda: Natantia).* US Government Printing Office, Washington, DC, 179pp.

Williams, A. B. 1984. *Shrimps, Lobsters and Crabs of the Atlantic Coast.* Smithonian Institution, 550pp.

General

Barnes, R. D. 1974. *Invertebrate Zoology.* W. B. Saunders & Company, Philadelphia, 870pp.

Smith, R. I. (Ed.). 1964. *Keys to Marine Invertebrates of the Woods Hole Region.* Cont. No 11. Systematics—Ecology Program, MBL, Woods Hole, MA.

Fish Taxonomy

Eddy, S. and J. C. Underhill. 1979. *How to Know the Freshwater Fishes.* William C. Brown and Company. Dubuque, IA, 215pp.

(manuscript).

Gilbert, C. R. *Key to Species of Florida Centrarchidae* (manuscript).

Lee, D. S., C. R. Gilbert, C. H. Hocutt, R. E. Jenkins, D. E. McAllister, and J. R. Stauffer. 1980. *Atlas of North American Freshwater Fishes.* North Carolina St. Museum of Natural History, 865pp.

Pfliegh, W. L. *The Fishes of Missouri.* Missouri Department of Conservation, 1975, 343pp.

Smith-Vaniz, W. *Freshwater Fishes of Alabama.* Auburn University Agricultural Exp. St.

Chao, L. N. 1977. FAO Species Identification Sheets. #13., 1877pp.

Dohlberg, M. 1975. *Coastal Fishes of Georgia.* University of Georgia Press, 187pp.

Hoese, H. D. and R. H. Moore. 1977. *Fishes of the Gulf of Mexico Texas, Louisiana, and Adjacent Water.* Texas A&M University Press, 327pp.

Parker, J. C. (Ed.). 1974. *Key to the Estuarine and Marine Fishes of Texas.* Texas A&M, 177pp.

QC CHECKS ROUTINES TO ASSESS PRECISION AND ACCURACY

Species Identification

An in-house specimen collection is always maintained.

A project voucher collection is kept for each project. This enables one to reference back to any particular specimen that might be in question or to verify any particular specimen.

Outside experts are used whenever needed to verify new species. As soon as a new specimen is obtained, corroboration is always sought.

REFERENCES FOR BIOLOGICAL COLLECTING METHODS

Livingston, R. J. 1976. Diurnal and seasonal fluctuations of estuarine organisms in a north Florida estuary: Sampling strategy, community structure, and species diversity. *Estuarine Coastal Mar. Sci.* 4: 373–400.

Livingston, R. J., R. S. Lloyd, and M. S. Zimmerman. 1976. Determination of adequate sample size for collections of benthic macrophytes in polluted and unpolluted coastal areas. *Bull. Mar. Sci.* 26: 569–575.

Stoner, A. W., H. S. Greening, J. D. Ryan, and R. J. Livingston. 1983. Comparisons of macrobenthos collected with cores and suction sampler in vegetated and unvegetated marine habitats. *Estuaries* 6: 76–82.

Nutrient Limitation Experiments

Field Work

Water was pumped from 0.5 m depths into plastic tanks stations that were established in the first phase of the analysis. The field program continued during warm weather periods. The water was pumped through a 64 μm plankton net to remove the meroplankton and zooplankton. The plankton material was preserved for identification and enumeration according to procedures given in the original study plan

(Livingston, 1988). Approximately 400 L per station (1200 L, total) was taken to fill the microcosm tanks and provide renewal water. A submersible pump (Ruhl model 1500) was used to pump water from the bay through the 64 μm plankton net, which was placed into a standpipe system with an overflow. Water moved through the overflow pipe into plastic buckets for transport to the Tallahassee Microcosm Facility of EP&A., Inc. Sampling was conducted during late afternoon and samples were transported during the early evening to eliminate heating from daytime peak temperatures.

Microcosm Facility

Water from each station was run through 25 μm plankton nets into a 1000 L Plexiglas collecting tank. Plankton taken in the nets was preserved and saved for quantification and identification to species. Water was gently stirred in the collecting tank by a rubber paddle and poured sequentially into microcosms (containing about 18 L) to ensure homogeneous filling requirements. At that time, samples were taken for chlorophyll *a* analysis by spectrophotometric and fluorometric methods. The analysis was carried out with a Beckman DU-64 spectrophotometer and a model 112 Turner fluorometer according to methods described in QA/AC/SOP reports generated for this project. At the end of the experiment, the samples were run for chlorophyll *a* in the same manner as described earlier.

Three sets of 12 microcosms were placed in the established (temperature-controlled) water baths. Water from one estuarine station and two gulf stations were used for each group of microcosms. There were three sets of four microcosms in each water bath for a given station. This microcosm setup followed a randomized block experimental plan with three replicates for each of the four treatments. This allowed an ANOVA statistical model to be applied to the results of the experiments for each of the three (station-specific) experiments.

Before each experiment, all culture microcosms were acid washed (10% HCl) and rinsed twice with freshwater. Microcosms were placed in the temperature control (outdoor) tanks with the ambient temperatures maintained throughout each experiment. Filtered air was gently sparged into each microcosm to ensure mixing and to prevent settling of the live cells and the nonliving particles.

The microcosms were pulse diluted with bay water filtered through a Gelman type A/E glass fiber filter (nominal porosity of 0.7 μm) immediately after filling the microcosms. Prefiltering through 5 μm Whatman paper filter was required. The dilution water was stored in the dark at 4.0°C to minimize any residual microbial activity. Beginning on the second day, each microcosm received 1.8 L of dilution water taken from the respective bay stations and this procedure was followed for 5 days. A total of 9 L of dilution water was required for each microcosm. On day 7, the microcosm experiments were terminated.

Prior to filling the microcosms but after filtration with the 64 and 25 μm nets, water from each source station was sampled for the following parameters (using established protocols; Livingston, 1988):

Phytoplankton (total counts to species)
Chlorophyll *a*
Particulate organic matter (POM)
Dissolved organic carbon (DOC)
Particulate organic nitrogen (PON)
Dissolved organic nitrogen (DON)
Ammonia-N
Nitrite-N
Nitrate-N
Inorganic (ortho) phosphate-P
Dissolved organic phosphorus
Particulate phosphorus
Color
Turbidity

These measurements will represent time-zero conditions of the experimental water. Chlorophyll *a* was measured daily before dilution water was added to each culture tank according to methods described previously. The full suite of samples was taken and analyzed at the termination of the experiment.

Nutrient enrichment included the following treatment: three control cultures; three P-enriched cultures to achieve a composite input level of 10 micromol PO_4–P L^{-1} and ambient dissolved nitrogen; three nitrate N-enriched cultures to achieve a composite input level of 50 micromole NO_3–N L^{-1} and ambient dissolved P; and three cultures that received a combined N and P level as described for the single additions. These experiments were run with filtered and unfiltered water to determine the possible effects of zooplankton on the productivity results.

Experimental Procedures

We will add the following as added treatments for each of the stations:

3 replicates—control 3 replicates—Si (Na_2SiO_3, 30 μm)
3 replicates—EDTA (0.6 μm)
3 replicates—N (NO_3, 25 μm above ambient), P, (PO_4, 2.0 μm above ambient)

Si, EDTA (concentrations as indicated previously)

These experiments were run with the nutrient experiments (present protocol— N, P, NP, control) during monthly periods. The existing nutrient protocols can be used for a comparison with the new treatments. As part of this research, whole bay water treatments will be used to allow us to start to analyze the effects of the experimental setup on whole water samples (net phytoplankton, zooplankton). Such experiments will evaluate the effects of changes in zooplankton numbers on the utilization of phytoplankton blooms induced by addition of various factors such as N/P as indicated earlier.

At each station, four centrifugal pumps will be used to pump 220 gallons (832 L) of water into four (4) 55 gallon containers. We will need 114 gallons (432 L) at each station to set up the experiment and another 80 gallons (303 L) for dilution water (which will also be used for the whole water treatment). The total water needed for each station was 194 gallons (734 L). The water was transferred to the microcosms by simultaneously pumping from the four containers to achieve homogeneity. Dilution water was stored in cubitainers in two horizontal refrigeration units as before. Experiments were run monthly from April 1992 through March 1993.

Experimental Design

A randomized block design was used with statistical significance (å = 0.05) within and among stations. The four treatments include the controls (3), P-amended (3), N-amended (3), and P- and N-amended (3). The four treatments in each row were randomly selected so that each of the three station-specific experiments will have 12 treatments (randomized four treatments by three replicates). Data analysis was carried out using the computer system maintained by E.P.&A., Inc.

Ancillary Environmental Data

Daily solar radiation values were obtained using a spectroradiometer and additional data were taken by maintained weather stations in Tallahassee. The monthly field data (water quality, phytoplankton to species) were taken as usual according to established protocols (Livingston, 1988). These field data were taken monthly through March 1993. Additional data were taken by the experimental weather stations in Apalachee Bay (e.g., solar insolation, wind speed and direction, rainfall, water temperature, water level, and barometric pressure). Freshwater discharge data were available from the U.S. Geological Survey (Tallahassee, FL).

Appendix B: Statistical Analyses Used in the Long-Term Studies of the NE Gulf of Mexico (1970–2012)

DIFFERENCE TESTING

A series of statistical methods were used to analyze the long-term data. Scattergrams of the long-term field data were examined and either logarithmic or square root transformations were made, where necessary, to approximate the best fit for a normalized distribution. These transformations were used in all statistical tests of significance.

To examine spatial differences in mean values of various field characteristics, the nonparametric Wilcoxon signed-rank test for paired differences (i.e., applied to the differences between paired observations through time at the sites being compared) was used. The Wilcoxon test (Wilcoxon, 1949) is well suited for these types of field data comparisons because the differencing used tends to remove much of the serial correlation in the data sets, and the test does not require the normality assumption that is often difficult to meet with field data. In all such comparisons, the null hypothesis ($\mu_1 = \mu_2$) was tested against the alternative ($\mu_1 \neq \mu_2$) using a two-tailed test with $\alpha = 0.05$. Variances of the field characteristics were tested for significant differences using a modified Levene's statistic (Levene, 1960) when the underlying populations were nonnormal. Both W_{10} and W_{50} statistics were examined using the recommendations of Brown and Forsythe (1974). Nonparametric testing, while avoiding problems with normality, is still subject to the assumptions of independence. Dependence was examined in both the Wilcoxon and Levene tests with an overall Q statistic (Ljung and Box, 1978).

We used a number of basic statistical tests regarding comparisons of two or more data sets. These included the following:

Independence Tests

Random, independent observations are integral to the comparison of two samples. The analysis can check to see if there are certain forms of numeric dependence in the data; if there is dependence due to problems of design or sampling methodology, for example, it cannot be detected. The independence tests used for preliminary analyses examine each sample for autocorrelation at a number of lags. For each sample, autocorrelation is computed for the lesser of 24 or n/4 lags, where n is the number of observations. The program then computes the Q statistic of Ljung and Box (1978) for an overall test of the autocorrelations as a group where the null hypothesis is that correlation at each lag is equal to zero. This statistic is approximately chi-square distributed under the null hypothesis with K-m degrees of freedom, where K is the number of lags examined and m is the number of estimated parameters (1 if an AR1 filter has been applied to the data, 0 otherwise). The p-value of the chi-square test is checked, and if it remains less than 0.05, the null hypothesis is rejected.

The output sheet also contains plots of the autocorrelation functions (ACFs) for the two samples. These graphs are bar charts showing the values of autocorrelation coefficients at all computed lags and are overlain with dashed lines at the critical value (value for which any individual correlation may be considered significant) for each series. Critical values are equal to $2/\sqrt{n}$ where n is the number of observations.

Normality Tests

The parametric F-test for comparison of variances and t-test for comparison of means assume that the data come from normally distributed populations. In support of those tests, the two-sample spreadsheet provides a (chi-square) goodness-of-fit test that compares distributions of the samples to normal distributions having the same means and variances as the samples. Because the number of observations is generally fairly small, the program is designed to divide the data into seven equally sized intervals (frequency classes) for comparison with expected (normally distributed) frequencies. The program also examines the expected frequencies for the leftmost and rightmost classes and expands (doubles) the width of these classes if the initial expected frequency is less than one. These procedures both help to minimize the problem of having extremely low expected frequencies in the denominator of the chi-square computation. However, very low sample sizes and sample distributions with extreme departures from normality can still interfere with the test. For this reason, the program prints a warning when sample size is below 25. The chi-square test should not be considered reliable for such low sample sizes.

The spreadsheet contains a graph of the frequency distribution (with a normal curve overlain for visual comparison purposes) for each of the two samples. Additionally, the sheet contains calculated values for skewness, kurtosis, and Geary's g statistic (Geary, 1936), which is another measure of kurtosis and which may be compared with the value 0.7979 (g value for a normal distribution). All of these values may be used by an investigator to assist in their consideration of the normality issue.

Variance Testing

For independent, random samples from normally distributed populations, the F-test can be used to compare the variances of two samples. F is computed as the ratio of the variances of the samples:

$F = \frac{s_1^2}{s_2^2}$, where the null hypothesis is $H_0: \frac{\sigma_1^2}{\sigma_2^2} = 1$ and where s_1^2 is the larger of the two-sample variances. The program prints the value of F, the associated p-value from an F-table with $n_1 - 1$ numerator degrees of freedom and $n_2 - 1$ denominator degrees of freedom. It also prints a decision to reject or not to reject the null hypothesis (using an alpha level of 0.05) and a warning if it seems that one or more test assumptions have been violated. If one or both of the data series violate the independence assumption due to serial dependence at lag 1, one can use the AR(1) filter option in the two-sample menu. Running the filter will prompt the program to use a modified formula for the calculation of F, which computes F in the presence of lag 1 dependence.

For those cases where the samples do not satisfy the assumption of normality, two modifications to Levene's statistic (Levene, 1960) for variance comparisons are included in the spreadsheet. The statistics are W_{10} and W_{50}, which are both robust to departures from normality in the underlying populations (Brown and Forsythe, 1974). They use robust measures of central location (W_{10} uses the mean of the middle 80% of the observations; W_{50} uses the median observation) and thereby decrease or eliminate the influence of extreme values in testing variance. Brown and Forsythe (1974) recommend using W_{50} for testing the variances of asymmetric distributions and using W_{10} for testing long-tailed distributions. Both formulas involve the generation of a new data series z_{ij} that also must be free of serial dependence. If a warning message is printed with the robust tests, it applies to the new series, not the original series, and it is possible to have dependence in the z_{ij} series even if the original data were independent. Dependence is examined with an overall Q statistic (Ljung and Box, 1978) and chi-square test developed in a fashion similar to that described earlier in the independence testing section. The output of the variance tests should not be used if a transformation (log, square root, arcsin, normalization) has been applied to the data. Also, performing a sequential average tends to "smooth" the data (i.e., variation is lowered in both data sets) and may have an effect on the variance test.

Means Testing

For independent, random samples from normally distributed populations, the parametric t-test can be used to compare the sample means. For unpaired samples, the spreadsheet provides either a t-test with equal variance or a t-test with unequal variance (the results of the aforementioned variance test are used as the criterion for deciding which t-test to output). For paired samples, the sheet provides the results of a paired t-test. For all tests, the output contains the computed value of the t-statistic, a probability value (two-sided test, alpha = 0.05) from the t distribution, a reject or do not reject decision regarding the null hypothesis, and a warning if it seems that one or more test assumptions have been violated. If one or both of the data series violate the independence assumption due to serial dependence at lag 1, one can use the AR(1) filter option from the two-sample menu. Running the filter will prompt the program to use a modified formula for the calculation of t, which computes t in the presence of lag 1 dependence.

For cases where one or both of the data sets violate the assumption of normality, a data transformation can be made to bring the data into normality (although variance testing is affected) or use one of the nonparametric tests (see appendix) that have been included in the spreadsheet. The tests are the rank sum test for unpaired comparisons (Wilcoxon, 1945) and the signed-rank test for paired differences (Wilcoxon, 1949). Both tests are subject to the assumption of random independent samples. For the rank sum test, both of the data series should be independent, while for the signed-rank test, only the series of differences needs to be independent. This can make the signed-rank test a valuable tool for two-sample testing because, in practice, serial correlation often disappears when dependent data sets are differenced. Wilcoxon test results should not be used if an AR1 filter has been applied to the data.

Output for the Wilcoxon tests includes a description of the rejection region, the computed value for the test statistic T for small samples and z for larger samples ($n > 25$ for signed-rank test, $n > 10$ for rank sum), a reject or do not reject decision regarding the null hypothesis (with a probability value based on a two-tailed test and alpha level of 0.05), and a warning if the independence assumption has been violated. Output for the signed-rank Wilcoxon test also includes for the differenced data series an overall Q statistic and chi-square probability value, along with a graph of the ACF.

Distributional Shape Testing

The two-sample spreadsheet also includes a chi-square test, which compares the shapes of the distributions for the two samples. Although the samples may individually exhibit dependence, the assumption is made that the two samples are independent of each other. There are no assumptions regarding the form of the underlying distributions. In order to examine shape only, the program adjusts the two data sets by subtracting the respective mean value from each observation such that both series then have a common mean value of zero. Then the overall range of data in the two series is divided into the lesser of 9 or n/4 classes and a frequency distribution is produced for each sample. The frequency distributions are compared with a chi-square test.

Output for the shape test includes the computed values for the chi-square statistic and its associated probability value, a reject or do not reject decision regarding the null hypothesis (based on a one-tailed test and alpha level of 0.01), and a warning if the sample size is too small for a reliable test ($n < 25$) or if the independence assumption has been violated.

Principal Component Analysis and Regressions

Principal component analysis (PCA) and associated correlation matrices were determined using monthly data and the SAS™ statistical software package. A PCA was carried out as a preliminary review of the data using river flow information, sediment data, and water quality variables. The PCA was used to reduce the physical–chemical variables into a smaller set of linear combinations that could account for most of the total variation of the original set. For the physical–chemical variables in this study, the series were stationary after appropriate transformations; thus, the sample correlation matrix of the variables was a good estimate of the population correlation matrix. Therefore, the standard PCA could be carried out based on the sample correlation matrix.

A matrix of the water and sediment quality data associated with the sampling stations in the Perdido system was prepared. The data were then grouped by station by year. Values for the dependent variables (total biomass m^{-2}, herbivore biomass m^{-2}, omnivore biomass m^{-2}, primary carnivore biomass m^{-2}, secondary carnivore biomass m^{-2}, tertiary carnivore m^{-2}) were then paired with the water/sediment quality data (independent variables) taken for stations within each sector. Unless otherwise defined, these statistics were run using SAS, SYSTAT™, and SuperANOVA™. Data analyses were run on the three independent data sets. Significant principal components were then used to run a series of regression models with the biological factors as dependent variables. Residuals were tested for independence using serial correlation (time series) analyses and the Wald–Wolfowitz (Wald and Wolfowitz, 1940) runs test. A chi-square test was run to evaluate normality.

SPATIAL AUTOCORRELATION

The nonparametric Wilcoxon signed-rank test for paired differences (Wilcoxon, 1949) was used to examine spatial differences in various field characteristics. The test was applied to the differences between paired observations through time at the sites being compared. The Wilcoxon test is well suited for field data comparisons because differencing tends to remove much of the serial correlation in the data sets, and the test does not require normally distributed data. In all such comparisons, the null hypothesis ($\mu_1 = \mu_2$) was tested against the alternative ($\mu_1 \neq \mu_2$) using a two-tailed test with $p = 0.05$. Variances of the field characteristics were tested for significant differences using a modified Levene's statistic when the underlying populations were nonnormal. Both W_{10} and W_{50} statistics were examined using the recommendations of Brown and Forsythe (1974). Nonparametric testing, while avoiding problems with normality, was still subject to the assumptions of independence. Dependence was examined in both the Wilcoxon and Levene tests with an overall Q statistic (Ljung and Box, 1978). Despite attempts to satisfy the independence assumption, it was often violated, particularly in the weekly field analyses.

Spatial autocorrelation analyses were run independently on each of the nine 100-core data sets collected at stations 3, 5a, and ML to examine within-site variability. Spatial autocorrelation coefficients (ρ) were calculated based on a generalized form of Moran's statistic (Moran, 1950) for the four community characteristics (i.e., total numbers of individuals, numbers of taxa, and numbers of the top two dominant species). This procedure is described in detail in Appendix C. Correlation coefficients were calculated between each and all pairs of cores of increasing distance (h) apart, where h is referred to as the order of the coefficient. Correlation coefficients were calculated between each core and all surrounding cores up to order 8. Sample autocorrelation coefficients for each order were then averaged over the three 100-core replicates at each site. If $\rho_{kl}(h)$ denotes the average order-h spatial autocorrelation at the kth station for a given community characteristic l, then the average sample spatial autocorrelation $\rho_{kl}(h)$ has a mean $-1/(n-1)$ and variance $s^2(h)/3$. The sample Z-score for $\rho_{kl}(h)$ is

$$Z_{kl}(h) = \frac{\left[\hat{\rho}_{kl}(h) + \frac{1}{n-1}\right]}{\sqrt{\sigma^2(h)/3}}$$

which has an approximate standard normal distribution. For testing the null hypothesis $\rho_{kl}(h) = 0$ versus the alternative hypothesis $\rho_{kl}(h) \neq 0$, the rejection region occurs where the sample Z-score falls outside of the interval ±1.96, when $p = 0.05$ (i.e., when $Z_{kl}(h) > 1.96$ or $Z_{kl}(h) < -1.96$, we conclude that $\rho_{kl}(h)$ is significantly different from zero).

To examine spatial autocorrelation in community characteristics among cores in each of the 100-core arrays, correlation coefficients were calculated and tested for significance using the following methodology. Suppose that a spatial process $\{Y(s)\}$ is observed at locations $s_1,\ldots,s_n$. Let $\gamma(s_i,s_j)$ denote the covariance between $Y(s_i)$ and $Y(s_j)$. In this study, we assume that $\{Y(s)\}$ is an isotropic process (i.e., the mean value and variance of Y(s) are independent of location s, and in addition, $\gamma(s_i,s_j)$ depends only on the distance $\| s_i - s_j \|$ between the two spatial locations s_i and s_j). When the observations $\{Y(s_i),\ldots,Y(s_n)\}$ are available, the spatial autocovariance function $\{\gamma(h), h \geq 0\}$ can be estimated by

$$\hat{\gamma}(h) = \frac{1}{|N(h)|} \sum_{(i,j)\in N(h)} \left(Y(s_i) - \bar{Y}\right)(Y(s_j) - \bar{Y})$$

where

$\bar{Y} = (1/n)\sum_{i=1}^{n} Y(s_i),$

$N(h) = \{(i,j) : \| s_i - s_j \| = h\}$ is the neighborhood set with distance h

$|N(h)|$ denotes the number of elements of N(h)

A typical spatial neighbor scheme for regular lattices indicates different order neighbors are specified relative to a reference point.

For $h > 0$, let $w_{ij}(h) = 1$ for $(i,j) \in N(h)$ and $w_{ij}(h) = 0$ for $(i,j) \notin N(h)$. Since $w_{ij}(h) = w_{ij}(h)$, the weighting matrix $W(h) = [w_{ij}(h)]$ is symmetrical. The sample order-h spatial autocorrelation can be expressed as

$$\hat{\rho}(h) = \frac{n}{\sum_{i,j=1}^{n} w_{ij}(h)} \left[\frac{\sum_{i,j=1}^{n} w_{ij}(h)\left(Y(s_i) - \bar{Y}\right)\left(Y(s_j) - \bar{Y}\right)}{\sum_{i=1}^{n} \left(Y(s_i) - Y\right)^2} \right]$$

which is a generalization of Moran's statistic (Moran, 1950).

When $\{Y(s_i),\ldots,Y(s_n)\}$ are independent identically distributed normal variables, the spatial ACF is $\rho(h) = 0$ for $h > 0$. In this case, Cliff and Ord (1981) showed that the first two moments of the sample autocorrelation $\hat{\rho}(h)$ are

$$E\left(\hat{\rho}(h)\right) = -\frac{1}{n-1} \quad \text{and} \quad E\left(\hat{\rho}^2(h)\right) = \left(n^2 S_1(h) - nS_2(h)\right) + 3S_0(h) / \left[\left(n^2 - 1\right) S_0^2(h)\right]$$

where

$$S_0(h) = \sum_{i,j=1}^{n} w_{ij}(h) \quad S_1(h) = 2\sum_{ij=1}^{n} w_{ij}^2(h) = 2S_0(h) \quad S_2(h) = 4\sum_{i=1}^{n} w_{i\cdot}^2(h)$$

$$w_{i\cdot}(h) = \sum_{j=1}^{n} w_{ij}(h)$$

The variance of $\hat{\rho}(h)$, denoted by $\sigma^2(h)$, can be calculated by the formula

$$\sigma^2\left((h)\right) = \mathrm{Var}\left(\hat{\rho}(h)\right) = E\left(\hat{\rho}^2(h)\right) - \left[E\left(\hat{\rho}(h)\right)\right]^2$$

For large sample size n, $\hat{\rho}(h)$ is approximately normally distributed and can be defined as

$$Z(h) = \frac{\left[\hat{\rho}(h) - E(\hat{\rho}(h))\right]}{\sqrt{\sigma^2(h)}}$$

Z(h) is called the sample Z-score for the spatial autocorrelation coefficient $\rho(h)$ and can be used to test the null hypothesis $\rho(h) = 0$ versus the alternative hypothesis $\rho(h) \neq 0$.

DATA TRANSFORMATIONS

Scattergrams of the long-term field data were examined and either logarithmic or square root transformations were made, where necessary, to approximate the best fit for a normalized distribution. These transformations were used in all statistical tests of significance.

Analysis of Variance

Analysis of variance (ANOVA) models were run using SYSTAT and SuperANOVA. The ANOVA determinations were run using year-by-season data sets for the infauna, macroinvertebrates, and fishes separately. Three-month intervals, starting with March of each year, were used to define the seasonal patterns. These seasonal definitions were based on established seasonal temperature patterns in the study area. The basic model followed that outlined by Winer (1971) as

$$Y_{ij} = m + a_i + b_j + ab_{ij} + e_{ij}$$

where

Y_{ij} is the response associated with the ith year and the jth season
m represents an overall effect
a_i is the effect of the ith year
b_j is the effect of the jth season
ab_{ij} is the interaction between the ith year and jth season
e_{ij} represents the random error

Specific hypotheses were tested (e.g., years and seasons have no effect on dependent variables). A two-factor analysis was carried out with years and seasons as the two factors.

Major assumptions of the model were tested by studying residual distributions. The three monthly values within each season were used as replicates even though these factors were not replicates in the true sense. Therefore, an ACF was run to check for serial correlation among the monthly residual values (positive correlations would invalidate the ANOVA test). No positive serial correlations of residuals for sequential months were noted for the various dependent variables. In fact, negative autocorrelations were usually found at lags 1 and 2, which indicated generally conservative ANOVA results (i.e., error variance was overestimated). Residuals were plotted against fitted values in addition to carrying out normal probability plots. Unless otherwise stated, ANOVA results showed random distributions of the residuals in normal probability plots. The Wald–Wolfowitz runs test (with cutoff = 0) was run on residuals to determine the possibility of clumping of the positive and negative residuals (Wald and Wolfowitz, 1940). A lack of significance in the Wald–Wolfowitz runs tests on residuals was usually found. F-values and their associated p-values were calculated for the final determinations with pairwise comparison of the effects/interactions via post hoc tests for further comparisons. Post hoc tests included Tukey's for the comparisons of means and Scheffe's for all comparisons of between factors only.

Principal Component Analysis

PCA and associated correlation matrices were determined using monthly data and the SAS statistical software package. A PCA was carried out as a preliminary review of the data using the river and rainfall information and the water quality variables. The PCA was used to reduce the physical–chemical variables into a smaller set of linear combinations that could account for most of the total variation of the original set. For the physical–chemical variables in this study, the series were stationary after appropriate transformations; thus, the sample correlation matrix of the variables was a good estimate of the population correlation matrix. Therefore, the standard PCA could be carried out based on the sample correlation matrix.

TIME SERIES ANALYSIS AND DYNAMIC REGRESSION

Univariate time series and dynamic regression models were developed for the five biological series, herbivores {Hb(t)}, omnivores {Om(t)}, primary carnivores $\{C_1(t)\}$, secondary carnivores $\{C_2(t)\}$, and tertiary carnivores $\{C_3(t)\}$, following the well-known Box–Jenkins modeling procedures (Box and Jenkins, 1976; Pankratz, 1983, 1991). All parameters in the models were estimated using the maximum likelihood method.

Appendix C: Lists of Species of Plants and Animals Taken in the Various River–Bay Systems in the NE Gulf of Mexico

Data are averages over the time spans of the respective studies (Table 1).

APALACHEE BAY (ECONFINA AND FENHOLLOWAY ESTUARIES)

Phytoplankton

Species	E06	E07	E08	E09	E10	E11	E25	Econ Total
Undetermined nanoflagellates	1,189,476	1,130,202	833,999	811,522	668,498	590,077	499,334	817,587
Undetermined cryptophyte	407,925	844,156	693,640	350,982	1,104,062	648,352	285,048	619,166
Chaetoceros fragilis	7,326	6,660					2,126,205	305,742
Undetermined nanococcoids	400,266	411,422	374,126	202,797	170,663	293,373	109,224	280,267
Thalassionema nitzschioides	82,917	172,328	353,147	83,084	120,214	65,268	10,323	126,754
Navicula spp.	334,332	218,115	54,114	90,743	58,275	78,588	18,982	121,878
Cyclotella sp. 17 (5–10 μm)					822,510			117,501
Pyramimonas spp.	136,863	203,131	182,152	32,968	61,605	144,856	29,637	113,030
Undetermined pennate diatoms	201,465	251,748	53,946	66,767	73,260	73,926	10,490	104,515
Chaetomorpha spp.	16,317	79,754	171,662	57,942	110,556	75,591	58,442	81,466
Gymnodinium spp.	60,939	71,762	155,345	82,252	47,453	55,944	47,952	74,521
Skeletonema sp. 1 (2-celled)	23,310	436,896	1,332			4,995	666	66,743
Chaetoceros compressus	1,665	4,329	41,625	16,983	304,362	6,327	81,585	65,268
Nitzschia section *pseudonitzschia*	666	14,319	10,989	270,729	39,960	1,665	11,489	49,974
Cylindrotheca closterium	113,220	101,066	23,477	31,469	8,159	61,272	3,497	48,880
Thalassiosira spp.	48,951	73,094	83,917	40,793	30,969	40,293	18,482	48,071
Amphidinium spp.	666	15,152	109,891	5,994	161,339	26,308	4,996	46,335
Amphora spp.	82,917	90,077	25,808	44,456	32,634	39,294	8,992	46,311
Skeletonema costatum (narrow)	114,885	125,541			23,976	17,982		40,341
Rhizosolenia fragilissima	333	37,962	134,199	44,622	30,636	5,661	25,974	39,912
Cyclotella choctawhatcheeana	75,924	78,588	30,969	16,317	25,308	35,964	8,991	38,866
Chaetoceros diversus	14,985	54,945	62,937	27,306	81,585	21,978		37,677
Cocconeis scutellum	43,290	54,280	35,465	45,122	59,275	19,315	3,830	37,225
Nitzschia spp.	99,567	59,442	20,147	34,133	20,480	19,980	5,661	37,059
Cyclotella spp.	37,962	67,599	46,454	12,654	28,639	54,446	8,826	36,654
Cocconeis spp.	45,954	76,923	20,647	33,301	27,973	32,135	13,487	35,774
Undetermined centric diatoms	46,287	45,456	63,271	15,818	19,148	35,798	13,487	34,181
Minutocellus scriptus	26,973	33,300	16,650	1,332	58,941	94,239	1,332	33,252
Bacteriastrum hyalinum	666	26,640	59,940	28,971	69,930	5,328	833	27,473
Skeletonema costatum	54,279	52,947	4,995	35,298	8,658	2,664	1,998	22,977
Rhizosolenia stolterfothii	27,972	3,996	18,981	87,912	6,993	8,159	5,329	22,763
Skeletonema costatum (2–5 μm)			66,933	17,982	43,956	19,647	2,331	21,550

Thalassiosira sp. (3–5 μm)	6,993	15,984	18,648	666	64,269	35,964	2,331	20,694
Asterionellopsis japonica	1,665	15,318	65,435	4,662	41,126	333	8,991	19,647
Eutreptiella spp.	14,985	36,964	15,318	1,499	167	57,942	333	18,173
Nitzschia pungens	999	333	77,256	11,322	999	333	26,973	16,888
Nitzschia sp. (heteropolar and hyaline)	5,994	29,970	25,974	18,648	20,646	5,661	666	15,366
Chaetoceros gracilis	333				89,910		333	12,939
Chaetoceros sp. 3 (E08 3/93)			88,911					12,702
Rhizosolenia setigera	33,300	11,656	8,658	5,661	10,657	6,660	6,660	11,893
Amphidinium sp. 4 (E08 12/92)			78,255					11,179
Achnanthes manifera	13,320	23,644	7,660	4,662	8,492	13,154	1,998	10,419
Chaetoceros perpusillus			33,633		38,295			10,275
Entomoneis alata	19,314	18,482	333	999	333	30,969		10,061
Johannesbaptistia sp.			1,332	27,306			30,636	8,468
Undetermined cuboidal colony		1,332	31,968	13,653	4,995	333	1,998	7,754
Entomoneis-like sp.	23,310	9,492	4,662	1,998	5,661	8,658	333	7,731
Entomoneis spp.	21,312	17,150	3,996	4,496	2,165	2,997	999	7,588
Chaetoceros coarctatus	999	1,998	4,329	25,641	11,822	333	2,331	6,779
Synedra tabulata	2,664	17,484	6,328	6,327	11,988	667	500	6,565
Proboscia alata cf. *gracillima*				45,621				6,517
Leptocylindrus danicus	333	8,159	5,328	6,327	1,998	12,321	10,989	6,494
Hemiaulus hauckii		4,329	333	35,298	666	1,332		5,994
Gonyaulax spp.	31,635	1,665	999	999		333	3,330	5,566
Chaetoceros brevis			8,658	2,664	23,310		4,329	5,566
Paralia sulcata	6,660	2,165	3,996	21,312	999	1,665	833	5,376
Johannesbaptistia pellucida	6,327	3,996	2,331	9,990	6,660	666	4,995	4,995
Melosira nummuloides	31,635	666						4,614
Cerataulina pelagica			20,646	1,332	9,657			4,519
Minidiscus sp.	1,332	11,322				18,648		4,472
Undetermined coccoids	31,302							4,472
Cryptomonas spp.	11,322	5,994	999	4,329	999	2,165	2,997	4,115
Proboscia alata		1,332	9,324	2,331	4,662	1,998	7,327	3,853
Chaetoceros affinis		7,659	6,327	2,997	8,658			3,663
Undetermined cyanophyte	1,332			16,317	6,660			3,473

(Continued)

Species	E06	E07	E08	E09	E10	E11	E25	Econ Total
Cymatosira lorenziana	10,323	3,663	3,664	1,832	2,997	333	999	3,402
Prorocentrum micans	1,665	2,997	3,830	3,830	5,329	1,998	3,996	3,378
Bacillaria paxillifer	6,993	4,829		8,658	2,664			3,306
Undetermined dinoflagellate cysts	7,992	666	1,665	3,663	3,330	1,665	2,831	3,116
Cocconeis disculus	3,330	3,330	1,998	2,997	1,665	2,332	5,828	3,069
Grammatophora marina	4,662	9,990	1,499	2,332	1,499	333	833	3,021
Prorocentrum minimum	999	999	4,329	2,331	8,325	1,665	333	2,712
Undetermined filamentous cyanophyte				18,648				2,664
Chaetoceros cf. *fragilis*		16,650						2,379
Chaetoceros subsecundus		7,992	5,328			1,998		2,188
Chaetoceros didymus		666	999	6,660	4,662	333	666	1,998
Gyrosigma spp.	3,330	4,163	1,332	1,332	1,665	1,332	167	1,903
Dictyocha speculum			3,996	1,665	6,327		999	1,855
Undetermined dinoflagellates	6,993	1,332	2,331	333	666		1,332	1,855
Chaetoceros constrictus			9,324	2,664	999			1,855
Chaetoceros sp. 1 (1 seta)	5,994	3,996	999	333	999	333	333	1,855
Euglena spp.	3,330	2,664	333		167	6,327		1,832
Chaetoceros wighami		12,654						1,808
Bellerochea malleus	666	7,992			3,663			1,760
Pleurosigma spp.	3,330	4,829		2,664	500	333	500	1,737
Navicula gracilis	666	4,329	999	3,330	1,332			1,522
Ceratium hircus	666	1,665	2,997	1,665	2,331	333	500	1,451
Rhizosolenia imbricata			333	1,332	1,665		6,328	1,380
Chaetoceros dichaeta							8,991	1,284
Chaetoceros throndsenii	1,665	1,665	999		999	3,330		1,237
Chaetoceros lorenzianus			2,997		4,995			1,142
Rhizosolenia delicatula	1,332	1,665		3,330	666	999		1,142
Chaetoceros subtilis		333	1,998		2,664	2,997		1,142
Licmophora spp.	999	3,996	500	833	666	333	167	1,071
Ardissonia crystalina	999	4,995			333	999		1,047
Actinoptychus undulatus	1,665	2,997	167		1,665	333	333	1,023
Guinardia flaccida	333	333	333	4,662			1,332	999

Rhizosolenia styliformis		1,332	2,664	999	999	333		904
Chrysochromulina spp.	1,998	1,665	1,998					809
Nitzschia sigmoidea	1,998	2,665	333			333		761
Rhopalodia gibba	1,665	666		999	1,999			761
Tropidoneis lepidoptera	1,332	666		2,997		333		761
Cyclotella sp. 6 (5 μm)	1,998	3,330						761
Delphineis livingstonii		333		2,165	666	333	1,499	714
Protoperidinium spp.	1,332	999	333	1,998	333			714
Navicula lyra	999	1,166	333	1,998			333	690
Streptotheca thamensis	1,332	1,998				999	333	666
Dactyliosolen antarcticus			4,329		333			666
Dictyocha sp.				4,662				666
Thalassiosira incerta	3,996					666		666
Cymatosira belgica	666		999	2,664				618
Dactyliosolen mediterraneus			999		3,330			618
Falcula hyalina	1,998	500	333		1,166	167		595
Hemiaulus sinensis	1,998			1,665			333	571
Prymnesium spp.		666				1,998	1,332	571
Chaetoceros sp. 2 (3 setae)		1,332	1,332		1,332			571
Lyrella clavata		666	3,330					571
Achnanthes spp.	2,664				333	666		523
Gyrodinium spp.	666	2,664			333			523
Chaetoceros simplex		1,332	1,332		666			476
Nitzschia panduriformis	666	666		1,332	333			428
Asteromphalus spp.						2,997		428
Odontella mobiliensis		666	333	833	999			404
Nitzschia amphicephala		1,332		1,332				381
Ceratium fusus		666	333	333	999		333	381
Lyngbya spp.	333			2,331				381
Chaetoceros atlanticus					2,664			381
Chaetoceros eibonii		2,664						381
Haslea wawrikae		666	333	666			666	333
Chaetoceros danicus			999	1,332				333

(*Continued*)

Species	E06	E07	E08	E09	E10	E11	E25	Econ Total
Lithodesmium undulatum		1,332		666				285
Diploneis spp.		666		1,332				285
Eucampia cornuta			666	333			999	285
Nitzschia longissima v. *reversa*		666		1,332				285
Chaetoceros costatum				666	1,332			285
Chaetoceros-like sp.		1,998						285
Pedinomonas spp.		1,998						285
Melosira moniliformis	1,665					167		262
Corethron criophilum	333	333		666	333			238
Chaetoceros peruvianus			333	999			333	238
Gyrosigma balticum	333	333		333		666		238
Nitzschia tryblionella		999		666				238
Nitzschia longissima	333	666	333		167			214
Triceratium reticulatum	666			333	333			190
Synedra tabulata v. *acuminata*	666	666						190
Actinoptychus senarius	666	666						190
Dimeregramma marina	1,332							190
Nitzschia circumsuta		1,332						190
Prorocentrum gracile			333			333	666	190
Chaetoceros curvisetus	1,332							190
Chaetoceros pseudocurvisetus							1,332	190
Melosira spp.	1,332							190
Thalassiothrix longissima			333	999				190
Grammatophora spp.		999	167					167
Striatella interrupta				999	167			167
Cyclotella striata	333			333			333	143
Nitzschia gracilis		666					333	143
Chaetoceros decipiens					999			143
Fragilaria sp.		999						143
Amphora coffeaeformis	666	333						143
Cocconeis dirupta		666					333	143
Thalassiosira eccentrica		999						143

Actinocyclus ehrenbergii		999						143
Amphora obtusa	666					333		143
Cyclotella stylorum	999							143
Hemidiscus sp.							999	143
Navicula yarrensis	333				333	333		143
Synedra spp.	999							143
Pleurosigma angulatum	333	500						119
Actinocyclus spp.						666	167	119
Undetermined euglenoid	333	333						95
Diplopsalis lenticula						333	333	95
Fallacia forcipata	666							95
Plagiotropis lepidoptera				666				95
Spirulina subsalsa		666						95
Eunatogramma sp.		666						95
Eutreptia spp.							666	95
Fallacia nummularia				333			333	95
Nitzschia constricta	0	666						95
Odontella rhombus	666							95
Actinocyclus cf. *normanii*			666					95
Ardissonia spp.	333			333				95
Dimeregramma minor	666							95
Eupodiscus radiatus		333	333					95
Grammatophora oceanica	666							95
Mastogloia spp.			666					95
Oscillatoria spp.				333	333			95
Petroneis plagiostoma				333	333			95
Rhabdonema adriaticum		666						95
Striatella unipunctata					167	333		71
Pachysphaera spp.			333				167	71
Undetermined curved cell with long bristle			500					71
Gyrosigma fasciola	333							48
Undetermined chrysophyte					333			48
Achnanthes hauckiana	333							48

(*Continued*)

Species	E06	E07	E08	E09	E10	E11	E25	Econ Total
Coscinodiscus spp.	333							48
Nitzschia fusoides				333				48
Auliscus pruinosus				333				48
Amphora arenaria	333							48
Amphora ocellata					333			48
Amphora ovalis						333		48
Auliscus caelatus				333				48
Ceratium trichoceros							333	48
Chaetoceros pendulus				333				48
Cocconeis disculoides				333				48
Fallacia pseudony					333			48
Hemiaulus membranaceus							333	48
Mastogloia crucicula					333			48
Navicula placenta	333							48
Prorocentrum lima						333		48
Rhaphoneis amphiceros	333							48
Rhizosolenia hebetata v. *semispina*	333							48
Stenopterobia delicatissima	333							48
Prorocentrum sphaeroideum							167	24
Achnanthes curvirostrum							167	24
Amphora bigibba				167				24
Nitzschia punctata							167	24
Gyrosigma parkerii		167						24
Nitzschia lanceolata					167			24
Prorocentrum spp.				167				24
Protoperidinium cf. *aciculiferum*					167			24
Synedra hennedyana					167			24
Synedra undulata				167				24
Total	**4,077,918**	**5,250,266**	**4,360,318**	**2,984,526**	**4,781,398**	**2,823,516**	**3,586,760**	**3,980,672**

Species	F06	F09	F10	F11	F14	F16	F25	Fen Total
Undetermined nanoflagellates	779,700	832,833	513,153	507,659	467,366	716,752	351,815	545,780
Skeletonema costatum (narrow)		3,240,756						463,155
Skeletonema costatum	440,226	19,980	1,051,947	475,524	958,041	55,231	1,998	431,204
Undetermined cryptophyte	325,974	393,606	741,924	265,068	343,823	374,639	156,844	349,909
Undetermined nanococcoids	261,126	286,713	223,943	197,969	340,160	268,165	227,441	226,795
Podocystis spathulata						333		172,188
Navicula spp.	273,873	266,400	108,059	73,427	111,555	107,895	17,816	134,458
Pyramimonas spp.	36,963	103,896	103,730	95,904	237,096	65,861	12,322	93,087
Thalassiosira spp.	21,645	449,883	22,311	48,452	24,975	12,987	3,663	83,060
Nitzschia section *pseudonitzschia* spp.		6,327	34,299	384,615	1,332	43,374	42,458	67,373
Cylindrotheca closterium	22,644	184,149	52,614	43,124	63,603	49,118	5,329	59,869
Undetermined ochromonadaceae	149,850	252,414						57,847
Nitzschia spp.	59,094	69,597	20,646	21,978	55,112	22,923	2,997	35,621
Gymnodinium spp.	9,990	37,296	42,458	30,470	53,280	60,786	54,446	34,372
Minutocellus scriptus	50,283	55,944	13,653	14,652	88,578	6,662	1,332	32,967
Rhizosolenia stolterfothii			24,143	6,494	333	182,651	33,633	30,517
Undetermined pennate diatoms	88,878	27,306	11,655	22,811	26,307	30,554	18,315	29,930
Cyclotella choctawhatcheeana	43,328	42,291	24,642	36,630	26,640	14,848	4,329	28,767
Cocconeis spp.	27,099	32,967	34,466	18,149	28,805	36,938	17,483	25,608
Bacteriastrum hyalinum	371		76,923	45,621	666	53,392	1,998	25,591
Thalassionema nitzschioides	17,982	20,979	31,968	18,648	16,817	64,436	29,804	24,690
Undetermined centric diatoms	21,858	23,976	13,653	9,158	64,935	11,822	8,160	20,772
Amphora spp.	20,798	18,648	23,310	19,980	20,813	32,691	9,491	20,176
Johannesbaptistia sp.			92,241	35,964	333	999		18,981
Rhizosolenia setigera	7,659	13,653	14,985	64,103	7,326	19,647	3,830	18,196
Chaetoceros lorenzianus				91,242			2,664	14,129
Chaetomorpha spp.	7,992	11,988	7,659	23,310	1,332	37,796	24,975	13,011
Cocconeis scutellum	14,727	16,317	15,485	12,821	10,989	19,647	1,166	12,855
Gymnodinium sp. 10						85,914		12,273
Entomoneis spp.	14,652	25,641	2,831	2,997	37,796	1,832	833	12,250
Skeletonema sp. 1 (2-celled)	10,323	53,280	13,986		4,329	666		12,083
Eutreptiella spp.	666	6,660	8,492	13,320	54,279	500	1,000	11,988

(Continued)

Species	F06	F09	F10	F11	F14	F16	F25	Fen Total
Cryptomonas spp.	23,219	37,296	2,498	1,166	5,994	3,996		10,833
Cyclotella spp.	7,992	23,976	15,152	8,991	5,994	9,326	1,665	10,204
Entomoneis-like sp.	6,993	22,311	1,998	8,159	2,331	1,332		8,825
Bacillaria paxillifer	999	12,654	13,320	5,661	14,319	10,656	999	8,539
Asterionellopsis japonica	999	5,661	10,323	15,984	1,332	22,201	1,998	8,381
Chaetoceros diversus	5,328	9,990	4,329	9,324	2,997	22,644	1,665	7,802
Proboscia alata cf. *gracillima*			17,649			35,964		7,659
Amphidinium spp.	333	333	9,824	28,305	2,664	9,491	1,998	7,279
Johannesbaptistia pellucida	4,329	6,327	1,332	1,332	1,832	21,978	3,331	5,923
Nitzschia sp. (heteropolar and hyaline)	9,324	6,327	3,996	3,663	13,653	2,083	4,995	5,863
Chaetoceros fragilis						7,326		5,851
Bellerochea malleus	333	10,656	2,331	6,660	7,659	12,987		5,804
Chaetoceros compressus				31,635		7,992	36,963	5,661
Thalassiosira sp. (3–5 μm)		8,658	11,988	999	16,650			5,637
Merismopedia thermalis	26,895	9,990	999					5,507
Navicula sp. 46 (F11 9/92)				30,969	6,660			5,471
Nitzschia amphicephala	34,632	1,998						5,233
Rhizosolenia fragilissima	333	9,657	11,988	999	1,998	7,438	6,327	4,773
Paralia sulcata		4,662	3,663	2,997	2,331	17,982	1,166	4,519
Nitzschia pungens	1,665		1,665	999		26,973	999	4,496
Leptocylindrus danicus		5,994	5,328	4,496	6,327	7,326	2,165	4,210
Carteria-like sp.					28,305			4,139
Cymatosira lorenziana	666	999	1,665	6,660	999	15,569		3,842
Cocconeis disculus	2,664	2,664	1,665	2,664	2,997	12,321	7,660	3,568
Prorocentrum minimum	666	999	2,331	666	13,653	999		3,568
Merismopedia punctata	14,985	6,327		167	333			3,116
Euglena spp.		14,985			5,162			3,092
Chaetoceros didymus				666	1,665	17,982	18,648	2,997
Synedra tabulata	1,998	666	2,165	4,829	4,995	1,998	167	2,950
Prorocentrum micans		666	11,655	1,998	1,665	4,526	2,165	2,930
Melosira nummuloides	19,314							2,759
Chaetoceros affinis				6,660		10,989	666	2,593

Chaetoceros coarctatus				1,332		9,324		2,521
Cerataulina pelagica		666	13,653	2,331	333	666		2,521
Gyrosigma spp.	9,042	1,665	500	333		4,662	6,993	2,315
Achnanthes manifera	666	3,330	1,998	2,664	2,331	4,829	2,166	2,260
Hemiaulus hauckii				1,998		12,321	1,665	2,046
Chaetoceros subsecundus	1,998	9,990			1,332		666	1,903
Streptotheca thamensis	2,664	4,662	1,665	1,665	1,332	1,332		1,903
Odontella mobiliensis		1,665	1,332	1,499		2,664		1,879
Chaetoceros cf. *fragilis*				11,988				1,713
Gonyaulax spp.	666	5,661			1,332	3,996		1,665
Undetermined dinoflagellate cysts	1,370	2,664	1,332	999	1,332	3,663	3,996	1,623
Pleurosigma spp.	1,998	1,665	1,665	1,332	1,499	3,164	500	1,618
Undetermined filamentous cyanophyte						9,990		1,427
Gyrosigma fasciola		5,328	999	666	2,664	167		1,403
Bellerochea-like sp.	7,992							1,308
Skeletonema costatum (2–5 μm)			8,658				12,987	1,237
Undetermined cuboidal colony				666		7,992	1,998	1,237
Actinoptychus undulatus		333	1,998	333	1,332	2,997		1,237
Chaetoceros subtilis		4,995	333	2,664				1,189
Delphineis livingstonii	999	333	333	1,665	333	3,830		1,070
Proboscia alata		4,995		333	333	1,611	4,329	1,039
Chaetoceros vixvisibilis				6,993				999
Navicula sp. (epiphytic)			6,993					999
Entomoneis alata	1,665	999			3,996			951
Grammatophora marina	333		833	2,331	1,332	1,665	333	928
Pseudanabaena spp.			4,329			1,998		904
Hemiaulus sinensis			5,328					856
Chaetoceros brevis			1,332					809
Ceratium hircus	666		3,330	333		1,332	666	809
Chrysochromulina spp.		666	1,998					761
Striatella unipunctata			666		666	3,164		737
Dictyocha speculum			333	666		3,996	4,329	714
Rhizosolenia styliformis				1,665		3,330	666	714

(Continued)

Species	F06	F09	F10	F11	F14	F16	F25	Fen Total
Licmophora spp.		333	1,166	666	1,332	1,166		690
Prymnesium spp.			666	3,164	666	333		690
Protoperidinium spp.							666	618
Undetermined chrysophyte	4,076							582
Rhizosolenia delicatula		2,664			666	666	5,661	571
Corethron criophilum	666		333	2,664		333	1,665	571
Rhizosolenia imbricata		333			666	2,331	1,499	523
Undetermined euglenoid	666	333		333	333	1,332	999	500
Pleurosigma angulatum		1,665			666	1,083		488
Synedra tabulata v. *acuminata*	2,664				666			476
Cymatosira belgica		333				2,664		428
Falcula hyalina		333	1,998		666			428
Triceratium reticulatum	999	666	333			999	666	428
Cyclotella striata	666	333	999	333	666		333	428
Campylosira spp.					2,997		999	428
Undetermined dinoflagellates	1,332	1,332					667	381
Chaetoceros throndsenii		999		333	1,332			381
Achnanthes spp.		999	666		999			381
Lithodesmium undulatum					1,499	666		357
Tropidoneis lepidoptera			333			999		333
Nitzschia sigmoidea	333	999	333		500		333	309
Nitzschia gracilis	999	999						285
Pseudosolenia calcar-avis						1,998	333	285
Cyclotella sp. 17 (5–10 μm)								238
Chaetoceros constrictus			1,332	333				238
Ceratium fusus						1,332	1,166	238
Achnanthes hauckiana	1,665							238
Peridinium spp.		666				666	666	238
Cocconeis sp.	1,665							238
Nitzschia fusoides	333		666		333			238
Chaetoceros decipiens						1,334		191
Undetermined cyanophyte				1,332				190

Haslea wawrikae			333				2,664	190
Gyrodinium spp.						1,332		190
Lyngbya spp.	333		333					190
Chaetoceros peruvianus				666		666		190
Diplopsalis lenticula			333		333	666	500	190
Coscinodiscus spp.	666			333	333			190
Auliscus pruinosus						666		190
Gyrosigma balticum			333	333		500		167
Navicula gracilis			333			666		143
Guinardia flaccida						333	333	143
Rhopalodia gibba	666		333				167	143
Eunatogramma spp.						999		143
Licmophora remulus				333		666		143
Navicula cf. *clavata*			666	333				143
Nitzschia lorenziana			666	333				143
Stylochrysalis sp. 2	999							143
Chaetoceros dichaeta				666				119
Navicula lyra	333		333			167	167	119
Diploneis spp.						833		119
Fallacia forcipata			666			167		119
Amphidinium sp. 4 (E08 12/92)								95
Minidiscus sp.					666			95
Chaetoceros sp. 2 (3 setae)		333	333					95
Lyrella clavata		333			333			95
Nitzschia panduriformis	333				333		333	95
Nitzschia tryblionella		666						95
Fragilaria sp.	666							95
Amphora coffeaeformis						333		95
Spirulina subsalsa	666							95
Coscinodiscus sp.				666			333	95
Eutreptia spp.					333			95
Protoperidinium depressum			333	333			333	95
Calothrix crustacea			666					95

(Continued)

Species	F06	F09	F10	F11	F14	F16	F25	Fen Total
Coccolithophore sp.						666		95
Gymnodinium variabile						666		95
Tryblionella littoralis	666							95
Dinophysis caudata			333					95
Gyrosigma acuminatum	333							71
Cocconeis placentula						445		64
Delphineis surirella						417		60
Gonyaulax polygramma						417		60
Chaetoceros gracilis					333			48
Chaetoceros sp. 1 (1 seta)					333			48
Ardissonia crystalina				333				48
Hemiaulus hauckii–like sp.							5,994	48
Chaetoceros simplex				333				48
Chaetoceros danicus								48
Eucampia cornuta						333		48
Nitzschia longissima v. *reversa*	333							48
Actinoptychus senarius	0		333					48
Dimeregramma marina	333							48
Nitzschia circumsuta		333						48
Cocconeis dirupta					333			48
Plagiotropis lepidoptera					333		333	48
Thalassiosira eccentrica				333				48
Thalassiothrix longissima								48
Eunatogramma sp.				333				48
Nitzschia constricta					333			48
Odontella rhombus	333							48
Amphora arenaria						333		48
Ceratium tripos						333	333	48
Achnanthes curvirostrum		333						48
Amphora bigibba						333		48
Nitzschia punctata	333							48
Amphora robusta				333				48

Coscinodiscus asteromphalus						333		48
Diploneis crabro						333		48
Donkinia recta						333		48
Fallacia pygmaea		333						48
Gomphonema spp.		333						48
Hantzschia virgata	333							48
Odontella regia				333				48
Oxytoxum sp. 1						333		48
Plagiogramma pulchellum v. *pygmaea*				333				48
Pseudoscourfieldia sp.			333					48
Triceratium sp.				333				48
Tryblionella calida	333							48
Undetermined colonial alga				333				48
Grammatophora spp.					167			24
Toxarium rostrata			167					24
Total	**2,925,150**	**6,762,564**	**3,538,300**	**2,827,346**	**3,237,433**	**2,791,000**	**1,198,153**	**3,358,223**

Zooplankton (Numbers)

Species	E06	E07	E09	Econave
Acartia tonsa	11,115	11,745	3,163	8,674
Calanoid copepodids	2,784	660	10,071	4,505
Oithona colcarva	2,188	2,017	6,063	3,423
Spionidae larvae			4,882	1,627
Crustacean nauplii	2,112	330	866	1,103
Pseudodiaptomus pelagicus	132	1,775	377	761
Bivalve larvae (umbone stage)	66	881	843	597
Euterpina acutifrons			1,513	504
Gastropod larvae (shelled stage)	330		876	402
Copepod nauplii		1,166		389
Paradactylopodia brevicornis		660		220
Caridean larvae		132	414	182
Uca spp. zoeae	303	132	99	178
Barnacle larvae (cypris)	66	308	136	170
Centropages hamatus			495	165
Hydromedusae			407	136
Temora turbinata		66	224	97
Sagitta sp.			286	95
Diarthrodes nobilis		198	35	78
Tortanus setacaudatus	132	88		73
Barnacle larvae (nauplii)	105		103	70
Caridean prozoeae			203	68
Scottolana canadensis		198		66
Paracalanus parus			129	43
Penilia avirostris			96	32
Callinectes sapidus zoeae			66	22
Harpacticus sp.		66		22
Metis sp.			35	12
Labidocera aestiva			35	12
Menippe mercenaria zoeae			35	12
Corycaeus americanus			32	11
Total	**19,334**	**20,422**	**31,486**	**23,747**

Species	F06	F09	F16	Fenave
Acartia tonsa	7,172	14,676	2,519	8,122
Crustacean nauplii	10,558	6,370	1,246	6,058
Calanoid copepodids	429	3,752	7,068	3,750
Oithona colcarva	1,602	3,407	4,614	3,208
Gastropod larvae (shelled stage)			6,061	2,020
Uca spp. zoeae	4,112	182		1,431
Centropages hamatus		132	1,882	671
Bivalve larvae (umbone stage)			1,688	563
Pseudodiaptomus pelagicus	297	1,047	197	514
Barnacle larvae (nauplii)	132	936	358	475
Copepod nauplii	610	546		385
Barnacle larvae (cypris)	650	222	70	314
Spionidae larvae		132	804	312
Paracalanus parus			897	299
Euterpina acutifrons			814	271
Temora turbinata			660	220
Corycaeus americanus			339	113
Caridean larvae			218	73
Tortanus setacaudatus		198		66
Callinectes sapidus zoeae	81		103	62
Rhithropanopeus harrisii zoeae	122			41
Sagitta sp.			117	39
Paradactylopodia brevicornis	66			22
Saphirella sp.		66		22
Hydromedusae			41	14
Mesocyclops sp.	41			14
Metis sp.			35	12
Labidocera aestiva			33	11
Harpacticus sp.			32	11
Total	**25,872**	**31,667**	**29,796**	**29,112**

Submerged Aquatic Vegetation (g m^{-2})

Species	E07	E08	E09	E10	E11	E12	E13	Eav
Syringodium filiforme	0.5	1276.3	655.0	1023.8	0.4	258.5	1135.4	621.4
Halimeda incrassata	0.1	39.3	20.4	1562.2		232.3	229.6	297.7
Thalassia testudinum	0.3	42.1	285.4	126.7		1125.3	105.6	240.8
Digenea simplex	31.2	347.8	2.9	389.8	33.4	16.0	723.0	220.5
Spyridia filamentosa	71.3	326.4	408.0	213.1	31.8	106.0	171.0	189.6
Laurencia poitei	19.2	191.1	185.9	240.6	37.1	1.5	277.0	136.0
Laurencia intricata	77.8	209.8	32.3	102.9	40.1	14.7	235.1	101.8
Caulerpa prolifera	231.9	45.6	7.5	128.6	236.1	15.4	12.1	96.7
Gracilaria debilis	0.2	11.1		8.9	5.0		278.4	43.4
Halodule wrightii	3.3	22.4	50.7	28.9	49.4	21.7	38.1	30.6
Gracilaria cylindrica	5.0	24.6	0.1	12.6	4.3	0.1	139.1	26.5
Halophila engelmannii	13.9	18.8	15.3	14.6	48.4	8.2	34.5	22.0
Ruppia maritima	26.3	11.3		0.1	110.8		1.1	21.4
Penicillus capitatus	0.2	13.8	4.4	37.2		27.0	7.6	12.9
Chondria spp.	8.4	5.9	13.6	9.6	7.4	0.8	26.7	10.3
Anadyomene stellata		27.6		7.2			7.3	6.0
Polysiphonia spp.	0.8	2.0	8.9	0.2	0.5	0.2	12.8	3.6
Jania sp.		6.6	0.8	8.9			6.3	3.2
Syringodium filiforme (with flowers)		7.4	8.3	1.1		0.2	5.3	3.2
Gracilaria compressa	0.3	0.7		3.3	0.3	0.1	14.7	2.8
Unknown green filamentous algae		2.2				13.4	1.1	2.4
Caulerpa cupressoides			15.6					2.2
Derbesia sp.		0.6	0.2	0.2	0.1	13.4	0.7	2.1
Gracilaria cervicornis	6.5	0.9		0.3		0.1	4.0	1.7
Sargassum pteropleuron	1.2	0.6	0.8	3.0		2.1	0.4	1.1
Vaucheria spp.	0.2	4.0		1.2	0.1		1.9	1.0
Polysiphonia hapalacantha	0.3	3.7		0.1	0.8	0.8	0.4	0.8
Dictyota dichotoma	0.4	1.1	0.1	0.1	1.9		0.9	0.6
Vaucheria sp.	0.1	3.3		0.2			0.7	0.6
Spyridia filamentosa (yellow)	3.0				0.8			0.5
Solieria tenera	0.1	2.7					1.0	0.5
Champia parvula	0.7	0.4		0.1	1.6		0.6	0.5
Gracilaria foliifera	0.3			0.1	0.1		2.8	0.5
Gracilaria verrucosa	0.1	1.5	0.1		0.4	0.4	0.4	0.4
Wrightiella tumanowiczii							2.0	0.3
Cladosiphon occidentalis	0.5	0.3			0.2		0.7	0.2
Penicillus sp.				1.4				0.2
Halophila engelmannii (with flowers)	1.1	0.2			0.1			0.2
Ceramium byssoideum		1.3						0.2
Cladophora sp.		0.1			0.2	0.3	0.4	0.1
Avrainvillea levis			0.8					0.1
Dictyota cervicornis	0.1	0.3			0.2	0.1	0.2	0.1
Derbesia sp. (dark)		0.2		0.4			0.1	0.1
Caulerpa ashmeadii			0.1			0.3	0.3	0.1
Sargassum fluitans	0.1	0.1	0.2		0.2		0.1	0.1

(*Continued*)

Species	E07	E08	E09	E10	E11	E12	E13	Eav
Jania adherens		0.2				0.4		0.1
Codium isthmocladum			0.3				0.2	0.1
Chaetomorpha spp.		0.3	0.1				0.2	0.1
Ceramium spp.	0.1	0.1					0.3	0.1
Enteromorpha spp.	0.1	0.2	0.1		0.1		0.1	0.1
Penicillus stalk		0.1		0.1		0.2		0.1
Sporochnus sp.	0.2		0.1		0.1			0.0
Padina sp.		0.1					0.3	0.0
Sargassum filipendula				0.3				0.0
Unknown pink hair ball		0.3						0.0
Hypnea spinella							0.2	0.0
Unknown red algae	0.2							0.0
Penicillus lamourouxii				0.2				0.0
Polysiphonia denudata							0.2	0.0
Stictyosiphon subsimplex					0.2			0.0
Sargassum sp.							0.1	0.0
Enteromorpha prolifera							0.1	0.0
Caulerpa mexicana						0.1		0.0
Padina vickersiae		0.1						0.0
Hypnea spp.							0.1	0.0
Acetabularia crenulata		0.1						0.0
Laurencia papillosa			0.1					0.0
Total	**505**	**2,654**	**1,717**	**3,927**	**611**	**1,859**	**3,480**	**2,108**

Species	**F09**	**F10**	**F11**	**F12**	**F14**	**F15**	**F16**	**Fav**
Syringodium filiforme	0.1	0.3	32.1	1454.1	0.3	8.5	521.8	288.1
Halimeda incrassata	0.1		1.7	319.9		0.1	18.8	48.6
Halophila engelmannii	0.7	24.9	32.0	3.5	4.8	141.3	68.6	39.4
Halodule wrightii	0.2	2.0	5.3	29.3	0.6	16.0	193.6	35.3
Laurencia poitei	16.0	9.7	9.6	9.4	24.1	35.0	83.9	26.8
Spyridia filamentosa	0.5	3.9	4.7	127.1	6.8	1.5	31.4	25.1
Thalassia testudinum	0.1	0.1		159.7			5.0	23.5
Jania sp.	2.6	5.3	91.1	0.2	0.3	16.1	0.1	16.5
Laurencia intricata	4.2	6.3	8.1	36.6	2.3	6.0	40.9	14.9
Gracilaria cylindrica	6.6	20.1	5.5	6.0	19.9	44.9	0.2	14.7
Gracilaria compressa	20.5	7.4	5.2		18.8	22.2		10.6
Gracilaria debilis	8.1	6.4	2.5	8.1	24.0	22.4	0.3	10.2
Ulva lactuca	11.9	2.2	1.1	0.3	34.6	6.6		8.1
Caulerpa prolifera	0.6	1.0	1.2	13.8	0.3	19.0	4.7	5.8
Chondria spp.	1.7	1.1	2.8	17.0	1.0	3.0	4.6	4.5
Codium isthmocladum	1.9	18.9	0.6	1.0	1.8	1.9	0.7	3.8
Gracilaria foliifera	3.3	6.0	1.1		7.3	7.2	0.1	3.6
Cladosiphon occidentalis		0.3	0.3	17.3		0.9	0.9	2.8
Gracilaria cervicornis	1.1	0.4	1.3	0.2	6.3	0.3	5.7	2.2
Hypnea spinella	2.5	1.8	0.4		4.6	0.7	0.1	1.4
Ruppia maritima	0.2	1.1	0.7		0.7	4.8		1.1
Halophila engelmannii (with flowers)		1.5	0.7			4.3		0.9
Caulerpa ashmeadii		0.3	0.5	0.7		1.4	3.3	0.9
Polysiphonia spp.	0.3	0.7	1.3	0.1	0.3	1.6	1.3	0.8
Avrainvillea levis							4.7	0.7
Gracilaria foliifera v. *angustissima*		0.3	0.2		3.4	0.8		0.7
Champia parvula	0.2	0.8	1.1	0.1	0.5	1.8		0.6
Digenea simplex	0.9	0.5	0.2	1.1	0.2	0.5	1.1	0.6
Solieria tenera		0.1	3.8		0.1			0.6
Caulerpa mexicana	0.1	0.6	2.4	0.1		0.1	0.6	0.5
Sargassum pteropleuron		0.8	0.2	1.0	0.3	0.7	0.7	0.5
Anadyomene stellata	0.1			3.3			0.3	0.5
Enteromorpha spp.	0.2	0.1	0.1		2.9			0.5
Syringodium filiforme (with flowers)				0.6			2.1	0.4
Penicillus capitatus				0.1		0.1	2.1	0.3
Hypnea cornuta			0.2		2.0			0.3
Sporochnus sp.		0.4		0.1		1.2		0.2
Derbesia sp.			0.1	0.6		0.9		0.2
Polysiphonia hapalacantha					0.2		1.4	0.2
Sargassum fluitans				0.2		0.4	0.9	0.2
Dictyota dichotoma			0.2	0.2	0.2	0.5	0.2	0.2
Bryopsis plumosa						1.0		0.1
Jania sp. 2					0.8			0.1
Udotea sp.							0.8	0.1
Daysa sp.	0.7					0.1		0.1
Padina vickersiae				0.6				0.1
Cladophora sp.				0.5				0.1
Chaetomorpha spp.		0.2	0.3			0.1		0.1
Gracilaria spp.						0.5		0.1

(*Continued*)

Species	F09	F10	F11	F12	F14	F15	F16	Fav
Jania adherens				0.2		0.3		0.1
Hypnea cervicornis	0.4							0.1
Eucheuma nudum		0.2						0.0
Gracilaria damaecornis						0.2		0.0
Gracilaria verrucosa							0.2	0.0
Hypnea spp.	0.2							0.0
Heterosiphonia sp.		0.2						0.0
Ceramium byssoideum							0.1	0.0
Sargassum sp.				0.1				0.0
Wrangelia penicillata			0.1					0.0
Acetabularia crenulata						0.1		0.0
Laurencia papillosa				0.1				0.0
Unknown green filamentous algae								0.0
Caulerpa cupressoides								0.0
Vaucheria spp.								0.0
Vaucheria sp.								0.0
Spyridia filamentosa (yellow)								0.0
Wrightiella tumanowiczii								0.0
Penicillus sp.								0.0
Total	**85.2**	**125.0**	**218.2**	**2212.3**	**168.7**	**374.4**	**1000.6**	**597.7**

Fish (Numbers)

Species	E07	E08	E09	E10	E11	E12	E13	Eave
Lagodon rhomboides	2,620	1,528	212	2,223	1,079	1,375	558	1,371
Diplodus holbrooki		242	627	202	1	4,234	168	782
Micrognathus criniger	218	173	25	26	1,678	12	271	343
Orthopristis chrysoptera	224	60	12	48	109	137	63	93
Eucinostomus argenteus	261	52	74	98	57	41	13	85
Syngnathus floridae	35	38	36	44	37	173	34	57
Bairdiella chrysoura	70	27	51	60	51	50	48	51
Paraclinus fasciatus	9	55	37	10	28	14	89	35
Monacanthus ciliatus	1	26	49	33	3	96	22	33
Gobiosoma robustum	14	16	1	4	116	1	66	31
Haemulon plumieri	1	32	39	37	1	55	32	28
Sciaenidae sp.	96	2	5	19	59	15		28
Centropristis striata	6	42	23	43	3	37	27	26
Opsanus beta	11	15	2	4	95	3	33	23
Syngnathus scovelli	74	1	3	3	60	1		20
Chasmodes saburrae	34			1	63		4	15
Chilomycterus schoepfi	7	8	16	12	10	30	12	14
Lactophrys quadricornis	3	6	1	5	4	18	4	6
Cynoscion nebulosus	13	1		2	6	6	2	4
Leiostomus xanthurus	3	17			1	6		4
Hippocampus zosterae	6	1	3	1	2	2	6	3
Sphoeroides nephelus	6	5	1	2	2		3	3
Sphoeroides parvus	8	2		1	2	1	1	2
Synodus foetens	4	3		1	3	1	2	2
Paralichthys albigutta	4	2	1		2	1	4	2
Syngnathus louisianae	8		2		4			2
Haemulon sp.		2		1	11			2
Blenniidae		3			10			2
Paraclinus marmoratus		3	6		1		1	2
Calamus arctifrons		3	1			1	3	1
Lutjanus synagris			2	1		4	1	1
Chloroscombrus chrysurus	3		1			2	1	1
Aluterus schoepfi	2			1	1	3		1
Paralichthys lethostigma	3				2	1		1
Urophycis floridana	2	1	2					1
Alosa chrysochloris				4		1		1
Hippocampus erectus	1		1		1	1		1
Sphyraena borealis			2			2		1
Lucania parva	1				1		2	1
Microgobius gulosus	1				2			0
Menidia peninsulae					3			0
Fundulus heteroclitus					3			0
Anchoa mitchilli			1		1			0
Leptocephalus larvae	1		1					0
Chasmodes bosquianus		2						0
Lutjanus griseus	1		1					0
Mycteroperca microlepis						1	1	0

(*Continued*)

Species	E07	E08	E09	E10	E11	E12	E13	Eave
Brevoortia patronus					1			0
Prionotus tribulus	1							0
Hypleurochilus bermudensis	1							0
Nicholsina usta			1					0
Lachnolaimus maximus						1		0
Chaetodipterus faber							1	0
Dasyatis sabina					1			0
Diplectrum formosum		1						0
Peprilus burti						1		0
Ancylopsetta quadrocellata	1							0
Ophidion holbrooki		1						0
Prionotus scitulus				1				0
Archosargus probatocephalus					1			0
Bothidae sp.					1			0
Echeneis naucrates		1						0
Gambusia holbrooki	1							0
Halichoeres bivittatus			1					0
Micropogonias undulatus		1						0
Scorpaena plumieri		1						0
Total	**3,773**	**2,384**	**1,252**	**2,890**	**3,526**	**6,350**	**1,484**	**3,094**

Species	F09	F10	F11	F12	F14	F15	F16	Fave
Lagodon rhomboides	148	195	164	1,382	269	683	563	486
Leiostomus xanthurus	820	215	8	1	306	239	18	230
Anchoa mitchilli	270	141	20	6	546	9		142
Orthopristis chrysoptera	2	65	81	217	14	227	251	122
Diplodus holbrooki			8	513		34	252	115
Eucinostomus argenteus	301	139	56	25	30	46	39	91
Micrognathus criniger	8	86	79	58	81	171	34	74
Syngnathus floridae	5	27	18	135	7	59	60	44
Monacanthus ciliatus	1		1	105		16	94	31
Bairdiella chrysoura	17	4	20	77	14	20	55	30
Sciaenidae sp.	37	5	5	83	8	18	17	25
Paraclinus fasciatus	4	24	27	15	20	35	23	21
Gobiosoma robustum	24	5	40	29	26	20	3	21
Syngnathus scovelli	26	39	17	1	47	13	1	21
Brevoortia patronus	102				24			18
Haemulon plumieri		1	1	52		10	57	17
Centropristis striata		1	1	34		10	59	15
Opsanus beta	1	19	19	5	15	31	10	14
Chilomycterus schoepfi	5	6	6	34	1	14	15	12
Cynoscion nebulosus	20		9	12	3	5	4	8
Lactophrys quadricornis	1	10	8	4	1	4	4	5
Gobiidae sp.	28							4
Urophycis floridana	1	3		4		4	13	4
Chasmodes saburrae	1				20	2		3
Chloroscombrus chrysurus	6	2			2	13		3
Sphoeroides nephelus	7	2	2	2	2	5	2	3
Paraclinus marmoratus	1	3	1	3			13	3
Calamus arctifrons		1	1				18	3
Hippocampus zosterae		1	1	6		3	3	2
Prionotus tribulus	13							2
Synodus foetens	1	1		7			3	2
Paralichthys albigutta	4	1	2		4			2
Syngnathus louisianae	5	1	3			2		2
Aluterus schoepfi				8			2	1
Microgobius gulosus	8		1		1			1
Pogonias cromis	2	3			2	3		1
Blenniidae		8	1					1
Leptocephalus larvae		1		3		3	2	1
Hippocampus erectus			2		1	5		1
Lutjanus synagris		1		3			2	1
Chasmodes bosquianus		2				4		1
Hypleurochilus bermudensis					6			1
Nicholsina usta							6	1
Sphoeroides parvus	1		1	1		2		1
Alosa chrysochloris							5	1
Lachnolaimus maximus				1			3	1
Etropus crossotus	2		1			1		1
Sciaenops ocellatus	2	1				1		1
Menidia peninsulae					3			0

(*Continued*)

Species	F09	F10	F11	F12	F14	F15	F16	Fave
Chaetodipterus faber	1				1	1		0
Dasyatis sabina	1	1			1			0
Lutjanus griseus				1	1			0
Diplectrum formosum							2	0
Peprilus burti		1	1					0
Selene vomer	1					1		0
Haemulon sp.							1	0
Paralichthys lethostigma				1				0
Sphyraena borealis			1					0
Ancylopsetta quadrocellata						1		0
Ophidion holbrooki							1	0
Prionotus scitulus		1						0
Anchoa hepsetus		1						0
Arius felis							1	0
Caranx hippos	1							0
Gobiesox strumosus	1							0
Gobiosoma bosci	1							0
Ophidion welshi					1			0
Serranus subligarius						1		0
Symphurus plagiusa	1							0
Trinectes maculatus	1							0
Total	**1,882**	**1,031**	**614**	**2,845**	**1,457**	**1,728**	**1,676**	**1,605**

Invertebrate (Numbers)

Species	E07	E08	E09	E10	E11	E12	E13	Eave
Tozeuma carolinense		225	1,621	540	1	4,727	471	1083.6
Anachis avara	243	1,354	217	1,310	242	244	1,921	790.1
Hippolyte zostericola	14	211	1,420	305	58	374	1,015	485.3
Periclimenes longicaudatus	10	511	1,056	347	18	360	801	443.3
Neopanope texana	293	401	23	409	464	31	408	289.9
Pagurus sp.	64	271	217	521	119	217	540	278.4
Palaemon floridanus	319	160	122	176	232	251	404	237.7
Periclimenes americanus	159	367	21	99	39	9	140	119.1
Neopanope packardii	53	79	172	122	98	42	73	91.3
Thor dobkini	18	279	28	108	23	14	108	82.6
Farfantepenaeus duorarum	29	53	36	150	25	117	54	66.3
Brachidontes exustus	12	4	44	10	14	4	370	65.4
Alpheus normanni	24	26	236	14	54	11	11	53.7
Palaemonetes intermedius	227	10	5	5	69		10	46.6
Nassarius vibex	10	58	2	160	76	2	18	46.6
Metoporhaphis calcarata		13	146	39	1	53	52	43.4
Libinia dubia	32	20	10	35	16	25	68	29.4
Callinectes sapidus	45	16		14	71	5	6	22.4
Ophiothrix angulata	2	15	80	21	8	3	16	20.7
Lysmata wurdemanni	5	10	28	29	4	8	60	20.6
Penaeidae sp.	2	38			57		17	16.3
Menippe mercenaria		12	21	8	7	6	27	11.6
Petrolisthes armatus	4	7	13	32	4		13	10.4
Podochela riisei		2	48	7		9	5	10.1
Echinaster sp.	10	8	1	21	3	15	10	9.7
Haminoea succinea		48		2	1		13	9.1
Latreutes fucorum		1	41			15	1	8.3
Crassostrea virginica	1		1	8	37		4	7.3
Pelia mutica			22	8		2	15	6.7
Latreutes parvulus		11	1	1	5		24	6.0
Crepidula fornicata	3	3		27	1		8	6.0
Palaemonetes vulgaris	11		3		22			5.1
Prunum apicinum		10	4	13		1	6	4.9
Portunus gibbesii	2	8		3	9	1	8	4.4
Argopecten irradians			3	2		14	4	3.3
Epialtus dilatatus		3	9	1	1	2	3	2.7
Crepidula plana		1		2			15	2.6
Pilumnus sayi			15					2.1
Holothuroidea			15					2.1
Lolliguncula brevis	4	2	2	1	3	1	1	2.0
Palaemonetes pugio	8	1			3		1	1.9
Ophioderma brevispinum			2	4			7	1.9
Pilumnus dasypodus			10	1				1.6
Synalpheus fritzmuelleri		3	7					1.4
Eurypanopeus depressus	2	1			1		4	1.1
Pitho anisodon		1		1		1	5	1.1
Aplysia willcoxi			4			3	1	1.1

(Continued)

Species	E07	E08	E09	E10	E11	E12	E13	Eave
Fasciolaria hunteria		1	2	1	1	1	1	1.0
Megalobrachium soriatum			5				1	0.9
Synalpheus minus			5					0.7
Urosalpinx tampaensis	1	1		1		2		0.7
Petrolisthes galathinus		1	3					0.6
Columbella rusticoides			2		1		1	0.6
Eupleura sulcidentata			1			3		0.6
Macrocoeloma camptocerum			2	1				0.4
Processa hemphilli							2	0.3
Trachypenaeus similis	1			1				0.3
Trachypenaeus constrictus				1		1		0.3
Mithrax pleuracanthus			1		1			0.3
Rhithropanopeus harrisii	1				1			0.3
Luidia clathrata		1	1					0.3
Erato maugeriae							2	0.3
Melongena corona			1			1		0.3
Chione cancellata			2					0.3
Polinices duplicatus	1				1			0.3
Geukensia demissa							2	0.3
Mellita quinquiesperforata							1	0.1
Diadema antillarum			1					0.1
Pisidium sp.							1	0.1
Alpheus heterochaelis	1							0.1
Diodora cayenensis							1	0.1
Luidia sp.		1						0.1
Ebalia cariosa							1	0.1
Urosalpinx cinerea				1				0.1
Anadara sp.						1		0.1
Ensis minor					1			0.1
Laevicardium mortoni					1			0.1
Leptogorgia sp.							1	0.1
Macroma sp.	1							0.1
Pinnixa sp.					1			0.1
Pinnotheres spp.			1					0.1
Total	**1,612**	**4,248**	**5,733**	**4,562**	**1,794**	**6,576**	**6,772**	**4471.0**

Species	F09	F10	F11	F12	F14	F15	F16	Fave
Tozeuma carolinense		1	6	6,553		1,251	3,120	1561.6
Pagurus sp.	43	287	600	1,168	218	411	30	393.9
Anachis avara	33	141	247	1,707	183	350	89	392.9
Periclimenes longicaudatus	1	133	164	1,198	16	497	422	347.3
Palaemon floridanus	33	145	165	1,059	25	251		239.7
Hippolyte zostericola	5	295	147	765	14	204	171	228.7
Neopanope texana	222	135	91	212	184	196	13	150.4
Ophiothrix angulata	1	34	3	176	2	19	226	65.9
Palaemonetes intermedius	262	2	54	1	23	15		51.0
Neopanope packardii	1	14	10	228	11	24	35	46.1
Metoporhaphis calcarata	2	65	33	74	2	37	100	44.7
Libinia dubia	12	29	53	23	103	26	1	35.3
Farfantepenaeus duorarum	99	7	9	38	45	11	10	31.3
Mellita quinquiesperforata	5	14	47	1		139		29.4
Nassarius vibex	1	19	57	16	30	44		23.9
Menippe mercenaria		11	14	68	1	21	50	23.6
Callinectes sapidus	101	3	4	5	25	4	1	20.4
Brachidontes exustus	9	10	17	58	32	7	4	19.6
Portunus gibbesii		31	54	4	1	25	21	19.4
Thor dobkini	6	2	2	81		8	22	17.3
Petrolisthes armatus		15	1	59	3	7	35	17.1
Pelia mutica		7	1	41	1	19	29	14.0
Latreutes parvulus		57	23	2	3	7	2	13.4
Epialtus dilatatus		14	4	47	9	14	5	13.3
Lolliguncula brevis		15	41		5	21	5	12.4
Crassostrea virginica	6	1	1	1	57			9.4
Periclimenes americanus	16	5	20	12	3	5	1	8.9
Podochela riisei		1	4	21		2	34	8.9
Eurypanopeus depressus	3	2	1		56			8.9
Penaeidae sp.	9			47	2			8.3
Latreutes fucorum			2	22		2	29	7.9
Crepidula plana	2	13	2	3	25	3		6.9
Palaemonetes vulgaris	26				18			6.3
Megalobrachium soriatum		1		22		1	20	6.3
Alpheus normanni		5	4	4	1	7	21	6.0
Synalpheus minus		2		16		1	21	5.7
Pilumnus dasypodus		1	2	12			25	5.7
Synalpheus fritzmuelleri		2		21		1	15	5.6
Crepidula fornicata	1			30	1	4		5.1
Argopecten irradians				30		3		4.7
Lysmata wurdemanni		1	3	21		3	4	4.6
Palaemonetes pugio	20	1	6		4			4.4
Petrolisthes galathinus			1	9		1	20	4.4
Prunum apicinum		3	2	16		2	2	3.6
Lytechinus variegatus				13			11	3.4
Pilumnus sayi			2	6			15	3.3
Trachypenaeus constrictus		3	2		10	3	1	2.7
Pitho anisodon		2	2	6		3	2	2.1
Aplysia willcoxi		1	2	6		2	2	1.9

(Continued)

Species	F09	F10	F11	F12	F14	F15	F16	Fave
Ophioderma brevispinum				11	1	1		1.9
Rhithropanopeus harrisii	11					1		1.7
Echinaster sp.		2		3		2	3	1.4
Pisidium sp.		3	2	3		1	1	1.4
Calliostoma euglyptum				9			1	1.4
Haminoea succinea		2			4	3		1.3
Fasciolaria hunteria				5				0.7
Eupleura sulcidentata			1	3		1		0.7
Xanthidae sp.				2			3	0.7
Processa bermudensis							4	0.6
Alpheus heterochaelis	1			1	2			0.6
Luidia clathrata			1			2	1	0.6
Pisania sp.		2	1					0.4
Holothuroidea				3				0.4
Urosalpinx tampaensis			1	1	1			0.4
Processa hemphilli			1			1		0.3
Leptochela serratorbita							2	0.3
Macrocoeloma camptocerum				2				0.3
Columbella rusticoides				2				0.3
Chione cancellata				2				0.3
Luidia sp.		2						0.3
Microphyrs bicornutus		2						0.3
Mithrax pleuracanthus				1				0.1
Erato maugeriae				1				0.1
Diodora cayenensis							1	0.1
Melongena corona					1			0.1
Polinices duplicatus				1				0.1
Astropecten articulatus		1						0.1
Sicyonia typica							1	0.1
Ebalia cariosa						1		0.1
Persephona mediterranea		1						0.1
Branchiostoma caribaeum							1	0.1
Dissodactylus mellitae						1		0.1
Pilaria sp.							1	0.1
Sicyonia laevigata				1				0.1
Total	**931**	**1,546**	**1,910**	**13,953**	**1,122**	**3,665**	**4,633**	**3965.7**

PERDIDO MARSH PONDS

List of Plankton Species Taken in Marsh Ponds Monthly from August 2004 to July 2007

Species	PM1	PM2	PM3	PM4	PM5	PM6
Heterosigma akashiwo	557,519	195,947	1,553,367	1,014,369	612,341	41,827
Synechococcus cf. *elongatus*	1,054,500	554,333	514,303	146,550	439,758	48,331
Cyclotella choctawhatcheeana	192,674	146,780	55,857	41,950	36,027	9,779
Merismopedia tenuissima	95,342	24,582	40,485	17,498	16,044	4,924
Chaetoceros compressus	8,793	42,140	3,976	414	384	
Merismopedia elegans			646		39,657	2,099
Prorocentrum cordatum	20,282	5,293	1,998	2,044	2,165	2,573
Ceratoneis closterium	6,004	5,462	4,059	3,130	2,382	4,662
Cryptomonas sp. (15/10 μm)	24,528					
Akashiwo cf. *splendens* (in mucus)	1,436	2,432	656	1,160	7,296	7,699
Chaetoceros cf. *compressus*				11,383	8,355	636
Bacillaria paxillifer	4,205	1,872	2,917	3,583	3,048	3,653
Gymnodinium sp. (*Heterocapsa*-like sp.)						11,110
Undetermined nanoflagellates (<10 μm)	1,832	2,765	1,080	2,079	1,100	1,796
Scenedesmus spp.	4,735	1,897	828	1,342	787	666
Alexandrium-like *Gymnodinium* sp.			555	495	5,631	2,210
Chattonella cf. *subsalsa*	2,092	3,925	172	1,342	939	363
Chaetoceros subtilis	2,695	3,189	313	1,337	601	232
Nitzschia reversa	1,332	838	984	964	1,464	2,614
Amphora spp.	1,342	1,600	1,413	959	974	1,494
Merismopedia glauca						5,167
Navicula gregaria	187	101	1,373	656	969	1,342
Lepocinclis sp. (papillate)	167	151	1,241	1,050	222	1,453
Gymnodinium sp. (*Akashiwo*-like)	1,207	1,039	111	232	404	313
Gymnodinium sp.		3,259				
Navicula spicula	1,259	606	354	288	444	232
Akashiwo sp. (in mucus)	593	252	303	283	535	1,161
Undetermined flagellate (cf. cryptophyte)			303	262	1,927	414
Urosolenia eriensis	916	384	424	424	424	283
Chaetoceros throndsenii	1,852	162	121	222	167	172
Diploneis spp.	270	404	429	349	626	555
Cyclotella spp.	375	338	359	581	460	323
Pseudonitzschia pseudodelicatissima	957	1,009	51	313	51	
Gyrosigma macrum var.	62	20	1,019	404	212	585
Undetermined euglenoid flagellate	42	515	283	666	384	263
Chaetoceros affinis	489	949	91	273	51	40
Nitzschia sigmoidea	146	172	343	353	268	565

(Continued)

Species	PM1	PM2	PM3	PM4	PM5	PM6
Lyngbya sp. (10–15 μm with calyptra)	1,322	252	20	172	51	
Achnanthes spp.	114	192	187	182	464	666
Entomoneis sp. 8–like	947	560		121	81	
Spermatozopsis exsultans	166	81	172	373	293	605
Amphora pediculus	1,603	51		10		
Chaetoceros cf. *rigidus*	604	928				
Chaetoceros diversus	427	918	20	151		
Prorocentrum compressum	354	182	121	252	182	262
Gyrosigma beaufortiana	375	192	177	91	202	263
Gyrosigma macrum	21	111	253	343	283	283
Tryblionella levidansis	125	162	313	152	258	273
Prorocentrum sp. (with spine)		212	91	353	303	222
Crucigenia tetrpedia	458	121	121	121	192	121
Skeletonema costatum	416	414	71	20	192	20
Desikaneis gessneri	312	273	253	91	71	51
Fallacia pygmaea	156	40	283	81	147	283
Fragilaria construens v. *venter*			30	71		838
Peridinium quinquecorne	42	91	40	81	212	434
Thalassionema nitzschioides	302	394	30	40		10
Nitzschia dissipata	437	91	71	20	71	40
Achnanthes curvirostrum	208	162	162	86	81	30
Guinardia striata	177	333		71		
Selenastrum sp.	42	192	61	111	81	61
Alexandrium sp.	31	131		51	131	202
Tryblionella debilis	62	101	91	71	96	121
Pseudanabaena spp.			121		273	101
Frustulia krammeri		40	30	76	197	131
Prorocentrum micans	156	101	91	40	30	51
Entomoneis ornata	104	71	91	61	51	91
Amphora ocellata	302	81	51	20		10
Gymnodinium sp. (wide)					182	262
Nitzschia filiformis	21	61	51	71	91	131
Chaetoceros rigidus	83		293	40		
Thalassiosira oestrupii	135	86	30		71	81
Plagiotropis lepidoptera	21	20	116	101	61	61
Dinobryon divergens	208	61		81		
Cerataulina pelagica	125	172	40	10		
Ankistrodesmus falcatus	10			30	71	232
Fragilaria construens elliptica					30	313
Entomoneis costata	73	51	51	51	51	61
Paralia sulcata	104	111	71	20	20	
Ceratoneis gracilis	104	101	20	86	10	
Stauroneis pachycephala	31	10	40		136	81
Diploneis suborbicularis	42	10	86	20	61	71
Eunotia pectinalis	42	40	111	10	40	40
Ankistrodesmus spiralis	187	61		10	25	
Skeletonema menzellii		20		262		
Johannesbaptistia pellucida		283				

(*Continued*)

Species	PM1	PM2	PM3	PM4	PM5	PM6
Navicula sp. (on Nit. sig.)					192	91
Nitzschia obtusa	42	111	20	10	10	71
Rhopalodia gibba	21	40	20	10	61	111
Chaetoceros fragilis					262	
Opephora schwartzii	42	40	111	40	25	
Dactyliosolen fragilissimus	187	20		51		
Entomoneis alata	10	30	61	51	30	71
Diploneis bombus	10	61	30	51	51	51
Crucigenia crucifera	83	81	40	40		
Gymnodinium sp. (>15 μm)	146	20	10		51	10
Synedra radians	21	91		10	71	40
Aulacoseira granulata	21	91	10	101	10	
Cyclotella meneghiniana	62	20		51	71	20
Synedra cf. *parasitica* (on *Nit*. sig.)		101				121
Cocconeis pensacolae	62	40	30	10	56	20
Nitzschia sp. (tube-dwelling)	52	20	10	30	51	51
Pseudonitzschia-like *Bacillaria* sp.		81	131			
Entomoneis paludosa (var.)	115	20	30	20	20	
Calycomonas gracilis	21	10	20	20	51	81
Cyclotella atomus	10		51	141		
Brachysira serians		46	76	20	51	10
Total	**1,999,185**	**1,010,900**	**2,194,920**	**1,261,712**	**1,191,939**	**166,783**

Epipelic Microalgae

Species	PM1	PM2	PM3	PM4	PM5	PM6
Achnanthes apiculata	65	—	—	—	—	—
Achnanthes brevipes–like sp.	—	—	—	63	—	—
Achnanthes curvirostrum	16,220	2,747	708	1,401	565	639
Achnanthes exigua	—	65	73	—	—	—
Achnanthes hauckiana	925	194	—	—	108	328
Achnanthes hungarica	51	—	—	—	—	—
Achnanthes lanceolata	43	—	—	—	—	—
Achnanthes lanceolata v. *apiculata*	—	—	125	—	—	—
Achnanthes lanceolata v. *dubia*	86	—	—	—	—	125
Achnanthes longipes	—	—	—	—	54	52
Achnanthes spp.	3,833	1,231	2,302	1,350	5,731	33,194
Actinella punctata	54	—	188	333	4,961	667
Actinella spp.	—	—	—	—	581	125
Actinocyclus cf. *normanii*	—	—	—	188	—	—
Actinocyclus ehrenbergii	215	65	115	172	43	89
Actinocyclus ehrenbergii v. *ralfsii*	—	—	—	—	22	—
Actinocyclus normanii	97	70	—	68	70	—
Actinocyclus octanarius	65	387	505	1,292	1,683	563
Actinocyclus spp.	285	827	1,573	1,071	731	892
Actinoptychus spp.	—	65	—	—	—	—
Actinoptychus undulatus	151	204	693	238	184	386
Alexandrium spp.	—	—	—	135	—	—
Amphidinium spp.	65	—	—	—	—	—
Amphora arenaria (without transverse bar)	54	—	—	—	—	—
Amphora arenaria	1,065	54	63	156	70	—
Amphora bigibba	—	—	63	83	274	203
Amphora binoides	51	—	—	—	140	—
Amphora coffeaeformis	1,038	569	130	125	134	—
Amphora decussata	43	—	21	—	—	—
Amphora ocellata	1,918	898	203	271	54	—
Amphora ovalis	1,156	129	—	—	—	—
Amphora sp.	—	—	—	—	—	271
Amphora sp. (3 μm wide; PM6 12/05)	—	—	—	—	—	4,230
Amphora sp. (small)	—	—	—	—	—	313
Amphora sp. 8–like	2,903	206	68	63	—	—
Amphora spp.	87,115	59,476	67,138	45,854	45,839	80,367
Anomoeoneis sphaerophora	—	—	—	—	—	21
Anomoeoneis spp.	—	65	—	—	—	—
Anorthoneis spp.	—	—	—	83	—	—
Ardissonia crystalina	—	—	—	104	—	—
Ardissonia spp.	—	—	68	—	—	68
Asterionellopsis glacialis	258	—	—	—	—	68
Aulacoseira ambigua	22	3,226	7,875	6,094	3,161	5,109

(*Continued*)

Species	PM1	PM2	PM3	PM4	PM5	PM6
Aulacoseira granulata	—	65	135	—	70	—
Aulacoseira spp.	—	54	172	—	—	—
Auliscus pruinosus	—	—	63	—	—	—
Bacillaria paxillifer	355,638	243,332	193,514	223,263	174,271	190,238
Bacillaria paxillifer (longer cells)	—	—	—	—	2,516	—
Bacillaria sp. (giant)	—	—	130	—	—	63
Bacillaria spp.	—	—	3,313	—	—	—
Biddulphia alternans	—	—	—	—	—	21
Biddulphia biddulphianum	—	—	—	63	—	—
Biddulphia spp.	—	—	63	—	—	—
Biddulphiopsis sp.	—	65	—	63	—	—
Biremis lucens	—	101	—	—	—	—
Biremis spp.	—	—	—	—	108	—
Brachysira follis	—	54	—	—	—	—
Brachysira follis v. *hannae*	—	—	—	—	65	—
Brachysira fossilis	—	97	—	—	—	188
Brachysira serians	215	376	328	188	3,475	821
Brachysira spp.	255	1,043	266	427	2,307	1,198
Caloneis amphisbaena	187	70	—	68	—	—
Caloneis lewisii	—	172	63	125	70	625
Caloneis liber	—	—	208	42	—	42
Caloneis oregonica	145	939	3,255	1,224	312	125
Caloneis spp.	140	1,032	4,391	1,375	1,011	693
Caloneis weilesii	—	—	—	—	—	130
Calycomonas gracilis	442	417	880	745	1,328	1,693
Calycomonas sp.	—	65	—	—	65	—
Calycomonas spp.	215	—	135	1,036	140	1,355
Calycomonas-like sp.	32	204	1,630	1,453	1,634	5,354
Campylodiscus clypeus	—	—	63	—	70	—
Campylodiscus echeneis	481	1,730	8,599	11,870	9,777	5,491
Campylodiscus spp.	—	179	194	625	204	363
Capartogramma crucicula	—	54	135	68	—	—
Centritractus belanophorus	—	—	—	—	—	21
Cerataulina pelagica	38	—	—	—	—	—
Ceratoneis closterium	179,426	14,419	4,147	3,339	2,737	8,453
Ceratoneis fusiformis	—	—	—	68	—	—
Ceratoneis gracilis	1,524	1,509	795	52	140	233
Chaetoceros affinis	—	—	188	—	—	—
Chaetoceros cf. *compressus*	—	—	—	203	—	—
Chaetoceros curvisetus	129	—	—	—	—	—
Chaetoceros decipiens	194	—	—	—	—	—
Chaetoceros diversus	387	57	—	166	—	—
Chaetoceros rigidus	65	—	—	—	—	—
Chaetoceros subtilis	—	—	203	—	—	—
Chroococcus sp.	430	—	—	—	—	—
Climaconeis scopularum	—	65	—	42	—	—
Cocconeis cf. *pediculus*	—	—	—	—	—	2,358
Cocconeis pediculus	38	—	—	63	—	—

(*Continued*)

Species	PM1	PM2	PM3	PM4	PM5	PM6
Cocconeis pensacolae	156	461	396	328	328	571
Cocconeis placentula	—	375	193	63	231	505
Cocconeis scutellum	—	226	188	240	280	255
Cocconeis sp.	—	—	—	—	—	677
Cocconeis sp. (8/5 μm PM2 7/07)	—	909	—	—	—	—
Cocconeis sp. (small)	—	—	—	—	—	365
Cocconeis sp. (wide)	—	69	—	—	54	104
Cocconeis spp.	1,580	3,630	1,531	1,432	1,731	2,824
Coscinodiscus centralis	—	—	255	63	—	188
Coscinodiscus granii	—	65	229	—	134	63
Coscinodiscus jonesianus	—	—	68	—	—	—
Coscinodiscus radiatus	—	613	2,063	1,261	715	542
Coscinodiscus spp.	—	538	1,813	474	387	313
Cosmioneis delawarensis	—	—	63	—	65	—
Cosmioneis pulchellum	—	140	—	—	—	—
Cosmioneis pusilla	—	122	333	63	108	130
Cryptomonas spp.	65	—	—	—	54	—
Ctenophora pulchella	—	32	—	—	70	—
Cyclotella choctawhatcheeana	6,489	6,485	589	625	1,016	—
Cyclotella meneghiniana	65	172	130	281	285	21
Cyclotella sp. (marsh)	65	129	292	208	258	—
Cyclotella sp. (with granulated center)	—	—	—	—	—	68
Cyclotella spp.	1,491	3,317	4,333	2,792	3,705	3,943
Cyclotella spp. (*Actinocyclus*-like)	—	—	—	—	108	—
Cyclotella striata	—	—	—	—	—	115
Cymatosira lorenziana	656	2,140	2,594	474	172	1,094
Cymbella aspera	—	—	63	—	—	—
Cymbella cuspidata	—	—	—	63	—	—
Cymbella spp.	—	269	63	52	—	63
Decussata placenta	38	—	—	—	—	271
Delphineis livingstonii	48	—	—	—	—	—
Delphineis surirella	97	—	68	—	—	—
Denticula sp.	—	—	—	63	—	—
Desikaneis gessneri	1,359	1,401	776	482	344	500
Desikaneis howelli	—	54	—	—	—	—
Desmogonium spp.	—	—	—	68	—	—
Diadesmis confervacea	323	452	688	906	194	—
Diadesmis contenta	—	—	—	—	—	125
Diploneis bombus	355	527	2,005	625	683	193
Diploneis didyma	—	—	68	339	—	135
Diploneis elliptica	—	—	104	—	—	21
Diploneis intermedia	—	—	—	68	—	—
Diploneis interrupta	172	1,334	844	115	312	188
Diploneis obliqua	—	—	104	—	65	—
Diploneis oblongella	—	65	63	63	—	—
Diploneis ovalis	—	97	—	610	140	203
Diploneis smithii	—	—	—	135	—	—

(Continued)

Species	PM1	PM2	PM3	PM4	PM5	PM6
Diploneis spp.	2,908	9,654	18,835	10,935	11,022	14,172
Diploneis suborbicularis	574	3,235	7,615	5,443	5,704	15,551
Donkinia spp.	129	—	—	—	70	—
Entomoneis alata	645	950	5,347	4,105	3,681	5,630
Entomoneis costata	8,328	2,613	583	505	183	740
Entomoneis gigantea	97	—	—	68	—	68
Entomoneis ornata	23,108	13,500	3,623	3,375	2,457	2,471
Entomoneis paludosa	—	—	73	188	—	—
Entomoneis paludosa (var.)	9,260	3,156	68	135	70	135
Entomoneis sp. 8	1,156	—	—	—	—	—
Entomoneis sp. 8–like	4,596	495	63	135	172	—
Entomoneis spp.	3,167	1,408	913	1,699	1,462	1,281
Eucampia zodiacus	—	—	250	—	—	—
Eunotia alpina	—	—	—	—	456	—
Eunotia bidentula	—	—	—	—	30,640	63
Eunotia bigibba	—	—	—	—	10,226	55
Eunotia diodon	—	—	63	125	184	—
Eunotia flexuosa	32	925	828	276	1,358	370
Eunotia flexuosa v. *eurycephala*	—	1,521	526	495	21,373	1,946
Eunotia formica	—	—	—	—	65	—
Eunotia lunaris	—	—	118	115	1,314	73
Eunotia monodon	113	721	724	703	20,600	1,604
Eunotia monodon v. *constricta*	43	70	380	287	5,775	2,271
Eunotia naegeli	—	43	—	—	—	—
Eunotia parallela	—	—	—	—	204	68
Eunotia pectinalis	223	4,791	1,943	2,326	4,955	781
Eunotia pectinalis v. *undulata*	188	2,219	724	485	127	188
Eunotia punctinatis v. *undulata*	—	—	—	—	54	—
Eunotia serra	—	127	177	118	65	198
Eunotia spp.	6,188	4,063	4,724	3,302	7,528	1,349
Eunotia tautoniensis	86	1,025	286	250	269	125
Eunotia tautoniformis	—	—	52	—	—	—
Fallacia forcipata	—	—	—	63	—	—
Fallacia nummularia	65	—	73	—	—	63
Fallacia pygmaea	789	1,414	2,474	1,745	2,114	1,740
Fallacia spp.	108	70	68	198	—	542
Fragilaria brevistriata	1,027	782	258	—	—	—
Fragilaria capucina	—	—	221	—	—	—
Fragilaria construens	258	—	—	—	1,355	28,006
Fragilaria construens v. *pumilia*	—	—	—	745	—	—
Fragilaria construens v. *venter*	1,548	—	1,240	1,354	677	3,850
Fragilaria leptostauron	97	—	—	—	—	—
Fragilaria spp.	105,286	33,945	23,737	42,126	53,290	220,592
Fragilaria spp. (chain-forming)	559	731	667	1,666	2,752	4,124
Fragilaria virescens	129	—	—	—	—	—
Frustulia krammeri	215	1,184	661	1,156	169,217	13,787
Frustulia rhomboides	38	791	185	172	19,471	3,384

(*Continued*)

Species	PM1	PM2	PM3	PM4	PM5	PM6
Frustulia rhomboides v. *saxonica*	—	—	—	—	65	—
Frustulia rhomboides var.	—	—	—	—	50,844	—
Frustulia spp.	156	409	844	63	645	604
Gomphonema acuminatum	—	22	—	68	22	—
Gomphonema acuminatum v. *coronata*	—	69	—	130	—	—
Gomphonema angustatum	54	—	—	—	—	—
Gomphonema intricatum	—	—	—	63	—	—
Gomphonema spp.	129	667	125	323	145	266
Grammatophora marina	—	—	130	104	—	68
Grammatophora spp.	—	—	—	63	—	63
Guinardia striata	—	—	—	948	—	—
Gymnodinium spp.	145	—	125	—	54	—
Gyrosigma balticum	151	570	3,271	719	833	1,875
Gyrosigma beaufortiana	10,173	3,713	3,498	3,758	3,280	6,546
Gyrosigma cf. *eximium*	—	—	—	52	—	—
Gyrosigma closterioides	—	—	—	—	57	—
Gyrosigma eximium	54	43	—	—	—	—
Gyrosigma fasciola	65	22	243	206	65	118
Gyrosigma hummii	22	—	—	—	—	68
Gyrosigma macrum	436	4,782	16,815	14,872	8,168	8,532
Gyrosigma macrum var.	137	773	19,619	5,246	1,988	2,677
Gyrosigma obscurum	—	—	42	—	65	63
Gyrosigma parkerii	—	—	63	55	22	63
Gyrosigma scalproides	54	65	—	68	—	—
Gyrosigma sp.	—	—	63	63	—	—
Gyrosigma spp.	448	665	1,786	901	715	968
Haslea spp.	54	—	130	—	129	—
Hemiaulus hauckii	—	—	21	—	—	—
Hemiaulus sinensis	—	—	63	—	—	—
Hemidiscus spp.	—	—	—	21	—	—
Heterosigma akashiwo	—	161	—	—	—	—
Hyalodiscus scoticus	—	—	68	—	—	—
Hyalosynedra spp.	—	114	188	271	172	144
Licmophora spp.	—	—	21	—	—	—
Lyngbya sp.	1,581	—	948	—	—	—
Lyngbya sp. (10–15 μm with calyptra)	6,467	559	271	339	559	68
Lyngbya sp. (constricted)	—	—	—	—	16,129	5,813
Lyngbya spp.	1,624	172	3,573	875	495	21
Lyrella clavata	65	—	135	—	—	—
Lyrella diffluens	—	—	63	—	—	—
Lyrella lyra	65	376	620	198	70	130
Lyrella spp.	65	574	1,089	604	328	193
Lyrella sulcifera	502	526	193	193	129	—
Mastogloia binotata	—	—	375	—	—	—
Mastogloia elliptica	—	32	68	—	—	—
Mastogloia smithii	—	234	193	68	—	—
Mastogloia sp.	—	—	—	63	—	—

(Continued)

Species	PM1	PM2	PM3	PM4	PM5	PM6
Mastogloia spp.	32	183	130	276	237	118
Melosira moniliformis	—	65	542	250	608	573
Melosira nummuloides	1,220	4,606	3,302	3,776	3,656	4,177
Melosira pensacolae	—	15,469	5,986	4,490	7,979	3,287
Melosira spp.	108	—	135	375	398	417
Melosira varians	—	—	313	—	301	188
Merismopedia elegans	58,581	220,259	398,987	142,875	132,096	—
Merismopedia glauca	6,483	—	39,125	—	—	1,167
Merismopedia sp.	—	0	—	—	—	—
Merismopedia spp.	38,760	—	141,939	240,493	44,740	14,418
Merismopedia tenuissima	2,065	—	—	—	—	—
Mougeotia spp.	581	—	—	—	—	—
Navicula capitata	—	65	63	—	—	63
Navicula capitata v. *hungarica*	161	70	—	219	65	271
Navicula cf. *pelliculosa*	202	—	—	—	—	13,194
Navicula cruciculoides	—	—	—	—	—	68
Navicula cruciculoides–like sp.	—	70	—	—	—	—
Navicula cuspidata	—	139	63	63	22	—
Navicula gregaria	863	2,379	8,304	5,079	16,567	9,333
Navicula halophila	—	—	—	63	—	—
Navicula hasta	51	509	860	1,189	3,380	6,889
Navicula hungarica v. *capitata*	—	—	21	—	—	21
Navicula incerta	339	—	—	—	—	—
Navicula lacustris	—	—	—	21	—	—
Navicula lanceolata	—	—	—	63	—	—
Navicula peregrina	425	968	2,714	2,917	3,532	7,880
Navicula placenta	305	129	255	198	597	698
Navicula pseudocomoides	—	65	—	—	—	—
Navicula radiosa	—	32	—	—	—	—
Navicula rhyncocephala	—	—	—	—	—	125
Navicula sovereignae	1,281	993	110	396	70	68
Navicula sp. (bow tie)	1,645	554	135	—	—	—
Navicula sp. (septate)	253	54	—	—	—	—
Navicula sp. (tube-dwelling)	2,774	—	—	—	—	—
Navicula sp. with polar septa	258	—	—	—	—	—
Navicula spicula	157,821	28,572	3,844	4,287	4,536	6,209
Navicula spp.	94,069	117,705	181,990	157,346	187,891	253,113
Navicula viridula	—	—	—	—	70	—
Navicula yarrensis	586	989	5,740	2,045	1,925	4,243
Navicula yarrensis v. *americana*	454	323	2,083	597	344	641
Navicula zeta	142	175	—	—	—	120
Neidium affine	—	—	177	—	65	—
Neidium iridis v. *amphigomphus*	—	75	—	—	—	—
Neidium productum	—	—	—	63	70	—
Neidium spp.	—	366	302	73	328	193
Nitzschia acicularis	—	645	68	125	86	104
Nitzschia angustata	—	—	—	—	—	63

(Continued)

Species	PM1	PM2	PM3	PM4	PM5	PM6
Nitzschia dissipata	15,539	4,077	271	406	70	224
Nitzschia dubia	70	498	1,526	662	452	172
Nitzschia dubia–like sp.	—	129	—	—	—	—
Nitzschia epithemioides	—	—	—	—	—	68
Nitzschia fasciculata	—	204	385	214	242	438
Nitzschia filiformis	65	387	609	339	253	1,089
Nitzschia hungarica	86	—	42	109	—	219
Nitzschia insignis–like sp.	—	—	—	—	—	42
Nitzschia lanceola	65	—	63	—	156	—
Nitzschia longissima	866	1,290	641	531	140	21
Nitzschia lorenziana	—	—	—	—	65	—
Nitzschia marginulata	—	—	21	—	—	—
Nitzschia nana	—	—	—	68	—	—
Nitzschia obtusa	183	1,339	2,641	1,135	2,374	2,646
Nitzschia reversa	49,727	24,215	11,183	7,103	3,855	4,021
Nitzschia scalaris	661	2,979	9,654	5,663	55,500	8,123
Nitzschia scalaris–like sp.	—	—	135	—	—	—
Nitzschia sigma	—	—	21	—	—	—
Nitzschia sigmoidea	4,506	48,962	150,957	91,240	86,899	147,750
Nitzschia sigmoidea–like sp.	—	—	115	—	—	—
Nitzschia sp.	—	54	—	—	—	—
Nitzschia sp. (beaded)	—	194	—	63	—	—
Nitzschia sp. (curved with capitate ends)	—	65	—	—	—	—
Nitzschia sp. (giant *Nit. epi.*–like)	—	—	—	—	—	500
Nitzschia sp. (long cell with central keel)	—	—	—	135	70	—
Nitzschia sp. (tube-dwelling)	—	—	—	—	—	636
Nitzschia sp. (with central space)	—	—	—	—	—	63
Nitzschia spathulata	38	—	—	—	—	—
Nitzschia spp.	46,569	73,891	131,636	86,011	81,568	107,580
Nitzschia stagnorum	—	—	—	—	5,522	—
Odontella aurita	—	—	57	—	—	—
Odontella reticulata	65	—	—	—	—	21
Odontella rhombus	—	65	188	63	—	—
Odontella sp.	—	—	—	—	70	—
Opephora marina	32	32	—	—	—	—
Opephora schwartzii	645	187	—	135	—	—
Opephora spp.	2,995	764	—	490	194	333
Orthoseira roseana	—	65	—	63	237	188
Orthoseira sp.	65	—	130	—	—	—
Orthoseira spp.	32	546	1,162	3,261	2,283	766
Oscillatoria sp. (curved tip)	—	—	125	250	129	—
Oscillatoria spp.	226	65	542	1,092	140	21
Paralia sulcata	1,634	9,985	32,850	12,419	4,767	6,110
Pediastrum duplex	—	—	583	—	—	—
Petrodictyon gemma	22	—	482	417	151	797
Petroneis granulata	38	—	68	—	—	—

(*Continued*)

Species	PM1	PM2	PM3	PM4	PM5	PM6
Petroneis maculata	231	108	—	52	65	292
Petroneis maculata v. *orbiculata*	554	312	—	—	—	68
Petroneis marina	178	—	—	68	—	118
Petroneis monilifera	32	—	—	—	—	—
Petroneis plagiostoma	27	65	—	—	70	—
Petroneis spp.	952	920	998	1,396	651	646
Pinnularia boyeri	—	—	125	—	—	—
Pinnularia braunii	—	—	63	—	—	—
Pinnularia major	—	197	125	177	350	526
Pinnularia major v. *pulchra*	—	22	—	—	—	—
Pinnularia spp.	516	2,769	2,149	1,188	4,180	1,604
Pinnularia viridis	—	448	673	310	258	214
Plagiogramma moretonii	—	65	—	—	—	—
Plagiogramma pulchellum	32	—	—	—	—	—
Plagiotropis lepidoptera	2,118	1,274	1,326	1,677	1,586	7,083
Plagiotropis spp.	425	247	373	146	527	667
Plagiotropis zebra	32	258	232	807	1,339	4,031
Plagiotropis zebra–like sp.	—	—	—	—	—	125
Pleurosigma angulatum	—	86	406	63	—	68
Pleurosigma elongatum	237	2,376	4,948	1,417	570	1,583
Pleurosigma intermedium	86	219	1,797	774	1,086	1,125
Pleurosigma spp.	2,249	8,780	23,424	8,718	7,555	7,867
Pleurosira laevis	505	398	557	370	457	146
Pleurosira laevis–like sp.	—	—	—	250	—	—
Pleurosira spp.	—	—	—	—	22	—
Proboscia alata	—	—	21	—	—	—
Prorocentrum compressum	151	—	—	—	—	63
Prorocentrum cordatum	344	—	—	125	—	—
Prorocentrum mexicanum	65	—	—	—	—	—
Prorocentrum micans	—	70	—	—	—	—
Prorocentrum spp.	32	—	—	—	—	—
Psammodictyon panduriforme	65	445	490	234	306	73
Psammodictyon spp.	43	—	68	—	—	—
Pseudauliscus peruvianus	—	—	—	—	—	52
Pseudendoclonium sp.	54	—	—	—	—	—
Pseudauliscus radiatus	—	—	68	—	—	—
Pseudonitzschia spp.	65	140	286	—	—	—
Pseudoscourfieldia spp.	22	—	—	—	—	—
Rhabdonema adriaticum	—	97	52	—	—	—
Rhaphoneis spp.	—	—	—	68	—	—
Rhizosolenia imbricata	—	—	63	—	—	—
Rhizosolenia setigera	—	—	125	—	—	—
Rhopalodia gibba	387	290	932	797	210	891
Rhopalodia rupestris	—	140	—	—	70	68
Rhopalodia rupestris–like sp.	—	—	—	—	70	68
Rhopalodia spp.	—	43	135	313	280	772
Scenedesmus quadricauda	—	—	—	—	—	250
Scenedesmus spp.	194	258	—	250	—	—

(*Continued*)

Species	PM1	PM2	PM3	PM4	PM5	PM6
Scillatoria spp.	828	—	2,209	—	—	—
Selenastrum spp.	—	22	—	—	—	125
Sellaphora pupula	—	54	—	21	134	—
Sellaphora spp.	48	194	63	—	199	—
Seminavis spp.	43	1,194	188	63	—	—
Skeletonema costatum	1,108	1,544	333	271	—	—
Skeletonema potamos	—	—	—	—	753	—
Spirogyra spp.	—	65	—	—	—	—
Staurastrum spp.	—	—	—	—	—	63
Stauroneis anceps	65	129	—	63	784	266
Stauroneis pachycephala	—	247	68	516	1,592	672
Stauroneis phoenicenteron	129	333	21	120	489	240
Stauroneis spicula	—	108	—	—	—	125
Stauroneis spp.	129	183	490	193	1,681	451
Stenopterobia densestriata	—	—	—	—	54	—
Stenopterobia spp.	—	—	—	—	129	—
Surirella bilinearis	4,486	711	406	—	489	406
Surirella biseriata	194	65	—	—	—	—
Surirella fastuosa	113	209	771	73	118	214
Surirella febigeri	—	—	68	—	65	63
Surirella gemma	—	—	52	104	—	—
Surirella ovata	194	—	125	—	—	—
Surirella sp.	—	—	52	—	—	—
Surirella sp. (giant)	—	168	755	391	452	865
Surirella spp.	9,355	4,242	8,248	5,574	5,885	8,859
Surirella tenera	—	—	—	—	140	83
Synedra acus	—	—	—	21	22	—
Synedra acus v. *radians*	43	—	21	—	—	—
Synedra cf. *parasitica*	1,871	2,129	20,388	11,377	6,011	17,983
Synedra cf. *parasitica* (on *Nit. sig.*)	—	4,302	2,667	2,412	8,840	708
Synedra delicatissima	—	—	—	—	—	21
Synedra pulchella	—	—	—	—	22	135
Synedra radians	301	65	63	130	323	130
Synedra sp. (curved)	—	—	—	—	22	—
Synedra spp.	264	258	313	354	75	896
Synedra tabulata	—	—	—	104	54	—
Synedra tenera	1,290	—	—	—	—	—
Synedra ulna v. *oxyrhynchus*	—	129	—	—	—	—
Tabellaria fenestrata	226	532	740	260	4,962	844
Tabellaria spp.	—	—	—	63	—	125
Tabularia fasciculatum	51	—	—	—	—	—
Tabularia tabulata	65	608	—	188	65	52
Talaroneis furcigerum	—	—	68	—	—	—
Terpsinöe americana	129	129	1,896	2,427	3,199	4,610
Terpsinöe musica	362	1,143	11,614	22,862	35,537	23,823
Thalassionema nitzschioides	—	22	—	—	—	—
Thalassiosira eccentrica	108	129	208	83	129	73
Thalassiosira gessneri	—	—	—	—	—	125

(*Continued*)

Species	PM1	PM2	PM3	PM4	PM5	PM6
Thalassiosira oestrupii	75	129	135	63	258	—
Thalassiosira spp.	86	151	776	511	118	63
Toxarium hennedyanum	—	—	—	—	—	52
Toxarium undulatum	—	54	—	—	—	—
Triceratium dubium	—	—	—	—	140	—
Triceratium favus	—	—	—	—	—	68
Tryblionella circumsuta	108	269	646	427	866	333
Tryblionella debilis	427	681	1,380	1,290	851	1,932
Tryblionella dubia	—	275	55	68	350	—
Tryblionella granulata	483	648	1,573	594	683	821
Tryblionella hungarica	32	118	52	—	—	68
Tryblionella levidansis	919	1,204	2,516	2,182	1,603	2,185
Tryblionella littoralis	—	194	—	125	65	—
Tryblionella navicularis	—	—	—	21	—	—
Tryblionella perversa	—	—	—	—	—	68
Tryblionella plana	27	—	—	—	—	—
Tryblionella spp.	1,307	4,497	9,597	11,158	7,633	9,354
Undetermined centric diatoms	742	656	563	870	946	2,092
Undetermined cryptophyte	919	—	—	—	—	—
Undetermined diatom cells in chains	8,107	—	—	4,875	1,613	21,624
Undetermined dinoflagellate cysts	—	—	—	21	—	21
Undetermined epiphytic diatom	1,871	—	—	—	—	—
Undetermined nanococcoids	129	—	—	—	—	—
Undetermined nanoflagellates	387	—	—	—	—	—
Undetermined pennate diatoms	14,640	13,344	16,457	9,572	19,923	30,977
Undetermined pennate diatoms (in chains)	3,871	774	—	3,875	1,032	11,188
Urosolenia eriensis	—	—	—	—	54	—
Total	**1,422,808**	**1,091,539**	**1,725,123**	**1,314,764**	**1,523,195**	**1,494,796**

PERDIDO BAY SYSTEM

Phytoplankton

Species	18	22	23	25	26	31	37	40
Merismopedia tenuissima	4,257	24,172,440	3,044,624	121,867	475,323	51,187	8,736	8,571
Undetermined cryptophyte	572,896	448,493	1,236,344	1,305,937	1,482,455	1,074,569	996,051	758,599
Undetermined nanoflagellates	330,011	588,758	579,820	651,382	651,339	613,596	679,893	620,779
Cyclotella choctawhatcheeana	85,177	33,400	411,718	628,772	846,312	629,353	568,064	469,368
Chaetoceros throndsenii	13,765	633	77,853	186,093	251,806	164,456	613,180	198,879
Chaetoceros spp.	8,035	1,337	74,155	152,451	160,440	194,100	419,466	411,962
Heterosigma akashiwo	143,936	15,805	415,519	169,731	354,966	101,609	60,807	25,421
Cryptomonas spp.	204,593	235,120	172,290	97,386	141,670	100,895	28,671	21,304
Undetermined nanococcoids	50,419	97,232	90,419	101,159	98,968	87,130	94,204	86,573
Gymnodinium spp.	28,023	16,657	77,562	89,714	91,035	68,880	73,775	67,334
Pyramimonas spp.	14,294	7,370	60,717	95,468	74,233	77,198	74,341	60,073
Urosolenia eriensis	35,305	2,519	24,925	97,908	93,441	74,089	44,169	26,197
Chlamydomonas spp.	41,241	129,718	61,179	39,101	60,124	31,800	19,307	14,521
Chrysochromulina spp.	3,869	959	29,636	59,390	56,545	68,734	69,956	62,374
Synedropsis karsteteri	1,206	367	78,795	41,438	30,867	18,286	195,321	3,492
Chaetoceros sp.	5,210	372	30,675	37,237	73,101	42,964	41,221	37,199
Pseudoscourfieldia spp.	8,045	1,594	39,275	50,401	49,640	24,856	15,870	12,304
Chaetoceros subtilis	1,750	282	18,403	29,032	32,391	32,069	41,379	43,476
Skeletonema costatum	7,641	784	6,023	11,594	10,522	26,234	50,196	58,750
Scenedesmus spp.	56,028	34,077	12,695	18,615	19,381	10,250	7,426	5,882
Prorocentrum cordatum	3,003	344	24,185	30,273	39,647	22,804	8,750	8,377
Ankistrodesmus sp.	8,119	1,706	8,229	29,950	24,621	29,741	17,495	8,993
Ankistrodesmus spp.	39,811	11,056	6,839	12,463	11,730	8,173	7,060	5,011
Spermatozopsis exsultans	2,745	69,260	13,731	5,668	7,941	2,513	2,354	1,225
Heterocapsa pygmaea	3,914	498	14,993	17,803	20,247	9,970	8,943	7,992
Nitzschia reversa	1,237	571	6,048	11,813	23,646	7,287	6,251	5,338
Chattonella cf. *subsalsa*	297	128	24,840	16,837	6,088	3,621	83	2
Pseudonitzschia pseudodelicatissima	725	143	5,895	5,586	6,780	8,002	14,795	16,047

Tetraedron spp.	1,436	413	3,895	10,140	9,964	8,921	10,879	7,857
Undetermined pennate diatoms	5,706	9,583	7,162	7,348	5,537	5,704	7,152	8,160
Selenastrum spp.	7,549	17,471	5,501	5,692	5,347	3,813	2,868	2,123
Undetermined curved cell with long bristle	2,080	129	6,970	1,341	27,217	1,730	493	29
Chaetoceros wighami	442	32	6,067	3,666	3,118	4,583	11,301	10,555
Leptocylindrus danicus	36	13	1,544	109	17,324	631	2,957	5,265
Cyclotella spp.	600	19,110	2,364	1,584	4,420	2,689	1,509	4,091
Cerataulina pelagica	609	231	991	3,462	4,870	3,961	7,848	8,080
Mallomonas spp.	5,528	8,212	5,978	3,448	4,929	1,550	212	320
Dictyosphaerium pulchellum	8,444	7,796	1,823	4,221	2,988	1,205	199	195
Dactyliosolen fragilissimus	228	63	823	2,546	2,946	3,427	8,497	9,462
Eutreptiella spp.	1,871	255	5,002	4,205	5,600	2,053	1,048	917
Skeletonema potamos	103	21,287	566	419	995	277	647	579
Stylochrysalis spp.	1,023	5,611	1,884	3,557	2,993	2,771	1,852	1,316
Pedinophora sp. (4 fl. in mucus)	83	22,715	727	16	56	7		
Cryptomonas sp.	862	47	22,643					
Dictyosphaerium ehrenbergianum	7,574	4,329	1,570	1,738	2,796	1,051	357	162
Gyrodinium spp.	106	81	2,238	4,254	3,483	3,501	2,713	1,981
Chaetoceros diversus	60	5	1,990	2,333	1,634	3,254	4,259	5,749
Scenedesmus quadricauda	7,890	1,832	1,803	2,243	2,746	810	504	409
Gymnodinium cf. *splendens*	893	119	3,575	3,848	5,179	1,486	495	214
Chaetoceros simplex	208	46	543	1,751	1,929	2,845	4,750	4,833
Chaetoceros affinis	150	48	1,105	1,272	1,531	2,223	4,925	5,848
Dactylococcopsis–like sp.	67	32	1,632	4,095	2,792	3,506	651	514
Nitzschia spp.	1,856	4,000	2,602	1,976	2,615	880	618	687
Cyclotella meneghiniana	597	13,792	1,002	224	66	18		188
Undetermined green algae in mucus	3,866	7,242	1,267	800	893	394		333
Kirchneriella spp.	4,007	3,946	693	863	1,706	616	231	316
Crucigenia quadrata	2,844	2,088	1,169	1,153	1,315	777	1,239	981
Undetermined oxytoxoid flagellate	657	54	1,547	2,673	2,032	2,495	996	491
Pandorina morum		11,310	936	128		110		
Chaetoceros sp. (in mucilage)				752	594	749	4,754	4,732
Thalassiosira spp.	54	54	379	973	1,255	2,117	2,542	2,757

(*Continued*)

Species	18	22	23	25	26	31	37	40
Chaetoceros ceratosporum	21		738	1,448	1,239	1,613	2,430	2,196
Euglena spp.	679	6,025	1,266	628	1,209	254	97	26
Thalassionema cf. *frauenfeldii*	55	8	351	688	813	1,542	3,253	3,403
Pachysphaera spp.	86	27	524	1,073	1,269	1,830	2,482	2,422
Undetermined dinoflagellates	689	462	2,402	1,384	1,603	867	906	942
Cryptomonas sp. (*Teleaulax*-like)			10,204					
Prymnesium spp.			284		19	8,868		381
Gymnodinium sp. D	396	6,668	868	325	399	248	77	164
Undetermined filamentous cyanophyte		8,310	593			5		
Closterium spp.	442	179	720	2,536	1,450	1,298	422	131
Leptocylindrus minimus	13		58	80	138	745	2,833	4,624
Oocystis spp.	1,184	1,536	1,317	1,158	770	438	440	448
Dinobryon spp.	593	673	345	451	476	672	1,981	2,128
Golenkinia spp.	1,026	2,997	319	1,202	695	171	73	62
Navicula spp.	417	2,100	996	827	399	887	473	583
Anabaena spp.	3,427	971	427	712	717	325	218	
Cyclotella sp. 6 (5 μm)			1,545			3,473		1,919
Crucigenia tetrpedia	675	2,778	624	662	813	570	104	106
Asterionellopsis glacialis	137	43	122	351	528	750	2,226	2,277
Synura spp.	96	6,202	388	29	24			2
Peridinium quinquecorne	5,510	16	201	30	49			2
Thalassionema spp.	16		146	618	651	881	1,470	1,415
Bacteriastrum hyalinum	70	4	237	120	185	614	1,809	2,093
Dinobryon divergens	2,421	173	732	729	350	232		
Prorocentrum micans	3		187	725	579	808	1,170	1,033
Synedra spp.	2,768	777	410	388	346	125	20	49
Actinastrum cf. *hantzschii*	78	4,608	366	75				
Undetermined centric diatoms	349	1,078	397	432	495	232	717	870
Crucigenia spp.	869	713	526	383	1,034	347	336	38
Pseudonitzschia spp.	23		71	293	124	669	1,817	1,860
Undetermined prymnesiophyte			2,915			904		645
Gymnodinium cf. *elongatum*	143	11	145	461	534	524	893	826

Dictyosphaerium spp.	330	1,362	450	651	449	137	16	29
Chaetoceros didymus	29		16	48	177	308	1,647	1,614
Micractinium spp.	93	742	113	1,344	724			48
Cyclotella cf. *atomus*	287	333	320	259	281	449	886	845
Amphidinium spp.	5	5	90	64	491	272	779	1,654
Guinardia striata			83	61	290	273	1,157	1,223
Scenedesmus dimorphus	629	1,606	146	272	388	194	10	11
Dinobryon sertularia	292	88	150	1,088	493	135	3	105
Nephroselmis spp.			495	328	1,928	282	26	41
Chaetoceros throndsenii v. *trisetosa*	31		433	464	790	356	399	343
Achnanthes spp.	242	2,043	454	177	15	35	83	55
Pedinophora sp.		2,915	76					
Aulacoseira spp.	645	885	209	291	324	208	42	
Achnanthes exigua	21	2,540	245	19	25	9	12	10
Pedinomonas spp.			1,193			792		886
Merismopedia spp.	481	1,798	517					
Carteria spp.	10	2,341	368	19	39			
Scrippsiella spp.	58	43	416	620	409	182	291	247
Synedra cf. *delicatissima*	368	98	48	676	621	133	8	10
Ceratium hircus	16		231	208	346	504	579	396
Heterocapsa spp.	30		150	760	190	295	550	368
Chlorogonium spp.	193	1,613	320	56	70	18	13	
Bacillaria paxillifer	85	481	604	418	397	92	61	19
Ankistrodesmus falcatus	1,113	141	297	123	55	55	236	
Undetermined chrysophyte flagellate				43	1,986			
Undetermined blue-green algae	1,984							
Cosmarium spp.	529	114	46	417	219	121		55
Heterocapsa rotundata	5		8			30	696	1,185
Aulacoseira granulata	1,338	221	140	93			10	24
Staurastrum spp.	190	134	209	458	331	93	15	9
Gyrosigma spp.	6	49	485	188	498	202	170	201
Micractinium pusillum	186	1,189	24	224	19	18		
Acanthoceras magdeburgense	595	148	119	320	227	10	10	2

(*Continued*)

Species	18	22	23	25	26	31	37	40
Cymbella spp.	102	119	780	129	124	99	54	100
Acanthoceras magdeburgense (spore)			301		678			
Leucocryptos marina			535			699		286
Gyrosigma sp. (with stauros)	5	43	230	635	208	128	21	42
Amphidinium crassum	39	11	223	178	282	76	184	265
Undetermined prasinophytes			240			281		914
Navicula spicula	84		90	231	52	217	366	347
Schroederia setigera	241	157	121	312	220	65	15	17
Lepocinclis spp.	3	1,242	66		34			
Hillea spp.			335	78	468	171	126	
Chaetoceros spp.	238	151				9		930
Gymnodinium sanguineum			180	160	419	167	94	115
Chaetoceros compressus	13		48		41	156	242	798
Microcoleus spp.		1,286						
Plagioselmis spp.			394			593		269
Komma spp.			884			226		94
Chattonella spp.	311	228	279	29	260	27	42	
Amphora spp.	57	39	152	175	75	173	128	346
Gomphonema spp.	45	985	116	13	7	5		7
Treubaria setigerum	66	62	151	248	270	118	10	
Opephora spp.	39	462	164	86	78	98	106	86
Coelastrum spp.	42	215	65	213	155	7		114
Ceratoneis closterium	56	13	86	144	101	66	145	256
Hemiaulus hauckii	10	4	32	91	53	65	327	308
Eucampia cornuta	13		14	32	65	67	275	468
Chaetoceros simplex v. *calcitrans*			18		150			
Cocconeis placentula	49	790	50	37	15	19	5	
Entomoneis spp.	40	14	83	228	146	152	149	102
Elakatothrix gelatinosa	452	49	46	48	102	7		34
Centritractus belanophorus	5	163	161	200	181	37		7
Dinobryon bavaricum	111	16	26	154	229	5		
Pontosphaera sp.		839	1					

Undetermined colonial alga (2-celled)			48	288	340	107		
Hemiselmis spp.						490		346
Phacus spp.	21	754	30	3	14			2
Cyclotella-like sp.		258	567					
Eudorina elegans		817						
Chaetoceros decipiens	5		16	45	62	121	149	242
Anabaenopsis sp.							630	171
Undetermined euglenoid		342	188			224		25
Dictyosphaerium sp.	325			123	10	96	84	
Arthrodesmus spp.	289	24	52	152	72	45	20	
Mallomonas cf. *schwemmeki*			694			40		2
Protoperidinium spp.	2		152	84	65	76	162	127
Thalassionema nitzschioides			8	5	91	92	161	261
Crucigenia fenestrata	253	234	32		155			
Diploneis spp.	65	52	67	153	88	75	40	50
Falcomonas spp.			369			253		40
Entomoneis alata	119	11	135	145	89	35	10	27
Chaetoceros minimus–like sp.	8		48	144	51	56	45	97
Rhizosolenia longiseta	115	41	400	14	46	4		
Rhizosolenia spp.	25		38	40	62	11	249	124
Oscillatoria spp.	19	148	75	129	87	11	18	10
Synedra ulna	248	157	50	55	39	27		10
Diatoma spp.	310			8				
Cocconeis spp.	93	44	109	109	38	52	37	65
Gonium pectorale	44	193	324					
Cryptomonas acuta			396			71		82
Chrysochromulina sp. (st. 23 10/97)			536					
Dunaliella spp.			375			121		38
Prorocentrum compressum			8			7	234	268
Cyclotella striata			5	53	50	73	141	115
Microcystis aeruginosa	357	145		8				
Lioloma pacifica	8		8		10	39	215	212
Eunotia spp.	269	61	54	38	11	12	3	28

(*Continued*)

Species	**18**	**22**	**23**	**25**	**26**	**31**	**37**	**40**
Undetermined coccoids						355		150
Gymnodinium sp.							501	
Pinnularia spp.	38	267	39	72	9	20		7
Pediastrum duplex	83	32	32					
Undetermined wing-shaped alga				186	72	148	10	7
Nitzschia sigmoidea	12	42	106	184	31	7	1	3
Climaconeis sp.				40	2	64	210	96
Coscinodiscus spp.			22	29	24	55	85	121
Paralia sulcata			8	16	48	61	115	183
Scenedesmus smithii	140	22			87	55		
Undetermined chrysophyte	10	394	22					
Gonyaulax spp.	29		33	11	111	77	47	68
Nitzschia sigma	114	56	41	35	79	18	18	
Pedinomonas mikron			127			202		77
Ankistrodesmus spiralis	154	3	6	62	51	14	89	
Plagioselmis sp. A (sensu *Throndsen*)			167			233		
Brachymonas-like sp.		392						
Chlamydomonas sp. (with 2 pyrenoids in mucus)	17	286	88					
Merismopedia spp. (colonies of 8 cells)		387						
Diploneis bombus	5	5	58	119	25	54		
Minidiscus trioculatus			43			84		230
Crucigenia crucifera	31	323						
Protoperidinium bipes	16		16	45	61	48	24	35
Proboscia alata	3		14	5	12	21	54	223
Prorocentrum spp.	5	5	24	21	17	48	110	59
Undetermined coccoid cells in mucus				131	101			
Amphidinium amphidinioides			75			215		19
Undetermined dinoflagellate cysts	15	1	108	5	4	113		61
Rhizosolenia styliformis v. *longispina*			105		203			
Entomoneis-like sp.		1	303					
Gymnodinium splendens			145	12		78	13	56

Cryptomonas acerosa			160			82		61
Merismopedia punctata		250	48					
Prorocentrum gracile			2	16	19	58	109	85
Mallomonas sp. (bottle-shaped)	15		105	104	14			
Cryptomonas spp. (median swelling)			294					
Scenedesmus intermedius	73	22	24	43		55	21	38
Surirella spp.	23	29	22	56	78	27		10
Ceratium fusus			48	8	19	38	63	84
Chaetoceros socialis							24	249
CYTCLO			54			126		82
Chaetoceros simplex v. *calcitrans* (spore)			2		198			
Thalassiosira cedarkeyensis		5	50	48	39	25	24	29
Stylochrysalis cf. *libera*		247						
Plagioselmis sp. B (sensu *Throndsen*)			191			53		
Peridinium spp.	28		103	27	14	31	3	2
Fallacia pygmaea	10		21	57	17	33	40	50
Pedinomonas sp.						233		
Synedra sp. 4 (PD 18 6/93)	231							
Cyclotella "*wolfeana*"	3		14	32	29	66	34	31
Spermatozopsis sp.			195			28		1
Nitzschia longissima	3		8	74	57	36	10	12
Climaconeis scopularum							105	112
Aphanothece spp.	214							
Rhizosolenia fragilissima			10		19	22		30
Rhizosolenia setigera	3	11	4	8	17	32	52	79
Heterocapsa triquetra	10	11	54	19	48	14	18	
Anabaena macrosporum			85			117		
Nitzschia sp. (linear with capitate ends)	103	78	20					
Treubaria spp.	26	129	16	11	7	7		
Micromonas spp.	6	73	82			7		28
Nitzschia palea			190	3				
Oxyphysis oxytoxoides			8	11	28	62	42	10
Chaetoceros subsecundus			20			11	29	123

(*Continued*)

Species	18	22	23	25	26	31	37	40
Undetermined coccoid blue-green algae		183						
Merismopedia elegans				181				
Tryblionella spp.	3		34	54	12	11	21	2
Mastogloia spp.	5	8	2			14	44	93
Pleurosigma spp.	6	3	24	28	2	17	18	62
Chaetoceros fragilis					99			
Fragilaria spp.	83	48		5		5	5	5
Nitzschia pungens			38			115		6
Anabaenopsis spp.							52	105
Dictyocha fibula			14	13	10	63		50
Pseudanabaena spp.		56				33		47
Pediastrum simplex		148						
Heterosigma spp.			137					
Bacteriastrum spp.	3					9	21	107
Cymatosira lorenziana	10		5				71	54
Desikaneis gessneri		134		5				
Undetermined green flagellates			137					
Selenastrum-like sp.			19	24	14	29	24	7
Akashiwo sp. (in mucus)			85	8	26	7		
Chroococcus spp.		43		85				
Pandorina spp.	41	84	2					
Chromulina spp.			48			73		
Eurosolenia eriensis	1		45		12	61		2
Kirchneriella-like sp. (terminal spines)	90	8		16				
Hemiaulus sinensis						14	31	64
Trachelomonas spp.	6	92	11					
Pseudosolenia calcar-avis						18	30	41
Dictyocha speculum			21		2	8	37	20
Eunotia lunaris	37	5	14	19	2			
Entomoneis paludosa		102						
Chaetoceros curvisetus								95
Pseudoscourfieldia sp.			10	8		22		50

Scenedesmus serratus	93		1					
Actinocyclus spp.	3		4		12	12	32	24
Aulacodiscus spp.	83			8				
Gloeocapsa-like sp.		74	17					
Chroomonas spp.						70		20
Coelastrum microporum			32		58			
Chaetoceros laciniosus			16	21		14		38
Diploneis smithii	5		14		10	23		7
Gloeocystis planktonica								
Undetermined colony				85				
Pediastrum tetras	31	52						
Tabellaria spp.	44		6		4	4		
Anabaena spiroides			65			16		
Prorocentrum cf. *minimum*						81		
Pterosperma sp.			81					
Chaetoceros brevis					10	19	10	40
Frustulia rhomboides	26	5	40		2	5		
Treubaria triappendiculata	50		12		0			
Diatoma vulgare	78							
Stylochrysalis sp.		75						
Leucocryptos spp.			63			11		
Stauroneis spp.	21	5		4		11	8	15
Asterionella formosa	57		12					
Synedra parasitica					68			
Mougeotia spp.	42	25						
Tetraselmis spp.			13			19		33
Nostoc spp.		65						
Nitzschia acicularis	3	6	54					
Pachysphaera marshalliae					16		5	32
Spondylosium cf. *secedens*				48				
Undetermined green algae		43			19			
Gomphonema acuminatum		59						
Melosira spp.	10		24		22			

(*Continued*)

Species	18	22	23	25	26	31	37	40
Chaetoceros atlanticus								55
Navicula capitata		38	16			1		
Oxytoxum gracile					10		16	29
Eunotia zazuminensis	20	19					10	5
Frustulia spp.	22	8	4	6		4		5
Campylosira spp.								53
Pachysphaera pelagica			14		12			5
Undetermined centric diatoms (hyaline)		51						
Guinardia delicatula								50
Eunotia pectinalis	13	5	12	13	2			
Tabellaria sp.	49							
Fragilaria crotonensis	32		16					
Gyrosigma sp. (S-shaped)			12		19			
Phacus longicauda		48						
Micromonas sp.			39			8		
Kirchneriella subsolitaria		46						
Polykrikos spp.			4		34	7		
Chaetoceros peruvianus						14		31
Chroococcus sp.				43				
Undetermined chroococcales	3							39
Dinobryon crenulatum			34		7			
Nephrochloris spp.						40		
Teleaulax sp.			38			2		
Navicula pupula	3	35						
Thalassiosira oestrupii							24	14
Gymnodinium sp. (curved)						34	3	
Scenedesmus denticulatus	10	11	16					
Ditylum brightwellii							16	20
Kirchneriella contorta		24	12					
Calycomonas ovalis			12					22
Dinophysis spp.						3	5	26
Psammodictyon spp.			16			18		

Crucigenia apiculata		33						
Synedra tabulata	9	7	6	11	0			
Undetermined straight cell with long bristle	3			6		7	13	4
Undetermined loricate green algae			18	3	7	2		
Nitzschia dubia			10	5	14			
Thalassiosira minima/proschikinae complex								31
Lyngbya spp.		22	8					
Karenia brevis								29
Mallomonas akrokomas	15		14					
Neidium spp.	8	13		6		2		
Diploneis elliptica			28					
Emiliania huxleyi			2			2		24
Actinoptychus undulatus							3	
Alexandrium spp.			12	8	6			
Fallacia spp.	5			3			8	5
Kirchneriella lunaris	26							
Rhizosolenia delicatula			2		2	2		20
Chaetoceros subtilis v. *trisetosa*	3			21				
Euastrum spp.	14				10			
Delphineis surirella				8		2	3	10
Gyrosigma distortum	3			11	2	2	3	
Nitzschia section *pseudonitzschia* spp.						13		10
Semiorbis-like sp.	10			13				
Synedra pulchella	15	5		3				
Thalassiothrix mediterranea v. *pacifica*			2		12			
Quadrigula chodatii			22					
Synedra acus		4	18					
Thalassiothrix frauenfeldii			1	3			5	13
Amphidinium sp. (top-shaped)						7	13	1
Campylosira cymbelliformis								21
Haslea spp.						9	11	1
Lithodesmium undulatum							21	

(*Continued*)

Species	18	22	23	25	26	31	37	40
Selenastrum westii	21							
Thalassiosira minima					5	11		
Eunotia flexuosa	10		7	3				
Melosira nummuloides						8		7
Schizostauron crucicula		4	16					
Ceratium spp.						7	10	2
Coscinodiscus radiatus	3		4		5	5		2
Undetermined paired cells				19				
Cocconeis pensacolae				3			8	2
Navicula yarrensis			7			11		
Selenastrum bihraianum	18							
Spirogyra spp.			18					
Dactyliosolen antarcticus					7		5	5
Fallacia forcipata				16		1		
Navicula hungarica v. *capitata*		13	4					
Pyramimonas cf. *adriatica*								17
Achnanthes manifera						2	11	3
Bleakeleya notata			16					
Lagerheimia chodatii		16						
Nitzschia tryblionella			4					7
Prorocentrum lima	3							13
Calycomonas gracilis						5		10
Cocconeis distans							8	5
Rhizosolenia styliformis			2			2	5	1
Scenedesmus bijuga			6		2			
Tabellaria fenestrata	9		4		2			
Thalassiosira rotula						5		10
Cyclotella atomus								14
Actinoptychus senarius			6				3	5
Amphidinium musicolum						1		13
Chaetoceros costatum								14
Cymatosira belgica				3				11

Gomphonema angustatum		8	6					
Nitzschia panduriformis			1			9		4
Oxyrrhis marina					7			
Pyramimonas cf. *propulsar*								14
Pyramimonas grossii						14		
Amphora coffeaeformis		8					5	
Entomoneis sp. (with knots)			2	11				
Monoraphidium sp. (curved)	10			3				
Rhizosolenia stolterfothii						5		8
Spondylosium spp.	8			5				
Tryblionella levidansis			8		5			
Akashiwo splendens					7			
Hemiaulus sp. (long curved p.v. axis)								12
Licmophora spp.				8		2		2
Chattonella sp.		11						
Dictyocha sp.							5	6
Falcula hyalina			2			2	5	2
Gymnodinium simplex			11					
Navicula cf. *distans*				11				
Skeletonema spp.		11						
Tryblionella hungarica		11						
Coscinodiscus granii			2			5	3	
Crucigenia-like sp.						5	5	
Cyclotella sp. 7 (9 μm)			6			4		
Dinobryon sp.					10			
Diploneis obliqua							3	7
Dunaliella sp.			10					
Gymnodinium cf. *agiliforme*							10	
Hemiaulus sp. (long straight p.v. axis)								10
Odontella aurita							5	5
Prorocentrum mexicanum								
Prorocentrum sphaeroideum			1					9
Sphaerocystis-like sp.			10					

(*Continued*)

Species	18	22	23	25	26	31	37	40
Synedra radians	5	5						
Chaetoceros lorenzianus					5			4
Dinobryon faculiferum	5		4					
Eunotia bidentula	9							
Haslea cruciculoides						7		2
Myxosarcina sp.						9		
Odontella mobiliensis						9		
Tetraselmis wettsteinii								9
Heterocapsa rotundata–like sp.							8	
Achnanthes exigua–like sp.		8						
Amphidinium cf. *sphaeroides*							8	
Chaetoceros danicus							8	
Cymatosira belgica (fusiform)							8	
Dinophysis caudata								8
Euglena acus		1	2			5		
Falcula-like sp.			4			4		
Gephyrocapsa oceanica			8					
Lagerheimia subsalsa			8					
Melosira pensacolae			8					
Monoraphidium spp.	8							
Navicula sovereignae				3			5	
Nitzschia dissipata						3	5	
Nitzschia sp. 6			8					
Nitzschia spp. (S-shaped)				5	2			1
Petrodictyon gemma			8					
Amphora obtusa								7
Anorthoneis spp.					2			5
Cerataulus smithii							5	2
Chloromonas spp.			4			3		
Chloronephris sp.			7					
Chroomonas cf. *lateralis*			7					
Fragilaria sp.								7

Guinardia flaccida						7		
Gymnodinium sp. (with acuminate tip)							5	2
Gyrosigma macrum var.			2		2		3	
Haslea sp. (with stauros)							5	2
Lyrella clavata				5				
Martyana martyi			2			5		
Plagiogramma cf. *tenuissima*					2		5	
Pyramimonas cf. *grossii*								7
Pyramimonas cf. *nephroides*								7
Selenastrum sp.					7			
Thalassiothrix longissima						4	3	
Calycomonas wulffii								6
Chaetoceros coarctatus								6
Chaetoceros constrictus			6					
Dimeregramma marina							6	
Sellaphora pupula				6				
Staurastrum ionatum v. *tortum*			6					
Striatella unipunctata			5					1
Thalassiosira eccentrica	3			3				
Polykrikos schwartzii							3	
Oxytoxum spp.							5	
Actinella punctata	1		4					
Biremis spp.								5
Caloneis westii								
Capartogramma crucicula		3	2					
Closterium cf. *lineatum*		5						
Dimeregramma fulvum							5	
Dimeregramma minor							5	
Dinophysis caudata cf. *acutiformis*								5
Entomoneis gigantea				5				
Gymnodinium sp. 3			5					
Leptocylindrus danicus (spore)								
Meridion circulare	5							

(*Continued*)

Species	18	22	23	25	26	31	37	40
Navicula cf. *cruciculoides*								5
Nitzschia gracilis	1		1					3
Nitzschia lanceola				3				
Nitzschia sp. 23			5					
Octacanthium spp.	5							
Oxytoxum sp.						5		
Prorocentrum sp.(with spine)				3		2		
Protoperidinium depressum					5			
Skeletonema menzellii		5						
Skeletonema sp. 1 (2-celled)						5		
Skeletonema-like sp.					5			
Spiniferomonas spp.	3				2			
Spirulina spp.		5						
Trachelomonas hispida	5							
Cocconeis scutellum	1							2
Achnanthes curvirostrum	3							1
Amphidinium cf. *extensum*								4
Cymbomonas sp.						4		
Eudorina spp.			4					
Eutreptiella marina			4					
Gymnodinium filum								4
Gymnodinium varians								4
Licmophora flabellata						3		1
Protoperidinium divergens			4					
Surirella amphioxys			4					
Synura uvella			2			2		
Tribonema sp.			4					
Amphidinium lanceolatum							3	
Anorthoneis sp.		3						
Biddulphia spp.	3							
Brachysira spp.	3							
Caloneis oregonica	3							

Caloneis spp.	3							
Closterium setaceum		3						
Cymbella subcuspidata				3				
Diploneis didyma			1		2			
Euglena viridis			3					
Eunatogramma marina							3	
Eunotia diodon	3							
Eunotia pectinalis v. *undulata*			3					
Gambierdiscus sp.								3
Gyrosigma beaufortiana	3							
Kirchneriella lunaris v. *dianae*	3							
Lyrella sulcifera							3	
Mastogloia apiculata								3
Navicula gregaria		3						
Navicula hasta	3							
Navicula yarrensis v. *americana*							3	
Ophiocytium capitatum v. *longispinum*	3							
Oscillatoria anguina			3					
Oscillatoria sp. (colony)		3						
Prorocentrum cf. *lima*	3							
Prorocentrum concavum							3	
Spirulina subsalsa		3						
Staurodesmus indentatus			3					
Stauroneis pachycephala	3							
Stenopterobia curvula			1			1		1
Tabellaria flocculosa	1		2					
Tetraedron caudatum	3							
Tetraedron pentaedrium	3							
Tetraedron verrucosum	3							
Akashiwo spp.					2			
Alexandrium sp.								
Amphidinium sp. 3			2					
Apiochloris sp.			2					

(*Continued*)

Species	**18**	**22**	**23**	**25**	**26**	**31**	**37**	**40**
Calycomonas spp.								
Campylodiscus clypeus						2		
Cerataulina bergonii								2
Ceratium tripos								2
Chlorogonium sp. 1 (St.22 5/89)			2					
Chromophysomonas spp.					2			
Cryptomonas ovata						2		
Cryptomonas salina			2					
Cyclotella meneghiniana complex			2					
Dinophysis punctata			2					
Eucampia zodiacus						2		
Eunotia formica			2					
Eunotia monodon			2					
Gyrosigma balticum			2					
Gyrosigma parkerii			2					
Hantzschia virgata			2					
Licmophora remulus						2		
Navicula lyra			1					1
Nitzschia scalaris						2		
Plagiogramma pulchellum						2		
Pleurosigma elongatum			2					
Pleurosigma reversum								2
Psammodictyon panduriforme								
Rhizosolenia imbricata			2					
Scenedesmus armatus			1					1
Stylochrysalis libera					2			
Surirella fastuosa			1					1
Tryblionella compressa								
Tryblionella debilis						2		
Undetermined pennate diatoms (in chains)								2
Achnanthes brevipes								1
Actinocyclus sp.			1					

Amphidinium longum							1	
Amphora ovalis			1					
Anorthoneis excentrica								1
Astasia sp.						1		
Aulacoseira italica		1						
Centritractus spp.		1						
Ceratium furca						1		
Chaetoceros gracilis								1
Chlamydocapsa maxima	1							
Closterium moniliferum		1						
Cyclotella meneghiniana–like sp. (20–30 μm)	1							
Cymbella sp.			1					
Diploneis crabro			1					
Dunaliella acidophila			1					
Euastrum insulare			1					
Gymnodinium cf. *aeruginosum*			1					
Haslea wawrikae								1
Neidium affine		1						
Neosynedra provincialis			1					
Nitzschia longissima v. *reversa*			1					
Phacus orbicularis							1	
Pinnularia gibba			1					
Rhaphoneis amphiceros								1
Staurastrum iversenii			1					
Stauroneis anceps	1							
Streptotheca thamensis								1
Surirella robusta			1					
Synedra undulata			1					
Tetmemorus brebissonii			1					
Protoperidinium sp. (St. 42B 8/07)								
Protoperidinium sp. (elongated with spine)								

(*Continued*)

Species	**18**	**22**	**23**	**25**	**26**	**31**	**37**	**40**
Undetermined cyanophyte								
Peridinium sp. (smooth)								
Schroederia judayi								
Gonatozygon spp.								
Opephora schwartzii								
Rhopalodia gibba								
Monoraphidium sp.	0							
Pseudanabaena sp.			0					
Scenedesmus ecornis					0			
Scenedesmus obliquus			0					
Synedra tabulata v. *acuminata*			0					
Tetraedron regulare			0					
Total	**1,782,382**	**26,115,882**	**6,800,152**	**4,232,796**	**5,362,779**	**3,699,780**	**4,295,307**	**3,187,866**

Infauna Biomass (g m^{-2})

Species	18	22	23	25	26	31	33	37	40	42B	Total
Mediomastus ambiseta	16.73	4.41	363.01	1423.86	131.81	206.77	82.79	20.59	34.29	212.66	2496.92
Streblospio benedicti	118.08	162.58	283.33	411.01	128.18	119.85	109.81	76.81	57.51	272.95	1740.11
Rangia cuneata	2.79	0.86	200.59	170.71	11.57	1.24	0.95	0.30	0.42	59.56	448.99
Phoridae	0.19	—	—	—	—	—	0.10	—	—	—	0.29
Copepoda	1.83	0.77	25.54	27.40	0.86	2.01	1.43	2.62	7.89	13.15	83.50
Nemertean sp.	2.98	0.67	18.17	51.32	6.89	7.65	5.61	1.82	11.63	12.06	118.80
Immature tubificid without cap setae	0.58	0.86	6.79	12.11	0.10	7.17	—	—	0.31	1.46	29.39
Scottalana canadensis	—	0.19	5.55	4.21	0.10	0.57	0.10	0.30	0.21	—	11.22
Assiminea succinea	3.17	0.19	27.74	29.56	2.39	0.19	0.19	—	—	—	63.43
Hobsonia florida	4.62	20.49	21.81	35.51	2.01	0.19	0.10	0.30	—	3.65	88.67
Macoma mitchelli	0.19	0.19	35.58	35.97	11.00	2.01	2.09	0.50	1.66	1.83	91.03
Chironomus sp.	0.96	14.82	46.97	38.25	3.06	—	—	—	—	2.56	106.62
Leitoscoloplos fragilis	0.29	—	7.56	18.27	9.57	13.50	9.61	7.77	12.77	12.79	92.13
Parandalia americana	0.19	0.10	7.56	31.60	0.10	0.96	0.57	—	0.62	0.37	42.06
Heteromastus filiformis	0.10	0.10	7.56	67.72	0.57	0.10	0.29	0.10	0.73	10.23	87.49
Laeonereis culveri	0.19	0.10	6.41	21.49	—	0.38	—	0.10	0.10	1.83	30.60
Polydora ligni	0.67	2.91	4.88	16.74	0.48	1.53	0.29	3.13	14.22	32.52	77.37
Oligochaeta	1.83	2.20	3.92	56.67	0.10	0.10	—	—	0.31	17.17	82.30
Paraprionospio pinnata	0.38	—	0.10	0.48	0.48	9.85	14.65	19.08	19.55	—	64.57
Grandidierella bonnieroides	1.25	0.38	4.97	52.01	—	—	0.19	0.10	—	—	58.91
Glycinde solitaria	0.48	0.10	1.63	4.97	2.20	4.78	4.85	5.35	9.79	2.92	37.08
Chironomus decorus group	2.31	10.42	7.37	1.63	1.05	—	—	—	—	—	22.78
Capitella capitata	0.29	0.19	2.68	12.72	0.86	0.10	0.19	—	0.10	14.25	31.38
Eteone heteropoda	0.48	—	4.97	10.52	0.57	0.67	0.67	0.10	0.93	4.02	22.94
Calliphoridae	—	—	7.94	32.81	—	—	—	—	0.10	—	40.85
Cryptochironomus fulvus group	0.38	0.96	2.77	5.57	0.10	—	—	—	—	—	9.78
Cossura soyeri	—	—	—	—	—	0.19	0.57	9.59	22.66	—	33.02
Sigambra tentaculata	—	—	—	0.38	0.10	1.43	1.81	4.85	15.47	—	24.03
Lepidactylus sp.	—	—	—	—	—	0.48	0.10	—	—	—	0.57
Paranais litoralis	—	—	0.19	1.43	0.29	0.10	—	—	—	—	2.01

(*Continued*)

Species	18	22	23	25	26	31	33	37	40	42B	Total
Tagelus plebeius	0.10	—	0.19	11.29	—	0.10	—	—	—	0.37	12.04
Nereis succinea	0.48	—	1.05	2.01	—	—	—	—	0.73	2.19	6.46
Crab zoeae	0.10	0.19	—	0.19	0.67	0.19	0.48	1.31	1.97	9.87	14.97
Oligochaeta 1	—	—	—	—	—	—	—	—	—	—	0.00
Gyptis vittata	—	—	0.29	0.67	0.19	1.43	2.38	0.71	4.67	0.37	10.70
Edotea sp. (cf. *montosa*)	0.19	—	2.20	1.63	—	0.10	—	—	0.10	1.83	6.04
Tubificoides heterochaetus	0.10	—	8.70	2.97	—	0.10	—	—	—	—	11.86
Pectinaria gouldii	0.29	—	1.43	3.54	0.10	0.19	0.29	—	0.31	1.46	7.61
Cumacea	—	—	0.96	2.20	0.10	0.10	—	0.40	0.42	0.73	4.90
Chaoborus punctipennis	1.54	3.54	0.29	0.29	0.19	—	—	—	—	—	5.84
Corophium louisianum	0.67	—	0.48	5.26	0.10	—	—	—	0.73	—	7.23
Tanytarsus sp.	0.48	0.19	3.25	0.67	0.10	—	0.19	—	—	0.37	5.25
Procladius sublettei	0.19	0.77	0.19	0.10	0.10	—	—	—	—	—	1.34
Polydora sp.	0.10	0.19	0.57	1.05	—	0.19	—	0.30	0.42	3.29	6.11
Nemertean sp. 1	—	—	—	0.48	—	0.10	—	—	—	—	0.57
Corophium sp.	—	—	0.38	3.44	—	—	—	—	0.73	0.73	5.28
Procladius sp.	0.87	1.63	1.24	0.57	—	—	—	—	—	—	4.32
Procladius bellus	0.10	—	—	—	0.10	—	—	—	—	—	0.19
Dicrotendipes sp.	—	—	0.48	0.48	—	—	—	—	—	—	0.96
Polypedilum scalaenum	0.10	1.53	0.19	0.29	0.10	—	—	—	—	2.56	4.76
Boccardia ligerica	0.10	—	0.48	1.63	0.10	—	0.38	0.40	—	—	3.08
Armandia agilis	—	—	—	0.10	—	0.19	—	0.30	3.81	—	4.40
Hargeria rapax	0.19	—	0.10	3.44	—	—	—	—	0.21	—	3.94
Nuculana acuta	—	—	—	—	—	—	—	0.30	3.43	—	3.73
Piromis roberti	—	—	—	—	—	—	—	0.30	2.80	—	3.11
Tharyx annulosus	—	—	—	—	—	0.10	—	—	2.91	—	3.00
Polydora aggregata	—	—	—	—	—	—	—	—	—	2.92	2.92
Armandia maculata	—	—	—	0.10	—	0.10	—	0.10	2.60	—	2.89
Monoculodes edwardsi	—	—	0.10	1.72	—	—	—	—	—	—	1.82
Polypedilum illinoense	0.10	0.29	0.67	0.38	0.29	—	—	—	0.10	—	1.83
Callianassa sp.	—	—	—	0.67	—	0.10	0.10	—	—	—	0.86
Eurythoe complanata	—	—	—	—	—	—	0.48	0.50	1.66	—	2.64

Minuspio perkinsi	—	—	—	—	—	—	—	—	2.60	—	2.60
Turbellarian sp.	—	—	0.48	1.24	—	—	—	0.10	0.21	—	2.03
Goeldichironomus carus	—	—	—	—	—	—	—	0.10	—	—	0.10
Paralauterborniella nigrohalterale	0.29	—	0.10	0.10	—	—	0.10	—	—	—	0.57
Mysidopsis bahia	0.38	0.29	0.57	0.29	0.19	—	—	0.10	—	—	1.82
Prionospio perkinsi	—	—	—	—	—	—	0.10	0.10	1.66	—	1.86
Polypedilum sp.	—	0.67	1.05	0.10	0.10	—	—	—	—	0.37	2.28
Edotea montosa	—	0.29	0.38	0.48	—	—	—	—	—	1.10	2.24
Haminoea succinea	—	—	—	0.19	—	0.10	—	1.21	0.73	—	2.22
Prionospio sp.	—	—	—	—	—	—	—	0.10	1.14	—	1.24
Chironomus sp. pupae	—	0.29	1.05	0.38	—	—	—	—	—	—	1.72
Monopylephorus sp.	—	—	—	—	—	—	—	—	—	—	0.00
Annelida (unident.)	—	—	—	0.19	—	0.29	—	—	—	—	0.48
Terebellidae	—	—	—	—	—	—	—	0.20	1.66	—	1.86
Ensis minor	—	—	—	0.38	—	—	—	—	—	0.37	0.75
Dero sp.	—	0.77	0.77	—	—	—	0.10	—	—	—	1.63
Chironomidae	—	0.19	0.48	0.30	—	—	—	—	0.10	—	1.07
Collembola	—	0.10	—	—	—	—	—	—	—	—	0.10
Callianassa atlantica	—	—	—	0.10	—	—	—	—	—	—	0.10
Tubificoides sp.	—	—	1.43	0.10	—	—	—	—	—	—	1.53
Tellina sp.	—	—	—	—	—	—	—	—	—	—	0.00
Prostoma sp.	—	—	0.29	0.19	—	—	—	—	—	—	0.48
Tribelos jucundum	0.38	0.57	0.10	0.10	—	0.10	—	—	—	—	1.25
Holothuroidea	—	—	—	0.19	—	—	—	0.10	0.42	—	0.71
Notomastus latericeus	—	—	—	—	—	—	0.10	—	0.52	—	0.61
Dero digitata	—	0.10	—	—	—	—	—	—	—	—	0.10
Polydora socialis	—	—	0.10	0.10	—	—	0.10	—	0.73	—	1.01
Nanocladius sp.	—	—	—	—	—	—	—	—	—	—	0.00
Oxyurostylis smithi	—	—	—	—	—	—	—	—	—	—	0.00
Enchytraeidae	0.19	—	—	0.29	—	—	—	—	—	—	0.48
Glycera americana	—	—	—	0.10	—	0.10	—	0.10	0.62	—	0.92
Aulodrilus pigueti	—	0.77	—	—	—	—	—	—	—	—	0.77
Culicoides sp.	—	0.19	0.48	0.10	—	—	—	—	—	—	0.77

(Continued)

Species	18	22	23	25	26	31	33	37	40	42B	Total
Goeldichironomus sp.	—	—	0.29	0.48	—	—	—	—	—	—	0.77
Ampelisca sp.	—	—	—	0.10	—	—	0.10	—	0.21	0.37	0.76
Mysidopsis bigelowi	—	—	—	0.10	—	—	0.10	0.10	0.10	0.37	0.76
Acetes americanus	—	—	—	0.10	—	—	—	—	0.62	—	0.72
Gyptis brevipalpa	—	—	—	0.10	—	—	—	0.20	0.42	—	0.71
Mysidopsis almyra	—	—	—	0.29	0.19	—	—	0.10	—	—	0.58
Gyptis sp.	—	—	0.10	0.38	—	—	—	0.10	—	—	0.58
Tharyx sp.	—	—	—	—	—	—	—	—	0.31	0.37	0.68
Ceratopogonidae sp.	0.19	0.10	0.29	0.10	—	—	—	—	—	—	0.67
Neritina reclivata	—	—	—	0.67	—	—	—	—	—	—	0.67
Callinectes sapidus	—	0.19	0.19	0.29	—	—	—	—	—	—	0.67
Polychaete (unident.)	—	0.10	—	0.10	—	0.10	—	—	—	0.37	0.65
Bullidae sp.	—	—	—	—	—	—	—	0.61	—	—	0.61
Lepidoptera	0.10	—	—	—	—	—	—	—	—	—	0.10
Brachidontes exustus	—	—	—	0.57	—	—	—	—	—	—	0.57
Polypedilum halterale group	—	0.10	0.38	0.10	—	—	—	—	—	—	0.57
Naididae	—	0.29	0.10	0.10	—	—	0.10	—	—	—	0.57
Prionospio cristata	—	—	—	—	—	—	—	—	0.52	—	0.52
Podarke obscura	—	—	—	—	—	—	—	0.30	0.21	—	0.51
Lucifer faxoni	—	—	—	—	—	—	0.10	0.20	0.21	—	0.50
Coelotanypus sp.	—	—	—	—	—	—	—	—	—	—	0.00
Diopatra cuprea	—	—	—	—	—	—	—	—	—	—	0.00
Monoculodes sp.	—	—	—	—	—	—	—	—	—	—	0.00
Nais communis	—	—	—	—	—	—	—	—	—	—	0.00
Onuphidae	—	—	—	—	—	—	—	—	—	—	0.00
Paracladopelma sp.	—	—	—	—	—	—	—	—	—	—	0.00
Tanaidacea	—	—	—	—	—	—	—	—	—	—	0.00
Scolelepis texana	—	—	—	0.48	—	—	—	—	—	—	0.48
Pristina sp.	—	—	—	—	—	—	—	—	0.42	—	0.42
Ampelisca verrilli	—	—	—	—	—	—	—	—	0.42	—	0.42
Schistomeringos rudolphi	—	—	—	—	—	—	—	—	0.42	—	0.42
Chironomid pupae	—	—	0.19	0.10	—	—	—	—	—	—	0.29

Hesionidae	—	—	—	0.19	—	—	—	0.10	0.10	—	0.40
Gammarus sp.	0.19	—	—	0.19	—	—	—	—	—	—	0.38
Bezzia/*Palpomyia* group	—	0.29	0.10	—	—	—	—	—	—	—	0.38
Diptera	—	0.38	—	—	—	—	—	—	—	—	0.38
Harnischia complex	—	0.10	0.29	—	—	—	—	—	—	—	0.38
Periclimenes americanus	—	—	—	0.38	—	—	—	—	—	—	0.38
Tanytarsini sp.	—	0.10	0.29	—	—	—	—	—	—	—	0.38
Prionospio cirrifera	—	—	—	—	—	—	—	—	0.31	—	0.31
Ancistrosyllis sp.	—	—	—	—	—	—	—	—	0.31	—	0.31
Maldanidae	—	—	—	—	—	—	—	—	0.31	—	0.31
Chaetognatha sp.	—	—	—	—	—	—	—	0.20	0.10	—	0.31
Bivalvia (pelecypoda)	—	—	—	—	—	0.10	—	—	0.21	—	0.30
Listriella barnardi	—	—	0.10	—	—	—	—	0.10	0.10	—	0.30
Cladocera	0.10	0.19	—	—	—	—	—	—	—	—	0.29
Pristina synclites	0.10	0.19	—	—	—	—	—	—	—	—	0.29
Tanypodinae	0.10	0.10	0.10	—	—	—	—	—	—	—	0.29
Ablabesmyia mallochi	—	0.29	—	—	—	—	—	—	—	—	0.29
Penaeidae	—	—	0.10	—	0.10	—	0.10	—	—	—	0.29
Taphromysis bowmani	—	—	—	—	—	—	—	—	—	—	0.00
Ancistrosyllis jonesi	—	—	—	—	—	—	—	—	0.21	—	0.21
Ancistrosyllis sp.	—	—	—	—	—	—	—	—	0.21	—	0.21
Magelona sp.	—	—	—	—	—	—	—	—	0.21	—	0.21
Sthenelais boa	—	—	—	—	—	—	—	—	0.21	—	0.21
Armandia sp.	—	—	—	—	—	—	—	0.10	0.10	—	0.20
Hirudinea	—	—	—	0.10	—	—	—	—	0.10	—	0.20
Mulinia lateralis	—	—	—	—	—	0.10	—	—	0.10	—	0.20
Phyllodoce arenae	—	—	—	—	—	—	0.10	—	0.10	—	0.20
Bowmaniella brasiliensis	—	—	—	0.10	—	—	—	0.10	—	—	0.20
Gammarus daiberi	0.19	—	—	—	—	—	—	—	—	—	0.19
Stenoninereis martini	0.19	—	—	—	—	—	—	—	—	—	0.19
Sphaeriidae	0.10	—	—	0.10	—	—	—	—	—	—	0.19
Chironomus stigmaterus	—	0.19	—	—	—	—	—	—	—	—	0.19
Tribelos sp.	—	0.19	—	—	—	—	—	—	—	—	0.19

(*Continued*)

Species	18	22	23	25	26	31	33	37	40	42B	Total
Myrophis punctatus	—	—	—	0.10	—	0.10	—	—	—	—	0.19
Cylichnella bidentata	—	—	—	—	—	0.10	0.10	—	—	—	0.19
Thienemannimyia group	—	0.10	—	—	—	—	0.10	—	—	—	0.19
Leitoscoloplos robustus	—	—	—	—	—	—	—	—	—	—	0.00
Farfantepenaeus aztecus	—	—	—	—	—	—	—	—	—	—	0.00
Scaleworm (unident.)	—	—	—	—	—	—	—	—	—	—	0.00
Ampelisca vadorum	—	—	—	—	—	—	—	—	0.10	—	0.10
Amphitrite sp.	—	—	—	—	—	—	—	—	0.10	—	0.10
Ancistrosyllis papillosa	—	—	—	—	—	—	—	—	0.10	—	0.10
Bivalve sp. A	—	—	—	—	—	—	—	—	0.10	—	0.10
Branchiostoma caribaeum	—	—	—	—	—	—	—	—	0.10	—	0.10
Crab megalopae	—	—	—	—	—	—	—	—	0.10	—	0.10
Glycera sp.	—	—	—	—	—	—	—	—	0.10	—	0.10
Lumbrineris sp.	—	—	—	—	—	—	—	—	0.10	—	0.10
Lyonsia hyalina	—	—	—	—	—	—	—	—	0.10	—	0.10
Macoma tenta	—	—	—	—	—	—	—	—	0.10	—	0.10
Nereidae	—	—	—	—	—	—	—	—	0.10	—	0.10
Ogyrides alpharostris	—	—	—	—	—	—	—	—	0.10	—	0.10
Parahesione luteola	—	—	—	—	—	—	—	—	0.10	—	0.10
Paramphinoe sp.	—	—	—	—	—	—	—	—	0.10	—	0.10
Paramphinome pulchella	—	—	—	—	—	—	—	—	0.10	—	0.10
Paraonis fulgens	—	—	—	—	—	—	—	—	0.10	—	0.10
Pinnotheridae sp.	—	—	—	—	—	—	—	—	0.10	—	0.10
Pogonophora	—	—	—	—	—	—	—	—	0.10	—	0.10
Retusa canaliculata	—	—	—	—	—	—	—	—	0.10	—	0.10
Rheotanytarsus sp.	—	—	—	—	—	—	—	—	0.10	—	0.10
Spiophanes bombyx	—	—	—	—	—	—	—	—	0.10	—	0.10
Trachypenaeus similis	—	—	—	—	—	—	—	—	0.10	—	0.10
Spiochaetopterus costarum	—	—	—	—	—	—	—	0.10	—	—	0.10
Immature tubificid with cap setae	0.10	—	—	—	—	—	—	—	—	—	0.10
Ablabesmyia rhamphe group	—	—	0.10	—	—	—	—	—	—	—	0.10
Ablabesmyia sp.	—	—	0.10	—	—	—	—	—	—	—	0.10

Asheum sp.	—	—	—	0.10	—	—	—	—	—	—	0.10
Barnacle	—	—	—	0.10	—	—	—	—	—	—	0.10
Bivalve	—	—	—	0.10	—	—	—	—	—	—	0.10
Capitellidae	—	—	—	—	—	0.10	—	—	—	—	0.10
Chaoborus punctipennis pupa	—	0.10	—	—	—	—	—	—	—	—	0.10
Chironomid adult	—	—	—	0.10	—	—	—	—	—	—	0.10
Chironominae	—	—	0.10	—	—	—	—	—	—	—	0.10
Chironomus decorus group pupa	—	0.10	—	—	—	—	—	—	—	—	0.10
Coleoptera	—	—	—	—	0.10	—	—	—	—	—	0.10
Cryptotendipes sp.	—	0.10	—	—	—	—	—	—	—	—	0.10
Cymadusa compta	—	—	—	—	—	0.10	—	—	—	—	0.10
Erpobdellidae	—	—	—	0.10	—	—	—	—	—	—	0.10
Gammarus mucronatus	—	—	—	0.10	—	—	—	—	—	—	0.10
Goeldichironomus holoprasinus	—	—	—	0.10	—	—	—	—	—	—	0.10
Hydracarina	—	0.10	—	—	—	—	—	—	—	—	0.10
Hydropsyche sp.	—	—	0.10	—	—	—	—	—	—	—	0.10
Limnodrilus hoffmeisteri	—	0.10	—	—	—	—	—	—	—	—	0.10
Mysidopsis sp.	—	—	—	0.10	—	—	—	—	—	—	0.10
Mytilopsis leucophaeata	—	0.10	—	—	—	—	—	—	—	—	0.10
Nais variabilis	—	0.10	—	—	—	—	—	—	—	—	0.10
Parachironomus sp.	—	—	0.10	—	—	—	—	—	—	—	0.10
Sigambra bassi	—	—	—	—	—	0.10	—	—	—	—	0.10
Spiochaetopterus oculatus	—	—	—	0.10	—	—	—	—	—	—	0.10
Thienemanniella sp.	—	—	0.10	—	—	—	—	—	—	—	0.10
Argulus sp.	—	—	—	—	—	—	0.10	—	—	—	0.10
Notomastus sp.	—	—	—	—	—	—	0.10	—	—	—	0.10
Ophiuroidea	—	—	—	—	—	—	0.10	—	—	—	0.10
Penaeidae zoeae	—	—	—	—	—	—	0.10	—	—	—	0.10
Total	**169.24**	**239.54**	**1139.15**	**2630.99**	**317.19**	**386.13**	**242.84**	**162.51**	**258.06**	**705.95**	**6251.61**

Invertebrate Biomass (g)

Species	18	22	23	25	26	31	33	37	40	Total
Farfantepenaeus aztecus	0.35	0.78	7.39	2.86	5.07	6.50	4.41	1.64	1.03	30.03
Penaeidae	0.09	1.28	7.07	2.19	2.09	5.29	7.59	1.04	0.15	26.79
Callinectes sapidus	0.22	1.76	5.30	1.29	1.72	0.77	0.63	0.28	0.24	12.21
Haminoea succinea	—	—	—	0.01	—	0.01	0.06	11.07	0.30	11.45
Farfantepenaeus duorarum	0.06	0.14	0.62	0.31	0.47	0.98	0.71	0.64	0.43	4.36
Acetes americanus	0.03	—	0.02	0.04	0.01	1.15	0.95	0.65	0.83	3.68
Litopenaeus setiferus	0.02	0.43	0.42	0.06	0.71	0.16	0.04	0.03	—	1.87
Callinectes similis	—	—	—	—	0.06	0.47	0.40	0.36	0.58	1.87
Lolliguncula brevis	—	—	—	0.01	0.01	0.24	0.24	1.05	0.64	2.19
Squilla empusa	—	—	—	—	—	0.02	0.09	0.34	0.25	0.70
Rhithropanopeus harrisii	0.03	—	0.12	0.10	—	—	—	—	—	0.25
Neritina reclivata	0.03	0.15	0.27	0.09	—	—	—	—	—	0.54
Trachypenaeus constrictus	—	—	—	0.01	—	0.05	0.05	0.03	0.21	0.35
Rangia cuneata	0.01	—	0.16	0.08	0.07	—	0.01	—	—	0.33
Aplysia willcoxi	—	—	—	—	—	0.01	0.21	—	—	0.22
Macoma mitchelli	—	—	0.04	—	0.05	0.05	—	—	—	0.14
Palaemonetes pugio	0.02	0.02	0.01	0.01	0.04	—	—	—	—	0.10
Brachidontes exustus	—	—	0.02	0.07	—	—	—	—	—	0.09
Ischadium recurvum	—	—	—	—	—	—	—	—	—	0.00
Callibaetis sp.	—	0.01	0.05	—	—	—	—	—	—	0.06
Leander tenuicornis	—	—	—	—	—	—	—	—	0.06	0.06
Nassarius vibex	—	—	—	—	—	0.02	0.04	—	—	0.06
Bursatella leachi plei	—	—	—	—	—	—	—	0.05	—	0.05
Palaemonetes sp.	—	—	—	—	—	—	—	—	0.01	0.01
Trachypenaeus similis	—	—	—	—	—	0.01	0.03	—	0.01	0.05
Aplysia sp.	—	—	—	—	—	0.01	—	0.02	—	0.03
Mulinia lateralis	—	—	—	—	—	—	—	—	0.03	0.03
Eurypanopeus depressus	—	—	—	0.02	—	—	—	—	—	0.02
Pagurus sp.	—	—	—	—	—	—	—	—	0.02	0.02
Alpheus heterochaelis	—	—	—	—	—	0.01	—	—	—	0.01
Astropecten sp.	—	—	—	—	—	—	—	—	0.01	0.01
Crassostrea virginica	—	—	—	0.01	—	—	—	—	—	0.01
Libinia dubia	—	—	—	—	—	—	—	0.01	—	0.01
Luidia clathrata	—	—	—	—	—	—	—	—	0.01	0.01
Luidia sp.	—	—	—	—	—	—	—	—	0.01	0.01
Metoporhaphis calcarata	—	—	—	—	—	—	—	—	0.01	0.01
Myropsis quinquespinosa	—	—	—	—	—	—	—	—	0.01	0.01
Nuculana acuta	—	—	—	—	—	—	—	0.01	—	0.01
Palaemonetes vulgaris	—	—	—	—	—	—	—	—	0.01	0.01
Polinices duplicatus	—	—	—	—	—	—	—	—	0.01	0.01
Portunus gibbesii	—	—	—	—	—	—	—	0.01	—	0.01
Total	**0.86**	**4.57**	**21.49**	**7.16**	**10.30**	**15.75**	**15.46**	**17.23**	**4.86**	**97.68**

Fish Biomass (g)

Species	18	22	23	25	26	31	33	37	40	Total
Leiostomus xanthurus	21.18	268.64	620.05	457.22	783.23	1068.10	758.04	302.93	80.70	4360.09
Anchoa mitchilli	44.62	130.57	141.27	109.72	194.04	103.19	91.35	97.26	43.85	955.87
Brevoortia patronus	3.15	10.72	36.09	20.06	67.48	46.84	79.72	75.91	53.26	393.23
Micropogonias undulatus	3.42	28.89	35.14	10.79	27.72	60.88	48.47	10.51	2.72	228.53
Lagodon rhomboides	1.24	2.71	31.03	36.80	13.70	19.18	18.05	10.25	3.26	136.22
Arius felis	0.01	0.29	0.76	2.08	1.86	1.15	1.39	0.98	0.09	8.62
Eucinostomus argenteus	3.18	5.41	3.59	3.83	0.54	0.54	0.15	0.19	0.22	17.65
Sciaenidae	0.35	0.99	0.36	1.32	0.52	0.30	0.89	4.34	0.41	9.49
Gobionellus boleosoma	0.04	0.32	0.84	0.98	0.58	1.14	0.45	0.40	0.81	5.56
Anchoa hepsetus	—	—	0.02	0.39	0.05	1.01	0.45	1.52	7.42	10.86
Anchoa nasuta	1.51	0.34	0.05	0.08	0.15	1.42	0.98	0.99	5.67	11.19
Trinectes maculatus	0.44	0.16	1.53	1.34	0.27	0.16	0.07	0.02	0.01	4.00
Chloroscombrus chrysurus	0.01	—	0.10	0.03	0.41	1.21	1.15	2.85	1.11	6.87
Menidia beryllina	0.19	0.01	0.35	0.09	0.02	0.01	—	—	—	0.67
Cynoscion arenarius	0.09	0.13	0.42	0.10	0.23	0.18	0.13	0.04	0.04	1.37
Citharichthys spilopterus	—	—	0.28	0.29	0.14	0.37	0.33	0.49	0.25	2.15
Polydactylus octonemus	—	—	0.01	—	0.12	0.33	0.25	0.38	0.13	1.22
Paralichthys lethostigma	0.03	0.05	0.31	0.40	0.08	0.26	0.13	0.02	0.01	1.29
Microgobius gulosus	0.39	0.29	0.37	0.18	0.06	—	—	—	—	1.30
Symphurus plagiusa	—	—	0.01	0.05	0.07	0.15	0.16	0.26	0.52	1.21
Dorosoma petenense	0.04	0.04	0.10	0.05	0.43	0.30	0.42	0.34	0.04	1.77
Caranx hippos	—	0.01	0.18	0.39	0.06	0.16	0.02	0.13	—	0.95
Synodus foetens	—	—	0.01	0.09	0.06	0.35	0.18	0.16	0.44	1.30
Gobiosoma bosci	0.18	0.11	0.45	0.16	0.08	0.02	—	—	—	1.01
Peprilus burti	0.03	—	—	—	0.27	0.15	0.37	0.15	0.09	1.06
Urophycis floridana	—	—	—	0.01	0.01	0.13	0.13	0.10	0.25	0.63
Sphoeroides parvus	—	—	0.03	0.05	0.02	0.15	0.08	0.03	0.01	0.37
Peprilus paru	0.02	—	—	—	0.02	0.15	0.16	0.36	0.10	0.81
Bairdiella chrysoura	0.02	0.01	0.15	0.02	0.05	0.08	0.03	0.05	0.02	0.43
Lepomis macrochirus	0.07	0.60	0.05	0.02	—	—	—	—	—	0.74

(Continued)

Species	18	22	23	25	26	31	33	37	40	Total
Sphoeroides nephelus	—	—	—	0.10	0.06	0.32	0.12	0.04	0.02	0.67
Notropis sp.	0.73	—	—	0.01	—	—	—	—	—	0.74
Prionotus tribulus	—	—	0.08	0.06	0.05	0.13	0.06	0.07	0.02	0.47
Dasyatis sabina	—	0.01	0.12	0.02	0.02	0.05	0.03	0.01	—	0.26
Leptocephalus larvae	—	0.07	0.02	—	—	0.03	0.01	0.37	0.13	0.63
Archosargus probatocephalus	—	0.02	0.20	0.05	0.02	—	0.01	—	—	0.30
Mugil cephalus	—	0.13	0.03	0.01	0.01	0.03	0.03	0.01	—	0.25
Lepisosteus osseus	—	0.03	0.06	—	0.09	0.03	0.04	—	—	0.25
Lutjanus griseus	0.01	0.17	0.11	0.09	—	0.01	—	—	—	0.39
Myrophis punctatus	—	—	0.06	0.01	0.01	0.01	0.01	0.01	—	0.11
Bagre marinus	—	—	—	0.01	0.07	0.11	0.05	0.02	0.01	0.27
Anchoa lyolepis	—	—	—	—	—	0.02	0.02	0.16	0.07	0.27
Ictalurus catus	0.01	0.10	0.04	0.02	0.01	—	—	—	—	0.18
Pomoxis nigromaculatus	0.01	0.25	—	—	—	—	—	—	—	0.26
Gobionellus hastatus	—	0.01	—	0.04	0.01	0.02	0.01	0.03	0.04	0.16
Microgobius thalassinus	—	—	—	—	0.01	—	0.05	0.06	0.08	0.20
Gobiidae	—	0.04	—	0.04	—	0.10	—	0.01	—	0.19
Etropus crossotus	—	—	—	0.02	—	0.04	0.05	0.04	0.03	0.18
Pogonias cromis	—	0.17	—	—	—	—	—	—	—	0.17
Trichiurus lepturus	—	—	—	—	—	—	0.01	0.12	0.03	0.16
Harengula jaguana	—	—	—	—	—	—	0.01	0.09	0.01	0.11
Lepisosteus oculatus	—	0.12	0.02	0.01	—	—	—	—	—	0.15
Micropterus punctulatus	0.09	—	—	—	0.01	—	—	—	—	0.10
Cynoscion nebulosus	—	0.01	0.04	0.02	—	0.02	0.01	—	—	0.10
Bothidae	—	—	0.03	0.02	0.02	0.04	—	0.03	—	0.14
Cyprinella venusta	0.13	—	—	—	—	—	—	—	—	0.13
Syngnathus scovelli	—	0.01	0.07	0.03	—	—	0.01	0.01	—	0.13
Orthopristis chrysoptera	—	—	—	—	—	—	—	0.01	0.06	0.07
Sciaenops ocellatus	—	0.08	—	0.02	—	—	—	—	—	0.10
Notropis petersoni	0.04	—	0.04	—	—	—	—	—	—	0.08
Pomoxis annularis	—	0.07	—	—	—	—	—	—	—	0.07
Atractosteus spatula	—	0.06	0.01	—	—	—	—	—	—	0.07

Syngnathus louisianae	—	—	—	—	—	0.02	—	0.01	0.04	0.07
Sphoeroides spengleri	—	—	0.02	—	0.03	0.01	—	0.01	—	0.07
Scomberomorus maculatus	—	—	—	0.01	—	0.01	—	—	—	0.02
Selene vomer	—	—	—	—	—	0.01	0.01	0.04	—	0.06
Ameiurus nebulosus	—	0.01	—	—	—	—	—	—	—	0.01
Lepomis gulosus	0.01	0.04	—	—	—	—	—	—	—	0.05
Peprilus sp.	—	—	—	—	—	0.03	—	—	0.02	0.05
Paralichthys albigutta	—	—	—	0.01	—	0.01	0.01	0.02	—	0.05
Trachinotus carolinus	—	—	—	—	—	—	—	—	—	0.00
Sardinella aurita	—	—	0.01	—	—	0.02	—	—	0.01	0.04
Sphyraena guachancho	—	—	—	—	—	0.01	—	0.03	—	0.04
Dorosoma cepedianum	—	—	—	—	0.01	—	0.03	—	—	0.04
Paralichthys sp.	—	—	—	0.02	—	—	0.01	—	—	0.03
Pomatomus saltatrix	—	—	0.02	0.01	—	—	—	—	—	0.03
Sphoeroides sp.	—	—	—	—	—	0.02	0.01	—	—	0.03
Gobionellus shufeldti	—	—	—	—	0.02	—	—	—	0.01	0.03
Etrumeus teres	—	—	—	—	—	—	—	—	0.02	0.02
Eucinostomus gula	—	—	—	—	—	—	—	—	0.02	0.02
Ancylopsetta quadrocellata	—	—	—	—	—	0.01	—	0.01	—	0.02
Chaetodipterus faber	—	—	—	—	—	0.01	—	0.01	—	0.02
Micropterus salmoides	—	—	0.02	—	—	—	—	—	—	0.02
Oligoplites saurus	—	—	—	—	0.01	—	0.01	—	—	0.02
Gambusia affinis	—	—	—	0.01	—	—	—	—	—	0.01
Echeneis naucrates	—	—	—	—	—	—	—	—	0.01	0.01
Ericymba buccata	0.01	—	—	—	—	—	—	—	—	0.01
Gymnura micrura	—	—	—	—	—	—	—	—	0.01	0.01
Ictalurus punctatus	0.01	—	—	—	—	—	—	—	—	0.01
Lucania parva	0.01	—	—	—	—	—	—	—	—	0.01
Morone americana	0.01	—	—	—	—	—	—	—	—	0.01
Morone saxatilis	0.01	—	—	—	—	—	—	—	—	0.01
Prionotus scitulus	—	—	—	—	—	0.01	—	—	—	0.01
Prionotus sp.	—	—	—	—	—	0.01	—	—	—	0.01
Sphyraena borealis	—	—	—	—	—	—	—	—	0.01	0.01

(Continued)

Species	18	22	23	25	26	31	33	37	40	Total
Stephanolepis hispidus	—	—	—	—	—	—	—	—	0.01	0.01
Syngnathus sp.	—	—	—	—	—	—	—	—	0.01	0.01
Eucinostomus lefroyi	—	—	—	—	0.01	—	—	—	—	0.01
Eucinostomus sp.	—	—	—	—	0.01	—	—	—	—	0.01
Gambusia holbrooki	—	—	0.01	—	—	—	—	—	—	0.01
Gobiosoma robustum	—	—	—	—	0.01	—	—	—	—	0.01
Gobiosoma sp.	—	—	0.01	—	—	—	—	—	—	0.01
Hemicaranx amblyrhynchus	—	—	—	—	—	—	—	0.01	—	0.01
Menticirrhus americanus	—	—	—	—	0.01	—	—	—	—	0.01
Syngnathus floridae	—	—	—	—	—	—	—	0.01	—	0.01
Total	**81.31**	**451.71**	**874.58**	**647.19**	**1092.77**	**1309.10**	**1004.16**	**511.89**	**202.12**	**6174.83**

Appendix D: Personnel Involved in the Perdido Study (1988–2007)

Birkholz, Dr. Detlef A. (Toxicology, residue analyses)
Environmental Toxicologist
EnviroTest Laboratories
9936 67th Avenue
Edmonton, Alberta T6E OP5
Canada

Brush, Dr. Grace S. (Long-core analysis, long-term changes in estuaries)
Department of Geography and Environmental Engineering
The Johns Hopkins University
313 Ames Hall
Baltimore, Maryland 21218

Cifuentes, Dr. Luis A. (Nutrient studies, isotope analyses)
Department of Oceanography
Texas A&M University
Texas A&M Research Foundation
Box 3578
College Station, Texas 77843

Coffin, Dr. Richard A. (Nutrient studies, isotope analyses)
U.S. Environmental Protection Agency
Environmental Research Laboratory
Sabine Island
Gulf Breeze, Florida 32561-5299

Davis, Dr. William P. (Biology of fishes)
U.S. Environmental Protection Agency
Environmental Research Laboratory
Sabine Island
Gulf Breeze, Florida 32561-5299

Epler, Dr. John H. (Aquatic insects)
Rt. 3, Box 5485
Crawfordville, Florida 32327

Flemer, Dr. David A. (Nutrients and productivity in aquatic systems)
U.S. Environmental Protection Agency
Environmental Research Laboratory
Sabine Island
Gulf Breeze, Florida 32561-5299

Franklin, Dr. Marvin (Riverine hydrology)
U.S. Geological Survey
Tallahassee, Florida

Gallagher, Dr. Tom (Hydrological modeling)
HydroQual, Inc.
1 Lethbridge Plaza
Mahwah, New Jersey 07430

Gilbert, Dr. Carter R. (Fish systematics)
Curator in Fishes
Florida Museum of Natural History
Department of Natural Sciences
Museum Road, University of Florida
Gainesville, Florida 32611-2035
Homann, Mr. Phillip (Data processing)
Environmental Planning & Analysis, Inc.
933 1/2 West Tharpe Street
Tallahassee, Florida 32303
Howell, Mr. Robert L. (Fishes and invertebrates)
Environmental Planning & Analysis, Inc.
933 1/2 West Tharpe Street
Tallahassee, Florida 32303
Isphording, Dr. Wayne C. (Marine geology)
Tierra Consulting
P.O. Box 2243
Mobile, Alabama 36688
Kalke, Dr. Richard D. (Estuarine zooplankton)
Marine Science Institute
The University of Texas
P/O/Box 1267
Port Aransas, Texas 78373-1267
Karsteter, Mr. William R. (Aquatic macroinvertebrates)
Environmental Planning & Analysis, Inc.
933 1/2 West Tharpe Street
Tallahassee, Florida 32303
Klein, Dr. C. John, III (Aquatic engineering, estuarine modeling)
Klein Engineering & Planning Consultants
3011 North Branch Lane
Baltimore, Maryland 21234
Koenig, Dr. Christopher C. (Biology of fishes)
Environmental Planning & Analysis, Inc.
933 1/2 West Tharpe Street
Tallahassee, Florida 32303
Pensacola, Florida 32406
Livingston, Dr. Robert J. (Aquatic ecology)
Environmental Planning & Analysis, Inc.
933 1/2 West Tharpe Street
Tallahassee, Florida 32303
Niedoroda, Dr. Alan W. (Oceanographic modeling)
Environmental Science & Engineering, Inc.
P.O. Box 1703
Gainesville, Florida 32602-1703
Prasad, Dr. Alshinthala K. (Aquatic algae)
Environmental Planning & Analysis, Inc.
933 1/2 West Tharpe Street
Tallahassee, Florida 32303

Price, Ms. Janet H. (Project management)
Champion International Corporation
375 Muscogee Road
Cantonment, Florida 32533-0087

Ray, Dr. Gary L. (Aquatic macroinvertebrates)
Environmental Planning & Analysis, Inc.
933 1/2 West Tharpe Street
Tallahassee, Florida 32303

Rhew, Dr. Kyeongsik (Aquatic algae)
Environmental Planning & Analysis, Inc.
933 1/2 West Tharpe Street
Tallahassee, Florida 32303

Woodsum, Mr. Glenn C. (Computer programming, data management)
Environmental Planning & Analysis, Inc.
933 1/2 West Tharpe Street
Tallahassee, Florida 32303

Appendix E: Final Recommended Order, Lane et al. vs International Paper and FDEP, Bram D. E. Canter, Presiding (2007)

State of Florida Division of Administrative Hearings
Mellita A. Lane; Jacqueline M. Lane; Zachary P. Lane; Sarah M. Lane;
Peter A. Lane; Friends of Perdido Bay, Inc.; and James

LANE,		)			
			)		
		Petitioners,	)		
			)		
vs.			) Case Nos. 05-1609		
			)		05-1610
DEPAR	TMENT OF	ENVIRONMENTAL	)		05-1611
PROT	ECTION AND	INTERNATIONAL	)		05-1612
PAPER	COMPANY,		)		05-1613
				)	05-1981
	Respondents.	)			
			)		

RECOMMENDED ORDER

The final hearing in this case was held on May 31, June 1, 2, and 26 through 30, and July 17, 27, and 28, 2006, in Pensacola, Florida, before Bram D. E. Canter, an Administrative Law Judge of the Division of Administrative Hearings (DOAH).

APPEARANCES

For Petitioners Friends of Perdido Bay, Inc. and James Lane:

Howard K. Heims, Esquire Littman, Sherlock & Heims, P.A. Post Office Box 1197 Stuart, Florida 34995 and Marcy I. LaHart, Esquire 711 Talladega Street West Palm Beach, Florida 33405-1143

For Petitioners Mellita A. Lane, Jacqueline M. Lane, Zachary P. Lane, Peter A. Lane, and Sarah M. Lane:

Jacqueline M. Lane, pro se, and Qualified Representative 10738 Lillian Highway Pensacola, Florida 32506

For Respondent International Paper Company (IP):

Terry Cole, Esquire Patricia A. Renovitch, Esquire Oertel, Fernandez, Cole, & Bryant, P.A. Post Office Box 1110 Tallahassee, Florida 32302-1110

For Respondent Florida Department of Environmental Protection (Department):

W. Douglas Beason, EsquireDavid K. Thulman, Esquire The Douglas Building, Mail Station 35 3900 Commonwealth Boulevard Tallahassee, Florida 32399-3000

STATEMENT OF THE ISSUES

The issues in this case are whether IP is entitled to issuance of National Pollutant Discharge Elimination System (NPDES) Permit Number FL0002526-001/001-IW1S ("the proposed permit"), Consent Order No. 04-1202, Authorization for Experimental Use of Wetlands Order No. 04-1442, and Waiver Order No. 04-0730 (collectively, "the Department authorizations"), which would authorize IP to discharge treated industrial wastewater from its paper mill in Cantonment, Escambia County, Florida, into wetlands which flow to Elevenmile Creek and Perdido Bay.

PRELIMINARY STATEMENT

On April 12, 2005, the Department published notice of its intent to issue the proposed permit, Consent Order, exemption, and waiver. The Department authorizations would allow IP to change its industrial wastewater treatment system at the Cantonment paper mill, construct an effluent distribution system on approximately 1500 acres of wetlands owned by IP that are located near the mill site, construct a 10-mile pipeline to transport its treated wastewater to the wetlands, and discharge the treated wastewater into the wetlands.

In April 2005, Mellita A. Lane, Jacqueline M. Lane, Zachary P. Lane, Peter A. Lane, and Sarah M. Lane ("Lane Petitioners") filed identical petitions challenging the Department authorizations on numerous grounds. The Department forwarded the petitions to DOAH for assignment of an Administrative Law Judge and to conduct an evidentiary hearing. The Lane Petitioners subsequently amended their petitions.

In May 2005, Friends of Perdido Bay, Inc. (FOPB), and James Lane filed a petition for hearing to challenge the Department authorizations. The FOBP petition was forwarded to DOAH by the Department, and the pending cases were consolidated for the final hearing. The FOPB petition was subsequently amended.

In October 2005, while the cases were pending, IP applied for a revision to its NPDES permit renewal application. The cases were abated so that the Department could review and act on the permit revision. In January 2006, DEP issued a proposed revised NPDES permit and a corresponding First Amendment to Consent Order.

At the final hearing, the parties' Joint Exhibits 1 through 18 were admitted into evidence. IP presented the testimony of Kyle Moore, accepted as an expert in environmental engineering and management; Dr. Glenn Daigger, accepted as an expert in civil engineering, environmental engineering and wastewater treatment, with a specialty in biological wastewater treatment; Thomas Gallagher, accepted as an expert in environmental engineering with a specialty in water quality modeling; Dr. Robert Livingston, accepted as an expert in aquatic ecology and ecosystem ecology, with a specialty in pollution biology; and Dr. Wade Nutter, accepted as an expert in hydrology, soils, and forested wetlands. IP's Exhibits 1, 3, 17, 19, 21 through 33, 35 through 37, 39 through 41, 43 through 46, 49 through 52, 54, 55, 57, 58, 60 through 64, 65A, 65B, 66 through 75, 78 through 81, 82B through 82D, 83A through 83C, 84 through 86, 87A, 87B, 88A through 88D, 89A, 89B, 90A through 90E, 93, 94, 95A, 95B, 100, 101, and 104A through 104J were admitted into evidence.

The Department presented the testimony of William Evans, who was accepted as an expert in environmental engineering. The Department's Exhibit 38 was admitted into evidence.

Petitioners presented the testimony of Barry Sulkin, an environmental scientist; William Evans; William Chandler, a Department chemist; Donald Ray, a Department biologist; Laurence Donelan, a Department biologist; Glenn Butts, a Department marine biologist; Dr. Steven Schropp, an environmental scientist; Dr. Jacqueline Lane, a marine biologist; Dr. Kevin White, a civil engineer;

Dr. Richard Weickowitz, a Department scientist who works with water quality models; and fact witnesses James Lane, Alexander Zelius, Pompelio Ucci, Ray Barkowski, Winford Howell, Bob DeGraff, and Matthew Dimitroff. Petitioners also offered the depositions of five persons, which were admitted into evidence: Cynthia Arnold, a civil engineer and IP consultant on landfill matters; Jeff Hilleke, an IP chemical engineer; Dr. Michael Steltenkamp, IP's Environmental Health and Safety Manager; Dr. Leslee Williams, a Department oceanographer; and Joel Bolduc, a former IP environmental engineer. Petitioners' Exhibits 5, 5A through 5C,[1] 6 through 8, 27, 35, 43, 51 through 56, 59, 62, 67, 71, 79, 80A, 81 through 83, 91, 92A, 92B, 100, 101, 103A, and 103B were admitted into evidence.

J.D. Brown was allowed to present comments on the record as a private citizen and member of the public.

The undersigned took official recognition of applicable sections of the Florida Statutes (2005) and Florida Administrative Code, the "cluster rule" published in the Code of Federal Regulations, and three prior administrative orders involving the Cantonment mill.

On July 26, 2006, the Department filed without objection a revision to the Consent Order. On July 31, 2006, the Department filed Joint Trial Exhibit 18 that integrated the Consent Order dated April 12, 2005, the First Amendment to Consent Order dated January 11, 2006, and the Department's Notice of Minor Revision to Consent Order filed on July 26, 2006.

The 21-volume Transcript of the hearing was filed with DOAH. At the request of Petitioners, the time for filing proposed findings of fact and conclusions of law was extended. The parties timely filed Proposed Recommended Orders (PROs). The parties were allowed to file supplemental argument on the interpretation of Florida Administrative Code Rule 62-660.300(1). All parties filed supplemental argument. The PROs and supplemental argument have been carefully considered in the preparation of this Recommended Order.

FINDINGS OF FACT

I. Introduction

A. The Parties

1. 1. 1. The Department is the state agency authorized under Chapter 403, Florida Statutes (2006),[2] to regulate discharges of industrial wastewater to waters of the state. Under a delegation from the United States Environmental Protection Agency (EPA), the Department administers the NPDES permitting program in Florida.
2. 2. 2. IP owns and operates the integrated bleached kraft paper mill in Cantonment, Escambia County, Florida.
3. 3. 3. FOPB is a non-profit Alabama corporation[3] established in 1988 whose members are interested in protecting the water quality and natural resources of Perdido Bay. FOPB has approximately 450 members. About 90% of the members own property adjacent to Perdido Bay. James Lane is the President of FOPB.
4. 4. 4. Mellita A. Lane, Zachary P. Lane, Peter A. Lane, and Sarah M. Lane are the adult children of Dr. Jacqueline Lane and James Lane. Dr. Lane and James Lane live on property adjacent to Perdido Bay with their son Peter.

B. The Adjacent Waters

1. 5. The mill's wastewater effluent is discharged into Elevenmile Creek, which is a tributary of Perdido Bay. The creek flows southwest into the northeastern portion of Perdido Bay. Elevenmile Creek is a freshwater stream for most of its length but is sometimes tidally affected 1–2 miles from its mouth. Elevenmile Creek is designated as a Class III water.

2. 6. Perdido Bay is approximately 28 square miles in area and is bordered by Escambia County on the east and Baldwin County, Alabama on the west. The dividing line between the states runs north and south in the approximate middle of Perdido Bay. U.S. Highway 90 crosses the Bay, going east and west, and forms the boundary between what is often referred to as the "Upper Bay" and "Lower Bay." The Bay is relatively shallow, especially in the Upper Bay, ranging in depth between 5 and 10 ft. Perdido Bay is designated as a Class III water.

. 7. Sometime around 1900, a manmade navigation channel was cut through the narrow strip of land separating Perdido Bay from the Gulf of Mexico. The channel, called Perdido Pass, allowed the salt waters of the Gulf to move with the tides up into Perdido Bay. Depending on tides and freshwater inflows, the tidal waters can move into the most northern portions of Perdido Bay and even further, into its tributaries and wetlands.

3. 8. The Perdido River flows into the northwest portion of Perdido Bay. It is primarily a freshwater river but it is sometimes tidally influenced at and near its mouth. The Perdido River was designated an Outstanding Florida Water (OFW) in 1979.

9. At the north end of Perdido Bay, between Elevenmile Creek and the Perdido River, is a large tract of land owned by IP called the Rainwater Tract. The northern part of the tract is primarily freshwater wetlands. The southern part is a tidal marsh. Tee and Wicker Lakes are small (approximately 50 acres in total surface area) tidal ponds within the tidal marsh. Depending on the tides, the lakes can be as shallow as 1 ft, or several feet deep. A channel through the marsh allows boaters to gain access to Tee and Wicker Lakes from Perdido Bay.

C. The Mill

1. Production

. 10. Florida Pulp and Paper Company first began operating the Cantonment paper mill in 1941. St. Regis Paper Company (St. Regis) acquired the mill in 1946. In 1984, Champion International Corporation (Champion) acquired the mill. Champion changed the product mix in 1986 from unbleached packaging paper to bleached products such as printing and writing grades of paper. In 2001, Champion merged with IP, and IP took over operation of the mill. The primary product of the mill continues to be printing and writing paper.

4. 11. The mill is integrated, meaning that it brings in logs and wood chips, makes pulp, and produces paper. The wood is chemically treated in cookers called digesters to separate the cellulose from the lignin in the wood because only the cellulose is used to make paper. Then the "brown stock" from the digesters goes through the oxygen delignification process, is mixed with water, and is pumped to paper machines that make the paper products.

5. 12. There are two paper machines located at the mill. The larger paper machine, designated P5, produces approximately 1000 tons per day of writing and printing paper. The smaller machine, P4, produces approximately 400–500 tons per day of "fluff pulp."

2. The Existing Wastewater Treatment Plant

. 13. The existing wastewater treatment plant (WWTP) at the mill is described in the revised NPDES permit as a "multi-pond primary and secondary treatment system, consisting of a primary treatment system (primary settling basin, polymer addition, two solids/sludge dewatering basins, and a floating dredge), and secondary treatment system (four ponds in series; two aerated stabilization basins with approximately 2200 horsepower (HP) of aeration capacity, a nutrient feed system, two non-aerated polishing ponds and a final riffle section to re-aerate the effluent)."

2. 14. The WWTP is a system for reducing the biological oxygen demand (BOD) of the mill's wastewater by bacteria. IP's wastewater is nutrient deficient when it enters the WWTP. Nutrients in the form of phosphorus and nitrogen must be added for the growth of bacteria.

3. 15. The WWTP begins with a primary settling basin in which suspended solids settle to the bottom. The solids form a sludge that is pumped by hydraulic dredge into two dewatering basins. The dewatering basins are used alternately so that, as one pond is filled, water is removed from the other pond.

After being dewatered, the sludge is removed and allowed to dry. Then, it is transported to a landfill located about 5 miles west of the mill on land owned by IP.

. 16. The water removed from the dewatering basins moves into to the first aeration basin. The aeration basin has floating aerator devices that add oxygen to facilitate biological conversion of the wastewater. The wastewater then flows sequentially through three more basins where there is further oxygenation and settling of the biological solids. The discharge from the fourth settling basin flows through a riffle section where the effluent is aerated using a series of waterfalls. This is the last element of the treatment process from which the mill's effluent enters waters of the state.

4. 17. Chemicals are added during the treatment process to control phosphorus and color. Chemicals are also added to suppress foam.

5. 18. Sanitary wastewater from the mill, after pretreatment in an activated sludge treatment system, is "sewered" to the mill's WWTP and further treated in the same manner as the industrial wastewater. A separate detention pond collects and treats stormwater from onsite and offsite areas and discharges at the same point as the wastewater effluent from the WWTP. Stormwater that falls on the industrial area of the mill is processed through the WWTP.

6. 19. The discharge point from the WWTP, and the point at which the effluent is monitored for compliance with state effluent limitations, is designated D-001, but is also called the Parshall Flume. The effluent is discharged from the Parshall Flume through a pipe to an area of natural wetlands. After passing through the wetlands, the combined flow runs through a pipe that enters Elevenmile Creek from below the surface. This area is called the "boil" because the water can be observed to boil to the surface of Elevenmile Creek. From the boil, the mill effluent flows approximately 14 miles down (apparently misnamed) Elevenmile Creek to upper Perdido Bay.

D. Regulatory History of the Mill

1. 20. Before 1995, the mill had to have both state and federal permits. The former Florida Department of Environmental Regulation (DER) issued St. Regis an industrial wastewater operating permit in 1982 pursuant to Chapter 403, Florida Statutes. The EPA issued St. Regis an NPDES permit in 1983 pursuant to the Clean Water Act. When it acquired the facility in 1984, Champion continued to operate the mill under these two permits.

2. 21. In 1986, Champion obtained a construction permit from DER to install the oxygen delignification technology and other improvements to its WWTP in conjunction with the conversion of the production process from an unbleached to a modified bleached kraft production process.

3. 22. In 1987, Champion applied to DER for an operating permit for its modified WWTP and also petitioned for a variance from the Class III water quality standards in Elevenmile Creek for iron, specific conductance, zinc, and transparency. DER's subsequent proposal to issue the operating permit and variance was formally challenged.[4]

4. 23. In 1988, while the challenges to the DER permit and variance were still pending, Champion dropped its application for a regular operating permit and requested a temporary operating permit (TOP), instead.

5. 24. In December 1989, DER and Champion entered into Consent Order No. 87-1398 ("the 1989 Consent Order"). The 1989 Consent Order included an allegation by DER that the mill's wastewater discharge was causing violation of state water quality standards in Elevenmile Creek for dissolved oxygen (DO), un-ionized ammonia, and biological integrity.

25. The 1989 Consent Order authorized the continued operation of the mill, but established a process for addressing the water quality problems in Elevenmile Creek and Perdido Bay and bringing the mill into compliance in the future. Champion was required to install equipment to increase the DO in its effluent within a year. Champion was also required to submit a plan of study and, 30 months after DER's approval of the plan of study, to submit a study report on the impacts of the mill's effluent on DO in Elevenmile Creek and Perdido Bay and recommended measures for reducing or eliminating adverse impacts. The study report was also supposed to address the other water quality violations caused by Champion.

6. 26. A comprehensive study of the Perdido Bay system was undertaken by a team of 24 scientists lead by Dr. Robert Livingston, an aquatic ecologist and professor at Florida State University.

The initial 3-year study by Dr. Livingston's team of scientists was followed by a series of related scientific studies, which will be referred to collectively in this Recommended Order as "the Livingston studies."

7. 27. The 1989 Consent Order had no expiration date, but it was tied to the TOP, which had an expiration date of December 1, 1994. Champion was to be in compliance with all applicable water quality standards by that date.
8. 28. The TOP established the following specific effluent discharge limitations for the mill:

Monthly Average	Maximum Biological Oxygen Demand (BOD)
(Mar–Oct) 4500	6885 lb/day
(Nov–Feb) 5100	6885 lb/day

Total Suspended Solids (TSS)	
(Mar–Oct) 8000 lb/day	27,000 lb/day
(Nov–Feb) 11,600 lb/day	27,000 lb/day

Iron	Specific Conductance	Zinc
3.5 mg/L	2,500 micromhos/cm	0.75 mg/L

0. 1. 29. The limits stated above for iron, specific conductance, and zinc were derived from the variance granted to Champion. Champion was also granted variances from the water quality standards for biological integrity, un-ionized ammonia, and DO.
1. 2. 30. The 1989 Consent Order, TOP, and variance were the subject of the Recommended Order and Final Order issued in Perdido Bay Environmental Association, Inc. v. Champion International Corporation, 89 ER FALR 153 (DER Nov. 14, 1989). Champion's deviation from the standards for iron, zinc, and specific conductance pursuant to the variance was determined to present no significant risk of adverse effect on the water quality and biota of Elevenmile Creek and Perdido Bay. The mill effluent's effect on transparency (reduced by color in the mill effluent) was considered a potentially significant problem. However, because it was found that there was no practicable means known or available to reduce the color, and there was insufficient information at that time to determine how Champion's discharge of color was affecting the biota, Champion was allowed to continue its discharge of color into Elevenmile Creek pending the results of the Livingston studies.

31. In the administrative hearing, the petitioners argued that it was unreasonable to put off compliance for 5 years, but the hearing officer determined that 5 years was reasonable under the circumstances. One finding in the Recommended Order and a reason for recommending approval of the TOP and Consent Order was:
After the studies referred to in the consent order, the Department will not allow Champion additional time to study problems further. Significant improvements will be required within the 5-year period and at the end of that period, the plant will be in compliance with all water quality standards or will be denied an operating permit, with related enforcement action.

0. 32. The requirement of the 1989 Consent Order that Champion be in compliance with all applicable standards by December 1994, was qualified with the words "unless otherwise agreed." In considering this wording, the hearing officer opined that any change in the compliance deadline "would require a new notice of proposed agency action and point of entry for parties who might wish to contest any modification in the operational requirements, or changes in terms of compliance with water quality standards."
2. 2. 33. The mill was not in compliance with all water quality standards in December 1994. No enforcement action was taken by the Department and no modification of the 1989 Consent Order or TOP was formally proposed that would have provided a point of entry to any members of the public who might have objected. Instead, the Department agreed through correspondence

with Champion to allow Champion to pursue additional water quality studies and to investigate alternatives to its discharge to Elevenmile Creek.

3. 3. 34. In 1994 and 1995, Champion applied to renew its state and federal wastewater permits, which were about to expire. The Department and EPA notified Champion that its existing permits were administratively extended during the review of the new permit applications. Today, the Cantonment mill is still operating under the 1989 TOP which, due to the administrative extension, did not terminate in December 1994, as stated on its face.
4. 4. 35. In November 1995, following EPA's delegation of NPDES permitting authority to the Department, the Department issued an order combining the state and federal operating permits into a single permit identified as Wastewater Permit Number FL0002526-002-IWF/MT.
5. 5. 36. In summary, the permit requirements currently applicable to the operation of the Cantonment paper mill are contained in the following documents:

January 3, 1983, EPA NPDES Permit
December 13, 1989, DER Temporary Operating Permit (TOP) December 13, 1989, DER Consent Order December 12, 1989, DER Variance November 15, 1995, DEP Order (combining the NPDES permit and the State-issued wastewater permit) April 22, 1996, DEP Letter (clarifying November 15, 1995, Order regarding 1983 NPDES Permit)

0. 1. 37. During the period from 1992 to 2001, more water quality studies were conducted and Champion investigated alternatives to discharging into upper Elevenmile Creek, including land application of the effluent and relocation of the discharge to lower Elevenmile Creek or the Escambia River.
1. 38. In 2001, IP and Champion merged and IP applied to the Department to have the mill permit and related authorizations transferred to IP. Dr. Lane formally challenged the proposed transfer, but she was determined to lack standing. One conclusion of law in the Recommended Order issued in the 2001 administrative case was that the mill was in compliance with the consent order, TOP, and
2. variance. That conclusion was not based on a finding that Champion was in compliance with all applicable water quality standards, but that the deadline for compliance (December 1, 1994) had been extended indefinitely by the pending permit renewal application.
3. 2. 39. In 2001, Dr. Lane twice petitioned the Department for a declaratory statement regarding the Department's interpretation of certain provisions of the 1989 Consent Order. The first petition was denied by the Department because Dr. Lane failed to adequately state her interests and because she was a party in a pending case in which the Consent Order was at issue. Dr. Lane second petition was denied for similar reasons.
4. 3. 40. Over 14 years after the deadline established in the 1989 TOP for the mill to be in compliance with all applicable standards in Elevenmile Creek, IP is still not meeting all applicable standards. However, the combination of (1) Consent Order terms that contemplated unspecified future permit requirements based on yet-to-be-conducted studies, (2) the wording in the TOP that tied the deadline for compliance to the expiration of the TOP, and (3) the administrative extension of the TOP, kept the issue of Champion's and IP's compliance in a regulatory limbo. It increased the Department's discretion to determine whether IP was in compliance with the laws enacted to protect the State's natural resources, and reduced the opportunity of interested persons to formally disagree with that determination.

II. The Proposed Authorizations

A. In General

0. 1. 41. In September 2002, while Champion's 1994 permit renewal application was still pending at DEP, IP submitted a revised permit renewal application to upgrade the WWTP and relocate its discharge. The WWTP upgrades consist of converting to a modified activated sludge treatment process, increasing aeration, constructing storm surge ponds, and adding a process for pH adjustment.

The new WWTP would have an average daily effluent discharge of 23.8 million gallons per day (mgd). IP proposes to convey the treated effluent by pipeline 10.7 miles to a 1464 acre wetland tract owned by IP,[5] where the effluent would be distributed over the wetlands as it flows to lower Elevenmile Creek and upper Perdido Bay.

1. 2. 42. IP revised its permit application again in October 2005, to obtain authorization to reconfigure the mill to produce unbleached brown paper for various grades of boxes. If the mill is reconfigured, only softwood (pine) would be used in the new process.
2. 3. 43. On April 12, 2005, the Department issued a Notice of Intent to Issue the proposed NPDES permit, together with Consent Order No. 04-1202, Authorization for Experimental Use of Wetlands Order No. 04-4442, and Waiver Order No. 04-0730.
3. 4. 44. An exemption from water quality criteria in conjunction with the experimental use of wetlands for wastewater treatment is provided for in Florida Administrative Code Rule 62-660.300(1). The proposed exemption order would exempt IP from Class III water quality criteria for pH, DO, transparency, turbidity, and specific conductance.
4. 5. 45. The proposed waiver order is associated with the experimental use of wetlands exemption and relieves IP of the necessity to comply with two exemption criteria related to restricting public access to the area covered by the exemption. The Department and IP contend that restricting public access to Tee and Wicker Lakes is unnecessary.
5. 6. 46. The proposed Consent Order is an enforcement document that is necessary if the mill is to be allowed to operate despite the fact that its wastewater discharge is causing violations of water quality standards. A principal purpose of the proposed Consent Order is to impose a time schedule for the completion of corrective actions and compliance with all state standards. The proposed Consent Order would supersede the 1989 Consent Order.

B. The Proposed NPDES Permit

1. WWTP Upgrades

0. 1. 47. IP's primary objective in upgrading the WWTP was to reduce the nitrogen and phosphorus in the mill's effluent discharge. The upgrades are designed to reduce un-ionized ammonia, total soluble nitrogen, and phosphorus. They are also expected to achieve a modest reduction of BOD and TSS.
1. 2. 48. Upgraded pond 1 is expected to convert soluble BOD to suspended solids and to accomplish other biological conversions seven or eight times faster than the current pond 1.
2. 3. 49. The modification of pond 3 to an activated sludge system is expected to more rapidly remove and recycle the solids back into pond 1. Pond 3 will have a much larger bacterial population to treat the effluent. There would also be additional pH control at the end of pond 3.
3. 4. 50. IP would continue to use its Rock Crossing Landfill for disposal of wastewater sludge removed from the WWTP. Authorization for the landfill is part of the proposed NPDES permit. Groundwater monitoring beneath the landfill is required.
4. 5. 51. The WWTP upgrades would include increased storm surge capacity by converting two existing aeration and settling basins (ponds 2 and 4) to storm surge basins. The surge basins would allow the mill to manage upsets and to withstand a 25-year, 24 h storm event of 11 in. of rain. Rainfall that falls into the production areas would flow to the WWTP, and be impounded in ponds 2 and 4. After the storm event this impounded water would flow back through the WWTP where it would be treated before flowing through the compliance point and into the pipeline to the wetland tract.
5. 6. 52. The Department required IP to monitor for over 129 pollutants in its stormwater runoff from the mill's manufacturing facility, roads, parking lots, and offsite nonpoint sources. No pollutants were found in the stormwater at levels of concern.
53. The average volume of mill discharge would be 23.8 mgd. IP plans to obtain up to 5 mgd of treated municipal wastewater from a new treatment facility planned by the Emerald Coast Utility Authority (ECUA), which would be used in the paper production process and would reduce the need for groundwater withdrawals by IP for this purpose. The treated wastewater would enter the WWTP along with other process wastewater, be treated in the same manner in the WWTP, and become part of the effluent conveyed through the pipeline to the wetland tract.

2. *Effluent Limitations*

1. 1. 54. The effluent limitations required by the proposed permit include technology-based effluent limits (TBELs) that apply to the entire pulp and paper industry. TBELs are predominantly production-based and are designed to limit the amount of pollutants that may be discharged per ton of product produced. The Cantonment mill has not had a problem in meeting TBELs.
2. 2. 55. The TBELs that IP must meet are in the "Cluster Rule" promulgated by the EPA and adopted by the Department. The mill already meets the TBELS applicable to its current bleaching operation. In fact, EPA determined that the mill was performing in the top 5% of similar mills in the nation. The mill would have to meet the TBELs for a brown kraft operation if that conversion is made by IP.
3. . 56. The proposed permit also imposes water quality-based effluent limits (WQBELs) that are specific to the Cantonment mill and the waters affected by its effluent discharge. The WQBELs for the mill are necessary for certain constituents of the mill's effluent because the TBELs, alone, would not be sufficient to prevent water quality criteria in the receiving waters from being
4. . violated. For example, the TBEL for BOD for similar pulp and paper mills is 15,943 pounds per day (ppd) on a monthly average, but the WQBEL for BOD for the Cantonment mill would be 4500 ppd in summer and 5100 ppd in winter.
5. 3. 57. Dr. Livingston developed an extensive biological and chemical history of Perdido Bay and then evaluated the nutrient loadings from Elevenmile Creek over a 12-year period to correlate mill loadings with the biological health of the Bay. Because Dr. Livingston determined that the nutrient loadings from the mill that occurred in 1988 and 1989 did not adversely impact the food web of Perdido Bay, he recommended effluent limits for ammonia nitrogen, orthophosphate, and total phosphorous that were correlated with mill loadings of these nutrients in those years. The Department used Dr. Livingston's data, and did its own analyses, to establish WQBELs for orthophosphate for drought conditions and for nitrate–nitrite. WQBELs were ultimately developed for total ammonia, orthophosphate, nitrate–nitrite, total phosphorus, BOD, color, and soluble inorganic nitrogen.
6. . 58. The WQBELs in the proposed permit were developed to assure compliance with water quality standards under conditions of pollutant loadings at the daily limit (based on a monthly average) during low flow in the receiving
7. . waters. The proposed permit also establishes daily maximum limits (the most that can be discharged on any single day). For BOD, the daily maximum limit is 9000 ppd. William Evans, the Department employee with primary responsibility for the technical review of the proposed Department authorizations, said that setting the daily maximum limit at twice the monthly average was a standard practice of the Department. The maximum daily limits are not derived from the Livingston studies.
8. . 59. Dr. Glen Daigger, a civil and environmental engineer, designed a model for the WWTP and determined the modifications necessary to enable the WWTP's discharge to meet all TBELs and WQBELs. Petitioners did not dispute that the proposed WWTP is capable of achieving the TBELs and WQBELs. Their main complaint is that the WQBELs are not adequate to protect the receiving waters.

9.3. Discharge to the Wetland Tract

10. . 60. IP proposes to relocate its discharge to the wetland tract as a means to end decades of failure by the mill to meet water quality standards in Elevenmile Creek. Discharging to the wetland tract, which flows to the marine waters of lower Elevenmile Creek and Perdido Bay, avoids many of the problems associated with trying to meet the
11. . more stringent water quality standards applicable in a freshwater stream.
12. 4. 61. An effluent distribution system is proposed for the wetland tract to spread the effluent out over the full width of the wetlands so that their full assimilative capacity is utilized. This would be accomplished by a system of berms running perpendicular to the flow of water through the wetlands, and gates and other structures in and along the berms to gather and redistribute the flow as it moves in a southerly

direction toward Perdido Bay and lower Elevenmile Creek. The design incorporates four existing tram roads that were constructed on the wetland tract to serve the past and present silviculture activities there. The tram roads, with modifications, would serve as the berms in the wetland distribution system.

13. 5. 62. As the effluent is discharged from the pipeline, a point designated D-003, it would be re-aerated[6] and distributed across Berm 1 through a series of adjustable, gated openings. Mixing with naturally occurring waters, the effluent would move by gravity to the next lower berm. The water will re-collect behind each of the vegetated berms and be distributed again through each berm. The distance between the berms varies from a quarter to a half mile.
14. 6. 63. Approximately 70% of the effluent discharged at D-003 would flow by gravity a distance of approximately 2.3 miles to Perdido Bay. The remaining 30% of the effluent would flow a shorter distance to lower Elevenmile Creek.
15. 7. 64. A computer simulation performed by Dr. Wade Nutter, an expert in hydrology, soils, and forested wetlands, indicated that the effluent discharged at D-003 will move through the wetland tract at a velocity of approximately a quarter-of-a-foot per second and the depth of flow across the wetland tract will be about one-half inch. It would take 4 or 5 days for the effluent to reach lower Elevenmile Creek and Perdido Bay. As the treated effluent flows through the wetland tract, there will be some removal of nutrients by plants and soil. Nitrogen and phosphorous are expected to be reduced approximately 10%. BOD in the effluent is expected to be reduced approximately 90%.
16. . 65. Construction activities associated with the effluent pipeline and berm modifications in the wetland tract were permitted by the Department in 2003 through issuance of a Wetland Resource Permit to IP. The United States Army Corps of Engineers has also permitted this
17. . work. No person filed a petition to challenge those permits.
18. 8. 66. A wetland monitoring program is required by the proposed permit. The stated purpose of the monitoring program is to assure that there are no significant adverse impacts to the wetland tract, including Tee and Wicker Lakes, and is referred to as the No Significant Adverse Impact (NSAI) analysis. A year of "baseline data" on the wetlands and Tee and Wicker Lakes was collected and submitted to the Department for use in developing the NSAI analysis, but was not made a part of the record in this case. After the discharge to the wetland tract commences, the proposed permit requires IP to submit wetland monitoring reports annually to the Department.
19. 9. 67. A monitoring program was also developed by Dr. Livingston and other IP consultants to monitor the impacts of the proposed discharge on Elevenmile Creek and Perdido Bay. It was made a part of the proposed permit.

C. The Exemption for Experimental Use of Wetlands

0. 68. Florida Administrative Code Rule 62-660.300(1) provides an exemption from water quality criteria for the experimental use of wetlands. The proposed Authorization for Experimental Use of Wetlands Order would exempt IP from
1. Class III water quality criteria for pH, DO, transparency, turbidity, and specific conductance.
2. 2. 69. The proposed exemption order sets forth "interim limits" for pH, DO, color, turbidity, and specific conductance. The proposed exemption order also states that IP may petition for alternative water quality criteria pursuant to Florida Administrative Code Rule 62-66D.300(1)(b)(c) and (d).
3. 3. 70. The exemption is for 5 years beginning with the commencement of discharge into the wetland tract at D-003. The exemption it can be renewed by IP by application to the Department.

D. The Waiver

0. 1. 71. To qualify for the experimental use of wetlands exemption, Florida Administrative Code Rules 62-660.300(1)(a)3 and 4 require, respectively, that the public be restricted from the exempted wetland area and that the waters not be used for recreation. IP proposes to prevent public access to

the area of the wetland tract where the effluent distribution system is located. This is the freshwater area of the wetland tract and includes the four berms.

1. 2. 72. However, IP does not want, nor believe it is necessary, to prevent public access and recreation on Tee and Wicker Lakes within the tidal marsh below berm 4. These lakes are accessible by boat from Perdido Bay and are used now by the public for boating and fishing.

E. The Proposed Consent Order

1. 1. 73. The proposed Consent Order establishes a schedule for the construction activities associated with the proposed WWTP upgrades and the effluent pipeline and for incremental relocation of the mill's discharge form Elevenmile Creek to the wetland tract. IP is given 24 months to complete construction activities and begin operation of the new facilities. At least 25% of the mill's effluent must be diverted to the wetland tract. At least 25% of the effluent is to be diverted to the wetland tract when the new facilities begin operations. The volume of effluent diverted to the wetlands is to increase another 25% every 3 months thereafter so that 3 years after issuance of the permit 100% of the effluent is being discharged into the wetland tract and there is no longer a discharge at D-001 into Elevenmile Creek.[7]
2. . 74. The proposed Consent Order establishes interim effluent limitations that would apply immediately upon the effective date of the Consent Order and continue during the 24-month construction period when the mill will continue to
3. . discharge into Elevenmile Creek. Other interim effluent limits would apply during the 12-month period following construction when the upgraded WWTP would be operating and the effluent would be incrementally diverted from Elevenmile Creek to the wetland tract. A third set of interim effluent limits would apply at D-003 when 100% of the discharge is into the wetland tract. They include the interim limits for specific conductance, pH, DO, color, and turbidity established through the experimental use of wetland exemption.
4. 2. 75. The proposed Consent Order requires IP to submit a report within 6 months with the results of the 2004 transparency study. The Department must be satisfied that the study shows the transparency standard will not be violated before the wetlands can be used for the discharge. This report has already been submitted to the Department, but the Department has not yet completed its review of the report. Nevertheless, it was admitted into the record as IP Exhibit 79.
5. . 76. The proposed Consent Order provides that, in the event IP's does not receive treated sanitary wastewater from the planned ECUA facility, IP will notify the Department and submit an alternate compliance plan to the Department for the Department's approval. The submittal
6. . and approval of an alternate compliance plan would extend the time for compliance with water quality standards by another 6 months. The Department amended the proposed Consent Order at the conclusion of the hearing to provide for notice to the public and an opportunity for persons to object to the Department's action on any alternate compliance plan.
7. 3. 77. The Consent Order requires a "Plan of Action" to determine "whether there remains a critical period for ortho-phosphate loading to lower Elevenmile Creek and Perdido Bay."
8. 4. 78. The proposed Consent Order requires IP to submit within 97 months (which would allow for 5 years of discharge to the wetland tract) a final report on whether there has been significant adverse impacts in the wetlands and Tee and Wicker Lakes resulting from the discharge of effluent pursuant to the interim limits for pH, DO, specific conductance, turbidity, and color. If the NSAI analysis shows no significant adverse impact has occurred, the proposed Consent Order contemplates that IP or the Department would establish alternative water quality criteria that would apply permanently in the wetland tract.
9. . 79. IP is required by the Consent Order to submit quarterly progress reports of its progress toward
10. . compliance with the required corrective actions and deadlines.
11. 5. 80. The Consent Order imposes a "stipulated penalty" of $500 per day for noncompliance with its terms. It also contains a statement that a violation of its terms may subject IP to civil penalties up to $10,000 per day.

III. The Principal Factual Disputes

A. The Evidence in General

0. 1. 81. Much of the water quality and biological data presented by Petitioners were limited in terms of the numbers of samples taken, the extent of the area sampled, and the time period covered by the sampling. Much of the expert testimony presented by Petitioners was based on limited data, few field investigations, and the review of some, but not all relevant permit documents.[8]
1. 82. On the other hand, the Livingston studies represent perhaps the most complete scientific evaluation ever made of a coastal ecosystem. Even Dr. Lane called the Livingston studies "huge" and "amazing." Therefore, with regard to the factual issues raised by Petitioners that involved scientific subjects investigated in the Livingston studies, Petitioners' data and the expert opinions based on those data were generally of much less weight than the data and conclusions of the Livingston studies. However, the
2. Livingston studies did not address all of the factual issues in dispute.
3. 2. 83. Some of the evidence presented by Petitioners regarding historical water quality conditions in Perdido Bay and Elevenmile Creek was lay testimony. The lay testimony was competent and sufficient to prove the existence of environmental conditions that are detectable to the human senses, such as an offensive smell, a dark color, or a sticky texture.

B. Historical Changes in Perdido Bay

0. 1. 84. Petitioners claim that, before the Cantonment mill began operations in the 1940s, Perdido Bay was a rich and diverse ecosystem and a beautiful place for swimming, fishing, boating, and other recreational activities. Petitioners blame the mill effluent for all the adverse changes they say have occurred in Perdido Bay.
1. 2. 85. Petitioners claim that the water in Perdido Bay was much clearer before the mill was built. James Lane, who has lived on the Bay for 65 years, said he began to notice in the late 1940s that the water was becoming dark and filled with wood fibers.
2. 86. Mr. Lane recalls that there used to be an abundance of fish in the Perdido Bay, including croakers, pinfish, flounder, redfish, minnows, and catfish. Now
3. Mr. Lane sees few of these fish in the Bay and he believes the remaining fish are unfit to eat because they look diseased to him.
4. 3. 87. Mr. Lane said there were extensive areas of seagrasses in the Bay which supported large numbers of shrimp, crabs, and mussels, but these grasses are now gone.
5. 4. 88. The Lane family used to enjoy swimming in Perdido Bay but stopped swimming years ago because the water felt sticky and often had a brown foam or scum on the surface. Mr. Lane and others members of FOPB claim to have gotten infections from swimming in the Bay.
6. 5. 89. Mr. Lane and other witnesses described the odor of Elevenmile Creek near the mill as unpleasant and, at times, offensive. They consider the Creek to be too polluted for swimming.
7. 6. 90. Donald Ray, who has been a Department biologist for 30 years, said he has received many complaints from citizens about the conditions in Perdido Bay. He said the foam that occurs in Perdido Bay is not natural foam, but one that persists and leaves a stain on boats.
8. 7. 91. On the other hand, it is Dr. Livingston's opinion that the ecological problems of Perdido Bay are due primarily to the opening of Perdido Pass around 1900. The opening of the pass allowed Gulf waters to enter Perdido Bay and caused salinity stratification in the Bay, with marine waters on the bottom and freshwater from the Perdido River, Elevenmile Creek, and other tributaries on the top. The stratification occurs regularly in the lower Bay, but only during low flow conditions in most of the upper Bay, Perdido River, and Elevenmile Creek. It restricts DO exchange between the upper and lower water layers and results in low DO levels in the lower layer. Low DO, or "hypoxia," is the primary cause of reduced biological diversity and productivity in Perdido Bay.

92. Dr. Livingston's initial study of the Perdido Bay system (1988–1991) included an investigation of historical conditions, using documents and maps, anecdotal statements of area residents, as well as historic water quality and sediment data. Dr. Livingston found general agreement from most sources that:

> [P]rior to the 1940s, the various rivers and the bay in the Perdido Basin were quite different from what they are today. Eyewitness accounts from 1924 indicate a bay that was clear and "bluish" in color; the bottom could be seen at depths of five feet. According to resident' accounts, seagrasses grew from Garth Point to Witchwood; the grassbeds provided cover for many shrimp that were taken at the time. Flounder were taken with gigs and crabs were taken with hand nets. According to these accounts, the water from the various rivers and creeks in the area was relatively clear, and white sand/gravel bottoms were dominant forms of habitat in the freshwater and estuarine systems. The water was tea-colored but clear. Redfish, trout, blue crabs, shrimp, and mullet were abundant.
>
> * * *
>
> [T]hrough the early 1900s, the Elevenmile Creek was said to be crystal clear with soft white sand and good fishing.
>
> * * *
>
> According to various reports, in the early 1950s, the waters of Elevenmile Creek turned black, with concentrations of foam observed floating on the surface. By 1986, more than 28 million gallons of largely untreated effluent was flowing into the Elevenmile Creek-Perdido Bay system each day. Experiments by the Florida Game and Freshwater Fish Commission had shown that the creek waters were lethal. The Florida Board of Health reported that Elevenmile Creek was "grossly polluted" and that Perdido Bay had been "greatly degraded within the 1.5 mile radius of where Elevenmile Creek dumped into the bay."

Nevertheless, Dr. Livingston discounted much of this historical record, especially with regard to the belief that the mill's effluent had adversely affected Perdido Bay, because it was not based on what he considers reliable scientific data. He found "little evidence in the long-term sediment record of a direct response to historical activities of the pulp and paper mill, suggesting that the flushing capacity of Perdido Bay quickly diluted effluents that enter Perdido Bay from Elevenmile Creek."

1. 1. 93. The evidence is persuasive that the salinity stratification in Perdido Bay is a major cause of low DO in the Bay.[9] However, the stratification does not explain all of the observed changes in water quality, biological productivity, and recreational values. The stratification does not account for the markedly better conditions in the Bay that existed before the Cantonment paper mill began operations. The Livingston studies confirmed that when nutrient loadings from the mill were high, they caused toxic algae blooms and reduced biological productivity in Perdido Bay. As recently as 2005, there were major toxic blooms of heterosigma in Tee and Wicker Lakes caused by increased nutrient loading from the mill. Other competent evidence showed that the mill's effluent has created nuisance conditions in the past, such as foam and scum, which adversely affected the recreational values of these public waters.
2. . 94. Some of the adverse effects attributable to the mill effluent were most acute in the area of the Bay near the Lanes' home on the northeastern shore of the Bay, because the flow from the Perdido River tends to push the
3. . flow from Elevenmile Creek toward the northeastern shore. Petitioners were justified in feeling frustrated in having their concerns about the adverse impacts of the mill's effluent discounted for many years, and in having to wait so long for an effective regulatory response.
4. 2. 95. However, with regard to many of their factual disputes, Petitioners' evidence lacked sufficient detail regarding the dates of observations, the locations of observations, and in other respects, to distinguish the relative contribution of the mill effluent from other factors that contributed to the adverse impacts in the Bay, such as salinity stratification, natural nutrient loading from the Perdido River and other tributaries, and anthropogenic sources of pollution other than the paper mill.[10]
5. 3. 96. Petitioners generally referred to the mill effluent and its impacts to Perdido Bay as if they have been relatively constant for 65 years. The Livingston studies, however, showed clearly that the mill effluent and its impacts, as well as important factors affecting the impacts, such as drought, have frequently changed.

6. . 97. Focusing on the fact that the average daily BOD loading allowed under the proposed permit would be same as under the 1989 TOP (4500 ppd), Petitioners remarked
7. . several times at the final hearing that the proposed permit for the mill was no different than the existing permit. According to Petitioners, if the mill is allowed to operate under the proposed permit, one can predict that the future adverse impacts to Perdido Bay will be the same as the past adverse impacts. However, the 1989 TOP and the proposed permit are very different. Therefore, it cannot be assumed that the impacts would be the same.
8. . 98. Petitioners' evidence was generally insufficient to correlate past adverse impacts to Perdido Bay with the likely impacts that would occur under the proposed permit. In contrast, that was the focus of the Livingston studies.

9. . C. Development of the WQBELs

10. . 1. Whether Perdido Bay is an Alluvial System and Whether Elevenmile Creek is a Blackwater Stream

11. 4. 99. Alluvial systems are generally characterized by relatively high nutrient inputs from tributaries and associated wetlands that provide for high biological productivity in the receiving bay or estuary. Petitioners disagree with Dr. Livingston's characterization of the Perdido Bay system as an alluvial system. Petitioners presented the testimony of Donald Ray, a Department biologist, who said that the Perdido River is not an alluvial river and the natural nutrient loadings to Perdido Bay are less than would occur in an alluvial system.

100. Although it is curious that two experienced biologists cannot agree on whether Perdido Bay is part of an alluvial system, the dispute is immaterial because it was not shown by Petitioners that any of the four proposed Department authorizations is dependent on the applicability of the term "alluvial." The WQBELs developed by Dr. Livingston, for example, were not dependent on a determination that Perdido Bay meets some definition of an alluvial system, but were based on what the data indicated about actual nutrient loadings into Perdido Bay and the Bay's ecological responses to the loadings. If the dispute is not immaterial, then Dr. Livingston's opinion that Perdido Bay is part of an alluvial system is more persuasive, because he has greater experience and knowledge of the coastal bay systems on the Florida Panhandle than does Mr. Ray.

101. Petitioners also take exception to Dr. Livingston's characterization of Elevenmile Creek as a blackwater creek. Petitioners claim Elevenmile Creek is naturally clear to "slightly tannic" stream. This dispute, however, is also immaterial because the proposed permit calls for the termination of the mill's discharge to Elevenmile Creek, including its contribution of color to the Creek.

102. Petitioners assert that Dr. Livingston's characterizations of Perdido Bay as an alluvial system and Elevenmile Creek as a blackwater creek show he is biased and that his "overall analysis" lacks credibility. Dr. Livingston's opinions on these points do not show bias nor compromise the credibility of his overall analysis of the Perdido Bay system, which is actually the product of many scientists and based on 18 years of data.[11]

2. Selection of 1988 and 1989 Mill Loadings as a Benchmark for the WQBELs

103. Generally, the Department establishes effluents limits for nutrients based on Chlorophyl A analysis. However, the Livingston studies showed that Chlorophyl A was not significantly associated with plankton blooms in Perdido Bay. Therefore, the Department accepted Dr. Livingston's recommendation to base the WQBELs for nutrients on the nutrient loading from the mill in 1988 and 1989, which the Livingston studies showed were good years for Perdido Bay with respect to its biological health.

104. Phytoplankton are a fundamental component of the food web in Perdido Bay. The number of phytoplankton species is a sensitive indicator of the overall ecological health of the Bay. The Livingston studies showed that the loadings of ammonia and orthophosphate from the mill had a direct effect on the number of phytoplankton species. In the years when the mill discharged high loadings of

ammonia and orthophosphate, there were toxic algae blooms and reduced numbers of phytoplankton species. In 1988 and 1989, when the loadings of ammonia and orthophosphate were lower, there were no toxic algae blooms, and there were relatively high numbers of phytoplankton species.

105. Petitioners dispute that 1988 and 1989 are appropriate benchmarks years for developing the WQBELs because Petitioners claim there were high nutrient loadings and algae blooms in those years. Mr. Ray testified that the Department received citizen complaints about algae blooms in those years. Dr. Livingston's analysis was more persuasive, however, because it distinguished types of algae blooms according to their harmful effect on the food web and was based on considerably more water quality and biological data.

106. Petitioners also presented water quality data collected from 1971 to 1994 by the Bream Fishermen Association at one sampling station in the northeastern part of Perdido Bay, which indicate that in 1988 and 1989, the concentrations of nutrients were sometimes high. The proposed nutrient WQBELs were derived from data about the actual response of the Perdido Bay ecosystem over time to various inputs. The sampling data from the Bream Fishermen Association were not correlated to ecosystem response and, therefore, are insufficient to refute Dr. Livingston's evidence that 1988 and 1989 were years of relatively high diversity and productivity in Perdido Bay. Furthermore, nutrients loadings would be reduced under the proposed permit.

3. DO and Sediment Oxygen Demand

107. The parties agreed that sediment oxygen demand (SOD) is a major reason for the low DO in Perdido Bay in areas where there is salinity stratification. SOD is caused by the bacterial degradation of particulate organic matter that settles to the bottom. SOD decreases DO in the lower water layer, but also can cause a reduction of DO in the surface layer. Low DO has substantially reduced the biological productivity of Perdido Bay.

108. Thomas Gallagher, an environmental engineer and water quality modeling expert, showed that even without the mill discharge, DO in the bottom waters of Perdido Bay would fall below the applicable Class III water quality standard of 5 mg/L. Low DO conditions are now a "natural" characteristic of the Bay, usually occurring during summer and early fall when freshwater flows are low and temperatures are high. At these times, surface water DO levels are usually above the state standard, but DO in the bottom waters usually range between 1.0 and 2.0 mg/L.

109. Petitioners claim that the dominant source of the sediment in Perdido Bay is the carbon and nutrient loading in the mill's effluent that flows into the Bay from Elevenmile Creek. Mr. Ray, who sampled sediments in Perdido Bay over several years for the Department, believes that the mill effluent is the main source of the sediment and, consequently, the sediment oxygen demand.

110. Dr. Livingston did extensive sediment analyses in Perdido Bay. He compared the data with sediment data from other bays on the Florida Panhandle. It is Dr. Livingston's opinion that the mill effluent contributes little to the sediments or SOD in Perdidio Bay. His initial 3-year study concluded:

> [T]he hypoxic conditions of Elevenmile Creek are due, in part, to mill discharges. However, low dissolved oxygen conditions at depth in Perdido Bay are not due to the release of mill effluents from Elevenmile Creek, and can actually be attributed to a long history of human activities that include alteration of the hydrological interactions at the gulfward end of the estuary. The entry of saline water from the Gulf and the resulting stratification have been coupled with various forms of human development that release carbon, nitrogen, and phosphorus compounds into the estuary. The landward movement of high-salinity water from the Gulf of Mexico, laden with various types of oxygen-consuming compounds from various sources, together with oxygen demand from sediments to the lower water column that is isolated from reaeration due to salinity stratification, are thus responsible for a large portion of the observed hypoxic conditions at depth in Perdido Bay. [The paper mill] is responsible for a relatively small amount of these oxygen-consuming effects.

111. In East Bay, which is a part of Escambia Bay and a relatively pristine system, there was SOD that caused DO to fall below standards in the lower water layer.

Dr. Livingston also found severe oxygen deprivation at times in the lower waters of the Styx River and Perdido River, which do not receive mill effluent. Dr. Livingston believes the low DO that occasionally occurs in these rivers is due to agricultural runoff, urban discharges, and natural organic loading from adjacent wetlands.

112. There was extensive evidence, some of which was presented by Petitioners, showing that the mill loadings of carbon and nutrients are less than the loadings from the Perdido River. Mr. Gallagher concluded that the sediment in the Bay is mostly "terrestrial carbon," and not from the mill's effluent. His water quality modeling work determined that the mill's effluent reduced bottom layer DO by about 0.1 mg/L.
113. Dr. Lane believes that the organic solids in the mill's effluent are accumulating in Perdido Bay sediments, but Mr. Gallagher pointed out that degrading solids cannot accumulate because they are degrading. In addition, Mr. Gallagher said that logic dictates that solids that have not settled out after spending several days in the settling basins of IP's WWTP are not going to readily settle in the more turbulent environment of Perdido Bay. Some of the solids are oxidizing or being transported into the Gulf.
114. Mr. Gallagher determined that in summer and late fall, 60% of the water in the bottom layer in the upper Bay is from the Gulf and almost all the rest is from the Perdido River. He believes only 0.1%–2.0% of the water in the bottom layer is mill effluent.
115. Dr. Livingston responded to the BOD and carbon issues that "these Petitioners raised over the years" by investigating them as part of the Livingston studies. He found no relationship between loading and DO. Dr. Livingston concluded that the mill was not having much effect on SOD.
116. Dr. Livingston and Mr. Gallagher referred to a carbon isotope study of the sediment in Perdido Bay by Coffin and Cifuentes. The isotope study was a part of the initial 3-year Livingston study entitled "Ecological Study of the Perdido Bay Drainage System." The study identified a unique carbon isotope in the mill's effluent and looked for traces of the isotope in the sediments of Perdido Bay. Very little of the carbon isotope was found in the sediments, suggesting that the mill's effluent was not contributing much to the sediments. The carbon isotope study was not offered into evidence.
117. Petitioners assert that the isotope study is hearsay and cannot be used to support a finding of fact.[12] However, Dr. Livingston's opinion about the sources of the sediment was not based solely on the isotope study. The isotope study was consistent with his other studies and with Mr. Gallagher's water quality modeling analysis. Therefore, the conclusions of the isotope study serve to support and explain Dr. Livingston's expert opinion that the mill effluent is not the primary source of the sediment and low DO in Perdido Bay.
118. Dr. Livingston summarized his opinion regarding DO and SOD as follows: "all of these lines of evidence, from all the bays that I have worked in and from them scientific literature and from our own studies, every line of evidence simply eliminated the pulp mill as the primary source of the low dissolved oxygen in the bay."

4. Long-term BOD

119. BOD is a measurement of the oxygen demand exerted by the oxidation of carbon, nitrogen, and the respiration of algae. A 5-day BOD analysis is the standard test used in the regulatory process. The use of the standard 5-day BOD measurement is not restricted to organic material that is expected to completely degrade in 5 days. Five days is simply the time period selected to standardize the measurement. For example, the 5-day BOD analysis is used in the regulation of domestic wastewater even though most of the organic material in domestic wastewater takes about 60 days to degrade and would exert an oxygen demand throughout the 60 days.
120. It was undisputed that paper mill effluent will continue to consume DO after 5 days. One estimate given was that it would take 100 days to completely degrade. Some of the naturally occurring organic material flowing into Perdido Bay from the Perdido River and Gulf of Mexico would also include material with long-term BOD.

121. Petitioners claim that long-term BOD analysis is essential to determine the true impacts of the mill's effluent on Perdido Bay, but they failed to show that the Livingston studies did not consider long-term BOD.[13] The evidence shows that Dr. Livingston's studies accounted for DO demand in all its forms and for any duration. Dr. Livingston's studies focused on the response of Perdido Bay's food web to nutrients and various other inputs as they changed over time. If long-term BOD was having an adverse effect on the food web, the Livingston studies were designed to detect that effect. Dr. Livingston's opinion is that long-term BOD is not a significant problem for Perdido Bay because the Bay is part of a dynamic system and the sediments are regularly flushed out or otherwise recycled in a matter of a few months, not years.[14]

5. Carbon

122. Dr. Lane, who is a marine biologist, believes a major reason for low DO in Perdido Bay is "organic carbonaceous BOD." However, Dr. Lane presented no evidence other than statements of the theoretical process by which carbon from the mill would cause low DO in the Bay. She presented no scientific data from Perdido Bay to prove her theory.[15] Dr. Livingston said that 16 years of studies in the Bay have found DO and carbon to be "totally uncorrelated."

D. Other Water Quality Issues

1. Toxicity

123. Petitioners allege that the mill effluent has had occasional problems passing toxicity tests. Un-ionized ammonia is the likely cause, and the reduction of un-ionized ammonia in the proposed permit and the distribution of the effluent over the wetland tract should prevent toxicity problems from recurring.
124. Dr. Livingston examined tissue samples from various fish and invertebrates and found low levels of bioconcentrating chlorine compounds in Perdido Bay that he believes were "probably associated with discharges from the Pensacola mill." Although they are toxic substances, Dr. Livingston found no diseased organisms and no evidence of food web magnification of these potentially bioaccumulable compounds.
125. Mr. Ray testified that Perdido Bay was the worst of all the bays he has studied in terms of high sediment metals. Most of his sediment sampling was done in 1977–1983, years before the Livingston studies got started. His knowledge about subsequent years was based on only two samples, one in 1988 and another in 2005.[16] Dr. Lane did an analysis of 12 sediment samples in Perdido Bay, Perdido River, and Elevenmile Creek in 1999 and concluded that "Eleven Mile Creek appears to be the source of all elevated levels [of metals] except silver."
126. The Livingston studies included toxics analysis of Perdido Bay sediments, including metals, dioxin, and other chlorinated organic compounds. Dr. Livingston testified that metal concentrations in the sediments of Elevenmile Creek did not differ from the metal concentrations in the Perdido River and other streams in the area. The concentrations were not significantly different from concentrations in other bays he has studied that do not have a paper mill discharge.

2. Mutagenic Compounds

127. Petitioners claim that there are chemicals in paper mill effluent that are mutagenic and are causing changes in the sex of fish. They introduced an exhibit from the Department's exhibit list (DEP Exhibit 38) that discussed investigations of effluent from the Cantonment mill and other Florida paper mills which found abnormally high testosterone levels and related mutations in female Gambusia fish. The most recent such study[17] implicates androgens produced by the microbial degradation of natural chemicals in the trees pulped at the mills, especially softwood trees (pines), as the cause.
128. Petitioners believe IP's proposal to begin using 100% pine at the Cantonment mill could cause mutations in fish and other animals exposed to the mill's effluent. Although IP and the Department are

aware of the sex change studies, there was no evidence presented that the subject was investigated or addressed by them in the permitting process.

129. DEP Exhibit 38 is hearsay and no non-hearsay evidence was presented on the issue of mutagenic compounds in the mill's effluent. Therefore, no finding of fact in this Recommended Order can be based on the data and analysis in DEP Exhibit 38.[18] Furthermore, Petitioners did not raise the issue of mutagenic compounds in the mill's effluent discharge in their petitions for hearing or in the pre-hearing stipulation.[19]

E. Antidegradation Policy

130. Petitioners claimed the proposed permit violated the antidegradation policy for surface waters established in Florida Administrative Code Rule 62-302.300(1). An element of that policy is to require, for any discharge that degrades water quality, a demonstration that the degradation is necessary or desirable under circumstances which are clearly in the public interest. Florida Administrative Code Rule 62-4.242(1)(a) contains a list of factors to be considered and balanced in applying the antidegradation policy. These include consideration of whether the proposed project would be beneficial to public health, safety, or welfare and whether the discharge would adversely affect the, conservation of fish and wildlife, and recreational values.

131. The greater weight of the evidence supports the position of IP and the Department that the proposed discharge to the wetland tract would be an improvement over the existing circumstances. However, as discussed below, there was an insufficient demonstration that the discharge would not cause significant adverse impact to the biological community within the wetland tract, and there was an insufficient demonstration that the Perdido River OFW would not be significantly degraded. Without sufficient demonstrations on these points, it is impossible to find that the degradation has been minimized.

132. Petitioners did not prove that the proposed project was not in the public interest, but the burden was on IP to show the opposite. Because IP did not make a sufficient demonstration regarding potential adverse impacts on the biological community within the wetland tract and on the Perdido River OFW, IP failed to prove compliance with Florida's antidegradation policy.

F. Perdido River OFW

133. Florida Administrative Code Rule 62-302.300(2) contains the standards applicable to OFWs and prohibits a discharge that significantly degrades an OFW unless the proposed discharge is clearly in the public interest or the existing ambient water quality of the OFW would not be lowered.[20] Petitioners contend that the water quality of the Perdido River would be significantly degraded by the mill's effluent under the authorizations.

134. Mr. Gallagher's modeling analysis predicted improved water quality in the Perdido River for DO and several other criteria over the conditions that existed in 1979, the year the river was designated as an OFW. However, the modeling also predicted that the discharge would reduce the DO in the river (as it existed in 1979) by.01 mg/L under unusual conditions of effluent loading at the daily limit (based on a monthly average) during a drought. Mr. Gallagher's modeling indicated that a very small (less than 0.1 mg/L) reduction in DO in the surface water of the lower Perdido River would occur as a result of the proposed project. He considered that to be an "insignificant" effect and it was within the model's range of error.

135. However, IP made the wrong comparisons in its modeling analysis to determine compliance with the OFW rule, Florida Administrative Code Rule 62-4.242(2). Mr. Gallagher used the model to compare the DO levels in the Perdido River that would result from the mill's discharge of BOD at the proposed permit limit of 4500 ppd with the predicted DO levels that would have existed in 1979 if St. Regis was discharging 5100 ppd of BOD. IP should have compared the DO levels resulting from the proposed permit with the actual DO levels in 1979, or at least the DO levels that the model would have simulated using actual BOD loadings by St. Regis in 1979.

136. The DO levels that would have existed in 1979 if St. Regis had discharged 5100 ppd of BOD are irrelevant. No DO data from 1979 were presented at the hearing and no explanation was given for why DO data for 1979 were not used in the analysis. No evidence was presented that St. Regis discharged 5100 ppd of BOD as a monthly average in 1979.[21] It might have discharged substantially less.[22]
137. Petitioners did not prove that the proposed permit would significantly degrade the Perdido River, but the burden was on IP to show the opposite. Because the wrong anti-degradation comparison was made, IP failed to provide reasonable assurance that the Perdido River would not be significantly degraded by the proposed discharge.

G. The Experimental Use of Wetlands Exemption

138. Petitioners claim that IP did not demonstrate compliance with all the criteria for the experimental use of wetlands exemption. There are seven criteria set forth in Florida Administrative Code Rule 62-660.300(1)(a) that must be met to qualify for the exemption. IP is seeking a waiver from two of the criteria and those will be discussed later in this Recommended Order.

1. Impact on the Biological Community

a. In General

139. Florida Administrative Code Rule 62-660.300(1)(a)1 requires a demonstration that "the wetlands ecosystem may reasonably be expected to assimilate the waste discharge without significant adverse impact on the biological community within the receiving waters."
140. Dr. Nutter used a "STELLA" wetland model to predict the effects of discharging mill effluent to the wetland tract. The STELLA model was programmed to evaluate the "water budget" for the wetland tract, as well as simulate the fate of nitrogen, phosphorus, and total dissolved solids (TDS). Petitioners contend that the STELLA model is too limited to adequately assess potential adverse impacts on the biological community, but the model was not the sole basis upon which Dr. Nutter formed his opinions. He also relied on relevant scientific literature, his general knowledge of wetland processes, and on his 40 years of experience in land treatment of wastewater.
141. The STELLA model predicted that there would be about a 10% reduction in nitrogen and phosphorus. Dr. Nutter testified that that figure was a conservative prediction and the scientific literature suggests there could be a greater reduction.
142. Wetlands are effective in processing TSS and BOD. Dr. Nutter ran the model with the proposed permit limits and the model predicted 90%–95% BOD removal before the effluent reached berm 4.
143. Dr. Nutter expected pH levels to be in the range of background levels in the wetlands, which vary between 6.5 and 8.0.[23]
144. Dr. Nutter predicts that in high flow conditions, there will be more DO in the water flowing from the wetlands into Tee and Wicker Lakes. During low flow conditions, he predicts no change in the DO level. Background DO levels in the wetland tract now range between 0.0 and 5.0 mg/L. Mr. Gallagher's water quality modeling for Perdido Bay assumed that the water flowing from the wetland tract would have a DO level of 2.0 mg/L, which Dr. Nutter believes this is a conservative estimate, meaning it could be higher.

b. Specific Conductance

145. A fundamental premise of the relocated discharge is that it solves the mill's decades-long failure to meet the stricter water quality standards applicable in the freshwaters of Elevenmile Creek because the new receiving waters would be marine waters. However, the majority (about 70%) of the wetland tract is a freshwater wetland. The tidal influence does not reach above berm 4 in the wetland tract. Before the mill's effluent reaches marine waters, it would be distributed over the entire freshwater portion of the wetland tract.

146. Dr. Livingston explained that, but for the mill's discharge, minnows and other small "primary" freshwater fish species would be found in Elevenmile Creek. The primary fish cannot tolerate the mill's discharge because the high levels of sodium chloride and sulfide (specific conductance) cause osmoregulatory problems, disrupting their blood metabolism and ion regulation. High conductivity also eliminates sensitive microinvertebrates.
147. Because Tee and Wicker Lakes are in the tidally influenced, southern portion of the wetland tract, the fish and other organisms in the lakes are polyhaline, which means they are adapted to rapid changes in salinity, temperature and other habitat features. That is not true of the organisms in the freshwater area of the wetland tract.
148. A constructed wetlands pilot project was built in 1990 at the Cantonment mill. The initial operational phase of the pilot project was July 1991 through June 1993. A second phase was conducted for just 3 months, from September 1997 through December 1997. The pilot project generated some information about "benthic macroinvertebrate diversity," which was "low to moderate." In addition, there were "observations" made of "three amphibian species, three reptile species, approximately 31 bird species, three fish species that were introduced, and two mammal species."
149. The information generated by the pilot project is ambiguous with respect to the effect of the effluent on fish and other organisms attributable to the specific conductance of the effluent, indicating both successes and failures in terms of survival rates. Moreover, the data presented from the pilot wetland project lacks sufficient detail, both with respect to the specific conductivity of the effluent applied to the wetlands and with respect to the response of salt-intolerant organisms to the specific conductivity of the effluent, to correlate the findings of the pilot project with the proposed discharge to the wetland tract.
150. Freshwater wetlands do not have naturally high levels of specific conductance. The specific conductance in the wetland tract is 100 micromhos/cm or less.[24] The proposed interim limit for specific conductance for the discharge into the wetland tract is "2500 micromhos/cm or 50% above background, whichever is greater."
151. Using total dissolved solids (TDS) as a surrogate for analyzing the effects on specific conductance, Dr. Nutter predicted that average TDS effluent concentrations would only be reduced by 1.0%.[25] His prediction is consistent with the literature on the use of wetlands for wastewater treatment, which indicates wetlands are not effective in reducing TDS and specific conductance.
152. The wetland tract would not assimilate TDS in mill's effluent. The potential exists, therefore, for the discharge to cause specific conductance in the freshwater area of the wetland tract to reach levels that are too high for fish and other organisms which can only live, thrive, and reproduce in waters of lower specific conductance. It was the opinion of Barry Sulkin, an environmental scientist, that the "freshwater community" would be adversely impacted by the salts in the effluent.
153. Although the freshwater area of the wetland tract is not dominated by open water ponds, creeks, and streams,[26] the evidence shows that it contains sloughs, creeks, and other surface water flow. No evidence was presented about the biological community associated with the sloughs, creeks, and other waters in the wetland tract, other than general statements about the existing plants and the trees that are being planted.
154. Petitioners did not prove that granting the exemption would cause significant adverse impact to the biological community in the freshwater area of the wetland tract, but it was IP's burden to affirmatively demonstrate the opposite. Because IP did not adequately address the impact of increased specific conductance levels on fish and other organisms in the freshwater area of the wetland tract, IP did not provide reasonable assurance that the proposed discharge would be assimilated so as not to cause significant adverse impact on the biological community within the wetland tract.

c. Tee and Wicker Lakes

155. When the Department issued the proposed exemption order, it did not have sufficient data and analyses regarding Tee and Wicker Lakes to determine with reasonable confidence that these waterbodies would not be adversely impacted by the proposed discharge. A transparency study of the lakes, which

IP introduced as an exhibit at the final hearing, had not previously been reviewed by Department staff. Dr. Livingston is still developing data and analyses for the lakes to use in the NSAI analysis.

156. The proposed NSAI monitoring plan states that one of its objectives is to determine the "ecological state" of the tidal ponds, including whether the ponds "could comprise an important nursery area for estuarine populations." In addition, the monitoring is to determine "the normal distributions of salinity, temperature, color, and dissolved oxygen" in the tidal ponds. These are data that must be known before a determination is possible that the discharge would not have a significant adverse impact on the biological community associated with the lakes.

157. Petitioners did not prove that granting the exemption would cause significant adverse impact to the biological community of Tee and Wicker Lakes, but it was IP's burden to affirmatively demonstrate the opposite. Because insufficient data exists regarding baseline conditions in Tee and Wicker Lakes, IP did not provide reasonable assurance that the proposed discharge would not cause significant adverse impact on the biological community within the wetland tract.

2. *Public Interest and Public Health*

158. Florida Administrative Code Rule 62-660.300(1)(a)2. requires the applicant to demonstrate that "granting the exemption is in the public interest and will not adversely affect public health or the cost of public health or other related programs."

a. *Public Interest*

159. Petitioners made much of a statement by Mr. Evans that the public interest consideration in this permit review was "IP's interest." Petitioners claimed that this statement was an admission by the Department that it gave no consideration to the public interest. However, in context, Mr. Evan's statement was not such an admission. Moreover, Florida Administrative Code Rule 62-302.300(6) expressly provides that the public interest is not confined to activities conducted solely for public benefits, but can also include private activities conducted for private purposes.

160. The proposed exemption order does not directly address the public interest criterion, but it notes that "existing impacted wetlands will be restored." In IP's application for the exemption, it states that the exemption would "contribute to our knowledge of wetlands in general and to the refinement of performance guidelines for the application of pulp mill wastewater to wetlands."

161. Petitioners dispute that the wetland tract is being restored. The evidence shows that some restoration would be accomplished. The natural features and hydrology of the tract have been substantially altered by agriculture, silviculture, clearing for pasture, ditching, and draining. The volume of flow in the discharge would offset the artificial drainage that occurred. A mixture of hardwood tree species would be planted, which would restore more of the diversity found in a natural forested wetland.

162. However, an aspect of the project that could substantially detract from the goal of restoration is the transformation of the freshwater wetlands to an unnatural salty condition. Dr. Nutter said that the salt content of the mill's effluent was equivalent to Gatorade, but for many freshwater organisms, that is too salty.

163. Another public benefit of the exemption that was discussed at the final hearing is that it would allow IP to relocate its discharge from Elevenmile Creek and thus end its adverse impacts to the Creek. That public benefit is not given much weight because IP has not shown that its adverse impacts to Elevenmile Creek cannot be eliminated or substantially reduced by decreasing its production of paper products. The evidence shows only that IP has attempted to solve its pollution problems through environmental engineering.[27]

164. A sufficient public interest showing for the purpose of obtaining the experimental use of wetlands exemption should not be a rigorous challenge if all the other exemption criteria are met, because that means the proposed wetland discharge was shown to have no harmful consequences. The public interest showing in this proceeding was insufficient, however, because the other exemption criteria were not met and there is a reasonable potential for harmful consequences.

b. Public Health

165. Petitioners raised the issue of the presence of Klebsiella bacteria, which can be a public health problem when they occur at high levels. The more detection of Klebsiella, however, does not constitute a public health concern. Petitioners did not show that Klebsiella bacteria exist in the mill's effluent at levels that exceed applicable water quality standards. Petitioners also did not present competent evidence about the likely fate of Klebsiella bacteria in the proposed effluent distribution system. Dr. Lane's statement that Klebsiella bacteria might be a problem is not sufficient to rebut IP's prima facie showing that the proposed permit will not cause or contribute to a violation of water quality standards applicable to pathogenic bacteria.
166. Petitioners also point to past incidents of high total coliform concentrations in Elevenmile Creek in support of their contention that the proposed exemption poses a risk to public health. However, these past incidents in Elevenmile Creek are not sufficient to prove that fecal coliform in the effluent discharged to the wetland tract will endanger the public health. IP proposes to restrict access to the wetland distribution system. Furthermore, the fate of bacteria in the wetlands is much different than in the Creek. The more persuasive evidence is that the wetland tract would destroy the bacteria by solar radiation and other mechanisms so that bacteria concentrations in waters accessible by the public would not be at levels which pose a threat to public health.

3. Protection of Potable Water Supplies and Human Health

167. Florida Administrative Code Rule 62-660.300(1)(a)5. requires the applicant for the exemption to demonstrate that "the presently specified criteria are unnecessary" to protect potable water supplies and human health, which presupposes that the applicant has applied for an exemption from water quality criteria applicable to human health. IP has not requested such an exemption and, therefore, this particular criterion appears to be inapplicable. Even if it were applicable, the evidence does not show that the effluent would cause a problem for potable water supplies or human health.

4. Contiguous Waters

168. Florida Administrative Code Rule 62-660.300(1)(a)6. requires a showing that "the exemption will not interfere with the designated uses of contiguous waters."
169. Contiguous waters, for the purpose of this criterion, would be Elevenmile Creek, Perdido Bay, and the Perdido River. Petitioners argue that Tee and Wicker Lakes should be considered contiguous waters for the purpose of this criterion of the exemption rule. However, Tee and Wicker Lakes are within the exempted wetland tract so they are not contiguous waters.
170. Petitioners contend that IP failed to account for the buildup of detritus in the wetlands and its eventual export to Perdido Bay. Their contention is based primarily on the opinion of Dr. Kevin White, a civil engineer, that treatment wetlands must be scraped or burned to remove plant buildup. However, Dr. Nutter explained that periodic removal of plant material is needed for the relatively small "constructed wetland" treatment systems that Dr. White is familiar with, but should not be needed in the 1464-acre wetland tract.
171. Nevertheless, because IP did not provide reasonable assurances that the proposed permit and related authorizations would not significantly degrade the Perdido River OFW, IP failed to meet this particular exemption criterion regarding interference with contiguous waters.

5. Scientifically Valid Experimental Controls

172. Florida Administrative Code Rule 62-660.300(1)(a)6. requires a showing that "scientifically valid environmental controls are provided ... to monitor the long-term effects and recycling efficiency."
173. Petitioners' argument about this particular criterion was largely misplaced. The term "environmental controls" modifies the term "monitor" and connotes only that the experiment would be monitored in a manner that will generate reliable information about long-term effects and performance. For monitoring purposes, IP's proposed NSAI protocol is an innovative and comprehensive plan that complies with this exemption criterion.

174. Petitioners' objections to the lack of sufficient information about Tee and Wicker Lakes is more appropriately an attack on the sufficiency of IP's showing that its discharge would not cause a significant adverse impact on the biological community within the wetland tract. That issue was discussed above.

6. Duration of the Exemption

175. Petitioners argue that the exemption cannot exceed 5 years in duration, but the time schedules established by the proposed Consent Order and proposed permit would allow the exemption to be in effect for 9 years. The Department's exemption order states that the 5 years does not begin to run until IP begins to discharge effluent at D-003 into the wetland tract. The possibility that IP might seek to renew the exemption after 5 years does not make the exemption something other than a 5-year exemption. The Department's action on the request to renew the exemption would be subject to public review and challenge by persons whose substantial interests are affected.

H. The Waiver

176. The proposed waiver order would excuse IP from compliance with the criteria in Florida Administrative Code Rule 62-660.300(1)(a)3. and 4., which require that public access and recreation be restricted in the area covered by the exemption for experimental use of wetlands. Without the waiver, the public would have to be excluded from Tee and Wicker Lakes.
177. Section 120.542, Florida Statutes, requires a showing by the person seeking the waiver that the purpose of the underlying statute will be achieved by other means and the application of a rule would create a substantial hardship or would violate principles of fairness. Petitioners contend that IP failed to demonstrate substantial hardship. However, Petitioners do not want public access to Tee and Wicker Lakes restricted. The sole reason for their objection to the proposed waiver is apparently to thwart the issuance of the exemption.
178. Section 120.542, Florida Statutes, defines "substantial hardship" as a demonstrated economic, technological, legal, or other type of hardship to the person requesting the waiver. In the proposed waiver order, the Department identifies IP's hardship as the possibility that denial of the waiver could result in denial of IP's NPDES permit and closure of the mill. The proposed waiver order then describes the number of jobs and other economic benefits of the mill that would be lost if the mill were closed.
179. As discussed in the Conclusions of Law below, the Department's interpretation of Section 120.542, Florida Statutes, to accept a demonstration of hardship that is associated with denial of the waiver is mistaken. The statute requires that the hardship arise from the application of the rule. In this case, IP must demonstrate that it would suffer substantial hardship if it were required to restrict public access and recreation on Tee and Wicker Lakes.
180. Petitioners claimed that IP has no authority to restrict the public from gaining access to Tee and Wicker Lakes because those are public waterbodies which the public has a right to enter and use. A substantial legal hardship for IP in complying with the exemption rule, therefore, is that compliance is impossible.

I. The Consent Order

1. Compliance Schedule

181. Subsections 403.088(2)(d) and (e), Florida Statutes, provide that no permit shall be issued unless a reasonable schedule for constructing, installing, or placing into operation of an approved pollution abatement facility or alternative waste disposal system is in place. Petitioners claim the time

schedules for compliance are not reasonable. Petitioners presented no competent evidence, however, that the WWTP upgrades, pipeline construction, and other activities required by the proposed permit can be accomplished in a shorter period of time.

182. One recurring theme in the Petitioners' case was that the adverse impacts associated with the continued discharge to Elevenmile Creek should not be allowed to continue, even for an interim period associated with construction of the WWTP upgrades and effluent pipeline. However, Petitioners also advocated the relocation of the discharge to the Escambia River, or to a "constructed wetlands." Both of these alternatives would have required a transition period during which the discharge to Elevenmile Creek would likely have continued. Furthermore, the Consent Order imposes interim limits on the discharge to Elevenmile Creek that would apply immediately upon issuance of the proposed permit. Although altered by the mill's effluent discharge, Elevenmile Creek is now a relatively stable biological system. The proposed permit would effectuate some improvement in the creek and Perdido Bay even during the construction phase.

2. Contingency Plan

183. The proposed Consent Order includes a contingency plan in the event that the NSAI monitoring analysis shows adverse impacts to the biological community within the wetland tract. The plan provides for alternative responses including relocating all or part of the wetland discharge to Elevenmile Creek. Petitioners object to the plan, primarily because they contend it is vague.
184. The provisions in the contingency plan for relocating all or part of the discharge from the wetland tract to Elevenmile Creek, appear to reflect a presumption that the negatives associated with continued discharge to the wetlands would outweigh the negatives associated with returning the discharge to Elevenmile Creek. However, it is not difficult to imagine scenarios where the harm to the biological community of the wetland tract is small in relationship to the harm to the biological community that might have reestablished itself in Elevenmile Creek.
185. Because the selection of an alternative under the contingency plan requires the consideration of data and analyses associated with future events, it is impossible to know at this time whether future action taken by the Department and IP pursuant to the contingency plan would be reasonable.
186. If the contingency plan is intended by the Department and IP to authorize future action when circumstances described in the plan are present, then the plan is too vague. On the other hand, there is adequate detail in the plan if the purpose of the plan is merely to establish a framework for future decision-making that would be subject to permit modification, public review and challenge. Clarification is needed.

3. Penalties

187. Petitioners complained that the stipulated of $500 per day for violations of the proposed Consent Order is too small to provide a deterrent to a company of the size of IP. Petitioners are correct, but did not present evidence to show what size penalty would be appropriate.

CONCLUSIONS OF LAW

188. DOAH has jurisdiction over the parties and subject matter of this proceeding pursuant to Sections 120.569 and 120.57(1), Florida Statutes.
189. The Department has authority to issue NPDES permits under Federal delegation pursuant to Section 403.0885(2), Florida Statutes. The NPDES program is implemented through rules found in Florida Administrative Code Chapters 62-4, 62-302, 62-620, 62-650, and 62-660, but primarily in Chapter 62-620.

STANDING

190. In order to establish their standing, the Lanes must establish that they have a substantial interest that would be affected by the proposed agency action. § 120.569(1), Fla. Stat.
191. For organizational standing, FOPB must prove that a substantial number of its members, but not necessarily a majority, have a substantial interest that would be affected; that the subject matter of the proposed project is within the general scope of the interests and activities for which the organization was created; and that the relief requested is of the type appropriate for the organization to receive on behalf of its members. See Florida League of Cities, Inc. v. Department of Environmental Regulation, 603 So. 2d 1363 (Fla. 1992); Friends of the Everglades, Inc. v. Board of Trustees of the Internal Improvement Trust Fund, 595 So. 2d 186 (Fla. 1st DCA 1992).
192. IP contends that Petitioners lack standing because they failed to demonstrate that the Department authorizations would result in substantial harm to the waters of the State, or that Petitioners will suffer any injury in fact as a result of the Department authorizations. However, Petitioners presented evidence of their interest in fishing, swimming, boating, and making other uses of the waters into which IP's wastewater effluent flows (primarily Perdido Bay), and there is no dispute that these waters are affected by IP's effluent. Therefore, Petitioners proved their standing in this case.
193. Petitioners' standing is not dependent on proving their claim that the proposed permit and related authorizations would cause substantial adverse impacts to water quality and natural resources. Standing and the merits of a claim are different concepts. See, for example, Village Park Mobile Home Ass'n., Inc. v. State Dept. of Business Regulation, 506 So. 2d 426, 433 (Fla. 1st DCA 1987); St. Martin's Episcopal Church v. Prudential-Bache Securities, 613 So. 2d 108, 109, n. 4 (Fla. 4th DCA 1993). If standing was based on whether a claim was proved, every losing petitioner would lack standing.

BURDEN AND STANDARD OF PROOF

194. As the applicant for the permit and the related Department authorizations, IP has the ultimate burden of proving its entitlement by a preponderance of the evidence. Dept. of Transp. v. J.W.C. Co., Inc., 396 So. 2d 778 (Fla. 1st DCA 1981).
195. Florida Administrative Rule 62-4.070(1) states that a permit shall be issued only if the applicant affirmatively provides the Department with reasonable assurance based on plans, test results, installation of pollution control equipment, or other information, that the construction, expansion, modification, operation, or activity of the installation will not discharge, emit, or cause pollution in contravention of Department standards or rules.
196. "Reasonable assurance," in this context means a demonstration that there is a substantial likelihood of compliance with standards, or "a substantial likelihood that the project will be successfully implemented." Metropolitan Dade County, v. Coscan Florida, Inc., 609 So. 2d 644, 648 (Fla. 3d DCA 1992). See also City of Newberry v. Watson Constr. Co., 19 F.A.L.R. 2067, 2080 (DER 1996). It does not mean absolute guarantees. Save our Suwannee v. Florida Dep't of Envtl. Protection and Piechocki, 18 F.A.L.R. 1467, 1472 (DEP 1996).
197. Competent substantial evidence based upon detailed site plans and engineering studies, coupled with credible expert engineering testimony is a sufficient basis for a finding of reasonable assurances. Hamilton County Board of County Commissioners v. FDEP, 587 So. 2d 1378 (Fla. 1st DCA 1991).
198. Reasonable assurance standard requires the applicant to address "reasonably foreseeable contingencies" in establishing entitlement. See Putnam County Envtl. Council, Inc. v. Florida Dep't of Envtl. Protection and Georgia-Pacific Corp., 24 F.A.L.R. 4674, RO at 4714, (Fla. DEP 2002), app. den., Case Nos. 1D02-3673 and 1D02-3674 (Fla. 1st DCA, Nov. 26, 2003).

199. The applicant bears the ultimate burden of providing reasonable assurance that all applicable criteria and standards will be met. J.W.C. Co., supra. If the applicant presents prima facie evidence of entitlement to the permit, the burden shifts to the objecting party to present "contrary evidence of equivalent quality." Id. at 789. This burden cannot be satisfied with speculative concerns about potential or possible adverse environmental effects. See Rowe v. Oleander Power Project, L.P., 22 F.A.L.R. 1173, 1185 (DEP 1999); Chipola Basin Protective Group, Inc. v. Fla. Chapter Sierra Club, 11 F.A.L.R. 467, 481 (DER 1988); J.T. McCormick v. City of Jacksonville, 12 F.A.L.R. 960, 971 (DER 1990).

THE PROPOSED PERMIT

200. IP's water quality modeling indicated that the discharge allowed under the proposed permit would cause a 0.1 mg/L decrease in DO levels in Perdido Bay under circumstances of maximum BOD loading and low flow conditions. Because the DO levels in Perdido Bay sometimes fall below the Class III water quality standard of 5.0 mg/L, Petitioners assert that it is wrong to allow IP to contribute to the existing violation of the water quality standard for DO.
201. The Amendment to the Fact Sheet that accompanies the proposed permit states: Although Elevenmile Creek and Perdido Bay are not expected to meet the applicable Class III DO criteria with or without the [IP] discharge, the modeling indicates the [IP] discharge will not cause DO levels to be significantly decreased below background DO levels. As such, the [IP] discharge at the proposed permitted BOD_5 level is not expected to cause or contribute to a violation of the DO standard.
202. The salinity stratification and sediment oxygen demand in Perdido Bay, cause DO levels in the bottom water layer of Perdido Bay to fall below the Class III water standard of 5.0 mg/L even without the mill's discharge. The modeled decrease in DO of 0.1 mg/L that would be caused by the mill's effluent under unusual conditions is an insignificant impact and does not prevent approval of the proposed permit. See Pacetti v. Florida Dep't of Envtl. Regulation, DOAH Case Nos. 84-3810 and 84-3811 (DER FO, April 18, 1986) (a lowering of DO by 0.1 mg/L in waters that are naturally below the DO standard is a de minimus impact and does not prevent issuance of the permit).
203. IP provided reasonable assurance that the proposed permit would not result in a discharge of pollution in contravention of Department standards or rules, except as specifically noted below.
204. IP did not provide reasonable assurance that the proposed permit would not cause the Perdido River to be significantly degraded.
205. IP did not provide reasonable assurance that the mill's effluent would be assimilated so as not to cause significant adverse impact on the biological community within the wetland tract.
206. Section 403.088(1)(e), Florida Statutes, sets forth special conditions that must be met to obtain an operation permit for a discharge that will not meet permit conditions or applicable statutes and rules. The statute requires an applicant to show, among other things, that issuing the permit will be in the public interest and the discharge will not be unreasonably destructive to the quality of the receiving waters. IP did not make an adequate showing on these two criteria.
207. Because IP did not make a sufficient demonstration that the discharge to the wetlands would not cause significant adverse impact to the biological community within the wetland tract, and IP did not make a sufficient showing that the discharge would not significantly degrade the Perdido River OFW, it cannot be determined that issuance of the proposed permit would be in the public interest. Furthermore, a discharge is unreasonably destructive if its destructive effect can be reduced or eliminated by reasonable means. Because IP did not make a sufficient showing with regard to the potential adverse impact of the proposed discharge on the biological community within the wetland tract, it is impossible to know whether any destructive effects could be reduced or eliminated by reasonable[28] means. Therefore, IP did not meet the special permit conditions required by Section 403.088(1)(e), Florida Statutes.
208. Because IP did not provide reasonable assurance that the proposed permit would not result in a discharge of pollution in contravention of Department standards or rules, Florida Administrative Code Rule 62-4.070(2) requires the Department to deny the permit.

THE EXEMPTION

209. Petitioners argue that it was improper to include Tee and Wicker Lakes in the exemption because they are not wetlands. Although Florida Administrative Code Rule 62-660.300(1) is entitled "Exemptions to Provide for the Experimental Use of Wetlands for Low-Energy Water and Wastewater Recycling," the rule text refers to "wetlands and freshwaters," "waters affected," "receiving waters," and "waters." The purpose of the rule and the use of terms other than "wetlands" provide support for the Department's interpretation of the rule to allow associated lakes or ponds to be included within the area of the exemption when, as in this case, the lakes or ponds are relatively small and located within a much larger wetland system.
210. However, IP failed to demonstrate that granting the exemption would not cause significant adverse impact on the biological community within the wetland tract as required by Florida Administrative Code Rule 62.660.300(1)(a). With regard to the freshwater area of the wetland tract, IP did not provide reasonable assurance that the specific conductance of the mill's effluent would not cause significant adverse impact to fish and other organisms. With regard to Tee and Wicker Lakes, IP does not have adequate baseline information about the biological community associated with the lakes to provide reasonable assurance that the proposed discharge would not cause significant adverse impact. Ongoing studies being conducted for the NSAI monitoring program would generate the needed information, but reasonable assurances must be based on current information. See Metropolitan Dade County v. Coscan Florida, Inc. 609 So. 2d 644 (Fla. 3d DCA 1992).
211. The Department intends to apply Florida Administrative Code Rule 62-660.300(1) in two stages, the first stage being the issuance of the proposed exemption order that "provides the opportunity to develop the information necessary to amend the water quality standards by Secretary Order," and the second stage being the issuance of a "permanent exemption with alternative criteria." The undersigned provided the parties an opportunity to supplement their PROs with argument on the validity of the Department's application of the exemption rule in this bifurcated manner. The Department's interpretation of the rule, so that permanent alternative criteria would not be established until the results of the experiment (i.e., actual use of the wetland tract to assimilate the mill's effluent) are known, is reasonable.

THE WAIVER

212. The proposed Waiver Order is based on an erroneous analysis of the substantial hardship showing required by Section 120.542, Florida Statutes. The analysis focused on the consequences to IP of the denial of the waiver, which was assumed to be closure of the mill. However, the statute requires that the hardship arise from the application of the rule that the petitioner seeks to have waived. In this case, IP must demonstrate that it would suffer substantial hardship if it were required to restrict public access and recreation on Tee and Wicker Lakes.
213. The hardship that must be shown to qualify for a waiver should be commensurate with the potential environmental harm or other negative consequences of granting the waiver. The record in this case does not show there would be public health problems or other negative consequences associated with the public's continued access to and use of Tee and Wicker Lakes. Granting the waiver preserves the public interest in public access and recreation on public waters. Granting the waiver would also be consistent with Petitioners' desire to maintain public access and recreation on Tee and Wicker Lakes. IP's legal hardship in complying with the public access and recreation criteria would be substantial because restricting public access to public waters would require the approval of the Board of Trustees of the Internal Improvement Trust Fund, the difficulty of which is widely known and was acknowledged by Petitioners. The record evidence that Tee and Wicker Lakes are public waters is sufficient to establish IP's substantial hardship for the purpose of the waiver.
214. However, it is recommended that IP's petition for exemption be denied and the waiver would serve no independent purpose. Therefore, IP's petition for waiver should also be denied.

THE CONSENT ORDER

215. IP provided reasonable assurance that the proposed Consent Order complies with all applicable Department statutes and rules, except as specifically noted below.
216. If an applicant for an operation permit involving a discharge that will not comply with all applicable statutes and rules is able to meet the special conditions of Section 403.088(1)(e), Florida Statutes, an order must be issued to accompany the permit which establishes a schedule for achieving compliance with all permit conditions. § 403.088(1)(f), Fla. Stat. That is the purpose of the proposed Consent Order. However, because IP did not comply with all the special conditions of Section 403.088(1)(e), Florida Statutes, the proposed Consent Order should not be approved.
217. The Contingency Plan for Management of the Combined Effluent Distribution Project Wetland Site, which is Attachment I to the proposed Consent Order, states that it is intended "to safeguard the biological integrity" of the wetland tract, including Tee and Wicker Lakes. Alternative management options are discussed in the contingency plan that would be implemented if the NSAI monitoring program results in determination that the biological integrity of the wetland tract has been adversely affected by the effluent. The contingency plans are described as "built-in," suggesting that they are pre-authorized and would not require new agency action. This interpretation is supported by a statement that the Department would be contacted for assistance if problems are encountered that are "not anticipated or accounted for in the contingency plan."
218. However, there is insufficient detail in the contingency plan to determine now that the relocation of 50% of the discharge to Elevenmile Creek, for example, would be an appropriate response to future circumstances. Any relocation of the discharge would have to include an assessment of whether the adverse impact to the biological community of the wetland tract is worse than the adverse impact that might occur to the biological community of Elevenmile Creek. The contingency plan should be amended to clarify what agency action, if any, is required to implement the management alternatives. Because the contingency plan appears now to pre-authorize future action without reasonable assurance that the future action would be appropriate, it creates another ground for disapproving the proposed Consent Order.
219. The $500 per day stipulated penalty provision of the proposed Consent Order is too low to be a deterrent to noncompliance or an incentive for quick correction of noncompliance. However, the record contains no evidence upon which to base a higher penalty figure. A stipulated penalty amount, such as $2500 per day, might be too large for noncompliance with some of the less important provisions of the proposed Consent Order. A penalty range such as "up to $10,000 per day" might grant too much discretion to the Department. The penalty provision of the proposed Consent Order should be revised to increase the stipulated penalty amount, but to also distinguish between major and minor problems associated with noncompliance.
220. Finally, to the extent that the proposed Consent Order incorporates or authorizes those specific provisions of the proposed permit and exemption that are determined in this Recommended Order to be in conflict with Department statutes and rules, the proposed Consent Order is also in conflict with Department statutes and rules.

SUMMARY

221. The more persuasive evidence presented at the final hearing strongly indicates that the proposed Department authorizations would likely effectuate a significant improvement to the Perdido Bay system over the current discharge to Elevenmile Creek. However, IP must demonstrate more than improvement. It must demonstrate compliance with all applicable Department standards and rules. Improving overall environmental conditions is in the public interest, but the statutes and rules that require consideration of the public interest also contain other criteria that make it impossible to ignore the areas where IP's showing was insufficient.

RECOMMENDATION

Based on the foregoing Findings of Fact and Conclusions of Law, it is RECOMMENDED that the Florida Department of Environmental Protection enter a final order:

0. 1. Denying proposed revised NPDES Permit Number FL0002526-001/001-IW1S;
1. 2. Disapproving revised Consent Order Number 04-1202;
2. 2. 3. Denying IP's petition for authorization for the experimental use of wetlands; and
4. Denying IP's petition for waiver.

DONE AND ENTERED this 11th day of May, 2007, in Tallahassee, Leon County, Florida.

Bram D.E. Canter Administrative Law Judge Division of Administrative Hearings The DeSoto Building 1230 Apalachee Parkway Tallahassee, Florida 32399-3060 (850) 488-9675 SUNCOM 278-9675 Fax Filing (850) 921-6847 www.doah.state.fl.us

Filed with the Clerk of the Division of Administrative Hearings this 11th day of May, 2007.

COPIES FURNISHED

Michael W. Sole, Secretary Department of Environmental Protection The Douglas Building, Mail Station 35 3900 Commonwealth Boulevard Tallahassee, Florida 32399-2400 W. Douglas Beason, Esquire Department of Environmental Protection The Douglas Building, Mail Station 35 3900 Commonwealth Boulevard Tallahassee, Florida 32399-3000.

Terry Cole, Esquire Oertel, Fernandez, Cole & Bryant, P.A. 301 South Bronough Street, Fifth Floor Post Office Box 1110 Tallahassee, Florida 32302-1110

Jacqueline M. Lane 10738 Lillian Highway Pensacola, Florida 32506

Howard K. Heims, Esquire Littman, Sherlock & Heims, P.A. Post Office Box 1197 Stuart, Florida 34995-1197

Marcy I. LaHart, Esquire 711 Talladega Street West Palm Beach, Florida 33405-1443

David K. Thulman, Esquire Department of Environmental Protection The Douglas Building, Mail Station 35 3900 Commonwealth Boulevard Tallahassee, Florida 32399-3000

Stacey D. Cowley, Esquire Department of Environmental Protection The Douglas Building, Mail Station 35 3900 Commonwealth Boulevard Tallahassee, Florida 32399-2400

Lea Crandall, Agency Clerk Department of Environmental Protection The Douglas Building, Mail Station 35 3900 Commonwealth Boulevard Tallahassee, Florida 32399-2400 Tom Beason, General Counsel Department of Environmental Protection The Douglas Building, Mail Station 35 3900 Commonwealth Boulevard Tallahassee, Florida 32399-2400

NOTICE OF RIGHT TO SUBMIT EXCEPTIONS

All parties have the right to submit written exceptions within 15 days from the date of this Recommended Order. Any exceptions to this Recommended Order should be filed with the agency that will issue the Final Order in this case.

ENDNOTES

1 Two exhibits were introduced as Petitioners' Exhibit 5, the curriculum vita of Mr. Sulkin and a similar document for Dr. White. They have been re-marked for the record, respectively, as Petitioners' Exhibits 5 and 5A.

2 All references to the Florida Statutes are to the 2006 codification, unless otherwise indicated.

3 FOPB's Articles of Incorporation (Petitioners' Exhibit 6) state that it is an Alabama corporation. James Lane testified that FOPB is "registered" in Florida.

4 Petitioners Jacqueline Lane and James Lane were among the challengers, but they dropped out of the case before the hearing based on their belief that DER's requirement for nutrient studies would lead to acceptable modifications to the mill's effluent that would eliminate its adverse effects on Perdido Bay.

5 The parties often referred to the wetland tract as the Rainwater Tract. However, because the wetland tract is only a portion of the entire Rainwater Tract, the latter name is not being used in this Recommended Order.

6 Because DO in the effluent is expected to be depleted during the effluent's 10-mile conveyance through the pipeline, there is a passive aeration system at the end of the pipeline.

7 The proposed permit allows for discharges to Elevenmile Creek from D-001 in an emergency situation, such as a pipeline break.

8 Remarkably, some Department witonesses whose testimony was presented by Petitioners had little or no familiarity with the data and analyses generated by the Livingston studies, even though their employment responsibilities require them to be knowledgeable about the biology and water quality of the Perdido Bay system.

9 Dr. Lane appeared to adopt the opinion that salinity stratification is the major cause of hypoxia in her examination of Dr. Stephen Schropp. See Transcript at 2180.

10 For example, Dr. Livingston found no infauna (organisms living in the sediments) in Wolf Bay, which is southwest of, but connected to Perdido Bay, and which indicates that significant adverse impacts are being caused by sources other than the mill.

11 Although Champion and IP funded the Livingston studies, they had no say in the study results that were published. Dr. Livingston was never paid a salary or fee for his work on the studies. IP paid him for his time as an expert witness at the hearing.

12 In her cross-examination, Dr. Lane asked Mr. Gallagher to relate the results of the Coffin and Cifuentes study, and Petitioners' Exhibit 59 includes a critique of the study. However, the failure of a party to object to hearsay evidence does not convert hearsay evidence into non-hearsay evidence.

13 Petitioner's own witness, Dr. Richard Weikowitz, said the long-term BOD assumption in the Mr. Gallagher's model was reasonable.

14 Dr. Livingston said, "Everywhere I've gone, I've found that BOD has been overrated as a variable that's of importance to these systems."

15 Petitioners presented the testimony of Glen Butts, a marine biologist with the Department, but he disagrees with Petitioners' theory that the problem in Perdido Bay is organic loading. He thinks the problem is toxicity.

16 Mr. Ray testified that mercury concentrations were high, but another witness for Petitioners, Dr. Steven Schropp, testified that mercury concentrations were low.

17 The study that was introduced into evidence as DEP Exhibit 38 is not dated. However, there is information in the study that indicates it was completed in 2001 or later.

18 The hearsay study concludes that the mutagenic compounds in the effluent are probably not chlorinated compounds, so the study does not support or explain Dr. Livingston's tissue analysis which implicated chlorine compounds.

19 In their petitions, Petitioners cited Florida Administrative Code Rule 62.302.500, which prohibits discharges that cause mutagenic effects. However, the same rule also prohibits discharges that cause nuisance conditions such as scum and odor, and Petitioners specifically alleged that IP's discharge was causing nuisance scum and odor.

20 The OFW rule's contemplation of a discharge that would significantly degrade an OFW but not lower ambient water quality is just one of several ambiguities in the rule.

21 There was also no evidence presented that 5100 ppd for BOD was a permit limit in 1979. The record does not identify any permit issued prior to Champion's 1982 permit.

22 For example, the mill has a permit limit of 4500 ppd, but the long-term average BOD loading has only been 2,544 ppd.

23 Curiously, Petitioners repeatedly pointed out that IP would be deviating from the Class III water quality standards for pH, transparency, and DO. Deviating from existing standards is the idea behind the exemption.

24 Water samples taken in the wetland tract showed specific conductance of 37.2 micromhos/cm in one place and 60 micromhos/cm in another. Mr. Ray's samples from "little creeks" in the wetland tract showed "about" 100 micromhos/cm.

25 In its antidegradation analysis, IP states that the ECUA wastewater would have a diluting effect that could reduce by 20% the specific conductance in the combined discharge to the wetland tract.

26 Aerial photography of the wetland track shows some stream-like features below berm 1. Petitioners contend these are streams, but the more persuasive evidence indicates they are firebreaks that were cut in 2001 by Florida Division of Forestry personnel to control a fire in the area.

27 IP stated that the Cantonment mill employs 650 people and spends millions annually in Escambia County for goods and services and these public benefits would terminate if the mill closed down. However, this worse case scenario for IP was not correlated with any particular set of regulatory controls. IP did not show that it was impossible to meet water quality standards except by closing down the mill.

28 Dilution of the effluent, for example, might be a reasonable means to reduce or avoid the problems associated with the specific conductance of the discharge.

REFERENCES

Abood, K.A. and S.G. Metzger. 1996. Comparing impacts to shallow-water habitats through time and space. *Estuaries* 19: 220–228.

Adams, D.A., J.S. O'Conner, and S.B. Weisberg. 1998. Sediment quality of the NY/NJ harbor system—An investigation under the Regional Environmental Monitoring and Assessment Program (REMAP). US Environmental Protections Agency, Division of Environmental Science and Assessment, Region 2. Edison, NJ. Final report, EPA/902-R-98-001, 255pp.

Adams, S.M. 1976. Feeding ecology of eelgrass fish communities. *Trans. Am. Fish. Soc.* 105: 514–519.

Adler, R.W. 1996. Model watershed protection. *Natl. Wetlands Newslet.* 1996: 7–12.

Alabama Water Improvement Commission. 1979. Agriculture runoff management plan. Montgomery, AL, 223pp.

Alexander, C., R. Smith, B. Loganathan, J. Ertel, H.L. Windom, and R.F. Lee. 1999. Pollution history of the Savannah River Estuary and comparisons with Baltic Sea pollution history. *Limnologica* 29: 267–273.

Allan, R.P. and J.S. Soden. 2008. Atmospheric warming and the amplification of precipitation extremes. *Science* 2008: 1481–1484.

Ambrose, R.B., Jr. and T.O. Barnwell. 1989. Environmental software at the U.S. Environmental Protection Agency's Center for exposure assessment modeling. *Environ. Software* 4: 76–93.

Ambrose, R.B., Jr., S.B. Vandergriff, and I.A. Wool. 1996. WASP3, a hydrodynamic and water quality model—Model theory, user's manual, and programmers guide. USEPA Publication EPA/600/3-86/034, Athens, GA, 379pp.

American Public Health Association (APHA). 1989. *Standard Methods for the Examination of Water and Wastewater*, 17th edn. American Public Health Association, Washington, DC, 1268pp.

Anderson, D.A. and D.J. Garrison. 1997. The ecology and oceanography of harmful algal blooms. *Limnol. Oceanogr.* 42: 1–1305.

Anderson, D.M. 1996. Control and mitigation of red tides. Manuscript prepared for Project Start (Solutions to Avoid Red Tide). Sarasota, FL, 59pp, December 4.

Anderson, D.M. 1997a. Bloom dynamics of toxic *Alexandrium* species in the northeastern U.S. *Limnol. Oceanogr.* 42: 1009–1022.

Anderson, D.M. 1997b. Physiology and bloom dynamics of toxic *Alexandrium* species, with emphasis on life cycle transitions. In: D.M. Anderson, A.D. Cembella, and G.M. Hallagraeff (Eds.), *Physiological Ecology of Harmful Algal Blooms*. Springer-Verlag, Heidelberg, Germany, pp. 13–28.

Andersson, L. and L. Rudberg. 1988. Trends in nutrient and oxygen conditions within the Kattegar: Effects on local nutrient supply. *Estuarine Coastal Shelf Sci.* 26: 559–579.

Appleton, E. 1995. A cross-media approach to saving the Chesapeake Bay. *Environ. Sci. Technol.* 29: 550–555.

Archer, C.L. and K. Caldeira. 2008. Historical trends in the jet streams. Department of Global Ecology, Carnegie Institution of Washington, Stanford, California. *Geophys. Res. Lett.* 35: L0880, 6pp.

Arditi, R. and L.R. Ginzberg. 1989. Coupling in predator-prey dynamics: Ratio-dependence. *J. Theor. Biol.* 139: 311–326.

Armstrong, N.E. 1982. Responses of Texas estuaries to freshwater inflows. In: V.S. Kennedy (Ed.), *Estuarine Comparisons*. Academic Press, New York, pp. 103–120.

Babcock, S. 1976. Lake Jackson studies, annual report for research project. Florida Game and Freshwater Fish Commission, Tallahassee, FL, 120pp.

Bacchus, S.T. 2002. The "Ostrich" component of the multiple stressor model: Undermining South Florida. In: J.W. Porter and K.G. Porter (Eds.), *The Everglades, Florida Bay and Coral Reefs of the Florida Keys: An Ecosystem Sourcebook*. CRC Press, Boca Raton, FL, pp. 677–748.

Badylak, S. and E.J. Phlips. 2004. Spatial and temporal patterns of phytoplankton composition in subtropical coastal lagoon, the Indian River Lagoon, Florida, USA. *J. Plankton Res.* 26: 1229–1247.

Bahr, L.M. and W.P. Lanier. 1981. The ecology of intertidal oyster reefs of the South Atlantic coast: A community profile. US Fish and Wildlife Service, Office of Biological Services, Washington, DC. FWS/OBS-81/15, 105pp.

Baird, D. and R.E. Ulanowicz. 1989. The seasonal dynamics of the Chesapeake Bay ecosystem. *Ecol. Monogr.* 59: 329–364.

Baird, R.C. 1996. Toward new paradigms in coastal resource management: Linkages and institutional effectiveness. *Estuaries* 19: 320–335.

Balls, P.W., A. MacDonald, K. Pugh, and A.C. Edwards. 1995. Long-term nutrient enrichment of an estuarine system: Ythan, Scotland (1958–1993). *Environ. Pollut.* 90: 311–321.

Barbour, M.T., J. Gerritsen, and J.S. White. 1996. *Development of the Stream Condition Index (SCI) for Florida*, Vols. I and II. Florida Department of Environmental Protection, Tallahassee, FL, 36pp.

Barnett, E., J. Lewis, J. Marx, and D. Trimble (Eds.). 1995. *Ecosystem Management Implementation Strategy*. Florida Department of Environmental Protection, Tallahassee, FL.

Barry, J.P., M.M. Yorklavich, G.M. Cailliet, D.A. Ambrose, and B.S. Antrim. 1996. Trophic ecology of the dominant fishes in Elkhorne Slough, California, 1974–1980. *Estuaries* 19: 115–138.

Bass, D.G., Jr. 1990. Stability and persistence of fish assemblages in the Escambia River, Florida. *Rivers* 1: 296–306.

Bass, D.G., Jr. 1991. Riverine fishes of Florida. In: R.J. Livingston (Ed.), *The Rivers of Florida*. Springer-Verlag, New York, pp. 65–84.

Bass, D.G., Jr. and V. Hitt. 1977. Ecology of the backwater river system, Florida. Northwest streams research project. Florida Fish and Wildlife Conservation Commission, Tallahassee, FL, 348pp.

Bates, S.S., C.J. Bird, A.S.W. Frietas, R. de Foxall, and M. Gilgan. 1989. Pennate diatom *Nitzschia pungens* a the primary source of domoic acid, a toxin in shellfish from eastern Prince Edward Island, Canada. *Can. J. Fish. Aquat. Sci.* 48: 1203–1215.

Bay Area Resource Council. 1988. Environment of the Pensacola Bay System. Pensacola, FL, 28pp. (unpublished report).

Bearon, R.N., D. Grunbaum, and R.A. Cattolico. 2006. Effects of salinity structure on swimming behavior and harmful algal bloom formation in *Heterosigma akashiwo*, a toxic raphidophyte. *Mar. Ecol. Prog. Ser.* 306: 153–163.

Beaugrand, G. and P.C. Reid. 2003. Long-term changes in phytoplankton, zooplankton and salmon related to climate. *Global Change Biol.* 9: 801–817.

Becker, M.E. 1996. The effect of nutrient loading on benthic microalgal biomass and taxonomic composition. MS thesis, University of North Carolina at Wilmington, Wilmington, NC.

Beers, J.R. 1978. Pump sampling. In: A. Sournia (Ed.), *Phytoplankton Manual*, Vol. 6, Monographs on Oceanographic Methodology. UNESCO Press, Paris, France, pp. 41–49.

Bell, G.R. 1961. Penetration of spines from a marine diatom into the gill tissue of lingcod. *Nature* 192: 279–280.

Bell, T. and J. Kalff. 2001. The contribution of picophytoplankton in marine and freshwater systems of different trophic status and depth. *Limnol. Oceanogr.* 46: 1243–1248.

Belton, T.J., R. Hazen, B.E. Ruppel, K. Lockwood, R. Mueller, E. Stevenson, and J.J. Post. 1985. A study of dioxin (2,3,7,8-tetrachlorodibenzo-*p*-dioxin) contamination in select finfish, crustaceans, and sediments of New Jersey waterways. Office of Science and Research, New Jersey Department of Environmental Protection, CN-409, Trenton, NJ (unpublished report).

Bergqusit, A.M. and S.R. Carpenter. 1986. Limnetic herbivory: Effects on phytoplankton populations and primary production. *Ecology* 67(5): 1351–1360.

Bernhard, A.E. and E.R. Peele. 1997. Nitrogen limitation of phytoplankton in a shallow embayment in northern Puget Sound. *Estuaries* 20: 759–769.

Berrigan, M. 1990. Biological and economical assessment of an oyster resource development project in Apalachicola Bay, Florida. *J. Shellfish Res.* 9: 149–158.

Bevis, T.H. 1995. The response of fish assemblages to stormwater discharge into a north Florida solution lake. MS thesis, Department of Biological Sciences and Center for Aquatic Research and Resource Management, Florida State University, Tallahassee, FL.

Bidagare, R.R. and M.E. Ondrusek. 1996. Spatial and temporal variability of phytoplankton pigment distributions in the central equatorial Pacific Ocean. *Deep Sea Res.* 43: 809–833.

Biere, A., F. Rienks, L.S. Galdon, and K. Soetaert (Eds.). 1999. Progress report 1998. Netherlands Institute of Ecology, Nieuwersluis, the Netherlands, 80pp.

Biggs, R.B., T.B. Demos, M.M. Carter, and E.L. Beasley. 1989. Susceptibility of U.S. estuaries to pollution. *Rev. Aquat. Sci.* 1: 189–207.

Biggs, R.B. and D.A. Flemer. 1972. The flux of particulate carbon in an estuary. *Mar. Biol.* 12: 11–17.

Birkholz, D.A. 1999. Chemical analysis report. Enviro-Test Laboratories, Edmonton, Alberta, Canada.

Bishop, S.S., K.A. Emmanuele, and J.A. Yoder. 1984. Nutrient limitation of phytoplankton growth in Georgia nearshore waters. *Estuaries* 7: 506–512.

Blanchet, R.H. 1979. The distribution and abundance of ichthyoplankton in the Apalachicola Bay, Florida area. MS thesis, Florida State University, Tallahassee, FL.

Bloesch, J. and H. Burgi. 1989. Changes in phytoplankton and zooplankton biomass and taxonomic composition reflected by sedimentation. *Limnol. Oceanogr.* 34: 1048–1061.

Blumberg, A.F. and G.L. Mellor. 1980. A coastal ocean numerical model. In: J. Sundermann and K.P. Holz (Eds.), *Mathematical Modeling of Estuarine Physics, Proceedings of the International Symposium*, Hamburg, Germany, August 24–26, 1978. Springer-Verlag, Berlin, Germany, pp. 203–214.

Blumberg, A.F. and G.L. Mellor. 1987. A description of a three-dimensional coastal ocean circulation model. In: N.S. Heaps (Ed.), *Three-Dimensional Coastal Ocean Models*. American Geophysical Union, Washington, DC, pp. 1–16.

Bobbie, R.J., S.J. Morrison, and D.C. White. 1978. Effects of substrate biodegradability on the mass and activity of the associated estuarine microbiota. *Appl. Environ. Microbiol.* 35: 179–184.

Boesch, D.F., N.E. Armstrong, C.F. D'Elia, N.G. Maynard, H.W. Paerl, and S.L. Williams. 1993. Deterioration of the Florida Bay ecosystem: An evaluation of the scientific evidence. Report to the Interagency Working Group on Florida Bay. Everglades National Park, Homestead, FL. www.aoml.noaa.gov/flbay (unpublished report, 1999).

Bohnsack, J.A. 1983. Resilience of reef fish communities in the Florida Keys following a January 1977 hypothermal fish kill. *Environ. Biol. Fish.* 9: 41–53.

Bolgiano, R.W. 1980. Mercury contamination of the floodplains of the South and South Fork Shenandoah Rivers. State Water Control Board, Richmond, VA. *Basic Data Bull.* 47: 75pp.

Bologna, P., R. Lathrop, P. Bowers, and K. Able. 2000. Assessment of submerged aquatic vegetation in Little Egg Harbor, New Jersey. Institute of Marine and Coastal Sciences, Rutgers University, New Brunswick, NJ. Technical report 2000–2011, 30pp.

Bonin, D.J., M.R. Droop, S.Y. Maestrini, and M.C. Bonin. 1986. Physiological features of six microalgae to be used as indicators of seawater quality. *Cryptogram. Algol.* 7: 23–83.

Bopp, R.F., S.N. Chillrud, E.L. Shuster, and H.J. Simpson. 1998. Trends in chlorinated hydrocarbon levels in Hudson River basin sediments. *Environ. Health Perspect.* 106: 1075–1081.

Bopp, R.F., M.L. Gross, H. Tong, H.J. Simpson, S.J. Monson, B.L. Deck, and F.C. Moser. 1991. A major incident of dioxin contamination: Sediments of New Jersey estuaries. *Environ. Sci. Technol.* 25: 951–956.

Borja, A., J. Bald, J. Franco, J. Larreta, I. Muxika, M. Revilla, J.G. Rodríguez, O. Solaun, A. Uriarte, and V. Valencia. 2009. Using multiple ecosystem components, in assessing ecological status in Spanish (Basque Country) Atlantic marine waters. *Mar. Pollut. Bull.* 59: 54–64.

Borja, A., D.M. Dauer, M. Elliott, and C.A. Simenstad. 2010. Medium- and long-term recovery of estuarine and coastal ecosystems: Patterns, rates and restoration effectiveness. *Estuaries Coasts* 33: 1249–1260.

Borsuk, M.E., C.A. Stow, and K.H. Reckhow. 2004. Confounding effect of flow on estuarine response to nitrogen loading. *J. Environ. Eng.* 130: 605–614.

Bortone, S. 1991. Seagrass mapping of Perdido Bay. Florida Department of Environmental Protection, Tallahassee, FL (unpublished report).

Boschen, C.J. 1996. The feeding ecology of the dominant fishes that occur in a eutrophicated north Florida solution lake. MS thesis, Department of Biological Sciences and Center for Aquatic Research and Resource Management, Florida State University, Tallahassee, FL.

Botton, M.L. 1979. Effects of sewage sludge on the benthic invertebrate community of the inshore New York Bight. *Estuarine Coastal Shelf Sci.* 8: 169–180.

Box, G.E.P. and G.M. Jenkins. 1978. *Time Series Analysis: Forecasting and Control*. Holden-Day, Inc., San Francisco, CA, 575pp.

Boyer, J.N., J.W. Fourqurean, and R.D. Jones. 1999. Seasonal and long-term trends in the water quality of Florida Bay. In: J.W. Fourqurean, M.B. Robblee, and L.A. Deegan (Eds.), Florida Bay: A Dynamic Subtropical Estuary. *Estuaries* 22: 417–430.

Boyer, J.N. and R.D. Jones. 2002. A view from the bridge: External and internal forces affecting the ambient water quality of the Florida Keys National Marine Sanctuary (FKNMS). In: J.W. Porter and K.G. Porter (Eds.), *The Everglades, Florida Bay and Coral Reefs of the Florida Keys: An Ecosystem Sourcebook*. CRC Press, Boca Raton, FL, pp. 609–628.

Boynton, W.R. 1997. Estuarine ecosystem issues of the Chesapeake Bay. In: R.D. Simpson and N.J.L. Christensen, Jr. (Eds.), *Ecosystem Function and Human Activities*. Chapman and Hall Publishers, New York, pp. 71–93.

Boynton, W.R., J.H. Garber, R. Summers, and W.M. Kemp. 1995. Inputs, transformations, and transport of nitrogen and phosphorus in Chesapeake Bay and selected tributaries. *Estuaries* 18: 285–314.

Boynton, W.R., W.M. Kemp, and C.W. Keefe. 1982. A comparative analysis of nutrients and other factors influencing estuarine phytoplankton production. In: V.S. Kennedy (Ed.), *Estuarine Comparisons*. Academic Press, New York, pp. 69–90.

Boynton, W.R., W.M. Kemp, and L.G. Osborne. 1980. Nutrient fluxes across the sediment water interface in the turbid zone of a coastal plain estuary. In: V.S. Kennedy (Ed.), *Estuarine Perspectives*. Academic Press, New York, pp. 93–109.

Brady, K. 1982. Larval fishes of Apalachee Bay. MS thesis, Department of Biological Sciences, Florida State University, Tallahassee, FL.

Brand, L.E. 2000. An evaluation of the scientific basis for "restoring" Florida Bay by increasing freshwater runoff from the Everglades. Reef Relief, Key West, FL.

Brand, L.E. 2002. The transport of terrestrial nutrients to South Florida coastal waters. In: J.W. Porter and K.G. Porter (Eds.), *The Everglades, Florida Bay and Coral Reefs of the Florida Keys: An Ecosystem Sourcebook*. CRC Press, Boca Raton, FL, pp. 361–413.

Brank, C.W.C. and P.A. Senna. 1994. Factors influencing the development of *Cylindrospermopsis raciborskii* and *Microcystis aeruginosa* in the Paranoa Reservoir, Brasilia, Brazil. *Algolog. Stud.* 75: 85–96.

Breitburg, D.L. 1990. Near-shore hypoxia in the Chesapeake Bay: Patterns and relationships among physical factors. *Estuarine Coastal Shelf Sci.* 30: 593–609.

Breitburg, D.L. 1992. Episodic hypoxia in Chesapeake Bay: Interacting effects of recruitment, behavior, and physical disturbance. *Ecol. Monogr.* 62: 525–546.

Bricelj, V.M. and S. Kuenstner. 1989. Effects of the "brown tide" on the feeding physiology and growth of juvenile and adult bay scallops and mussels. In: E.M. Cosper, E.J. Carpenter, and V.M. Bricelj (Eds.), *Novel Phytoplankton Blooms: Causes and Impacts of Recurrent Brown Tides and Other Unusual Blooms*. Lecture Notes on Coastal and Estuarine Studies. Springer-Verlag, Berlin, Germany, pp. 491–509.

Bricker, S.B., C.G. Clement, D.E. Pirhalla, S.P. Orlando, and D.R.G. Farrow. 1999. National estuarine eutrophication assessment. Effects of nutrient enrichment in the nation's estuaries. National Oceanic and Atmospheric Administration, National Ocean Service, Special Projects Office and the National Centers for Coastal Ocean Science, Silver Spring, MD, 71pp.

Brim, M.S. 1993. Toxics characterization report for Perdido Bay, Alabama, and Florida. US Fish and Wildlife Service Publications PCFO-EC-93-94, Washington, DC, 137pp.

Brinson, S.T. and J.M. Keltner, Jr. 1981. Water quality and biological assessment of Bayou Chico, Florida: A two-part study. Florida Department of Environmental Regulation, Pensacola, FL (unpublished report).

Brockman, U.H., P.M. Laane, and H. Postma. 1990. Cycling of nutrient elements in the North Sea. *Neth. J. Sea Res.* 26: 239–264.

Brook, I.M. 1976. Trophic relationships in a *Thalassia* community: Fish diets in relation to abundance and biomass. In: *Thirty-Ninth Annual Meeting*. American Society of Limnology and Oceanography, Savannah, GA.

Brook, I.M. 1977. Trophic relationships in a seagrass community (*Thalassia testudinum*), in Card Sound, Florida-fish diets in relation to macrobenthic and cryptic faunal abundance. *Trans. Am. Fish. Soc.* 106: 219–229.

Brouard, D., C. Demers, R. Lalumiere, R. Schetagne, and R. Verdon. 1990. Evolution of mercury levels in fish of the La Grande Hydroelectric Complex. Hydro Quebec (Montreal) and Schooner, Inc. (Quebec), Canada. Summary report, 97pp.

Brown, L.C. and T.O. Barnwell, Jr. 1987. The enhanced stream water quality models QUAL2E and QUAL2E-UNCAS. USEPA Publication EPA/600/3-87/607, Athens, GA, 189pp.

Brown, M.B. and A.B. Forsythe. 1974. Robust tests for the equality of variances. *J. Am. Stat. Assoc.* 69: 364–367.

Brown, R.C., R.H. Pierce, and S.A. Rice. 1985. Hydrocarbon contamination in sediments from urban stormwater runoff. *Mar. Pollut. Bull.* 16: 236–240.

Brush, G.S. 1984. Stratigraphic evidence of eutrophication in an estuary. *Water Resour. Res.* 20: 531–541.

Brush, G.S. 1991. Long-term trends in Perdido Bay: A stratigraphic study. Florida Department of Environmental Regulation, Pensacola, FL (unpublished report).

Bureau of Economic and Business Research. 1995. *1995 Florida Statistical Abstract*, 29th edn. University Press of Florida, Gainesville, FL, 800pp.

Bureau of Economic and Business Research. 1996. *Florida Estimates of Population, 1995*. University Press of Florida, Gainesville, FL, 809pp.

Burgess, N.M., D.C. Evers, J.D. Kaplan, M. Duggan, and J.J. Kerekes. 1998. Ecological impacts of mercury. In: N. Burgess, S. Beauchamp, G. Brun, T. Chair, C. Roberts, L. Rutherford, R. Tordon, and O. Vaidya (Eds.), Mercury in the Atlantic: A progress report. Regional Science Coordinating Committee, Environment Canada, Sackville, New Brunswick, Canada.

Burkholder, J.A. and H.B. Glasgow, Jr. 1997. *Pfiesteria piscicidia* and other *Pfiesteria*-dinoflagellates: Behaviors, impacts, and environmental controls. *Limnol. Oceanogr.* 42: 1052–1075.

Burkholder, J.M., E.J. Noga, C.H. Hobbs, and H.B. Glasgow, Jr. 1992. New "phantom" dinoflagellate is the causative agent of major estuarine fish kills. *Nature* 358: 407–410.

Buskey, E.J., P.A. Montagna, A.F. Amos, and T.E. Whitledge. 1997. Disruption of grazer populations as a contributing factor to the initiation of the Texas brown tide algal bloom. *Limnol. Oceanogr.* 42: 1215–1222.

Buskey, E.J. and D.A. Stockwell. 1993. Effects of a persistent "brown tide" on zooplankton populations in the Laguna Madre of South Texas. In: T.J. Samyda and Y. Shimizu (Eds.), *Toxic Phytoplankton Blooms in the Sea: Proceedings of the Fifth International Conference on Toxic Marine Phytoplankton*, Newport, RI, October 28–November 1, 1991. Elsevier, New York, pp. 659–665.

Butts, G. and D. Ray. 1986. An investigation of Upper Escambia Bay during drought conditions July–August 1986. Florida Department of Environmental Regulation, Northwest District, Pensacola, FL, 11pp. (unpublished report).

Byrne, C.J. 1980. The geochemical cycling of hydrocarbons in Lake Jackson, Florida. PhD dissertation, Department of Oceanography, Florida State University, Tallahassee, FL.

Cairns, D.J. 1981. Detrital production and nutrient release in a southeastern flood-plain forest. MS thesis, Florida State University, Tallahassee, FL.

Cairns, J., Jr. and K.L. Dickson. 1977. Recovery of streams and spills of hazardous materials. In: J. Cairns, Jr., K.L. Dickson, and E.E. Herricks (Eds.), *Recovery and Restoration of Damaged Ecosystems*. University of Virginia Press, Charlottesville, VA, pp. 24–42.

Cake, E.W., Jr. 1983. Habitat suitability index models: Gulf of Mexico American oyster. US Fish and Wildlife Service, Office of Biological Services, Washington, DC. FWS/OBS-82/10.57, 37pp.

Campbell, K.R., C.J. Ford, and D.A. Levine. 1998. Mercury distribution in Poplar Creek, Oak Ridge, Tennessee, USA. *Environ. Toxicol. Chem.* 17: 1191–1198.

Canuel, E.A. and A.R. Zimmerman. 1999. Composition of particulate organic matter in the southern Chesapeake Bay: Sources and reactivity. *Estuaries* 22: 980–994.

Caperon, J., S.A. Cattell, and G. Krasnick. 1971. Phytoplankton kinetics in a subtropical estuary: Eutrophication. *Limnol. Oceanogr.* 26: 599–607.

Caraco, N., A. Tamse, O. Boutros, and I. Valiela. 1987. Nutrient limitation of phytoplankton growth in brackish coastal ponds. *Can. J. Fish. Aquat. Sci.* 4: 473–476.

Cardwell, R.D., S. Olsen, M.I. Carr, and E.W. Sanborn. 1979. Causes of oyster mortality in South Puget Sound. Washington Department of Fisheries, Washington, DC. NOAA Technical Memorandum ERL MESA-39, 79pp.

Carlsson, P. and E. Graneli. 1993. Availability of humic-bound nitrogen for coastal phytoplankton. *Estuarine Coastal Shelf Sci.* 36: 433–447.

Carr, W.E.S. and C.A. Adams. 1972. Food habits of juvenile marine fishes: Evidence of the cleaning habit in the leatherjacket, *Oligoplites saurus*, and the spottail pinfish, *Diplodus holbrooki*. *Fish. Bull.* 70: 1111–1120.

Carr, W.E.S. and C.A. Adams. 1973. Food habits of juvenile marine fishes occupying seagrass beds in the estuarine zone near Crystal River, Florida. *Trans. Am. Fish. Soc.* 102: 511–540.

Carvalho, L.R., E.J. Cox, S.C. Fritz, S. Juggins, P.A. Sims, F. Gasse, and R.W. Battarbee. 1995. Standardizing the taxonomy of saline lake *Cyclotella* spp. *Diatom Res.* 10: 229–240.

Cassie-Cooper, V. 1996. *Microaglae Marvels*. Arrow Press, Morrinsville, New Zealand, 164pp.

Causey, B.D. 2002. The role of the Florida Keys National Marine Sanctuary in the South Florida ecosystem restoration initiative. In: J.W. Porter and K.G. Porter (Eds.), *The Everglades, Florida Bay and Coral Reefs of the Florida Keys: An Ecosystem Sourcebook*. CRC Press, Boca Raton, FL, pp. 883–894.

Cerco, C.F. and S.P. Seitzinger. 1997. Measured and modeled effects of benthic algae on eutrophication in Indian River-Rehoboth Bay, Delaware. *Estuaries* 20: 231–248.

Chaalali, A., G. Beaugrand, P. Boët, and B. Sautour. 2013. Climate-caused abrupt shifts in a European macrotidal estuary. *Estuaries Coasts* 36: 1193–1205.

Chaky, D.A. 2003. Polychlorinated biphenyls, polychlorinated dibenzo-P-dioxins, and furans in the New York metropolitan area: Interpreting atmospheric deposition and sediment chronologies. PhD dissertation, Rensselaer Polytechnic Institute, Troy, NY.

Chang, E.H. 1988. Distribution, abundance and size composition of phytoplankton off Westland, New Zealand, February 1982. *N. Z. J. Mar. Freshw. Res.* 22: 345–367.

Chang, F.J., C. Anderson, and N.C. Boustead. 1990. First record of *Heterosigma* (Raphidophyceae) bloom with associated mortality of cage-reared salmon in Big Glory Bay, New Zealand. *N. Z. J. Mar. Freshw. Res.* 24: 461–469.

Chanton, J. and F.G. Lewis. 1999. Plankton and dissolved inorganic carbon isotopic composition in a river-dominated estuary: Apalachicola Bay, Florida. *Estuaries* 22: 575–583.

Chanton, J. and F.G. Lewis. 2002. Examination of coupling between primary and secondary production in a river-dominated estuary: Apalachicola Bay, Florida, U.S.A. *Limnol. Oceanogr.* 47: 683–697.

Chao, L.N. and J.A. Musick. 1977. Life history, feeding habits, and functional morphology of juvenile sciaenid fishes in the York River Estuary, Virginia. *Fish. Bull.* 75: 657–702.

Charpy-Roubaud, C.J., L.J. Charpy, and S.Y. Maestrini. 1983. Nutrient enrichment of waters of Golfo de San Jose (Argentina, 42°S), growth and species selection of phytoplankton. *Mar. Ecol.* 4: 1–18.

Christensen, H. and E. Kanneworff. 1985. Sedimenting phytoplankton as a major food source for suspension and deposit feeders in the Oresund. *Ophelia* 24: 223–244.

Christensen, J.D., M.E. Monaco, R.J., Livingston, G. Woodsum, T.A. Battista, C.J. Klein, B. Galperin, and W. Huang. 1998. Potential impacts of freshwater inflow on Apalachicola Bay, Florida oyster (*Crassostrea virginica*) populations: Coupling hydrologic and biological models. NOAA/NOS Strategic Environmental Assessments Division Report, Silver Spring, MD, 58pp.

Christensen, V. and D. Pauly. 1992. *A Guide to the Econpathii Software System (Version 2.1)*. International Center for Living Aquatic Resources Management (ICLARM), Manila, Philippines, 72pp.

Clarke, T.A. 1978. Diel feeding patterns o f 16 species of mesopelagic fishes from Hawaiian waters. *Fish. Bull.* 76: 495–514.

Clement, A. and G. Lembeye. 1993. Phytoplankton monitoring program in the fish farming of south Chile. In: T.J. Smayda and Y. Shimizu (Eds.), *Toxic Phytoplankton Blooms in the Sea: Proceedings of the Fifth International Conference on Toxic Marine Phytoplankton*, Newport, RI, October 28–November 1, 1991. Elsevier, New York, pp. 223–228.

Clement, B. and G. Merlin. 1995. The contributions of ammonia and alkalinity to landfill leachate toxicity to duckweed. *Sci. Total Environ.* 170: 71–79.

Clements, W.H. and R.J. Livingston. 1983. Overlap and pollution-induced variability in the feeding habits of filefish (Pisces: Monacanthidae) from Apalachee Bay, Florida. *Copeia* 1983: 331–338.

Clements, W.H. and R.J. Livingston. 1984. Prey selectivity of the fringed filefish *Monacanthus ciliatus* (Pisces: Monacanthidae): Role of prey accessibility. *Mar. Ecol. Prog. Ser.* 16: 291–295.

Clewell, A.F. 1977. Geobotany of the Apalachicola River region. In: R.J. Livingston (Ed.), *Proceedings of the Conference on the Apalachicola Drainage System*. Florida Department of Natural Resources, St. Petersburg, FL. Florida Marine Resources Publication 26, pp. 6–15.

Cliff, A.D. and J.K. Ord. 1981. *Spatial Processes*. Pion Ltd., London, U.K.

Cloern, J.E. 1979. Phytoplankton ecology of the San Francisco Bay system: The status of our current understanding. In: J.T. Conomos (Ed.), *San Francisco Bay: The Urbanized Estuary*. Pacific Division AAAS, San Francisco, CA, pp. 247–264.

Cloern, J.E. 1996. Phytoplankton bloom dynamics in coastal ecosystems: A review with some general lessons from sustained investigations of San Francisco Bay, California. *Rev. Geophys.* 34: 127–168.

Cloern, J.E. 2001. Our evolving conceptual model of the coastal eutrophication problem. *Mar. Ecol. Prog. Ser.* 210: 223–253.

Cloern, J.E., A.E. Alpine, B.E. Cole, R.L.J. Wong, J.F. Arthur, and M.D. Ball. 1983. River discharge controls phytoplankton dynamics in the northern San Francisco Bay estuary. *Estuarine Coastal Shelf Sci.* 16: 415–429.

Cloern, J.E. and R. Dufford. 2005. Phytoplankton community ecology: Principles applied in San Francisco Bay. *Mar. Ecol. Prog. Ser.* 285: 11–28.

Cloern, J.E., K.A. Hieb, T. Jacobson, B. Sanso, E. Di Lorenzo, M.T. Stacey, J.L. Largier et al. 2010. Biological communities in San Francisco Bay track large-scale climate forcing over the North Pacific. *Geophys. Res. Lett.* 37: L21602, 6pp.

Codd, G.A., C. Edwards, K.A. Beattle, L.A. Lawton, D.L. Campbell, and S.G. Bell. 1995. Toxins from cyanobacteria (bluegreen algae). In: W. Wiessner, E. Schnept, and R.C. Starr (Eds.), *Algae, Environment and Human Affairs*. Biopress Ltd., Bristol, U.K., pp. 1–17.

Coffin, R.B. and L.A. Cifuentes. 1992. A stable isotope study of carbon sources and microbial transformations in the Perdido Estuary, FL. In: R.J. Livingston (Ed.), Ecological study of the Perdido drainage system. Florida Department of Environmental Regulation, Tallahassee, FL (unpublished report).

Coffin, R.B. and L.A. Cifuentes. 1999. A stable isotope analysis of carbon cycling in the Perdido Estuary, Florida. *Estuaries* 22: 917–926.

Collard, S.B. 1989. Benthic macroinvertebrate species indicator list. Department of Environmental Regulation, Tallahassee, FL. STAR grant final report, 850pp.

Collard, S.B. 1991a. The Pensacola Bay System: Biological trends and current status. Northwest Florida Water Management District, Havana, FL. Water Resources Special Report 91-3, 184pp.

Collard, S.B. 1991b. Management Options for the Pensacola Bay System: The potential value of seagrass transplanting and oyster bed refurbishment programs. Northwest Florida Water Management District, Havana, FL. Water Resources Special Report 91-44.

Collette, B.B. and F.H. Talbot. 1972. Activity patterns of coral reef fishes with emphasis on diurnal–nocturnal changeover. *Bull. Natl. Hist. Mus. Los Angeles County* 14: 98–125.

Conley, D.J., J. Carstensen, G. Aertebjerg, P.B. Christensen, T. Dalsgaard, J.L.S. Hansen, and A.B. Josefson. 2007. Long-term changes and impacts of hypoxia in Danish coastal waters. *Ecol. Appl.* 17: S165–S184.

Cooley, N.R. 1978. An inventory of the estuarine fauna in the vicinity of Pensacola, Florida. Department of Natural Resources, Marine Research Laboratory, St. Petersburg, FL. Florida Marine Resources Publication 31, pp. 1–119.

Cooper, S.R. 1995a. Diatoms in sediment cores from the mesohaline Chesapeake Bay, U.S.A. *Diatom Res.* 10: 39–89.

Cooper, S.R. 1995b. An abundant, small brackish water *Cyclotella* species in Chesapeake Bay, U.S.A. In: J.P. Kociolek and M.J. Sullivan (Eds.), *A Century of Diatom Research in North America: A Tribute to the Distinguished Careers of Charles W. Reimer and Ruth Patrick*. Koeltz Scientific Books, Champaign, IL, pp. 133–140.

Corredor, J.E., R.W. Howarth, R.R. Twilley, and J.M. Morrell. 1999. Nitrogen cycling and anthropogenic impact in the tropical interamerican seas. *Biogeochemistry* 46: 163–178.

Cox, J., R. Kautz, M. MacLaughlin, and T. Gilber. 1994. Closing the gaps in Florida's wildlife habitat conservation system. Florida Fish and Wildlife Conservation Commission, Tallahassee, FL, 239pp. (unpublished report).

Craig, A., E.N. Powell, R.R. Fay, and J.M. Brooks. 1989. Distribution of *Perkinsus marinus* in Gulf coast oyster populations. *Estuaries* 12: 82–91.

Crosby, D.O. 1981. Environmental chemistry of pentachlorophenol. *J. Appl. Chem.* 53: 1051–1080.

Cross, R.D. and D.L. Williams (Eds.). 1981. *Proceedings of the National Symposium on Freshwater Inflow to Estuaries*. US Fish and Wildlife Service, Office of Biological Services, Washington, DC. FWS/OBS-81/04, 525pp.

Crossland, N.O. and C.J.M. Wolff. 1985. Fate and biological effects of pentacholorophenol in outdoor ponds. *Environ. Toxicol. Chem.* 4: 71–86.

Cullen, J.J., A.M. Ciotti, R.F. Davis, and M.L. Lewis. 1997. Optical detection and assessment of algal blooms. *Limnol. Oceanogr.* 42: 1223–1239.

Cullen, J.J., X. Yang, and H.L. MacIntyre. 1992. Nutrient limitation of marine photosynthesis. In: P.G. Falkolski and A.D. Woodhead (Eds.), *Primary Productivity and Biogeochemical Cycles in the Sea*. Plenum Press, New York, pp. 69–88.

Culliton, T.J., M.A. Warren, T.R. Goodspeed, D.G. Remer, C.M. Blackwell, and J.J. McDough, III. 1990. 50 years of population growth along the Nation's coasts, 1960–2010. Strategic Assessment Branch, Ocean Assessments Division, Office of Oceanography and Marine Assessment, National Ocean Survey, National Oceanic and Atmospheric Administration, Silver Spring, MD, 41pp.

Cupp, E.E. 1943. Marine plankton diatoms of the west coast of North America. *Bull. Scripps Inst. Oceanogr. Univ. Calif.* 5: 1–238.

Curl, H. 1962. Analysis of carbon in marine plankton organisms. *J. Mar. Res.* 20: 181–188.

D'Elia, C.F., W.R. Boynton, and J.G. Sanders. 2003. A watershed perspective on nutrient enrichment, science and policy in the Patuxent River, Maryland: 1960–2000. *Estuaries* 26: 171–185.

D'Elia, C.F., J.G. Sanders, and W.R. Boynton. 1986. Nutrient enrichment studies in a coastal plain estuary. Phytoplankton growth in large-scale, continuous cultures. *Can. J. Fish. Aquat. Sci.* 43: 397–406.

Dahl, E. and T. Hohannessen. 2001. Relationship between occurrence of *Dinophysis* species (dinophyceae) and shellfish toxicity. *Phycologia* 40: 223–227.

Danglade, E. 1917. Conditions and extent of the water level oyster beds and barren bottoms in the vicinity of Apalachicola, Florida. Appendix V, reports of the US commissioner of fishes for 1916. US Bureau of Fisheries 841, 75pp.

Dardeau, M.R., R.F. Modlin, W.W. Schroeder, and J.P. Stout. 1992. Estuaries. In: C.T. Hackney, S.M. Adams, and W.H. Martin (Eds.), *Biodiversity of the Southeastern United States*. John Wiley & Sons, New York, pp. 615–744.

Darnell, R.M. 1958. Food habits of fishes and larger invertebrates of Lake Pontchartrain, Louisiana, an estuarine community. *Publ. Inst. Mar. Sci. Univ. Tex.* 5: 353–416.

Darnell, R.M. 1961. Trophic spectrum of an estuarine community, based on studies of Lake Pontchartrain, Louisiana. *Ecology* 42: 553–568.

Darnell, R.M. 1967. Organic detritus in relation to the estuarine ecosystem. In: G.H. Lauff (Ed.), *Estuaries*. American Association for the Advancement of Science Publication 83, Washington, DC, pp. 376–388.

Darnell, R.M. and T.M. Soniat. 1981. Nutrient enrichment and estuarine health. In: B.J. Neilson and L.E. Cronin (Eds.), *Estuaries and Nutrients*. Human Press, Clifton, NJ, pp. 225–245.

Darst, M.R. and H.M. Light. 2007. Drying of floodplain forests associated with water-level decline in the Apalachicola River, Florida—Interim Results, 2006. US Geological Survey Open-File Report 2007-1019, Tallahassee, FL, 32pp.

Dauer, D.M., R.W. Ewing, G.H. Tourtellotte, W.T. Harlan, J.W. Sourbeer, and H.R. Baker. 1982. Predation, resource limitation and the structure of lower Chesapeake Bay. *Int. Rev. Gesamten. Hydrobiol.* 67: 477–489.

Dauer, D.M., J.A. Ranasinghe, and S.B. Weisberg. 2000. Relationships between benthic community condition, water quality, sediment quality, nutrient loads, and land use patterns in Chesapeake Bay. *Estuaries* 23: 80–96.

Dauer, D.M., A.J. Rodi, Jr., and J.A. Ranasinghe. 1992. Effects of low dissolved oxygen events on the macrobenthos of the lower Chesapeake Bay. *Estuaries* 15: 384–391.

Davies, T.T. and T.B. DeMoss. 1982. Chesapeake Bay Program technical studies: A synthesis. US Environmental Protection Agency, Washington, DC, 625pp.

Davis, J.C. 1975. Minimal dissolved oxygen requirements of aquatic life with emphasis on Canadian species: A review. *J. Fish. Res. Board Can.* 32: 2295–2332.

Davis, S.M. and J.C. Ogden (Eds.). 1994. *Everglades: The Ecosystem and Its Restoration*. St. Lucie Press, Delray Beach, FL, 826pp.

Davis, W.P. and S.A. Bortone. 1992. Effects of kraft mill effluents on the sexuality of fishes: An environmental early warning? In: T. Colborn and C. Clement (Eds.), *Chemically-Induced Alterations in Sexual and Functional Development: The Wildlife/Human Connections*. Princeton Scientific Publishing, Princeton, NJ, pp. 113–127.

Davis, W.P., M.R. Davis, and D.A. Flemer. 1999. Observations on the regrowth of subaquatic vegetation following transplantation: A potential method to assess environmental health of coastal habitats. In: S. Bortone (Ed.), *Seagrasses; Monitoring, Ecology, Physiology, and Management*. CRC Press, Boca Raton, FL, pp. 231–238.

Dawson, C.E. 1955. A contribution to the hydrograph of Apalachicola Bay. *Publ. Inst. Mar. Sci. Univ. Tex.* 4: 15–35.

Day, J.W., A. Yanez-Arancibia, J.H. Cowan, R.H. Day, R.R. Twilley, and J.R. Rybczyk. 2013. Global climate change impacts on coastal ecosystems of the Gulf of Mexico. Ecosystem-based management of the Apalachicola River-Apalachicola Bay system, Florida. In: J.W. Day and A. Yanez-Ararcibia (Eds.), *Gulf of Mexico, Origin, Waters and Biota*, Vol. 4, Ecosystem-Based Management. Texas A&M University, College Station, TX, pp. 253–271.

Day, K. and V. Ruttan. 1991. The deficit in natural resources research. *Bioscience* 41: 37–40.

Dayton, P.K. 1979. Ecology: A science and a religion. In: R.J. Livingston (Ed.), *Ecological Processes in Coastal and Marine Systems*. Plenum Press, New York, pp. 3–18.

Dayton, P.K., M.J. Tegner, P.E. Parnell, and P.B. Edwards. 1992. Temporal and spatial patterns of disturbance and recovery in a kelp forest community. *Ecol. Monogr.* 62: 421–445.

De Jong, V.N. 1995. The Ems estuary, the Netherlands. In: A.J. McComb (Ed.), *Eutrophic Shallow Estuaries and Lagoons*. CRC Press, New York, pp. 81–108.

De Jong, V.N. and W. van Raaphorst. 1995. Eutrophication of the Dutch Wadden Sea (Western Europe), an estuarine area controlled by the River Rhine. In: A.J. McComb (Ed.), *Eutrophic Shallow Estuaries and Lagoons*. CRC Press, New York, pp. 129–150.

Deegan, L.A., J.W. Day, Jr., J.G. Gosselink, A. Yanez-Arancibia, G. Soberon Chavez, and P. Sanchez-Gil. 1986. Relationships among physical characteristics, vegetation distribution, and fisheries yield in Gulf of Mexico estuaries. In: D.A. Wolfe (Ed.), *Estuarine Variability*. Academic Press, New York, pp. 83–100.

Deevey, E.S., Jr. 1988. Estimation of downward leakage from Florida lakes. *Limnol. Oceanogr.* 33: 1308–1320.

Dekshenieks, M.M., E.E. Hofmann, and E.N. Powell. 1993. Environmental effects on the growth and development of eastern oyster, *Crassostrea virginica* (Gmelin, 1791), larvae: A modelling study. *J. Shellfish Res.* 11: 399–416.

Delfino, J.J., D. Frazier, and J. Nepshinsky. 1984. Contaminants in Florida's coastal zone. University of Florida, Gainesville, FL. Florida Sea Grant Report 62, 176pp.

Demas, G.P., M.C. Rabinhorst, and J.C. Stevenson. 1996. Subaqueous soils: A pedological approach to the study of shallow-water habitats. *Estuaries* 19: 229–237.

DeMort, C.L. 1991. The St. Johns river system. In: R.J. Livingston (Ed.), *The Rivers of Florida*. Springer-Verlag, New York, pp. 97–120.

Denman, K. and T. Platt. 1977. Time series analysis in marine ecosystems. In: H.H. Shugart, Jr. (Ed.), *Time Series and Ecological Processes*. Siam Institute for Mathematics and Society, Philadelphia, PA, pp. 227–242.

Dennison, W.C., R.J. Orth, K.A. Moore, J.C. Stevenson, V. Carter, S. Kollar, P.W. Bergstrom, and R.A. Batuik. 1993. Assessing water quality with submersed aquatic vegetation. *Bioscience* 43: 86–94.

Deyrup, M. and R. Franz (Eds.). 1994. *Rare and Endangered Biota of Florida*, Vol. IV, Invertebrates. University Press of Florida, Gainesville, FL, 798pp.

Diaz, R.J., M. Luckenbach, S. Thornton, M.H. Roberts, Jr., R.J. Livingston, C.C. Koenig, G.L. Ray, and L.E. Wolfe. 1985. Field validation of multi-species laboratory test systems for estuarine benthic communities. Office of Research and Development, US Environmental Protection Agency, Gulf Breeze, FL. Report CR B12053, 81pp.

Diaz, R.J. and R. Rosenberg. 1995. Marine benthic hypoxia: A review of its ecological effects and the behavioral responses of benthic macrofauna. *Oceanogr. Mar. Biol. Annu. Rev.* 33: 245–305.

Dissmeyer, G.E. 1994. *Evaluating the Effectiveness of Forestry Best Management Practices in Meeting Water Quality Goals or Standards*. US Department of Agriculture, US Forest Service, Atlanta, GA, 166pp.

DiTuillio, G.R., D.A. Hutchins, and K.W. Bruland. 1993. Interaction of iron and major nutrients controls phytoplankton growth and species composition in the tropical North Pacific Ocean. *Limnol. Oceanogr.* 38: 495–508.

Dolan, D.M., A.K. Yui, and R.D. Geist. 1981. Evaluation of river load estimation methods for total phosphorus. *Great Lakes Res. Div. Publ.* 7: 207–214.

Dolbeth, M., P.G. Cardoso, S.M. Ferreira, T. Verdelhos, D. Raffaelli, and M.A. Pardal. 2007. Anthropogenic and natural disturbance effects on a macrobenthic estuarine community over a 10-year period. *Mar. Pollut. Bull.* 54: 576–585.

Dortch, Q. 1990. The interaction between ammonium and nitrate uptake in phytoplankton. *Mar. Ecol. Prog. Ser.* 61: 183–201.

Doudoroff, P. and D.L. Shumway. 1967. Dissolved oxygen criteria for the protection of fish. *Am. Fish. Soc. Spec. Publ.* 4: 13–19.

Downing, K.M. and J.C. Merkens. 1955. The influence of dissolved oxygen concentrations on the toxicity of unionized ammonia to rainbow trout (*Salmo gairdnerii* Richardson). *Ann. Appl. Biol.* 43: 243–246.

Drake, S.H. 1975. Effects of mercuric chloride on the embryological development of the Zebrafish (*Brachydanio rerio*). MS thesis, Florida State University, Tallahassee, FL, 148pp.

Droop, M.R. 1974. The nutrient status of algal cells in continuous culture. *J. Mar. Biol. Assoc. U.K.* 54: 825–855.

Duarte, C.M. 1995. Submerged aquatic vegetation in relation to different nutrient regimes. *Ophelia* 41: 87–112.

Duarte, C.M. and O. Piro. 2001. Interdisciplinary challenges and bottlenecks in the aquatic sciences. *Limnol. Oceanogr.* 10: 57–61.

Dugan, P.J. and R.J. Livingston. 1982. Long-term variation in macroinvertebrate communities in Apalachee Bay, Florida. *Estuarine Coastal Shelf Sci.* 14: 391–403.

Duke, T.W., J.I. Lowe, and A.J. Wilson. 1970. A polychlorinated biphenyl (Aroclor 1254) in the water, sediment, and biota of Escambia Bay, Florida. *Bull. Environ. Contam. Toxicol.* 5: 171–180.

Duncan, J.L. 1977. Short-term effects of storm water runoff on the epibenthic community of a north Florida estuary (Apalachicola, Florida). MS thesis, Florida State University, Tallahassee, FL.

Duxbury, A.C. 1975. Orthophosphate and dissolved oxygen in Pudget Sound. *Limnol. Oceanogr.* 20: 270–274.

Eadie, B.J., B.A. McKee, M.B. Lansing, and S. Metz. 1994. Records of nutrient-enhanced coastal ocean productivity in sediments from the Louisiana continental shelf. *Estuaries* 17: 754–765.

Ebeling, A. and R.N. Bray. 1976. Day versus night activity of reef fishes in a kelp forest off Santa Barbara, California. *Fish. Bull.* 74: 703–717.

Edmiston, H.L. 1979. The zooplankton of the Apalachicola Bay system. MS thesis, Florida State University, Tallahassee, FL.

Edmonson, W.T. 1970. Phosphorus, nitrogen, and algae in Lake Washington after diversion of sewage. *Science* 169: 690–691.

Edwards, N.C. 1976. A Study of the circulation and stratification of Escambia Bay, Florida, during the period of low freshwater inflow. MS thesis, Florida State University, Tallahassee, FL, 192pp.

Ehrlich, R., R.J. Wening, G.W. Johnson, S.H. Su, and D.J. Paustenbach. 1994. A mixing model for polychlorinated dibenzo-p-dioxins and dibenzofurans in surface sediments from Newark Bay, New Jersey, using polytopic vector analysis. *Arch. Environ. Contam. Toxicol.* 27: 486–500.

Ekino, S., T. Ninomiya, and M. Susa. 2004. Letter to the editor: More on mercury content in fish. *Science* 303: 764.

Elder, J.F. and D.J. Cairns. 1982. Production and decomposition of forest litter fall on the Apalachicola River floodplain, Florida. US Geological Survey, Tallahassee, FL. Water-Supply Paper 2195-B, 42pp.

Elder, J.F. and P.V. Dresler. 1983. Creosote discharge of the nearshore estuarine environment in Pensacola Bay, Florida: Preliminary assessment of effects, in movement and fate of creosote waste in ground water. US Geological Survey, Tallahassee, FL. Water-Supply Paper 228, 63pp.

Ellis, E.E. 1969. Some basic dynamics of the Pensacola Estuary. Florida Air and Water Pollution Control Committee, Tallahassee, FL, 19pp.

Elser, J.J., M.M. Elser, N.A. MacKay, and S.R. Carpenter. 1988. Zooplankton-mediated transitions between N- and P-limited algal growth. *Limnol. Oceanogr.* 33: 1–14.

Elser, J.J. and D.L. Frees. 1995. Microconsumer grazing and sources of limiting nutrients for phytoplankton growth: Application and complications of a nutrient-deletion/dilution-gradient technique. *Limnol. Oceanogr.* 40: 1–16.

Elsner, J.B. 1992. Predicting time series using a neural network as a method to distinguish chaos from noise. *J. Phys.* 25: 843–850.

Elton, C. 1927. *Animal Ecology*. Macmillan Company, New York, 256pp.

Elton, C. 1930. *Animal Ecology and Evolution*. Clarendon Press, Oxford, U.K., 296pp.

EnviroScience, Inc. 2005. Freshwater mussel and habitat surveys of the Apalachicola River, Chipola River, and selected sloughs/tributaries. Stow, OH (unpublished final report).

Eppley, R.W. 1972. Temperature and phytoplankton growth in the sea. *Fish. Bull.* 70: 1063–1085.

Eppley, R.W., E.H. Ringer, E.L. Venrick, and M.M. Mullin. 1973. A study of plankton dynamics and nutrient cycling in the central gyre of the North Pacific Ocean. *Limnol. Oceanogr.* 18: 534–551.

Erga, S.R. 1988. Phosphorus and nitrogen limitation of phytoplankton in the Inner Oslofjord (Norway). *Sarsia* 73: 229–243.

Ericksson, L.O. 1978. Nocturnalism vs. diurnalism: Dualism within fish individuals. In: J.E. Thorpe (Ed.), *Rhythmic Activity of Fishes*. Academic Press, New York, pp. 69–89.

Ernst, H.R. 2003. *Chesapeake Bay Blues*. Rowman & Littlefield Publishers, Oxford, U.K., 203pp.

Estabrook, R.H. 1973. Phytoplankton ecology and hydrography of Apalachicola Bay. MS thesis, Florida State University, Tallahassee, FL.

Estevez, E.D., L.K. Dixon, and M.F. Flannery. 1991. West-coastal rivers of peninsular Florida. In: R.J. Livingston (Ed.), *The Rivers of Florida*. Springer-Verlag, New York, pp. 187–222.

Eyre, B. and P. Balls. 1999. A comparative study of nutrient behavior along the salinity gradient of tropical and temperate estuaries. *Estuaries* 22: 313–326.

Ezer, T. and G.L. Mellor. 1997. Simulations of the Atlantic Ocean with a free surface sigma coordinate ocean model. *J. Geophys. Res.* 102: 647–657.

Fabien, J., S. Lefebvre, B.Ã. Vacron, and Y. Lagadeuc. 2006. Phytoplankton community structure and primary production in small intertidal estuarine-bay ecosystem (eastern English Channel, France). *Mar. Biol.* 151: 805–825.

Fahrny, S.A. et al. 2006. Submerged aquatic vegetation monitoring in Apalachicola Bay, Florida. Apalachicola National Estuarine Research Reserve, Eastpoint, FL, 24pp. (unpublished report to the National Oceanic and Atmospheric Administration).

Falconer, I.R. 2001. Toxic cyanobacterial bloom problems in Australian waters: Risks and impacts on human health. *Phycologia* 40: 228–233.

Farmer, J.D. and J.J. Sidorowich. 1987. Predicting chaotic time series. *Phys. Rev. Lett.* 59: 845–848.

Federle, T.W., M.A. Hullar, R.J. Livingston, D.A. Meeter, and D.C. White. 1983a. Spatial distribution of biochemical parameters indicating biomass and community composition of microbial assemblies in estuarine mud flat sediments. *Appl. Environ. Microbiol.* 45: 58–63.

Federle, T.W., R.J. Livingston, D.A. Meeter, and D.C. White. 1983b. Modifications of estuarine sedimentary microbiota by exclusion of epibenthic predators. *J. Exp. Mar. Biol. Ecol.* 73: 81–94.

Federle, T.W., R.J. Livingston, L.E. Wolfe, and D.C. White. 1986. A quantitative comparison of microbial community structure of estuarine sediments for microcosms in the field. *Can. J. Microbiol.* 32: 319–325.

Fenchel, T.M. and B.B. Jorgensen. 1978. Detritus food chains of aquatic ecosystems: The role of bacteria. *Adv. Microb. Ecol.* 1: 1–57.

Fernald, E.A. and D.J. Patton (Eds.). 1985. *Water Resources Atlas of Florida*. Institute of Science and Public Affairs, Florida State University, Tallahassee, FL, 291pp.

Fernald, E.A. and E.D. Purdum (Eds.). 1992. *Atlas of Florida*. University of Florida Press, Gainesville, FL, 280pp.

Fillos, J. and A.H. Molof. 1972. Effects of benthal deposits on oxygen and nutrient economy of flowing waters. *J. Water Pollut. Control Fed.* 44: 644–662.

Finlay, J.C. 2001. Stable-carbon-isotope ratios of river biota: Implications for energy flow in lotic food webs. *Ecology* 82: 1052–1064.

Finney, B.P., I. Gregory-Eaves, J. Sweetman, M.S.V. Douglas, and J.P. Smol. 2000. Impacts of climatic change and fishing on Pacific salmon abundance over the past 300 years. *Science* 290: 795–799.

Finney, B.P. and C. Huh. 1989. History of metal pollution in the Southern California Bight: An update. *Environ. Sci. Technol.* 23: 294–303.

Fisher, T.R., L.W. Harding, Jr., D.W. Stanley, and L.G. Ward. 1988. Phytoplankton, nutrients, and turbidity in the Chesapeake, Delaware, and Hudson Estuaries. *Estuarine Coastal Shelf Sci.* 27: 61–93.

Fisher, T.R., E.R. Peele, J.W. Ammerman, and L.W. Harding, Jr. 1992. Nutrient limitation of phytoplankton in Chesapeake Bay. *Mar. Ecol. Prog. Ser.* 82: 51–63.

Fisher, W.S., J.T. Winstead, L.M. Oliver, H.L. Edmiston, and G.O. Bailey. 1996. Physiologic variability of eastern oysters from Apalachicola Bay, Florida. *J. Shellfish Res.* 15: 543–553.

Fjerdingstad, E. 1965. Taxonomy and saprobic valency of benthic phytomicro-organisms. *Int. Rev. Gesamten. Hydrobiol.* 50: 475–604.

Flemer, D.A. 1970. Primary production in the Chesapeake Bay. *Chesap. Sci.* 11: 117–129.

Flemer, D.A. 1989. Perdido Bay as a long-term Gulf region estuarine ecosystem verification template. US Environmental Protection Agency, Environmental Research Laboratory, Gulf Breeze, FL. EPA/600/X-89/162, 72pp.

Flemer, D.A., R.J. Livingston, and S.E. McGlynn. 1998. Phytoplankton nutrient limitation: Seasonal growth responses of sub-temperate estuarine phytoplankton to nitrogen and phosphorus—An outdoor microcosm experiment. *Estuaries* 21: 145–159.

Fletcher, R. 1990. Impact of cultural eutrophication on aquatic invertebrates in Lake Jackson (Tallahassee, Florida). Honors and Scholars Program, Department of Biological Sciences and Center for Aquatic Research and Resource Management, Florida State University, Tallahassee, FL (unpublished final report).

Flint, R.W. 1985. Long-term estuarine variability and associated biological response. *Estuaries* 8: 158–169.

Florida Department of Environmental Protection (FDEP). 2012. Site-specific information in support of establishing numeric nutrient criteria in Apalachicola Bay. Division of Environmental Assessment and Restoration, Standards and Assessment Section, Tallahassee, FL, 135pp. http://www.dep.state.fl.us/water/wqssp/nutrients/docs/new/apalachicola_bay_121112.pdf, 2012.

Freeman, D.B. 1989. The distribution and trophic significance of benthic microalgae in Masonboro Sound, North Carolina. MS thesis, University of North Carolina at Wilmington, Wilmington, NC.

Friedland, K.D., D.W. Ahrenholz, and J.F. Guthrie. 1996. Formation and seasonal evolution of Atlantic menhaden juvenile nurseries in coastal estuaries. *Estuaries* 19: 105–114.

Fritz, L., M.A. Quilliam, J.L.C. Wright, A.M. Beale, and T.M. Work. 1992. An outbreak of domoic acid poisoning attributed to the pennate diatom *Psuedonitzschia australis*. *J. Phycol.* 28: 439–442.

Frost, B.W. 1980. Grazing. In: I. Morris (Ed.), *The Physiological Ecology of Phytoplankton*. University of California Press, Berkeley, CA, pp. 465–492.

Fry, B. 1983. Fish and shrimp migrations in the northern Gulf of Mexico analyzed using stable C, N, and S isotope ratios. *Fish. Bull.* 81: 789–801.

Fryxell, G.A. 1976. The position of the labiate process in the diatom genus *Skeletonemia*. *Br. Phycol. J.* 11: 93–99.

Fryxell, G.A. and M.C. Villac. 1999. Toxic and harmful marine diatoms. In: E.F. Stoermer and J.P. Smol (Eds.), *The Diatoms: Applications for the Environmental and Earth Sciences*. Cambridge University Press, Cambridge, U.K., pp. 419–428.

Fucik, K.W. 1974. The effect of petroleum operations on the phytoplankton ecology of the Louisiana coastal waters. MS thesis, Texas A&M University, College Station, TX.

Fuhs, G.W., S.D. Demmerle, E. Canelli, and M. Chen. 1972. Characterization of phosphorus-limited plankton algae. *Limnol. Oceanogr. Spec. Symp.* 1: 113–133.

Fukuyo, Y., H. Takano, M. Chihara, and K. Matsuoka. 1990. *Red Tide Organisms in Japan: An Illustrated Taxonomic Guide*. Uchida Rokakuho, Tokyo, Japan, 407pp.

Fulmer, J.M. 1997. Nutrient enrichment and nutrient input to Apalachicola Bay, Florida. MS thesis, Florida State University, Tallahassee, FL, 59pp.

Funicelli, N.A. 1984. Assessing and managing effects of reduced freshwater inflow to two Texas estuaries. In: V.S. Kennedy (Ed.), *The Estuary as a Filter*. Academic Press, New York, pp. 435–446.

Galehouse, J.S. 1971. Sedimentary analysis. In: R.E. Carver (Ed.), *Procedures in Sedimentary Petrology*. Wiley-Interscience, New York, pp. 69–94.

Gallagher, J.C. 1980. Population genetics of *Skeletonema costatum* (Bacillariophyceae) in Narragansett Bay. *J. Phycol.* 16: 464–474.

Gallegos, C.L. 1989. Microzooplankton grazing on phytoplankton in the Rhode River, Maryland: Non-linear feeding kinetics. *Mar. Ecol. Prog. Ser.* 57: 23–33.

Gallegos, C.L., T.E. Jordan, and D.L. Correll. 1992. Event-scale response of phytoplankton to watershed inputs in a subestuary: Timing, magnitude, and location of blooms. *Limnol. Oceanogr.* 37: 813–828.

Galperin, B., A.F. Blumberg, and R.H. Weisberg. 1992. The importance of density driven circulation in well-mixed estuaries: The Tampa Bay experience. In: M.L. Spaulding and A. Blumberg (Eds.), *Estuarine Coastal Model*. ASCE, Tampa, FL, pp. 332–341.

Galperin, B., L.H. Kantha, S. Hassid, and A. Rosati. 1988. A quasi-equilibrium turbulent energy model for geophysical flows. *J. Atmos. Sci.* 45: 55–62.

Galperin, B. and G.L. Mellor. 1990a. A time-dependent, three-dimensional model of the Delaware Bay and River. Part 1: Description of the model and tidal analysis. *Estuarine Coastal Shelf Sci.* 31: 231–253.

Galperin, B. and G.L. Mellor. 1990b. A time-dependent, three-dimensional model of the Delaware Bay and River. Part 2: Three-dimensional flow fields and residual circulation. *Estuarine Coastal Shelf Sci.* 31: 255–281.

Galtsoff, P.S. (Ed.). 1954. Gulf of Mexico, its origin, waters and marine life. US Fish and Wildlife Service. *Fish. Bull.* 64: 604pp.

Galvan, K., J.W. Fleeger, and B. Fry. 2008. Stable isotope addition reveals dietary importance of phytoplankton and microphytobenthos to saltmarsh infaunal. *Mar. Ecol. Prog. Ser.* 359: 37–49.

Gameson, A.L.H., N.J. Barrett, and J.S. Strawbridge. 1973. The aerobic Thames Estuary. In: S.H. Jenkins (Ed.), *Advances in Water Pollution Research. Sixth Information Congress*, Jerusalem. Pergamon Press, New York, pp. 853–850.

Gaston, G.R. 1985. Effects of hypoxia on macrobenthos of the inner shelf off Cameron, Louisiana. *Estuarine Coastal Shelf Sci.* 20: 600–613.

Gastrich, M.D. and C.E. Wazniak. 2002. A brown-tide bloom index based on the potential harmful effects of the brown tide alga, *Aureococcus anophagefferens*. *Aquat. Ecosyst. Health Manag.* 5: 435–441.

Geary, R.C. 1936. Moments of the ratio of the mean deviation to the standard deviation for normal samples. *Biometrika* 28: 295–307.

George, S.M. 1988. The sedimentology and mineralogy of the Pensacola Bay system. MS thesis, University of Southern Mississippi, Hattiesburg, MS.

Gerhart, D.Z. 1975. Nutrient limitation in a small oligotrophic lake in New Hampshire. *Int. Ver. Theor. Ansew. Limnol. Verh.* 19: 1013–1022.

Gerhart, D.Z. and G.E. Likens. 1975. Enrichment experiments for determining nutrient limitation: Four methods compared. *Limnol. Oceanogr.* 20: 649–653.

Gihara, G. and R.M. May. 1990. Nonlinear forecasting as a way of distinguishing chaos from measurement error in time series. *Nature* 344: 734–741.

Gilbert, P.M., D.J. Conley, T.R. Fisher, L.W. Harding, and T.C. Malone. 1995. Dynamics of the 1990 winter/spring bloom in Chesapeake Bay. *Mar. Ecol. Prog. Ser.* 122: 27–43.

Gilbert, P.M., R. Magnien, M.W. Lomas, J. Alexander, C. Fan, E. Haramoto, M. Trice, and T.M. Kana. 2001. Harmful algal blooms in the Chesapeake and coastal bays of Maryland, USA: Comparison of 1997, 1998, and 1999 events. *Estuaries* 24: 875–883.

Gillis, C.A., N.L. Bonnevie, S.H. Su, J.G. Ducy, S.L. Huntley, and R.J. Wenning. 1995. DDT, DDD, and DDE contamination of sediment in the Newark Bay Estuary, New Jersey. *Arch. Environ. Contam. Toxicol.* 28: 85–92.

Gillis, C.A., N.L. Bonnevie, and R.J. Wenning. 1993. Mercury contamination in the Newark Bay. *Ecotoxicol. Environ. Saf.* 25: 214–226.

Glass, N.R. 1971. Computer analysis of predation energetics in the largemouth bass. In: B.C. Patten (Ed.), *Systems Analysis and Simulation in Ecology*, Vol. 1. Academic Press, New York, pp. 325–363.

Glassen, R.C., J.E. Armstrong, J.A. Calder, R.W.G. Carter, P.A. La Rock, J.O. Pilotte, and J.W. Winchester. 1977. Bayou Chico Restoration Study. Florida State University, Florida Resources and Environmental Analysis Center, Tallahassee, FL (unpublished report).

Goldman, J.C., J.J. McCarthy, and D.G. Peavey. 1979. Growth rate influence on the chemical composition of phytoplankton in oceanic waters. *Nature* 279: 210–215.

Gorham, P.R. 1964. Toxic algae. In: D.F. Jackson (Ed.), *Algae and Man*. Plenum Press, New York, pp. 301–336.

Gorsline, D.S. 1963. Oceanography of Apalachicola Bay. In: *Essays in Marine Geology in Honor of K. O. Emery*. University of Southern California Press, Los Angeles, CA, pp. 145–176.

Gossett, R.W., D.A. Brown, and D.R. Young. 1983. Predicting the bioaccumulation of organic compounds in marine organisms using octanol/water partition coefficients. *Mar. Pollut. Bull.* 14: 387–392.

Goudie, A. 1994. *The Human Impact on the Natural Environment*, 4th edn. The MIT Press, Cambridge, MA, 454pp.

Graco, M., L. Farias, V. Molina, and D. Gutierrez. 2001. Massive developments of microbial mats following phytoplankton blooms in a naturally eutrophic bay: Implications of nitrogen cycling. *Limnol. Oceanogr.* 46: 821–832.

Graham, M.H. 1942. Scientific results of cruise VII of the Carnegie during 1928–1929 under command of Captain JP Ault—Biology: 3. Studies in the morphology, taxonomy and ecology of the Peridiniales. Carnegie Institution of Washington Publication 542, Washington, DC, pp. 1–129.

Graham, M.H. and P.K. Dayton. 2002. On the evolution of ecological ideas: Paradigns and scientific progress. *Ecology* 83: 1481–1489.

Graneli, E. 1978. Algal assay of limiting nutrients for phytoplankton production in the Oresund. *J. Water Manag. Res.* (*Vatten*) 2: 117–128.

Graneli, E. 1987. Nutrient limitation of phytoplankton biomass in a brackish water bay highly influenced by river discharge. *Estuarine Coastal Shelf Sci.* 25: 555–565.

Graneli, E. and P. Carlsson. 1997. The ecological significance of phagotrophy in photosynthetic flagellates. In: D.M. Anderson, A.D. Cembella, and G.M. Hallagraeff (Eds.), *Physiological Ecology of Harmful Algal Blooms*. Springer-Verlag, Berlin, Germany, pp. 540–557.

Graneli, E., D. Gediziorowska, and U. Nyman. 1985. Influence of humic and fluvic acids on *Prorocentrum minimum* (Pav.) J. Shiller. In: D.M. Anderson, A.W. White, and D.G. Baden (Eds.), *Toxic Dinoflagellates*. Elsevier, Amsterdam, the Netherlands, pp. 201–206.

Graneli, E., W. Graneli, and L. Rydberg. 1986. Nutrient limitation at the ecosystem and the phytoplankton community level in the Laholm Bay, Southeast Kaftagaf. *Ophelia* 26: 181–194.

Graneli, E., K. Wallstrom, U. Larsson, W. Graneli, and R. Elmgren. 1990. Nutrient limitation of primary production in the Baltic Sea area. *Ambio* 19: 142–151.

Greening, H.S. 1980. Seasonal and diel variations in the structure of macroinvertebrate communities: Apalachee Bay, Florida. MS thesis, Florida State University, Tallahassee, FL.

Greening, H.S. 1995. *Resource-Based Watershed Management in Tampa Bay*. Tampa Bay National Estuary Program, St. Petersburg, FL.

Greening, H.S. and R.J. Livingston. 1982. Diel variations in the structure of epibenthic macroinvertebrate communities of seagrass beds (Apalachee Bay, Florida). *Mar. Ecol. Prog. Ser.* 7: 147–156.

Greve, W. and I.R. Parsons. 1977. Photosynthesis and fish production. Hypothetical effects of climate change and pollution. *Helgol. Wissenschaftliche Meeresunters* 30: 666–672.

Grimm, N.B. (Ed.). 2013. *Front. Ecol. Environ.* 11: 455–510 (Ecological Society of America).

Grimm, N.B., F.S. Chapin, III, B. Bierwagin, P. Gonzalez, P.M. Groffman, Y. Luo, F. Melton et al. 2013a. Climate change impacts on ecological systems: Introduction to a US assessment. In: N.B. Grimm (Ed.). *Front. Ecol. Environ.* 11: 474–482 (Ecological Society of America).

Grimm, N.B., M.D. Staudinger, A. Staudt, S.L. Carter, F.S. Chapin, III, P. Kareiva, M. Ruckelhaus, and B.A. Stein. 2013b. Climate change impacts on ecological systems: Introduction to a US assessment. In: N.B. Grimm (Ed.). *Front. Ecol. Environ.* 11: 456–464 (Ecological Society of America).

Grontved, J. 1952. Investigations on the phytoplankton in the southern North Sea in May 1947. *Medd. Kom. Dan. Fisk. Havunders. Plankton* 5: 1–49.

Grossman, G.D., R. Coffin, and P.B. Moyle. 1980. Feeding ecology of the bay goby (Pisces: Gobiidae). Effects of behavioral, ontogenetic, and temporal variation in diet. *J. Exp. Mar. Biol. Ecol.* 44: 47–59.

Guildford, S.J. and R.E. Hecky. 2000. Total nitrogen, total phosphorus, and nutrient limitation in lakes and oceans: Is there a common relationship? *Limnol. Oceanogr.* 45: 1213–1223.

Gunster, D.G., N.L. Bonnevie, C.A. Gillis, and R.J. Wenning. 1993. Assessment of chemical loadings to Newark Bay, New Jersey from petroleum and hazardous chemical accidents occurring from 1986 to 1991. *Ecotoxicol. Environ. Saf.* 25: 202–213.

Guo, G.Z. and T. Pratt. 1993. Stormwater assessment of the Palafox Watershed. Northwest Florida Water Management District, Havana, FL. Water Resources Special Report 93-8, 86pp.

Haas, L.W., S.J. Hastings, and K.L. Webb. 1981. Phytoplankton response to a stratification mixing cycle in the York River estuary during late summer. In: B.J. Neilson and L.E. Cronin (Eds.), *Estuaries and Nutrients*. Humana Press, Clifton, NJ, pp. 619–636.

Hackney, C.T. and S.M. Adams. 1992. Aquatic communities of the southeastern United States: Past, present, and future. In: C.T. Hackney, S.M. Adams, and W.J. Martin (Eds.), *Biodiversity of the Southeastern United States*. John Wiley & Sons, New York, pp. 747–760.

Hackney, C.T. and E.B. Haines. 1980. Stable carbon isotope composition of fauna and organic matter collected in a Mississippi estuary. *Estuarine Coastal Mar. Sci.* 10: 703–708.

Hague, P.M., T.J. Belton, B.E. Ruppel, K. Lockwood, and R.T. Mueller. 1994. 2,3,7,8-TCDD and 2,3,7,8-TCDF in blue crabs and American lobsters from the Hudson-Raritan Estuary and the New York Bight. *Bull. Environ. Contam. Toxicol.* 52: 734–741.

Hagy, J.D. 1996. Residence times and net ecosystem processes in Patuxent River estuary. MS thesis, University of Maryland, College Park, MD, 205pp.

Hagy, J.D., W.R. Boynton, C.W. Keefe, and K.V. Wood. 2004. Hypoxia in Chesapeake Bay (1950–2001): Long-term change in relation to nutrient loading and river flow. *Estuaries* 27: 634–658.

Hagy, J.D., III, J.C. Lehrter, and M.C. Murrell. 2006. Effects of Hurricne Ivan on water quality in Pensacola Bay, Florida. *Estuaries Coasts* 29: 919–925.

Haigh, R. and F.J.R. Taylor. 1991. Mosaicism of microplankton communities in the northern Strait of Georgia, British Columbia. *Mar. Biol.* 110: 301–314.

Haines, E.B. 1979. Interactions between Georgia salt marshes and coastal waters: A changing paradigm. In: R.J. Livingston (Ed.), *Ecological Processes in Coastal and Marine Systems*. Plenum Press, New York, pp. 35–46.

Haines, E.B. and C.L. Montague. 1979. Food sources of estuarine invertebrates analyzed using $^{13}C/^{12}C$ ratios. *Ecology* 60: 48–56.

Hall, D.J., E.E. Werner, J.F. Gilliam, G.G. Mittelbach, D. Howard, C.G. Doner, J.A. Dickerman, and A.J. Stewart. 1979. Diel foraging behavior and prey selection in the golden shiner (*Notemigonus crysoleucas*). *J. Fish. Res. Board Can.* 39: 1029–1039.

Hallegraeff, G.M. 1993. A review of harmful algal blooms and their apparent global increase. *Phycologia* 32: 79–99.

Hallegraeff, G.M. 1995. Harmful algal blooms: A global overview. In: G.M. Hallegraeff, D.M. Anderson, and A.D. Cembella (Eds.), *Manual on Harmful Marine Microalgae*, IOC Manuals and Guides No. 33. International Oceanographic Commission, United Nations Educational, Scientific and Cultural Organization, Paris, France, pp. 1–22.

Hallegraeff, G.M., D.M. Anderson, and A.D. Cembella (Eds.). 2003. *Manual on Harmful Marine Microalgae*, Vol. 11, Monographs on Oceanographic Methodology. UNESCO Publishing, Paris, France, 793pp.

Hallegraeff, G.M. and S. Fraga. 1997. Bloom dynamics of the toxic dinoflagellate *Gymnodinium catenatum*, with emphasis on Tasmanian and Spanish coastal waters. In: D.M. Anderson, A.D. Cembella, and G.M. Hallagraeff (Eds.), *Physiological Ecology of Harmful Algal Blooms*. Springer-Verlag, Berlin, Germany, pp. 59–80.

Hallegraeff, G.M., M.A. MCausland, and R.K. Brown. 1995. Early warning of toxic dinoflagellate blooms of *Gymnodinium catenatum* in southern Tasmanian waters. *J. Plankton Res.* 17: 1163–1176.

Hand, J., J. Cole, and L. Lord. 1996. Florida water quality assessment, 305(b) technical appendix. Florida Department of Environmental Protection, Tallahassee, FL.

Hanes, N.B. and T.M. White. 1968. Effects of seawater concentration on oxygen uptake of a benthal system. *J. Water Pollut. Control Fed.* 40: 272–280.

Hannah, R.P. 1972. Primary productivity and certain limiting factors in a bayou estuary. MS thesis, University of West Florida, Pensacola, FL.

Hannah, R.P., A.T. Simmons, and G.A. Moshiri. 1973. Nutrient–productivity relationships in a bayou estuary. *J. Water Pollut. Control Fed.* 45: 2508–2520.

Hansen, P.J. 1997. Phagotrophic mechanisms and prey selection in mixotrophic phytoflagellates. In: D.M. Anderson, A.D. Cembella, and G.M. Hallagraeff (Eds.), *Physiological Ecology of Harmful Algal Blooms*. Springer-Verlag, Berlin, Germany, pp. 523–537.

Hara, Y. and M. Chihara. 1987. Morphology, ultrastructure, and taxonomy of the Raphidophycean alga *Heterosigma akashiwo*. *The Botanical Magazine* (Tokyo, Japan) [Shokubutsu-gaku-zasshi] 100: 151–163.

Harding, L.W. 1994. Long-term trends in the distribution of phytoplankton in the Chesapeake Bay: Role of light, nutrients and stream-flow. *Mar. Ecol. Prog. Ser.* 104: 267–291.

Hargraves, P.H. 1976. Studies on marine plankton diatoms. II. Resting spore morphology. *J. Phycol.* 12: 118–129.

Hargraves, P.H. 1990. Studies on marine planktonic diatoms: V. Morphology and distribution of *Leptocylindrus minimus* Gran. *Beihefte zur Nova Hedwigia* 100: 47–60.

Harlow, V.H. and R.W. Alden. 1997. Analysis of organomercury from fish collected in conjunction with the Shenandoah River mercury monitoring: Risk analysis of organomercury fish contamination in the South and South Fork Shenandoah Rivers. Applied Marine Research Laboratory, Old Dominion University, Norfolk, VA. Unpublished report for Chesapeake Bay Foundation.

Harmelin-Vivien, M.L. and C. Bouchon. 1976. Feeding behavior of some carnivorous fishes (Serranidae and Scorpaenidae) from Tulear (Madagascar). *Mar. Biol.* 37: 329–340.

Harris, H.H., I.J. Pickering, and G.N. George. 2003. The chemical form of mercury in fish. *Science* 301: 1203.

Harrison, G.W. 1979. Stability under environmental stress: Resistance, resilience, persistence, and variability. *Am. Nat.* 113: 659–669.

Harrison, P.J., M.H. Hu, Y.P. Yang, and X. Lu. 1990. Phosphate limitation in estuarine and coastal waters of China. *J. Exp. Mar. Biol. Ecol.* 140: 79–87.

Harriss, R.C. and R.R. Turner. 1974. Job completion report: Lake Jackson investigations, Job 5: Nutrients, water quality, and phytoplankton productivity, Job 7: Data analysis and reporting. Florida Game and Fresh Water Fish Commission, Tallahassee, FL, 231pp.

Hart, T.J. 1934. On the phytoplankton of the southwest Atlantic and Bellingshausen Sea, 1929–31. Cambridge University Press, London, U.K. Discovery Report 8, pp. 1–268.

Hartmann, K.J. 1993. Striped bass, bluefish, and weakfish in the Chesapeake Bay: Energetics, trophic linkages, and bioenergetics model applications. PhD dissertation, University of Maryland, College Park, MD.

Harvey, R. and K.E. Havens. 1999. Lake Okeechobee action plan. Lake Okeechobee Issue Team for the South Florida Ecosystem Restoration Working Group. South Florida Water Management District, West Palm Beach, FL.

Hasle, G.R. 1978a. The inverted microscope. In: A. Sournia (Ed.), *Phytoplankton Manual*, Vol. 6, Monographs on Oceanographic Methodology. UNESCO Press, Paris, France, pp. 88–96.

Hasle, G.R. 1978b. Diatoms. In: A Sournia (Ed.). *Phytoplankton Manual*, Vol. 6, Monographs on Oceanographic Methodology. UNESCO Press, Paris, France, pp. 136–142.

Hasle, G.R. 1978c. Using the inverted microscope. In: A Sournia (Ed.). *Phytoplankton Manual*, Vol. 6, Monographs on Oceanographic Methodology. UNESCO Press, Paris, France, pp. 191–196.

Hasle, G.R. and G.A. Fryxell. 1970. Diatoms: Cleaning and mounting for light and electron microscopy. *Trans. Am. Microsc. Soc.* 89: 469–474.

Hasle, G.R. and G.A. Fryxell. 1995. Taxonomy. Diatoms. In: G.M. Hallegraeff, D.M. Anderson, and A.D. Cembella (Eds.), *Manual on Harmful Marine Microalgae*, IOC Manuals and Guides No. 33. International Oceanographic Commission, United Nations Educational, Scientific and Cultural Organization, Paris, France, pp. 339–364.

Hasle, G.R. and E.E. Syvertsen. 1996. Marine Diatoms. In: C.R. Tomas (Ed.), *Identifying Marine Diatoms and Dinoflagellates*. Academic Press, San Diego, CA, pp. 5–385.

Hatanaka, M. and K. Iizuka. 1962. Studies and community structure of the *Zostera* area. Trophic order in a fish group living outside of the *Zostera* area. *Bull. Jpn. Soc. Sci. Fish.* 28: 155–161.

Haugh, L.D. and G.E.P. Box. 1977. Identification of dynamic regression (distributed lag) models connecting two time series. *J. Am. Stat. Assoc.* 72: 121–130.

Havens, K., M. Allen, E. Camp, T. Irani, A. Lindsey, J.G. Morris, A. Kane et al. 2013. Apalachicola Bay oyster situation report. University of Florida, Gainesville, FL. Florida Sea Grant Report TP 200, 32pp.

Hayes, P.F. and R.W. Menzel. 1981. The reproductive cycle of early setting *Crassostrea virginica* (Gmelin) in the northern Gulf of Mexico, and its implications for population recruitment. *Biol. Bull.* 160: 80–88.

Hayward, R.S. and F.J. Margraf. 1987. Eutrophication effects on prey size and food available to yellow perch in Lake Erie. *Trans. Am. Fish. Soc.* 116: 210–223.

Heard, W.H. 1977. Freshwater mollusca of the Apalachicola drainage. In: R.J. Livingston (Ed.), *Proceedings of the Conference on the Apalachicola Drainage System*, Gainesville, FL. Florida Department of Natural Resources, St. Petersburg, FL. Florida Marine Resources Publication 26, pp. 20–21.

Heck, K.L., Jr. 1973. The impact of pulp mill effluents on species assemblages of epibenthic marine invertebrates in Apalachee Bay, Florida. MS thesis, Florida State University, Tallahassee, FL.

Heck, K.L., Jr. 1979. Some determinants of the composition and abundance of motile macroinvertebrate species in tropical and temperate turtlegrass (*Thalassia testudinum*) meadows. *J. Biogeogr.* 6: 183–200.

Hecky, P.E. 1998. Low N:P ratios and the nitrogen fix: Why watershed nitrogen removal will not improve the Baltic. In: T. Hellstrom (Ed.), *Effects of Nitrogen in the Aquatic Environment*. Swedish National Committee for IAWQ (International Association of water Quality), Royal Swedish Academy of Sciences, Stockholm, Sweden, pp. 85–115.

Hecky, R.E. and P. Kilham. 1988. Nutrient limitation of phytoplankton in freshwater and marine environments. *Limnol. Oceanogr.* 33: 796–822.

Heffernan, J.B., P.A. Soranno, M.J. Angilletta, Jr., L.B. Buckley, D.S. Gruner, T.H. Keitt, J.R. Kellner et al. 2014. Macrosystems ecology: Understanding ecological patterns and processes at continental scales. *Front. Ecol. Environ.* 12: 1, 5–14.

Heil, C.A., M. Revilla, and P.M. Gilbert. 2007. Nutrient quality drives differentia phytoplankton community composition on the southwest Florida shelf. *Limnol. Oceanogr.* 52: 1067–1078.

Hein, M., M.F. Pedersen, and K. Sand-Jensen. 1995. Size-dependent nitrogen uptake in micro- and macroalgae. *Mar. Ecol. Prog. Ser.* 118: 247–253.

Helfman, G.S. 1978. Patterns of community structure in fishes: Summary and overview. *Environ. Biol. Fish.* 3: 129–148.

Hellstrom, T. 1996. An empirical study of nitrogen dynamics in lakes. *Water Environ. Res.* 68: 55–65.

Henderson-Sellers, B. and H.R. Markland. 1988. *Decaying Lakes: The Origins and Control of Cultural Eutrophication*. John Wiley & Sons, New York, 991pp.

Hershberger, P.K., J.E. Rensei, A.L. Matter, and F.B. Taub. 1997. Vertical distribution of the chloromonad flagellate *Heterosigma carterae* in columns: Implications for bloom development. *Can. J. Fish. Aquat. Sci.* 54: 2228–2234.

Hiatt, R.W. and D.W. Strasburg. 1960. Ecological relationships of fish fauna on coral reefs of the Marshal Islands. *Ecol. Monogr.* 30: 65–127.

Hillebrand, H. and U. Sommer. 1997. Response of epilithic microphytobenthos of the western Baltic Sea to in situ experiments with nutrient enrichment. *Mar. Ecol. Prog. Ser.* 160: 35–46.

Hitchcock, G.L. 1982. A comparative study of the size-dependent organic composition of marine diatoms and dinoflagellates. *J. Plankton Res.* 4: 363–377.

Hobson, E.S. 1965. Diurnal–nocturnal activity of some inshore fishes in the Gulf of California. *Copeia* 1965: 291–302.

Hobson, E.S. 1968. Predatory behavior of some shore fishes in the Gulf of California. US Fish and Wildlife Service Research Report 73, Washington, DC, 92pp.

Hobson, E.S. 1973. Diel feeding migrations in tropical reef fishes. *Helgol. Wissenschaftliche Meeresunters.* 24: 361–370.

Hobson, E.S. 1974. Feeding relationships of teleostean fishes on coral reefs in Kona, Hawaii. *Fish. Bull.* 72: 915–1031.

Hobson, E.S. 1975. Feeding patterns among tropical reef fishes. *Am. Sci.* 63: 382–392.

Hobson, E.W.S. 1978. Aggregating as a defense against predators in aquatic and terrestrial environments. In: S. Reese and F.J. Lighter (Eds.), *Contrasts in Behavior*. John Wiley, New York, pp. 219–234.

Hobson, E.S. and J.R. Chess. 1976. Trophic interactions among fishes and zooplankters near shore at Santa Catalina Island, California. *Fish. Bull.* 74: 567–598.

Hodgkiss, I.J. and W.S. Yim. 1995. A case study of Tolo Harbour, Hong Kong. In: A.J. McComb (Ed.), *Eutrophic Shallow Estuaries and Lagoons*. CRC Press, Boca Raton, FL, pp. 41–58.

Hoehn, T. 2002. Apalachicola–Chattahoochee–Flint River water control plan technical assistance document. Florida Department of Environmental Protection, Tallahassee, FL.

Hoese, H.D., B.J. Copeland, F.N. Moseley, and E.D. Lane. 1968. Fauna of the Aransas Pass Inlet, Texas: Diel and seasonal variations in trawlable organisms of the adjacent area. *Tex. J. Sci.* 20: 33–60.

Holland, A.F., N.K. Mountford, M.H. Hiefel, K.R. Kaunmeyer, and J.A. Mihursky. 1980. Influence of predation on infaunal abundance in upper Chesapeake Bay, USA. *Mar. Biol.* 57: 221–235.

Holland, A.F., N.K. Mountford, and J.A. Mihursky. 1977. Temporal variation in upper bay mesohaline benthic communities: 1. The 9-m mud habitat. *Chesap. Sci.* 18: 370–378.

Holling, C.S. 1973. Resilience and stability of ecological systems. *Annu. Rev. Ecol. Syst.* 4: 1–24.

Holmes, R.W. 1970. The Secchi disk in turbid coastal waters. *Limnol. Oceanogr.* 15: 688–694.

Holmguist, J.G. 1997. Disturbance and gap formation in a marine betnic mosaic influence of shifting macro algal patches on seas structure and invertebrates, *Mar. Ecol. Prog. Ser.* 158:121–130.

Honjo, T. 1993. Overview on bloom dynamics and physiological ecology of *Heterosigma akashiwo*. In: T.J. Smayda and Y. Shimizu (Eds.), *Toxic Phytoplankton Blooms in the Sea. Proceedings of the Fifth International Conference on Toxic Marine Phytoplankton*, Newport, RI. Elsevier, New York, pp. 33–41.

Honjo, T. 1994. The biology and prediction of representative red tides associated with fish kills in Japan. *Rev. Fish. Sci.* 1: 225–253.

Hood, M.A. and G.A. Moshiri. 1978. Environmental assessment of potential restoration programs for Bayou Texar, Pensacola, Florida. University of West Florida, Pensacola, FL (unpublished report).

Hooks, T.A., K.L. Heck, and R.J. Livingston. 1976. An inshore marine invertebrate community: Structure and habitat associations in the N.E. Gulf of Mexico. *Bull. Mar. Sci.* 26: 99–109.

Hopkins, T.S. 1969. The Escambia river and Escambia bay during summer 1969: A report in two parts. US Bureau of Commercial Fisheries, Gulf Breeze, FL. Contract No. 14-17-002-177, 54pp.

Hopkins, T.S. 1973. Marine ecology in Escarosa. Florida Department of Natural Resources, Coastal Coordinating Council, University of West Florida, Pensacola, FL, 100pp.

Hopkinson, C.S., I. Buffam, J. Hobbie, J. Vallino, M. Perdue, B. Eversmeyer, F. Prahl et al. 1998. Terrestrial inputs of organic matter to coastal ecosystems: An intercomparison of chemical characteristics and bioavailability. *Biogeochemistry* 43: 211–234.

Hopkinson, C.S., A.E. Giblin, J. Tucker, and R.H. Garritt. 1999. Benthic metabolism and nutrient cycling along an estuarine salinity gradient. *Estuaries* 22: 863–881.

Hopkinson, C.S. and R. Wetzel. 1982. In situ measurements of nutrient and oxygen fluxes in a coastal marine benthic community. *Mar. Ecol. Prog. Ser.* 10: 29–35.

Horner, R.A., D.L. Garrison, and E.G. Plumley. 1997. Harmful algal blooms and red tide problems on the U.S. west coast. *Limnol. Oceanogr.* 42: 1076–1088.

Horner, R.A., J.R. Postel, and J.E. Rensel. 1990. Noxious phytoplankton blooms in western Washington waters: A review. In: E.B. Graneli, B. Sundstrom, L. Elder, and D.M. Anderson (Eds.), *Toxic Marine Phytoplankton*. Elsevier, New York, pp. 171–176.

Horvath, G.J. 1968. The sedimentology of the Pensacola Bay system, Pensacola, Florida. MS thesis, Florida State University, Tallahassee, FL, 68pp.

Howarth, R.W. 1988. Nutrient limitation of net primary production in marine ecosystems. *Annu. Rev. Ecol. Syst.* 19: 89–110.

Howarth, R.W., D.M. Anderson, T.M. Church, H. Greening, C.S. Hopkinson, W.C. Huber, N. Marcus et al. 2000. *Clean Coastal Waters: Understanding and Reducing the Effects of Nutrient Pollution*. Ocean Studies Board and Water Science and Technology Board, National Academy Press, Washington, DC, 391pp.

Howarth, R.W., H.S. Jensen, R. Marino, and H. Postma. 1995. Transport to and processing of phosphorus in near-shore and oceanic waters. In: H. Tiessen (Ed.), *Phosphorus in the Global Environment*. John Wiley & Sons, Chichester, U.K., pp. 323–345.

Howarth, R.W. and R. Marino. 1998. A mechanistic approach to understanding why so many estuaries and brackish waters are nitrogen limited. In: T. Hellstrom (Ed.), Effects of nitrogen in the aquatic environment. KVA Report 1998. Royal Swedish Academy of Sciences, Stockholm, Sweden, pp. 117–136.

Howell, P. and D. Simpson. 1994. Abundance of marine resources in relation to dissolved oxygen in Long Island Sound. *Estuaries* 17: 394–402.

Huang, W. and W.K. Jones. 1997. Three-dimensional modeling of circulation and salinity for the low river flow season in Apalachicola Bay, Florida. Northwest Florida Water Management District, Havana, FL. Water Resources Special Report 97-10, 77pp.

Huber, W.C. and R.E. Dickinson. 1988. Storm water management model, Version 4, User's manual. Environmental Research Laboratory, Office of Research and Development. US Environmental Protection Agency, Athens, GA. EPA/600/3-99/001a, 76pp.

Hudson, P.L., D.R. Lenat, B.A. Caldwell, and D. Smith. 1990. Chironomidae of the southeastern United States: A checklist of species and notes on biology, distribution, and habitat. U.S. Fish and Wildlife Service. *Fish Wildl. Res.* 7: 1–46.

Hudson, R.J.M. and C.W. Shade. 2003. Letter-to-the-editor: More on mercury content in fish. *Science* 303: 763.

Hudson, T.J. and D.R. Wiggins. 1996. Comprehensive shellfish harvesting area survey of Pensacola Bay System, Escambia and Santa Rosa Counties, Florida. Department of Environmental Protection, Tallahassee, FL.

Hughes, T.P., A.M. Szmant, R. Steneck, R. Carpetner, and S. Miller. 1999. Algal blooms on coral reefs: What are the causes? *Limnol. Oceanogr.* 44: 1583–1586.

Hulburt, E.M. and N. Corwin. 1970. Relation of the phytoplankton to turbulence and nutrient renewal in Casco Bay, Maine. *J. Fish. Res. Board Can.* 27: 2081–2090.

Hulburt, E.M. and J. Rodman. 1963. Distribution of phytoplankton species with respect to salinity between the coast of southern New England and Bermuda. *Limnol. Oceanogr.* 8: 263–269.

Hunner, W.C., F.B. Ard, and J. Edmonds. 1994. Preliminary nonpoint source loading rate analysis of the Pensacola Bay System. Northwest Florida Water Management District, Havana, FL. Water Resources Special Report 92-2.

Hunter, M.D. and P.W. Price. 1992. Playing chutes and ladders: Heterogeneity and the relative roles of bottom-up and top-down forces in natural communities. *Ecology* 73: 724–732.

Huntley, S.L., N.L. Bonnevie, and R.J. Wenning. 1995. Polycyclic aromatic hydrocarbon and petroleum hydrocarbon contamination in sediment from the Newark Bay Estuary, New Jersey. *Arch. Environ. Contam. Toxicol.* 28: 93–107.

Huntley, S.L., R.J. Wenning, D.J. Paustenbach, A.S. Wong, and W.J. Luksemburg. 1994. Potential sources of polychlorinated dibenzothiophenes in the Passaic River, New Jersey. *Chemosphere* 29: 257–272.

Hutchinson, G.E. 1957. *A Treatise on Limnology*, Vol. 1, Geography, Physics, and Chemistry. John Wiley & Sons, New York, 1015pp.

Hutchinson, G.E. 1973. Eutrophication. *Am. Sci.* 61: 269–279.

Hydroqual, Inc. 1992. Elevenmile Creek and Perdido Bay modeling analysis. Hydroqual, Inc., Mahwah, NJ (unpublished report).

Iannuzzi, T.J., S.L. Huntley, N.L. Bonnevie, B.L. Finley, and R.J. Wenning. 1995. Distribution and possible sources of polychlorinated biphenyls in dated sediments from the Newark Bay Estuary, New Jersey. *Arch. Environ. Contam. Toxicol.* 28: 108–117.

Imai, I. 2003. Life histories of raphidophycean flagellates. In: T. Okaichi (Ed.), *Red Tides*. Terra Scientific Publishing Co., Tokyo, Japan, pp. 111–127.

Imai, I. and S. Iakura. 1999. Importance of cysts in the population dynamics of the red tide flagellate *Heterosigma akashiwo* (Raphicophyceae). *Mar. Biol.* 133: 755–762.

Imai, I., M. Kim, K. Nagasaki, S. Iakura, and Y. Ishida. 1998. Relationships between dynamics of red tide-causing raphidophycean flagellates and algicidal micro-organisms in the coastal Sea of Japan. *Phycol. Res.* 46: 139–146.

Imai, I., M. Yamaguchi, and M. Watanabe. 1997. Ecophysiology, life cycle, and bloom dynamics of *Chattonella* in the Seto inland sea, Japan. In: D.M. Anderson, A.D. Cembella, and G.M. Hallagraeff (Eds.), *Physiological Ecology of Harmful Algal Blooms*. Springer-Verlag, Berlin, Germany, pp. 93–112.

Ingle, R.M. 1951. Spawning and setting of oysters in relation to seasonal environmental changes. *Bull. Mar. Sci. Gulf Carib.* 1: 111–135.

Ingle, R.M. and C.E. Dawson. 1952. Growth of the American oyster *Crassostrea virginica* (Gmelin) in Florida waters. *Bull. Mar. Sci. Gulf Carib.* 2: 393–404.

Ingle, R.M. and C.E. Dawson. 1953. A survey of Apalachicola Bay. Florida State Board of Conservation, Tallahassee, FL. Technical Service 10, 38pp.

Inoue, Y. and K. Nozawa. 1989. Separation of toxin from harmful red tides occurring along the coast of Kagoshima Prefecture. In: T. Okaichi, D.M. Andersen, and T. Nemato (Eds.), *Red Tides: Biology, Environmental Science, and Toxicology*. Elsevier, New York, pp. 371–374.

Isphording, W.C., F.D. Imsand, and G.C. Flowers. 1987. Storm-related rejuvenation of a northern Gulf of Mexico estuary. *Trans. Gulf Coast Assoc. Geol. Soc.* 37: 357–370.

Isphording, W.C., F.D. Imsand, and G.C. Flowers. 1989. Physical characteristics and aging of gulf coast estuaries. *Trans. Gulf Coast Assoc. Geol. Soc.* 39: 387–401.

Isphording, W.C. and R.J. Livingston. 1989. Report on synoptic analyses of water and sediment quality in the Perdido Drainage System. Florida Department of Environmental Regulation, Tallahassee, FL (unpublished report).

Iverson, R.L. and H.F. Bittaker. 1986. Seagrass distribution and abundance in eastern Gulf of Mexico coastal waters. *Estuarine Coastal Shelf Sci.* 22: 577–602.

Iverson, R.L., W. Landing, B. Mortazawi, and J. Fulmer. 1997. Nutrient transport and primary productivity in the Apalachicola River and Bay. In: F.G. Lewis (Ed.), Apalachicola River and Bay freshwater needs assessment. Report to the ACF/ACT Comprehensive Study. Northwest Florida Water Management District, Havana, FL (unpublished report).

Ivlev, V.S. 1961. *Experimental Ecology of the Feeding Fishes*. Yale University Press, New Haven, CT, 302pp.

Jackson, J.B.C., M.X. Kirby, W.H. Wolfgang, H. Berger, K.A. Bjorndal, L.W. Botsford, B.J. Bourque et al. 2001. Historical overfishing and the recent collapse of coastal ecosystems. *Science* 293: 629–638.

Jaworski, N.A. 1981. Sources of nutrients and the scale of eutrophication problems in estuaries. In: B.J. Neilson and L.E. Cronin (Eds.), *Estuaries and Nutrients*. Humana Press, Clifton, NJ, pp. 81–100.

Jaworski, N.A., D.W. Lear, Jr., and O. Villa. 1972. Nutrient management in the Potomac Estuary. In: G.E. Likens (Ed.), *Nutrients and Eutrophication*, Vol. 1, American Society of Limnology and Oceanography. Special Symposia. Allen Press, Lawrence, KS, pp. 246–272.

Jay, D.A., R.J. Uncles, L. Largier, W.R. Geyer, J. Vallino, and W.R. Boynton. 1997. A review of recent developments in estuarine scalar flux estimation. *Estuaries* 20: 262–280.

Jenkin, P.M. 1937. Oxygen production by the diatom *Gloacinodiscus excentricus* in relation to submarine illumination in the English Channel. *J. Mar. Biol. Assoc. U.K.* 22: 1–301.

Jenkins, E.W. 2003. Environmental education and the public understanding of sciences. *Front. Ecol. Environ.* 1: 437–443.

Jones, K.C., J.A. Stratford, K.S. Waterhouse, and N.B. Vogt. 1989. Organic contaminants in welsh soils: Polynuclear aromatic hydrocarbons. *Environ. Sci. Technol.* 23: 540–550.

Jones, W.K. and W. Huang. 1996. Modeling changing freshwater delivery to Apalachicola Bay, Florida. In: M.L. Spaulding and R.T. Cheng (Eds.), *Estuarine and Coastal Modeling. Proceedings of the Fourth International Conference*, Apalachicola Bay, FL. American Society of Civil Engineering, New York, pp. 116–127.

Jones, W.K., J.H. Cason, and R. Bjorklund. 1992. A literature-based review of the physical, sedimentary, and water quality aspects of the Pensacola Bay System. Northwest Florida Water Management District, Havana, FL. Water Resources Special Report 92-5 (revision of Water Resources Special Report 90-3).

Jordan, T.E., D.L. Correll, J. Miklas, and D.E. Weller. 1991. Nutrients and chlorophyll *a* at the interface of a watershed and an estuary. *Limnol. Oceanogr.* 36: 251–267.

Jordan, T.E., D.E. Weller, and D.L. Correll. 2003. Sources of nutrient inputs to the Patuxent River Estuary. *Estuaries* 26: 226–243.

Jorgensen, B.B. and K. Richardson (Eds.). 1996. *Eutrophication in Coastal Marine Systems*, Vol. 52, Coastal and Estuarine Studies. American Geophysical Union, Washington, DC, 273pp.

Jouenne, F., S. Lefebvre, B. Véron, and Y. Lagadeuc. 2007. Phytoplankton community structure and primary production in small intertidal estuarine-bay ecosystem (eastern English Channel, France). *Mar. Biol.* 151: 805–825.

Justic, D. 1987. Long-term eutrophication of the northern Adriatic Sea. *Mar. Pollut. Bull.* 89: 1–5.

Kaiser, J. 2000. Mercury report backs strict rules. *Science* 289: 371–372.

Kankaanpaa, H.T., V.O. Sipia, J.S. Kuparinen, J.L. Ott, and W.W. Carmichael. 2001. Nodularin analyses and toxicity of a *Nodularia spumigena* (Nostocales, Cyanobacteria) water-bloom in the western Gulf of Finland, Baltic Sea, in August 1999. *Phycologia* 40: 268–274.

Karouna-Renier, N.K., R.A. Snyder, and K.R. Rao. 2005. Assessing fisheries as vectors for toxic materials from the environment to humans. An assessment of potential health risks posed by shellfish collected in estuarine waters near Pensacola, Florida. Center for Environmental Diagnostics and Bioremediation, University of West Florida, Pensacola, FL.

Kaul, L.W. and P.N. Froelich, Jr. 1984. Modeling estuarine nutrient geochemistry in a simple system. *Geochim. Cosmochem. Acta* 48: 1417–1433.

Keast, A. 1979. Patterns of predation in generalist feeders. In: H. Clepper (Ed.), *Predator-Prey Systems in Fisheries Management*. Sports Fishing Institute, Washington, DC, pp. 243–255.

Keefe, C.W., D.A. Flemer, and D.H. Hamilton. 1976. Seston distribution in the Patuxent River Estuary. *Chesap. Sci.* 17: 56–59.

Keister, J.E., E.D. Houde, and D.L. Breitburg. 2000. Effects of bottom-layer hypoxia on abundance and depth distributions of organisms in Patuxent River, Chesapeake Bay. *Mar. Ecol. Prog. Ser.* 205: 43–59.

Keller, B.D. and A. Itkin. 2002. Shoreline nutrients and chlorophyll *a* in the Florida Keys, 1994–1997: A preliminary analysis. In: J.W. Porter and K.G. Porter (Eds.), *The Everglades, Florida Bay and Coral Reefs of the Florida Keys: An Ecosystem Sourcebook*. CRC Press, Boca Raton, FL, pp. 649–658.

Kemp, W.M. and W.R. Boynton. 1980. Influence of biological and physical factors on dissolved oxygen dynamics in an estuarine system. *Estuarine Coastal Mar. Sci.* 11: 407–431.

Kemp, W.M. and W.R. Boynton. 1984. Spatial and temporal coupling of nutrient inputs to estuarine primary production: The role of particulate transport and decomposition. *Bull. Mar. Sci.* 35: 522–535.

Kemp, W.M. and W.R. Boynton. 1992. Benthic–pelagic interactions: Nutrient and oxygen dynamics. In: D.E. Smith, M. Leffler, and G. Mackiernan (Eds.), *Oxygen Dynamics in the Chesapeake Bay: A Synthesis of Recent Research*, Maryland Sea Grant Publication UM-SG-TS-92-01. University of Maryland, College Park, MD, pp. 149–209.

Kennedy, V.S. 1996. Biology of larvae and spat. In: V.S. Kennedy, R.E.I. Newell, and A.F. Eble (Eds.), *The Eastern Oyster: Crassostrea virginica*. University of Maryland, College Park, MD, Maryland Sea Grant Publication UM-SG-TS-96-01, pp. 371–421.

Kennedy, V.S., R.I.E. Newell, and A.F. Eble (Eds.). 1996. *The Eastern Oyster: Crassostrea virginica*. University of Maryland, College Park, MD. Maryland Sea Grant Publication UM-SG-TS-96-01, 734pp.

Kennish, M.J. 1997. *Estuarine and Marine Pollution*. CRC Press, Boca Raton, FL, 524pp.

Kennish, M.J. 1999. *Estuary Restoration and Maintenance*. CRC Press, Boca Raton, FL, 359pp.

Kennish, M.J. (Ed.). 2001a. Barnegat Bay-Little Egg Harbor, New Jersey: Estuary and watershed assessment. *J. Coastal Res. Spec. Iss.* 32: 280pp.

Kennish, M.J. 2001b. Physical description of the Barnegat Bay-Little Egg Harbor estuarine system. *J. Coastal Res.* 32: 13–27.

Kennish, M.J. 2001c. Benthic communities of the Barnegat Bay-Little Egg Harbor Estuary. *J. Coastal Res.* 32: 167–177.

Kennish, M., R.J. Livingston, D. Raffaelli, and K. Reise. 2003. The future of estuaries. *Review Paper for the International Conference on Environmental Future*, Zurich, Switzerland.

Kercher, J.R. and H.H. Shugart, Jr. 1975. Trophic structure, effective trophic position and connectivity in food webs. *Am. Nat.* 109: 191–206.

Ketchen, G.H. and R.C. Staley. 1979. *A Hydrographic Survey of Pensacola Bay*. Department of Oceanography, Florida State University, Tallahassee, FL.

Kiefer, D.A. 1973. Fluorescence properties of natural phytoplankton populations. *Mar. Biol.* 22: 263–269.

Kikuchi, T. 1974. Japanese contributions on consumer ecology in eelgrass (*Zostera marina*) beds with special reference to trophic relationships and resources in inshore fisheries. *Aquaculture* 4: 15–160.

Kikuchi, T. and J.M. Peres. 1977. Consumer ecology of seagrass beds. In: C.P. McRoy and C. Helffrich (Eds.), *Seagrass Ecosystems: A Scientific Perspective*. Marcel Dekker, New York, pp. 147–193.

Kim, M.K., I. Yoshinaga, I. Imai, K. Nagasaki, S. Itakura, and Y. Ishida. 1998. A close relationship between algicidal bacteria and termination of *Heterosigma akashiwo* (Raphidophyceae) blooms in Hiroshima Bay, Japan. *Mar. Ecol. Prog Ser.* 170: 25–32.

King, R.J. and B.R. Hodgson. 1995. Tuggerah Lakes system, New South Wales, Australia. 1995. In: A.J. McComb (Ed.), *Eutrophic Shallow Estuaries and Lagoons*. CRC Press, Boca Raton, FL, pp. 19–30.

Kirk, J.T.O. 1983. *Light and Photosynthesis in Aquatic Ecosystems*. Cambridge University Press, New York, 401pp.

Kirk, J.T.O. 1986. Optical limnology—A manifesto. In: P. De Deckker and W.D. Williams (Eds.), *Limnology in Australia*. CSIRO & Dr. W. Junk Publishers, Dordrecht, the Netherlands, pp. 33–62.

Kirk, R.E. 1995. *Experimental Design: Procedure for the Behavioral Sciences*. Brooks/Cole Publishing Company, Belmont, CA, 921pp.

Kivi, K., S. Kaitala, H. Kuosa, J. Kuparinen, E. Leskinen, R. Lignell, B. Marcussen, and T. Tamminen. 1993. Nutrient limitation and grazing control of the Baltic plankton community during annual succession. *Limnol. Oceanogr.* 38: 893–905.

Klein, C.J., III and J.A. Galt. 1986. A screening model framework for estuarine assessment. In: D.A. Wolfe (Ed.), *Estuarine Variability*. Academic Press, New York, pp. 483–501.

Koditschek, L.K. and P. Guyre. 1974. Antimicrobial-resistant coliforms in New York Bight. *Mar. Pollut. Bull.* 5: 71–74.

Koenig, C.C., R.J. Livingston, and C.R. Cripe. 1976. Blue crab mortality: Interaction of temperature and DDT residues. *Arch. Environ. Contam. Toxicol.* 4: 119–128.

Komarek, J. 1991. A review of water-bloom forming *Microcystis* species, with regard to populations from Japan. *Arch. Hydrobiol. Algolo. Stud.* 64: 115–127.

Komarek, J. and K. Anagnostidis. 1998. Cyanoprokaryota. Part 1. Chroococcales. In: H. Ettl, G. Gartner, H. Heynig, and D. Mollenhaeur (Eds.), *Süßwasserflora von Mitteleuropa*. Gustav Fischer Verlag, Stuttgart, Germany, pp. 1–548.

Kórner, S., S.K. Das, S. Veenstra, and J.E. Vermaat. 2001. The effect of pH variation at the ammonium-ammonia equilibrium in wastewater and its toxicity to *Lemna gibba*. *Aquat. Bot.* 71: 71–78.

Kroncke, I., J.W. Dippner, H. Heyen, and B. Zeiss. 1998. Long-term changes in macrofaunal communities off Norderney (East Frisia, Germany) in relation to climate variability. *Mar. Ecol. Prog. Ser.* 167: 25–36.

Kuo, A.Y. and B.J. Neilson. 1987. Hypoxia and salinity in Virginia estuaries. *Estuaries* 10: 277–283.

Kuo, A.Y., K. Park, and M.Z. Moustafa. 1991. Spatial and temporal variabilities of hypoxia in the Rappahannock River, Virginia. *Estuaries* 14: 113–121.

Lalli, C.M. (Ed.). 1990. *Enclosed Experimental Marine Ecosystems: A Review and Recommendations—A Contribution of the Scientific Committee on Oceanic Research Working Group 85*. Springer-Verlag, New York, 218pp.

Lam, C.W.Y. and K.C. Ho. 1989. Red tides in Tolo Harbor, Hong Kong. In: T. Okaichi, D.M. Anderson, and T. Nemoto (Eds.), *Red Tides: Biology, Environmental Science and Toxicology*. Elsevier, New York, pp. 49–52.

Landry, M.R. and R.P. Hassett. 1982. Estimating the grazing impact of marine micro-zooplankton. *Mar. Biol.* 67: 283–288.

Lapointe, B.E. 1997. Nutrient thresholds for bottom-up control of macroalgal blooms on coral reefs in Jamaica and southeast Florida. *Limnol. Oceanogr.* 42: 1583–1586.

Lapointe, B.E. 1999. Simultaneous top-down and bottom-up forces control macroalgal blooms on coral reefs. (Reply to the comment by Hughes et al.) *Limnol. Oceanogr.* 44: 1586–1592.

Lapointe, B.E. and P.J. Barile. 1999. Seagrass die-off in Florida Bay: An alternative interpretation. *Estuaries* 22: 460–470.

Lapointe, B.E. and P.J. Barile. 2004. Comment on J.C. Zieman, J.W. Fourqurean, and T.A. Frankovich. 1999. Seagrass die-off in Florida Bay: Long-term trends in abundance and growth of Turtlegrass, *Thalassia testudinum*. (*Estuaries* 22: 460–470) *Estuaries* 27: 157–164.

Lapointe, B.E., W.R. Matzie, and P.J. Barile. 2002. Biotic phase-shifts in Florida Bay and fore reef communities of the Florida Keys: Linkages with historical freshwater flows and nitrogen loading from Everglades runoff. In: J.W. Porter and K.G. Porter (Eds.), *The Everglades, Florida Bay and Coral Reefs of the Florida Keys: An Ecosystem Sourcebook*. CRC Press, Boca Raton, FL, pp. 629–648.

Lassus, P. and J.P. Berthome. 1988. Status of 1987 algal blooms in IFREMER. International Council for the Exploration of the Sea/Annex III C, Copenhagen K, Denmark, pp. 5–13.

Latasa, M. and R.R. Bidagare. 1998. Comparison of phytoplankton populations of the Arabian Sea during the spring inter-monsoon and southwest monsoon of 1995 as described by HPLC-analyzed pigments. *Deep Sea Res.* 11: 2133–2170.

Lathrop, R.G., R.M. Styles, S.P. Seitzinger, and J.A. Bognar. 2001. Use of GIS mapping and modeling approaches to examine the spatial distribution of sea grasses in Barnegat Bay. *Estuaries* 24: 904–916.

Laughlin, R.A. 1979. Trophic ecology and population distribution of the blue crab, *Callinectes sapidus*, Rathbun, in the Apalachicola estuary (North Florida, U.S.A.). PhD dissertation. Florida State University, Tallahassee, FL.

Laughlin, R.A. and R.J. Livingston. 1982. Environmental and trophic determinants of the spatial/temporal distribution of the brief squid (*Lolliguncula brevis*) in the Apalachicola Estuary (North Florida, USA). *Bull. Mar. Sci.* 32: 489–497.

Leach, J.H., M.G. Johnson, J.R.M. Kelso, J. Hartmann, W. Numann, and B. Entz. 1977. Responses of percid fishes and their habitats to eutrophication. *J. Fish. Res. Board Can.* 34: 1964–1971.

Leadbeater, B.S.C. 1972. Identification, by means of electron microscopy, of nanoplankton flagellates from the coast of Norway. *Sarsia* 49: 107–124.

Leber, K.M. 1983. Feeding ecology of decapod crustaceans and the influence of vegetation on foraging success in a subtropical seagrass meadow. PhD dissertation, Florida State University, Tallahassee, FL.

Leber, K.M. 1985. The influence of predatory decapods, refuge, and microhabitat selection on seagrass communities. *Ecology* 66: 1951–1964.

Lee, R.F., R. Sauerheber, and G.H. Dobbs. 1972. Uptake, metabolism and discharge of polycyclic aromatic hydrocarbons by marine fish. *Mar. Biol.* 17: 201–208.

Lee, T.N., E. Williams, E. Johns, D. Wilson, and N.P. Smith. 2002. Transport processes linking south Florida coastal ecosystems. In: J.W. Porter and K.G. Porter (Eds.), *The Everglades, Florida Bay and Coral Reefs of the Florida Keys: An Ecosystem Sourcebook*. CRC Press, Boca Raton, FL, pp. 309–342.

Legendre, L. and S. Demers. 1985. Auxiliary energy, ergoclines, and aquatic biological production. *Nat. Can.* 112: 5–14.

Lehman, P.W. 1992. Environmental factors associated with long-term changes in chlorophyll concentration in the Sacramento-San Joaquin Delta and Suisun Bay, California. *Estuaries* 15: 335–348.

Lehrter, J.C. 2008. Regulation of eutrophication susceptibility in oligohaline regions of a northern Gulf of Mexico estuary, Mobile Bay, Alabama. *Mar. Pollut. Bull.* 56: 1446–1460.

Leitman, H.M., J.E. Sohm, and M.A. Franklin. 1982. Wetland hydrology and tree distribution of the Apalachicola River floodplain, Florida. US Government Printing Office, Washington, DC. US Geological Survey Report 82, 92pp.

Leitman, S.F. 2003a. An evaluation of evapo-precipitation loses from impoundments in the ACF basin. Florida Department of Environmental Protection. Tallahassee, FL (unpublished report).

Leitman, S.F. 2003b. Review of cumulative monthly deficits during three major drought events. Florida Department of Environmental Protection, Tallahassee, FL (unpublished report).

Leitman, S.F. 2003c. Overview of consumptive demands in the Apalachicola–Chattahoochee–Flint drainage basin. Prepared for the nature conservancy. Florida Field Office, Altamonte Springs, FL.

Leitman, S.F. 2003d. A review of reservoir management in the ACF basin. Prepared for the nature conservancy. Florida Office, Tallahassee, FL.

Leitman, S.F., L. Ager, and C. Mesing. 1991. The Apalachicola experience: Environmental effects of physical modifications to a river for navigational purposes. In: R.J. Livingston (Ed.), *The Rivers of Florida*. Springer-Verlag, New York, pp. 223–246.

Leitman, S.F., J. Dowd, and S. Hombeck-Pelham. 2003. An evaluation of observed and unimpaired flow and precipitation during drought events in the ACF basin. In: K.J. Hatcher (Ed.), *Proceedings of the 2003 Georgia Water Resources Conference*, April 23–24, 2003. University of Georgia, Institute of Ecology, Athens, GA.

Levene, H. 1960. Robust tests for equality of variances. In: I. Olkin (Ed.), *Contributions to Probability and Statistics*. Stanford University Press, Palo Alto, CA, pp. 278–292.

Levin, L.A. 1984. Life history and dispersal patterns in a dense infaunal polychaete assemblage: Community structure and response to disturbance. *Ecology* 65: 1185–1200.

Levinton, J.S. 1982. *Marine Biology*. Prentice Hall, Englewood Cliffs, NJ, 526pp.

Lewis, F.G., III. 1974. Avoidance reactions of two species of marine fishes to kraft pulp mill effluent. MS thesis, Florida State University, Tallahassee, FL.

Lewis, F.G., III. 1982. Habitat complexity in a subtropical seagrass meadow. PhD dissertation, Florida State University, Tallahassee, FL.

Lewis, F.G., III. 2010. East Bay/Blackwater Bay/Lower Yellow River: Preliminary baseline resource characterization with a discussion of flow-dependent habitats and species. Northwest Florida Water Management District, Havana, FL. Special Report 2010-02, 101pp.

Lewis, F.G., III and R.J. Livingston. 1977. Avoidance of bleached kraft pulp mill effluent by pinfish (*Lagodon rhomboides*) and gulf killifish (*Fundulus grandis*). *J. Fish. Res. Board Can.* 34: 568–570.

Lewis, F.G., III and A.W. Stoner. 1983. Distribution of macro-fauna within seagrass beds: An explanation for patterns of abundance. *Bull. Mar. Sci.* 33: 296–304.

Light, H.M., M.R. Darst, and J.W. Grubbs. 1998. Aquatic habitats in relation to river flow in the Apalachicola River floodplain, Florida. US Geological Survey Professional Paper 1594, Tallahassee, FL, pp. 1–59.

Light, H.M., K.R. Vincent, M.R. Darst, and F.D. Price. 2006. Water-level decline in the Apalachicola River, Florida, from 1954 to 2004, and effects on floodplain habitats. US Geological Survey Scientific Investigations Report 2006-5173, Tallahassee, FL, 83pp.

Light, W.J. 1978. *Spionidae (Polychaeta: Annelida). Invertebrates of the San Francisco Bay Estuary System.* Boxwood Press, Pacific Grove, CA, 211pp.

Lindeman, R.L. 1942. The trophic–dynamic aspect of ecology. *Ecology* 23: 399–418.

Lirdwitayaprasit, T., S. Nishio, S. Montani, and T. Okaichi. 1990. The biochemical processes during cyst formation in *Alexandrium catenella*. In: L.E. Graneli, B. Sundstorm, L. Edler, and D.M. Anderson (Eds.), *Toxic Marine Phytoplankton*. Elsevier, Amsterdam, the Netherlands, pp. 294–299.

Little, E.J. and J.A. Quick. 1976. Ecology, resource rehabilitation, and fungal parasitology of commercial oysters, *Crassostrea virginica* (Gmelin), in Pensacola Estuary, Florida. Florida Department of Natural Resources, St. Petersburg, FL. Florida Marine Research Laboratory Report 21, 89pp.

Litts, T., H.R.A. Thomas, and R. Welch. 2003. Mapping irrigated lands in the ACF river basin. *Proceedings of the 2001 Georgia Water Resources Conference*, Athens, GA.

Livingston, R.J. 1973. Analysis of Mulat-Mulatto Bayou. Florida Department of Transportation, Tallahassee, FL (unpublished report).

Livingston, R.J. 1975a. Impact of kraft pulp-mill effluents on estuarine and coastal fishes in Apalachee Bay, Florida, USA. *Mar. Biol.* 32: 19–48.

Livingston, R.J. 1975b. Resource management and estuarine function with application to the Apalachicola drainage system. *Estuarine Pollut. Control Assess.* 1: 3–17.

Livingston, R.J. 1975c. Long-term fluctuations of epibenthic fish and invertebrate populations in Apalachicola Bay, Florida. *Fish. Bull.* 74: 311–321.

Livingston, R.J. 1976a. Diurnal and seasonal fluctuations of organisms in a north Florida estuary. *Estuarine Coastal Mar. Sci.* 4: 373–400.

Livingston, R.J. 1976b. Environmental considerations and the management of barrier islands: St. George Island and the Apalachicola Bay system. In: *Barrier Islands and Beaches: Technical Proceedings of the Barrier Island Workshop*, Annapolis, MD, May 17–18, 1976. The Conservation Foundation, Washington, DC, pp. 86–102.

Livingston, R.J. 1976c. Dynamics of organochlorine pesticides in estuarine systems: Effects on estuarine biota. In: M. Wiley (Ed.), *Estuarine Processes: Uses, Stresses and Adapting to the Estuary*. Academic Press, New York, pp. 507–522.

Livingston, R.J. 1977. The Apalachicola dilemma: Wetlands development and management initiatives. *National Wetlands Protection Symposium*, Reston, VA. Environmental Law Institute and the US Fish and Wildlife Service, Washington, DC, pp. 163–177.

Livingston, R.J. 1979. Multiple factor interactions and stress in coastal systems: A review of experimental approaches and field implications. In: F.J. Vernberg (Ed.), *Marine Pollution: Functional Responses*. Academic Press, New York, pp. 389–413.

Livingston, R.J. 1980a. Ontogenetic trophic relationships and stress in a coastal seagrass system in Florida. In: V.S. Peterson (Ed.), *Estuarine Perspectives*. Academic Press, New York, pp. 423–435.

Livingston, R.J. 1980b. The Apalachicola experiment: Research and management. *Oceanus* 23: 14–28.

Livingston, R.J. 1981a. River-derived input of detritus into the Apalachicola Estuary. In: R.D. Cross and D.L. Williams (Eds.), *Proceedings of the National Symposium on Freshwater Inflow to Estuaries*. US Fish and Wildlife Service, Washington, DC, pp. 320–332.

Livingston, R.J. 1981b. Man's impact on the distribution and abundance of sciaenid fishes. In: H. Clepper (Ed.), *Sixth Annual Marine Recreational Fisheries Symposium; Sciaenids: Territorial Demersal Resources*. National Marine Fisheries Service, Houston, TX, pp. 189–196.

Livingston, R.J. 1982a. Trophic organization in a coastal seagrass system. *Mar. Ecol. Prog. Ser.* 7: 1–12.

Livingston, R.J. 1982b. Between the idea and the reality: An essay on the problems involved in applying scientific data to research management problems. In: A. Donovan and A.L. Berge (Eds.), *Working Papers in Science and Technology Studies*. Virginia Polytechnic Institution and State University, Blacksburg, VA, pp. 31–59.

Livingston, R.J. 1983a. Compendium of knowledge concerning the flint river system (Georgia). Science Advisory Committee for the Flint River Ecosystem Study, Montezeuma, GA. Final report.

Livingston, R.J. 1983b. Field and semi-field validation of laboratory-derived aquatic test systems. US Environmental Protection Agency, Gulf Breeze, FL (unpublished report).

Livingston, R.J. 1983c. Identification and analysis of sources of pollution in the Apalachicola River and Bay System. Florida Department of Natural Resources, Tallahassee, FL. Final report.

Livingston, R.J. 1983d. Resource atlas of the Apalachicola Estuary. University of Florida, Sea Grant College Program, Gainesville, FL. Florida Sea Grant Publication 55, 64pp.

Livingston, R.J. 1983e. Review and analysis of the environmental implications of the proposed development of the east point breakwater and associated dredging operations within the east point channel (Apalachicola Bay System). Franklin County Board of County Commissioners, Apalachicola, FL. Final report.

Livingston, R.J. 1983f. The Biota of St. George Island. Franklin County Board of County Commissioners, Apalachicola, FL. Final report.

Livingston, R.J. 1984a. The ecology of the Apalachicola Bay System: An estuarine profile. US Fish and Wildlife Service, Washington, DC. FWS/PBS 82/05, 148pp.

Livingston, R.J. 1984b. Trophic response of fishes to habitat variability in coastal seagrass systems. *Ecology* 65: 1258–1275.

Livingston, R.J. 1984c. River-derived input of detritus into the Apalachicola Estuary. In: R.D. Cross and D.L. Williams (Eds.), *Proceedings of the National Symposium on Freshwater Inflow to Estuaries*. US Fish and Wildlife Service, Washington, DC, pp. 320–332.

Livingston, R.J. 1984d. Field characterization study of a Gulf coastal seagrass system. US Environmental Protection Agency, Washington, DC (unpublished report).

Livingston, R.J. 1984e. Aquatic field monitoring and meaningful measures of stress. In: H.H. White (Ed.), *Concepts in Marine Pollution Measurements*. University of Maryland, College Park, MD, Maryland Sea Grant Publication UM-SG-TS-84-03, pp. 681–692.

Livingston, R.J. 1984f. Long-term effects of dredging and open-water disposal on the Apalachicola Bay System. National Oceanic and Atmospheric Administration, Washington, DC. Final report.

Livingston, R.J. 1985a. The relationship of physical factors and biological response in coastal seagrass meadows. Proceedings of a seagrass symposium. Estuarine Research Foundation. *Estuaries* 7: 377–390.

Livingston, R.J. 1985b. Application of scientific research to resource management: Case history, the Apalachicola Bay system. In: N.L. Chao and W. Kirby-Smith (Eds.), *Proceedings of the International Symposium on Utilization of Coastal Ecosystems: Planning, Pollution, and Productivity*. Fundacao Universidad do Rio Grande, Rio Grande, Brazil, pp. 103–125.

Livingston, R.J. 1986a. The Choctawhatchee River-Bay System. Northwest Florida Water Management District, Havana, FL. Final report (unpublished report).

Livingston, R.J. 1986b. Analysis of field data concerning Old Pass Lagoon (Choctawhatchee Bay, Florida: September, 1985–February, 1986). Northwest Florida Water Management District, Havana, FL (unpublished final report).

Livingston, R.J. 1986c. Ecological processes of recruitment in coastal epibenthic macrobiota. *Proceedings from a Workshop on Recruitment in Tropical Coastal Demersal Communities*. Intergovernmental Oceanographic Commission, UNESCO Press, Paris, France. Report 44, pp. 151–166.

Livingston, R.J. 1987a. Historic trends of human impacts on seagrass meadows in Florida (invited paper). *Symposium Proceedings: Subtropical–Tropical Seagrasses of the Southeastern U.S.*, Gainesville, FL. Florida Marine Resources Publication 42, pp. 139–151.

Livingston, R.J. 1987b. Field sampling in estuaries: The relationship of scale to variability. *Estuaries* 10: 194–207.

Livingston, R.J. 1987c. Scientific research and resource management: Relationships and inconsistencies. In: J.A. Kusler and P. Riexinger (Eds.), *Proceedings of the National Wetlands Assessment Symposium*, June 17–20, 1985, Portland, ME. Association of State Wetland Managers, Chester, VT.

Livingston, R.J. 1988a. Inadequacy of species-level designations for ecological studies of coastal migratory fishes. *Environ. Biol. Fish.* 22: 225–234.

Livingston, R.J. 1988b. Projected changes in estuarine conditions based on models of long-term atmospheric alteration. US Environmental Protection Agency, Washington, DC. Report CR-814608-01-0.

Livingston, R.J. 1989a. Historical overview and data review: Perdido River complex, Elevenmile Creek, Bayou Marcus, and the Perdido Bay system. Environmental Planning & Analysis, Inc., Tallahassee, FL (unpublished report).

Livingston, R.J. 1989b. The ecology of the Choctawhatchee river system. Northwest Florida Water Management District, Havana, FL. Final report.

Livingston, R.J. 1990a. Application of scientific data to the management of the Apalachicola Oyster Resource. National Oceanic and Atmospheric Administration and the Florida Department of Environmental Regulation, Washington, DC. Final report.

Livingston, R.J. 1990b. Inshore marine habitats. In: R.L. Myers and J.J. Ewel (Eds.), *Ecosystems of Florida*. University of Central Florida Press, Orlando, FL, pp. 549–573.

Livingston, R.J. (Ed.) 1991a. *The Rivers of Florida*, Vol. 83, Ecological Study. Springer-Verlag, New York, 289pp.

Livingston, R.J. 1991b. Medium sized rivers of the Gulf coastal plain. In: C.T. Hackney, S.M. Adams, and W.H. Martin (Eds.), *Biodiversity of the Southeastern United States: Aquatic Communities*. John Wiley & Sons, New York, pp. 351–358.

Livingston, R.J. 1991c. Historical relationships between research and resource management in the Apalachicola River-estuary. *Ecol. Appl.* 1: 361–382.

Livingston, R.J. 1992. Ecological study of the Perdido drainage system. Florida Department of Environmental Regulation, Tallahassee, FL (unpublished report).

Livingston, R.J. 1993a. River-Gulf study. Florida Department of Environmental Regulation, Tallahassee, FL (unpublished report).

Livingston, R.J. 1993b. Estuarine wetlands. In: J. Berry and M. Dennison (Eds.), *Wetlands*. Noyes Publications, Saddle River, NJ, pp. 128–153.

Livingston, R.J. 1994. Ecological study of Jack's Branch and Soldier Creek. Florida Department of Environmental Regulation, Tallahassee, FL (unpublished report).

Livingston, R.J. 1995. Nutrient analysis of upper Perdido Bay. Florida Department of Environmental Protection, Tallahassee, FL (unpublished report).

Livingston, R.J. 1997a. Update on the ecological status of Perdido Bay. Florida Department of Environmental Protection, Tallahassee, FL (unpublished report).

Livingston, R.J. 1997b. Nutrient analysis of the Perdido Bay system. Florida Department of Environmental Protection, Tallahassee, FL (unpublished report).

Livingston, R.J. 1997c. Final analyses of Perdido database: The Soldier Creek System. Florida Department of Environmental Protection, Tallahassee, FL (unpublished report).

Livingston, R.J. 1997d. Eutrophication in estuaries and coastal systems: Relationships of physical alterations, salinity stratification, and hypoxia. In: F.J. Vernberg, W.B. Vernberg, and T. Siewicki (Eds.), *Sustainable Development in the Southeastern Coastal Zone*. University of South Carolina Press, Columbia, SC, pp. 285–318.

Livingston, R.J. 1998. Perdido Bay analysis: 10/88-9/98. Florida Department of Environmental Protection. Tallahassee, FL (unpublished report).

Livingston, R.J. 1999. Pensacola Bay System environmental study; Ecology and Trophic Organization. Florida Department of Environmental Protection, Tallahassee, FL (unpublished report).

Livingston, R.J. 2000. *Eutrophication Processes in Coastal Systems: Origin and Succession of Plankton Blooms and Effects on Secondary Production in Gulf Coast Estuaries*. CRC Press, Boca Raton, FL, 327pp.

Livingston, R.J. 2001. Mercury distribution in sediments and mussels in the Penobscot River-Estuary. Natural Resources Defense Council, New York (unpublished report).

Livingston, R.J. 2002. *Trophic Organization in Coastal Systems*. CRC Press, Boca Raton, FL, 388pp.

Livingston, R.J. 2005. *Restoration of Aquatic Systems*. Taylor & Francis Group, Boca Raton, FL, 424pp.

Livingston, R.J. 2007a. Phytoplankton bloom effects on a Gulf estuary: Water quality changes and biological response. *Ecol. Appl.* 17: 110–128.

Livingston, R.J. 2007b. Perdido project: Final report; River, Bay and Marsh Analyses. Florida Department of Environmental Protection, Tallahassee, FL (unpublished report).

Livingston, R.J. 2007c. *Atlas Concerning the Effects of Urban Runoff on a Series of Sink Hole Lakes*. Lone Shark Publishers, Tallahassee, FL, 70pp.

Livingston, R.J. 2008. Water quality and seagrass survey in the Carrabelle River and St. George Sound. Apalachicola Riverkeepers and the City of Carrabelle (unpublished report).

Livingston, R.J. 2010. Impacts of anthropogenic nutrient loading on coastal systems of the Gulf of Mexico. Choctawhatchee River-Bay System. Florida Department of Environmental Protection, Tallahassee, FL (unpublished report).

Livingston, R.J. 2013. Ecosystem-based management of the Apalachicola River–palachicola Bay system, Florida. In: J.W. Day and A. Yanez-Arancibia (Eds.), *Gulf of Mexico, Origin, Waters and Biota*, Vol. 4, The Gulf of Mexico: Ecosystem-Based Management. Texas A&M University Press, College Station, TX, pp. 53–69.

Livingston, R.J., Cairns, J., Jr., and K. Dickson (Eds.). 1978. *Biological Data in Water Pollution Assessment: Quantitative and Statistical Analyses*. American Society for Testing and Materials, Conshohoken, PA.

Livingston, R.J., R.J. Diaz, and D.C. White. 1985. Field validation of laboratory-derived multispecies aquatic test systems. US Environmental Protection Agency, Office of Research and Development, Environmental Research Laboratory, Gulf Breeze, FL. EPA/600/SA-85/039.

Livingston, R.J. and J.L. Duncan. 1979. Climatological control of a north Florida coastal system and impact due to upland forestry management. In: R.J. Livingston (Ed.), *Ecological Processes in Coastal and Marine Systems*. Plenum Press, New York, pp. 339–382.

Livingston, R.J., R.L. Howell, X. Niu, F.G. Lewis, and G.C. Woodsum. 1999. Recovery of oyster reefs (*Crassostrea virginica*) in a Gulf estuary following disturbance by two hurricanes. *Bull. Mar. Sci. Gulf Carib.* 64: 75–94.

Livingston, R.J., R.L. Iverson, R.H. Estabrook, V.E. Keys, and J. Taylor, Jr. 1974. Major features of the Apalachicola Bay system: Physiography, biota, and resource management. *Fla. Sci.* 4: 245–271.

Livingston, R.J. and E.A. Joyce, Jr. (Eds.). 1977. *Proceedings of the Conference on the Apalachicola Drainage System*, April 23–24, 1976, Gainesville, FL. Florida Marine Resources Publication 26.

Livingston, R.J., G. Kobylinski, F.G. Lewis, and P. Sheridan. 1976a. Analysis of long-term fluctuations of estuarine fish and invertebrate populations in Apalachicola Bay. *Fish. Bull.* 74: 311–321.

Livingston, R.J., S. Leitman, G.C. Woodsum, B. Galperin, P. Homann, J.D. Christensen, and M.E. Monaco. 2003. Relationships of river flow and productivity of the Apalachicola River-bay system. National Oceanic and Atmospheric Administration, Silver Spring, MD. Final report, NOS/CCMA Biogeography Program.

Livingston, R.J., F.G. Lewis, III, G.C. Woodsum, X. Niu, R.L. Howell, IV, G.L. Ray, J.D. Christensen et al. 2000. Coupling of physical and biological models: Response of oyster population dynamics to freshwater input in a shallow Gulf estuary. *Estuarine Coastal Shelf Sci.* 50: 655–672.

Livingston, R.J., R.S. Lloyd, and M.S. Zimmerman. 1976b. Determination of adequate sample size for collections of benthic macrophytes in polluted and unpolluted coastal areas. *Bull. Mar. Sci.* 26: 569–575.

Livingston, R.J. and O.L. Loucks. 1978. Productivity, trophic interactions, and food-web relationships in wetlands and associated systems. In: P.E. Greeson, J.R. Clark, and J.E. Clark (Eds.), *Wetland Functions and Values: The State of Our Understanding*. American Water Resources Association, Lake Buena Vista, FL, pp. 101–119.

Livingston, R.J., S.E. McGlynn, and X. Niu. 1998a. Factors controlling seagrass growth in a Gulf coastal system: Water and sediment quality and light. *Aquat. Bot.* 60: 135–159.

Livingston, R.J., A.W. Niedoroda, T.W. Gallagher, and A. Thurman. 1998b. Environmental studies of Perdido Bay. Florida Department of Environmental Protection, Tallahassee, FL (unpublished report).

Livingston, R.J., X. Niu, F.G. Lewis, and G.C. Woodsum. 1997. Freshwater input to a Gulf estuary: Long-term control of trophic organization. *Ecol. Appl.* 7: 277–299.

Livingston, R.J., A.K. Prasad, X. Niu, and S.E. McGlynn. 2002. Effects of ammonia in pulp mill effluents on estuarine phytoplankton assemblages: Field descriptive and experimental results. *Aquat. Bot.* 74: 343–367.

Livingston, R.J. and A.K.S.K. Prasad. 1989. A photographic atlas of north Florida estuarine phytoplankton and a summarization of life history relationships and associations with various environmental conditions. Florida Department of Environmental Regulation, Tallahassee, FL (unpublished report).

Livingston, R.J. and G.L. Ray. 1989. A simplified and rapid method for assessing the biological disturbances resulting from stormwater and marina discharges in estuaries. Florida Institute of Government and Florida Department of Environmental Regulation, Tallahassee, FL. Final report (unpublished report).

Livingston, R.J., P.S. Sheridan, B.G. McLane, F.G. Lewis, III, and G.G. Kobylinski. 1977. The biota of the Apalachicola Bay system: Functional relationships. In: R.J. Livingston (Ed.), *Proceedings of the Conference on the Apalachicola Drainage System*. Florida Department of Natural Resources, St. Petersburg, FL. Florida Marine Resources Publication 26, pp. 75–100.

Livingston, R.J., K.R. Smith, and W.H. Clements. 1984. Distribution of macroinvertebrates in the Flint River-Lake blackshear system. Science Advisory Committee for the Flint River Ecosystem Study, Montezeuma, GA. Final report.

Livingston, R.J. and H. Swanson. 1993. Project overview: The ecology of the lakes of Leon County. Leon County Board of County Commissioners, Florida State University and Department of Growth and Environmental Management of Leon County, Tallahassee, FL (unpublished report).

Livingston, R.J., N.P. Thompson, and D.A. Meeter. 1978. Long-term variation of organochlorine residues and assemblages of epibenthic organisms in a shallow north Florida (USA) estuary. *Mar. Biol.* 46: 355–372.

Livingston, R.J. and S.H. Wolfe. 1988. Restoration and preservation ranks of water bodies within the Northwest Florida Water Management District. Report for the Northwest Florida Water Management District, Havana, FL.

Livingstone, D.A. 1963. Chemical composition of rivers and lakes. In: M. Fleischer (Ed.), *Data of Geochemistry*, 6th edn. Tallahassee, FL, US Geological Survey Professional Paper 4406, pp. 1–61.

Ljung, R. and G.E.P. Box. 1978. On a measure of lack of fit in time series models. *Biometrika* 65: 297–303.

Locarnini, S.J.P. and B.J. Presley. 1996. Mercury concentrations in benthic organisms from a contaminated estuary. *Mar. Environ. Res.* 41: 225–239.

Lockwood, J.L., M.C. Ross, and J.P. Sah. 2003. Smoke on the water: The interplay of fire and water flow on Everglades restoration. *Ecol. Environ.* 1: 462–468.

Loesch, H. 1960. Sporadic mass shoreward migrations of demersal fish and crustaceans in Mobile Bay, Alabama. *Ecology* 41: 292–298.

Lohman, R., E. Nelson, S.J. Eisenreich, and K.C. Jones. 2000. Evidence for dynamic air–water exchange of PCDD/Fs: A study in the Raritan Bay/Hudson River Estuary. *Environ. Sci. Technol.* 34: 3086–3093.

Loneragan, N.R. 1999. River flows and estuarine ecosystems: Implications for coastal fisheries from a review and a case study of the Logan River, southeast Queensland. *Aust. J. Ecol.* 24: 431–441.

Long, E.R., G.M. Sloane, R.S. Carr, K.J. Scott, G.B. Thursby, E. Crecelius, C. Peven, and H.L. Windom. 1997. Magnitude and extent of sediment toxicity in four bays of the Florida panhandle: Pensacola, Choctawhatchee, St. Andrew, and Apalachicola. National Oceanic and Atmospheric Administration, Silver Spring, MD. Technical Memorandum NOS ORCA 117.

Lotze, H.K., H.S. Lenihan, B.J. Bourque, R.H. Bradbury, R.G. Cooke, M.C. Kay, S.M. Kidwell, M.X. Kirby, C.H. Peterson, and J.B.C. Jackson. 2006. Depletion, degradation, and recovery potential of estuaries and coastal seas. *Science* 312: 1806–1809.

Lovegrove, T. 1960. An improved form of sedimentation apparatus for use with an inverted microscope. *J. Cons.* 25: 279–284.

Lucas, L.V., J.R. Koseff, J.E. Cloern, S.G. Monismith, and J.K. Thompson. 1999a. Processes governing phytoplankton blooms in estuaries. I: The local production-loss balance. *Mar. Ecol. Prog. Ser.* 187: 1–15.

Lucas, L.V., J.R. Koseff, S.G. Monismith, J.E. Cloern, and J.K. Thompson. 1999b. Processes governing phytoplantion blooms in estuaries. II: The role of horizontal transport. *Mar. Ecol. Prog. Ser.* 187: 17–30.

Luczkovich, J.J. 1986. The patterns and mechanisms of selective feeding on seagrass-meadow epifauna by juvenile pinfish, *Lagodon rhomboids*. Ph.D Dissertation, State University, Tallahassee, FL.

Luczkovich, J.J. 1988. The role of prey detection in the selection of prey by the pinfish, *Lagodon rhomboides* (Linnaeus). *J. Exp. Mar. Biol. Ecol.* 123: 15–30.

Luczkovich, J.J., S.P. Borgatti, J.C. Johnson, and M.G. Everett. 2003. Defining and measuring trophic role similarity in food webs using regular coloration. *J. Theor. Biol.* 220: 303–321.

Luczkovich, J.J., G.P. Ward, R.R. Christian, J.C. Johnson, D. Baird, H. Neckels, and W. Rizzo. 2002. Determining the trophic guilds of fishes and macroinvertebrates in a seagrass food web. *Estuaries* 25: 1143–1164.

Macauley, J.M., V.D. Engle, J.K. Summers, J.R. Clark, and D.A. Flemer. 1995. An assessment of water quality and primary productivity in Perdido Bay, a northern Gulf of Mexico estuary. *Environ. Monit. Assess.* 78: 1–15.

MacDonald, D.D. 1994. Approach to the assessment of sediment quality in Florida Coastal Waters. Florida Department of Environmental Protection, Tallahassee, FL.

MacIntre, H.L., R.J. Geider, and D.C. Miller. 1996. Microphytobenthos: The ecological role of the "secret garden" of unvegetated, shallow-water marine habitats. 1. Distribution, abundance and primary production. *Estuaries* 19: 186–201.

MacKenzie, C.L., Jr. 1996. Management of natural populations. In: V.S. Kennedy, R.I.E. Newell, and A.F. Eble (Eds.), *The Eastern Oyster: Crassostrea virginica.* University of Maryland, College Park, MD. Maryland Sea Grant Publication UM-SG-TS-96-01, pp. 707–721.

Mackiernan, G.B. (Ed.). 1987. *Dissolved Oxygen in the Chesapeake Bay: Processes and Effects.* University of Maryland, College Park, MD. Maryland Sea Grant Publication UM-SG-TS-87-03, 174pp.

MacPherson, E. 1981. Resource partitioning in a Mediterranean demersal fish community. *Mar. Ecol. Prog. Ser.* 4: 183–193.

Maestrini, S.Y., D.J. Bonin, and M.R. Droop. 1984a. Phytoplankton as indicators of seawater quality: Bioassay approach and protocols. In: L.E. Shubert (Ed.), *Algae as Ecological Indicators.* Academic Press, New York, pp. 71–131.

Maestrini, S.Y., M.R. Droop, and D.J. Bonin. 1984b. Test algae as indicators of sea water quality: Prospects. In: L.E. Shubert (Ed.), *Algae as Ecological Indicators.* Academic Press, New York, pp. 132–188.

Mahoney, B.M.S. 1982. Seasonal fluctuations of benthic macrofauna in the Apalachicola estuary, Florida. The role of predation and larval availability. PhD dissertation, Florida State University, Tallahassee, FL.

Mahoney, B.M.S. and R.J. Livingston. 1982. Seasonal fluctuations of benthic macrofauna in the Apalachicola Estuary, Florida, U.S.A.: The role of predation. *Mar. Biol.* 69: 207–213.

Mahoney, J.B., F.H. Midlige, and D.G. Deuel. 1973. A fin rot disease of marine and euryhaline fishes in the New York Bight. *Trans. Am. Fish. Soc.* 102: 596–605.

Main, K.L. 1983. Behavioral response of acaridean shrimp to a predatory fish. PhD dissertation. Department of Biological Science, Florida State University, Tallahassee, FL.

Maine Department of Environmental Protection. 1998. Mercury in Maine. Land and Water Resources Council, 1997 Annual report, Appendix A. Augusta, ME.

Mallin, M.A. 1994. Phytoplankton ecology in North Carolina estuaries. *Estuaries* 17: 561–574.

Mallin, M.A., H.W. Paerl, J. Rudek, and P.W. Bates. 1993. Regulation of estuarine primary production by watershed rainfall and river flow. *Mar. Ecol. Prog. Ser.* 93: 199–203.

Mallin, M.S. and H.W. Paerl. 1994. Planktonic trophic transfer in an estuary: Seasonal, diel, and community structure effects. *Ecology* 75: 2168–2184.

Malone, T.C. 1977. Environmental regulation of phytoplankton productivity in the lower Hudson Estuary. *Estuarine Coastal Mar. Sci.* 13: 157–172.

Malone, T.C. 1992. Effects of water column processes on dissolved oxygen, nutrients, phytoplankton and zooplankton. In: D.E. Smith, M. Leffler, and G. Mackiernan (Eds.), *Oxygen Dynamics in the Chesapeake Bay: A Synthesis of Recent Research.* University of Maryland, College Park, MD. Maryland Sea Grant Publication UM-SG-TS-92-01, pp. 61–112.

Malone, T.C., D.J. Conley, T.R. Fisher, P.M. Gilbert, L.W. Harding, and K.G. Sellner. 1996. Scales of nutrient-limited phytoplankton productivity in Chesapeake Bay. *Estuaries* 19: 371–385.

Malone, T.C., L.H. Crocker, S.E. Pike, and B.A. Wendler. 1988. Influences of river flow on the dynamics of phytoplankton in a partially stratified estuary. *Mar. Ecol. Prog. Ser.* 32: 149–160.

Marcomini, A., A. Sfriso, B. Pavoni, and A.A. Orio. 1995. Eutrophication of the Lagoon of Venice: Nutrient loads and exchanges. In: A.J. McComb (Ed.), *Eutrophic Shallow Estuaries and Lagoons.* CRC Press, Boca Raton, FL, pp. 59–80.

Marino, D., G. Giuffre, M. Montressor, and A. Zingone. 1991. An electron microscope investigation on *Chaetoceros minimus* (Levande) and new observations *Chaetoceros throndsenii* (Marino et al. comb). *Diatom Res.* 6: 317–326.

Marino, D., G. Giuffre, M. Montresor, and A. Zingone. 1992. An electron microscope investigation on *Chaetoceros minimus* (Levander) Comb. nov. and new observations on *Chaetoceros throndsenii* (Marineo, Montresor & Zingone) comb. nov. *Diatom Res.* 6: 317–326.

Marino, D., M. Montresor, and A. Zingone. 1987. *Chaetoceros throndsenii* gen. et sp. nov., a planktonic diatom from the Gulf of Naples. *Diatom Res.* 2: 205–211.

Markley, S.M., D.K. Valdes, and R. Menge. 1990. Sanitary sewer contamination of the Miami River. Metro-Dade County Department of Environmental Resource Management, Miami, FL. DERM Technical Report 90-9.

Marmer, H.A. 1954. Tides and sea level in the Gulf of Mexico. In: P.S. Galtsoff (Coord.), Gulf of Mexico: Its origin, waters, and marine life. US Fish and Wildlife Service. *Fish. Bull.* 89: 101–118.

Marsh, G.P. 1964. *Man and Nature: Physical Geography as Modified by Human Actions.* Belknap Press, Cambridge, MA, 472pp.

Marsh, O.T. 1966. Geology of Escambia and Santa Rosa Counties, western Florida Panhandle. *Fla. Geol. Surv. Bull.* 46: 140pp.

Marshall, H.G. 1982. Phytoplankton distribution along the eastern coast of the USA. Part IV. Shelf waters between Cape Lookout, North Carolina, and Cape Canaveral, Florida. *Proc. Biol. Soc. Wash.* 95: 99–113.

Marshall, H.G. 1984a. Meso-scale distribution patterns for diatoms over the northeastern continental shelf of the United States. In: D.G. Mann (Ed.), *Proceedings of the Seventh International Diatom Symposium*, Philadelphia, PA, August 22–27, 1982. Koeltz Publications, Koenigstein, Germany, pp. 393–400.

Marshall, H.G. 1984b. Phytoplankton distribution along the eastern coast of the USA. Part V. Seasonal density and cell volume patterns for the northeastern continental shelf. *J. Plankton Res.* 6: 169–193.

Marshall, H.G. 1988. Distribution and concentration patterns of ubiquitous diatoms for the northeastern continental shelf of the United States. In: F.E. Round (Ed.), *Proceedings of the Ninth International Diatom Symposium*, August 24–30, 1986. Biopress Ltd., Bristol, England, U.K., pp. 75–85.

Marshall, H.G. and J.A. Ranasinghe. 1989. Phytoplankton distribution along the eastern coast of the U.S.A. Part VII. Mean cell concentrations and standing crop. *Cont. Shelf Res.* 9: 153–164.

Marshall, N.B. 1954. *Aspects of Deep Sea Biology*. Hutchinson, London, U.K., 380pp.

Martens, J. 1931. Beaches of Florida. Florida Geological Survey, Tallahassee, FL. 21st & 22nd annual report, pp. 67–119.

Martin, B.J. 1980. Effects of petroleum compounds on estuarine fishes. US Environmental Protection Agency, Washington, DC. EPA-600/3-80-019.

Martin, D.B. and W.A. Hartman. 1985. Arsenic, cadmium, lead, mercury, and selenium in sediments of riverine and pothole wetlands of the north central United States. *J. Assoc. Off. Anal. Chem.* 67: 1141–1146.

Martin, D.F., M.T. Doig, III, and R.H. Pierce, Jr. 1971. Distribution of naturally occurring chelators (humic acids) and selected trace metals in some Florida streams, 1968–1969. Florida Department of Natural Resources, St. Petersburg, FL. Professional Paper Series 12.

Martin, D.F. and W.H. Taft. 1998. Management of the Florida red tide-revisited. *Fla. Sci.* 61: 10–16.

Mason, C.F. 1987. A survey of mercury, lead, and cadmium in muscle of British freshwater fish. *Chemosphere* 16: 901–906.

Mattraw, H.C. and J.F. Elder. 1984. Nutrient and detritus transport in the Apalachicola River, Florida. US Geological Survey, Tallahassee, FL. Water-Supply Paper 2196-C, 62pp.

May, E.B. 1973. Extensive oxygen depletion in Mobile Bay, Alabama. *Limnol. Oceanogr.* 18: 353–366.

McAfee, R.O. 1984. Pensacola Bay Water Quality Monitoring Program review and assessment, June 30, 1983 to July 1, 1984. Escambia County Utilities Authority, Pensacola, FL.

McCain, B.B., S. Chan, M.M. Krahn, D.W. Brown, M.S. Myers, J.T. Landahl, S. Pierces, R.C. Clark, and U. Varanasi. 1992. Chemical contamination and associated fish diseases in San Diego Bay. *Environ. Sci. Technol.* 26: 725–733.

McCarthy, J.J., W.R. Taylor, and J.L. Taft. 1977. Nitrogenous nutrition of the plankton in the Chesapeake Bay. I. Nutrient availability and phytoplankton preferences. *Limnol. Oceanogr.* 22: 996–1011.

McComb, A.J. 1995. *Eutrophic Shallow Estuaries and Lagoons*. CRC Press, Boca Raton, FL, 240pp.

McComb, A.J., R.P. Atkins, P.B. Birch, D.M. Gordon, and R.J. Lukatelich. 1981. Eutrophication in the Peel-Harvey estuarine system, Western Australia. In: B.J. Neilson, and L.E. Cronin (Eds.), *Estuaries and Nutrients*. Humana Press, Clifton, NJ, pp. 223–342.

McComb, A.J. and R.J. Lukatelich. 1995. The Peel-Harvey estuarine system, Western Australia. In: A.J. McComb (Ed.), *Eutrophic Shallow Estuaries and Lagoons*. CRC Press, Boca Raton, FL, pp. 5–18.

McCormick, P.V., S. Newman, S. Miao, D.E. Gawlik, and D. Marley. 2002. Effects of anthropogenic phosphorus inputs on the Everglades. In: J.W. Porter and K.G. Porter (Eds.), *The Everglades, Florida Bay and Coral Reefs of the Florida Keys: An Ecosystem Sourcebook*. CRC Press, Boca Raton, FL, pp. 83–126.

McEachran, J.D., D.F. Boesch, and J.A. Musick. 1976. Food division within two sympatric species-pairs of skates (Pisces: Rajidae). *Mar. Biol.* 35: 301–317.

McElroy, A.D. and S.Y. Chiu. 1974. Water pollution investigation: Duluth-Superior area. Great Lakes Initiative Contract Program Report. EPA 905/9-74-014, 99pp.

McGlynn, S.E. 1995. Polynuclear aromatic hydrocarbons: Sediment and plant interactions. PhD dissertation, Department of Biological Sciences and Center for Aquatic Research and Resource Management, Florida State University, Tallahassee, FL.

McGlynn, S.E. and R.J. Livingston. 1997. The distribution of polynuclear aromatic hydrocarbons between aquatic plants and sediments. *Int. J. Quantum Chem.* 64: 1–13.

McKeown, J., A.H. Benedict, and G.M. Locke. 1968. Studies on the behavior of benthal deposits of wood origin. *J. Water Pollut. Control Fed.* 40: 333–353.

McLane, B.G. 1980. An investigation of the infauna of East Bay-Apalachicola Bay. MS thesis, Florida State University, Tallahassee, FL.

McManus, G.B. and M.C. Ederington-Cantrell. 1992. Phytoplankton pigments and growth rates, and microzooplankton grazing in a large temperate estuary. *Mar. Ecol. Prog. Ser.* 87: 77–85.

McNulty, J.K., W.N. Lindall, Jr., and E.A. Anthony. 1972. Cooperative Gulf of Mexico estuarine inventory and study, Florida: Phase 1 area description. National Oceanic and Atmospheric Administration, Silver Spring, MD. National Marine Fisheries Service Technical Report CIRC-368, 126pp.

McPherson, B.F. and K.M. Hammett. 1991. Tidal rivers of Florida. In: R.J. Livingston (Ed.), *The Rivers of Florida*. Springer-Verlag, New York, pp. 31–46.

McQueen, D.J., M.R.S. Johannes, J.R. Post, T.J. Stewart, and D.R.S. Lean. 1989. Bottom-up and top-down impacts on freshwater pelagic community structure. *Ecol. Monogr.* 59: 289–309.

Means, D.B. 1977. Aspects of the significance to terrestrial vertebrates of the Apalachicola River drainage basin, Florida. In: R.J. Livingston (Ed.), *Proceedings of the Conference on the Apalachicola Drainage System*. Florida Department of Natural Resources, St. Petersburg, FL. Florida Marine Resources Publication 26, pp. 37–67.

Mearns, A.J., M. Matta, G. Shigenaka, D. MacDonald, M. Buchman, H. Harris, J. Golas, and G. Lauenstein. 1991. Contaminant trends in the Southern California Bight: Inventory an assessment. National Oceanic and Atmospheric Administration, Seattle, WA. Technical Memorandum NOS ORCA 62, 404pp.

Meeter, D.A. and R.J. Livingston. 1978. Statistical methods applied to a four-year multivariate study of a Florida estuarine system. In: J. Cairns, Jr., K. Dickson, and R.J. Livingston (Eds.), *Biological Data in Water Pollution Assessment: Quantitative and Statistical Analyses*. American Society for Testing and Materials, Conshohoken, PA. ASTM Special Technical Publication 652, pp. 53–67.

Meeter, D.A., R.J. Livingston, and G.C. Woodsum. 1979. Short and long-term hydrological cycles of the Apalachicola drainage system with application to Gulf coastal populations. In: R.J. Livingston (Ed.), *Ecological Processes in Coastal and Marine Systems*. Plenum Press, New York, pp. 315–338.

Meinesz, A. 1999. *Killer Algae*. University of Chicago Press. Chicago, IL, 360pp.

Mellor, G.L. and T. Yamada. 1982. Development of a turbulence closure model for geophysical fluid problems. *Rev. Geophys. Space Phys.* 20: 851–875.

Menge, B.A. 1992. Community regulation: Under what conditions are bottom-up factors important on rocky shores. *Ecology* 73: 755–765.

Menzel, R.W. 1955a. Effects of two parasites on the growth of oysters. *Proc. Natl. Shellfish. Assoc.* 45: 184–186.

Menzel, R.W. 1955b. The growth of oysters parasitized by the fungus *Dermocystidium marinum* and by the trematode *Bucephalus cuculus*. *J. Parasitol.* 41: 333–342.

Menzel, R.W., N.C. Hulings, and R.R. Hathaway. 1957. Causes of depletion of oysters in St. Vincent Bar, Apalachicola Bay, Florida. *Proc. Natl. Shellfish. Assoc.* 48: 66–71.

Menzel, R.W., N.C. Hulings, and R.R. Hathaway. 1966. Oyster abundance in Apalachicola Bay, Florida, in relation to biotic associations influenced by salinity and other factors. *Gulf Res. Rep.* 2: 73–96.

Menzel, R.W. and F.E. Nichy. 1958. Studies of the distribution and feeding habits of some oyster predators in Alligator Harbor, Florida. *Bull. Mar. Sci. Gulf Carib.* 8: 125–145.

Menzie, C.A., B.B. Potocki, and J. Santodonato. 1992. Exposure to carcinogenic PAHs in the environment. *Environ. Sci. Technol.* 26: 1278–1284.

Merrett, N.R. and H.J.S. Roe. 1974. Patterns and selectivity in the feeding of certain mesopelagic fishes. *Mar. Biol.* 28: 115–126.

Meyer, M.S., D.C. Evers, J.J. Hartigan, and P.S. Rasmussen. 1998. Patterns of common loon (*Gavia immer*) mercury exposure, reproduction, and survival in Wisconsin, USA. *Environ. Toxicol. Chem.* 17: 184–190.

Meyers, V.B. and R.I. Iverson. 1981. Phosphorus and nitrogen limited phytoplankton productivity in northwestern Gulf of Mexico coastal estuaries. In: B.G. Neilson and L.E. Cronin (Eds.), *Estuaries and Nutrients*. Humana Press, Clifton, NJ, pp. 569–582.

Micheli, F. 1999. Eutrophication, fisheries, and consumer-resource dynamics in marine pelagic ecosystems. *Science* 285: 1396–1398.

Miller, J.M. and M.L. Dunn. 1980. Feeding strategies and patterns of movement in juvenile estuarine fishes. In: V.S. Kennedy (Ed.), *Estuarine Perspectives*. Academic Press, New York, pp. 437–448.

Miller, R. 1979. Relationships between habitat and feeding mechanisms in fishes. In: H. Clepper (Ed.), *Predator-Prey Systems in Fisheries Management*. Sports Fishery Institute, Washington, DC, pp. 269–288.

Miller, R.R., J.D. Williams, and J.E. Williams. 1989. Extinctions of North American fishes during the past century. *Fisheries* 14: 22–38.

Mitsch, W.J. and J.G. Gosselink. 1993. *Wetlands*, 2nd edn. John Wiley & Sons, New York, 722pp.

Mittelbach, G.G. 1988. Competition among refuging surf-fishes and effects of fish density on littoral zone invertebrates. *Ecology* 69: 614–623.

Mittelbach, G.G., C.W. Osenberg, and M.A. Leibold. 1988. Trophic relations and ontogenetic niche shifts in aquatic ecosystems. In: B. Ebenman and L. Persson (Eds.), *Size-Structured Populations*. Springer-Verlag, Berlin, Germany, pp. 219–235.

Moe, S.J., K. De Schamphelaere, W.H. Clements, M. Sorensen, P. van den Brink, and M. Liess. 2013. Interaction of global climate change and toxicant impacts on populations and communities. *Environ. Toxicol. Chem.* 32: 49–61.

Monaco, M.E. 1995. Comparative analysis of estuarine biophysical characteristics and trophic structure: Defining ecosystem function to fishes. PhD dissertation, University of Maryland, College Park, MD.

Monaco, M.E. and R.E. Ulanowicz. 1997. Comparative ecosystem trophic structure of three U.S. mid-Atlantic estuaries. *Mar. Ecol. Prog. Ser.* 161: 239–254.

Monselise, B.E. and D. Kost. 1993. Different ammonium uptake, metabolism, and detoxification efficiencies in two Lemnaceae. *Planta* 189: 167–173.

Montagna, P.A. and R.D. Kalke. 1992. The effect of freshwater inflow on meiofaunal and macrofaunal populations in the Guadalupe and Nueces estuaries, Tex. *Estuaries* 15: 307–326.

Montagna, P.A. and W.B. Yoon. 1991. The effect of freshwater inflow on meiofaunal consumption of sediment bacteria and microphytobenthos on San Antonio Bay, Texas. *Estuarine Coastal Shelf Sci.* 33: 599–547.

Montgomery, R.T., B.F. McPherson, and E.T. Emmons. 1991. Effects of nitrogen and phosphorus additions on phytoplankton productivity and chlorophyll a in a subtropical estuary, Charlotte Harbor, Florida. Denver, CO. US Geological Survey—Water-Resources Investigations Report 91-4077, 88pp.

Moore, K.A., D.J. Wilcox, and R.J. Orth. 2000. Analysis of abundance of submersed aquatic vegetation communities in Chesapeake Bay. *Estuaries* 23: 115–127.

Moran, P.A.P. 1950. Notes on continuous stochastic phenomena. *Biometrika* 47: 17–23.

Morang, A. 1992. A study of geologic and hydraulic processes at East Pass, Destin, Florida, Vol. I. US Army Corps of Engineers, Waterways Experiment Station, Coastal Engineering Research Center, Vicksburg, MS. Technical Report CERC-92-5, 28pp.

Morgan, E. (coordinator). 1998. Mercury in Maine. Land & water resources council, 1997 Annual report, Appendix A. Maine Department of Environmental Protection, Augusta, ME.

Morgan, J.P. and G.W. Stone. 1989. Recommendations for the dredging at the mouth of Bayou Texar, Pensacola, Escambia County, Florida. University of West Florida, Pensacola, FL.

Morrison, S.J., J.D. King, R.J. Bobbie, R.E. Bechtold, and D.C. White. 1977. Evidence of microfloral succession on allochthonous plant litter in Apalachicola Bay, Florida, USA. *Mar. Biol.* 41: 229–240.

Mortazavi, B., R.L. Iverson, W.M. Landing, and W. Huang. 2000a. Phosphorus budget of Apalachicola Bay: A river-dominated estuary in the northeastern Gulf of Mexico. *Mar. Ecol. Prog. Ser.* 198: 33–42.

Mortazavi, B., R.L. Iverson, W.M. Landing, F.G. Lewis, and W. Huang. 2000b. Control of phytoplankton production and biomass in a river-dominated estuary: Apalachicola Bay, Florida, USA. *Mar. Ecol. Prog. Ser.* 198: 19–31.

Mortazavi, B., R.L. Iverson, and W. Huang. 2000c. Dissolved organic nitrogen and nitrate in Apalachicola Bay, Florida: Spatial distributions and monthly budgets. *Mar. Ecol. Prog. Ser.* 214: 79–91.

Moshiri, G.A. 1976. Interrelationships between certain microorganisms and some aspects of sediment-water nutrient exchange in two bayou estuaries. Phases I and II. University of Florida, Gainesville, FL. Water Resources Research Center Publication 37, 45pp.

Moshiri, G.A. 1978. Certain mechanisms affecting water column-to-sediment phosphate exchange in a bayou estuary. *J. Water Pollut. Control Fed.* 50: 392–394.

Moshiri, G.A. 1981. Study of selected water quality parameters in Bayou Texar. University of West Florida, Pensacola, FL. Final report on Contract No. DACW01-80-0252.

Moshiri, G.A., N.G. Aumen, and W.G. Swann, III. 1980. Water quality studies in Santa Rosa Sound, Pensacola, Florida. US Environmental Protection Agency, Gulf Breeze Environmental Research Laboratory, Gulf Breeze, FL.

Moshiri, G.A. and W.G. Crumpton. 1978. Some aspects of redox trends in the bottom muds of a mesotrophic bayou estuary. *Hydrobiologia* 57: 155–158.

Moshiri, G.A., W.G. Crumpton, and N.G. Aumen. 1979. Dissolved glucose in a bayou estuary, possible sources and utilization by bacteria. *Hydrobiologia* 62: 71–74.

Moshiri, G.A., W.G. Crumpton, and D.A. Blaylock. 1978. Algal metabolites and fish kills in a bayou estuary: An alternative explanation to the low dissolved oxygen controversy. *J. Water Pollut. Control Fed.* 50: 2043–2046.

Moshiri, G.H., N.G. Aumen, and W.G. Crumpton. 1987. Reversal of the eutrophication process: A case study. In: B.G. Neilson and L.E. Cronin (Eds.), *Estuaries and Nutrients*. Humana Press, Clifton, NJ, pp. 370–390.

Moy, L.D. and L.A. Levin. 1991. Are *Spartina* marshes a replaceable resource? A functional approach to evaluation of marsh-creation efforts. *Estuaries* 14: 1–16.

Muessig, P.H. 1974. Acute toxicity of mercuric chloride and the accumulation and distribution of mercury chloride and methyl mercury in channel catfish (*Ictalurus punctatus*). PhD dissertation, Florida State University, Tallahassee, FL.

Muller, K. 1970. Phasenwechsel der lokomotorischen aktivitat bei der Quappe *Lota lota* L. *Oikos Suppl.* 13: 122–129.

Murdoch, W.W. 1966. Community structure, population control and competition—A critique. *Am. Nat.* 100: 219–226.

Murdock, J.F. 1955. An evaluation of pollution conditions in the Lower Escambia River. University of Miami, Coral Gables, FL. Florida Board of Conservation. Florida Marine Laboratory Report 55-28, 7pp.

Murphy, T.P., D.R.S. Lean, and C. Nalewajko. 1976. Blue-green algae: Their excretion of iron-selective chelators enables them to dominate other algae. *Science* 192: 900–902.

Musgrove, R.H., J.T. Barraclough, and R.G. Graham. 1965. Water resources of Escambia and Santa Rosa Counties, Florida. Florida Geological Survey, Tallahassee, FL. Report of Investigations No. 40.

Myers, V.B. 1977. Nutrient limitation of phytoplankton productivity in north Florida coastal systems: Technical considerations; spatial patterns; and wind mixing effects. PhD dissertation, Florida State University, Tallahassee, FL.

Myers, V.B. and R.I. Iverson. 1977. Aspects of nutrient limitation of phytoplankton productivity in the Apalachicola Bay system. In: R.J. Livingston (Ed.), *Proceedings of the Conference on the Apalachicola Drainage System*. Florida Department of Natural Resources, St. Petersburg, FL. Florida Marine Resources Publication 26, pp. 68–74.

Myers, V.B. and R.I. Iverson. 1981a. Phosphorus and nitrogen limited phytoplankton productivity in northeastern Gulf of Mexico coastal waters. In: B.J. Neilson and L.E. Cronin (Eds.), *Estuaries and Nutrients*. Humana Press, Clifton, NJ, pp. 569–582.

Myers, V.B. and R.I. Iverson. 1981b. Phosphorus and nitrogen limited phytoplankton productivity in northeastern Gulf of Mexico coastal waters. In: B.J. Neilson and L.E. Cronin (Eds.), *Estuaries and Nutrients*. Humana Press, Clifton, NJ, pp. 569–582.

Naito, K., M. Smatsui, and I. Imai. 2005. Influence of iron chelation with organic ligands on the growth of red tied phytoplankton. *Plankton Biol. Ecol.* 52: 14–26.

Nakazima, M. 1965. Studies on the source of shellfish poison in Lake Hamana. III. Poisonous effects on shellfishes feeding on *Prorocentrum* sp. *Bull. Jpn. Soc. Sci. Fish.* 31: 281–285.

National Commission on Water Quality. 1976. Environmental impact statement: Water quality analysis, Escambia River and Bay. Atlantic Scientific, Tallahassee, FL.

National Oceanic and Atmospheric Administration (NOAA). 1978. MESA New York Bight project annual report for fiscal year 1977. NOAA, Boulder, CO.

National Oceanic and Atmospheric Administration (NOAA). 1985. *National Estuarine Inventory: Data Atlas*, Vol. 1, Physical and Hydrologic Characteristics. NOAA Strategic Assessment Branch, Ocean Assessment Division, Rockville, MD, 103pp.

National Oceanic and Atmospheric Administration (NOAA). 1987. *National Estuarine Inventory: Data Atlas*, Vol. 2, Land-Use Characteristics. NOAA Strategic Assessment Branch, Ocean Assessment Division, Rockville, MD, 39pp.

National Oceanic and Atmospheric Administration (NOAA). 1995. Magnitude and extent of sediment toxicity in the Hudson-Raritan Estuary. National Ocean Service. Technical Memorandum NOS ORCA 88, Silver Spring, MD.

National Oceanographic and Atmospheric Administration (NOAA). 1997. *Estuarine Eutrophication Survey*, Vol. 4, Gulf of Mexico Region. Office of Ocean Resources Conservation and Assessment, Silver Spring, MD, 77pp.

National Oceanic and Atmospheric Administration (NOAA). 2003. Watershed database and mapping projects/Newark Bay. National Ocean Service, Silver Spring, MD.

National Research Council (NRC). 1983. *Proceedings of the Eutrophication Workshop*. May 1993, Marine Sciences Research Center at SUNY, Stony Brook, NY. National Research Council Marine Board, 54pp.

National Research Council (NRC). 1999. *From Monsoons to Microbes: Understanding the Ocean's Role in Human Health*. National Academy Press, Washington, DC, 178pp.

Neilson, B.G. and L.E. Cronin (Eds.). 1981. *Estuaries and Nutrients*. Humana Press, Clifton, NJ, 643pp.

Nelson, E.J., P. Kareiva, M. Ruckelhaus, K. Arkema, G. Geller, E. Gervetz, D. Goodrich et al. 2013. Climate change impact on key ecosystem services and the human well-being they support in the US. In: N.B. Grimm (Ed.). *Front. Ecol. Environ.* 11: 483–493 (Ecological Society of America).

Neutel, A., J.A.P. Heesterbeek, and P.C. De Rulter. 2002. Stability in real food webs: Weak links in long loops. *Science* 296: 1120–1123.

Newell, R.I.E. 1988. Ecological changes in Chesapeake Bay: Are they the result of over-harvesting the American oyster, *Crassostrea virginica*? Understanding the estuary: Advances in Chesapeake Bay Research. *Proceedings of a Conference*, March 29–31, 1988, Chesapeake Research Consortium Publication 129, Baltimore, MD, pp. 536–546.

Nichols, F.H. 1985. Increased benthic grazing: An alternative explanation for low phytoplankton biomass in northern San Francisco Bay during the 1976–1977 drought. *Estuarine Coastal Shelf Sci.* 21: 379–388.

Niedoroda, A.W. 1989. A study of the hydrography and oceanography of Perdido Bay. Florida Department of Environmental Regulation. Environmental Science & Engineering, Inc., Tallahassee, FL (unpublished report).

Niedoroda, A.W. 1990. A bathymetric survey of Perdido Bay. Florida Department of Environmental Regulation, Environmental Science & Engineering, Inc., Tallahassee, FL (unpublished report).

Niedoroda, A.W. 1992. Hydrography and oceanography of Perdido Bay. In: R.J. Livingston (Ed.), Ecological study of the Perdido Drainage system. Florida Department of Environmental Regulation, Tallahassee, FL, pp. 536–546 (unpublished report).

Niedoroda, A.W. 1999. Pensacola Bay system environmental study; Physical processes and oceanography. Florida Department of Environmental Protection, Environmental Science & Engineering, Inc., Tallahassee, FL (unpublished report).

Nihoul, J.C.J. 1986. *Marine Interfaces: Ecohydrodynamics*, Vol. 42, Elsevier Oceanography Series. Elsevier, New York, 670pp.

Nilsson, P., B. Jonsson, I. Swanberg, and K. Sundback. 1991. Response of a marine shallow-water sediment system to an increased load of inorganic nutrients. *Mar. Ecol. Prog. Ser.* 71: 275–290.

Niu, X.-F. 1995. Asymptotic properties of maximum likelihood estimates in a class of space-time models. *J. Multivar. Anal.* 55: 82–104.

Niu, X.-F., H.L. Edmiston, and G.O. Bailey. 1998. Time series models for salinity and other environmental factors in the Apalachicola National Estuarine Research Reserve. *Estuarine Coastal Shelf Sci.* 46: 549–563.

Nixon, S.W. 1980. Between coastal marshes and coastal waters: A review of twenty years of speculation and research on the role of salt marshes in estuarine productivity and water chemistry. In: P. Hamilton and K.B. MacDonald (Eds.), *Estuarine and Wetland Processes*. Plenum Press, New York, pp. 438–525.

Nixon, S.W. 1981a. Freshwater inputs and estuarine productivity. In: R.D. Cross and D.L. Williams (Eds.), *Proceedings of the National Symposium on Freshwater Inflow to Estuaries*. US Fish and Wildlife Service, Office of Biological Services, Washington, DC, FWS/OBS-81/04, pp. 31–57.

Nixon, S.W. 1981b. Remineralization and nutrient cycling in coastal marine ecosystems. In: B.G. Neilson and L.E. Cronin (Eds.), *Estuaries and Nutrients*. Humana Press, Clifton, NJ, pp. 111–138.

Nixon, S.W. 1988a. Comparative ecology of freshwater and marine systems. *Limnol. Oceanogr.* 33: 1–1025.

Nixon, S.W. 1988b. Physical energy inputs and the comparative ecology of lake and marine ecosystems. *Limnol. Oceanogr.* 33: 1005–1025.

Nixon, S.W. 1995. Coastal marine eutrophication: A definition, social causes, and future concerns. *Ophelia* 41: 199–219.

Nixon, S.W. and M.E.Q. Pilson. 1983. Nitrogen in estuaries and coastal marine ecosystems. In: E.J. Carpenter and D.G. Capone (Eds.), *Nitrogen in the Marine Environment*. Academic Press, New York, pp. 565–648.

Nixon, S.W., M.E.Q. Pilson, C.A. Oviatt, P. Donaghy, B. Sullivan, S. Seitzinger, D. Rudnick, and J. Frithsen. 1984. Eutrophication of a coastal marine ecosystem: An experimental study using the MERL microcosms. In: M.J.R. Fasham (Ed.), *Flows of Energy and Materials in Marine Ecosystems: Theory and Practice*. Plenum Press, New York, pp. 105–135.

Nordlie, F.G. 1990. Rivers and springs. In: R.L. Myers and J.J. Ewel (Eds.), *Ecosystems of Florida*. University of Central Florida Press, Orlando, FL, pp. 392–428.

Norin, L.L. 1977. ^{14}C-bioassays with the natural phytoplankton in the Stockholm Archipelago. *Ambio Spec. Rep.* 5: 15–21.

Northwest Florida Water Management District (NWFWMD). 1978. Evaluation of the sedimentation and hydraulic characteristics of Bayou Texar and Carpenters Creek, Escambia County, Florida. Report to Office of Water Resource and Restoration, Division of Environmental Program, Florida Department Environmental Regulation, Havana, FL.

Northwest Florida Water Management District (NWFWMD). 1988a. Stormwater evaluation for the restoration of Bayou Texar. Havana, FL. Water Resources Special Report 88-3.

Northwest Florida Water Management District (NWFWMB). 1988b. Surface water improvement and management program, Pensacola Bay System. Havana, FL. Program Development Series 89-2 and Technical Appendix.

Northwest Florida Water Management District (NWFWMD). 1990. Pensacola Bay system circulation project description. Unpublished Florida Coastal Management Grant Application. Havana, FL.

Northwest Florida Water Management District (NWFWMD). 1992. Characterization of karst development in Leon County, Florida, for the delineation of wellhead protection areas. Havana, FL (unpublished report).

Northwest Florida Water Management District (NWFWMD). 1993. Review of the Pensacola Bay system nonpoint source 208 plan methodologies. Havana, FL. Water Management District Program Development Series 93.2.

O'Brien, W.J. and F. DeNoyelles, Jr. 1974. Relationship between nutrient concentration, phytoplankton density, and zooplankton density in nutrient enriched experimental ponds. *Hydrobiologia* 44: 105–125.

O'Connor, D.J. 1981. Modeling of eutrophication in estuaries. In: B.G. Neilson and L.E. Cronin (Eds.), *Estuaries and Nutrients*. Humana Press, Clifton, NJ, pp. 182–223.

O'Connor, D.J., R.V. Thomann, and D.M. DiToro. 1977. Water quality analysis of estuarine systems. In: National Research Council (U.S.) (Ed.), *Geophysics of Estuaries Panel, Estuaries, Geophysics, and the Environment*. National Academy of Sciences, Washington, DC, pp. 71–83.

O'Connors, H.B. and I.W. Duedall. 1975. The seasonal variation in sources, concentrations, and impacts of ammonium in the New York Bight apex. In: T.M. Church (Ed.), *Marine Chemistry in the Coastal Environment: A Special Symposium Sponsored by the Middle Atlantic Region at the 169th Meeting of the American Chemical Society*, Philadelphia, PA, April 8–10, 1975, pp. 636–663.

Oczkowski, A.J., F.G. Lewis, S.W. Nixon, H.L. Edmiston, R.S. Robinson, and J.P. Chanton. 2011. Fresh water inflow and oyster productivity in Apalachicola Bay, FL (USA). *Estuaries Coasts* 34: 993–1005.

Odum, E.P. 1953. *Fundamentals of Ecology*. W.B. Saunders, Philadelphia, PA, 384pp.

Odum, E.P. and A.A. de la Cruz. 1967. Particulate organic detritus in a Georgia salt marsh-estuarine ecosystem. In: G.H. Lauff (Ed.), *Estuaries*. American Association for the Advancement of Science Publication 83, Washington, DC, pp. 383–388.

Odum, H.T. and R.F. Wilson. 1962. Further studies on reaeration and metabolism of Texas bays, 1958–1960. *Publ. Inst. Mar. Sci. Univ. Tex.* 8: 23–55.

Odum, W.E., J.S. Fishes, and J.C. Pickral. 1979. Factors controlling the flux of particulate organic carbon from estuarine wetlands. In: R.J. Livingston (Ed.), *Ecological Processes in Coastal and Marine Systems*. Plenum Press, New York, pp. 69–82.

Odum, W.E. and E.J. Heald. 1972. Trophic analyses of an estuarine mangrove community. *Bull. Mar. Sci.* 22: 671–738.

Odum, W.E., C.C. McIvor, and T.G. Smith. 1982. The Florida mangrove zone: A community profile. US Fish and Wildlife Service, Office of Biological Services, Washington, DC. FWS/OBS-82/24, 144pp.

Oey, L.-Y., G.L. Mellor, and R.I. Hires. 1985a. A three-dimensional simulation of the Hudson-Raritan Estuary. Part I: Description of the model and model simulations. *J. Phys. Oceanogr.* 15: 1676–1692.

Oey, L.-Y., G.L. Mellor, and R.I. Hires. 1985b. A three-dimensional simulation of the Hudson-Raritan Estuary. Part II: Comparison with observation. *J. Phys. Oceanogr.* 15: 1693–1709.

Office of Program Policy Analysis and Government Accountability. 1995. Review of the implementation of the surface water improvement and management program. Tallahassee, FL. OPPAGA Report 95-20.

Officer, C.B., R.B. Biggs, J.L. Taft, L.E. Cronin, M.A. Tyler, and W.R. Boynton. 1984. Chesapeake Bay anoxia: Origin, development, and significance. *Science* 223: 22–27.

Ogren, L.H. and H.A. Brusher. 1977. The distribution and abundance of fishes caught with a trawl in the St. Andrews Bay System, Florida. *Northeast Gulf Sci.* 1: 83–105.

Okaichi, T. 1997. Red tides in the Seto Inland Sea. In: T. Okaichi, and T. Yanagi (Eds.), *Sustainable Development in the Seto Inland Sea, Japan: From the Viewpoint of Fisheries*. Terra Scientific Publishing Company, Tokyo, Japan, pp. 251–304.

Okaichi, T. and Y. Imatomi. 1979. Toxicity of *Prorocentrum minimum* var. Mariae-Lebouriae assumed to be a causative agent of short-necked clam poisoning. In: D.L. Taylor and H.H. Seliger (Eds.), *Toxic Dinoflagellate Blooms*. Elsevier-North-Holland, Amsterdam, the Netherlands, pp. 385–388.

Okey, T.A. and D. Pauly. 1999. A mass-balanced model of trophic flows in Prince William Sound: De-compartmentalizing ecosystem knowledge. In: S. Keller (Ed.), *Ecosystem Approaches for Fisheries Management*. University of Alaska Sea Grant, Fairbanks, AK, pp. 621–635.

Olinger, L.W., R.G. Rogers, P.L. Force, T.L. Todd, B.L. Mullings, F.T. Bisterfeld, and L.A. Wise, II. 1975. Environmental and recovery studies of Escambia Bay and the Pensacola Bay System, Florida. US Environmental Protection Agency, Washington, DC, 471pp.

Oliver, L.M., W.S. Fisher, S.E. Ford, L.M. Ragone-Calvo, E.M. Burreson, E.B. Sutton, and J. Gandy. 1998. *Perkinsus marinus* tissue distribution and seasonal variation in oysters *Crassostrea virginica* from Florida, Virginia, and New York. Dis. *Aquat. Organ.* 34: 51–61.

Olsen, L.A. 1973. Food and feeding in relation to the ecology of two estuarine clams, *Rangia cuneata* (Gray) and *Polymesoda caroliniana* (Bosc). MS thesis, Florida State University, Tallahassee, FL, 103pp.

Olsen, P.S. and J.B. Mahoney. 2001. Phytoplankton in the Barnegat Bay-Little Egg Harbor estuarine system: Species composition and picoplankotn bloom development. In: M.J. Kennieh (Ed.), Barnegat Bay-Little Egg Harbor, New Jersey: Estuary and watershed assessment. *J. Coast. Res. Spec. Issue* 32: 113–143.

O'Neill, R.V. 2001. Is it time to bury the ecosystem concept? (With full military honors, of course!). *Ecology* 82: 3275–3284.

Orlando, S.P., Jr., L.P. Rozas, G.H. Ward, and C.J. Klein. 1993. *Salinity Characteristics of Gulf of Mexico Estuaries*. National Oceanic and Atmospheric Administration, Office of Ocean Resources Conservation and Assessment, Silver Spring, MD, 209pp.

Orth, R.J. and K.L. Heck, Jr. 1980. Structural components of eelgrass (*Zostera marina*) meadows in the lower Chesapeake Bay. II. Fishes. *Estuaries* 3: 278–288.

O'Shea, M.L. and T.M. Brosnon. 2000. Trends in indicators of eutrophication in western Long Island Sound and the Hudson-Raritan Estuary. *Estuaries* 23: 877–901.

Oshima, Y., C.J. Bolch, and G.M. Hallegraeff. 1992. Acute effects of the cyanobacterium cysts of *Alexandrium tamarense* (Dinophyceae). *Toxicon* 30: 1539–1544.

Oviatt, C.A., P. Doering, B. Nowicki, I. Reed, J. Cole, and J. Frithsen. 1995. An ecosystem level experiment on nutrient limitation in temperate coastal marine environments. *Mar. Ecol. Prog. Ser.* 116: 171–179.

Paasche, E.E. and S.R. Erga. 1988. Phosphorus and nitrogen limitation of phytoplankton in the Inner Oslofjord (Norway). *Sarsia* 73: 229–243.

Paerl, H.W. 1997. Coastal eutrophication and harmful algal blooms: The importance of atmospheric and groundwater as "new" nitrogen and other nutrient sources. *Limnol. Oceanogr.* 42: 1154–1165.

Paerl, H.W. 2009. Controlling eutrophication along the freshwater-marine continuum: Dual nutrient N and P reductions are essential. *Estuaries Coasts* 32: 593–601.

Paerl, H.W., K.L. Rossignol, S.N. Hall, B.L. Peierls, and M.S. Wetz. 2009. Phytoplankton community indicators of short- and long-term ecological change in the anthropogenically and climatically impacted Neuse River Estuary, North Carolina, USA. *Estuaries Coasts* 33: 485–497.

Paerl, H.W., L.M. Valdes, A.R. Joyner, B.L. Peirls, M.F. Piehler, S.R. Riggs, R.R. Christian et al. 2006. Ecological response to hurricane events in the Pamlico Sound System, North Carolina and implications for assessment and management in a regime of increased frequency. *Estuaries Coasts* 29: 1033–1045.

Paerl, H.W., L.M. Valdes, B.L. Peierls, J.E. Adolf, and L.W. Harding. 2006. Anthropogenic and climatic influences on the eutrophication of large estuarine ecosystems. *Limnol. Oceanogr.* 51: 448–462.

Paine, R.T., M.J. Tegner, and E.A. Johnson. 1998. Compounded perturbations yield ecological surprises. *Ecosystems* 1: 535–545.

Palmer, C.M. and T.E. Maloney. 1954. A new counting slide for nannoplankton. *Am. Soc. Limnol. Oceanogr. Spec. Publ.* 21: 6pp.

Palmer, S.L. 1984. Surface water. In: E.A. Fernald and D.J. Patton (Eds.), *Water Resources Atlas of Florida*. Institute of Science and Public Affairs, Florida State University, Tallahassee, FL, pp. 54–67.

Pankratz, A. 1983. *Forecasting with Univariate Box-Jenkins Models: Concepts and Cases*. John Wiley & Sons, New York, 562pp.

Pankratz, A. 1991. *Forecasting with Dynamic Regression Models*. John Wiley & Sons, New York, 400pp.

Parker, C.A. and J.E. O'Reilly. 1991. Oxygen depletion in Long Island Sound: A historical perspective. *Estuaries* 14: 248–264.

Parks, J.W., A. Lutz, and J.A. Sutton. 1989. Water column methylmercury in the Wabigoon/English River-Lake system: Factors controlling concentrations, speciation, and net production. *Can. J. Fish. Aquat. Sci.* 46: 2184–2202.

Parry, G.D., J.S. Langdon, and J.M. Huisman. 1989. Toxic effects of a bloom of the diatom *Rhizosolenia chunii* on shellfish in Port Philip Bay. *Mar. Biol.* 102: 25–41.

Parsons, K.C. 2003. Chemical residues in cormorants from New York Harbor and control location. Contract C003868. New York State Department of Environmental Conservation, Division of Fish, Wildlife and Marine Resources, Albany, NY (unpublished report).

Parsons, T.R., Y. Maita, and C.M. Lalli. 1984. *A Manual of Chemical and Biological Methods for Seawater Analysis*. Pergamon Press, New York, 173pp.

Patrick, L. 1991. Land use characterization report: Perdido River and Bay. US Fish and Wildlife Service, Panama City, FL (unpublished report).

Patrick, R.M. 1967. Diatom communities in estuaries. *Estuaries* 83: 311–315.

Patrick, R.M. 1971. Diatom communities. In: J. Cairns, Jr. (Ed.), *The Structure and Function of Freshwater Microbial Communities*, Vol. 3, Research Division Monograph. Virginia Polytechnic Institute and State University, Blacksburg, VA, pp. 151–164.

Patrick, R.M. and C.W. Reimer. 1966. *The Diatoms of the United States*, Vol. 13, Monograph. Academy of Natural Sciences of Philadelphia, Philadelphia, PA, 688pp.

Paulic, M. and J. Hand. 1994. Florida water quality assessment 1994 305(b) report. Florida Department of Environmental Protection, Tallahassee, FL, 261pp. (unpublished report).

Paulic, M., J. Hand, and L. Lord. 1996. Florida water quality assessment, 1996 305 (b) report. Florida Department of Environmental Protection, Tallahassee, FL, 318pp. (unpublished report).

Pauly, D.V., V. Christensen, A. Dalsgaard, R. Froese, and J. Torres. 1998. Fishing down marine food webs. *Science* 279: 860–863.

Pauly, D.V., V. Christensen, and C. Walters. 2000. Ecopath, ecosim, and ecospace as tools for evaluating ecosystem impact of fisheries. *ICES J. Mar. Sci.* 57: 697–706.

Pearce, J.W. 1977. Florida's environmentally endangered land acquisition program and the Apalachicola River system. In: R.J. Livingston (Ed.), *Proceedings of the Conference on the Apalachicola Drainage System*. Florida Department of Natural Resources, St. Petersburg, FL. Florida Marine Resources Publication 26, pp. 141–145.

Pearson, T.H. 1980. Marine pollution effects of pulp and paper industry wastes. *Helgol. Wissenschaftliche Meeresunters.* 33: 340–365.

Pearson, T.H. and R. Rosenberg. 1978. Macrobenthic succession in relation to organic enrichment and pollution of the marine environment. *Oceanogr. Mar. Biol. Annu. Rev.* 16: 229–311.

Pennock, J.R. 1985. Chlorophyll distribution in the Delaware estuary: Regulation by light-limitation. *Estuarine Coastal Shelf Sci.* 21: 711–725.

Pennock, J.R. 1987. Temporal and spatial variability in plankton, ammonium, and nitrate uptake in the Delaware River. *Esturaine Coastal Shelf Sci.* 24: 841–857.

Pennock, J.R. and J.H. Sharp. 1994. Temporal alteration between light- and nutrient-limitation of phytoplankton production in a coastal plain estuary. *Mar. Ecol. Prog. Ser.* 111: 275–288.

Pensacola Bay System Technical Symposium. 1997. Foundation of sustainability status of Pensacola Bay system water quality, Pensacola, FL.

Peters, R.H. 1977. The unpredictable problems of trophodynamics. *Environ. Biol. Fish.* 2: 97–101.

Peterson, B.J. and R.W. Howarth. 1987. Sulfur, carbon, and nitrogen isotopes used to trace organic matter flow in the salt-marsh estuaries of Sapelo Island, Georgia. *Limnol. Oceanogr.* 32: 1195–1213.

Peterson, C.H. 1979. Predation, competitive exclusion, and diversity in the soft-sediment benthic communities of estuaries and lagoons. In: R.J. Livingston (Ed.), *Ecological Processes in Coastal and Marine Systems.* Plenum Press, New York, pp. 233–264.

Peterson, C.H. 1982. The importance of predation and intra- and interspecific competition in the population biology of two infaunal suspension feeding bivalves, *Protothaca staminea* and *Chione undatella. Ecol. Monogr.* 52: 437–475.

Peterson, C.H. 1991. Intertidal zonation of marine invertebrates in sand and mud. *Am. Sci.* 79: 236–249.

Peterson, C.H. 1992. Competition for food and its community-level implications. *Benthos Res.* (*Bull. Jpn. Assoc. Benthol.*) 42: 1–11.

Philippart, C.J.M., G.C. Cadee, W. van Raaphorst, and R. Riegman. 2000. Long-term phytoplankton-nutrient interactions in a shallow coastal sea: Algal community structure, nutrient budgets, and denitrification potential. *Limnol. Oceanogr.* 45: 131–144.

Philips, E.J., M. Cichra, F.J. Aldridge, and J. Jembeck. 2000. Light availability and variations in phytoplankton standing crops in a nutrient-rich blackwater river. *Limnol. Oceanogr.* 45: 916–929.

Pinckney, J.L. and A.R. Lee. 2008. Spatiotemporal patterns of subtidal benthic microalgal biomass and community composition in Galveston Bay, Texas, USA. *Estuaries Coasts* 31: 444–454.

Plafkin, J.L., M.T. Barbour, K.D. Porter, S.K. Gross, and R.M. Hughes. 1989. Rapid bioassessment protocols for use in streams and rivers: Benthic macroinvertebrates and fish. Office of Water, US Environmental Protection Agency, Washington, DC. EPA/440/4-89-001.

Platt, T. and A.D. Jassby. 1976. The relationship between photosynthesis and light in natural assemblages of coastal marine phytoplankton. *J. Phycol.* 12: 421–430.

Platt, T.L., M. Dickie, and R.W. Trites. 1970. Spatial heterogeneity of phytoplankton in a near-shore environment. *J. Fish. Res. Board Can.* 27: 1453–1473.

Platt, T.L. and D.V. Subba Rao. 1975. Primary production of marine microphytes. In: J.P. Cooper (Ed.), *Photosynthesis and Productivity in Different Environments*, Vol. 3, International Biology Program. Cambridge University Press, Cambridge, MA, pp. 249–280.

Porter, J.S., S.K. Lewis, and K.G. Porter. 1999. The effect of multiple stressors on the Florida Keys coral reef ecosystem. A landscape hypothesis and a physiological test. *Limnol. Oceanogr.* 44: 941–939.

Porter, J.W., V. Kosmynin, K.L. Patterson, K.G. Porter, W.C. Jaap, J.L. Wheaton, K. Hackett et al. 2002. Detection of coral reef change by the Florida Keys Coral Reef monitoring project. In: J.W. Porter and K.G. Porter (Eds.), *The Everglades, Florida Bay and Coral Reefs of the Florida Keys: An Ecosystem Sourcebook.* CRC Press, Boca Raton, FL, pp. 749–769.

Porter, J.W. and K.G. Porter (Eds.). 2002. *The Everglades, Florida Bay and Coral Reefs of the Florida Keys: An Ecosystem Sourcebook.* CRC Press, Boca Raton, FL, 1000pp.

Portnoy, J.W. 1991. Summer oxygen depletion in a diked New England estuary. *Estuaries* 14: 122–129.

Posey, M.H., T.D. Alphin, L. Cahoon, D. Lindquist, and M.E. Becker. 1999. Interactive effects of nutrient additions and predation on infaunal communities. *Estuaries* 22: 785–792.

Postma, H. 1985. Eutrophication of Dutch coastal waters. *Neth. J. Zool.* 35: 348–359.

Potts, M. 1980. Blue-green algae (Cyanophyta) in marine coastal environments of the Sinai Peninsula: Distribution, zonation, stratification and taxonomic diversity. *Phycologia* 19: 60–73.

Powell, E.N., J.D. Gauthier, E.A. Wilson, A. Nelson, R.R. Fay, and J.M. Brooks. 1991. Oyster disease and climate change. Are yearly changes in *Perkinsus marinus* parasitism in oysters (*Crassostrea virginica*) controlled by climate cycles in the Gulf of Mexico? *Mar. Ecol.* 12: 82–91.

Powell, E.N., J.M. Klinck, E.E. Hofmann, and S.M. Ray. 1994. Modeling oyster populations IV. Rates of mortality, population crashes, and management. *Fish. Bull.* 92: 347–373.

Power, M.P. 1992. Top-down and bottom-up forces in food webs: Do plants have primacy? *Ecology* 73: 733–746.

Powers, C.F., D.W. Schults, K.W. Malueg, R.M. Brice, and M.D. Schuldt. 1972. Algal responses to nutrient additions in natural waters. II. Field experiments. *Limnol. Oceanogr. Spec. Symp.* 1: 141–156.

Prasad, A.K.S.K. and G.A. Fryxell. 1991. Habit, frustule morphology and distribution of the Antarctic marine benthic diatom *Entopyla australis* var. *gigantea* (Greville) Fricke (Entopylaceae). *Br. Phycol. J.* 26: 101–122.

Prasad, A.K.S.K. and R.J. Livingston. 1987. An atlas of diatoms and other algal forms from selected drainage areas in central and north Florida. Florida Department of Environmental Regulation, Tallahassee, FL (unpublished report).

Prasad, A.K.S.K. and R.J. Livingston. 1995. A microbiological and systematic study of *Coscinodiscus jonesianus* (Bacillariophyta) from Florida coastal waters. *Nova Hedw. Beih.* 112: 247–263.

Prasad, A.K.S.K. and R.J. Livingston. 1998. Fine structure and taxonomy of *Fryxelliella*, a new genus of centric diatom (Triceratiaceae: Bacillariophyta) with a new valve feature, a circumferential marginal tube, and descriptions of *F. floridana* sp. nov. from the Atlantic coast of Florida and *F. inconspicua* (Rattray) comb. nov. from the Miocene. *Phycologia* 36: 305–323.

Prasad, A.K.S.K., J.A. Nienow, and R.J. Livingston. 1990. The genus *Cyclotella* (Bacillariophyta) in Choctawhatchee Bay, Florida, with special reference to *C. striata* and *C. choctawhatcheeana* sp. nov. *Phycologia* 29: 418–436.

Prasad, M., M.R.P. Sapiano, C.R. Anderson, W. Long, and R. Murtugudde. 2010. Long-term variability of nutrients and chlorophyll in the Chesapeake Bay: A retrospective analysis, 1985–2008. *Estuaries Coasts* 33: 1128–1143.

Pratt, D.M. 1965. The winter-spring diatom flowering in Narragansett Bay. *Limnol. Oceanogr.* 10: 173–184.

Pratt, T.R., P.F. McGinty, G.Z. Guo, P.E., W.C. Hunner, and E.F. Songer. 1993. Stormwater assessment of the Bayou Chico Watershed, Escambia County, Florida. Northwest Florida Water Management District, Havana, FL. Water Resources Special Report 93-7.

Pratt, T.R., P.G. Weiland, J.H. Cason, J.M. Starnes-Smith, W.K. Jones, D.J. Cairns, and L.P. Simoneaux. 1990. The Pensacola Bay System surface water improvement and management plan: A comprehensive plan for the restoration and preservation of the Pensacola Bay System. Northwest Florida Water Management District, Havana, FL. Program Development Series 91-2.

Premila, V.E. and M.U. Rao. 1977. Distribution and seasonal abundance of *Oscillatoria nigroviridis* Thwaites ex. Gomant in the waters of Visakhapatnam Harbour. *Indian J. Mar. Sci.* 3: 79–91.

Prescott, G.W. 1962. *Algae of the Western Great Lakes*. Wm. C. Brown Company, Dubuque, IA, 977pp.

Prescott, G.W. 1980. *How to Know the Aquatic Plants*. Wm. C. Brower Company, Dubuque, IA, 158pp.

Price, K.S., D.A. Flemer, J.L. Taft, G.B. Mackiernan, W. Nehlsen, R.B. Biggs, N.H. Burger, and D.A. Blaylock. 1985. Nutrient enrichment of Chesapeake Bay and its impact on the habitat of striped bass: A speculative hypothesis. *Trans. Am. Fish. Soc.* 114: 97–106.

Pritchard, D.W. and J.K. Schubel. 1981. Physical and geological processes controlling nutrient levels in estuaries. In: B.G. Neilson and L.E. Cronin (Eds.), *Estuaries and Nutrients*. Humana Press, Clifton, NJ, pp. 47–69.

Purcell, B.H. 1977. The ecology of the epibenthic fauna associated with *Vallisneria americana* beds in a north Florida estuary. MS thesis, Florida State University, Tallahassee, FL.

Pyke, G.H., H.R. Pulliam, and E.L. Charnou. 1977. Optimal foraging: A selective review of theory and tests. *Q. Rev. Biol.* 52: 137–154.

Quinlan, E.L. and E.J. Phlips. 2007. Phytoplankton assemblages across the marine to low-salinity transition zone in a blackwater dominated estuary. *J. Plankton Res.* 29: 401–416.

Rabalais, N.N. 1992. An updated summary of status and trends in indicators of nutrient enrichment in the Gulf of Mexico. Report to Gulf of Mexico Program, Nutrient Enrichment Subcommittee. US Environmental Protection Agency, Office of Water, Gulf of Mexico Program, Stennis Space Center, MS. EPA/800-R-92-004, 421pp.

Rabalais, N.N., J. Berg, and E. Hagmeier. 1990. Long-term changes of the annual cycles of meteorological, hydrographic, nutrient, and phytoplankton time series at Helgoland and at LV. ELBE 1 in the German Bight. *Cont. Shelf Res.* 10: 305–328.

Rabalais, N.N., R.E. Turner, D. Justic, Q. Dortch, W.J. Wiseman, Jr., and B.K. Sen Grupta. 1996. Nutrient changes in the Mississippi River and system responses on the adjacent continental shelf. *Estuaries* 19: 386–407.

Rabalais, N.N., R.E. Turner, D. Justic, Q. Dortch, and W.J. Wiseman, Jr. 1999. Characterization of hypoxia: Topic 1: Report for the integrated assessment of hypoxia in the Gulf of Mexico. National Oceanic and Atmospheric Administration, Coastal Ocean Program, Silver Spring, MD. Decision Analysis Series 15, 167pp.

Rada, R.G., J.E. Findley, and J.G. Wiener. 1986. Environmental fate of mercury discharged into the Upper Wisconsin River. *Water Air Soil Pollut.* 29: 57–76.

Raffaelli, D. 2002. From Elton to mathematics and back again. *Science* 296: 1035–1037.

Ragone, L.M. and E.M. Burreson. 1993. Effect of salinity on infection progression and pathogenicity of *Perkinsus marinus* in the eastern oyster, *Crassostrea virginica* (Gmelin). *J. Shellfish. Res.* 12: 1–7.

Rainville, R.P., B.J. Copeland, and W.T. McKean. 1975. Toxicity of Kraft mill wastes to an estuarine phytoplankton. *J. Water Pollut. Control Fed.* 47: 487–503.

Randall, J. 1967. Food habits of reef fishes of the West Indies. *Stud. Trop. Oceanogr.* 5: 665–847.

Randall, J.M. and J.W. Day. 1987. Effects of river discharge and vertical circulation on aquatic primary production in a turbid Louisiana (USA) estuary. *Neth. J. Sea Res.* 21: 231–242.

Raney, D.C. 1980. Study of methodology for evaluating water-quality problems at Bayou Texar, Florida. University of Alabama, Bureau of Engineering Research, Tuscoloosa, AL. 263-112, 155pp. (unpublished report).

Rappe, C., P. Bergqvist, L. Kjeller, S. Swanson, T. Belton, B. Ruppel, K. Lockwood, and P.C. Kahn. 1991. Levels and patterns of PCDD and PCDF contamination in fish, crabs and lobsters from Newark Bay and the New York Bight. *Chemosphere* 22: 239–266.

Reardon, J. 1999. Ecology of phytoplankton communities in Lake Jackson. MS thesis, Department of Biological Science, Florida State University, Tallahassee, FL.

Reddy, P.M. and V. Venkateswarlu. 1986. Ecology of algae in the paper mill effluents and their impact on the River Tungabhadra. *J. Environ. Biol.* 7: 215–223.

Redfield, A.C. 1934. On the proportions of organic derivatives in seawater and their relation to the composition of the plankton. In: R.J. Daniel (Ed.), *James Johnstone Memorial Volume*. Liverpool University Press, Liverpool, U.K., pp. 176–192.

Redfield, A.C., B.H. Ketchum, and F.A. Richards. 1963. The influence of organisms on the chemical composition of seawater. In: M.N. Hill (Ed.), *The Sea*, Vol. 2. Wiley-InterScience, New York, pp. 26–77.

Reeve, D.W. and P.F. Earl. 1988. Chlorinated organic compounds in bleached chemical pulp and pulp-mill effluents. Part 1: The potential for effluent regulation. *Canadian Pulp & Paper Association Branch Paper for Pacific Coast-Western Branches Joint Conference*.

Regnell, O. and G. Ewald. 1997. Factors controlling temporal variation in methyl mercury levels in sediment and water in a seasonally stratified lake. *Limnol. Oceanogr.* 42: 1784–1795.

Reich, C.D., E.A. Shinn, T.D. Hickey, and A.B. Tihansky. 2002. Tidal and meteorological influence on shallow marine groundwater flow in the upper Florida Keys. In: J.W. Porter and K.G. Porter (Eds.), *The Everglades, Florida Bay and Coral Reefs of the Florida Keys: An Ecosystem Sourcebook*. CRC Press, Boca Raton, FL, pp. 659–676.

Reid, G.K., Jr. 1967. An ecological study of the Gulf of Mexico fishes in the vicinity of Cedar Key, Florida. *Bull. Mar. Sci. Gulf Carib.* 4: 1–94.

Reidenauer, J. and C. Shambaugh. 1986. An analysis of estuarine degradations within the Pensacola Bay system and their relationship to land management practices. Florida Department of Community Affairs, Tallahassee, FL, 132pp.

Reise, K. 1978. Experiments on epibenthic predation in the Wadden Sea. *Helgol. Wissenschaftliche Meeresunters* 31: 51–101.

Rensel, J.E. 1993. Severe blood hypoxia of Atlantic salmon exposed to the marine diatom *Chaetoceros concavicornis*. In: T.J. Smayda and Y. Shimizu (Eds.), *Toxic Phytoplankton Blooms in the Sea*. Elsevier, New York, pp. 625–630.

Reyer, A.J., D.W. Field, J.E. Cassells, C.E. Alexander, and C.L. Holland. 1988. The distribution and areal extent of coastal wetlands in estuaries of the Gulf of Mexico. National Coastal Wetlands Inventory, National Oceanic and Atmospheric Administration, National Ocean Service, Office of Oceanography and Marine Assessment, Rockville, MD, 18pp.

Reyes, E. and M. Merino. 1991. Diel dissolved oxygen dynamics and eutrophication in a shallow well-mixed tropical lagoon. *Estuaries* 14: 372–381.

Rhee, G.-Y. 1978. Effects of N and P atomic ratios and nitrate limitation on algal growth, cell composition, and nitrate uptake. *Limnol. Oceanogr.* 23: 10–25.

Rhoads, D.C. 1974. Organism-sediment relations on the muddy sea floor. *Oceanogr. Mar. Biol. Annu. Rev.* 12: 263–300.

Rhodes, L.L., A.J. Haywood, W.J. Ballantine, and A.L. MacKenzie. 1993. Algal blooms and climate anomalies in northeast New Zealand, August–December 1992. *N. Z. J. Mar. Freshw. Res.* 27: 419–430.

Richey, J.E., R.C. Wissmar, A.H. Devol, G.E. Likens, J.S. Eaton, W.E. Odum, N.M. Johnson, O.L. Loucks, R.T. Prentke, and P.H. Rich. 1978. Carbon flow in four lake ecosystems: A structural approach. *Science* 202: 1183–1186.

Riegman, R. 1998. Species composition of harmful algal blooms in relation to macronutrient dynamics. In: D.M. Anderson, A.D. Cembella, and G.M. Hallagraeff (Eds.), *Physiological Ecology of Harmful Algal Blooms*. Springer-Verlag, Berlin, Germany, pp. 475–488.

Riegman, R., A.A.M. Noordeloos, and G.C. Cadee. 1992. *Phaeocystis* blooms and eutrophication of the continental coastal zones of the North Sea. *Mar. Biol.* 112: 479–484.

Rigler, F.H. 1975. The concept of energy flow and nutrient flow between trophic levels. In: W.H. van Dobben and R.H. Lowe-McConnell (Eds.), *Unifying Concepts in Ecology*. Junk/PUDOC, The Hague, the Netherlands, pp. 15–26.

Rizzo, W. 1990. Nutrient exchanges between the water column and a subtidal benthic microalgal community. *Estuaries* 13: 219–226.

Roaza, H.P. and T.R. Pratt. 1992. Analysis of the suitability of existing STORET data for loading rate calculations in the Pensacola Bay System. Northwest Florida Water Management District, Havana, FL. Technical File Report 92-1.

Robertson, B. 1982. Guardian of Apalachicola Bay. *Oceans* 5: 65–67.

Roegner, G.C. and A.L. Shanks. 2001. Coastally-derived chlorophyll *a* to South Slough, Oregon. *Estuaries* 24: 244–256.

Roelke, D.L., L.A. Cifuentes, and P.M. Eldridge. 1997. Nutrient and phytoplankton dynamics in a sewage-impacted Gulf Coast estuary: A field test of the PEG-model and equilibrium resource competition theory. *Estuaries* 20: 725–742.

Rogers, R. 1988. EIS for designation of a new ocean dredged material disposal site, Pensacola, Florida. US Environmental Protection Agency, Region IV, Atlanta, GA.

Rogers, R.G. and F.T. Bisterfield. 1975. Loss of submerged vegetation in the Pensacola Bay system, 1949–1974. In: R.R. Lewis (Ed.), *Proceedings of the Second Annual Conference on Restoration of Coastal Vegetation in Florida*. Hillsborough Community College, Tampa, FL, pp. 35–51.

Rosenberg, D.M., R.A. Bodaly, R. Hecky, and R.W. Newbury. 1987. The environmental assessment of hydroelectric impoundments and diversions in Canada. In: M.C. Healy and R.R. Wallace (Eds.), *Canadian Aquatic Resources*, Vol. 215, Canadian Bulletin of Fisheries and Aquatic Sciences. Fisheries and Oceans Canada, Ottawa, Ontario, Canada, pp. 71–104.

Rosenberg, R. 1977. Benthic macrofaunal dynamics, production, and dispersion in an oxygen-deficient estuary on west Sweden. *J. Exp. Mar. Biol. Ecol.* 26: 107–133.

Rosenberg, R., R. Elmgren, S. Fleischer, P. Jonsson, G. Persson, and H. Dahhm. 1990. Marine eutrophication case studies in Sweden. *Ambio* 19: 102–108.

Ross, L.T. and D.A. Jones (Eds.). 1979. Biological aspects of water quality in Florida. Part 1: Escambia-Perdido, Choctawhatchee, Apalachicola, Aucilla-Ochlockonee-St. Marks, and Suwannee Drainage Basins. Florida Department of Environmental Regulation, Tallahassee, FL. Technical Series 4.

Ross, S.T. 1977. Patterns of resource partitioning in searobins (Pisces: Triglidae). *Copeia* 1977: 561–571.

Ross, S.T. 1978. Trophic ontogeny of the leopard searobin, *Prionotus scitulus* (Pisces: Triglidae). *Fish. Bull.* 76: 225–234.

Round, F.E. 1981. *The Ecology of Algae*. Cambridge University Press, London, U.K., 653pp.

Rowe, G.T. and G.S. Boland. 1991. Benthic oxygen demand and nutrient regeneration on the Continental Shelf of Louisiana and Texas. *Abstract for LUMCON Meeting*.

Rudnick, D.T., Z. Chen, D.L. Childers, J.N. Boyer, and T.D. Fontaine, III. 1999. Phosphorus and nitrogen inputs to Florida Bay. The importance of the Everglades watershed. In: J.W. Fourqurean, M.B. Robblee, and L.A. Deegan (Eds.), *Florida Bay: A Dynamic Subtropical Estuary*; *Estuaries* 22: 398–416.

Ryan, J.R. 1981. Trophic analysis of nocturnal fishes in seagrass beds in Apalachee Bay, Florida. MS thesis, Florida State University, Tallahassee, FL.

Ryther, J.H. 1954. The ecology of phytoplankton blooms in Moriches Bay and Great South Bay, Long Island, New York. *Biol. Bull.* 106: 198–209.

Ryther, J.H. and W.M. Dunstan. 1971. Nitrogen, phosphorus, and eutrophication in the coastal marine environment. *Science* 171: 1008–1013.

Sabatier, R., J.D. Lebreton, and D. Chessel. 1989. Principal component analysis with instrumental variables as a tool for modeling composition data. In: R. Coppi and S. Bolasco (Eds.), *Multiway Data Analysis*. Elsevier Sciences, North Holland, the Netherlands, pp. 341–352.

Sabo, J.L., J.C. Finlay, T. Kennedy, and D.M. Post. 2010. The role of discharge variation in scaling of drainage area and food chain length in rivers. *Science* 330: 965–967.

Sakshaug, E. and S. Myklestad. 1973. Studies on the phytoplankton ecology of the Trondheimsland. III. Dynamics of phytoplankton blooms in relation to environmental factors, bioassay experiments, and parameters for the physiological state of the population. *J. Exp. Mar. Biol. Ecol.* 11: 157–188.

Sakshaug, E. and Y. Olsen. 1986. Nutrient status of phytoplankton blooms in Norwegian waters and algal strategies for nutrient competition. *Can. J. Fish. Aquat. Sci.* 43: 389–396.

Santos, S.L. and S.A. Bloom. 1980. Stability in an annually defaunated estuarine soft-bottom community. *Oecologia* 46: 290–294.

Santos, S.L. and J.L. Simon. 1980a. Marine soft-bottom community-establishment following annual defaunation: Larval or adult recruitment? *Mar. Ecol. Prog. Ser.* 2: 235–241.

Santos, S.L. and J.L. Simon. 1980b. Response of soft-bottom benthos to annual catastrophic disturbance in a south Florida estuary. *Mar. Ecol. Prog. Ser.* 3: 347–355.

Scavia, D., J.C. Field, D.F. Boesch, R.W. Buddemeier, V. Burrketts, D.R. Cayan, M. Fogarty et al. 2002. Climate change impacts on U.S. coastal and marine ecosystems. *Estuaries* 25: 149–164.

Schelske, C.L. 1974. In situ and natural phytoplankton assemblage bioassays. In: L.E. Shubert (Ed.), *Algae as Ecological Indicators*. Academic Press, New York, pp. 15–47.

Schelske, C.L. and E.F. Stoemer. 1972. Phosphorus, silica, and eutrophication of Lake Michigan. In: G.E. Likens (Ed.), *Nutrients and Eutrophication: The Limiting Nutrient Controversy*, Vol. 1, American Society of Limnology and Oceanography. Special Symposia. Allen Press, Lawrence, KS, pp. 157–171.

Schindler, D.W. 1977. Evolution of phosphorus limitation in lakes. *Science* 195: 260–262.

Schindler, D.W. and E.J. Fee. 1974. Experimental lakes area: Whole-lake experiments in eutrophication. *J. Fish. Res. Board Can.* 31: 937–953.

Schmidt-Gengenbach, J. 1991. A study of the effects of pollutants on the benthic macroinvertebrates in Lake Jackson, Florida: A descriptive and experimental approach. MS thesis, Department of Biological Sciences and Center for Aquatic Research and Resource Management, Florida State University, Tallahassee, FL.

Schnable, J.E. 1966. The evolution and development of part of the northwest Florida coast. PhD dissertation, Florida State University, Tallahassee, FL, 231pp.

Schoenly, K. and J.E. Cohen. 1991. Temporal variation in food web structure: 16 empirical cases. *Ecol. Monogr.* 61: 267–298.

Schroeder, W.W. 1978. Riverine influence on estuaries: A case study. In: V.S. Kennedy (Ed.), *Estuarine Interactions*. Academic Press, New York, pp. 347–364.

Schroeder, W.W., S.P. Dinnel, and W.J. Wiseman. 1990. Salinity stratification in a river-dominated estuary. *Estuaries* 213: 145–154.

Schroeder, W.W. and W.J. Wiseman, Jr. 1985. An analysis of the winds (1974–1984) and sea level elevations (1973–1983) in coastal Alabama. Mississippi-Alabama Sea Grant Consortium, Ocean Springs, MS. Publication MASGP-84-024, 102pp.

Schroeder, W.W. and W.J. Wiseman, Jr. 1986. Low-frequency shelf-estuarine exchange processes in Mobile Bay and other estuarine systems on the northern Gulf of Mexico. In: D.A. Wolfe (Ed.), *Estuarine Variability*. Academic Press, New York, pp. 355–366.

Schropp, S.J., F.D. Calder, G.M. Sloane, J.C. Carlton, G.L. Holcomb, H.L. Windom, F. Huan, and R.B. Taylor. 1991. A report on physical and chemical processes affecting the management of Perdido Bay: Results of Perdido Bay Interstate Project. Alabama Department of Environmental Management and Florida Department of Environmental Regulation, Tallahassee, FL.

Sciarotta, T.C. and D R. Nelson. 1977. Diel behavior of the blue shark, *Prionace glauca*, near Santa Catalina Island, California. *Fish. Bull.* 75: 519–529.

Science Applications International Corporation. 1986. Remote Sediment Profile Study of Pensacola Bay, United States Air Force Administration Record for George Air Force Base, Florida. Newport, RI. No. SAIC-86/7500.99.

Seal, T.L., F.D. Calder, G.M. Sloane, S.J. Schropp, and H.L. Windom. 1994. Florida coastal sediment contaminants atlas: A summary of coastal sediment quality surveys. Florida Department of Environmental Protection, Tallahassee, FL, 107pp.

Seliger, H.H. and J.A. Boggs. 1988. Long-term pattern of anoxia in the Chesapeake Bay. In: M.P. Lynch and E.C. Krome (Eds.), *Understanding the Estuary: Advances in Chesapeake Bay Research*. Chesapeake Research Consortium, Solomons, MD, pp. 570–583.

Seliger, H.H., M.E. Loftus, and D.V. Subba Rao. 1975. Dinoflagellate accumulations in Chesapeake Bay. In: V.R. LoCicero (Ed.), *Proceedings of the First International Conference on Toxic Dinoflagellate Blooms*. Boston Massachusetts Science and Technology Foundation, Wakefield, MA, pp. 181–205.

Sellner, K.G., S.E. Shumway, M.W. Luckenbach, and T.L. Cucci. 1995. The effects of dinoflagellate blooms on the oyster *Crassostrea virginica* in Chesapeake Bay. In: P. Lassus, G. Arzul, E. Erard, P. Gentien, and C. Marcalliou (Eds.), *Harmful Marine Algal Blooms*. Lavoisier, Intercept Ltd., Paris, France, pp. 505–511.

Semina, H.J. 1978. Treatment of an aliquot sample. In: A. Sournia (Ed.), *Phytoplankton Manual*, Vol. 6, Monographs on Oceanographic Methodology. UNESCO Press, Paris, France, p. 181.

Sharp, J.H., C.H. Culberson, and T.M. Church. 1982. The chemistry of the Delaware Estuary. General considerations. *Limnol. Oceanogr.* 27(6): 1015–1028.

Sheridan, P.F. 1978. Trophic relationships of dominant fishes in the Apalachicola Bay system (Florida). PhD dissertation, Florida State University, Tallahassee, FL.

Sheridan, P.F. 1979. Trophic resource utilization by three species of sciaenid fishes in a northwest Florida estuary. *Northeast Gulf Sci.* 3: 1–15.

Sheridan, P.F. and R.J. Livingston. 1979. Cyclic trophic relationships of fishes in an unpolluted, river-dominated estuary in North Florida. In: R.J. Livingston (Ed.), *Ecological Processes in Coastal and Marine Systems*. Plenum Press, New York, pp. 143–161.

Sheridan, P.F. and R.J. Livingston. 1983. Abundance and seasonality of infauna and epifauna inhabiting a *Halodule wrightii* meadow in Apalachicola Bay, Florida. *Estuaries* 6: 407–419.

Sherman, K., L.M. Alexander, and B.D. Gold. 1991. *Food Chains, Yields, Models and Management of Large Marine Ecosystems*. Westview Press, Boulder, CO, 320pp.

Shimada, H., T. Hayashi, and T. Mizushima. 1996. Spatial distribution of *Alexandrium tamarencse* "species complex" (Dinophyceae): Dispersal in the North American and West Pacific regions. *Phycologia* 34: 472–485.

Shimada, M., T.H. Murakami, T. Imahayashi, H.S. Ozaki, T. Toyoshima, and T. Okaichi. 1983. Effects of sea bloom, *Chattonella antiqua*, on gill primary lamellae of the young yellowtail, *Seriola quinqueradiata*. *Acta Histochem. Cytochem.* 16: 232–244.

Shoplock, B. 1999. Ecology of zooplankton in Lake Jackson. MS thesis, Department of Biological Science, Florida State University, Tallahassee, FL.

Shuba, P.J. 1981. Pensacola Bay Nutrient Monitoring Study, 15 July 1981 to 13 October 1981. Research report submitted to Pensacola Public Utilities Department, Pensacola, FL. Project Number A010.

Shulman, S. 2008. *Undermining Science: Suppression and Distortion in the Bush Administration*. University of California Press, Berkeley, CA.

Shumway, S.E. 1990. A review of the effects of algal blooms on shellfish and aquaculture. *J. World Aquac. Soc.* 21: 65–104.

Shumway, S.E., J. Barter, and S. Sherman-Caswell. 1990. Auditing the impact of toxic algal blooms on oysters. *Environ. Auditor* 2: 41–56.

Sikora, W.B., R.W. Heard, and M.D. Dahlberg. 1972. The occurrence and food habits of two species of hake, *Urophycis regius* and *U. floridanus* in Georgia estuaries. *Trans. Am. Fish. Soc.* 101: 513–525.

Simonsen, R. 1974. *The Diatom Plankton of the Indian Ocean Expedition of R/V "Meteor" 1964–1965*. "Meteor" Forsch. Ergebnisse, Riehe D. Gebruder Bontrager, Berlin, Stuttgart, Germany, 19, pp. 1–107.

Simmons, E.G. and W.H. Thomas. 1962. Phytoplankton of the eastern Mississippi Delta. *Publ. Inst. Mar. Sci. Univ. Tex.* 8: 269–298.

Simon, J.L. 1974. Tampa Bay estuarine system—A synopsis. *Fla. Sci.* 37: 217–244.

Simpson, R.L., R.E. Good, M.A. Leck, and D.F. Whigham. 1983. The ecology of freshwater tidal wetlands. *Bioscience* 33: 255–259.

Sin, Y., R.L. Wetzel, and I.C. Anderson. 1999. Spatial and temporal characteristics of nutrient and phytoplankton dynamics in the York River Estuary, Virginia: Analyses of long-term data. *Estuaries* 22: 260–275.

Sinclair, M., M. El-Sabh, and J. Brindle. 1976. Seaward nutrient transport in the Lower St. Lawrence estuary. *J. Fish. Res. Board Can.* 33: 1271–1277.

Sittig, M.H. 1985. *Handbook of Toxic and Hazardous Chemicals and Carcinogens*, 2nd edn. Noyes Data Corp., Park Ridge, NJ, 886pp.

Sklar, F., C. McVoy, R. VanZee, D.E. Gawlik, K. Tarboton, D. Rudnick, and S. Miao. 2002. The effects of altered hydrology on the ecology of the Everglades. In: J.W. Porter and K.G. Porter (Eds.), *The Everglades, Florida Bay and Coral Reefs of the Florida Keys: An Ecosystem Sourcebook*. CRC Press, Boca Raton, FL, pp. 39–82.

Skulberg, O.M., B. Underdal, and H. Utkilen. 1994. Toxic waterblooms with cyanophytes in Norway-current knowledge. *Algolog. Stud.* 75: 279–289.

Smar, D. 2010. An assessment of ecological processes in the Apalachicola Estuarine System, Florida. MS thesis, University of Central Florida, Orlando, FL, 259pp.

Smayda, T.J. 1978. From phytoplankters to biomass. In: A. Sournia (Ed.), *Phytoplankton Manual*, Vol. 6, Monographs on Oceanographic Methodology. UNESCO Press, Paris, France, pp. 273–279.

Smayda, T.J. 1980. Phytoplankton species succession. In: I. Morris (Ed.), *The Physiological Ecology of Phytoplankton*. University of California Press, Berkeley, CA, pp. 493–570.

Smayda, T.J. 1989. Primary production and the global epidemic of phytoplankton blooms in the sea: A linkage? In: E.M. Cosper, J. Carpenter, and V.M. Bricelj (Eds.), *Novel Phytoplankton Blooms: Causes and Impacts of Recurrent Brown Tides and Other Unusual Blooms*. Springer-Verlag, Berlin, Germany, pp. 449–483.

Smayda, T.J. 1990. Novel and nuisance phytoplankton blooms in the sea. Evidence for a global epidemic. In: E. Granneli, B. Sundstrom, R. Edler, and D.M. Anderson (Eds.), *Toxic Marine Phytoplankton: Proceedings of the Fourth International Conference*. Elsevier, New York, pp. 29–40.

Smayda, T.J. 1997a. What is a bloom? A commentary. *Limnol. Oceanogr.* 42: 1132–1136.

Smayda, T.J. 1997b. Harmful algal blooms. Their ecophysiology and general relevance to phytoplankton blooms in the sea. *Limnol. Oceanogr.* 42: 1137–1153.

Smayda, T.J. 1997c. Ecophysiology and bloom dynamics of *Heterosigma akashiwo* (Raphidophycae). Japan. In: D.M. Anderson, A.D. Cembella, and G.M. Hallagraeff (Eds.), *Physiological Ecology of Harmful Algal Blooms*, NATO ASI Series G, Ecological Sciences, Vol. 41, Physiological Ecology of Harmful Algal Blooms. Springer-Verlag, Heidelberg, Germany, pp. 113–131.

Smayda, T.J. and Y. Shimizu. 1993. Toxic phytoplankton blooms in the sea. In: *Proceedings of the Fifth International Conference on Toxic Marine Phytoplankton*, Newport, RI. Elsevier, New York, 952pp.

Smith, D.E., M. Laffler, and G. Mackiernan (Eds.). 1992. Oxygen dynamics in the Chesapeake Bay. University of Maryland, College Park, MD. Maryland Sea Grant Publication UM-SG-TS-92-01, 252pp.

Smith, G.A., J.S. Nickels, W.M. Davis, R.F. Martz, R.H. Findlay, and D.C. White. 1982. Perturbations in the biomass, metabolic activity, and community structure of the estuarine detrital microbiota: Resource partitioning in amphipod grazing. *J. Exp. Mar. Biol. Ecol.* 64: 125–143.

Smith, J.B. and D. Torpak. 1989. The potential effects of global climate change on the United States. Report to Congress. US Environmental Protection Agency, Office of Research and Development, Washington, DC, 413pp.

Smith, J.N. and E.M. Levy. 1990. Geochronology for polycyclic aromatic hydrocarbon contamination in sediments of the Saguenay Fjord. *Environ. Sci. Technol.* 24: 874–879.

Smith, N.P. and P.A. Pitts. 2002. Regional-scale and long-term transport patterns in the Florida Keys. In: J.W. Porter and K.G. Porter (Eds.), *The Everglades, Florida Bay and Coral Reefs of the Florida Keys: An Ecosystem Sourcebook*. CRC Press, Boca Raton, FL, pp. 343–360.

Smith, S.M. and G.L. Hitchcock. 1994. Nutrient enrichment and phytoplankton growth in the surface waters of the Louisiana Bight. *Estuaries* 17: 740–753.

Smith, S.V. 1981. Response of Kaneohe Bay, Hawaii, to relaxation of sewage stress. In: B.G. Neilson and L.E. Cronin (Eds.), *Estuaries and Nutrients*. Humana Press, Clifton, NJ, pp. 391–410.

Smith, S.V. 1991. Stoichiometry of C:N:P fluxes in shallow-water marine ecosystems. In: J.G. Cole, G. Lovett, and T. Finley (Eds.), *Comparative Analyses of Ecosystems: Patterns, Mechanisms, and Theories*. Springer-Verlag, New York, pp. 259–286.

Snedaker, S., D. de Sylva, and D. Cottrell. 1977. A review of the role of freshwater in estuarine ecosystems. Southwest Florida Water Management District, Brooksville, FL, Final report. 126pp.

Snowden, C. and D.J. Cairns. 1993. Surface water improvement and management program priority list for the Northwest Florida Water Management District. Northwest Florida Water Management District, Havana, FL.

Sobocinski, K.L., R.J. Orth, M.C. Fabrizio, and RJ. Latour. 2013. Historical comparison of fish community structure in Lower Chesapeake Bay Seagrass Habitats. *Estuaries Coasts* 36: 775–794.

Sommer, U. 1990. Phytoplankton nutrient competition—From laboratory to lake. In: J.B. Grace and D. Tilman (Eds.), *Perspectives on Plant Competition*. Academic Press, New York, pp. 193–213.

Soniat, T.M. 1996. Epizootiology of *Perkinsus marinus* disease of eastern oysters in the Gulf of Mexico. *J. Shellfish Res.* 15: 35–43.

Sorokin, Y.I., F. Dallocchio, F. Gelli, and L. Pregnollato. 1996. Phosphorus metabolism in anthropogenically transformed lagoon ecosystems: The Comacchio Lagoons (Ferrar, Italy). *J. Sea Res.* 35: 243–250.

Sournia, A. (Ed.). 1978. *Phytoplankton Manual*, Vol. 6, Monographs on Oceanic Methodology. UNESCO Press, Paris, France, 337pp.

Sournia, A., M.J. Chretiennot-Dinet, and M. Ricard. 1991. Marine phytoplankton: How many species in the world ocean? *J. Plankton Res.* 13: 1093–1099.

South Florida Water Management District (SFWMD). 1994. An update of the surface water improvement and management plan for Biscayne Bay. West Palm Beach, FL. Draft report.

Sowles, J.W. 1997a. Memo to Stacey Ladner. Subject: Interpretation of Hg in Penobscot sediments. Maine Department of Environmental Protection, Augusta, ME.

Sowles, J.W. 1997b. Water resources survey—HoltraChem (Part II). September 29, 1997, memorandum to Stacy Ladner. Maine Department of Environmental Protection, Augusta, ME.

Sowles, J.W. 1999. Mercury contamination in the Penobscot River Estuary at HoltraChem Manufacturing Company—An evaluation of monitoring data and interpretation of toxic potential and ecological implications. April 23, 1999, data summary report. Maine Department of Environmental Protection, Augusta, ME.

Squires, L.E. and N.A. Sinnu. 1982. Seasonal changes in the diatom flora in the estuary of the Damour River, Lebanon. In: D.G. Mann (Ed.), *Proceedings of the Seventh International Diatom Symposium*, Philadelphia, PA, August 22–27, 1982, pp. 359–372.

Stahl, R.G. Jr., M.J. Hooper, J. Balbus, W.H. Clements, A. Fritz, T. Gouin, R. Helm, C. Hickey, W. Landis and J. Moe. 2013. The influence of global climate change on the scientific foundations and applications of environmental toxicology and chemistry. *Environ. Toxicol. Chem.* 32: 13–19.

Stankelis, R.M., M.D. Naylor, and W.R. Boynton. 2003. Submerged aquatic vegetation in the mesohaline region of the Patuxent Estuary: Past, present and future status. *Estuaries* 26: 186–195.

Stanley, D.W. and S.W. Nixon. 1992. Stratification and bottom-water hypoxia in the Pamlico River Estuary. *Estuaries* 15: 270–281.

Starck, W.A., II and W.P. Davis. 1966. Night habits of fishes of Alligator Reef, Florida. *Icthyologia* 1966: 313–357.

Staudinger, M.D., S.L. Carter, M.S. Cross, N.S. Dubois, J.E. Duffy, C. Enquist, R. Griffis et al. 2013. In: N.B. Grimm (Ed.), *Front. Ecol. Environ.* 11: 465–473 (Ecological Society of America).

Staudt, A., A.K. Leidner, J. Howard, K.A. Braumann, J.S. Dukes, L.J. Hansen, C. Paukert, J. Sabo, and L.A. Solorzano. 2013. The added complications of climate change: Understanding and managing biodiversity and ecosystems. In: N.B. Grimm (Ed.). *Front. Ecol. Environ.* 11: 494–501 (Ecological Society of America).

Stefanou, P., G. Tsirtsis, and M. Karydis. 2000. Nutrient scaling for assessing eutrophication: The development of a simulated normal distribution. *Ecol. Appl.* 10: 303–309.

Steidinger, K.A. 1996. Dinoflagellates. In: C.R. Tomas (Ed.), *Identifying Marine Diatoms*. Academic Press, New York, pp. 387–598.

Steidinger, K.A., P. Carlson, D. Baden, C.D. Rodriguez, and J. Seagle. 1998a. Neurotoxic shellfish poisoning due to toxin retention in the clam *Chione cancellata*. In: B. Reguera, J. Blanco, M.L. Fernandez, and T. Wyatt (Eds.), *Proceedings of the Eighth International Conference on Harmful Algae*, Virgo, Spain. Xunta de Galicia and Intergovernmental Oceanographic Commission, UNESCO, pp. 457–458.

Steidinger, K.A., J.T. Davis, and J. Williams. 1966. Observations of *Gymnodinium breve* Davis and other dinoflagellates. In: *Observations of an Unusual Red Tide: A Symposium*, Vol. 8, Professional Paper Series. Florida Board of Conservation, Marine Laboratory, St. Petersburg, FL, pp. 8–15.

Steidinger, K.A., J.H. Landsberg, E.W. Truby, and B.S. Roberts. 1998b. First report of *Gymnodinium pulchellum* (Dinophyceae) in North America and associated fish kills in the Indian River, Florida. *J. Phycol.* 34: 431–437.

Steinberg, N., J. Way, D.J. Suszkowski, and L. Clark. 2002. Harbor Health/Human Health: An analysis of environmental indicators for the NY/NJ Harbor Estuary. Prepared for the New York/New Jersey Harbor Estuary Program by the Hudson River Foundation for Science and Environmental Research under a cooperative agreement with the US Environmental Protection Agency, Region II, New York.

Steinman, A.D., K.E. Havens, H.J. Carrick, and R. VanZee. 2002. The past, present, and future hydrology and ecology of Lake Okeechobee and its watersheds. In: J.W. Porter and K.G. Porter (Eds.), *The Everglades, Florida Bay and Coral Reefs of the Florida Keys: An Ecosystem Sourcebook*. CRC Press, Boca Raton, FL, pp. 19–37.

Stern, A.H. 2004. Letter to the editor: More on mercury content in fish. *Science* 303: 763.

Stevens, R.J. and M.A. Neilson. 1987. Response of Lake Ontario to reductions in phosphorus load, 1967–82. *Can. J. Fish. Aquat. Sci.* 44: 2059–2068.

Stewart, W.D.P. 1974. *Alga Physiology and Biochemistry*. Blackwell Scientific Publications, London, U.K., 989pp.

Stith, L., J. Barkuloo, and M.S. Brim. 1984. Fish and wildlife resource inventory for Escambia navigation project Escambia and Santa Rosa Counties, Florida. Division of Ecological Science, US Fish and Wildlife Service, Panama City, FL, 44pp.

Stockner, J.G. and D.D. Cliff. 1976. Effects of pulp mill effluent on phytoplankton production in coastal marine waters of British Columbia. *J. Fish. Res. Board Can.* 33: 2433–2442.

Stockner, J.G. and D.D. Costella. 1976. Marine phytoplankton growth in high concentrations of pulp mill effluent. *J. Fish. Res. Board Can.* 33: 2758–2765.

Stone, G.W., J.P. Morgan, G.A. Moshiri, and S. Elawad. 1990. Physical, biological and environmental studies of Bayou Texar Escambia County, Florida: Conclusions and recommendations for environmental improvement in Bayou Texar. Institute for Coastal and Estuarine Research, University of West Florida, Pensacola, FL. Report-02-90, 2pp.

Stone, G.W., D. White, J.P. Morgan, and G.A. Moshiri. 1991. Sedimentation rates and macroinvertebrate distribution in Bayou Texar Escambia County, Florida. Institute for Coastal and Estuarine Research, University of West Florida, Pensacola, FL. Report-02-91, 18pp.

Stone, G.W. and J.P. Morgan. 1992. Recent sediment flux and bathymetric changes in the entrance channel to Bayou Texar: Escambia County, Florida. Institute for Coastal and Estuarine Research, University of West Florida, Pensacola, FL. Report-02-92, 34pp.

Stone, G.W. and J.P. Morgan. 1993. Sedimentation rates and geochemistry of subsurface sediments 12th. Avenue, Bayou Texar Pensacola, Florida. Institute for Coastal and Estuarine Research, University of West Florida, Pensacola, FL. Report-02-93, 29pp.

Stoner, A.W. 1976. Growth and food conversion efficiency of pin-fish (*Lagodon rhomboides*) exposed to sublethal concentrations of bleached kraft mill effluents). MS thesis, Florida State University, Tallahassee, FL.

Stoner, A.W. 1979a. The macrobenthos of seagrass meadows in Apalachee Bay, Florida, and the feeding ecology of *Lagodon rhomboides* (Pisces: Sparidae). PhD dissertation, Florida State University, Tallahassee, FL.

Stoner, A.W. 1979b. Species-specific predation on amphipod Crustacea by the pinfish, *Lagodon rhomboides*: Mediation by macrophyte standing crop. *Mar. Biol.* 55: 201–207.

Stoner, A.W. 1980a. The role of seagrass biomass in the organization of benthic macrophyte assemblages. *Bull. Mar. Sci.* 30: 537–551.

Stoner, A.W. 1980b. Abundance, reproductive seasonality and habitat preferences of amphipod crustaceans in seagrass meadows of Apalachee Bay, Florida. *Contrib. Mar. Sci.* 23: 63–77.

Stoner, A.W. 1980c. Feeding ecology of the *Lagodon rhomboides* (Pisces: Sparidae): Variation and functional responses. *Fish. Bull.* 78: 337–352.

Stoner, A.W. 1982. The influence of benthic macrophytes on the foraging behavior of pinfish *Lagodon rhomboides* (Linnaeus). *J. Exp. Mar. Biol. Ecol.* 58: 271–284.

Stoner, A.W., H.S. Greening, J.D. Ryan, and R.J. Livingston. 1983. Comparison of macrobenthos collected with cores and suction dredge. *Estuaries* 6: 76–82.

Stoner, A.W. and R.J. Livingston. 1978. Respiration, growth and food conversion efficiency of pinfish (*Lagodon rhomboides*) exposed to sub-lethal concentrations of bleached Kraft mill effluent. *Environ. Pollut.* 17: 207–218.

Stoner, A.W. and R.J. Livingston. 1980. Distributional ecology and food habits of the banded blenny *Paraclinus fasciatus* (Clinidae): A resident in a mobile habitat. *Mar. Biol.* 56: 239–246.

Stoner, A.W. and R.J. Livingston. 1984. Ontogenetic patterns in diet and feeding morphology in sympatric sparid fishes from seagrass meadows. *Copeia* 1984: 174–187.

Strathmann, R.R. 1967. Estimating the organic carbon content of phytoplankton from cell volume or plasma volume. *Limnol. Oceanogr.* 12: 411–418.

Streckis, R.A., I.C. Potter, and R.C.J. Lenanton. 1995. The commercial fisheries in three southwestern Australian estuaries exposed to different degrees of eutrophication. In: A.J. McComb (Ed.), *Eutrophic Shallow Estuaries and Lagoons*. CRC Press, Boca Raton, FL, pp. 189–204.

Strickland, J.D.H. 1960. Measuring the production of marine phytoplankton. *Bull. Fish Res. Board Can.* 112: 1–172.

Strickland, J.D.H. 1965. Production of organic matter in the primary stages of the marine food chain. In: J.P. Riley and G. Skirrow (Eds.), *Chemical Oceanography*. Academic Press, New York, pp. 477–610.

Stumpf, H.P., M.L. Frayer, M.J. Durako, and J.C. Brock. 1999. Variations in water clarity and bottom albedo in Florida Bay from 1985 to 1997. In: J.W. Fourqurean, M.B. Robblee, and L.A. Deegan (Eds.), *Florida Bay: A Dynamic Subtropical Estuary*; *Estuaries* 22: 431–444.

Su, S.H., L.C. Pearlman, J.A. Rothrock, T.J. Iannuzzi, and B.L. Finley. 2002. Potential long-term ecological impacts caused by disturbance of contaminated sediments: A case study. *Environ. Manage.* 29: 234–249.

Sullivan, M.J. 1978. Diatom community structure: Taxonomical and statistical analysis of a Mississippi salt marsh. *J. Phycol.* 14: 468–475.

Summers, J.K. 1999. The ecological condition of estuaries in the Gulf of Mexico. US Environmental Protection Agency, Fulf Ecology Division, Gulf Breeze, FL, Publication EPA 620-R-98-004, 71pp.

Summers, J.K., T.T. Polgar, J.A. Tar, K.A. Rose, D.G. Heimbuch, J. McCurley, R.A. Cummins, G.F. Johnson, D.T. Yetman, and G.T. DiNardo. 1985. Reconstruction of long-term time series for commercial fisheries abundance and estuarine pollution loadings. *Estuaries* 8: 114–124.

Summerson, H.C. and C.H. Peterson. 1984. Role of predation in organizing benthic communities of a temperate-zone seagrass bed. *Mar. Ecol. Prog. Ser.* 15: 63–77.

Sundback, K., V. Enoksson, W. Graneli, and K. Pettersson. 1991. Influence of sublittoral microphytobenthos on the oxygen and nutrient flux between sediment and water: A laboratory continuous flow study. *Mar. Ecol. Prog. Ser.* 74: 263–279.

Surratt, D., J. Cherrier, L. Robinson, and J. Cable. 2008. Chronology of sediment nutrient geochemistry in Apalachicola Bay, Florida (U.S.A). *J. Coast. Res.* 24: 660–671.

Suttle, C.A. and P.J. Harrison. 1988. Ammonium and phosphate uptake rates, N:P supply ratios, and evidence for N and P limitation in some oligotrophic lakes. *Limnol. Oceanogr.* 33: 186–202.

Swanson, J.R. 1991. Quantification of environmental impact factors from various physiographic regions of Leon County using a Delphi Consensus Technique. MS thesis, Department of Geography, Florida State University, Tallahassee, FL.

Swift, F. 1896. Report of a survey of the oyster region of St. Vincent Sound, Apalachicola Bay, and St. George Sound, Florida. US Commission of Fish and Fisheries, US Government Printing Office. Report of the US Commission for 1896, Appendix 4, Washington, DC, pp. 187–221.

Sykes, J.E. and C.S. Manooch, III. 1979. Predator-prey systems in fisheries management. In: R.H. Clepper (Ed.), *International Symposium on Predator-Prey Systems in Fish Communities and Their Role in Fisheries Management*. Sport Fishing Institute, Washington, DC, pp. 93–101.

Taning, A.V. 1951. Olieforurening af havet og massedod af fugle. *Naturens Verden* 35: 34–43.

Tanner, W.F. 1960. Florida coastal classification. *Trans. Gulf Coast Assoc. Geol. Soc.* 10: 259–266.

Tansley, A.G. 1935. The use and abuse of vegetational concepts and terms. *Ecology* 16: 284–307.

Taylor, F.J.R. 1976a. Dinoflagellates from the International Indian Ocean Expedition. A report on material collected by the R.V. "Anton Bruun" 1965–1965. *Bibliotheca Botanica* 132: 1–226, 46pp.

Taylor, F.J.R. 1976b. Flagellates. In: H.F. Steedman (Ed.), *Zooplankton Fixation and Preservation*, Vol. 4. UNESCO Press, Paris, France, pp. 259–264.

Taylor, F.J.R. 1976c. Dinoflagellates. In: A. Sournia (Ed.), *Phytoplankton Manual*, Vol. 6, Monographs on Oceanographic Methodology. UNESCO Press, Paris, France, pp. 143–147.

Taylor, F.J.R. and R. Haigh. 1993. The ecology of fish-killing chloromonad flagellate *Heterosigma* in the Strait of Georgia and adjacent waters. In: T.J. Smayda and Y. Shimizu (Eds.), *Toxic Phytoplankton Blooms in the Sea. Proceedings of the Fifth International Conference on Toxic Marine Phytoplankton*, Newport, RI. Elsevier, New York, pp. 705–710.

Ter Braak, C.J.F. 1996. Unimodal models to relate species to environment. DLO-Agricultural Mathematics Group, Wageningen, the Netherlands, 266pp.

Tett, P., S.I. Heaney, and M.R. Droop. 1985. The Redfield ratio and phytoplankton growth rate. *J. Mar. Biol. Assoc. U.K.* 65: 487–504.

The Nature Conservancy. 1995. Research report: Eglin air force base, Florida 1995. A compilation of inventory, monitoring and research conducted in support of ecosystem management. The Nature Conservancy, Gainesville, FL.

Thienemann, A. 1918. Lebensgemeinschaft und Lebensraum. *Naturw. Wochenschrift N.F.* 17: 282–290, 297–303.

Thistle, D. 1981. Natural physical disturbances and communities of marine soft bottoms. *Mar. Ecol. Prog. Ser.* 6: 223–228.

Thomas, D.L. 1971. The early life history and ecology of six species of drum (Sciaenidae) in the lower Delaware River, a brackish tidal estuary. Ichthyology Association, Middletown, DE. Deleware Program Report 3, 247pp.

Thomas, W.H., D.L.R. Siebert, and A.N. Dodson. 1974. Phytoplankton enrichment experiments and bioassays in natural coastal seawater and in sewage outfall receiving waters of southern California. *Estuarine Coastal Mar. Sci.* 2: 191–206.

Thompson, F.G. 1984. *Freshwater Snails of Florida: A Manual for Identification.* University Press of Florida, Gainsville, FL, 94pp.

Thompson, P.A. 1998. Spatial and temporal patterns of factors influencing phytoplankton in a salt-wedge estuary, the Swan River, Western Australia. *Estuaries* 21: 801–817.

Thompson, R.L. 1996. Memorandum of January 30, 1996. Florida Department of Environmental Protection, Tallahassee, FL.

Thornton, J.A., H. Beekman, G. Boddington, R. Dick, W.R. Harding, M. Lief, I.R. Morrison, and A.J.R. Quick. 1995a. The ecology and management of Zandvlei (Cape Province, South Africa), an enriched shallow African estuary. In: A.J. McComb (Ed.), *Eutrophic Shallow Estuaries and Lagoons.* CRC Press, Boca Raton, FL, pp. 109–128.

Thornton, J.A., J. McComb, and S.O. Ryding. 1995b. The role of sediments. In: A.J. McComb (Ed.), *Eutrophic Shallow Estuaries and Lagoons.* CRC Press, Boca Raton, FL, pp. 205–224.

Thorpe, P., R. Bartel, P. Ryan, K. Albertson, T. Pratt, and D. Cairns. 1997. The Pensacola Bay system surface water improvement and management plan. Northwest Florida Water Management District, Havana, FL (unpublished report).

Throndsen, J. 1969. Flagellates of Norwegian coastal waters. *Nytt Mag. Bot.* 16: 161–216.

Throndsen, J. 1978. Preservation and storage. In: S. Sournia (Ed.), *Phytoplankton Manual*, Vol. 6. UNESCO Press, Paris, France, pp. 69–74.

Throndsen, J. 1993. The planktonic marine flagellates. In: C.R. Tomas (Ed.), *Marine Phytoplankton.* Academic Press, San Diego, CA, pp. 7–145.

Thrush, S.F., J.E. Heweitt, and R.D. Pridmore. 1989. Patterns in the spatial arrangement of polychaetes and bivalves in intertidal sandflats. *Mar. Biol.* 102: 529–536.

Thrush, S.F., R.D. Pridmore and J.E. Hewitt. 1994. Impacts on soft-sediment macrofauna: The effects of spatial variation on temporal trends. *Ecol. Appl.* 4: 31–41.

Tichken, E.J. 1991. Santa Monica Bay restoration project: Assessment of geophysical properties of ocean sediments off Palos Verdes Peninsula Los Angeles County, CA (unpublished report).

Tilman, D. 1982. *Resource Competition and Community Structure.* Princeton University Press, Princeton, NJ, 296pp.

Todd, G. 1980. Annual study of the mercury contamination of the fish and sediments in the South, South Fork Shenandoah, and Shenandoah Rivers. Virginia Water Control Board, Richmond, VA (unpublished report).

Tomas, C.R. (Ed.). 1993. *Marine Phytoplankton: A Guide to Naked Dinoflagellates and Coccolithophorids.* Academic Press, New York, 263pp.

Tomas, C.R. (Ed.). 1996. *Identifying Marine Diatoms and Dinoflagellates.* Academic Press, San Diego, CA, 598pp.

Tomas, C.R., B. Bendis, and D.K. Johns. 1999. Role of nutrients in regulating plankton blooms in Florida Bay. In: H. Kumpf, K. Steidinger, and K. Sherman (Eds.), *The Gulf of Mexico: Large Marine Ecosystem.* Blackwell Science, New York, pp. 323–337.

Tomoyuki, S., N. Sou, M. Tadashi, Y. Souta, Y. Yasuhiro, S. Yohei, O. Yuji, I.R. Jenkinson, and H. Tsuneo. 2008. Factors influencing the initiation of blooms of the raphidophyte *Heterosigma akashiwo* and the diatom *Skeletonema costatum* in a port in Japan. *Limmol. Oceanogr.* 53: 2503–2518.

Toner, W. 1975. Oysters and the good 'ol boys. *Planning* 41: 10–15.

Tong, H. 1990. *Non-Linear Time Series: A Dynamical System Approach.* Clarendon Press, Oxford, U.K., 564pp.

Tougas, J.I. and J.W. Porter. 2002. Differential coral recruitment patterns in the Florida Keys. In: J.W. Porter and K.G. Porter (Eds.), *The Everglades, Florida Bay and Coral Reefs of the Florida Keys: An Ecosystem Sourcebook.* CRC Press, Boca Raton, FL, pp. 789–311.

Tracey, G.A. 1985. Feeding reduction, reproductive failure, and mortality in *Mytilus edulis* during the 1985 "brown tide" in Narragansett Bay, Rhode Island. *Mar. Ecol. Prog. Ser.* 50: 73–81.

Trueblood, D.D. 1991. Spatial and temporal effects of terrebellid polychaete tubes on soft-bottom community structure in Phosphorescent Bay, Puerto Rico. *J. Exp. Mar. Biol. Ecol.* 149: 139–159.

Trush, W.J., S.M. McBain, and L.B. Leopold. 2000. Attributes of an alluvial river and their relation to water policy and management. *Proc. Natl. Acad. Sci. U.S.A.* 97: 11858–11863.

Tuovila, B.J., T.H. Johegen, P.A. LaRock, J.B. Outland, D.H. Esry, and M. Franklin. 1987. An evaluation of the Lake Jackson filter system and artificial marsh on nutrient and particulate removal from stormwater runoff. In: K.R. Reddy and W.H. Smith (Eds.), *Aquatic Plants for Water Treatment and Resource Recovery*. Magnolia Publishing Company, Pineville, LA, pp. 271–278.

Turner, R.E. 2000. *Coastal Ecosystems of the Gulf of Mexico and Climate Change*. US Global Change Research Project, Washington, DC, Chapter 6, pp. 83–104.

Turner, R.E., N.N. Rabalais, and Z.-N. Zhang. 1990. Phytoplankton biomass, production and growth limitations on the Huanghe (Yellow River) continental shelf. *Cont. Shelf Res.* 10: 545–571.

Turner, R.E., W.W. Schroeder, and W.J. Wiseman, Jr. 1987. The role of stratification in the deoxygenation of Mobile Bay and adjacent shelf bottom waters. *Estuaries* 10: 13–20.

Turner, S.J., S.F. Thrush, R.D. Pridmore, J.E. Hewitt, V.J. Cummings, and M. Maskery. 1995. Are soft-sediment communities stable? An example from a windy harbour. *Mar. Ecol. Prog. Ser.* 120: 219–230.

Tyrrell, T. 1999. The relative influences of nitrogen and phosphorus on oceanic primary production. *Nature* 368: 619–621.

US Army Corps of Engineers (USACOE). 1976a. Final environmental statement: Apalachicola, Chattahoochee, and Flint Rivers, Alabama, Florida, and Georgia (Operation and Maintenance). Mobile District, Mobile, AL.

US Army Corps of Engineers (USACOE). 1976b. Statement of findings: Perdido Pass Channel (maintenance dredging), Baldwin County, Alabama. Mobile District, Mobile, AL.

US Army Corps of Engineers (USACOE). 1977. Draft report on water pollution in Bayou Chico. Mobile District, Mobile, AL.

US Army Corps of Engineers (USACOE). 2008. Modification the current Interim Operations Plan (IOP) at Jim Woodruff Dam. Mobile District, Mobile, AL.

US Environmental Protection Agency (USEPA). 1971a. *Conference in the Matter of Pollution of the Interstate Waters of the Escambia River basin (Alabama–Florida) and the Intrastate Portions of the Escambia basin within the State of Florida*. Pensacola, FL.

US Environmental Protection Agency (USEPA). 1971b. Circulation and benthic characterization studies Escambia Bay, Florida. Federal Water Pollution Control Administration, US Department of the Interior, Athens, GA.

US Environmental Protection Agency (USEPA). 1972. *Proceedings of the Conference—Pollution of the Interstate Waters of the Escambia River Basin (Alabama–Florida) and the Interstate Portions of the Escambia Basin within the State of Florida: Third Session*, January 24–26, 1972. Gulf Ecology Division Laboratory, Gulf Breeze, FL.

US Environmental Protection Agency (USEPA). 1976. Quality criteria for water. Office of Water and Hazardous Materials, Washington, DC.

US Environmental Protection Agency (USEPA). 1979. St. Regis and Perdido Bay algal growth potential tests. Surveillance and Analysis Divisions, Athens, GA.

US Environmental Protection Agency (USEPA). 1983. Methods for chemical analysis of water and wastes. Environmental Monitoring and Support Laboratory, Office of Research and Development, Cincinnati, OH. EPA-600/4-79-020.

US Environmental Protection Agency (USEPA). 1989. Ambient water quality criteria for ammonia (saltwater). Office of Water Regulations and Standards, Criteria and Standards Division, Washington, DC. EPA 440/S88004.

US Environmental Protection Agency (USEPA). 1990. Analysis of the section 301(h) secondary treatment variance application by Los Angeles County Sanitation Districts for Joint Water Pollution Control Plant. Region 9, Water Management Division, San Francisco, CA.

US Environmental Protection Agency (USEPA). 1993. Water-quality protection program for the Florida Keys National Marine Sanctuary. Phase II report. Washington, DC.

US Environmental Protection Agency (USEPA). 1997a. Mercury Report to Congress. An assessment of exposure to mercury in the United States. Office of Air Quality Planning and Standards, and Office of Research and Development, Washington, DC. EPA452/R-97-006:

US Environmental Protection Agency (USEPA). 1997b. Unpublished data concerning toxic compounds in sediments of the Newark Bay Complex.

US Environmental Protection Agency (USEPA). 1997c. *Pensacola Bay System Technical Symposium, Foundations of Sustainability: Status of Pensacola Bay System Water Quality*, Pensacola, FL, September 19, 1997.

US Environmental Protection Agency (USEPA). 1998. National strategy for the development of regional nutrient criteria. Washington, DC.

US Environmental Protection Agency (USEPA). 2000. Fenholloway Nutrient Study, Perry, FL. Region 4, Athens, GA.

US Environmental Protection Agency (USEPA). 2003. River corridor and wetland restoration. Washington, DC.

US Department of the Interior (USDOI). 1970. Effects of pollution on water quality: Escambia River and Bay, Florida. Federal Water Pollution Control Administration, Southeast Water Laboratory, Technical Services Program, Athens, GA.

US Department of the Interior (USDOI). 1988. Statement for management. Gulf Islands National Seashore, Gulf Breeze, FL.

US Fish and Wildlife Service (USFWS). 1990. Aerial photography of the seagrass beds of Perdido Bay. Pensacola, FL (unpublished report).

US Navy. 1986. Draft environmental impact statement: U.S. Navy Gulf coast strategic homeporting, Appendix IV. Pensacola, Florida. Southern Division, Naval Facilities Engineering Command, Charleston, SC.

US Public Health Service (USPHS). 1962. *Proceedings of the Conference on the Interstate Pollution of the Conecuh—Escambia River*. US Department of Health, Education and Welfare, Pensacola, FL.

Ulanowicz, R.E. 1987. NETWRK4: A package of computer algorithms to analyze ecological flow networks. University of Maryland, Chesapeake Biological Laboratory, Solomons, MD.

Ulanowicz, R.E. and J.J. Kay. 1991. A package for the analysis of ecosystem flow networks. *Environ. Software* 6: 131–142.

Utermohl, H. 1931. Neue Wege in der quantitativen Erfassung des Planktons. *Verh. Int. Ver. Theor. Angew. Limnol.* 5: 567–596.

Utermohl, H. 1958. Zur vervollkommnung der quantitativen phytoplankton methodik. *Verh. Int. Ver. Theor. Angew. Limnol.* 9: 1–38.

Valiela, L., J. McClelland, J. Hauxwell, P.J. Behr, D. Hersh, and K. Foreman. 1997. Macroalgal blooms in shallow estuaries: Controls and ecophysiological and ecosystem consequences. *Limnol. Oceanogr.* 42: 1105–1118.

Valiella, I., J.M. Teal, and E.J. Carpenter. 1978. Nutrient and particulate fluxes in a salt marsh ecosystem: Tidal exchanges and inputs by precipitation and groundwater. *Limnol. Oceanogr.* 23: 798–812.

Van den Hoek, C., D.G. Mann, and H.M. Jahns (Eds.). 1995. *Algae: An Introduction to Phycology*. Cambridge University Press, Cambridge, U.K., 623pp.

Van Dolah, R.F. 1978. Factors regulating the distribution and population dynamics of the amphipod *Gammarus palustris* in an intertidal salt marsh community. *Ecol. Monogr.* 48: 191–217.

Van Es, F.B., M.A. Van Arkel, L.A. Bowman, and H.G.J. Schroder. 1980. Influence of organic pollution on bacterial, macrobenthic and meiobenthic populations in intertidal flats of the Dollard. *Neth. J. Sea Res.* 14: 288–304.

Van Hoose, M.S. 1987. Biological assessment of the Perdido Bay system. Alabama Department of Conservation and Natural Resources, Marine Resources Division, Dauphin Isalnd, AL (unpublished report).

Van Raalte, C.D., I. Valeila, and J.M. Teal. 1976. The effect of fertilization on the species composition of salt marsh diatoms. *Water Res.* 10: 1–4.

Van Sickle, V.R., B.B. Barrett, L.J. Gulick, and T.B. Ford. 1976. Barataria basin: Salinity changes and oyster distribution. Center for Wetland Resources, Louisiana State University. Baton Rouge, LA. Sea Grant Publication LSU-T-76-002, 22pp.

Van Valkenburg, S.D. and D.A. Flemer. 1974. The distribution and productivity of nannoplankton in a temperate estuarine area. *Esturaine Coastal Mar. Sci.* 2: 311–322.

Venrick, E.L. 1978. How many cells to count. In: A. Sournia (Ed.), *Phytoplankton Manual*, Vol. 6, Monographs on Oceanographic Methodology. UNESCO Press, Paris, France, pp. 167–180.

Verity, P.G., M. Alber, and S.B. Bricker. 2006. Development of hypoxia in well-mixed subtropical estuaries in the southeastern USA. *Estuaries Coasts* 29: 665–673.

Vinagre, C., J. Salgado, H.N. Cabral, and M.J. Costa. 2011. Food web structure and habitat connectivity in fish estuarine nurseries—Impact of river flow. *Estuaries Coasts* 34: 663–674.

Vince, S.I., I. Valiela, N. Backus, and J. Teal. 1976. Predation by the salt marsh killifish *Fundulus heteroclitus* (L.) in relation to prey size and habitat structure. Consequences of prey distribution and abundance. *J. Exp. Mar. Biol. Ecol.* 23: 255–266.

Virnstein, R.W. 1977. The importance of predation by crabs and fishes on benthic infauna in Chesapeake Bay. *Ecology* 58: 1199–1217.

Vollenweider, R.A. 1985. Elemental and biochemical composition of plankton biomass: Some comments and explorations. *Arch. Hydrobiol.* 105: 11–29.

Voynova, Y.G. and J.H. Sharp. 2012. Anomalous biogeochemical response to a flooding event in the Delaware Estuary: A possible typology shift due to climate change. *Estuaries Coasts* 35: 943–958.

Wagner, J.R. 1984. Hydrogeologic assessment of the October 1982 draining of Lake Jackson, Leon County, Florida. Northwest Florida Water Management District Havana, FL. Water Resources Special Report 84-1.

Wakeman, T.M. and D. Suszkowski. 1999. Contaminated sediment strategies at the Port of New York and New Jersey. In: N. Kraus and W. McDougal (Eds.), *Coastal Sediments '99*, Vol. 3. Hudson River Foundation, American Society of Civil Engineers, Reston, VA, pp. 2347–2354.

Wald, A. and J. Wolfowitz. 1940. On a test whether two samples are from the same population. *Ann. Math. Stat.* 11: 147–162.

Wales, D.J. 1991. Calculating the rate of loss of information from chaotic time series by forecasting. *Nature* 350: 485–488.

Wallin, J.M., M.D. Hattersley, D.F. Ludwig, and T.J. Iannuzzi. 2002. Historical assessment of the impacts of chemical contamination in sediments on benthic invertebrates in the tidal Passaic River, New Jersey. *Human Ecol. Risk Assess.* 8: 1156–1177.

Walsh, J.J. 1988. *On the Nature of Continental Shelves*. Academic Press, New York, 520pp.

Wang, W.C. 1980. Fractionation of sediment oxygen demand. *Water Res.* 14: 603–612.

Ward, C.H., M.E. Bender, and D.J. Reish (Eds.). 1979. The offshore ecology investigation: Effects of oil drilling and production in a coastal environment. *Rice Univ. Stud.* 65: 1–589.

Ware, D.H. 1972. Predation of rainbow trout (*Salmo gairdneri*): The influence of hunger, prey density, and prey size. *J. Fish. Res. Board Can.* 29: 1193–1201.

Waterbury, J.B., S.W. Watson, R.R.L. Guillard, and L.E. Brand. 1979. Wide spread occurrence of a unicellular, marine, planktonic, cyanobacterium. *Nature* 277: 293–294.

Weinstein, M. and K.L. Heck. 1979. Ichthyofauna of seagrass meadows along the Caribbean coast of Panama and in the Gulf of Mexico: Composition, structure and community ecology. *Mar. Biol.* 50: 97–107.

Weisberg, R.H. 1976. A note on estuarine mean flow estimation. *J. Mar. Res.* 34: 387–394.

Weisberg, R.H. 1989. Sikes Cut—A review of data and physical model studies by the COE on the salinity effects for Apalachicola Bay. Final report for the Florida Department of Environmental Regulation, Tallahassee, FL (unpublished report, revised).

Weiss, R.F. 1970. Helium isotope effect in solution in water and seawater. *Science* 168: 247–248.

Welsh, B.L. and F.C. Eller. 1991. Mechanisms controlling summertime oxygen depletion in western Long Island Sound. *Estuaries* 14: 265–278.

Welch, B.L., R.B. Whitlatch, and W.F. Bohlen. 1982. Relationship between physical characteristics and organic carbon sources as a basis for comparing estuaries in southern New England. In: V.S. Kennedy (Ed.), *Estuarine Comparisons*. Academic Press, New York, pp. 53–67.

Wenning, R.J., M.A. Harris, B. Finley, D.J. Paustenbach, and H. Bedbury. 1993. Application of pattern recognition techniques to evaluate polychlorinated dibenzo-p-dioxin and dibenzofuran distributions in surficial sediments from the lower Passaic River and Newark Bay. *Ecotoxicol. Environ. Saf.* 25: 103–125.

Werner, E.E. and D.J. Hall. 1974. Optimal foraging and the size selection of prey by bluegill sunfish (*Lepomis macrochirus*). *Ecology* 55: 1042–1052.

Westman, W.E. 1978. Measuring the inertia and resilience of ecosystems. *Bioscience* 28: 705–710.

Wetz, M.S. and H.W. Paerl. 2008. Estuarine phytoplankton responses to hurricanes and tropical storms with different characteristics (trajectory, rainfall, winds). *Estuaries Coasts* 31: 419–429.

Wetzel, R.G. 1984. Detrital dissolved and particulate organic carbon functions in aquatic ecosystems. *Bull. Mar. Sci.* 35: 503–509.

Wetzel, R.G. and G.E. Likens. 1990. *Limnological Analyses*, 2nd edn. Springer-Verlag, Berlin, Germany, 212pp.

White, D.C. 1983. Analysis of microorganisms in terms of quantity and activity in natural environments. In: J.H. Slater, R. Whittenbury, and J.W.T. Wimpenny (Eds.), *Microbes in Their Natural Environments. Society for General Microbiology Symposium*, Cambridge University Press, Vol. 34, pp. 37–66.

White, D.C., R.J. Bobbie, S.J. Morrison, D.K. Oesterhof, C.W. Taylor, and D.A. Meeter. 1977. Determination of microbial activity of lipid biosynthesis. *Limnol. Oceanogr.* 22: 1089–1099.

White, D.C., W.M. Davis, J.S. Nickels, J.D. King, and R.J. Bobbie. 1979a. Determination of the sedimentary microbial biomass by extractable lipid phosphate. *Oecologia* 40: 51–62.

White, D.C., R.J. Livingston, R.J. Bobbie, J.S. Nickels. 1979b. Effects of surface composition, water column chemistry, and time of exposure on the composition of the detrital microflora and associated macrofauna in Apalachicola Bay, Florida. In: R.J. Livingston (Ed.), *Ecological Processes in Coastal and Marine Systems*. Plenum Press, New York, pp. 53–67.

White, D.W., G.W. Stone, J.P. Morgan, and G.A. Moshiri. 1993. Relationship of Benthic invertebrates to sedimentation within the vicinity of two stormwater outfalls in Bayou Texar Escambia County, Florida. Wetlands Research Laboratory, Institute for Coastal and Estuarine Research, University of West Florida, Pensacola, FL. Rep-01.

White, M.E. and E.A. Wilson. 1996. Predators, pests, and competitors. In: V.S. Kennedy, R.I.E. Newell, and A.F. Eble (Eds.), *The Eastern Oyster Crassostrea virginica*. University of Maryland, College Park, MD. Maryland Sea Grant Publication UM-SG-TS-96-01, pp. 559–580.

Whitfield, M. 1974. The hydrolysis of ammonium ions in sea water—A theoretical study. *J. Mar. Biol. Assoc. U.K.* 54: 565–580.

Whitfield, W.K., Jr. and D.S. Beaumariage. 1977. Shellfish management in Apalachicola Bay: Past, present, and future. In: R.J. Livingston and E.A. Joyce, Jr. (Eds.), *Proceedings of the Conference on the Apalachicola Drainage System*, Gainsville, FL, April 23–24, 1976. Florida Department of Natural Resources, Marine Research Laboratory, St. Petersburg, FL. Florida Marine Resources Publication 26, pp. 130–140.

Whitlach, R.B. 1977. Seasonal changes in the community structure of the macrobenthos inhabiting the intertidal sand and mud flats of Barnstable Harbor, Massachusetts. *Biol. Bull.* 152: 275–294.

Whitlatch, R.B. and R.N. Zajac. 1985. Biotic interactions among estuarine infaunal opportunistic species. *Mar. Ecol. Prog. Ser.* 21: 299–311.

Wieland, R. 2003. Loss of credibility could undermine Chesapeake cleanup effort. Chesapeake Media Service, Seven Valleys, Pennsylvania. *Bay J.* 13: Comment 2.

Wiener, J.G. and P.J. Shields. 2000. Mercury in the Sudbury River (Massachusetts, U.S.A.): Pollution history and a synthesis of recent research. *Can. J. Fish. Aquat. Sci.* 57: 1053–1061.

Wikfors, G.H. and R.M. Smolowitz. 1993. Detrimental effects of a *Prorocentrum* isolate upon hard clams and bay scallops in laboratory feeding studies. In: T.J. Smayda and Y. Shimizu (Eds.), *Toxic Phytoplankton Blooms in the Sea. Proceedings of the Fifth International Conference on Toxic Marine Phytoplankton*, Newport, RI. Elsevier, New York, pp. 447–452.

Wilber, D.H. 1992. Associations between freshwater inflows and oyster productivity in Apalachicola Bay, Florida. *Estuarine Coastal Shelf Sci.* 35: 179–190.

Wilcoxon, F. 1945. Individual comparisons by ranking methods. *Biometrics* 1: 50–83.

Wilcoxon, F. 1949. *Some Rapid Approximate Statistical Procedures*. American Cyanamid Company, Stamford, CT, 16pp.

Wild, S.R., S.P. McGrath, and K.C. Jones. 1990. Organic contaminant in an agricultural soil with a known history of sewage sludge amendments: Polynuclear aromatic hydrocarbons. *Environ. Sci. Technol.* 24: 1706–1711.

Wiley, D., P. Weiland, and D.J. Cairns. 1990. Point source assessment of the Pensacola Bay System. Northwest Florida Water Management District, Havana, FL. Water Resources Special Report 90-5.

Willen, E. 1976. A simplified method of phytoplankton counting. *Br. Phycol. J.* 11: 265–278.

Williams, A.B. 1984. *Shrimps, Lobsters and Crabs of the Eastern United States, Maine to Florida.* Smithsonian Institution Press, Washington, DC, 550pp.

Williams, J.D., M.L. Warren, K.S. Cummings, J.L. Harris, and R.J. Neves. 1992. Conservation status of freshwater mussels of the United States and Canada. *Fisheries* 18: 6–22.

Williams, J.E., J.E. Johnson, D.A. Hendrickson, S. Contreras-Balderas, J.D. Williams, M. Navarro-Mendoza, D.E. McAllister, and J.E. Deacon. 1989. Fishes of North America, endangered, threatened, or of special concern: 1989. *Fisheries* 14: 2–20.

Williams, M., S. Filoso, B.J. Longstaff, and W.C. Dennison. 2010. Long-term trends of water quality and biotic metrics in Chesapeake Bay: 1986 to 2008. *Estuaries Coasts* 33: 1279–1299.

Williamson, R.B. 1985. Urban stormwater quality 1. Hillcrest, Hamilton, New Zealand. *N. Z. J. Mar. Freshw. Res.* 19: 413–427.

Williams, W.D. 1970. A revision of North American epigean species of *Asellus* (Crustacea: Isopoda). Smiths. *Contrib. Zool.* 49: 1–80.

Wilson, R.M., J. Chanton, F.G. Lewis, and D. Nowacek. 2010. Concentration-dependent stable isotope analysis of consumers in the upper reaches of a freshwater-dominated estuary: Apalachicola Bay, FL, USA. *Estuaries Coasts* 33: 1406–1419.

Wiltshire, K.H., A. Kraberg, I. Bartsch, M. Boersma, H. Franke, J. Freund, C. Gebühr, G. Gerdts, K. Stockmann, and A. Wichels. 2010. Helgoland Roads, North Sea: 45 years of change. *Estuaries Coasts* 33: 295–310.

Windsor, J.G., Jr. 1985. Nationwide review of oxygen depletion and eutrophication in estuarine and coastal waters: Florida Region. Department of Oceanography and Ocean Engineering. Florida Institute of Technology. Final report to Brookhaven National Laboratory, Upton, NY, and to the US Department of Commerce, NOAA, National Ocean Service, Office of Oceanography and Marine Assessment, Ocean Assessment Division, Rockville, MD.

Winemiller, K.O. 1990. Spatial and temporal variation in tropical fish trophic networks. *Ecol. Monogr.* 60: 331–367.

Winer, B.J. 1971. *Statistical Principles in Experimental Design*. McGraw-Hill, New York, 322pp.

Woelke, C.E. 1961. Pacific oyster *Crassostrea gigas* mortalities with notes on common oyster predators in Washington waters. *Proc. Natl. Shellfish. Assoc.* 50: 53–66.

Wolfe, D.A., M.A. Champ, D.A. Flemer, and A.J. Mearns. 1987. Long-term biological data sets: Their role in research, monitoring, and management of estuarine and coastal marine systems. *Estuaries* 10: 181–193.

Wolfe, D.A., E.R. Long, and G.B. Thursby. 1996. Sediment toxicity in the Hudson-Raritan Estuary: Distribution and correlations with chemical contamination. *Estuaries* 19: 901–912.

Wolfe, M.F., S. Schwarzbach, and R.A. Sulaiman. 1998. Effects of mercury on wildlife: A comprehensive review. *Environ. Toxicol. Chem.* 17: 146–160.

Wolfe, S.H., J.A. Reidenauer, and D.B. Means. 1988. An ecological characterization of the Florida panhandle. US Fish and Wildlife Service, Washington, DC. Biological Report 88, 277pp.

Wood, D.A. 1996. Florida's endangered species, threatened species, and species of special concern. Official lists. Bureau of Nongame Wildlife, Division of Wildlife, Florida Game and Fresh Water Fish Commission, Tallahassee, FL.

Wood, E.J.F. 1954. Dinoflagellates in the Australian region. *Aust J. Mar. Freshw. Res.* 5: 171–351.

Wood, K. and R.L. Bartel. 1994. Bayou Chico sediment and water quality data report, Escambia County, Florida. Northwest Florida Water Management District, Havana, FL. Technical File Report 94-3.

Woodhead, P.M.J. 1966. The behavior of fish in relation to light in the sea. *Oceanogr. Mar. Biol. Annu. Rev.* 4: 337–403.

Woodwell, G.M. and D.E. Whitney. 1977. Flax Pond ecosystem study: Exchanges of phosphorous between a salt marsh and the coastal waters of Long Island Sound. *Mar. Biol.* 41: 1–6.

Woodwell, G.M., D.E. Whitney, C.A.S. Hall, and R.A. Houghton. 1977. The Flax Pond ecosystem study: Exchanges of carbon in water between a salt marsh and Long Island Sound. *Limnol. Oceanogr.* 22: 833–838.

Wright, R.T. and B.J. Nebel. 2002. *Environmental Science: Toward a Sustainable Future*. Prentice Hall, Inc., Upper Saddle River, NJ, 682pp.

Wu, T.S. and W.K. Jones. 1992. Preliminary circulation simulations in Apalachicola Bay. In: M.L. Spaulding, K. Bedford, A. Blumberg, R. Cheng, and C. Swanson (Eds.), *Estuarine and Coastal Modeling: Proceedings of the Second International Conference of the American Society of Civil Engineers*, Tampa, FL, November 13–15, 1992. American Society of Civil Engineering, New York, pp. 344–356.

Yamochi, S. and T. Abe. 1984. Mechanisms to initiate a *Heterosigma akashiwo* red tide in Osaka Bay. *Mar. Biol.* 83: 255–261.

Yerger, R.W. 1977. Fishes of the Apalachicola River. In: R.J. Livingston (Ed.), *Proceedings of the Conference on the Apalachicola Drainage System*, Gainesville, FL, April 23–24, 1975. Florida Department of Natural Resources, St. Petersburg, FL, Florida Marine Resources Publication 26, pp. 22–33.

Young, D.L. and R.T. Barber. 1973. Effects of waste dumping in New York Bight on the growth of natural populations of phytoplankton. *Environ. Pollut.* 5: 237–252.

Young, P.H. 1964. Some effects of sewer effluent on marine life. *Calif. Fish Game* 50: 33–37.

Young, W.T. 1981. A biological and water quality survey of Blackwater, Yellow, and Shoal Rivers and Tributaries, and East Bay Estuaries. Florida Department of Environmental Regulation, Tallahassee, FL.

Young, W.T. 1985. Pensacola mainstreet sewage treatment plant discharge impact assessment survey. Florida Department of Environmental Regulation, Pensacola, FL.

Zajac, R.N. and R.B. Whitlatch. 1982. Responses of estuarine infauna to disturbance. II. Spatial and temporal variation of succession. *Mar. Ecol. Prog. Ser.* 10: 15–27.

Zaret, T.M. and A.S. Rand. 1971. Competition in tropical stream fishes: Support for the competitive exclusion principle. *Ecology* 52: 336–342.

Zeeman, C. 2001. Memo to S. Ladner regarding media protection goals for contaminants of concern released by the HoltraChem Manufacturing Facility, Orrington, Maine. Maine Department of Environmental Protection, Augusta, ME.

Zieman, J.C., J.W. Fourqurean, and T.A. Frankovich. 1999. Seagrass die-off in Florida Bay: Long-term trends in abundance and growth of turtle grass, *Thalassia testudinum. Estuaries* 22: 460–470.

Zieman, J.C., J.W. Fourqurean, and T.A. Frankovich. 2004. Reply to B.E. Lapointe and P.J. Barile. 2004. Comment on J.C. Zieman, J.W. Fourqurean, and T.A. Frankovich. 1999. Seagrass die-off in Florida Bay: Long-term trends in abundance and growth of turtle grass, *Thalassia testudinum.* (*Estuaries* 22: 460–470.) *Estuaries* 27: 165–178.

Zieman, J.C. and R.T. Zieman. 1989. The ecology of the seagrass meadows of the west coast of Florida: A community profile. Department of Environmental Sciences, University of Virginia, Charlottesville, VA. US Fish and Wildlife Service, Washington, DC. Publication No. BR-85 (7.25).

Ziewitz, J. 2003. Summary of relationships between flow regime and species protected under the Endangered Species Act on the Apalachicola River. US Fish and Wildlife Service, Washington, DC (unpublished report).

Zillioux, E.J., D.B. Porcella, and J.M. Benoit. 1993. Mercury cycling and effects in freshwater wetland ecosystems. *Environ. Toxicol. Chem.* 12: 2245–2264.

Zimmerman, M.S. 1974. A comparison of the benthic macrophytes of a polluted drainage system (Fenholloway River) with an unpolluted drainage system (Econfina River). MS thesis, Florida State University, Tallahassee, FL.

Zimmerman, M.S. and R.J. Livingston. 1976a. The effects of kraft mill effluents on benthic macrophyte assemblages in a shallow bay system (Apalachee Bay, North Florida, U.S.A.). *Mar. Biol.* 34: 297–312.

Zimmerman, M.S. and R.J. Livingston. 1976b. Seasonality and physico-chemical ranges of benthic macrophytes from a north Florida estuary (Apalachee Bay). *Contrib. Mar. Sci. Univ. Tex.* 20: 34–45.

Zimmerman, M.S. and R.J. Livingston. 1979. Dominance and distribution of benthic macrophyte assemblages in a north Florida estuary (Apalachee Bay, Florida). *Bull. Mar. Sci.* 29: 27–40.

Zongwei, C., V.M.S. Ramanujam, and M.L. Gross. 1994a. Levels of polychlorodibenzo-*p*-dioxins and dibenzofurans in crab tissues from the Newark/Raritan Bay System. *Environ. Sci. Technol.* 28: 1528–1534.

Zongwei, C., V.M.S. Ramanujam, and M.L. Gross. 1994b. Mass-profile monitoring in trace analysis: Identification of polychlorodibenzothiophenes in crab tissues collected from the Newark/Raritan Bay system. *Environ. Sci. Technol.* 28: 1535–1538.

Index

A

B

C

Q

R

W

Y

Z

For Product Safety Concerns and Information please contact our EU representative GPSR@taylorandfrancis.com Taylor & Francis Verlag GmbH, Kaufingerstraße 24, 80331 München, Germany

Batch number: 10731065

Printed by Printforce, the Netherlands